Respiratory Medicine

Series Editors

Sharon I.S. Rounds
Alpert Medical School of Brown University
Providence, RI
USA

Anne Dixon
University of Vermont, Larner College of Medicine
Burlington, VT
USA

Lynn M. Schnapp
University of Wisconsin - Madison
Madison, WI
USA

More information about this series at http://www.springer.com/series/7665

Kent E. Pinkerton • William N. Rom
Editors

Climate Change and Global Public Health

Second Edition

ATS We help the world breathe®
PULMONARY • CRITICAL CARE • SLEEP

Humana Press

Editors
Kent E. Pinkerton
Department of Pediatrics
School of Medicine
University of California, Davis
Davis, CA
USA

William N. Rom
NYU School of Global Public Health
New York University
New York, NY
USA

ISSN 2197-7372 ISSN 2197-7380 (electronic)
Respiratory Medicine
ISBN 978-3-030-54745-5 ISBN 978-3-030-54746-2 (eBook)
https://doi.org/10.1007/978-3-030-54746-2

This Humana imprint is published by the registered company Springer Nature Switzerland AG
The registered company address is: Gewerbestrasse 11, 6330 Cham, Switzerland

Foreword

This 2nd edition of *Climate Change and Global Public Health* provides a wealth of new and updated information about the impact of climate change on our planet, our communities, and our health. Since the 1st edition 5 years ago, the scientific understanding of climate change has progressed substantially. Unfortunately, so have the impacts. Climate change is rapidly becoming one of the most significant public health crises in human history. We face an increasingly urgent problem that could claim a quarter million lives annually by 2030 without concerted global action to rapidly cut greenhouse gas emissions.

As detailed in the most recent IPCC report, the buildup of climate forcing gases in the atmosphere is accelerating. Seventeen of the 18 hottest years on record have occurred since 2000. The visible impacts of climate change are increasingly apparent across the planet in the form of heatwaves, drought, catastrophic wildfires, rising seas, and destructive hurricanes. These increasingly frequent extreme weather events present acute public heath challenges. At the same time, climate change is exacerbating the deadly effects of air pollution, especially among vulnerable populations like children and the elderly.

The last few years have been a mixed bag in terms of our collective response to the climate crisis. President Obama established the USA as a global leader on climate action by setting an ambitious national carbon reduction target and put in place a range of strategies to cut emissions from the energy, transportation, oil and gas, and agriculture sectors. In 2016, nearly 200 countries signed onto the Paris Accord pledging to undertake ambitious action to begin driving down emissions in line with scientifically determined levels.

Unfortunately, much of this progress has slowed or altogether halted since the election of President Trump. The Trump Administration has waged an all-out assault on climate science and is actively working to undermine efforts to cut greenhouse gas emissions despite the overwhelming economic and scientific rationale to do so. In the absence of US leadership, a number of major global actors have also wavered in their commitments.

However, despite the Trump Administration's failure to take action, a growing number of states across the country are moving ahead. For example, the United

States Climate Alliance, a bipartisan coalition of states that represent more than half of the US population and a combined economy that is the third largest in the world, are working across a range of sectors to cut greenhouse gas emissions in line with the Paris Accord.

As we wait for renewed political leadership at the federal level, California continues to stand out as a leader in demonstrating ways to tackle climate change and grow our economy. Over the past decade, California's economic growth has outpaced the national average while at the same time cutting greenhouse gas emissions ahead of schedule. The Golden State is proof that smart climate policy is also smart economic policy.

This book presents a sobering assessment of the impact that climate change is already having on public health. By effectively making this connection, it provides a strong tool that we can use to mobilize action across the planet to take on this challenge.

Sacramento, CA, USA Mary D. Nichols

Contents

1 **Introduction: Consequences of Global Warming to Planetary and Human Health** 1
William N. Rom and Kent E. Pinkerton

2 **Climate Variability and Change Data and Information for Global Public Health** 35
Juli M. Trtanj and Tamara G. Houston

3 **Climate Change: Updates on Recent Global and United States Temperature Anomalies and Impacts to Water, Forests, and Environmental Health** 51
David H. Levinson and Christopher J. Fettig

4 **Eyewitness to Climate Change** 75
Will Steger and Nicole Rom

5 **Arctic Connections to Global Warming and Health** 91
M. Torre Jorgenson and Janet C. Jorgenson

6 **Assessing the Health Risks of Climate Change** 111
Kristie L. Ebi

7 **Heat Waves and Rising Temperatures: Human Health Impacts and the Determinants of Vulnerability** 123
Helene G. Margolis

8 **Climate, Air Quality, and Allergy: Emerging Methods for Detecting Linkages** 163
Patrick L. Kinney, Perry E. Sheffield, and Kate R. Weinberger

9 **The Human Health Co-benefits of Air Quality Improvements Associated with Climate Change Mitigation** 181
George D. Thurston and Michelle L. Bell

10 **Asthma, Hay Fever, Pollen, and Climate Change** 203
Anthony Szema, Jonathan Li, Ashlee Pagnotta, Malvika Singh, and Jo' Ale White

11 **California and Climate Changes: An Update** 237
Rupa Basu

12 **Vector-Borne Diseases in a Changing Climate and World** 253
Yesim Tozan, Ora Lee H. Branch, and Joacim Rocklöv

13 **Dengue Fever and Climate Change** 273
Lauren Cromar and Kevin Cromar

14 **Climate Variability and Change: Food, Water, and Societal Impacts** 311
Jonathan A. Patz

15 **Hurricanes and Health: Vulnerability in an Age of Climate Change** 339
Alexis A. Merdjanoff and Rachael Piltch-Loeb

16 **The Impact of Climate Change and Extreme Weather Conditions on Agricultural Health and Safety in California** 353
Heather E. Riden, Emily Felt, and Kent E. Pinkerton

17 **Household Air Pollution from Cookstoves: Impacts on Health and Climate** 369
William J. Martin II, Tara Ramanathan, and Veerabhadran Ramanathan

18 **Biomass Fuel and Lung Diseases: An Indian Perspective** 391
Rajendra Prasad, Rajiv Garg, and Nikhil Gupta

19 **Climate Change and Women's Health: Risks and Opportunities** ... 403
Cecilia J. Sorensen and John Balbus

20 **The Impact of Climate Change on Public Health in Small Island States and Caribbean Countries** 427
Muge Akpinar-Elci and Hugh Sealy

21 **Global Climate Change, Desertification, and Its Consequences in Turkey and the Middle East** 445
Hasan Bayram and Ayşe Bilge Öztürk

22 **Climate Change and the Risk of Desertification with a Focus in the United States** 459
Huda Asif and Mehdi Mirsaeidi

23 **Federal Programs in Climate Change and Health Research** 483
Cecilia J. Sorensen, Caitlin Rublee, and John Balbus

24 Management of Climate Change Adaptation at the United States Centers for Disease Control and Prevention 505
Jeremy J. Hess, Gino Marinucci, Paul J. Schramm, Arie Manangan, and George Luber

25 Rules, Rulings, and Repeals: The Shifting State of Climate Policy at U.S. EPA 525
Jack Lienke

26 California's Integrated Approach to Air Quality and Climate Change 541
John R. Balmes

27 Climate Change and Public Health Interventions 549
Kathryn C. Conlon and Chelsea M. Austin

28 Integrating Climate Change, the Environment, and Sustainability Themes Into Professional Health Sciences Courses: A Case Study Across a University System 565
Tammy Nicastro, Arianne Teherani, Helene G. Margolis, and Sheri Weiser

29 The Physician's Response to Climate Change 583
Mary B. Rice and Alexander S. Rabin

Index 593

Contributors

Muge Akpinar-Elci, MD, MPH Old Dominion University, Norfolk, VA, USA

Huda Asif, MD Division of Pulmonary and Critical Care, University of Miami, Miami, FL, USA

Chelsea M. Austin, MPH Department of Environmental Health Sciences, Yale School of Public Health, New Haven, CT, USA

John Balbus, MD, MPH National Institute of Environmental Health Sciences, Bethesda, MD, USA

John R. Balmes, MD University of California, San Francisco, CA, USA

Rupa Basu, PhD, MPH Office of Environmental Health Hazard Assessment, Air and Climate Epidemiology Section, Oakland, CA, USA

Hasan Bayram, MD, PhD Department of Pulmonary Medicine, Koc University School of Medicine, Koc University Research Center for Translational Medicine (KUTTAM), İstanbul, Turkey

Michelle L. Bell, PhD School of Forestry and Environmental Studies, Yale University, New Haven, CT, USA

Ora Lee H. Branch, PhD Department of Research, College of Health & Human Services, Concordia University, Portland, OR, USA

Kathryn C. Conlon, PhD, MPH Department of Public Health Sciences, University of California Davis, Davis, CA, USA

Kevin Cromar, PhD Marron Institute of Urban Management, New York University, New York, NY, USA

Lauren Cromar, MS Tahoe Consulting, LLC, North Salt Lake, UT, USA

Kristie L. Ebi, PhD, MPH University of Washington, Seattle, WA, USA

Emily Felt, MPP Center for Health and the Environment, University of California, Davis, Davis, CA, USA

Christopher J. Fettig, MS, PhD Pacific Southwest Research Station, USDA Forest Service, Davis, CA, USA

Rajiv Garg, MBBS, MD Department of Respiratory Medicine, K.G's. Medical University UP, Lucknow, India

Nikhil Gupta, MBBS, MD Department of General Medicine, Dr. Ram Manohar Lohia Institute of Medical Sciences, Lucknow, India

Jeremy J. Hess, MD, MPH, FACEP Emory Schools of Medicine and Public Health, Atlanta, GA, USA

Climate and Health Program, NCEH, CDC, Atlanta, GA, USA

Tamara G. Houston National Oceanic and Atmospheric Administration, National Climatic Data Center, Asheville, NC, USA

M. Torre Jorgenson, ScD (Honorary) Alaska Ecoscience, Fairbanks, AK, USA

Janet C. Jorgenson U.S. Fish and Wildlife Service, Retired Arctic National Wildlife Refuge, Fairbanks, AK, USA

Patrick L. Kinney, ScD Department of Environmental Health, Boston University School of Public Health, Boston, MA, USA

David H. Levinson, MS, PhD National Stream and Aquatic Ecology Center, Watershed, Fish, Wildlife, Air and Rare Plants Staff, USDA Forest Service, Fort Collins, CO, USA

Jack Lienke, JD Institute for Policy Integrity, New York University School of Law, New York, NY, USA

Jonathan Li, MD Three Village Allergy & Asthma, PLLC, South Setauket, NY, USA

Sidney Kimmel Medical College at Thomas Jefferson University, Philadelphia, PA, USA

George Luber, PhD Climate and Health Program, Division of Environmental Hazards and Health Effects, National Center for Environmental Health, Centers for Disease Control and Prevention, Atlanta, GA, USA

Arie Manangan, MA Climate and Health Program, Division of Environmental Hazards and Health Effects, National Center for Environmental Health, Centers for Disease Control and Prevention, Atlanta, GA, USA

Helene G. Margolis, MA, PhD Department of Internal Medicine, General Medicine Division, Center for Health Services Research in Primary Care, Women's Center for Health, University of California, Davis, Davis, CA, USA

Gino Marinucci, MPH Climate and Health Program, Division of Environmental Hazards and Health Effects, National Center for Environmental Health, Centers for Disease Control and Prevention, Atlanta, GA, USA

William J. Martin II, MD College of Public health, The Ohio state University, Columbus, OH, USA

Alexis A. Merdjanoff, PhD School of Global Public Health, New York University, New York, NY, USA

Mehdi Mirsaeidi, MD, MPH Division of Pulmonary and Critical Care, University of Miami, Miami, FL, USA

Tammy Nicastro, MS National Center of Excellence in Women's Health, Department of Obstetrics, Gynecology, and Reproductive Sciences, University of California San Francisco, San Francisco, CA, USA

Ayşe Bilge Öztürk, MD Department of Pulmonary Medicine, Koc University School of Medicine, Koc University Research Center for Translational Medicine (KUTTAM), İstanbul, Turkey

Ashlee Pagnotta, BS Three Village Allergy & Asthma, PLLC, South Setauket, NY, USA

Stony Brook University, Stony Brook, NY, USA

Jonathan A. Patz, MD, MPH UW-Madison School of Medicine and Public Health, Department of Population Health Science, Madison, WI, USA

Rachael Piltch-Loeb, PhD, MSPH T.H. Chan School of Public Health, Division of Policy Translation and Leadership Development, Harvard University, Boston, MA, USA

Kent E. Pinkerton, PhD Department of Pediatrics, School of Medicine, Department of Anatomy, Physiology and Cell Biology, School of Veterinary Medicine, Center for Health and the Environment, John Muir Institute of the Environment, University of California, Davis, Davis, CA, USA

Rajendra Prasad, MD, DTCD Department of Pulmonary Medicine, Era's Lucknow Medical College & Hospital, Era University, Lucknow, India

Alexander S. Rabin, MD Tufts Medical Center, Boston, MA, USA

Tara Ramanathan, MBA Nexleaf Analytics, Los Angeles, CA, USA

Veerabhadran Ramanathan, MS, PhD Scripps Institution of Oceanography, University of California at San Diego, La Jolla, CA, USA

Mary B. Rice, MD, MPH Harvard Medical School, Boston, MA, USA

Heather E. Riden, MA Western Center for Agricultural Health and Safety and the Department of Public Health Sciences, University of California, Davis, Davis, CA, USA

Joacim Rocklöv, PhD Department of Public Health and Clinical Medicine, Section of Sustainable Health, Umeå University, Umeå, Sweden

Nicole Rom, MS Climate Generation: A Will Steger Legacy, Minneapolis, MN, USA

William N. Rom, MD, MPH NYU Grossman School of Medicine, NYU School of Global Public Health, New York University, New York, NY, USA

Caitlin Rublee, MD, MPH Department of Emergency Medicine, University of Colorado School of Medicine, Aurora, CO, USA

Paul J. Schramm, MS, MPH Climate and Health Program, Division of Environmental Hazards and Health Effects, National Center for Environmental Health, Centers for Disease Control and Prevention, Atlanta, GA, USA

Hugh Sealy, PhD, MSc, BEng The University of the West Indies, Cave Hill, Barbados, West Indies

Perry E. Sheffield, MD, MPH Environmental Medicine & Public Health, Icahn School of Medicine at Mount Sinai, New York, NY, USA

Malvika Singh Three Village Allergy & Asthma, PLLC, South Setauket, NY, USA

Stony Brook University, Stony Brook, NY, USA

Cecilia J. Sorensen, MD Department of Emergency Medicine, University of Colorado School of Medicine, Aurora, CO, USA

University of Colorado School of Public Health, Aurora, CO, USA

Will Steger, MA Climate Generation: A Will Steger Legacy, Minneapolis, MN, USA

Anthony Szema, MD Three Village Allergy & Asthma, PLLC, South Setauket, NY, USA

Northwell Health International Center for Excellence in Deployment Health and Medical Geosciences, Manhasset, NY, USA

Department of Occupational Medicine, Epidemiology, and Prevention at Zucker School of Medicine at Hofstra/Northwell, Hempstead, NY, USA

Donald and Barbara Zucker School of Medicine at Hofstra/Northwell, Hempstead, NY, USA

Columbia University Global Psychiatric, Epidemiology Group CDC NIOSH U01 OH011308-01 9/11, Trauma and Toxicity in Childhood: Longitudinal Health and Behavioral Outcomes, New York, NY, USA

Faculty Committee on Health Professions (Pre-Med Committee) Stony Brook University, Stony Brook, NY, USA

RDS2 Solutions, Inc., Stony Brook, NY, USA

Division of Pulmonary and Critical Care, Northwell Health, Manhasset, NY, USA

Division of Allergy/Immunology, Northwell Health, Manhasset, NY, USA

Department of Occupational Medicine, Epidemiology and Prevention, Donald and Zucker School of Medicine at Hofstra/Northwell, Hempstead, NY, USA

Arianne Teherani, PhD Department of Medicine, University of California San Francisco, San Francisco, CA, USA

George D. Thurston, SM, ScD NYU School of Medicine, New York University, New York, NY, USA

Yesim Tozan, PhD Global Health and Environmental Public Health Sciences Program, College of Global Public Health, New York University, New York, NY, USA

Juli M. Trtanj National Oceanic & Atmospheric Administration, Climate Program Office, Silver Spring, MD, USA

Kate R. Weinberger, MA School of Population and Public Health, University of British Columbia, Vancouver, BC, Canada

Sheri Weiser, MD Department of Medicine, University of California, San Francisco, San Francisco, CA, USA

Jo'Ale White, MS Three Village Allergy & Asthma, PLLC, South Setauket, NY, USA

Stony Brook University, Stony Brook, NY, USA

Chapter 1
Introduction: Consequences of Global Warming to Planetary and Human Health

William N. Rom and Kent E. Pinkerton

Introduction to Greenhouse Gases and Temperature

Greenhouse gases (e.g., carbon dioxide [CO_2], methane, hydrofluorocarbons, nitrous oxide, and ozone) reflect infrared radiation back to the earth's surface causing a warming effect ([1–6]; Fig. 1.1). Global warming has a major effect on climate, meteorological patterns measured over long periods of time (e.g., months, years, millennia, or mega-anna), which can be differentiated from weather and atmospheric conditions measured over short terms (e.g., hours or days).

Measurements of climate changes began more than a half century ago with the International Geophysical Year (1958), when CO_2 was measured atop Mauna Loa in Hawaii [6]. At that time, CO_2 measurements were 316 ppm. However, annual averages have continually increased such that CO_2 levels reached 414 ppm in 2019 (Fig. 1.2). In March 2019, the annual increase was 2.72 ppm, a notable acceleration, since the original increases hovered between 0.5 and 1 ppm per year.

Currently, CO_2 is the major greenhouse gas, representing about 85% of the greenhouse gases and causing about 0.75 watts/m^2 imbalance in global heating [7]. In the recent fourth National Climate Assessment, global warming was emphatically shown to be anthropogenically derived and to bear no relation to solar sunspots or volcanic eruptions [3]. Anthropogenic sources of fossil fuels' use for 80% of energy use (electricity generation, burning oil and natural gas for transportation, cooking, and heating), and converting forest lands for agriculture have substantially increased CO_2

W. N. Rom (✉)
NYU Grossman School of Medicine, NYU School of Global Public Health, New York University, New York, NY, USA
e-mail: William.Rom@nyulangone.org

K. E. Pinkerton
Department of Pediatrics, School of Medicine, Department of Anatomy, Physiology and Cell Biology, School of Veterinary Medicine, Center for Health and the Environment, John Muir Institute of the Environment, University of California, Davis, Davis, CA, USA

K. E. Pinkerton, W. N. Rom (eds.), *Climate Change and Global Public Health*, Respiratory Medicine, https://doi.org/10.1007/978-3-030-54746-2_1

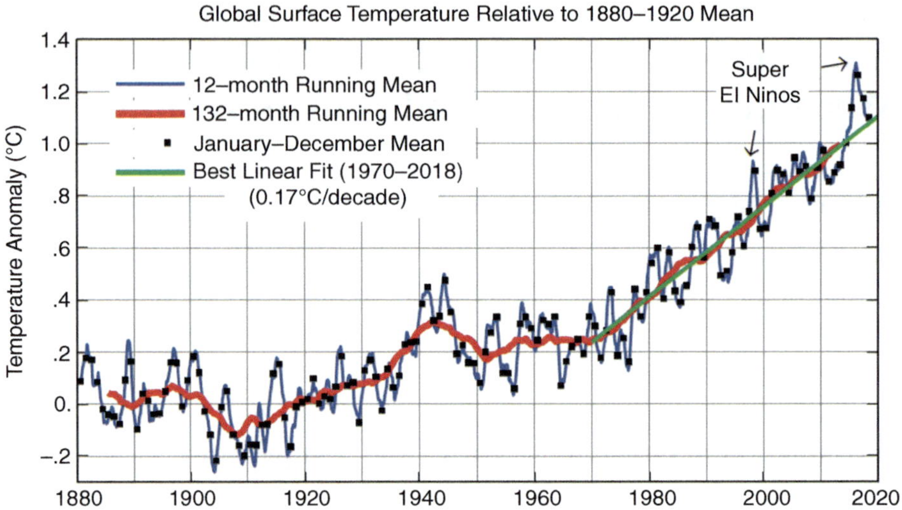

Fig. 1.1 Global Surface Temperature Relative to 1880–1920 Mean. Global surface temperature in 2018 was the fourth highest in the period of instrumental measurements in the Goddard Institute for Space Studies (GISS) analysis. The 2018 global temperature was +1.1 °C (~2 °F) warmer than in the 1880-01920 base period pre-industrial temperature. The four warmest years in the GISS record all occur in the past 4 years, and the 10 warmest years are all in the twenty-first century. Hansen J, Sato M, Ruedy R, Schmidt GA, and Lo K. Global temperature in 2018 and Beyond, 6 February 2019 (with permission)

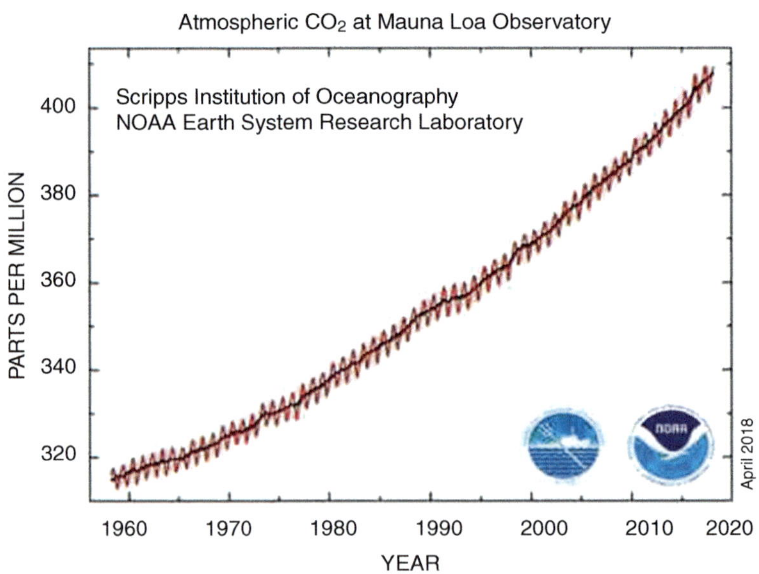

Fig. 1.2 CO_2 concentration (414.94 ppm May 1, 2019) on Mauna Loa, Hawaii. Reprinted with permission from The Scripps Institution of Oceanography

emissions since World War II. Only over the past 50 years have anthropogenic sources been prodigious enough to actually exceed natural CO_2 sinks (oceans and land, where it is used by plants in metabolism) and increase the global mean surface recordings of greenhouse gases and temperature. Given that the primary sources of CO_2 are the carbon fossil-fuel energy sources, which have been the basis of our modern lifestyle, curtailing these sources is inherently political and social; however, accomplishing a radical restructuring to a "green" or decarbonized lifestyle is essential.

Although economists optimistically suggest this can be accomplished with 1% of global gross world product, expenditures, such as in the Stern Report, green global infrastructure investments will need to be \$1 trillion annually [2]. The marketplace has been slow to respond because of global subsidies to the fossil fuel industry totaling ~\$500 billion/year, and the industry's support of politicians who are climate deniers. In the case of CO_2, the gas is long-lived, with 33% emitted per year remaining at 100 years and 19% at 1000 years, resulting in greater and greater cumulative emissions ([7]; Fig. 1.3). Therefore, as global CO_2 pollution emissions continue to exceed 36 gigatons annually, the consequences will take decades to centuries to reverse and return to levels <350 ppm, a level considered to keep the increase in mean surface temperature < 1.5 °C and avoid the worse consequences of climate change [8]. Measurements of CO_2 captured in air bubbles in the Antarctic ice show levels ~280 ppm going as far back as 800,000 years [9]. In this Introduction, we discuss the greenhouse gases' impacts on climate with consequences for planetary and human health. Computer models create predictions for the future, but

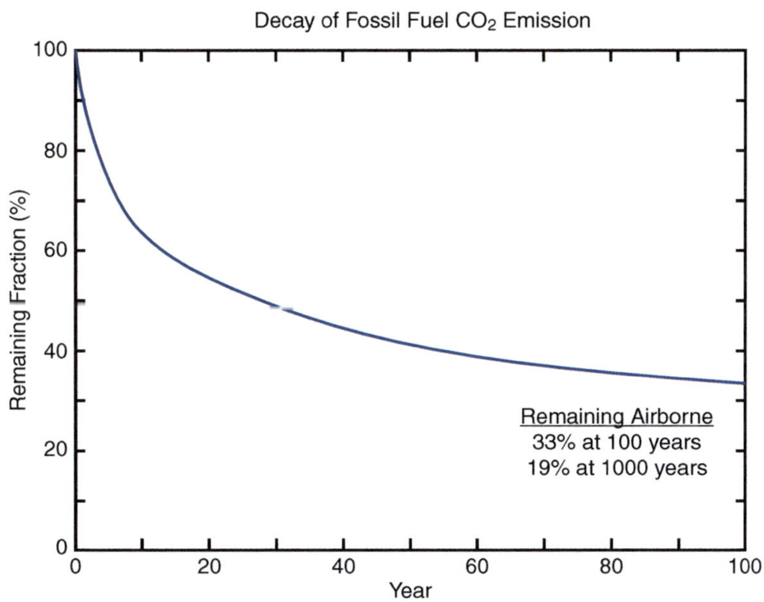

Fig. 1.3 Decay of fossil fuel CO_2 emissions. The fraction of CO_2 remaining in the air, after emission by fossil fuel burning, declines rapidly at first, but 1/3 remains in the air after a century and 1/5 after a millennium (Atmos Chem Phys 2007; 7: 2287–2312)

since there is little precedent for global warming at the rate and magnitude projected, there are uncertainties that only more observational data will resolve [10].

To add to this, there are other greenhouse gases (e.g., methane, hydrofluorocarbons, black carbon, nitrous oxide, ozone) that may have even greater radiative forcing than CO_2 [11–14]. The radiative forcing (heat-trapping ability) of methane, for example, is approximately 25 times greater than that of CO_2 over 100 years and 85 times greater over 20 years due to its comparatively shorter atmospheric lifespan compared to CO_2 [1–3]. Methane is emitted from fossil fuel and bovid livestock operations. Oil and natural gas production, processing, pipeline leaks, and storage are commonly associated with methane emissions, as are landfills and coal mines – methane from these latter two sources can be captured for natural gas generation of electricity. A rather exotic class of releases originates from the gastrointestinal tracts and manure of cattle, sheep, and goats and, interestingly, these releases can be mitigated with feed changes. From 1980 to 2018, methane measurements in the Earth's atmosphere increased from 1630 to 1860 ppb [14]. Because there is significant carbon in the frozen permafrost, thawing due to climate change could release stored methane and CO_2 that could further accelerate global warming trends [12]. Quantifying this risk is challenging [13].

Synthetic chemicals, including perfluorocarbons, hydrofluorocarbons (HFCs), and sulfur hexafluoride from fire extinguishers, refrigerants, and foam blowers have half-lives of 14 years (compared to >100 years for chlorofluorocarbons), and more than 2000-fold greater radiative forcing than CO_2 [14]. These chemicals have been synthesized to replace the chlorofluorocarbons (CFCs) that endanger the ozone layer in the stratosphere in the extreme cold of the Antarctic winter. CFCs have destroyed enough of the stratospheric ozone layer >50 kilometers above the surface of the Earth to result in a seasonal ozone hole over the entire continent of Antarctica during its extreme cold winter, when ice crystals can form. The stratospheric ozone layer filters out harmful ultraviolet light that causes skin cancer. Through the 1987 Montreal Protocol, 197 countries in the United Nations came together to ban CFCs and substitute hydrofluorocarbons (HFCs), which would not destroy ozone [15]. Unfortunately, HFCs are also greenhouse gases. HFC substitutes, such as isobutane and hydrofluoroolefins, retain the ability to be potent refrigerants, but are not climate forcing nor do they destroy ozone. The Kigali Agreement of 2016 amends the Montreal Protocol to move toward these substitutes by mid-century. This is especially important to equatorial countries that will need air conditioning to cool houses and apartments, as climate change increases ambient temperatures. Nitrous oxide, primarily from fertilizers and secondarily from coal- and gas-fired power plants, nylon production, and vehicle emissions, can contribute to radiative forcing, but is less potent. SO_2 aerosols and organic carbon can provide a small cooling effect, and black carbon from diesel emissions and biomass burning contribute to global warming [16, 17]. The latter lasts days to weeks providing an opportunity to mitigate warming trends by reducing emission of these small particles. SO_2 aerosols have been declining as fossil fuel plants have controlled this pollutant. Black carbon has been reduced by mandating clean diesel engines for new trucks, although biomass burning for cookstoves remains a formidable challenge. More than 3 billion women and children are exposed to wood, charcoal, and dung during cooking without proper ventilation, and mortality from respiratory diseases reaches three million annually. Clean cookstoves that burn concentrated wood pellets or solar pressure cookers may pave the way toward 100 million clean cookstoves in the near future. Ozone that forms on the Earth's surface (troposphere) from emissions of NO_2, volatile organic compounds, and catalyzed by sunlight is a surface pollutant and a greenhouse gas. Lastly, global warming increases evaporation, increasing water vapor and clouds, but the role of clouds role in causing or mitigating climate change is still poorly understood.

The primary and immediate consequence of greenhouse gas increase in the troposphere is rising global surface temperature ([1–3], Fig. 1.1). The National Oceanic and Atmospheric Administration (NOAA) within the United States Department of Commerce collates data on climate at its National Climatic Data Center in Asheville, NC. They have ~25,000 temperature stations around the world that, at present, cumulatively make >1.6 billion daily observations. Data from the World Meteorological Society show annual surface temperatures from 1861 deviate in a positive direction beginning in 1970–1980 and persist and increase from the norm until the present. Data collected from tree rings, corals, ice cores, and historical records corroborate the thermometer recordings. Temperature time series collected from NASA's Goddard Institute for Space Studies and United Kingdom's land series at the University of East Anglia show the same trends. Combining global land and ocean measurements, the trend is increasing to 0.215 °C/current decade with an increase of 1.1 °C warming over baseline calculated in 2019 ([1–3], Fig. 1.1). Since 2000, 17 of the 18 warmest years on record have occurred. Global warming is not spatially uniform and greater trends are seen in the Northern Hemisphere and in high Arctic latitudes, where the surface temperature increase is 2.5 °C.

The Intergovernmental Panel on Climate Change (IPCC) in its Fifth Assessment Report declared, "Warming of the climate system is unequivocal, and since the 1950s, many of the observed changes are unprecedented over decades to millennia [1]."They also recently reported that we have only 12 years (until 2030) to decarbonize our economy (by half) and halt current global warming trends before temperatures reach 1.5 °C (1 °C is 1.8 °F) above pre-industrial levels [2]. The report contends that warming beyond the1.5 °C threshold, which is expected around 2035, could expose tens of millions within the global population to life-threatening heat waves, water shortages, and coastal flooding. "At present, the atmosphere and ocean have warmed, the amounts of snow and ice have diminished, and sea level has risen. Human influence on the climate system is clear, and recent anthropogenic emissions of greenhouse gases are the highest in history. Recent climate changes have had widespread impacts on human and natural systems. Anthropogenic greenhouse gas emissions have increased since the pre-industrial era, driven largely by economic and population growth, and are now higher than ever. This has led to atmospheric concentrations of carbon dioxide, methane and nitrous oxide that are unprecedented in at least the last 800,000 years. Their effects, together with those of other anthropogenic drivers, have been detected throughout the climate system and are extremely likely to have been the dominant cause of the observed warming since the mid-twentieth century. Continued emission of greenhouse gases will cause further warming and long-lasting changes in all components of the climate system, increasing the likelihood of severe, pervasive and irreversible impacts for people and ecosystems. Limiting climate change would require substantial and sustained reductions in greenhouse gas emissions which, together with adaptation, can limit climate change risks. Surface temperature is projected to rise over the twenty-first century under all assessed emission scenarios. It is very likely that heat waves will occur more often and last longer, and that extreme precipitation events will become more intense and frequent in many regions. The ocean will continue to warm and acidify, and global mean sea level will continue to rise [2]."

Climate change has been called the leading public health priority for the twenty-first century [4, 5]. The rise in surface temperature has led to an increase in heat waves, with 157 million more people exposed to such events in 2017. During that time, labor capacity was saddled, with 153 billion hours of labor lost due to heat; vector-borne diseases increased as vector capacity for transmission increased almost 10%; and agricultural yield potential declined in more than 30 food producing countries [5]. Trends in climate change impacts suggest an unacceptably high risk for populations across the globe. The lack of progress in reducing emissions and building adaptive capacity means that the solutions to climate change will be more expensive, and accompanied by increased mortality.

Consequences of Climate Change on the Biosphere

Loss of Arctic Ice

Warming will significantly reduce the ice in the Arctic Ocean, where there is currently a 13% decrease in Arctic sea ice per decade, and a new autumnal minimum of 4.59 million square kilometers compared to seven million square kilometers in 1979, the first year that satellite measurements were available [18]. There is a further decline in the average multi-year ice (2.54 million square kilometers to 0.13 million square kilometers) to <5% over the past 32 years [18]. As more ice melts, more of the ocean is exposed to the 24-hour summer sun, decreasing its albedo or ability to reflect sunlight. Bright white ice reflects incoming sunlight, but dark ocean water absorbs it, heating the ocean and accelerating warming. As a result, the Arctic has been warming twice as fast as the rest of the globe, with temperatures in 2018, in latitudes above 60 degrees north, at 2.5 °C above the 1981–2010 average. Changes in sea ice influence ocean currents and the jet stream in ways that can affect weather in lower latitudes, including the United States. To personally visit the changing Arctic, I journeyed with the Thule Inuit by dogsled to witness the melting and receding glaciers (Fig. 1.4). When I found the Inuit hunter, Ikuo Oshima in Siorapaluk, the furthest north village occupied by the Thule Inuit in Greenland, he lamented the late freezing and early break-up of the sea ice. He was the only Inuit hunter that I encountered who spoke English. He told me that the sea ice would melt earlier in

Fig. 1.4 Calving and Collapse of the Tracy Glacier in Inglefield Gulf in Northwest Greenland. (Photo by dogsled with Thule Inuit by William N. Rom M.D., MPH)

the spring, moving from the traditional August to as early as May, making the late winter ice slushier and more dangerous to collapse under the sled. The ice-forming moved from October to December which made it difficult to hunt in the late fall because complete darkness had descended. These seasonal variations made it difficult for the Inuit people to compile sufficient meat to maintain their culture as hunters.

Resulting changes in the Arctic regions include warming that could melt permafrost, perennially frozen ground, sequestering CO_2 and methane gas [19]. Permafrost is mostly discontinuous, underlying 24% of the terrestrial Northern Hemisphere and 80% of Alaska. However, much of the known permafrost region is in disequilibrium with the current climate. Though melting of the permafrost could allow peat and attendant water to increase the number of plants taking up carbon, models predict a far greater release of carbon over time as frozen regions thaw and facilitate the microbial decomposition and conversion of organic carbon in soil to greenhouse gases (CO_2 and methane). Melting permafrost also disrupts forests and man-made structures including buildings, pipelines, roads, and other infrastructure.

Loss of Antarctic Ice

In Antarctica, there has been loss of ice shelves, such as the Larsen B in 2002 ([20]; Fig. 1.5). Western Antarctic regions, especially the Bellingshausen and Amundsen sea shelves, are at risk for increased melt from higher basal oceanic warming; if these ice shelves melt and their attendant glaciers including the Thwaites and Pine

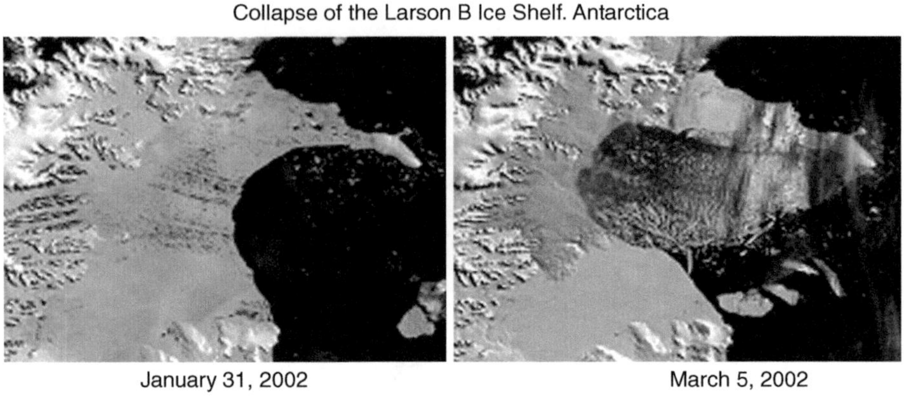

Fig. 1.5 Breakup of Larsen B Ice Shelf on Antarctic Peninsula January – March 2002 Provided by the National Snow and Ice Data Center

Island Glaciers advance, sea level would rise by >4.8 meters [21, 22]. Winters with extensive sea ice enhance krill abundance, so loss or reduction of the size of Antarctic ice shelves adversely affects the population of krill that are the basis of Antarctic biodiversity. Fish species that depend on krill and other crustaceans are the primary prey of Emperor penguins. The Adelie and Emperor penguins in Antarctica and the huge population of King penguins on South Georgia and other Antarctic islands are at risk because of potential declines in krill [23].

Loss of Glaciers

Glaciers are in retreat across the globe with Glacier National Park in Montana predicted to be glacier-free by mid-century [3]. Temperate glaciers near the equator are at immediate risk of complete loss. Importantly, glaciers on Kilimanjaro have declined from 12.5 to ~1.0 square kilometers from 1912 to 2017 (Fig. 1.6), and have lost a half-meter in height every year since 2000 [1]. The Kilimanjaro Northern Ice Field has been present for 11,700 years, and is expected to disappear in 15 years [1]. In Latin American cities, such as Lima, Peru, and La Paz, Bolivia, glaciers are the source of drinking water and hydropower, creating a potential cause for concern about future glacial demise. The seven great rivers arising out of the Himalayas and KunLun Ranges from glacial melt serve nearly 40% of the world's population. Increased glacial melting also produces lakes at the termini of their moraines; increased melt water can rupture these enlarging lakes and flood downstream communities. The Wrangell-St. Elias-Kluane-Alsek-Tatshenshini World Heritage Park is 85% covered by ice [24]. Figure 1.7 illustrates the diminution of the Alsek Glacier over the past century [24]. Photographs taken by the 1906 US-Canada Border Survey crew showed the Alsek Glacier towering over the river. By 2016 the Alsek

Fig. 1.6 Uhuru Point, Kilimanjaro Summit William N. Rom M.D., MPH in 1970 (left) and Daughter Nicole in 1999 (right). Notice disappearance of Kilimanjaro ice fields over 29 years

Glacier emptied into a large, iceberg-laden Alsek Lake with the glacier in full-scale retreat miles in the background. Neil Hartling, a Yukon rafting leader, was able to rescue these old photographs (Fig. 1.7) from the Canadian government and identify the locale of each photograph to compile this history [24].

Risks to Forests

The boreal and mountain forests are at risk due to climate change. The mountain pine bark beetle, lethal to spruce and pine trees, is normally killed by extreme cold 20 to 40 °F below zero [3]. With global warming, the mountain pine bark beetle is thriving, putting extensive forests in the sub-Arctic and US Rocky Mountain West at risk. Bark beetles and spruce budworms in Alaska's Kenai Peninsula have killed spruce across 1.2 million acres, nearly half of the peninsula's forest. An outbreak has consumed more than half of the merchantable pine in British Columbia with the outbreak spreading north and east into Alberta and higher altitudes. The mountain pine beetle has killed 85,000 square miles of ponderosa lodgepole pine trees in the western United States and 65,000 square miles of forest in British Columbia [3]. Unfortunately, there is evidence that it has spread to jack pine, which is a common species throughout the boreal forest. Complicating the mountain beetle infestation is the more rapid melting of the winter snowpack and drying of the climate leading to water-challenged forests, which leave root structures of such species as aspens unable to support the forest [3]. These forests are tinderboxes that serve to enable the forest fires becoming increasingly common in the western United States. Deforestation, usually to make way for agriculture, has been under way for decades, with Brazil and Indonesia being hotspots. The burning of tropical forests not only ends their ability to absorb carbon, but also produces an immediate flow of carbon back to the atmosphere, making it one of the leading sources of greenhouse gas emissions [1]. The world's forests cover 10 billion acres and absorb one-quarter of human emissions of CO_2. Deforestation of the Amazon is proceeding at a pace of 6000–7000 square kilometers per year (down from a peak of 29,000 km^2 in 1995)

Fig. 1.7 Alsek Glacier, Yukon. In 1906, the Canadian-American Border Survey took this photograph of the Alsek Glacier reaching the shore of the Alsek River on the Alaska-Yukon international border. These photographs were about to be discarded when Neil Hartling of Canadian River Expeditions serendipitously contacted the Canadian Boundary Commission about photographs that might illustrate glacier disappearance due to global warming. Below is a photograph from the same place where there is now Alsek Lake with the glacier retreating over 6 miles (second photograph by William N. Rom M.D., MPH on rafting expedition with Neil Hartling in 2016

due to roads, hydroelectric plants, forest burning, and soybean farming [1]. However, from July 2017 to August 2018, deforestation in the Amazon increased 13.7%, and more than 2300 illegal gold artisanal mining operations were observed. Second-growth forest may be able to keep pace with CO_2 absorption, but this needs study. Species may likely change as global warming proceeds, with southern species extending their range northward. In this regard, the sugar maple of Vermont may be at risk for replacement by oak, hickory, or pine; already maple sap is running up to 2 weeks earlier and at reduced amounts compared to previous years.

Mean Sea-Level Rise

The gradual rise of the sea level is of concern for low-lying nations and small island states such as the Maldives in the Indian Ocean, and Kiribati, Tuvalu, Fiji, Cook Islands, and Marshall Islands in the Pacific [1, 2]. The sea-level rise is based on two mechanisms: first, global warming exerts a steric force by thermal expansion of the volume of water ; second, increased mass of water (eustatic sea-level rise) from melting of glaciers, especially polar glaciers in Greenland and Antarctica. Both tide gauge sea-level reconstructions and satellite altimetry show that the current rate of global mean sea-level change is about 3 mm year, and the rate is accelerating [1, 2]. Usually, this change in the rate of sea-level rise is modeled as quadratic, but other functions (e.g., an exponential) may be equally valid; extrapolation of the quadratic fit to the altimeter record to 2100 finds 65 ± 12 cm of sea-level rise by 2100 relative to 2005, suggesting also that the rate of sea-level rise in 2100 could be ~10 mm per year [1–5]. The most likely glacial melting scenarios estimated a range of sea-level rise of 0.8 to 2.0 meters by 2100 [1, 2]. The Fourth National Climate Assessment found that relative sea-level rise along almost all US coastlines would make damaged coastal infrastructure more common during high tides nationwide [3]. Projected sea-level rise could reach 6 feet by 2100 and, under extreme conditions, as high as 8 feet along the upper Atlantic and western Gulf Coasts. Economic losses would be severe for cities such as Miami, but these losses would affect the entire national economy. Paleoclimate observations have found sea shells >30 feet higher than current sea levels, when past temperatures were 2 °C higher than present suggesting the possibility that much more sea-level rise could occur.

Bleaching of Coral Reefs

Coral reef ecosystems reflect symbiotic relationships between various coral species and algae, for example, crustose coralline algae, the latter of which impart red-pink color to the reefs. Marine algae are known as dinoflagellates and use irradiance for photosynthesis. In return for shelter and access to sunlight provided by the reefs, these algal zooxanthellae supply coral reefs with essential nutrients, especially carbon, produced by photosynthesis. However, coral reefs are sensitive to several of the factors associated with global warming (i.e., increased oceanic CO_2, temperature, and acidification, 25–29).

The oceans serve as sinks for CO_2 with >30% of CO_2 emitted to the atmosphere by human activities taken up by the ocean; the resultant carbonic acid has lowered the pH from 8.16 to 8.05 over the past two decades [28]. Among other things, acidification prevents calcium carbonate accretion by reef corals and disrupts the outer shell of Pteropods, tiny sea butterflies eaten by a variety of marine species, ranging from tiny krill to salmon to whales, setting the stage for wide-ranging ecological consequences. Oceans also serve as sinks for absorbing increased heat, perhaps 85% of the storage capacity of the Earth. Indeed, the North Atlantic has warmed 0.51 °C, with 3–4 °C warming in the fisheries near Maine and the Canadian Maritime provinces causing temperature sensitive cod species in their fisheries to move northward from Maine as the surrounding waters warmed over the past decade [30].

Increases in oceanic CO_2, acidification, and temperature cause coral reefs to starve, die, and turn white, a process known as coral bleaching ([28–29]; Fig. 1.8). Coral reef bleaching is occurring worldwide; although local pollution is another contributing factor, most experts attribute this to global warming from increased temperature and acidity of the oceans [25]. Experimental aquae on the Great Barrier Reef in Australia with several CO_2 and warming scenarios show striking bleaching up to 50% after 8 weeks of exposure to CO_2 520–1300 ppm [29]. Bleached corals are physiologically damaged and nutritionally compromised, and they can die if the bleaching is severe and recovery time of their symbionts is prolonged [26]. Bleaching and ocean acidification result in loss of reef structure, leading to lower fishery yields and loss of coastal protection and habitat, with impacts on tourism and food security. Coral reefs are critical to biodiversity of the ocean, for example, there are up to 800 types of coral, and 4000 fish species live and propagate on coral reefs. About one-fourth of salt water fish spend some of their life cycle on coral reefs. Seaside

Fig. 1.8 Coral reef at risk of warm ocean water bleaching. Photograph by Zack Rago of "Chasing Coral" at the Great Barrier Reef, Australia, where half of the reef has not recovered from serial bleaching events and has perished

communities in developing countries depend on coral fish populations for food, and as the coral reefs bleach and disappear, the health and survival of these people are at risk. Globally, 17% of our food derives from coral reefs, and this increases to 50% in Ghana, Sierra Leone, Bangladesh, and the Maldives, with significant contributions in the Philippines and Indonesia.

During the marine heat wave of 2016, the northern third of the Great Barrier Reef and much of the middle third were extensively bleached; 29% of the 3863 reefs had a catastrophic die-off of fast-growing staghorn and tabular corals [26]. Acidification was an additional effect, and any potential adaptation and acclimatization by coral reef organisms to thermal stress may be offset or overridden by CO_2 effects. Recent observations show that there is ecological memory for coral bleaching since there were more temperature-resistant corals found on the Great Barrier Reef following the 2017 versus 2016 warming bleaching episode consistent with selection of a more temperature resistant species [27].

Threat to Biodiversity

Biodiversity is threatened by global warming, and habitat loss and degradation threaten ~25% of mammalian species and ~ 13% of bird species with extinction [30]. There is a mismatch between the thermal preferences of many species and the new climate that they are experiencing in their present geographic distribution. Pika, a small mammal in the rabbit family, living in the high altitude of the Colorado Rockies, moves to higher altitudes as the climate warms. Migrating birds need to adapt to changes in flowering plants and insect arrival in northern latitudes. Japanese cherry blossoms flower 2 weeks earlier. Sardines disappear in the warming waters off the Galapagos Islands causing sea lions and blue-footed boobies to change their fish diets. Declines in algae cause marine iguanas to become smaller as they adapt to less food. Giant daisy trees, known as Scalesia, are flattened by El Niño storms, and invasive blackberry bushes predominate in the recovery [31].

Habitat loss affects over half of the Planet Earth due to human habitation and agriculture. Less than 4% of oceans and 15% of land are formally protected. Polar bears (n = 25,000) are a 'threatened' species due to loss of sea ice, because it is more difficult to hunt prey seals when they rest on top of the ice near their breathing holes [30, 31]. The United States Geological Survey predicts two-thirds of the polar bears will be lost by 2050 [30, 31]. At the 1.5 °C warming threshold marked by the IPCC, range loss is 6% for insects, 8% for plants, and 4% for vertebrates by 2100 [31]. The risks at 2 °C double, and at 4 °C excess warming, risks are ~10 times larger. The Great Extinction of 252 million years ago, between the Permian and Triassic ages, obliterated over 96% of all ocean species and 70% of land species as the temperature warmed 7–14 °F due to extreme volcanic activity and increased CO_2 [30, 31]. The United Nations in 2019 predicts that anthropogenic activities disruptive to nature will result in extinction of one million plant and animal species in the short term.

Consequences for Human Health

The World Health Organization estimates that the warming and precipitation trends due to anthropogenic climate change will claim 250,000 lives annually during the time period 2030–2050.

Heat waves

The first consequence of global warming will be increased heat stress, particularly in urban centers that already serve as heat islands [32, 33]. Increased temperature in urban heat islands will occur not only in the daytime but also at night preventing any nocturnal relief. They expose more than half of the global population to stagnant air, and impervious surfaces stressing the importance of the built environment to consider dissipating heat [34]. Heat waves will be accompanied by increased mortality due to cardiorespiratory diseases, diabetes, accidents, homicides, and suicides [35]. Mortality also goes up for heat stroke with its attendant dehydration. Heat stroke is defined clinically as core body temperature > 40.6 °C accompanied by hot, dry skin and central nervous system abnormalities. It is distinguishable from hyperthermia, a medical emergency due to failed thermoregulation by the body such that core body temperatures exceed 41–42 °C. Extreme heat events vary by region and adaptation. For example, a temperature of 102 °F would create a negative health outcome in Cleveland, whereas the same temperature would have little additional effect on people in Phoenix because of adaptation. The New York Times published a heat interactive on August 30, 2018, where one can find modeled heat projections following the carbon pollution commitments of the Paris Climate Agreement. For example, predictions indicate that in 2100, New York will experience 29 (range 16–40) more days of heat >90 °F compared to 8 days in 1960 [34]. Jakarta, Indonesia, would have temperatures >90 °F all year round compared to 5 months in 1960. Humidity and temperature thresholds causing dangerous climatic conditions occur about 20 days a year to 30% of the world's population [35]. By 2100, this scenario will increase to ~48% under Representative Common Pathway (RCP) 2.6 compared to ~74% under growing emissions, such as RCP 8.5. Under RCP 8.5, exposure to wet bulb temperatures above 35 °C – the theoretical limit for human tolerance – could exceed a million person-days per year by 2080 [36]. The wet bulb temperature is measured with a sling psychrometer incorporating the relative humidity; global warming will increase evaporation, increasing humidity impeding the ability of humans to cool themselves by sweat evaporation. Also, RCP 8.5 compared to RCP 2.6 for the year 2100, 85% of the land surface area will be three standard deviations beyond the mean temperature compared to ~20%, respectively. The IPCC Representative Common Pathways reflect the intensity of the heat on the surface of the Earth with RCP 2.6 watts/m^2 for a temperature of ~2 °C and RCP 8.5 watts/m^2 reflecting business as usual at ~4.8 °C [1].

The most famous heat wave occurred in Europe in August 2003, resulting in 32,000 excess deaths [37]. France experienced a loss of 15,000 individuals with 2000 heat-related deaths in 1 day [38, 39]. Hospitals, retirement facilities, and nursing homes without air conditioning were especially vulnerable as a positive association has been noted between heat waves and mortality in the elderly, especially elderly women in social isolation [40]. Heat effects may be lingering as in Lyon, France, where the 1-month and 2-year mortalities among 83 patients admitted for heat stroke August 1–20, were 58% and 71%, respectively. In the Assessment and Prevention of Acute Health Effects of Weather Conditions in Europe project, a 1 °C increase in maximum apparent temperature above the intersection of heat and mortality threshold, increased respiratory admissions by 4.5% (95%CI 1.9–7.3) in Mediterranean and North-Continental cities [41]. In the EuroHEAT project, heat wave-related mortality ranged from +7.6% in Munich to +33.6% in Milan from 1990 to 2004 [42]. The increase was up to three times greater during episodes of long duration and intensity. The highest effect was observed for respiratory diseases and among women aged 75–84 years in cities where the heat wave episode was characterized by unusual meteorological conditions.

Air Pollution

Higher surface temperatures, especially in urban areas, promote increased ground-level ozone with a synergistic effect on mortality. Biostatistical regression of temperature and ozone on mortality in nine French cities exhibited a significant effect of 1% per 10 μg/m^3 in ozone level [43]. US epidemiological studies show that a 10 °C increase in temperature on the same summer day increased cardiovascular mortality by 1.17%, and there was an 8.3% mortality difference comparing the highest level of ozone to the lowest among the 95 cities in the National Morbidity and Mortality Study [44]. Schwartz and colleagues found an association between elevated temperatures and short-term increases in cardiovascular-related admissions for 12 US cities [45, 46]. Increased cardiovascular and chronic pulmonary disease deaths have been associated with particulate matter size >10 microns (PM_{10}) in Wuhan, China, a city located in a deep valley susceptible to trapping air pollutants, where a dose response has been observed with the highest mortality on the days of extremely high temperature exceeding 33.1 °C [47]. Behavioral conditions (symptomatic mental disorders; dementia; mood (affective) disorders; neurotic, stress-related, and somatoform disorders; disorders of psychological development; and senility) have also been associated with higher temperatures in urban areas [48]. A recent study of 40 US cities projected extreme heat events to increase fivefold by mid-century resulting in 32,934 excess deaths, and eightfold by 2100, resulting in 150,322 excess deaths with business as usual (RCP 8.5, [49]). A recent study of Medicare hospital data from 1985 to 2006 for 135 cities evaluated mortality for congestive heart failure, myocardial infarction, COPD, and diabetes [50]. A Cox proportional hazard model for each cohort within each city was correlated to summer temperature variation. Mortality hazard ratios (comparison of heat-related day to non-heat-related day) ranged from 1.028 to 1.040 per 1 °C increase with higher associations for those >74 years of age. Hazard ratios were lower in cities with a higher percentage of land with green surface. Based on an average of 270,000 deaths per year across all four cohorts, a 5% increase in mortality would correspond to ~14,000 additional deaths per year due to an increase in temperature variability in the United States.

It has been postulated that allergic diseases, including hay fever and asthma, will increase in urban areas because global warming will increase pollen [51, 52]. Increases in CO_2 from 350 to 700 ppm in laboratory conditions can increase ragweed mass and pollen output from 40 to 60% [53]. The major ragweed allergen, Amb a 1, was also noted to increase in laboratory experiments [54]. The U.S. Department of Agriculture has performed field experiments with ragweed plots in Baltimore demonstrating a combined urban island heat and CO_2 effect on pollen release with less effect in suburban or rural plots. More than 40 million Americans complain of hay fever and 16 million have asthma defined by the Centers for Disease Control and Prevention, and the trend for asthma has been increasing over the past two decades [55].

Vector-Borne Diseases

Global warming may alter the distribution of vector-borne diseases with malaria and dengue fever expanding their ranges by moving north from tropical to mid-latitude regions, including the United States [56]. Malaria continues to plague African children with 445,000 deaths (90% are children) and approximately 216 million cases per year in 2017. The epidemic potential of malarial transmission has been projected to increase as a result of climate change [57]. The more compelling data comes from records of illnesses kept in health dispensaries on tea plantations stemming from the British colonial era in Kenya [58]. The cases of malaria were projected for the tea highlands with temperature and rainfall over three decades showing a nonlinear correlation with actual cases exceeding those predicted, suggesting an already existing effect from climate change [59].

The WHO chronicles a 30-fold increase in Dengue fever infections over recent decades to approximately 390 million in 2010 [60]. It is transmitted primarily by *Aedes (Stegomyia) aegypti* and secondarily by *Aedes albopictus*, and is characterized by high fever, headache, skin rash, and muscle and joint pains with the name break-bone fever. A more severe form, dengue hemorrhagic fever, which occurs in about 5% of cases, is characterized by shock with increased vascular permeability, internal bleeding, disseminated intravascular coagulation, and circulatory failure. Dengue is caused by an RNA flavivirus, and there are four distinct serotypes for which a multi-valent vaccine is being developed. These efforts are being funded by the Gates Foundation. The entire *Stegomyia* genome has been sequenced with 14,519 protein coding sequences arising from 1.38 billion base pairs; knowledge of the genetic background could enable scientists to understand the immune response to specific epitopes for efficient vaccine design. Computer modeling predicts 5–6 billion people at risk of dengue transmission by 2085, but if CO_2 were controlled at RCP 2.6, then only 3.5 billion (35% of the world's population) would be at risk [60]. Dengue fever time-series studies correlate outbreaks with temperature, rainfall, and humidity. *Aedes(Stegomyia) aegypti* and *albopictus* mosquitoes bite during the daytime making bed nets a less useful preventative compared to household spraying. Oviposition, or the number of eggs laid per female, increases dramatically with temperature, doubling with every 5 °C increase. Oviposition also increases when humidity climbs above 60%. The eggs need standing water to hatch, and increased rainfall will assist in the progression of the mosquito life cycle. Increasing temperature shortens the incubation time for the egg inside the mosquito and can increase mosquito abundance. At higher temperatures, there is a reduced size, weight, and wing span of the mosquito, which requires more frequent biting to complete one gonotrophic cycle. Higher temperatures require unfed females to feed sooner for the sake of their own survival than do lower temperatures. With increased mosquito abundance and more frequent feedings, the risk/incidence of dengue transmission is expected to increase as well.

Implications for Social Stability

Global warming leads to climate change, with potential effects on hurricane, cyclone, and storm intensity and frequency; drought and associated consequences related to food production and famine, population migrations, and potential war; precipitation increases with attendant flooding; and insurance companies and governments as adverse financial impacts diminish their abilities to respond to disasters [61]. The Bulletin of the American Meterological Society listed 15 extreme weather events in 2017 linked to climate change [62]. They concluded that extreme weather events are 2–3 times more likely to be related to climate change than without global warming, and that society is increasingly out of sync with the changing climate.

Drought, Forest Fires, Food Insecurity, and Migration

Below-average precipitation anomalies across the southern tier of the United States are indicative of ongoing major drought conditions [1–5]. It has also been exceptionally dry across the western United States, much of eastern and southern South America, particularly eastern Brazil, much of central Asia, including nearly all of Mongolia, and much of Australia. These hot, dry conditions exacerbate intensity and frequency of forest fires. Australian blazes have occurred after record heat waves and hot, dry winds in southern Victoria state. During the 2009 heat wave, the Black Saturday bush fire scorched 1.1 million acres, killed 173 people, and destroyed

>2000 houses [63, 64]. Smoke from the Australian bushfires increased overall mortality 5%, and increased hospital admissions for respiratory illnesses by 3–5% [63, 64]. In the western United States, since 1980, wildfire season has increased by >2 months, and the number of large wildfires has doubled. Temperatures have risen, snow melt is earlier, and the forests are drier for longer periods of time. The average duration of fires has increased fivefold. Epidemiological studies of fire smoke exposure show an increase of 10 μg/m^3 in PM_{10} from wildfires results in an approximately 1% increase in non-accidental mortality [65]. Time series analyses found California wildfires in 2015 were significantly associated with Emergency Department visits for ischemic heart disease, dysrhythmia, heart failure, pulmonary embolism, stroke, and respiratory conditions, especially in those >65 years of age [65]. In 2018, the Camp Fire wildfire obliterated the community of Paradise, California, and 89 lives were lost. This fire began in extremely dry forest, and 50 mph winds spread it at the rate of one football field per second. It covered 153,000 acres and destroyed >18,800 buildings. Former Governor Jerry Brown, in attributing this disaster to climate change, said this was the new abnormal.

Lack of rainfall over several seasons is the most immediate and most visible cause of the humanitarian crises in the Horn of Africa and sub-Saharan Sahel. Climate change is only one of several factors that have led to the crisis. Other factors include a very large population, especially children and youths that depend on rain-fed agriculture and pastoralism for their livelihood and sustenance. Environmental (soil, water) degradation and rapid population growth have compounded the problem. Much of sub-Saharan Africa has neglected agricultural development, and a recent phenomenon has been the purchase of large tracts of land to produce export food commodities. As a result, rural communities across Africa are trapped in worse poverty, vulnerability, and increasing dependence on external humanitarian assistance. Population growth and the desire for more wealth by the middle class in developing countries, has created the need for more energy sources, supplied largely from the burning of fossil fuels. Urban expansion has resulted in the net loss of agricultural land. Migration to urban centers from rural areas by rural denizens seeking a better life has increased the stress on food production. Some of the most fertile and productive farm lands are near cities. Agricultural lands have reduced production due to overdrawn and unreplenished aquifers. The combined effects of climate change, population increase, and expectations of a higher standard of living that lead to land and water scarcity for food production will affect the quantity of food and quality of the diet that can portend adverse effects on nutrition. Although increased CO_2 is expected to enhance crop growth, more likely there will be numerous other factors, including soil fertility and pests, that flourish in warmer climate, that will mitigate or eliminate any positive effects. Higher temperatures reduce wheat, rice, and maize 2.5 to 10% for each 1 °C increase in ambient temperature [66]. Wheat, corn, and cotton yields have statistical declines between 63% and 70% for high CO_2 emission scenarios, for example, RCP 8.5. Extreme high temperatures during the reproductive stage will affect pollen viability, fertilization, and grain or fruit formation. Free-air CO_2 enrichment experiments for 18 genetically diverse rice lines show declines in protein, iron, and zinc, with consistent declines in vitamins

B1, B2, B5, and B9 correlating with the fraction of nitrogen within the vitamin observed [67]. Insects may also reduce crop yields by 10–25% in future global warming scenarios [68].

Hurricanes and Extreme Weather Events

Hurricane Katrina highlighted the 2005 hurricane season that was the costliest in the United States at $145 billion in property damage. Katrina, a category 3 hurricane, was the fifth worse hurricane in the history of the United States, causing 1836 deaths from the hurricane and its attendant floods. Over 80% of New Orleans flooded after the failure of its levee system. Obradovich et al. published that increasing monthly temperatures to >30 °C increased mental health difficulties by 0.5% points; 1 °C warming over a multi-5-year period was associated with a 2% point increase; and exposure to Hurricane Katrina was associated with a 4% point increase in this metric [69]. In 2011, Americans experienced 14 record-breaking weather and climate disasters that each caused $1 billion or more in damages and contributed to a total cost of approximately $53 billion [70]. In 2012, Hurricane Sandy struck New York and New Jersey causing $65 billion in damages and killing over 65 people. Its >13-foot storm surge flooded subways, tunnels, basements and knocked out electrical power on Lower Manhattan [70]. At New York University's Langone Medical Center, patients were evacuated during the middle of the night as the surge shorted electrical power; the flood also inundated Bellevue Hospital's basement stopping the elevators and fuel pumps [70]. As the fuel pumps stopped, a brigade was formed to bring diesel fuel to emergency generators on the 13th floor until the hospital's patients were finally evacuated by the National Guard. Thousands of experimental mice at NYU Langone Health perished, and freezers lost power endangering many experiments until samples could be transferred to off-site locations. The hospitals were closed for 6 months requiring alternative plans to provide medical care and prescriptions to outpatients, train residents and fellows, and continue academic programs. One factor contributing to the storm's strength was abnormally warm sea surface temperatures offshore the East Coast of the United States – more than 3 °C (5 °F) above normal, to which global warming had contributed 0.6 °C (1 °F, 70). As the temperature of the atmosphere increases, the capacity to hold water increases, leading to stronger storms and higher rainfall amounts. Tropical cyclones get their energy from the warm surface layer of the ocean (which is getting warmer and deeper under climate change) and increasing water vapor in the atmosphere. A given cyclone will be more powerful in the presence of a warmer ocean and higher atmospheric water content than it would be otherwise. And the higher the local sea level is, the worse the storm. In 2013, Typhoon Haiyan hit the Philippines with winds estimated at 180 miles per hour, and killed 6300 people, injured 27,000, and destroyed or damaged 1.2 million homes. About 712 climate-related extreme events were responsible for $326 billion of losses in 2017, almost triple the losses of 2016 [5]. Importantly, 99% of the losses in low-income counties were uninsured. In 2017, three record-breaking hurricanes (Harvey, Irma, and Maria) cost an estimated >$300 billion in economic losses to the United States. Hurricane Harvey had near record-doubling rainfall accumulations leading to massive flooding and dam safety challenges in Houston resulting in $125 billion in losses. The hurricane stalled over the Houston area inundating it in rain; climate change increases the intensity of storms from warmer surface ocean waters, increases the amount of moisture in the clouds, and slows the speed of hurricanes increasing the likelihood of storms stalling upon landfall. Irma and Maria caused an additional $140 billion in damages and were category 4 hurricanes packing winds of 185 mph. Hurricane Maria destroyed much of the electrical infrastructure in Puerto Rico causing significant loss of life [71]. The Milken Institute/George Washington School of Public Health and the University of Puerto Rico Graduate School of Public Health estimated the total excess mortality using the migration displacement scenario to be 2975 (95% CI 2658-3290) due to Hurricane Maria for the total study period from September 2017 through February 2018. The risk of death was 45% higher and persistent until the end of the study period for populations living in low socioeconomic development municipalities, and older males (65+) experienced continuous elevated risk of death through February.

Hurricanes, rising sea levels, and increased precipitation are examples of compounding extreme events under climate change [72, 73]. Models predict up to 2.5 million displaced persons from South Florida northward to Orlando, Florida by 2100 due to rising sea levels. Climate change may result in increasing numbers of migrants as small island states and coastal cities experience increased flooding from rising sea levels and increased rainfall from the interior. Droughts will stimulate farmers' migration away from their failed crops, and the potential for water and resource conflict will increase.

Insurability for Extreme Events

Extreme weather events adversely affect the insurance and re-insurance industries, including Swiss Re, AIG, and others. Swiss Re estimates 3.4 billion people, primarily in the developing world, are at risk from storms, droughts, and floods creating a risk pool for innovative insurance solutions. Insured losses have jumped from an annual $5 billion 40 years ago to an annual $134 billion in 2017 (highest year ever). Over 70% of studies concluded that climate change has increased the risk of a given extreme event, such as heat, drought, rainfall, wildfires, and storms. Companies, such as Swiss Re, are offering commercial insurance solutions as pre-disaster planning for developing countries to off-set public budgets, but the countries must adopt climate-mitigation policies. At the World Economic Forum, it was estimated that moving to a low-carbon energy infrastructure and restricting warming to below 2 °C would require global investment in clean energy of roughly $500 billion per year by 2020 and ~ $1 trillion thereafter. However, public and private investment in clean energy in 2017 was only $333.5 billion, far below needed levels. Private sector investors are critical to global efforts to stimulate a low-carbon economy, adapt to the unavoidable impacts of climate change, and close the climate investment gap. They require risk-adjusted long-term certainty from governments and international institutions about the direction of clean energy and climate policies and financing. Capital is not flowing to low-carbon investments at the scale required because of a lack of investor confidence in their climate and clean energy policy framework.

United States Military

The United States' military is assessing risks for future conflict around the world relating to climate change. Recent war games and intelligence studies conclude that over the next 20 to 30 years, vulnerable regions, particularly sub-Saharan Africa, the Middle East, and South and Southeast Asia, will face the prospect of food shortages, water crises, and catastrophic flooding driven by climate change that could demand an American humanitarian relief or military response. As an example, Bangladesh will lose about 20% of its land mass, creating a major refugee population since it is already densely populated. There will be a spillover migration, or an exodus of people walking toward India. This will be one potential site for armed conflict with different religions, damage to infrastructure from flooding, and the spread of contagious diseases. The US military has seen 'catastrophic' damage to infrastructure such as Tyndall Air Force Base in Florida from 150 mph winds of Hurricane Michael in 2018 ($5 billion in losses), the potential loss of bases, such as Diego Garcia in the Indian Ocean, and new Arctic sea lanes to defend with the melting of the Arctic ice cap. They have been particularly innovative in creating fuel cells, solar panels for Afghan outposts, and alternative fuels for aircraft and vehicles since supply lanes are vulnerable to attack.

Efforts at Mitigation and Adaptation

Policy: Global

The United Nations has been the central focus on developing international consensus for climate change science and mitigation. Stockholm, Sweden, was the host for the first United Nations Conference on the Human Environment in 1972, and led to the establishment of the United Nations Environment Program (UNEP). The purpose of the conference was to unite the countries of the world against a common enemy, which was environmental degradation. Following this, the UN set up a commission on environment and development that issued a report using the term "sustainable development" as the way to ensure that economic development would not endanger the ability of future generations to enjoy the fruits of the earth. The twentieth anniversary of this conference was held in Rio de Janeiro in 1992 and called the "Earth Summit," which was attended by leaders of 105 nations demonstrating their commitment to sustainable development. The framework convention on climate change encouraged adoption of national policies that mitigate climate change by limiting anthropogenic emissions of greenhouse gases and protecting and enhancing their greenhouse gas sinks and reservoirs.

Since the 1992 agreement set no mandatory limits on greenhouse gas emissions for individual countries and contained no enforcement mechanisms, it was considered nonbinding. It did establish national greenhouse gas inventories of emissions and removals, and set up the Conferences of the Parties (COP). In 1997, the Kyoto Protocol established legally binding obligations for developed countries to reduce their greenhouse gas emissions. Most industrialized countries and some central European economies in transition agreed to legally binding reductions in greenhouse gas emissions of an average of 6 to 8% below 1990 levels between the years 2008 and 2012. The United States would be required to reduce its total emissions an average of 7% below 1990 levels. Despite the negotiations on behalf of the US government by Vice President Al Gore and the President's signature, the US Senate refused to consider ratification because developing countries, such as India, China, and Brazil were not bound to reduce their greenhouse gas emissions. The Byrd-Hagel Senate Resolution, agreed to by 95 senators, mandated that developing countries had to be included before the United States would ratify the treaty. In 2001, the Bush Administration rejected the Kyoto Treaty, and the United States was reduced to observer status. The Clean Development Mechanism (CDM) allowed industrialized countries to invest in renewable energy, energy efficiency, and fuel switching in developing countries to meet their CO_2 limits and invest more cheaply to achieve the target reduction of 1.5 billion tons of CO_2 equivalents. A Program of Activities was developed to bundle CDM efforts, such as distributing compact fluorescent lamps, efficient cook stoves, building refurbishment, or solar water heaters. The COP also agreed that credit would be granted for broad activities that absorb CO_2 from the atmosphere or store it, including forest and cropland management and re-vegetation.

Ministers and officials from 192 countries met at COP 15 in Copenhagen, Denmark, in 2009 to establish an ambitious global climate agreement for the period after Kyoto to begin in 2012. President Obama decided to put off the difficult task of reaching a climate change agreement and instead pursued a less specific political accord to limit the growth in CO_2 emissions with a temperature increase limited to 2 °C. The accord was notable in that it referred to a collective commitment by developed countries for $30 billion from 2010 to 2012 for forestry and investments through international institutions. In Cancún, Mexico, COP 16 confirmed the goal of limiting global warming to no more than 2 °C above pre-industrial levels, and agreed to set up a new Green Climate Fund to transfer money to developing countries. The agreement also noted that addressing climate change required a paradigm shift toward building a low-carbon society. The agreement, including the "Green Climate Fund," was for $100 billion a year by 2020 to assist poorer countries in financing emission reductions and adaptation. There was no specific agreement on how this fund would be raised, and the decisions of the legal form and level of emission reductions were once again deferred. They did develop a time frame for implementation of efforts to reduce emissions from deforestation and forest degradation (REDD) ; robust measurement, reporting, and verification to increase confidence in national climate policies; and support for the creation of well-functioning markets in developing countries for energy efficiency and renewable energy to accelerate the effective deployment and diffusion of these technologies at scale.

Policy : The United States

The United States was stymied to develop a national carbon policy with the Senate only mustering 44 votes for the first McCain-Lieberman Climate Bill that would set up a modified cap and trade program in 2003. The U.S. House of Representatives passed the first climate change bill in 2009 named after Representatives Waxman and Markey. It was based on cap-and-trade, with a goal of reducing greenhouse gas emissions 17% below 2005 levels by 2020, and 83% by 2050. This bill prohibited the EPA from regulating CO_2 under the Clean Air Act. In 2007 the U.S. Supreme Court decided that EPA had statutory authority under the Clean Air Act to regulate CO_2 (Massachusetts et al. vs EPA). The EPA also announced the Greenhouse Gas Reporting Rule, which affected entities with more than 25,000 tons/year (about 70% of all US emitters). The EPA also found that CO_2 endangered public health and welfare, allowing it to regulate CO_2 under the National Ambient Air Quality Standards (Endangerment Finding).

With lack of progress in the legislative branch (US Senate), President Obama developed his Climate Action Plan in 2013: cut carbon pollution in America, prepare the United States for the impacts of climate change, and lead international efforts to address climate change. The Environmental Protection Agency provided the lead on cutting carbon pollution by using the Clean Air Act. The Clean Power Plan carried out EPA's obligations under section 111(d) of the Clean Air Act and contained carbon dioxide emissions with performance rates for affected power plants that reflected the "best system of emission reduction" (BSER) . The Clean Power Plan included targets for reducing carbon pollution for each state based on each state's unique mix of power plants in 2012. The Clean Power Plan could result in a 32% reduction in carbon pollution from fossil fuel power plants by 2030. The projected annual benefits in 2030 were due primarily to the co-benefits of reducing power plant ozone and particle pollution: 1500–3600 premature deaths, 90,000 asthma attacks primarily in children, 180–1700 heart attacks, 1700 hospital admissions, and 300,000 missed school and work days for $34–54 billion in benefits including $20 billion in global climate benefits against $8.4 billion in costs. These benefits are what would be expected compared to not implementing the Clean Power Plan. The other EPA efforts were the methane rule to reduce methane leaks in oil and gas drilling and pipelines by 40%, and increasing fleet-wide corporate average fuel economy (CAFÉ) standards to 54.5 miles per gallon stepwise by 2025.

President Obama set the new target of reducing carbon pollution by 26–28% below 2005 levels by 2025. His team also met with Chinese leaders for a commitment to peak CO_2 emissions by 2030, while striving to peak earlier and boost its share of non-fossil fuel energy to 20%. After years of hard work, and thanks to President Obama's leadership, 195 countries came together in Paris and adopted the most ambitious agreement to combat climate change in history. Signed on Earth Day April 22, 2016, in New York, the goal was to keep global warming below 2 °C, and pursue efforts to limit temperature increase to 1.5 °C. The agreement set the world on a course to cut carbon pollution and other greenhouse gases. It ensured we can leave the planet a better place for our children. It was a clear sign that as citizens of the world, we fully recognized the science of climate change, we were already feeling its impacts, and we were ready to take ambitious, unified action. All countries set progressive climate targets for themselves – an approach for a long-term, durable system to ratchet down emissions over time. All countries were to communicate their climate targets every 5 years, starting in 2020. The agreement called for strong transparency and reporting of national emissions to build trust. It required countries to report on greenhouse gas inventories and on their mitigation progress. The national commitments fell short of even the 2 °C target, and are on a pathway to >3 °C by 2100; this requires negotiation to increase national commitments and leadership, particularly by the United States, China, and Europe. However, the election of Donald Trump to the US Presidency in 2016, in a faux populist revolt among the rural poor and working class, reversed these gains as his administration has rolled back all of the EPA regulations, including the Clean Power Plan, methane rule, corporate fuel economy standards, and announced US withdrawal from the Paris Climate Agreement by the end of his term. Furthermore, he attacked the scientific basis of public health standards, the social cost of carbon, and led climate deniers against the human cause of climate change rebutting the fact that 98% of climate scientists agree that climate change is happening and that humans are the cause [74]. The social cost of carbon was an agreed upon target ($45/ton CO_2) by federal agencies in considering federal projects, and this has also been adopted by several state Public Utilities Commissions in regulating electrical power generation. The social cost of carbon considered its global impact, which was rolled back to $1/ton by the Trump Administration by considering the impact only by and on the United States [75].

Withdrawal from the Paris Climate Agreement and the EPA rollbacks will increase future climate mortality and expenses to meet the requisite carbon mitigation and adaptation. If the future course were to continue along RCP 8.5 to 2100, one estimate of mortality is 106.5 million premature deaths versus 13.5 million for the RCP 2.6 pathway [76]. These estimates were based on mortality from heat waves, air pollution, droughts and food insecurity, loss of coastal reefs and fisheries, and extreme weather, including hurricanes, cyclones, and flooding [76]. Shindell et al. quantified the benefit of reducing carbon dioxide emissions, including the co-benefits of reducing particulate matter, ozone, and nitrogen oxides resulting in decreased air pollution, thus saving an estimated 153 ± 43 million lives worldwide with 40% of this benefit occurring in the next 40 years [77]. Another biostatistical estimate of increased premature mortality in the United States predicted an increase in 80,000 premature deaths per decade primarily from rolling back the Clean Power Plan [78].

Mitigation: Renewables

Mitigation efforts were strengthened by 157 Gigawatts (GW) of renewable energy installed globally in 2017 compared to 70 GW of fossil fuel capacity, reaching 10.3 million jobs in the renewable energy sector, a 5.7% increase over 2016 [5]. Pacala and Socolow [79] envisioned 15 stabilization options to solve the climate problem by 2050 with current technologies projecting mind-numbing options that humankind has not even approached considering; at least 7 of these stabilization wedges to reduce CO_2 would have to be completely accomplished [78]. More realistically, MacDonald et al. described future cost-competitive electricity systems capitalizing on the solar-rich Southwest and wind-rich Great Plains by transporting renewable energy to both coasts via high-voltage direct current transmission lines [80]. Their plan would result in an 80% reduction of CO_2emissions relative to 1990 with levelized costs. Using a low-cost renewable, high-cost natural gas scenario (2006–2008), US power consumers could save an estimated $47 billion annually with a national electrical power system versus the three regionally divided ones (this saves three times the cost of the new transmission lines). This electricity plan requires approximately 523 GW wind (~eightfold increase), 371 GW solar Photovoltaics (~62-fold increase); 100 GW nuclear; 74 GW hydroelectric; 461 GW natural gas for 1529 GW installed capacity compared to 2012 and to be online by 2030. This system would utilize natural gas rather than battery storage to fill gaps in renewable energy for electricity. The land required would be 6570 km^2 (460 km^2 for wind and 6110 km^2 for solar PV) constituting 0.08% of the United States; the amount of water would also be reduced by 65% due to fewer steam turbines. This plan would hasten the revolution to greener electricity with electric vehicles, heat pumps, electric stoves, etc., but would require investments into new renewable power plants and transmission lines plus surmounting hurdles in the legal, regulatory, commercial, and political worlds. Estimates are that to have at least a 50% chance of keeping warming below 2 °C throughout the twenty-first century, the cumulative carbon emissions between 2011 and 2050 need to be limited to around 1100 gigatons of carbon dioxide (Gt CO_2, [81]). Greenhouse gas emissions contained in current estimates of global fossil fuel reserves are around three times higher. Globally, a third of oil reserves, half of natural gas reserves, and over 80% of current coal reserves should remain unused from 2010 to 2050 in order to meet the 2 °C limit. Development of resources in the Arctic and any increase in unconventional oil production are incommensurate with efforts to limit average global warming to 2 °C [82].

Mitigation: Cap-and-Trade and Carbon Tax

Mitigation efforts for renewables, such as wind, solar, tidal, or fourth-generation nuclear are strengthened by policy approaches, such as cap-and-trade or a carbon fee or tax. The cap-and-trade approach was used for reducing SO_2 emissions successfully and abated acid rain. Cap-and-trade is best described as a system where emissions are capped producing a decline over time with affected industries reducing emissions steadily and trading excess emissions to those entities substantially below the cap until all polluting industries achieve a lower target. This has been unsuccessfully attempted legislatively, but has worked among some states as in the Regional Greenhouse Gas Initiative (RGGI) . RGGI includes nine New England and Mid-Atlantic states in a market-based carbon auction with a cap-and-trade program, and uses the proceeds to support renewable energy and energy efficiency. New Jersey is re-entering the program. The carbon fee or tax is supported by climate scientist and pioneer James Hansen with the carbon fee placed at the source of fossil fuel industries and the proceeds distributed on a per capita basis to individuals [83]. Economic forecasts suggest that Americans would get back more than what they would pay in higher energy prices. In a carbon fee, for example, \$50/ton rising at 2% per year, CO_2 emissions fall 39–46% below 2005 levels by 2025. None of the proposed taxes (including \$73/ton rising 1.5% per year) achieve 80% CO_2 below 2005 levels by 2050. Transportation emissions are stubbornly resistant to carbon taxes, for example, a \$50 carbon tax would reduce emissions from the transportation sector by only 2%. The macroeconomic effect is small, $<1\%$ of GDP in either direction. The tax proceeds could be used for per capita dividends (especially low and middle-income, vulnerable communities), reduce payroll or corporate taxes, reduce national debt, fund green energy, and/or infrastructure. In the long term, the benefits for humanity of a societal shift away from fossil fuels and toward cleaner sources of energy will far outweigh the costs. But the transition could have severe implications for some sectors, regions, and countries, for example, the 2018 gasoline tax in France that resulted in riots by the working class. Poorly managed, it could result in loss of income, opportunity, and future prospects for some workers and communities. Canada plans a carbon tax in 2019, where Prime Minister Justin Trudeau's government introduced a national "fee and dividend" scheme that will place a levy on the carbon emissions of fuels and other products, but then refund the money to individuals and companies through tax rebates. Most residents and businesses in Ontario, Saskatchewan, Manitoba, and New Brunswick,-the four provinces subject to the federal tax (other provinces have introduced their own versions) – will receive refunds that will be greater than the carbon tax paid by the average family. According to the Canadian government's estimates, some 70% of people will get back more in dividends than they pay in new tax. The most important reason for a carbon tax is to place a price on carbon pollution, and provide more opportunity for renewable energy to obtain investment and develop clean electricity.

Efforts to Combat Climate Change by Cities and States

As the US federal government announced its exit from the Paris Climate Agreement, grass roots organizers formed "We're Still In" to announce that they will continue with the Paris commitments. They include >3500 leaders in >200 cities and counties, 24 states in the U.S. Climate Alliance, > 2600 business leaders and lenders representing half of the US population and half of the nation's gross domestic product. The US government will re-join the Paris Climate Agreement on January 20, 2021, with a Democratic party in control of the government. It is critical to have mayors and leaders in cities to be active. Although cities represent only 3% of the land area of the Earth, they contain more than half of the population, consume 70% of the world's energy, and emit 75% of the CO_2 emissions. Globally, as many as 6225 companies headquartered in 120 countries have pledged to contribute to the Paris goals representing $36.5 trillion in revenue – more than the combined gross domestic products of the United States and China [84]. They expect as many as 65 million jobs in the low-carbon economy by 2030.

California is on the front line in the climate crisis and was the first state in the United States to adopt a cap-and-trade CO_2 regulation. Its target was to reduce CO_2 emissions by 15% by 2020 compared to a 1990 baseline implementing Assembly Bill (AB) 32, California's historic climate change law. The California Air Resources Board implemented regulations covering 360 businesses representing 600 facilities mandating caps or credits, and secondly, covered distribution of transportation fuel and natural gas. Under the program, companies were not given a specific limit on their greenhouse gas emissions, but supplied a sufficient number of allowances to cover their annual emissions. As a statewide cap declined annually, the total number of allowances issued also declined. The allowances given to electric utilities were to be sold at auction, with the proceeds distributed to ratepayers. Senate Bill (SB) 100 committed California to a 100% clean electricity grid by 2045 from 20% currently. In the future, California plans to import clean energy, for example, wind from Wyoming. Electricity was 16% of California's greenhouse gas emissions, and Executive order B-55-18 committed California to total economy-wide carbon neutrality by 2045. SB 32 mandated a 40% reduction of greenhouse gas emissions below 1990 levels by 2030 through continued use of cap-and-trade to achieve these reductions. They plan to have 50% of electricity to come from renewables by 2030, and to achieve this, solar panels are mandated on all new home construction in 2020 (building code). SB 350 and California's Public Utility Commission set a goal of five million zero-emission vehicles by 2030 and 250,000 electric charging stations by 2025. California has an exemption under the Clean Air Act to control its unique emissions, and is suing the federal government over fuel economy standards that the federal government is rolling back; 12 states follow the California regulations, potentially creating two markets for gasoline cars and trucks.

In June 2019, the New York State Climate Leadership and Community Protection Act passed, setting up the most ambitious climate goals in the United States:

requiring reductions in statewide greenhouse gas emissions to 60% of 1990 levels by 2030 and 15% of 1990 levels by 2050. The electricity generating sector would have to reach 70% renewable energy (from 60% from hydroelectric and nuclear currently) by 2030 and net zero CO_2 pollution by 2040. A 22-member Climate Action Council will develop scoping goals, including procurement of at least 9 GW of offshore wind electricity generation by 2035, 6 GW of solar electricity by 2025, and 3 GW of energy storage capacity by 2030. The law directs efforts to decarbonize industrial entities by switching from fossil fuels to renewable electricity, electrifying residential houses and office buildings for heating, cooking, and cooling, and electrifying transport where New York has ten million cars, trucks, and buses. The law specifically addresses low-income communities for environmental justice to provide funding for renewable energy.

Conclusion

The IPCC has reported that global citizens have only 12 years (by 2030) to decarbonize our economy by half and halt current global warming trends before temperatures reach 1.5 °C above pre-industrial levels [2]. The report contends that warming beyond the 1.5 °C threshold, which is expected around 2035, could expose tens of millions within the global population to life-threatening heat waves, water shortages, and coastal flooding. At 2 °C by 2050, 37% of the world's population would be exposed to extreme heat, and 411 million to extreme water scarcity. We are on a pathway to >3.2 °C by 2100 [2, 85]. Limiting the anthropogenic temperature anomaly to 1.5–2 °C is possible, but it requires transformational change across the board of modernity, especially massive development of renewable energy [86]. Why were 2 °C and especially 1.5 °C chosen? Impacts research indicates that unbridled anthropogenic climate change would be most likely to play out in a disruptive and irreparable way; key to understanding this is the non-linearity and irreversibility of the multiple tipping points ahead [86]. Risks from extreme precipitation events would increase dramatically with 2 °C warming, especially in eastern Asia and eastern North America; sea level would rise about 4 inches more with 2 °C warming than with 1.5 °C, affecting ten million more people; an extra 580,000 to one million square miles of permafrost would thaw at 2 °C compared to 1.5 C; and at 1.5 °C of warming, the Arctic is forecast to be ice-free once per century, but, at 2 °C warming, that would happen once every 10 years [2]. Carbon capture and sequestration by power plants to control CO_2 emissions, geoengineering to increase albedo or direct CO_2 removal from the atmosphere, new generation of modular nuclear power, reforestation, and enhancing carbon sequestration in soil are all emerging technologies. Political leaders, investors, and the public need to understand the importance of timing to act now to avoid the worst outcomes of climate change. A "Green New Deal" has been proposed in the U.S. House of Representatives to achieve decarbonization of the economy by 2030, and increase jobs and training for displaced fossil fuel workers toward the renewable energy economy. Lastly, there is a moral imperative to fight global warming: today's youth will suffer the future consequences of global heating both in cost and health; the poorest states in the United States, and the poorest countries in the Southern Hemisphere will suffer the most extreme weather, heat, and vector-borne disease; particulate air pollution ranks sixth and indoor air pollution ranks eighth in global burden of disease with developing countries' cities the most severely polluted; animals and plants will decline with one million species going extinct; and climate deniers will bear a moral responsibility for the delay in response to the challenge of climate change. As former Senator and Governor Gaylord Nelson, the Founder of Earth Day stated, "The ultimate test of man's conscience may be his willingness to sacrifice something today for future generations whose words of thanks will not be heard." The human health consequences of global warming disproportionately affect communities of color because of inadequate and crowded housing, greater sources of pollution, lack of air conditioning, and zoning restrictions.

References

1. IPCC. In: Core Writing Team, Pachauri RK, Meyer LA, editors. Climate Change 2014: Synthesis Report. Contribution of Working Groups I, II and III to the Fifth Assessment Report of the Intergovernmental Panel on Climate Change. Geneva: IPCC; 2014. 151 pp.
2. IPCC. Summary for policymakers. In: Global warming of 1.5°C. An IPCC Special Report on the impacts of global warming of 1.5°C above pre-industrial levels and related global greenhouse gas emission pathways, in the context of strengthening the global response to the threat of climate change, sustainable development, and efforts to eradicate poverty Masson-Delmotte V, Zhai P, Pörtner HO, Roberts D, Skea J, Shukla PR, Pirani A, Moufouma-Okia W, Péan C, Pidcock R, Connors S, Matthews JBR, Chen Y, Zhou X, Gomis MI, Lonnoy E, Maycock T, Tignor M, Waterfield T, editors. Geneva: World Meteorological Organization; 2018. 32 pp.
3. Fourth National Climate Assessment, U.S. Global Change Research Program, Vol I: Climate Science Special Report and Vol II: Impacts, Risks, and Adaptation in the United States, Washington, D.C. 2018.
4. Watts N, Adger WN, Agnolucci P, et al. Health and climate change: policy responses to protect public health. Lancet. 2015; published online June 23; https://doi.org/10.1016/S0140-6736(15)60854-62.
5. Watts N, Amann M, Arnell N, et al. The 2018 report of the Lancet Countdown on health and climate change: shaping the health of nations for centuries to come. Lancet. 2018; https://doi.org/10.1016/S0140-6736(18)32594-7.
6. Keeling CD. Climate change and carbon dioxide: an introduction. Proc Natl Acad Sci USA. 1997;94:8273–4. https://doi.org/10.1073/pnas.94.16.8273.
7. Hansen J, Sato M, Ruedy R, et al. Dangerous human-made interference with climate: a GISS model E study. Atmos Chem Phys. 2007;7:2287–312.
8. Lenton TM, Held H, Kriegler E, et al. Tipping elements in the Earth's climate system. Proc Natl Acad Sci. 2008;105:1786–93.
9. Luthi D, Le Floch M, Bereiter B, et al. High-resolution carbon dioxide concentration record 650,000-800,000 years before present. Nature. 2008;453:379–82.
10. Hansen J, Nazarenko L, Ruedy R, Sato M, Willis J, Del Genio A, et al. Earth's energy imbalance: confirmation and implications. Science. 2005;308:1431–5.
11. Montzka SA, Dlugokencky EJ, Butler JH. Non-CO_2 greenhouse gases and climate change. Nature. 2011;476:43–50.

12. Swindles GT, Morris PJ, Mullan D, et al. The long-term fate of permafrost peatlands under rapid climate warming. Sci Rep. 2015;5:17951. https://doi.org/10.1038/srep17951.
13. Oppenheimer M, O'Neill BC, Webster M, Agrawala S. The limits of consensus. Science. 2007;317:1505–6.
14. Hansen J, Sato M. Greenhouse gas growth rates. Proc Natl Acad Sci. 2004;101:16109–14.
15. Velders GJM, Andersen SO, Daniel JS, Fahey DW, McFarland M. The importance of the Montreal protocol in protecting climate. Proc Natl Acad Sci U S A. 2007;104:4814–9.
16. Hansen J, Nazarenko L. Soot climate forcing via snow and ice albedos. Proc Natl Acad Sci U S A. 2004;101:423–8.
17. Ramanathan V, Feng Y. On avoiding dangerous anthropogenic interference with the climate system: formidable challenges ahead. Proc Natl Acad Sci U S A. 2008;105:14245–50.
18. Notz D, Stroeve J. Observed Arctic sea-ice loss directly follows anthropogenic CO_2 emission. Science. 2016:aag2345. https://doi.org/10.1126/science.aag2345.
19. Jorgenson MT, Romanovsky V, Harden J, et al. Resilience and vulnerability of permafrost to climate change. Can J For Res. 2010;40:1219–36.
20. Steig EJ, Schneider DP, Rutherford SD, Mann ME, Comiso JC, Shindell DT. Warming of the Antarctic ice-sheet surface since the 1957 international geophysical year. Nature. 2009;457:459–62.
21. Witze A. Crucial West Antarctic glaciers are retreating unstoppably. Nature. 2014; https://doi.org/10.1038/nature.2014.15202.
22. Joughin I, Smith BE, Medley B. Marine ice sheet collapse potentially under way for the Thwaites Glacier Basin, West Antarctica. Science. 2014;344:735–8. https://doi.org/10.1126/science.1249055.
23. Jenouvrier S, Caswell H, Barbraud C, et al. Demographic models and IPCC climate projections predict the decline of an emperor penguin population. Proc Natl Acad Sci. 2009;106:1844–7. https://doi.org/10.1073/pnas.0806638106.
24. Rom W, Hartling N. Rafting the Alsek. North America's wildest river. Explorers J. 2017;95(3) Fall:34–43.
25. Carpenter KE, Abrar M, Aeby G, et al. One-third of reef-building corals face elevated extinction risk from climate change and local impacts. Science. 2008;321:560–3.
26. Hughes TP, Kerry JT, Baird AH, et al. Global warming transforms coral reef assemblages. Nature. 2018;556:492–6. https://doi.org/10.1038/s41586-018-0041-2.
27. Hughes TP, Kerry JT, Connolly SR, et al. Ecological memory modifies the cumulative impact of recurrent climate extremes. Nat Clim Chang. 2018; https://doi.org/10.1038/s41588-018-0351-2.
28. Anthony KRN, Kline DI, Diaz-Pulido G, Dove S, Hoegh-Guldberg O. Ocean acidification causes bleaching and productivity loss in coral reef builders. Proc Natl Acad Sci. 2008;105:17442–6.
29. Hoegh-Guldberg O, Mumby PJ, Hooten AJ, et al. Coral reefs under rapid climate change and ocean acidification. Science. 2007;318:1737–42.
30. Urban MC. Accelerating extinction risk from climate change. Science. 2015;348:571–3. https://doi.org/10.1126/science.aaa4984.
31. Warren R, Price J, Graham E, Forstenhaeusler N, Van Der Wal J. The projected effect on insects, vertebrates, and plants of limiting global warming to 1.5 °C rather than 2 °C. Science. 2018;360:791–5. https://doi.org/10.1126/science.aar3646.
32. Frumkin H, Hess J, Luber G, Malilay J, McGeehin M. Climate change: the public health response. Am J Public Health. 2008;98:435–45.
33. Rom WN, Pinkerton KE, Martin WJ, Forastiere F. Global warming: a challenge to all American Thoracic Society members. Am J Respir Crit Care Med. 2008;177:1053–7.
34. How much hotter is your hometown than when you were born? New York Times, 30 Aug 2018. https://www.nytimes.com/interactive/.../climate/how-much-hotter-is-your-hometown.htm... (Accessed 17 Dec 2018).

35. Mora C, Dousset B, Caldwell IR, et al. Global risk of deadly heat. Nat Clim Chang. 2017; https://doi.org/10.1038/NCLIMATE3322.
36. Coffel ED, Horton RM, Sherbinin AD. Temperature and humidity projections of a rapid rise in global heat stress exposure during the 21st century. Environ Res Lett. 2018;13:014001. https://doi.org/10.1088/1748-9326/aaa00e.
37. Fouillet A, Rey G, Laurent F, et al. Excess mortality related to the august 2003 heat wave in France. Int Arch Occup Environ Health. 2006;80:16–24.
38. Argaud L, Ferry T, Le Q-H, et al. Short- and long-term outcomes of heatstroke following the 2003 heat wave in Lyon, France. Arch Intern Med. 2007;167:2177–83.
39. Dhainaut JF, Claessens Y-E, Ginsberg C, Riou B. Unprecedented heat-related deaths during the 2003 heat wave in Paris: consequences on emergency departments. Crit Care. 2004;8:1–2.
40. Stafoggia M, Forastiere F, Berti G, et al. Factors associated with heat-related in-hospital mortality: a multicity case-crossover analysis. Epidemiology. 2006;17:S163–4.
41. Michelozzi P, Accetta G, De Sarlo M, et al. High temperature and hospitalizations for cardiovascular and respiratory causes in 12 European cities. Am J Respir Crit Care Med. 2009;179:383–9.
42. D'Ippoliti D, Michelozzi P, Marino C, et al. The impact of heat waves on mortality in 9 European cities: results from the EuroHEAT project. Environ Health. 2010;9:37.
43. Filleul L, Cassadou S, Medina S, et al. The relation between temperature, ozone and mortality in nine French cities during the heat wave of 2003. Environ Health Perspect. 2006;114:1344–7.
44. Ren C, Williams GM, Morawska L, Mengersen K, Tong S. Ozone modifies associations between temperature and cardiovascular mortality: analysis of the NMMAPS data. Occup Environ Med. 2008;65:255–60.
45. Schwartz J, Samet JM, Patz JA. Hospital admissions for heart disease: the effects of temperature and humidity. Epidemiology. 2004;15:755–61.
46. Medina-Ramon M, Zanobetti A, Cavanagh DP, Schwartz J. Extreme temperatures and mortality: assessing effect modification by personal characteristics and specific cause of death in a multi-city case-only analysis. Environ Health Perspect. 2006;114:1331–6.
47. Qian Z, He Q, Lin HM, et al. High temperatures enhanced actue mortality effects of ambient particle pollution in the "oven" city of Wuhan, China. Environ Health Perspect. 2008;116:1172–8.
48. Hansen A, Bi P, Nitschke M, Ryan P, Pisaniello D, Tucker G. The effect of heat waves on mental health in a temperate Australian City. Environ Health Perspect. 2008;116:1369–75.
49. Greene S, Kalkstein LS, Mills DM, Samenow J. An examination of climate change on extreme heat events and climate-change mortality relationships in large U.S. cities. Weather. Climate Soc. 2011;3:281–91.
50. Zanobetti A, O'Neill MS, Gronlund CJ, Schwartz JD. Summer temperature variability and long-term survival among elderly people with chronic disease. Proc Natl Acad Sci. 2012;109:6608–13.
51. Cecchi L, D'Amato G, Ayres JG, et al. Projections of the effects of climate change on allergic asthma: the contribution of aerobiology. Allergy. 2010;65:1073–81.
52. D'Amato G, Cecchi L. Effects of climate change on environmental factors in respiratory allergic diseases. Clin Exp Allergy. 2008;38:1264–74.
53. Wayne P, Forster S, Connelly J, Bazzaz FA, Epstein PR. Production of allergenic pollen by ragweed (*Ambrosia artemisiifolia)* is increased in CO_2 enriched atmospheres. Ann Allergy Asthma Immunol. 2002;88:279–82.
54. Singer BD, Ziska LH, Frenz DA, Gebhard DE, Straka JG. Increasing Amb a 1 content in common ragweed (*Ambrosia artemisiifolia*) pollen as a function of rising atmospheric CO_2 concentration. Funct Plant Biol. 2005;32:667–70.
55. Beggs PJ, Bambrick HJ. Is the global rise of asthma an early impact of anthropogenic climate change? Environ Health Perspect. 2005;113:915–9.
56. Haines A, Patz JA. Health effects of climate change. JAMA. 2004;291:99–103.
57. Patz JA, Olson SH. Malaria risk and temperature: influences from global climate change and local land use practices. Proc Natl Acad Sci. 2006;103:5635–6.

58. Pascual M, Ahumada JA, Chaves LF, Rodo X, Bouma M. Malaria resurgence in the east African highlands: temperature trends revisited. Proc Natl Acad Sci U S A. 2006;103:5829–34.
59. Alonso D, Bouma MJ, Pascual M. Epidemic malaria and warmer temperatures in recent decades in an East African highland. Proc R Soc B. 2010:1–9. https://doi.org/10.1098/rspb.2010.
60. Hales S, de Wet N, Maindonald J, Woodward A. Potential effect of population and climate changes on global distribution of dengue fever: an empirical model. Lancet. 2002;360:830–4.
61. Costello A, Abbas M, Allen A, et al. Managing the health effects of climate change. Lancet. 2009;373:1693–733.
62. Herring SC, et al. Explaining Extreme Events of 2017 from a Climate Perspective. 7th ed. Boston: Bulletin of the American Meterological Society; 2018.
63. Morgan G, Sheppeard V, Khalaj B, et al. Effects of bushfire smoke on daily mortality and hospital admissions in Sydney. Aust Epidemiol. 2010;21(1):47–55.
64. Johnston F, Hanigan I, Henderson S, et al. Extreme air pollution events from bushfires and dust storms and their association with mortality in Sydney, Australia 1994–2007. Environ Res. 2011;111(6):811–6.
65. Wettstein ZS, Hoshiko S, Harrison RJ, et al. Cardiovascular and cerebrovascular emergency department visits associated with wildfire smoke exposure in California in 2015. J Am Heart Assoc. 2018;7:7e007492.
66. Hatfield JL, Prueger JH. Temperature extremes: effect on plant growth and development. Weather Climate Extremes. 2015;104:4–10.
67. Zhu C, Kobayashi K, Loladze I, et al. Carbon dioxide (CO2) levels this century will alter the protein, micronutrients, and vitamin content of rice grains with potential health consequences for the poorest rice-dependent countries. Sci Adv. 2018;4:eaaq1012.
68. Deutsch CA, Tewksbury JJ, Tigchelaar M, et al. Increase in crop losses to insect pests in a warming climate. Science. 2018;361:916–9. https://doi.org/10.1126/science.aat3466.
69. Obradovich N, Migliorini R, Paulus MP, et al. Empirical evidence of mental health risks posed by climate change. Proc Natl Acad Sci USA. 2018;115:10953–8. https://doi.org/10.1073/pnas.1801528115.
70. Rom WN, Evans L, Uppal A. The sentinel event of climate change. Hurricane Sandy and its consequences for pulmonary and critical care medicine. Am J Respir Crit Care Med. 2013;187: iii–iv https://doi.org/10.1164/rccm.201212-2207OE.
71. Kishore N, Marques D, Mahmud A, et al. Mortality in Puerto Rico after Hurricane Maria. N Engl J Med. 2018;379:162–70. https://doi.org/10.1056/NEJMsa1803972.
72. Lehmann J, Mempel F, Coumou D. Increased occurrence of record-wet and record-dry months reflect changes in mean rainfall. Geophys Res Lett. 2018; https://doi.org/10.1029/2018GL079439.
73. Sarhadi A, Ausin MC, Wiper MC, et al. Multidimensional risk in a non-stationary climate: joint probability of increasing severe warm and dry conditions. Sci Adv. 2018;4:eaau3487.
74. Anderegg WR, Prall JW, Harold J, Schneider SH. Expert credibility in climate change. Proc Natl Acad Sci USA. 2010;107:12107–9. https://doi.org/10.1073/pnas.1003187107.
75. Ricke K, Drouet L, Caldeira K, Tavoni M. Country-level social cost of carbon. Nat Clim Chang. 2018; https://doi.org/10.1038/s41558-018-0282-y.
76. Rom WN. Estimated global mortality from present to 2100 from climate change. Int J Humanit Soc Sci Rev. 2018;4:1–8.
77. Shindell D, Faluveg G, Shindell C. Quantified, localized health benefits of accelerated carbon dioxide emissions reductions. Nat Clim Chang. 2018;8:291–5. https://doi.org/10.1038/s41558-018-0108-y.
78. Cutler D, Dominici F. A breath of bad air: cost of the trump environmental agenda may lead to 80000 extra deaths per decade. JAMA. 2018;319:2261–2. https://doi.org/10.1001/jama.2018.7351.
79. Pacala S, Socolow R. Stabilization wedges: solving the climate problem for the next 50 years with current technologies. Science. 2004;305:968–72.

80. MacDonald AE, Clack CTM, Alexander A, Dunbar A, Wilzak J, Xie Y. Future cost-competitive electricity systems and their impact on US CO_2 emissions. Nat Clim Chang. 2016; https://doi.org/10.1038/NCLIMATE2921.
81. McGlade C, Ekins P. The geographical distribution of fossil fuels unused when limiting global warming to 2°C. Nature. 2015;517:187–90. https://doi.org/10.1038/nature14016.
82. Rooney RC, Bayley SE, Schindler DW. Oil sands mining and reclamation cause massive loss of peatland and stored carbon. Proc Natl Acad Sci U S A. 2012;109:4933–7.
83. Hansen J, Kharecha P, Sato M, et al. Assessing "dangerous climate change": required reduction of carbon emissions to protect young people, future generations and nature. PLoS One. 2013;8(12):e81648.
84. Yale Data-Driven Environmental Solutions Group. Who's acting on climate change? Subnational and non-state global climate action. (Yale Data Driven 2017).
85. Bronselaer B, Winton M, Griffies SM, et al. Change in future climate due to Antarctic melt water. Nature. 2018;564:53–8. https://doi.org/10.1038/s41488-018-0712-z.
86. Schellnhuber HJ, Rahmstorf S, Winkelmann R. Why the right climate target was agreed in Paris. Nat Clim Chang. 2016;6:649–53.

Chapter 2
Climate Variability and Change Data and Information for Global Public Health

Juli M. Trtanj and Tamara G. Houston

In understanding and using climate data for health, the data challenges are not trivial—they are both technical and cultural. One of the great challenges in understanding the health consequences of climate variability and change is the paucity of temporally and spatially compatible data to underpin evidence based on scientifically sound knowledge and action. Robust results require data from many different disciplines, including from health, medical, social, and behavioral, to environmental, oceanographic, and climate sciences. Within each of those disciplines, there is yet greater granularity, variability, and quality of data. And within that are further challenges to access and availability—ranging from privacy concerns and private sector ownership surrounding some health data, making it altogether unavailable, or available but without the granularity needed for robust analysis, to accessing massive climate data sets in a usable way. And though it may sound like an oxymoron, while we may not have a robust temporally and spatially matched dataset for a given problem, as more and more data are gathered across disciplines, the challenge becomes how to integrate and use all this BIG data.

The key is to have a well-defined problem, ask the right questions to identify the most appropriate data, and find out as much as you can about the data. This done preferably by reaching the person who owns, collected or generated the data, but at the very least the metadata manager. This is an absolutely critical step to ensure that research in this field continues to develop, grow, and support greater knowledge about health consequences and adaptation options. Too often those in a specific discipline think their data are the most complex or difficult, and will think it

J. M. Trtanj (✉)
National Oceanic & Atmospheric Administration, Climate Program Office,
Silver Spring, MD, USA
e-mail: juli.trtanj@noaa.gov

T. G. Houston
National Oceanic and Atmospheric Administration, National Climatic Data Center,
Asheville, NC, USA

K. E. Pinkerton, W. N. Rom (eds.), *Climate Change and Global Public Health*,
Respiratory Medicine, https://doi.org/10.1007/978-3-030-54746-2_2

straightforward to just download or use data from another discipline, do their analysis and publish the results. The reality is that most data sets are complex and have significant strengths and weaknesses. Knowing how and when to use them appropriately is critical—even officially reported data have biases. Otherwise, the result is often erroneous conclusions about causality, or mechanism, which fundamentally detracts from the rigor of this whole community.

So What Is Climate Data, How Do I Know What I Need, How Do I Get It, and How Do I Use It?

The aim of this chapter is to provide a common understanding of climate terminology, climate data, and to highlight the major, long-standing data and modeling centers through which climate data and models are available. The secondary aim of this chapter is to provide a framework for how to *think* about and approach data such that data do not drive the process, but rather the question that needs answering drives the data. And the final goal is to highlight the massive opportunity for influence that the health community can have on the data providers simply by clarifying the decision that is to be made, or the research question being asked, and communicating that to those responsible for collecting data and turning it into useful information. With clarity of decision needs and articulation of a well-defined problem, those "requirements" can influence the environmental observations made and the information products produced.

Data Culture

One of the biggest differences between the climate and health communities is the approach to data—how we collect and generate the information we need to evaluate, understand, and take action in a given situation. This varies by observational mode, volume, scale, scope, frequency, and continuity. The climate community has a culture of data collection through targeted and sustained in situ and satellite observations, data management, archiving, reanalysis, and modeled data sets. Entire highly respectable careers are spent on data management, international cooperation is built around data sharing (see GEO), and supercomputer power is critical to manage and model it. In contrast, health data tend to be event- and illness-specific, often without the continuous collection so critical to understanding baseline conditions and trends, and sometimes without any geo-referenced environmental parameters. Actual individual health outcome data may even be sparser and, due to privacy issues, not available at all. So, while the health data collected may serve the immediate need for which they are obtained (i.e., an outbreak), it is often not particularly useful for climate and health analysis or for the prediction of future climate-sensitive health risks. The advent of syndromic surveillance, and other health proxies, along with Big Data, icloud storage, and increasingly accessible geospatial tools, are rapidly helping address this gap. Reliable and compatible data underpin and help establish the evidence base for credible actions that reduce climate-sensitive health risks. So, in the spirit of multidisciplinary collaboration, let us learn from each other and together tackle this data disconnect.

https://www.ncei.noaa.gov/news/noaa-expands-big-data-access

https://www.noaa.gov/big-data-project

Climate is what you expect, Weather is what actually happens

Defining Terms

In its most basic sense, this is the fundamental relationship between weather and climate. Yet, within that, there are a multitude of critically important differences. Understanding and using the correct terminology will greatly facilitate communication across disciplines, the development of a robust problem statement and identification of appropriate data to use in answering that problem. Climate is a continuum encompassing short term weather to seasonal, decadal, and long term changes in the climate system. On top of this is layered the operative functional capacity, i.e., forecast, early warning, prediction, scenario, with each having associated levels of uncertainty based on the lead time and model error. And even one step further, to really understand the complexity inherent in these coupled human and natural systems requires the consideration of other social and economic factors. Figure 2.1 provides an overview of time scales, function, and uncertainty.

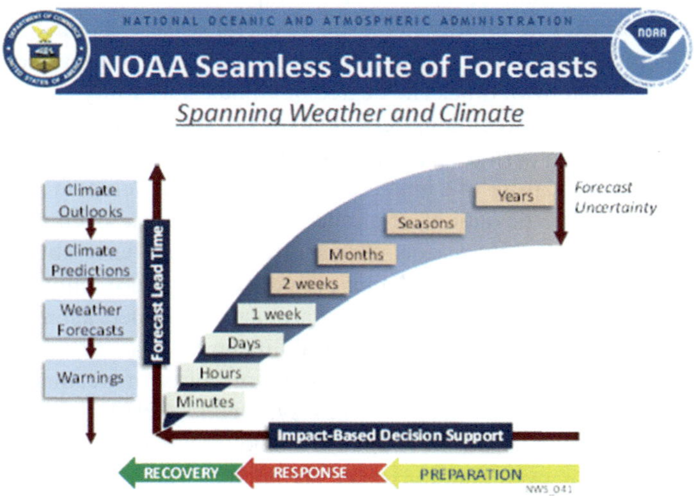

Fig. 2.1 NOAA seamless suite of forecasts

- **Weather** is the day-to-day state of the atmosphere, at some place and time, and its short-term (minutes to 7–8 days) variation. Weather is described as the combination of temperature, humidity, precipitation, cloudiness, visibility, and wind speed and direction. We talk about the weather in terms of "What will it be like today?" "How hot is it right now?" "When will that storm hit our section of the country?" Weather, over time, is what makes up climate—so if you have a weather-related health risk, that means there is a climate-related element as well. It does not mean that the knowledge about the interactions at weather time scales necessarily translate to the longer term climate time scales as there are other factors at play. (http://nsidc.org/arcticmet/basics/weather_vs_climate.html), http://www.ametsoc.org/amsedu/online/climateinfo/samplecourse/Ch01-1stEd.pdf
- **Climate** is the slowly varying aspect of the atmosphere-hydrosphere-land surface system, defined as statistical weather information that describes the variation of weather at a given place for a specified interval. It is typically characterized in terms of averages of specific states of the atmosphere, ocean, and land, including variables such as temperature (land, ocean, and atmosphere), salinity (oceans), soil moisture (land), wind speed and direction on (atmosphere), and current strength and direction on (oceans). In popular usage, it represents the synthesis of weather; more formally, it is the weather of a locality averaged over some period (usually 30 years) plus statistics of weather extremes (http://nsidc.org/arcticmet/basics/weather_vs_climate.html).
- Local or regional climate is in terms of the averages of weather elements, such as temperature and precipitation, derived from observations taken over a span of many years. In this empirically based context, climate is defined as weather (the state of the atmosphere) at some locality averaged over a specified time interval. Climate must be specified for a particular place and period because, like weather, climate varies both spatially and temporally http://www.ametsoc.org/amsedu/online/climateinfo/samplecourse/Ch01-1stEd.pdf
- **Climate (climatic) variability** In the most general sense, the term "climate variability" denotes the inherent characteristic of climate which manifests itself in changes of climate with time. The degree of climate variability can be described by the differences between long-term statistics of meteorological elements calculated for different periods. (In this sense, the measure of climate variability is the same as the measure of climate change). The term "climate variability" is often used to denote deviations of climate statistics over a given period of time (such as a specific month, season, or year) from the long-term climate statistics relating to the corresponding calendar period. (In this sense, climate variability is measured by those deviations, which are usually termed anomalies. (http://nsidc.org/arcticmet/glossary/climate_variability.html).
- **Climate change** is a change in the statistical distribution of weather over periods of time that range from decades to millions of years (APHA—Climate Change: Mastering the Public Health Role pp7). Climate change is expressed in terms of years, decades, or even centuries—but its impacts can be felt in the present. Scientists study climate to look for trends or cycles of variability (such as the changes in wind patterns, ocean surface temperatures, and precipitation over the

equatorial Pacific that result in El Niño and La Niña), and also to place cycles or other phenomena into the bigger picture of possible longer term or more permanent climate changes. (http://nsidc.org/arcticmet/basics/weather_vs_climate.html).

- **Global warming** is the gradual increase in the average temperatures of Earth's near-surface air and oceans since the mid-twentieth century and its projected continuation (APHA—Climate Change: Mastering the Public Health Role pp7).

Early Warning, Prediction, Forecast, Outlook, Projection, Scenario

In addition to the basic definitions, the application of those terms to a suite of predictive tools across time scales warrants similar clarification.

Early warning in its truest sense means information that makes it into the hands of a decision maker (individual or institutional) with sufficient lead time to allow preventive and protective action. Early warning can mean basic monitoring, forecasts, or predictions that provide advance warning to decision makers to allow preventive action to take place. This can cross time scales, ranging from things like tornado warnings where seconds count, to a risk map about potential pathogenic vibrio affecting shellfish, or using an El Nino forecast to help manage West Nile Virus risk 3 months ahead. The term "early warning" can be applied up to seasonal and annual time scales, but most commonly is used to refer to a weather event or time scale.

[vibrio risk map for Chesapeake Bay].

[link to Cal Serv and South Dakota WNV]

Forecasts are typically on weather time scales (daily and out 7 to 10 days). In cases of extreme weather events, such as hurricanes or tornados, the forecasts can be less than hourly with frequent updates. A forecast is related to a prediction in that the forecast is made by a particular person or with a particular technique or representation of current conditions that includes a prediction of those conditions. An example of a forecast is a statement by a weather forecaster that it will rain at 3:30 PM tomorrow—this reflects that individual's best judgment, perhaps drawn from a prediction that there is a 70% chance of rain tomorrow afternoon. For a decision maker, the credibility of the forecast depends critically on the credibility of the forecaster (or forecasting technique) as well as on the inevitability of the event.

A climate prediction is generally made on intraseasonal to seasonal to interannual time scales. A *prediction* is a probabilistic statement that something will happen in the future based on what is known today and is most influenced by the initial, or current, conditions. A prediction generally assumes that future changes in related conditions will not have a significant influence. For example, a weather prediction indicating whether tomorrow will be clear or stormy is based on the state of the atmosphere today (and in the recent past) and not on unpredictable changes in "boundary conditions" such as how ocean temperatures or even society may change between today and tomorrow. For decision makers, a prediction is a statement about an event that is likely to occur no matter what they do. http://sciencepolicy.colorado.edu/zine/archives/1-29/26/correspond.html

Climate predictions are usually expressed in probabilistic terms (e.g., probability of warmer or wetter than average conditions) for periods such as weeks, months, or seasons. A prediction is a probabilistic statement of something that could happen in the future based only on what is known today. Climate projections are long-range predictions of the future climate based on changing atmospheric conditions, such as increased or decreased pollutants due to emissions from the burning of fossil fuels (coal, oil, gas) http://www.nws.noaa.gov/om/csd/graphics/content/outreach/brochures/Weather&Climate_General_Public.pdf

Outlooks are probabilistic and typically made on climate time scales of 2 weeks, monthly, and seasonal and, often drawing on expert judgement. NOAA's Climate Prediction Center issues Extended Range Outlooks (out to 2 weeks) and monthly to seasonal Outlook maps showing probabilities of temperature and precipitation departing from normal, with an accompanying technical discussion. These outlooks are issued from 2 weeks to 13 months in advance, for the lower 48 states and Hawaii and other Pacific Islands. (https://www.cpc.ncep.noaa.gov/products/forecasts/).

Regional Climate Outlook Forums (RCOFs) produce consensus-based, user-relevant climate outlook products in real time in order to reduce climate-related risks and support sustainable development for the coming season in sectors of critical socioeconomic significance for the region in question. Regional Climate Outlooks are done globally through the National Hydrological and Meteorological Services and the World Meteorological Organization (Fig. 2.2). https://public.wmo.int/en/our-mandate/climate/regional-climate-outlook-products

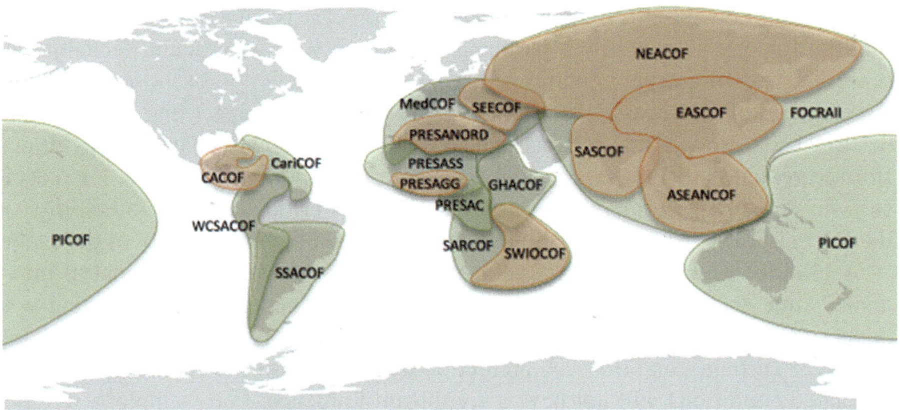

Fig. 2.2 Regional Climate Outlook Forums

Climate projections are generally decadal-to-centennial. In contrast to a prediction, a *projection* specifically allows for significant changes in the set of "boundary conditions" that might influence the prediction, creating "if this, then that" types of statements. Thus, a *projection is a probabilistic statement that it is possible that something will happen in the future if certain conditions develop*. The set of boundary conditions that is used in conjunction with making a projection is often called a scenario, and each scenario is based on assumptions about how the future will develop. For example, the Intergovernmental Panel on Climate Change (IPCC) would *project* a range of possible temperature changes that would likely occur for a range of plausible emissions scenarios and a range of model-derived estimates of climate sensitivity (the temperature change that would result from a CO2 doubling). This is clearly a projection of what *could* happen *if* certain assumed conditions prevailed in the future—it is neither a prediction nor a forecast of what will happen independent of future conditions. For a decision maker, a projection is an indication of a possibility, and normally of one that could be influenced by the actions of the decision maker or other policy actor. http://sciencepolicy.colorado.edu/zine/archives/1-29/26/correspond.html

A **climate scenario** is a coherent, internally consistent, and plausible description of a possible future state of the world. It is not a forecast; rather, each scenario is one alternative image of how the future can unfold. A projection may serve as the raw material for a scenario, but scenarios often require additional information (e.g., about baseline conditions). A set of scenarios is often adopted to reflect, as well as possible, the range of uncertainty in projections. Other terms that have been used as synonyms for scenario are "characterisation", "storyline", and "construction".

Scenarios are best thought of as "plausible alternative futures – each an example of what might happen under particular assumptions"; scenarios are not predictions or forecasts because they depend on assumed changes in key boundary conditions (like emissions) and scenarios are not fully projections of what is likely to happen because they have considered only a limited set of possible future boundary conditions (e.g., emissions scenarios). For the decision maker, scenarios provide an indication of possibilities, but not definitive probabilities. For instance, the Intergovernmental Panel on Climate Change will run several scenarios with different boundary conditions such as emissions and economic growth rates. WeatherZine, Number 26, 2001, NCAR http://sciencepolicy.colorado.edu/zine/archives/1-29/26/correspond.html

How to Think About Climate Data—Or When to Use What?

Climate data comprises many different types, scales, and resolution of data, derived from multiple sources (satellite or in situ), and made available through a number of products and service modes.

Scale

Climate data can be global, regional, or local in scale and comprises oceanic, atmospheric, and terrestrial data. Within that are mostly physical parameters such as precipitation, temperature (atmospheric and oceanic), sea level, waves, and winds. The data streams, while collected separately, but can be part of the same satellite or field collection effort. The different data streams are then combined to make climate data products and models. Scale is largely dependent on the means by which the data are collected (satellite or in situ observations), the area of coverage, and density of collection sites for in situ observations, or grid size for satellites.

Source

Data are collected or provided from multiple sources; satellite, in situ, modeled, reanalyzed, and projections. **Satellites** provide periodic but global coverage from polar orbital satellites or consistent coverage over parts of the globe through geostationary satellites. Polar orbital satellites provide total earth coverage, but will measure the same place twice each day at the same local time, every 12 hours, as part of their low earth orbit (approximately 500 miles altitude) moving from North Pole to South Pole. Because of their lower altitude, polar orbital satellites can use microwave radiometers which allow them to measure through clouds to sense precipitation, temperature in different layers of the atmosphere, and surface characteristics like ocean surface winds. Geostationary satellites are fixed high above the equator (approximately 22,000 miles altitude) providing continuous coverage of the same area, but the resolution is generally 1 km at best, and coverage is not global. In general, for climate and weather purposes, the National Aeronautics and Space Administration (NASA) launches research satellites mostly in polar orbital and in lower earth orbit. The National Oceanic and Atmospheric Administration (NOAA) operates the satellites needed for weather and climate predictions which include geostationary satellites.

In situ data are collected from ground, water-based, or airborne instruments and sensors. Availability varies by country, both in time and space, and access. The quality varies according to the instrumentation and human skill in collection and recording. Metadata may or may not be available, and upkeep, updates, and archiving are problematic for many countries. In situ data are useful alone, can be combined with other data into more comprehensive products, and can be used to validate and enhance satellite data. The networks and instruments for in situ data collection vary widely and include everything from permanent weather stations, to tide gauges, to drifting buoys in the ocean and ships of convenience, to the atmospheric radiation and temperature and Carbon dioxide measurements and Mauna Loa Observatory in Hawaii, which has tracked CO2 since the 1950s.

Products

Data can also be processed into products such as Sea Surface Temperature (SST) , SST Anomalies (commonly depicted during El Nino and La Nina events), Vegetation Indices, Sea Ice (see http://www.realclimate.org/index.php/data-sources/ for additional products). One of the most well-known data sets is the Global Historical Climatology Network (GHCN) Daily Dataset, which is a global, daily in situ dataset derived from multiple sources: approximately 25,000 temperature stations, 44,000 precipitation stations, 25,000 snowfall or snow depth stations, and currently ingests more than1.6 billion daily observations with the earliest value from January 2, 1833, and the latest value from yesterday. (see https://www.ncdc.noaa.gov/ghcn-daily-description).

The scientific community has established three global networks for terrestrial, oceanographic, and climate data. The Global Ocean Observing System (GOOS) is a permanent global system for observations, modeling, and analysis of marine and ocean variables to support operational ocean services worldwide (Fig. 2.3). GOOS comprises a network of ocean-based observations and satellite observations and, along with the Global Climate Observing System (GCOS) and the Global Terrestrial Observing System (GTOS) comprise a global network of monitoring to understand and predict climate, among other things.

Global Ocean Observing System (GOOS) in situ measurements

- 3000 Argo floats collect high-quality temperature and salinity profiles from the upper 2000 m of the ice-free global ocean and currents from intermediate depths.
- 1250 drifting buoys record the currents of surface, the temperature, and the atmospheric pressure.
- 350 embarked systems on commercial or cruising yachts which collect the temperature, salinity, the oxygen, and the carbon dioxide ($CO2$) in the ocean and the atmosphere, and the atmospheric pressure.
- 100 research vessels measure all the physical, chemical, and biological parameters, between the surface of the sea and the ocean floors every 30 nautical miles out of 25 transoceanic lines.
- 200 marigraphs and holographs which transmit information in quasi real time, thus providing the possibility of detecting tsunamis.
- 50 commercial ships which launch probes measuring the temperature and salinity between the surface and the ocean floor on their transoceanic ways.
- 200 moorings in open sea which are used as long-term observatories, recording weather, chemical, and biological parameters on a fixed site between the surface and the bottom.

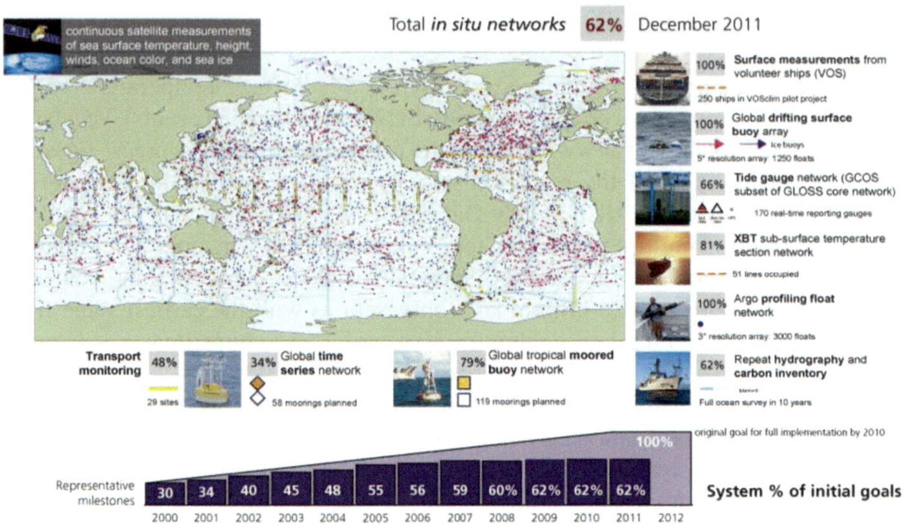

Fig. 2.3 Global Ocean Observing System (GOOS) in situ measurements

Reanalysis

In order to create consistent and comparable global datasets, major efforts are made by the community to create reanalysis datasets. Reanalysis is a scientific method for developing a comprehensive record of how weather and climate are changing over time. In it, observations and a numerical model that simulates one or more aspects of the Earth system are combined objectively to generate a synthesized estimate of the state of the system. (https://reanalyses.org/reanalysesorg-home-page).

These are weather models which have the real world observations assimilated into the solution to provide a "best guess" of the evolution of weather over time (although pre-satellite era estimates (before 1979) are less accurate). The newest as of this writing is the NCEP/NCAR reanalysis with 6-hour, daily, and monthly data available. https://www.esrl.noaa.gov/psd/data/reanalysis/reanalysis.shtml

Projections

Data are also generated through climate projections and scenarios. A climate projection is a model-derived estimate of the future and the pathway leading to it. When the certainty around a projection is branded "most likely" it becomes a forecast or prediction. A forecast is often obtained using deterministic models, possibly a set of these, outputs of which can enable some level of confidence to be attached to projections. General Circulation Models (GCMs), numerical models that represent the physical processes in the atmosphere, ocean, cryosphere, and land surface, are the most advanced tools currently available for simulating the response of the global climate system to increasing greenhouse gas concentrations. While simpler models have also been used to provide globally or regionally averaged estimates of the climate response, only GCMs, possibly in conjunction with nested regional models, have the potential to provide geographically and physically consistent estimates of regional climate change which are required in impact analysis. GCMs depict the climate using a three dimensional grid over the globe typically having a horizontal resolution of between 250 and 600 km, 10 to 20 vertical layers in the atmosphere and sometimes as many as 30 layers in the oceans. Many physical processes, and feedback mechanisms such as water vapor and warming, or clouds and radiation, occur at smaller scales and cannot be properly modelled. Instead, their known properties must be averaged over the larger scale in a technique known as parameterization, which are sources of uncertainty in GCM-based simulations of future climate.

Assessing Climate Data Partners

Atmosphere: surface	Atmosphere: Upper-air	Atmosphere: composition	Ocean: surface	Ocean: subsurface	Terrestrial
Air temperature	Earth rad'n budget	Carbon dioxide	Temperature	Temperature	Soil moisture
Precipitation	Temperature	Methane	Salinity	Salinity	Snow cover
Air pressure	Wind speed & dir	Ozone	Sea level	Current	Permafrost + seasonally frozen
SfcRad'n budget	Water vapor	Nitrous oxide	Sea state	Nutrients	Glaciers + ice caps
Wind speed & dir	Cloud properties	CFCs	Sea ice	Carbon	River discharge
Water vapor		Hydro CFCs	Current	Ocean tracers	Water use
		Hydrofluorocarbs	Ocean color	Phytoplankton	Ground water
		Sulfur hexafluorides	CO_2 partial pressure		Lake levels
		Perfluorocarbons			Albedo
		Aerosol properties			Land cover
					Percent absorbed photosynthetically active radiation
					Leaf area index
					Biomass
					Fire disturbance

NOAA houses much of the climate, weather, and ocean data not only for the United States but serves as the main repository for the World Meteorological Organization and other international bodies. In the United states, there is a three-tiered climate services support program. The partners of this program include NOAA's National Centers for Environmental Information—https://www.ncdc.noaa.gov/), six Regional Climate Centers (RCC—https://www.ncdc.noaa.gov/customer-support/partnerships/regional-climate-centers),and individual State Climate Offices (SCO—http://www.stateclimate.org/). NCEI is the world's largest active archive of weather data with over 150 years of in situ, radar, and satellite data available for use in a wide variety of applications. The Regional Climate Centers are a federal-university cooperative effort that is managed by NCEI. The RCCs are engaged in the timely production and delivery of useful climate data, information, and knowledge for decision makers and other users at the local, state, and national levels. The RCCs support NOAA's efforts to provide operational climate services, while leveraging improvements in technology and collaborations with partners to expand quality data dissemination capabilities. State Climatologists have the best understanding of the climate of their state, and the ability and knowledge to provide climate data and information to local users. Additional NOAA climate partners include the National Weather Service Climate Services Division https://www.weather.gov/climateservices/, the Climate Prediction Center https://www.cpc.ncep.noaa.gov/, the Climate Diagnostics Center (http://cires.colorado.edu/science/centers/cdc/), the Climate Program Office (https://cpo.noaa.gov/) and their Regional Integrated Sciences and Assessments (RISA) Program (https://cpo.noaa.gov/Meet-the-Divisions/Climate-and-Societal-Interactions/RISA/About-RISA), and six Regional Climate Service Directors (https://www.ncdc.noaa.gov/rcsd) that are located at the NWS Regional Headquarters (Fig. 2.4).

Some applications require data and information for areas outside of the United States. While the agencies mentioned above focus primarily at the national, regional, and local levels, some do participate in international activities as well. For example, NCEI operates a World Data Center for Meteorology (https://www.ncdc.noaa.gov/wdcmet) and a World Data Center for Paleoclimatology (https://www.ncdc.noaa.gov/data-access/paleoclimatology-data). The World Data Centers are part of a global network of discipline sub-centers that facilitate international exchange of scientific data. The World Meteorological Organization also maintains a list of member National Meteorological or Hydrometerological Services (https://public.wmo.int/en/about-us/members) in which users can go directly to the country of interest in order to obtain weather and climate data and information for their application (Fig. 2.5). WMO designated Regional Climate Centres are also being implemented to provide more regionally focused data and products to users (http://www.wmo.int/pages/prog/wcp/wcasp/rcc/rcc.php)

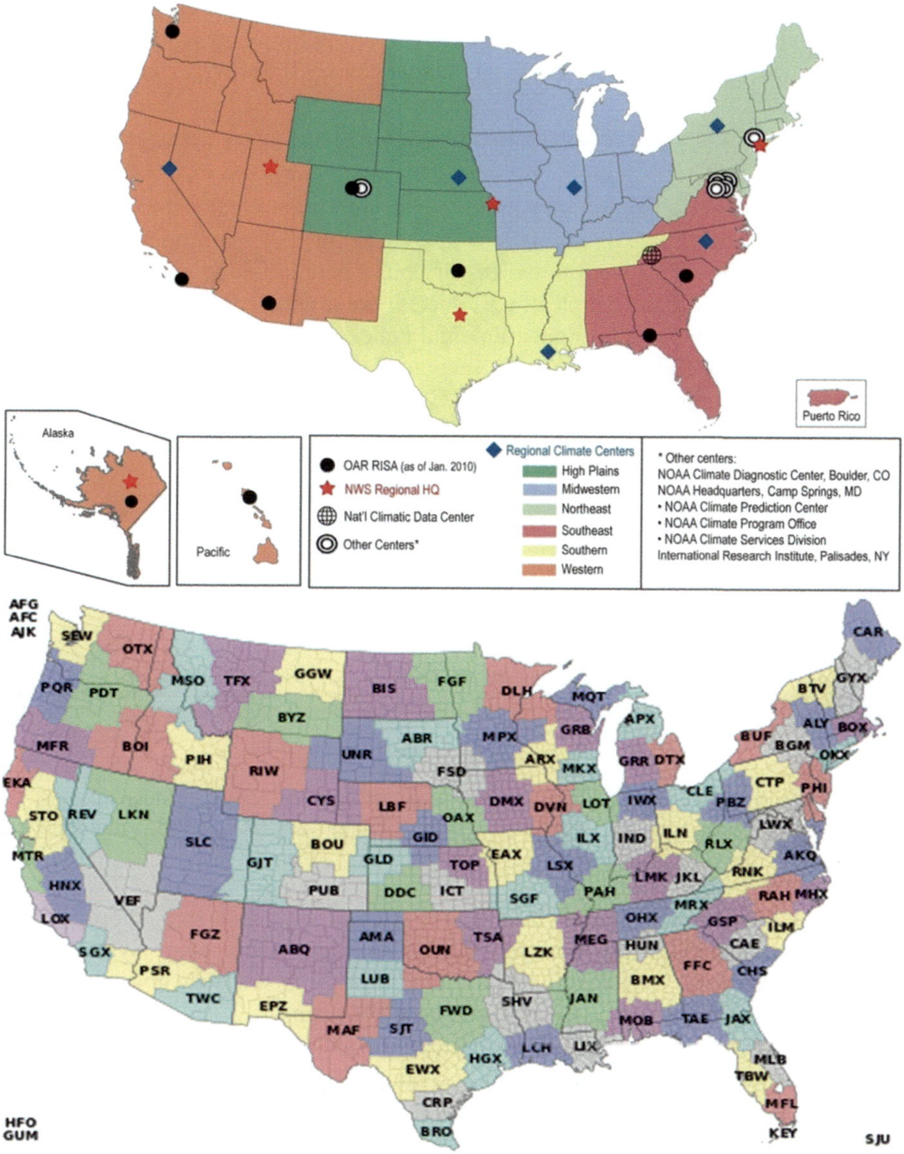

Fig. 2.4 Map depicting climate service partners throughout the United States

Fig. 2.5 World Meteorological Organization map of member National Meteorological or Hydrometeorological Services (http://www.wmo-dra.info/gmap/WMO_NMHS_regions/metservices.html)

Global Observing Systems Information Center is a one-stop shop for the Global Ocean Observing System, Global Climate Observing System, and Global Terrestrial Observing System. https://www.ncdc.noaa.gov/gosic/global-climate-observing-system-gcos/us-gcos-program.

Starting with the End in Mind: Integrated Information Systems

An Integrated Information System is a simple framework that can be used to ensure that appropriate sectors and decision makers are involved, together, at the beginning and throughout—this is how to craft an effective problem statement which leads to the appropriate data, products, and other information needed to answer the question. But it is fundamentally a sustained process—not a one-off meeting. An IIS basically starts with engagement of the decision maker as part of the team—*and that engagement is sustained over the life of the problem and sometimes beyond.* This leads to a clearer identification of the problem, the decision time frame, and the actionable options. This understanding then drives the development of actionable information needed. That in turn both draws from existing data—observations, surveillance, social, behavioral, etc., and helps define future

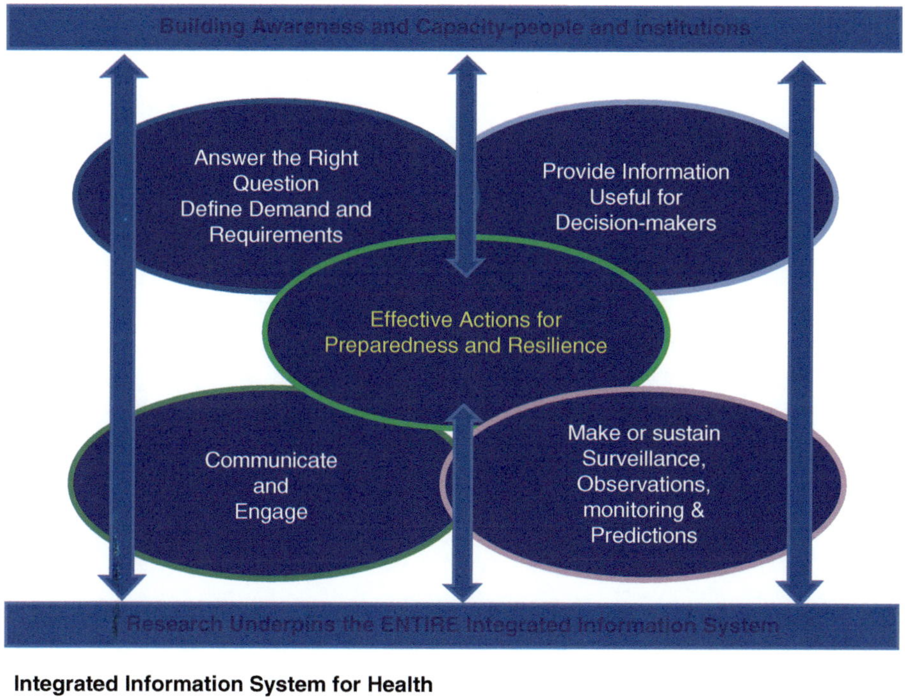

Fig. 2.6 Effective actions for preparedness and resilience

data collection and product development. The information provided throughout the process serves as the backbone for evidence-based action, and the *process itself*, of sustaining engagements over time, helps build the trust needed for a decision maker to take action (Fig. 2.6).

Conclusion

In summary, it is incumbent on the researchers in this interdisciplinary field to be informed enough to ask the right questions to find and understand the right data, to provide scientifically sound information to help people make the right decision. This requires active recognition of the need to really understand the caveats and best uses of a particular data set or product. In general, while some more widely used data and products such as those developed for the Intergovernmental Panel on Climate Change may have well-defined tutorials and use parameters, in general, it is wiser to find the owner or originator of the data and work with them to ensure appropriate use of the data and, therefore, robust scientific findings that both inform decisions, and move this interdisciplinary field forward in both science and policy contexts.

Chapter 3
Climate Change: Updates on Recent Global and United States Temperature Anomalies and Impacts to Water, Forests, and Environmental Health

David H. Levinson and Christopher J. Fettig

Recent Changes to Global and United States Surface Temperatures

A number of critical improvements have been implemented to global surface air temperature datasets over the past few years, allowing for increased statistical reliability when determining surface temperature anomalies and trends. For example, the surface air temperature data in the Global Historical Climatology Network-Monthly (GHCNm) dataset, developed and maintained by the National Oceanic and Atmospheric Administration (NOAA) National Centers for Environmental Information, was updated from Version 3.3.0 to Version 4.0 in October 2018 [57]. Previous improvements have been documented for past versions of GHCNm [77, 94], as well as the Extended Reconstructed Sea Surface Temperature (ERSST) dataset, most recently with Version 5 [41] that is combined with GHCNm to determine global land-ocean temperatures. GHCNm Version 4 includes data from approximately 26,000 surface stations, almost four times the number in the previous version. By including the additional surface stations, as well as estimates for missing base period (30-year) averages, NOAA has expanded the geographic coverage of surface temperature anomalies throughout the entire period of record (1880–present). Recent measurements show the continued rise in land surface temperatures, both globally and across the United States (Fig. 3.1 top-bottom, and Fig. 3.2), both of which show clear increasing trends since the 1970s. For global surface

D. H. Levinson (✉)
National Stream and Aquatic Ecology Center, Watershed, Fish, Wildlife, Air and Rare Plants Staff, USDA Forest Service, Fort Collins, CO, USA
e-mail: david.levinson@usda.gov

C. J. Fettig
Pacific Southwest Research Station, USDA Forest Service, Davis, CA, USA
e-mail: chris.fettig@usda.gov

K. E. Pinkerton, W. N. Rom (eds.), *Climate Change and Global Public Health*, Respiratory Medicine, https://doi.org/10.1007/978-3-030-54746-2_3

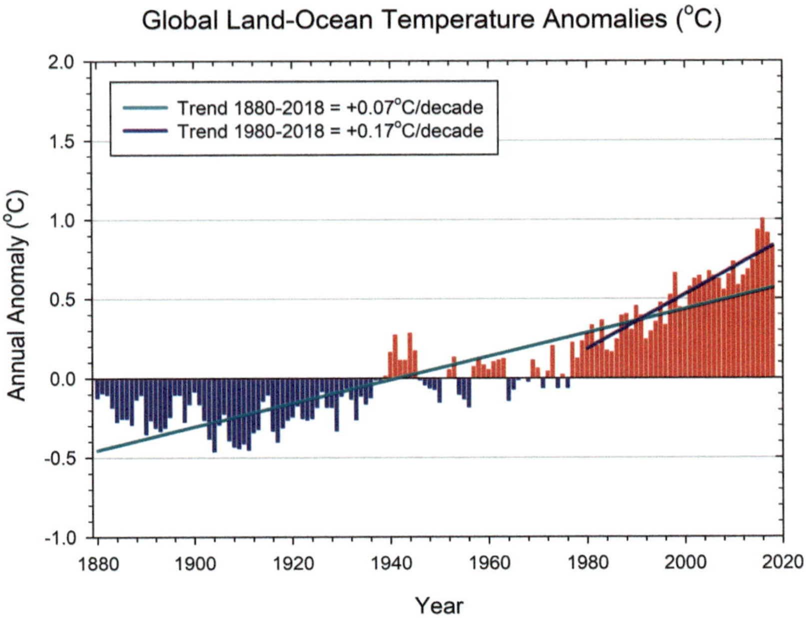

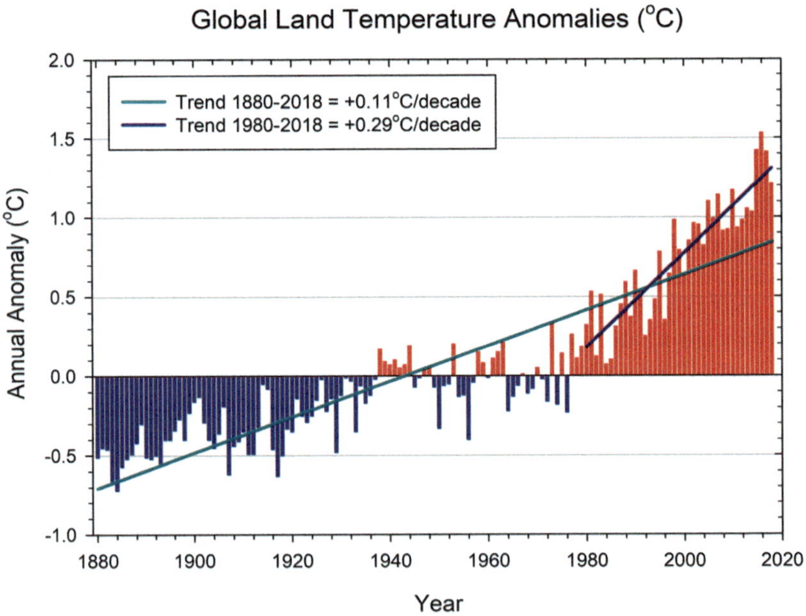

Fig. 3.1 (TOP) Combined land-ocean annual global surface temperature anomalies covering the period 1880–2018, relative to the 1901–2000 annual mean; (BOTTOM) Land-only annual global surface temperatures anomalies covering the period 1880–2018, relative to the 1901–2000 annual mean. Data for land surface temperatures are taken from NOAA's Global Historical Climatology Network-Monthly (GHCNm) dataset (Version 4; [57]), and data for ocean surface temperatures are from the Extended Reconstructed Sea Surface Temperature (ERSST) dataset (Version 5; [41])

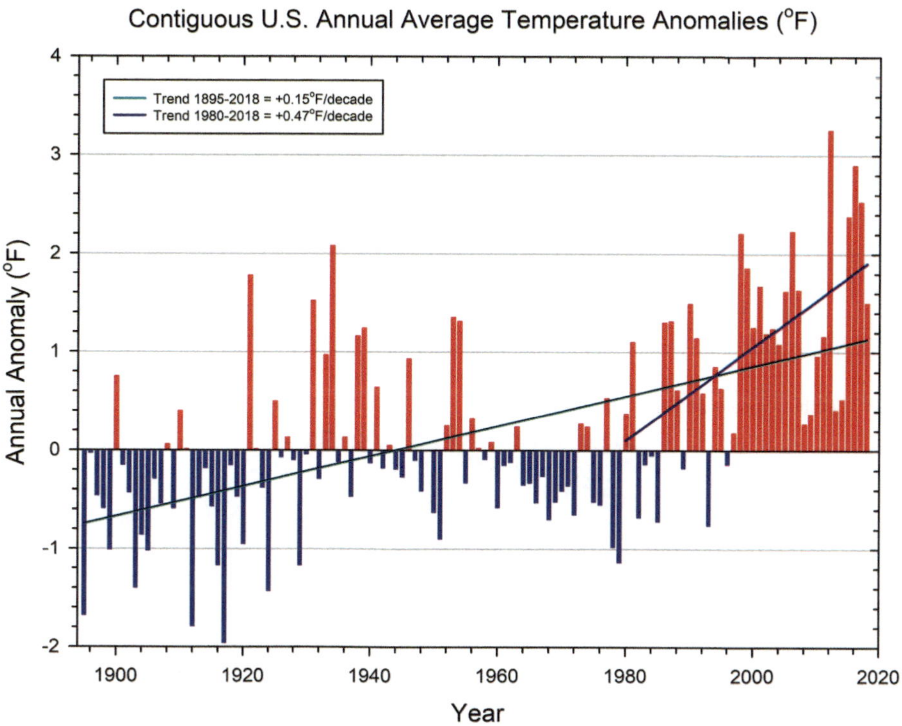

Fig. 3.2 Annual average surface temperatures (in °F) for the contiguous United States over the period 1895–2018, relative to the 1901–2000 mean. Data are from NOAA's Global Historical Climatology Network-Monthly (GHCNm) dataset (Version 4; [57])

temperatures, the observed trend over the century-scale, covering the period 1880–2018, is approximately +0.07 °C/decade. However, since 1980, the trend has increased significantly, and over the period 1980–2018 has risen to approximately +0.17 °C/decade. In addition, the observed surface temperatures for the contiguous United States have also increased steadily (Fig. 3.2) over the past several decades. Based on data covering the entire historical record from 1895 to 2018, the trend is just over +0.15 °F/decade, but, for the period 1980–2018, the trend has increased substantially to +0.47 °F/decade.

The observed increases in global surface temperatures has been larger over continental areas than oceans. As displayed spatially using gridded data in Fig. 3.3 (top-bottom), the observed trends over the 30-year period covering 1988–2017 exceeded +0.5 °C/decade across many Northern Hemisphere land areas, including most of North Africa, western parts of Asia, and higher latitudes of North America. The increased warming of global surface land temperatures is clearly shown when comparing the top 5 warmest and coolest years in the historical record. As shown in Table 3.1, the top 5 warmest years for global surface land temperatures have all occurred since 2015, with the top 5 coolest years all occurring in the late nineteenth and early twentieth centuries.

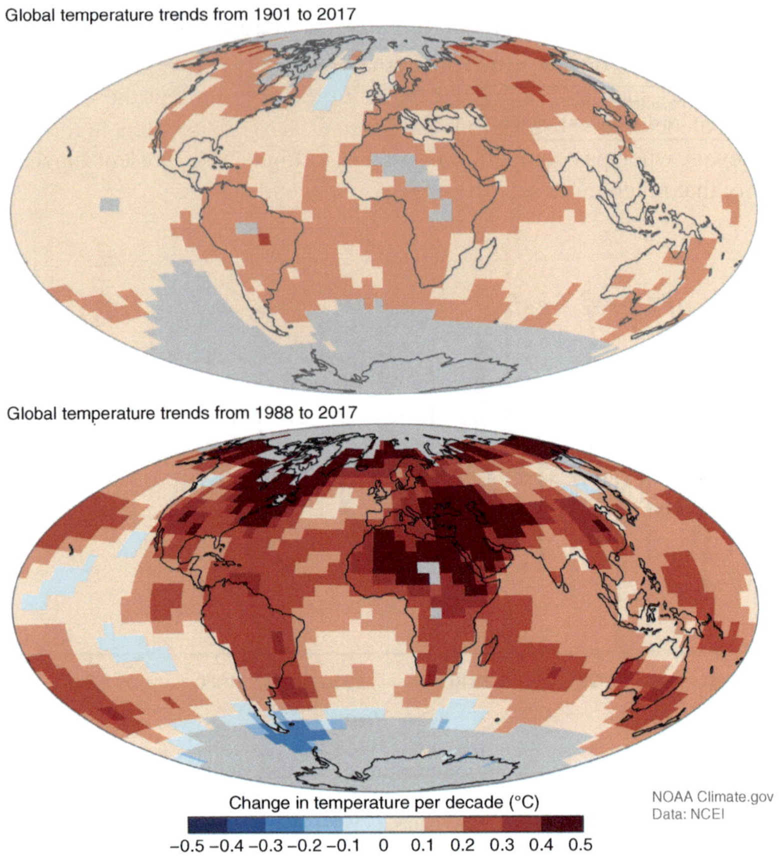

Fig. 3.3 (TOP) Observed trends (in °C/decade) of 5° × 5° gridded land and ocean annual global surface temperature anomalies covering the period 1901–2017, with anomalies relative to the 1901–2000 annual mean. (BOTTOM) Observed trends (in °C/decade) of 5°×°5° gridded land and ocean annual global surface temperature anomalies covering the period 1988–2017, with anomalies relative to the 1901–2000 annual mean. Data for land surface temperatures are taken from NOAA's Global Historical Climatology Network-Monthly (GHCNm) dataset (Version 4; [57]), and data for ocean surface temperatures are from the Extended Reconstructed Sea Surface Temperature (ERSST) dataset (Version 5; [41])

Climate Change Impacts on Water Resources and Water Quality

The impacts of climate change on the water cycle occur through changes in precipitation, stream and river flows, and water quality, and are felt across multiple sectors and regions [50]. Maintaining consistent, high-quality water resources is a primary concern for public and environmental health, and declining water supplies and degradation of water quality related to increasing surface air temperatures from climate change is of widespread concern. The environmental impacts of reduced water

Table 3.1 Top 5 warmest and coldest years for global surface air temperatures (land and ocean combined), based on data since 1880 (with annual temperature anomalies in °C relative to 1901–2000 mean). Data sources are the Global Historical Climatology Network-Monthly (GHCN-m) dataset for land temperatures and the Extended-Reconstructed Sea Surface Temperature Dataset (ERSST) for ocean surface temperatures, both developed and updated by NOAA

Warmest years	Coldest years
2016 (+1.55 °C)	1884 (−0.72 °C)
2019 (+1.51 °C)	1883 (−0.68 °C)
2015 (+1.43 °C)	1917 (−0.64 °C)
2017 (+1.42 °C)	1907 (−0.63 °C)
2018 (+1.23 °C)	1885 (−0.58 °C)

resources has been felt in regions around the world, with many drier regions having experienced increasing numbers and intensity of droughts, resulting in reductions in water supplies [84]. It is clear that the water supplies for many municipalities in the western United States are vulnerable to severe shortages [9], and this vulnerability increases in future climate simulations as surface temperatures increase [31]. Observed variations in river-flows indicate that the effects of increasing temperatures are dominated by fluctuations in precipitation [32, 37], although water management does have an effect where larger diversions, dams, and other water infrastructure play a role [50]. As shown by Brown et al. [16], water supplies across the contiguous United States are expected to decline in many areas due to the impacts of both population growth and climate change, specifically in the arid and semi-arid parts of the western United States (Fig. 3.4). Hydrologic model simulations of water supply and demand over the twenty-first century show that in the absence of further adaptation efforts, serious water shortages are likely in some regions. Specifically, those areas that currently experience water shortages are expected to experience increased water stress as surface air temperatures increase, while other areas will experience increasing streamflow related to increasing precipitation. For example, the observed runoff and streamflow at regional scales have declined during the last half-century in the northwestern United States [10], as well as in the Colorado River Basin [91], while streamflow is increasing in the Mississippi River Basin and the northeastern United States [95].

In addition, models of streamflow responses due to changes in climate show that many areas will see changes in the timing and magnitude of peak flows [39, 80]. For example, recent observations have shown a declining trend in annual runoff for the Colorado River Basin [97], the primary water supply for a vast area of the Southwest and Colorado Plateau. Significant changes have also been observed in the timing of winter-spring streamflows in river basins in eastern parts of North America [40]. Problems with water quality will also likely increase due to rising temperatures across stream and river systems, especially during severe drought events [68]. Lower and more persistent periods of low flows under drought conditions can worsen water quality [69]. The same is true for higher flows, which have increased due to extreme precipitation (flood) events [64]. The impacts of increasing

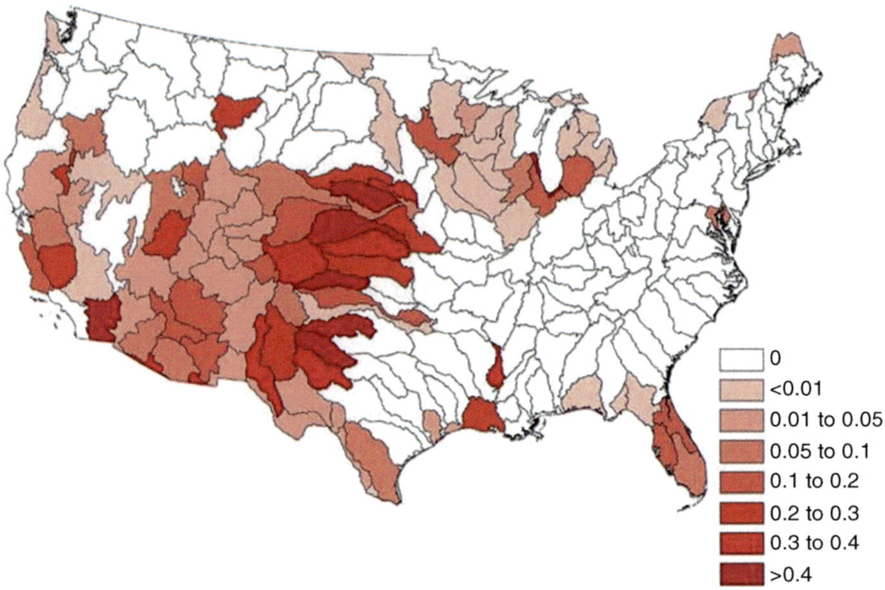

Fig. 3.4 Mean shortage frequency if relying only on renewable water supplies, determined using the mean of 14 future scenarios. The shortage frequency is shown by river basin for the period covering 2046–2070, which was determined using the mean of 14 future scenarios, which included the same number of both wet and dry future scenarios (originally from [16])

precipitation intensity, in conjunction with the effects of wildfires (see section "Wildfire", below) and widespread use of fertilizers, have been shown to increase sediment, nutrient, and contaminant loads in surface waters [55]. In addition, changing land cover, and increases in the frequency and magnitude of forest disturbances and floods will likely increase sediment loads in larger rivers [86, 88].

Significant rises in stream temperatures have been documented for some regions of the United States, which has implications for water quality and aquatic habitat. To address these concerns, the multi-agency NorWeST project combined stream temperature data from over 200,000,000 hourly temperature recordings at >20,000 unique stream monitoring sites across the entire western United States [42]. NorWeST uses observed data and numerical models to understand changes in stream and river temperatures, and current and potential impacts to aquatic habitat. They also utilized spatial statistical models to develop 36 historical and future climate scenarios at 1-km resolution for >1,000,000 km of streams, which are used comparatively with historical stream temperatures. Results from these simulations showed increasing water temperatures for all 36 scenarios. For many native fish species, cold headwater streams are needed for spawning and maintaining proper habitat conditions, and rising stream temperatures pose a threat to survival of some species, especially those adapted to colder stream conditions.

The impacts of climate change on groundwater storage vary significantly, and are not well understood due to inconsistent monitoring of large aquifers and a lack of understanding the connection between surface and sub-surface flows [50]. However, it is clear that precipitation is the primary source of aquifer recharge in water-limited ecosystems [14, 31], and that changes in recharge rates are amplified relative to changes in total precipitation, especially in drier environments [36]. In many forested and mountainous areas, groundwater recharge is primarily generated by snowmelt infiltration, which puts groundwater dependent ecosystems at risk due to observed reductions in snowpacks across much of the western United States [98].

Expected increases in water demand under climate change are projected in regions of the United States where groundwater aquifers are the primary water supply source [15], specifically the Great Plains and parts of the Southwest and Southeast. Mining groundwater is commonly used in many regions to supplement surface water supplies, especially in semi-arid and arid climates, but such mining is unsustainable in the long-term [14] and may significantly deplete storage in aquifers if not properly managed and regulated.

Climate Change Impacts on Watershed Health

Ecosystems across the globe are responding to the impacts of warming temperatures and increasingly variable precipitation events in a variety of ways. In the United States, forested watersheds are experiencing increasing numbers of large fires, droughts and floods, insect outbreaks, major land-use changes, and other environmental impacts that are detrimental to their proper function (see section "Climate Change Impacts on Forests and Forest Health"). To improve understanding of the health and function of forested watersheds, the USDA Forest Service developed the Watershed Condition Framework (WCF) to help monitor and assess the condition of over 15,000 watersheds managed by the agency [65].

To measure and evaluate how well watersheds are functioning, the Watershed Condition Classification (WCC) system [66] was developed as the initial step of the WCF to integrate numerous drivers of watershed condition into a combined rating (Functioning Properly, Impaired Function, Functioning At-Risk). To assess the condition of watersheds, WCC utilizes 12 indicators, and 24 sub-attributes, to measure the complex function of watersheds due to their physical and biological environments. The 12 indicator/24 attribute model is broken into four categories: Aquatic Physical, Aquatic Biological, Terrestrial Physical, and Terrestrial Biological (weighted of 30%, 30%, 30%, and 10%, respectively).

The initial evaluation of watershed condition using the WCC was conducted in 2011, establishing baseline conditions for all HUC 6 (12-digit) watersheds on National Forest System lands in the United States. Additional evaluations were conducted over multiple years in 2015–2018 on watersheds that experienced substantial disturbance or had large restoration projects completed. Figure 3.5 (left-right) shows both the initial WCC evaluation for watersheds in the western United States, along with those watersheds that changed their condition class (either improved or degraded) during subsequent evaluations. It is clear from the re-assessments that the vast majority of watersheds remained at the same condition class. However, of those that did have a change in their condition, the majority of those watersheds had a decline in their WCC rating and therefore a degradation of overall condition, especially those in the Southwest, where large fires and droughts have occurred.

To assess the condition of riparian zones, which have the highest biodiversity within forested and rangeland watersheds, the Western Riparian Threats Assessment was conducted in 2010 to provide an initial, coarse-scale assessment of historical, current, and future threats to streams and riparian areas in the western United States [83]. The effort supported the development of a strategic vision for the future of western wildland management that offers strategies for managing these important landscape elements and their watersheds, recognizing the need to balance sometimes conflicting interests and demands. Common indicators of stress in riparian ecosystems include reduced biodiversity, altered productivity, an increased prevalence of disease, shifts in species composition and "terrestrialization", reduced efficiency of nutrient cycling, and increased dominance of exotic and generalist species [58].

To develop this complex assessment, Theobold et al. [83] assessed threats to riparian ecosystems by examining their deviation from reference conditions in the

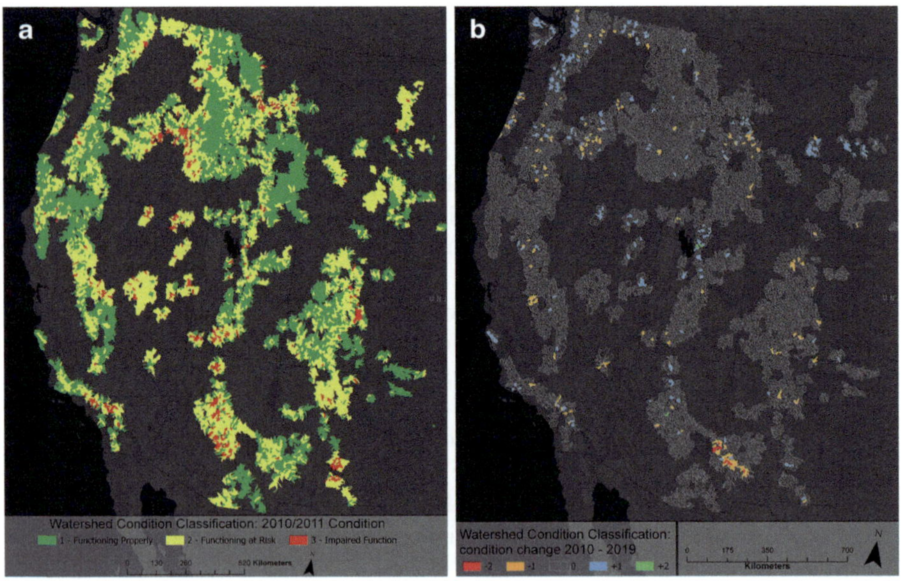

Fig. 3.5 (Left) Watershed Condition Classification (WCC) for the western United States, from initial condition assessments conducted on USDA Forest Service National Forest System lands in 2010–2011. (Right) The change in condition class for those watersheds reassessed in 2015–2018, with the observed change relative to the initial assessment in 2010–2011. (Source: data obtained from the USDA Forest Service's Watershed Condition Assessment and Tracking Tool)

physical processes that govern riparian pattern and process. These factors include the natural flow regime (and alterations to it), lateral connectivity to the landscape (and degree of anthropogenic confinement), and longitudinal connectivity. These were assessed using the most extensive geospatial information obtainable at the time, and used to model flow and sediment regimes at the 12 digit Hydrologic Unit Code (HUC) resolution. Comparisons of historical flow and sediment regimes to current and future regimes allowed for an assessment of threats to these systems as a result of human land uses and water development and the impacts of climate change on discharge and sediment yield from watersheds. Figure 3.6 is a map of the final threat assessment score on a scale of 0 to 100% (darker blue indicates a higher threat level). Based on this multi-parameter risk assessment, the riparian systems that are under the highest threat in the western United States include the central valley of California, the east slope of the Cascades in Washington, the Colorado River system, and the Southwest.

Fig. 3.6 Riparian threats score (scaled 0–100%) for reach catchment areas (~HUC12 scale) in the western United States, normalized by water resource region, with the highest threat levels to riparian zones represented by dark blue shading and the lowest threat levels by light blue shading. (From Theobald et al. 2010)

Climate Change Impacts on Forests and Forest Health

Forests cover ~42 million km^2 (~30%) of the earth's surface and are found in all regions at elevations and latitudes capable of sustaining tree growth, except where disturbances, whether natural or human-induced, are too frequent or too severe to enable establishment. Forests provide immeasurable ecological, economic, and social goods and services to both natural systems and humankind. These include, among others, purification of the air that we breathe; regulation of edaphic formation and water quantity and quality (see section "Climate Change Impacts on Watershed Health"); provision of fish and wildlife habitat, food, medicine, shelter, wood, and other forest products; provision of aesthetics, outdoor recreation and spiritual renewal; and regulation of climate through carbon storage and complex physical, chemical, and biological processes that affect planetary energetics [13]. In short, forests represent one of the earth's most important ecosystems, and are critical to the health, welfare, and survival of human societies across the globe.

Forests absorb the equivalent of ~2 billion tons of carbon dioxide (CO_2) each year [46] and are increasingly recognized for mitigation of atmospheric concentrations of CO_2 [20]. However, the sustainability of many forests is threatened by increases in the frequency and severity of disturbances exacerbated by climate change [5, 6, 25, 59, 86]. For example, a key finding of the recently-published Fourth National Climate Assessment is that "It is very likely that more frequent extreme weather events will increase the frequency and magnitude of severe ecological disturbances, driving rapid (months to years) and often persistent changes in forest structure and function across large landscapes" ([84], Table 3.2). Disturbances, such as wildfire and bark beetle epidemics, that cause large amounts of tree mortality may reverse the role of some forests from carbon sinks to carbon sources [7]. Alternatively, healthy forests assimilate, accumulate, and sequester large amounts of carbon from the atmosphere.

The effects of climate change on forests include both positive (e.g., increased growth through elevated water use efficiency and longer growing seasons) and negative impacts (e.g., increased frequency and severity of disturbances) with feedbacks that influence environmental health. Disturbances release growing space, alter nutrient cycling, and affect other key processes essential to the proper functioning of ecosystems [30]. For example, Schelhaas et al. [75] provided a quantitative overview of the role of natural disturbances in European forests, which they suggested was useful as a basis for modeling the future impacts of climate change by establishing a baseline. They reported storms were responsible for 53% of the net volume affected over a 40-year period, while biotic factors (e.g., bark beetle epidemics) contributed 16%. In managed forests, natural disturbance cycles have been altered by human interventions aimed at reducing the susceptibility of forests to disturbances. In some cases, these interventions have exacerbated the effects of other disturbances. For example, dry forests in portions of the western United States were once dominated by open and park-like stands of widely dispersed trees prior to Euro-American settlement. Frequent thinning of small-diameter and

Table 3.2 Key findings of the Fourth National Climate Assessment concerning the impacts of climate change on forests in the United States ([86], https://nca2018.globalchange.gov)

Ecological disturbances and forest health	It is very likely that more frequent extreme weather events will increase the frequency and magnitude of severe ecological disturbances, driving rapid (months to years) and often persistent changes in forest structure and function across large landscapes. It is also likely that other changes, resulting from gradual climate change and less severe disturbances, will alter forest productivity and health and the distribution and abundance of species at longer timescales (decades to centuries).
Ecosystem services	It is very likely that climate change will decrease the ability of many forest ecosystems to provide important ecosystem services to society. Tree growth and carbon storage are expected to decrease in most locations as a result of higher temperatures, more frequent drought, and increased disturbances. The onset and magnitude of climate change effects on water resources in forest ecosystems will vary, but are already occurring in some regions.
Adaptation	Forest management activities that increase the resilience of United States forests to climate change are being implemented, with a broad range of adaptation options for different resources, including applications in planning. The future pace of adaptation will depend on how effectively social, organizational, and economic conditions support implementation.

fire-intolerant tree species by low-intensity surface fires, and competitive exclusion of tree seedlings by understory grasses are believed to have maintained such conditions. Many of these forests are now denser, have more small trees and fewer large trees, and are dominated by more shade-tolerant and fire-intolerant tree species, primarily as a result of fire suppression activities and harvesting practices implemented in the twentieth century. These changes have led to heavy accumulations of forest fuels [92] that feed severe wildfires when natural or human-induced ignitions occur. Today, thinning and prescribed fire are commonly used to increase the resiliency of forests to wildfires, especially in the western United States (Fig. 3.7).

Climate has shaped the world's forests for millennia, and minor climatic shifts may have significant effects on community compositions [76]. A notable global assessment of forest health published in 2010 reported 88 unique episodes of tree mortality over the last 30 years [6]. Since then, numerous additional episodes have been identified. The common implicated causal factor in these examples is elevated temperatures and/or water stress, raising the concern that the world's forests are increasingly responding to ongoing warming and drying attributed to climate change. Reports of large tree mortality events occurring during "climate change-type droughts" (hotter, drier) are now common worldwide (e.g., [5, 6, 38, 89]) (Fig. 3.8).

Among other factors, the current distribution of trees results from climatic shifts dating back millions of years in addition to more recent recolonization of deglaciated lands [33]. These historical patterns perhaps foreshadow changes to current coniferous vegetation as climate change accelerates. For example, based on the best existing data for 130 tree species in North America and associated climate information, McKenney et al. [56] predicted that on average the geographic range for a

Fig. 3.7 Conditions of many seasonally dry forests in the western United States, especially those that once experienced low-to-moderate intensity fire regimes, leave them uncharacteristically susceptible to high-severity wildfire. Creating more fire-resilient stands generally requires treatment of surface and ladder fuels, reductions in crown density, and maintenance of large-diameter trees [1]. A combination of thinning (top) and prescribed burning (bottom) is commonly used. Most evidence suggests that these treatments are typically accomplished with few unintended consequences as most ecosystem components (e.g., carbon sequestration, soils, wildlife) exhibit subtle impacts or no measurable impacts [78]. Since increased wildfire activity is expected as a result of climate change and desired fuel treatment effects are transient, repeated applications of fuel reduction treatments are required to maintain resilient conditions. (Photo credits: top, C.J. Fettig, and bottom, S.R. McKelvey, USFS Pacific Southwest Research Station)

Fig. 3.8 Much of California experienced a "climate change-type drought" in 2012–2015, inciting a large tree mortality event in the central and southern Sierra Nevada. While droughts have had an important influence on this region for millennia, this drought was characterized by large precipitation deficits and abnormally high temperatures during both the wet and dry seasons [2, 90], and in some areas is thought to be the most severe in 1200 years [35]. In particular, 2014 is noted for the lowest Palmer Drought Severity Index recorded for 1895–2017, when instrumental records were widely available (www.ncdc.noaa.gov/cag/). The drought resulted in progressive canopy water stress of at least 888 million trees and severe canopy water stress of at least 58 million trees [8], substantial mortality of dominant and co-dominant trees, and impacts to many ecological goods and services [23]. Much of the tree mortality was attributed to western pine beetle (*Dendroctonus brevicomis*), a native species that readily colonizes drought-stressed ponderosa pine (*Pinus ponderosa*) [22], but other tree and shrub species were affected by other contributing factors [27]. Stephens et al. [79] concluded that a greater potential for "mass fires" exists in future decades as a result of the amount, size, and continuity of dry combustible woody fuels, which could produce large, severe, and uncontrollable wildfires. Climate change will further amplify evapotranspiration and moisture overdrafts, likely increasing levels of tree mortality in the Sierra Nevada during droughts by ~15–20% per °C increase in temperature [34]. (Photo credit: C.J. Fettig, USFS Pacific Southwest Research Station)

given tree species will decrease by 12% and shift northward 700 km during the twenty-first century. Under a scenario where survival only occurs in areas where anticipated climatic conditions overlap with current climatic conditions, niches for tree survival decrease by 58% and shift northward 330 km. In terms of tree species, there will be winners (e.g., ponderosa pine, *Pinus ponderosa*) and losers (e.g., Engelmann spruce, *Picea engelmanni*) [72]. The fate of any tree species will depend on genetic variation, phenotypic variation, fecundity and dispersal mechanisms, and their resistance and resilience to a multitude of disturbances. A number of recent improvements have been made in our understanding of the effects of climate change on forest disturbances. Below we highlight four examples focusing on impacts to coniferous forests in western North America.

Forest Insects

Insects have adapted multiple thermally dependent phenological strategies to survive and persist in harsh environments, including development rates and thermal thresholds, diapause (a dormant physiological state entered to survive harsh conditions and increase cohort synchrony) and cold-hardening (increased cold tolerance through acclimation and metabolic processes). These traits, in addition to interactions with host trees and community associates, determine species- and location-specific responses to climate change (Fig. 3.9). Overall, climate change is thought to increase levels of tree mortality attributed to insects (e.g., for bark beetles, [11]; for defoliators, [18]), but there are important exceptions to this trend, especially among defoliators (e.g., larch budworm (*Zeiraphera diniana*), [17]). Bark beetles are commonly recognized as a primary disturbance agent in coniferous forests. Frequently referred to as "aggressive" bark beetles, several species can kill healthy trees and have the capacity to cause landscape-scale tree mortality. In western North America, recent decades have seen elevated levels of tree mortality attributed to bark beetle epidemics in spruce forests of south-central Alaska and the Rocky Mountains, lodgepole pine (*P. contorta*) forests of western Canada and the Rocky Mountains, pinyon-juniper woodlands of the southwestern United States, and ponderosa pine forests of Arizona, California, Oregon, and South Dakota, among others [26]. Because bark beetles, like all insects, are highly sensitive to thermal conditions conducive to population survival and growth, and water stress can influence host tree vigor, epidemics have been correlated with recent shifts in temperature [67] and precipitation [47] attributed to climate change. Bentz et al. [11] predicted increases in thermal regimes conducive to population success for two economically important species in North America, spruce beetle (*Dendroctonus rufipennis*) and mountain pine beetle (*D. ponderosae*), although there was considerable spatial and temporal variability in their predictions. They suggested a northward and upward in elevation movement of temperature suitability, and identified regions with a high potential for epidemics to occur in the twenty-first century. In Europe, warming temperatures are increasing the area of spruce habitat that supports two rather than one generation per year of European spruce beetle (*I. typographus*) [61], the most economically important species in Europe, and a higher number of sister broods [19], both of which benefit population growth [12].

Forest Diseases

As with forest insects, forest diseases caused by native and introduced pathogens are generally predicted to become more severe as a result of climate change [71, 82]. However, diseases caused by pathogens directly affected by climate (e.g.,

Fig. 3.9 In western North America, recent epidemics of mountain pine beetle (*Dendroctonus ponderosae*) have been severe and long-lasting. Since 2001, >27 million hectares of forest have been impacted. The species ranges throughout British Columbia and Alberta, Canada, most of the western United States, into northern Mexico, and colonizes several pine species [60]. Episodic outbreaks are a common occurrence, but the magnitude of recent events have exceeded the range of historic variability and have occurred in areas where mountain pine beetle epidemics were once rare or previously unrecorded (e.g., jack pine forests (*Pinus banksiana*) in Canada). Historically, the range of mountain pine beetle was restricted by climatic conditions unfavorable to brood development, but is expanding due to climate change and other factors. Scientists are concerned that mountain pine beetle could expand eastward across the boreal forest of Canada and into the eastern United States ([74], but see [11]). Others have speculated that under continued warming the loss of whitebark pine (*Pinus albicaulis*) (bottom), and the unique ecological services that this species provides, is imminent in many areas. The U.S. Fish and Wildlife Service announced in 2011 that it determined whitebark pine warranted protection under the Endangered Species Act, but that adding the species to the Federal List of Endangered and Threatened Wildlife and Plants was precluded by the need to address other listing actions of higher priority. In 2015, the listing priority of whitebark pine was downgraded due to declines in mountain pine beetle populations [21]. (Photo credits: top, C.J. Fettig, and bottom, C.J. Hayes, USFS Pacific Southwest Research Station)

needle blights) are predicted to have a reduced impact. These groups of pathogens may cause disease in healthy hosts if the pathogen's environmental requirements are met, many of which require moist conditions. Relatedly, following a detailed meta-analysis, Jactel et al. [43] concluded that the direct effects of drought on forest pathogens is mainly negative, as many require high humidity for spore dispersal and germination. Despite this, forest diseases caused by pathogens indirectly affected by climate (e.g., root diseases) are generally predicted to have increased impacts [82]. While the ability of these pathogens to spread and infect new hosts is affected by moisture, factors associated with climate change that stress their hosts are generally considered to be more important to host invasion.

Wildfire

Increases in the annual area burned by wildfires and cost of suppression efforts have been observed since the mid-1980s (Fig. 3.10). Wildfires have been episodic, occurring during warm years and strongly associated with changes in timing of spring snowmelt in the western United States, which, in turn, is sensitive to changes in temperature and precipitation [87]. At the same time, rapid increases in human habitation of the wildland-urban interface, especially in the western and southeastern United States, further exacerbates wildfire risks [53, 85]. By 2100, the annual area burned in the United States is projected to increase 2–6 fold from present due to climate change [51, 63]. As a result, concerns regarding air quality [73], human

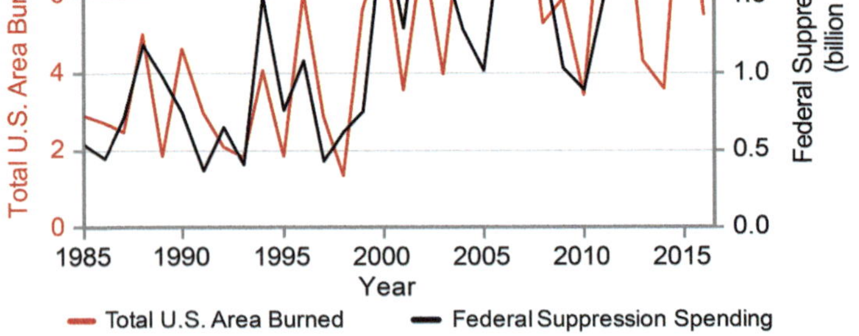

Fig. 3.10 Annual area burned by wildfire in the United States (1985–2016) and annual suppression costs (Consumer Price Index deflated, scaled to 2016 USD). (Source: USDA Forest Service, adapted from [86]). Wildfire-related concerns regarding air quality and human safety will become more important as a result of climate change

Fig. 3.11 A high-severity wildfire in a lodgepole pine (*Pinus contorta*) forest that was heavily impacted by mountain pine beetle (*Dendroctonus ponderosae*) a few years earlier. Dramatic changes are imposed on the forest fuel complex during and after a bark beetle epidemic. Associated anticipated changes in fire behavior include increased rates of fire spread and increased probabilities of torching, crowning, and spotting. As a result, fire fighters should expect increased difficulties with fire line construction and establishment of access, egress, and escape routes and safety zones [44, 45]. (Photo credit: C.J. Fettig, USFS Pacific Southwest Research Station)

safety, and protection of critical infrastructure will become increasingly important. As previously mentioned, bark beetles and wildfire are principal drivers of change in many North American forests, and both have increased in extent in recent years. As a result, these two disturbances are increasingly interacting on the landscape (Fig. 3.11).

Invasive Species

Finch et al. [29] provide an assessment of the effects of climate change on invasive species. They conclude that impacts are mediated primarily by direct effects on invaders, indirect effects that alter resource availability and interactions with other native and invasive species, and other factors such as human influences that alter the environment. Manipulative experiments, while uncommon, have shown that some invasive plants respond strongly to elevated CO_2. For example, growth of cheatgrass (*Bromus tectorum*), the notable invasive weed, is enhanced by elevated CO_2 [96] and increasing temperature [93], specifically during periods of high soil moisture. While desert plants like cheatgrass are likely to be among the most responsive to elevated CO_2 (i.e., due to increases in water use efficiency), similar relationships have been observed in many plants. While insects are not directly affected by elevated CO_2, increasing temperatures caused by elevated CO_2 can positively affect growth rates, phenology, dispersal, and survival. Conversely, rising temperatures also have the potential to negatively affect invasive insects by disrupting their synchrony with hosts and altering their overwintering environments. As such, climate change is also likely to modify competitive interactions to produce communities that are more or less susceptible to colonization by new invaders or expansion by existing invaders [29]. Of particular importance, many disturbances exacerbated by climate change often result in releases of large amounts of growing space which may increase the performance of some invasive species (Fig. 3.12). For example, a recent meta-analysis of relevant literature concluded that wildfires enhance invasive plant composition and performance, but have no effect on native species compositions [3]. Today, many invaders first arrive in new regions as stowaways on cargo ships. Climate change is reducing the extent and thickness of sea ice resulting in increases in shipping efficiencies [52, 81], which will increase survival rates of stowaways and enhance the likelihood of establishment in new regions [70].

Rapid and broad-scale tree mortality can have long-term impacts not only on forest health but human health (e.g., [4, 62]), with feedbacks that further influence climate and land use [49, 54]. For example, the annual amount of pollution removed by trees and forests in the conterminous United States is estimated at 17.4 million tons with a human health value of $6.8 billion USD [62]. Most of these health benefits come from reductions in the incidences of human mortality (850 cases), acute

Fig. 3.12 Forest disturbances often result in releases of large amounts of growing space. One potential consequence is subsequent invasion by exotic plants, in this case Canada thistle (*Cirsium arvense*) and bull thistle (*Cirsium vulgare*) in a lodgepole pine (*Pinus contorta*) forest that experienced high levels of tree mortality (>70%) attributed to mountain pine beetle (*Dendroctonus ponderosae*). (Photo by J.B. Runyon, USFS Rocky Mountain Research Station)

respiratory symptoms (670,000), asthma exacerbation (430,000) and lost school days (200,000) [62]. On the other hand, most states in the western United States have at least 10% of their forested landscapes at risk (defined as without remediation, at least 25% of standing live basal area greater than 2.54 cm in diameter will be killed in the next 15 years) to forest insects and diseases epidemics [48], the impacts of many of which are exacerbated by climate change. Seven trees species endemic to the western United States are expected to suffer losses of ≥25% in the next 15 years [48]. To that end, the recent loss of whitebark pine (*P. albicaulis*) stands due to interactions among climate change, mountain pine beetle, and white pine blister rust underscores the need for a greater understanding of climate change effects on complex interactions important to ecosystem resiliency and stability (Fig. 3.9). Characterizing thresholds for systems beyond which such changes are irreversible is important.

There are tools available to restore forest health and to increase the resistance and resilience of forests to climate change and disturbances exacerbated by climate change (e.g., [24, 78]). Resource managers can intervene and mitigate some of the negative effects [28, 64] (Fig. 3.13), but this requires knowledge of the effects of climate change on forests, forest-related enterprises, and resource-dependent communities, and of institutional, social, and environmental factors that influence adaptive capacity. Healthy forests have a vital role to play in combating climate change and its influence on environmental health, and are critical to the welfare and sustainability of human societies.

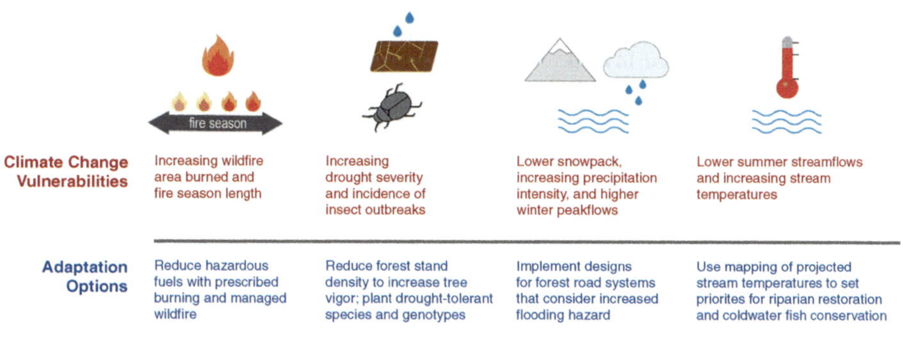

Fig. 3.13 In order to increase resistance and resilience to stressors and disturbances, adaptation practices have been developed in response to climate change vulnerabilities. Flexible approaches that promote learning and sharing among interested parties can help accelerate implementation. (Source: adapted from [86])

References

1. Agee JK, Skinner CN. Basic principles of forest fuel reduction treatments. For Ecol Manag. 2005;211:83–96.
2. Aghakouchak A, Cheng L, Mazdiyasni O, Farahmand A. Global warming and changes in risk of concurrent climate extremes: insights from the 2014 California drought. Geophys Res Lett. 2014;44:8847–52.
3. Alba C, Skálová H, McGregor KF, D'Antonio C, Pysek P. Native and exotic plant species respond differently to wildfire and prescribed fire as revealed by meta-analysis. J Veg Sci. 2015;26:102–13.
4. Allan BF, Keesing F, Ostfeld R. Effect of forest fragmentation on Lyme disease risk. Conserv Biol. 2003;17:267–72.
5. Allen CD, Breshears DD, McDowell NG. On underestimation of global vulnerability to tree mortality and forest die-off from hotter drought in the Anthropocene. Ecosphere. 2015;6:1–55.
6. Allen CD, Macalady AK, Chenchouni H, Bachelet D, McDowell N, Vennetier M, Kitzberger T, Rigling A, Breshears DD, Hogg EH, Gonzalez P, Fensham R, Zhangm Z, Castro J, Demidova N, Lim JH, Allard G, Running SW, Semerci A, Cobb N. A global overview of drought and heat-induced tree mortality reveals emerging climate change risks for forests. For Ecol Manag. 2011;259:660–84.
7. Arora VK, Peng Y, Kurz WA, Fyfe JC, Hawkins B, Werner AT. Potential near-future carbon uptake overcomes losses from a large insect outbreak in British Columbia, Canada. Geophys Res Lett. 2016;43:2590–8.
8. Asner GP, Brodrick PG, Anderson CB, Vaughn N, Knapp DE, Martin RE. Progressive forest canopy water loss during the 2012–2015 California drought. Proc Natl Acad Sci U S A. 2016;113:E249–E55.
9. Balling RC Jr, Gober P. Climate variability and residential water use in the city of Phoenix. Arizona. J Appl Meteor Climate. 2007;46:1130–7.
10. Barnett TP, Pierce DW, Hidalgo HG, Bonfils C, Santer BD, Das T, Bala G, Wood AW, Nozawa T, Mirin AA, Cayan DR, Dettinger MD. Human-induced changes in the hydrology of the western United States. Science. 2008;319:1080–3.
11. Bentz BJ, Régnière J, Fettig CJ, Hansen EM, Hayes JL, Hicke JA, Kelsey RG, Lundquist J, Negrón JF, Seybold S. Climate change and bark beetles of the western United States and Canada: direct and indirect effects. Bioscience. 2010;60:602–13.
12. Bentz BJ, Jönsson AM, Schroeder M, Weed A, Wilcke RAI, Larsson K. *Ips typographus* and *Dendroctonus ponderosae* models project thermal suitability for intra-and inter-continental establishment in a changing climate. Front For Global Chang. 2019;2 https://doi.org/10.3389/ffgc.2019.00001.
13. Bonan GD. Forests and climate change: forcings, feedbacks, and the climate benefits of forests. Science. 2009;320:1444–9.
14. Brown TC. Projecting US freshwater withdrawals. J Water Resour Res. 2000;36:769–80.
15. Brown TC, Foti R, Ramirez JA. Projected freshwater withdrawals in the United States under a changing climate. Water Resour Res. 2013;49:1259–76.
16. Brown TC, Mahat V, Ramirez JA. Adaptation to future water shortages in the United States caused by population growth and climate change. Earth's Future. 2019;7:1–16.
17. Büntgen U, Frank D, Liebhold A, Johnson D, Carrer M, Urbinati C, Grabner M, Nicolussi K, Levanic T, Esper J. Three centuries of insect outbreaks across the European Alps. New Phytol. 2009;182:929–41.
18. Carnicer J, Colla M, Ninyerolac M, Pons X, Sánchez G, Peñuelasa J. Widespread crown condition decline, food web disruption, and amplified tree mortality with increased climate change-type drought. Proc Natl Acad Sci U S A. 2011;108:1474–8.
19. Davídková M, Doležal P. Sister broods in the spruce bark beetle, *Ips typographus* (L.). For Ecol Manag. 2017;405:13–21.

20. Dugan AJ, Birdsey R, Mascorro VS, Magnan M, Smyth CE, Olguin M, Kurz WA. A systems approach to assess climate change mitigation options in landscapes of the United States forest sector. Carb Bal Manage. 2018;13:13.
21. Federal Register. Endangered and threatened wildlife and plants, vol. 80; 2015. p. 80584–614.
22. Fettig CJ. Native bark beetles and wood borers in Mediterranean forests of California. In: Lieutier F, Paine TD, editors. Insects and diseases of Mediterranean forest systems. Cham: Springer International; 2016.
23. Fettig CJ. Socioecological impacts of the western pine beetle outbreak in southern California: lessons for the future. J For. 2019;117:138–43.
24. Fettig CJ, Klepzig KD, Billings RF, Munson AS, Nebeker TE, Negrón JF, Nowak JT. The effectiveness of vegetation management practices for prevention and control of bark beetle infestations in coniferous forests of the western and southern United States. For Ecol Manag. 2007;238:24–53.
25. Fettig CJ, Reid ML, Bentz BJ, Sevanto S, Spittlehouse DL, Wang T. Changing climates, changing forests: a western North American perspective. J For. 2013;111:214–28.
26. Fettig CJ, Progar RA, Paschke J, Sapio FJ. Forest insects. In: Robertson G, Barrett T, editors. Implications of forest disturbance processes for sustainability in the western US. Portland: U.S. Department of Agriculture, Forest Service; 2020.
27. Fettig CJ, Mortenson LA, Bulaon BM, Foulk PB. Tree mortality following drought in the central and southern Sierra Nevada, California, US. For Ecol Manage. 2019;432:164–78.
28. Fettig CJ, Wuenschel A, Balachowski J, Butz RJ, Jacobsen AL, North MP, Ostoja SM, Pratt RB, Standiford RB. Drought management recommendations for California. In: Vose J, Patel-Weynand T, Peterson DL, Luce CH, editors. Drought impacts on U.S. forests and rangelands: translating science into management responses. Washington, DC: U.S. Department of Agriculture, Forest Service; 2019.
29. Finch DM, Butler JL, Runyon JB, Fettig CJ, Kilkenny FF, Jose S, Frankel SJ, Cushman A, Cobb RC, Dukes JS, Hicke JA, Amelon SK. Effects of climate change on invasive species. In: Poland TM, Patel-Weynand T, Finch D, Miniat CF, Lopez V, editors. Invasive species in forests and grasslands of the United States: a comprehensive science synthesis for the United States forest sector. Berlin: Springer; 2021.
30. Folke C, Carpenter S, Walker B, Scheffer M, Elmqvist T, Gunderson L, Holling CS. Regime shifts, resilience, and biodiversity in ecosystem management. Annu Rev Ecol Evol Syst. 2004;35:557–81.
31. Foti R, Ramirez JA, Brown TC. Vulnerability of U.S. water supply to shortage: a technical document supporting the Forest Service 2010 RPA assessment. Fort Collins: U.S. Department of Agriculture, Forest Service; 2011. 147 pp.
32. Fritze H, Stewart IT, Pebesma EJ. Shifts in Western North American snowmelt runoff regimes for the recent warm decades. J Hydrometeorol. 2011;12:989–1006.
33. Godbout J, Fazekas A, Newton CH, Yeh FC. Glacial vicariance in the Pacific northwest: evidence from a lodgepole pine mitochondrial DNA minisatellite for multiple genetically distinct and widely separated refugia. Mol Ecol. 2008;17:2463–75.
34. Goulden ML, Bales RC. California forest die-off linked to multi-year deep soil drying in 2012–2015 drought. Nat Geosci. 2019;12:632–7.
35. Griffin D, Anchukaitis KJ. How unusual is the 2012–2014 California drought? Geophys Res Lett. 2014;41:9017–23.
36. Groves DG, Yates D, Tebaldi C. Developing and applying uncertain global climate change projections for regional water management planning. Water Resour Res. 2008;44:W12413.
37. Hamlet AF, Mote PW, Clark MP, Lettenmaier DP. Effects of temperature and precipitation variability on snowpack trends in the western United States. J Clim. 2005;18:4545–61.
38. Hartmann H, Moura CF, Anderegg WR, Ruehr NK, Salmon Y, Allen CD, Arndt SK, Breshears DD, Davi H, Galbraith D, Ruthrof KX, Wunder J, Adams HD, Bloemen J, Cailleret M, Cobb R, Gessler A, Grams TE, Jansen S, Kautz M, Lloret F, O'Brien M. Research frontiers for improving our understanding of drought-induced tree and forest mortality. New Phytol. 2018;218:15–28.

39. Hidalgo HG, Das T, Dettinger MD, Cayan DR, Pierce DW, Barnett TP, Bala G, Mirin A, Wood AW, Bonfils C, Santer BD, Nozawa T. Detection and attribution of streamflow timing changes to climate change in the western United States. J Clim. 2009;22:3838–55.
40. Hodgkins GA, Dudley RW. Changes in the timing of winter–spring streamflows in eastern North America, 1913–2002. Geophys Res Lett. 2006;33:L06402.
41. Huang B, Thorne PW, Banzon VF, Boyer T, Chepurin G, Lawrimore JH, Menne MJ, Smith TM, Vose RS, Zhang H-M. Extended reconstructed sea surface temperature version 5 (ERSSTv5): upgrades, validations, and inter-comparisons. J Clim. 2017;30:8179–205.
42. Isaak D, Wenger S, Peterson E, Ver Hoef J, Nagel D, Luce C, Hostetler S, Dunham J, Roper B, Wollrab S, Chandler G, Horan D, Parkes-Payne S. The NorWeST summer stream temperature model and scenarios for the western U.S.: a crowd-sourced database and new geospatial tools foster a user community and predict broad climate warming of rivers and streams. Water Resour Res. 2017;53:9181–205. https://doi.org/10.1002/2017WR020969.
43. Jactel H, Petit J, Desprez-Loustau ML, Delzon S, Piou D, Battisti A, Koricheva J. Drought effects on damage by forest insects and pathogens: a meta-analysis. Global Change Bio. 2012;18:267–76.
44. Jenkins MJ, Page WG, Hebertson EG, Alexander ME. Fuels and fire behavior dynamics in bark beetle-attacked forests in western North America and implications for fire management. For Ecol Manag. 2012;275:23–34.
45. Jenkins MJ, Runyon JB, Fettig CJ, Page WG, Bentz BJ. Interactions among the mountain pine beetle, fires, and fuels. For Sci. 2014;60:489–501.
46. Köhl M, Lasco R, Cifuentes M, Jonsson Ö, Korhonen KT, Mundhenk P, de Jesus NJ, Stinson G. Changes in forest production, biomass and carbon: results from the 2015 UN FAO Global Forest Resource Assessment. For Ecol Manag. 2015;352:21–34.
47. Kolb TE, Fettig CJ, Ayres MP, Bentz BJ, Hicke JA, Mathiasen R, Stewart JE, Weed AS. Observed and anticipated impacts of drought on forests insects and diseases in the United States. For Ecol Manag. 2016;380:321–34.
48. Krist FJ, Ellenwood JR, Woods ME, McMahan AJ, Cowardin JP, Ryerson DE, Sapio FJ, Zweifler MO, Romero SA. 2013–2027 national insect and disease forest risk assessment. Washington, DC: U.S. Department of Agriculture, Forest Service, Forest Health Technology Enterprise Team; 2014.
49. Kurz WA, Dymond CC, Stinson G, Rampley GJ, Neilson ET, Carroll AL, Ebata T, Safranyik L. Mountain pine beetle and forest carbon feedback to climate change. Nature. 2008;452:987–90.
50. Lall U, Johnson T, Colohan P, Aghakouchak A, Brown C, McCabe G, Pulwarty R, Sankarasubramanian A. Water. In: Reidmiller DR, Avery CW, Easterling DR, Kunkel KE, KLM L, Maycock TK, Stewart BC, editors. Impacts, risks, and adaptation in the United States: fourth National Climate Assessment, vol. II. Washington, DC: U.S. Global Change Research Program; 2018.
51. Litschert SE, Brown TC, Theobald DM. Historic and future extent of wildfires in the southern Rockies ecoregion, USA. For Ecol Manage. 2012;269:124–33.
52. Liu JP, Song MR, Horton RM, Hu YY. Reducing spread in climate model projections of a September ice-free arctic. Proc Natl Acad Sci U S A. 2013;110:12571–6.
53. Liu Z, Wimberly MC, Lamsal A, Sohl TL, Hawbaker TJ. Climate change and wildfire risk in an expanding wildland–urban interface: a case study from the Colorado front range corridor. Landsc Ecol. 2015;30:1943–57.
54. McDowell NG, Pockman WT, Allen C, Breshears DD, Cobb N, Kolb T, Plaut J, Sperry J, West A, Williams D, Yepez EA. Mechanisms of plant survival and mortality during drought: why do some plants survive while others succumb to drought? New Phytol. 2008;178:719–39.
55. McIsaac GF, David MB, Gertner GZ, Goolsby DA. Relating net nitrogen input in the Mississippi River basin to nitrate flux in the lower Mississippi River: a comparison of approaches. J Environ Qual. 2002;31:1610–22.

56. McKenney DW, Pedlar JH, Lawrence K, Campbell K, Hutchinson MF. Potential impacts of climate change on the distribution of North American trees. Bioscience. 2007;57:939–48.
57. Menne MJ, Williams CN, Gleason BE, Rennie JJ, Lawrimore JH. The global historical climatology network – monthly temperature dataset, version 4. J Clim. 2018;31:9835–54.
58. Merritt DM, Poff NL. Shifting dominance of riparian *Populus* and *Tamarix* along gradients of flow alteration in western North American rivers. Ecol Appl. 2010;20:135–52.
59. Millar CI, Stephenson NL. Temperate forest health in an era of emerging mega disturbance. Science. 2015;349:823–6.
60. Negrón JF, Fettig CJ. Mountain pine beetle, a major disturbance agent in US western coniferous forests: a synthesis of the state of knowledge. For Sci. 2014;60:409–13.
61. Netherer S, Matthews B, Katzensteiner K, Blackwell E, Henschke P, Hietz P, Pennerstorfer J, Rosner S, Kikuta S, Schume H, Schopf A. Do water-limiting conditions predispose Norway spruce to bark beetle attack? New Phytol. 2015;205:1128–41.
62. Nowak DJ, Hirabayashi S, Bodine A, Greenfield E. Tree and forest effects on air quality and human health in the United States. Environ Pollut. 2014;193:119–29.
63. Ojima DS, Iverson LR, Sohngen BL, Vose JM, Woodall CW, Domke GM, Peterson DL, Littell JS, Matthews SN, Prasad AM, Peters MP, Yohe GW, Friggens MM. Risk assessment. In: Peterson DL, Vose JM, Patel-Weynand T, editors. Climate change and United States forests. Dordrecht: Springer; 2014.
64. Peterson DL, Millar CI, Joyce LA, Furniss MJ, Haolosky JE, Neilson RP, Morelli TL. Responding to climate change on national forests: a guidebook for developing adaptation options. Portland: U.S. Department of Agriculture, Forest Service; 2011.
65. Potyondy JP, et al. Watershed condition framework. FS-977. U.S. Department of Agriculture, Forest Service; 2011., 24 pp.
66. Potyondy JP, Geier TW. Watershed condition classification technical guide. FS-978. U.S. Department of Agriculture, Forest Service; 2011., 41 pp.
67. Powell JA, Logan JA. Insect seasonality-circle map analysis of temperature-driven life cycles. Theo Pop Bio. 2005;67:161–79.
68. Pruski FF, Nearing MA. Climate-induced changes in erosion during the 21st century for eight U.S. locations. Water Resour Res. 2002;38:34–11.
69. Pruski FF, Nearing MA. Runoff and soil-loss responses to changes in precipitation: a computer simulation study. J Soil Water Conserv. 2002;57:7–16.
70. Pyke CR, Thomas R, Porter RD, Hellmann JJ, Dukes JS, Lodge DM, Chavarria G. Current practices and future opportunities for policy on climate change and invasive species. Conserv Biol. 2008;22:585–92.
71. Ramsfield TD, Bentz BJ, Faccoli M, Jactel H, Brockerhoff EG. Forest health in a changing world: effects of globalization and climate change on forest insect and pathogen impacts. Forestry: Intl J For Res. 2016;89:245–52.
72. Rehfeldt GE, Crookston NL, Warwell MV, Evans JS. Empirical analyses of plant-climate relationships for the western United States. Int J Plant Sci. 2006;167:1123–50.
73. Reid C, Maestas M. Wildfire smoke exposure under climate change: impact on respiratory health of affected communities. Current Opinion Pul Med. 2019;25:179–87.
74. Safranyik L, Carroll AL, Régnière J, Langor DW, Riel WG, Shore TL, Peter B, Cooke BJ, Nealis VG, Taylor SW. Potential for range expansion of mountain pine beetle into the boreal forest of North America. Can Entomol. 2010;142:415–42.
75. Schelhaas MJ, Nabuurs GJ, Schuck A. Natural disturbances in the European forests in the 19th and 20th centuries. Global Change Bio. 2003;9:1620–33.
76. Shugart HH. Theory of forest dynamics: the ecological implications of forest succession models. New York: Springer; 2003.
77. Smith TM, Reynolds RW, Peterson TC, Lawrimore JH. Improvements to NOAA's historical merged land–ocean surface temperatures analysis (1880–2006). J Clim. 2008;21:2283–96.

78. Stephens SL, McIver JD, Boerner REJ, Fettig CJ, Fontaine JB, Hartsough BR, Kennedy P, Schwilk DW. Effects of forest fuel-reduction treatments in the United States. Bioscience. 2012;62:549–60.
79. Stephens SL, Collins BM, Fettig CJ, Finney MA, Hoffman CE, Knapp EE, North ME, Safford H. Drought, tree mortality, and wildfire in forests adapted to frequent fire. Bioscience. 2018;68:77–88.
80. Stewart IT, Cayan DR, Dettinger MD. Changes toward earlier streamflow timing across western North America. J Clim. 2005;18:1136–55.
81. Stroeve JC, Serreze MC, Holland MM, Kay JE, Malanik J, Barrett AP. The Arctic's rapidly shrinking sea ice cover: a research synthesis. Clim Chang. 2012;110:1005–27.
82. Sturrock RN, Frankel SJ, Brown AV, Hennon PE, Kliejunas JT, Lewis KJ, Worrall JJ, Woods AJ. Climate change and forest diseases. Plant Pathol. 2011;60:133–49.
83. Theobald DM, Merritt DM, Norman JB III. Assessment of threats to riparian ecosystems in the western U.S. A report presented to The Western Environmental Threats Assessment Center, Prineville, OR by the USDA Forest Service, Stream Systems Technology Center and Colorado State University, Fort Collins, CO, 61 pp, 2010.
84. USGCRP. Impacts, risks, and adaptation in the United States: fourth National Climate Assessment, vol. II. Washington, DC: U.S. Global Change Research Program; 2018.
85. Volker C, Radeloff DP, Helmers H, Kramer HA, Mockrin MH, Alexandre PM, Bar-Massada A, Butsic V, Hawbaker TJ, Martinuzzi S, Syphard AD, Stewart SI. Rapid growth of the US wildland-urban interface raises wildfire risk. Proc Natl Acad Sci U S A. 2018;115:3314–9.
86. Vose JM, Peterson DL, Domke GM, Fettig CJ, Joyce LA, Keane RE, Luce CH, Prestemon JP, Band LE, Clark JS, Cooley NE, D'Amato A, Halofsky JE. Forests. In: Reidmiller DR, Avery CW, Easterling DR, Kunkel KE, Lewis KLM, Maycock TK, Stewart BC, editors. Impacts, risks, and adaptation in the United States: fourth National Climate Assessment, volume II. Washington, DC: U.S. Global Change Research Program; 2018.
87. Westerling AL. Increasing western U.S. forest wildfire activity: sensitivity to changes in the timing of spring. Philos Trans R Soc Lond [Biol]. 2016;371:20150178.
88. Whitehead P, Wade A, Butterfield D. Potential impacts of climate change on water quality in six UK rivers. Hydrol Res. 2009;40:113–22.
89. Williams AP, Allen CD, Macalady AK, Griffin D, Woodhouse CA, Meko DM, Swetnam TW, Rauscher SA, Seager R, Grissino-Mayer HD. Temperature as a potent driver of regional forest drought stress and tree mortality. Nat Clim Chang. 2013;3:292–7.
90. Williams AP, Seager R, Abatzoglou JT, Cook BI, Smerdon JE, Cook ER. Contribution of anthropogenic warming to California drought during 2012–2014. Geophys Res Lett. 2015;42:6819–28.
91. Xiao M, Udall B, Lettenmaier DP. On the causes of declining Colorado River streamflows. Water Resour Res. 2018;54:6739–56.
92. Youngblood A, Max T, Coe K. Stand structure in eastside old-growth ponderosa pine forests of Oregon and northern California. For Ecol Manag. 2004;199:191–217.
93. Zelikova TJ, Hufbauer RA, Reed SC, Wertin T, Fettig C, Belnap J. Eco-evolutionary responses of *Bromus tectorum* to climate change: implications for biological invasions. Ecol Evol. 2013;3:1374–87.
94. Zhang H-M, et al. Recent development on the NOAA's Global Surface Temperature Dataset, Abstract GC53H-04 presented at 2016 Fall Meeting, American Geophysical Union, San Francisco, CA, 12–16 Dec 2016.
95. Zhang YK, Schilling KE. Increasing streamflow and baseflow in Mississippi River since the 1940s: effect of land use change. J Hydrol. 2006;324:412–22.
96. Ziska LH, Reeves JB III, Blank B. The impact of recent increases in atmospheric CO_2 on biomass production and vegetative retention of cheatgrass (*Bromus tectorum*): implications for fire disturbance. Global Change Bio. 2005;11:1325–32.
97. Xiao M, Udall B, Lettenmaier DP. On the causes of declining Colorado River streamflows. Water Resour Res. 2018;54:6739–56. https://doi.org/10.1029/2018WR023153.
98. Mote PW, Li S, Lettenmaier DP. et al. Dramatic declines in snowpack in the western US. npj Clim Atmos Sci. 1, 2 (2018). https://doi.org/10.1038/s41612-018-0012-1.

Chapter 4
Eyewitness to Climate Change

Will Steger and Nicole Rom

Eyewitness to Climate Change

Will Steger (Polar Explorer) and Nicole Rom

Climate change is a reality. It threatens both our society and life as we know it on earth. The overwhelming consensus of the scientific community for more than three decades has been that the planetary warming we are now experiencing, and the resulting climate change, is largely a human-induced phenomenon. Human-induced climate change is brought on mainly by the release of carbon dioxide through the burning of fossil fuels, which blankets our atmosphere, raising the earth's surface temperature.

The amount of carbon dioxide that is in the atmosphere today is the *minimum* level we are going to have to live with for the indefinite future. Once carbon dioxide is in the stratosphere above us, it will stay there for hundreds and hundreds of years. It is as though you gained the most weight in your life, and knew you would never weigh even a single pound less, ever. Carbon dioxide does eventually get pulled back out of the atmosphere by natural processes, but that happens very slowly. Climate scientists like to compare the atmosphere to a bathtub half-full of water, with a very slow drain and a slowly trickling faucet. If the drain and the trickle are balanced, the water level never changes—just as the trickle of natural carbon dioxide into the atmosphere and the drainage into trees, carbonate rocks, and other places have been in balance for at least 2000 years, and probably more. Atmospheric carbon dioxide hovered at around 270–290 parts per million that whole time, and the climate stayed more or less stable. Carbon dioxide levels have now reached over

W. Steger
Climate Generation: A Will Steger Legacy, Minneapolis, MN, USA

N. Rom (✉)
Climate Generation: A Will Steger Legacy, Minneapolis, MN, USA
e-mail: nicole@climategen.org

K. E. Pinkerton, W. N. Rom (eds.), *Climate Change and Global Public Health*, Respiratory Medicine, https://doi.org/10.1007/978-3-030-54746-2_4

400 ppm and are on course to reach 500 ppm, coinciding with the rise of the burning of fossil fuels after the industrial revolution. Climate scientists consider 350 ppm to be the safe level to avoid catastrophic changes.

For over 50 years, National Geographic explorer Will Steger has inspired thousands of people through his chronicles of arctic adventures, and, more recently, his remarkable eyewitness accounts of climate change on our polar regions. Steger is most known for his historical polar expeditions. In 1986, Steger led the first confirmed unsupported dogsled trip to the North Pole. With seven teammates and 49 dogs, the team traveled 500 miles in 56 days. They began their expedition at Canada's Ward Hunt ice shelf on Ellesmere Island, which has since disintegrated from climate change; over 30% of the polar ice pack the expedition team crossed is smaller and thinner, requiring teams that travel to the North Pole today to have some sort of flotation. The expedition included two other Minnesotans, including the first woman, Ann Bancroft.

In 1989–1990, together with five other men from six countries, Steger led an international expedition across Antarctica—3741 miles, farther than from New York to Los Angeles, to raise broad public awareness of the importance of the region for education and scientific research and to protect it from the threats of mineral exploration.

Trans-Antarctica Expedition team breaks for lunch. (Photo courtesy of Will Steger)

In a series of expeditions in the mid-1990s, Steger crossed the Arctic Ocean from Russia to Canada's Ellesmere Island on some of the most dynamic and moving surface on Earth. Steger returned to the Canadian Arctic in the 2000s, with a mission to draw worldwide attention to climate change and its impacts on the region through education and advocacy.

Starting in 1990, dramatic changes began happening in the polar regions. With intimate familiarity of its vast lands, wildlife, and climates, Steger has experienced firsthand observations about how the Earth's surface has changed, including vanishing glaciers, shattered ice shelves, melting permafrost, and displaced communities of people and animals. The resulting changes from climate change deeply affect Steger in a way a scientific study or a satellite image could. Today, *every* ice shelf Steger has crossed has disintegrated into the ocean as a result of climate change.

When Steger's Trans Antarctica expedition team first flew over Antarctica's Weddell Sea, near the Antarctic Peninsula, he recorded this: "July 26th 1989: And it's Antarctica that we are looking at that is going to be the main player in the destiny of the human race. It's this snow and ice here. If the atmosphere warms up, this ice right in this area is going to break off into the ocean." At the time, however, it did not seem possible that an ice mass this large could actually break up. It seemed that the Larsen, a long ice shelf jutting in to the Weddell Sea, was as permanent as the Antarctic continent itself.

But on March 2, 2002, Steger was thumbing through the Minneapolis-based Star Tribune newspaper and on page nine in bold print was the caption "Larsen B Ice Shelf Disintegrates." It seemed at first this was science fiction, and it took days before Steger could grasp the extent of this global environmental catastrophe. There is no way to comprehend the massiveness of the disintegration of the Larsen ice shelf unless you ski and walk every step of the way. It took Steger's expedition team 31 days, from July 27 to August 26 to cross the full length of this ice shelf. Every day, camp after camp, through storm, whiteouts, and clear weather, the team skied and pushed their sleds, becoming intimately familiar with the ice shelf that treated them with safe surface conditions. While crossing the Larsen, the ice shelf felt very stable to Steger's team. Scientists at Queen's University estimate the shelf could have been stable for as long as 12,000 years—that many years ago there were still mastodons, mammoths, and saber-toothed cats roaming the earth. Over the course of one month in 2002, however, a chunk of ice the size of the New England state of Rhode Island broke free from the Larsen B ice shelf. The speed of the collapse surprised even the scientists who were monitoring the shelf. Scientists link the collapse with climate change. As of 2010, both the Larsen A and B ice shelves have disintegrated, along with the Wilkins ice shelf. Scientists are watching the continent closely, paying particular attention to melting, calving, and complete disintegration of the Larsen C, the Ross (about the size of France), and the Ronne (about the size of Spain) ice shelves on Antarctica.

After Antarctica, Greenland's ice cap contains the second largest mass of frozen fresh water in the world. At 7000 feet, in July 2008, Steger was on a kite-ski expedition when his team came across something very unusual on Greenland: running water. Every year in the summer, on the coast of the ice cap, the temperature warms enough to melt out systems of rivers and lakes. Since 1992, the thawing levels during the summer season on Greenland have increased in elevation. Data from NASA's Gravity Recovery and Climate Experiment show Greenland lost 150 to 250 cubic kilometers (36 to 60 cubic miles) of ice per year between 2002 and 2006. In 2008, Steger literally ran into running water at 7000 feet, the highest point on Greenland. By 2018, the rising thaw levels on Greenland were showing signs of surging, melting at the fastest rate in 400 years.

Unlike Antarctica, which sits at the bottom, the Arctic sits at the top of the world. The Arctic is an ocean two miles deep, surrounded by the land of eight nations. The Arctic Ocean is covered with a layer of ice eight to twelve feet thick. It is like a bucket of water with a thin layer of dust on the surface—the bucket represents the Arctic Ocean, the layer of dust, the ice. In the spring and summer, the ice breaks up and the ice is in constant motion, moved by wind currents and the ocean's movements. Sea ice is frozen seawater that floats on the ocean surface. Blanketing millions of square kilometers, sea ice forms and melts with the polar seasons, affecting both humans and wildlife. In the Arctic, some sea ice persists year after year, whereas almost all Southern Ocean or Antarctic sea ice is "seasonal ice," meaning it melts away and reforms annually. Sea ice in the Arctic plays a unique role in regulating the Earth's climate because of its role in regulating global temperature.

On Steger's 1995 expedition from Russia to Ellesmere Island, traveling predominantly over the Arctic Ocean, Steger's team battled more open water in one day than his entire 1986 North Pole expedition just 9 years earlier; a harbinger of future changes to come. Steger tells the story: "We had just left the North Pole a few days after Earth Day in 1995 when I noticed the ice beneath our skis was dark, almost black. This is a sign of thin ice. Just ahead of me, it was too late for the sled and 10 dogs that had broken through the ice and tipped onto its side, half in the water, half on thin ice. It took several hours to get the sled back on sturdier ice. Over the course

Steger's expedition team attempts crossing open water on 1995 arctic expedition. (Photo courtesy of Will Steger)

of the expedition, we had to review our route, which changed daily because of varying ice conditions and shifting ice. It became clear that this expedition was behind schedule; not because of poor planning, but because of an unusual year in the Arctic. We had no idea at the time this would be the new normal for the Arctic. There was a lot of snow, and while the weather was cold, it was already seeing warmer than normal temperatures for the region, which means ice is thinne is more open water."

Climate Generation's Ellesmere Island Expedition travels through the ruins of the summer sea ice melt from 2007 a year later. (Photo courtesy of Climate Generation: A Will Steger Legacy)

Water can soak up a lot of heat. When the oceans get warmer, sea ice begins to melt in the Arctic and around Greenland. NASA's Earth satellites show us that every summer some Arctic ice melts and shrinks, getting smallest by September. Then, when winter comes, the ice grows again. But, since 1979, the September ice has been getting smaller and smaller and thinner and thinner. As a result of this change, the Arctic Ocean is also turning from a once reflective surface to an absorptive surface. Traditionally, the Arctic Ocean's layer of thick ice has reflected 90% of the sun's energy (the same amount of energy that hits the tropical regions near the equator) back into space, helping to keep the planet cool. Now that the ice is smaller and thinner, it melts more ice, creating a positive feedback loop, and at the same time, revealing darker ocean surfaces, which absorbs the once reflected energy into the ocean, melting even more ice.

On a large scale, what we are witnessing around the world are feedback loops that spur large and rapid changes to our environment. The Arctic sea ice is a great example of those changes. The Arctic sea ice has lost over half of its thickness and area in the last three decades. Its once reflective surface is now exposing the darker ocean surfaces; because darker surfaces absorb more light and energy than lighter surfaces, warmth is accelerated and leads to more melting of ice. As a result, without any additional greenhouse gases, the Arctic will soon be ice-free during the summer. If the summer sea ice disappears, animals like the polar bear and walrus will face probable extinction. Arctic warming is also especially problematic because of permafrost thaw, which releases both carbon and methane (both are greenhouse gases, with methane an even more potent gas) that has been trapped in the frozen ground, sometimes for over 1000 years. The loss of ice and release of carbon and methane that we are now experiencing worldwide is the fingerprint of climate change.

On Canada's Baffin Island, home of the Inuit, Steger set out to document how climate change was affecting the region, to meet with Inuit elders and students, to explore traditional ecological knowledge in the remote communities visited on the trail, and to put a human face and cultural voice on this complex issue with the support of his nonprofit, Climate Generation: A Will Steger Legacy in 2007.

Nowhere on earth is the climate changing more rapidly or more dramatically than in the Arctic's Baffin Island. Climate Generation's Baffin Island Expedition witnessed first-hand climate impacts as the team traveled by dog-team from Iqaluit to Pangnirtung. Warmer-than-normal temperatures made it difficult to simply walk from the land out on to the sea ice in Frobisher Bay; the tidal overflow along the shore was not refreezing and instead the water remained liquid or slushy. The team had to pick their way across the more-solidly frozen sections, even as their feet sank into the slush, which soaked their moose-hide mukluk boots.

During pre-trip planning, the team assumed only the American members would sleep in tents. The Inuit members planned to make an igloo every night. Different from normal snow conditions, however, made igloo-making impossible. In many places, there was simply not enough snow. In other places, the snow had weak and soft layers that made blocks cut from it collapse instead of stand up. Living conditions are much warmer inside an igloo than inside a tent, so it was a disappointment to the Inuit members to not be able to build igloos.

On the Hall Peninsula, as the dog-teams made their way overland from Iqaluit, the team crossed a small flowing creek that was completely open, unfrozen water. Theo Ikummaq, the Inuit team leader, said this time of year that creek should be frozen solid. The temperatures on South Baffin Island had been, however, as much as 40 degrees above normal during the weeks before the expedition's departure.

The 60-mile-wide Cumberland Sound stretches between the Hall Peninsula and Pangnirtung, the expedition's second village. Inuit elders recall a time when they would dogsled and snowmobile straight across Cumberland Sound to ice-fish for turbot and to reach camps for seal hunting on the other side. In 2007, however, the team heard reports of the worst ice conditions ever; even seal pups were reported to be falling through the ice. When the team reached Cumberland Sound, their fears were confirmed; open water stretched all the way to the top of the sound, adding 70 miles so they could skirt around the open water. In some places, large polynias, or open sections of water, separated them from the shore. Ikummaq said many of these polynias were larger than normal or in places where there had traditionally been only solid ice. The team safely arrived in Pangnirtung on March 10. The next day, however, the ice over which they had just traveled broke up.

Numerous glaciers carve their way down from the Penny Ice Cap and surrounding peaks in Auyuittuq National Park on Baffin Island. Ironically, the name of the park translates to "the land that never melts," but the glaciers are now receding rapidly, giving newfound irony to its name. Fifty years ago, the Fork Beard glacier reached all the way to the valley floor. It has now receded over 1000 vertical feet and is no longer even visible from the valley floor.

On an expedition a year later, traveling further north than Baffin Island, Steger and his youth-focused expedition team traveled 700 miles across the sounds and straights of Ellesmere Island. The 2008 expedition was unable to reach their original goal, to visit the last remaining ice shelves on northern Ellesmere, because they were stopped by thick rubbles of ice; ice that had been part of the Arctic Ocean, 500 miles away. As confirmed by the U.S. National Snow and Ice Data Center, the team traveled through the ruins of the Arctic Ocean, encountering the melt of multiyear ice from the top of the globe that had happened in 2007, just a year before. In over 50 years of Arctic exploration, Steger had never witnessed ice conditions like what he experienced on Ellesmere Island in 2008.

Unusual ice and snow conditions make travel difficult on expeditions, but they also make it possible to provide an eyewitness account of climate change and its impact. Swift loss of sea ice and ice shelves, permafrost thaw, and ice melt on the two largest ice caps are already considerably altering the landscape of the polar regions.

Debate among scientists is not about *if* but rather *by how much sea level will rise by a given date*. Without action, life in the Arctic faces extinction. With action, we can address the root causes and limit the impact. The latest findings by the Intergovernmental Panel on Climate Change state that the world has the capacity to reduce climate change in less than 30 years, using existing technology. They again stress the importance of taking action within the next 10 years to reduce the worst impacts to the world's most vulnerable populations.

So, how can we act to avert the worst consequences? Over the next 10 years, we must significantly reduce our emissions from today's levels. By the year 2050, we must have cut those emissions by 80%. Dramatic change is a personal and societal responsibility—it requires perseverance, courage, tenacity—the qualities of a polar explorer.

Addressing Climate Change at the International Level

Nicole Rom

The UNFCCC

The United Nations Framework Convention on Climate Change (UNFCCC) is an international environmental treaty that was agreed upon in 1992 at the United Nations Conference on Environment and Development, known as the "Earth Summit," in Rio de Janeiro, Brazil. The Parties to this treaty (i.e., the countries that have formally endorsed it) have been holding annual meetings since 1995. There are 197 Parties that have formally endorsed the UNFCCC—which is nearly all of the world's 203 sovereign states. Although the UNFCCC is technically considered a "treaty," it is most accurate to think of it as an "agreement to agree" to take action steps that prevent the worst impacts of climate change. Nothing in the UNFCCC itself requires countries to take such action steps. That is why there have been 24 UNFCCC conferences since 1995: the world still has a lot to do to adequately deal with the threat of climate change.

The Kyoto Protocol

The Kyoto Protocol, forged in 1997 in Kyoto, Japan, was the world's first attempt at a global agreement to address climate change. It called for mandatory cuts in greenhouse gas emissions from developed countries, but exempted developing countries—including China and India, which are the world's first and fourth largest emitters of greenhouse gases despite their status as developing countries. Because of this, the United States never ratified the Kyoto Protocol, although the United States is the world's second largest emitter of greenhouse gases after China.

The Bali Road Map

In December 2007, at the 13th Conference of Parties to the UNFCCC in Bali, Indonesia, the countries of the world agreed on a plan for producing a new agreement that would work alongside and eventually replace the Kyoto Protocol. In particular, this "Bali Road Map" called on Parties to develop strategies to deal with five challenges:

- Finding consensus on an overall "shared vision" for a post-Kyoto agreement
- Cutting greenhouse gas emissions, including those resulting from deforestation
- Adapting to those climate change impacts that are already guaranteed to occur as a result of past emissions
- Developing clean energy technologies, and transferring knowledge of these technologies to underdeveloped countries
- Forging financial agreements between countries to pay for the efforts above

Under the Bali Road Map, it was hoped that countries would agree on plans for "enhanced action" on these issues in 2008 and 2009, in time to roll the action plans together into a new international climate agreement by the end of the 15th conference in Copenhagen. Between 2007 and 2009, negotiators from around the world worked steadily to address the issues above, in hopes that their work would culminate with an agreement in Copenhagen. This hoped-for agreement was laden with expectations as a result of the failures of the Kyoto Protocol. This is why the COP 15 conference in Copenhagen received so much attention before it began, while it was going on, and after it ended.

The Copenhagen Accord

The outcome of the COP15 conference in 2009 was a three-page, non-binding "Copenhagen Accord" that, while not perfect, provided the beginnings of an agreement to tackle climate change. The Accord was agreed to in the final 48 hours of the conference by heads of state from the United States, China, India, Brazil, and South Africa. The other countries assembled at the conference agreed to "take note of" the Accord. What "taking note" means is open to interpretation; it was an indication that many of the other countries at the conference were unwilling to endorse a non-binding climate agreement, but were supportive of the agreement insofar as it leads to a binding agreement later on. The Copenhagen Accord was built on commitments to cut overall emissions by the United States and other developed countries, and commitments to cut emissions intensity (emissions per unit of economic output) by India, China, and other developing countries.

COP21: Paris Agreement

It was not until 20 years after the Rio Summit that in 2015 at COP21 in Paris, France that all Parties to the UNFCCC finalized and signed the Paris Agreement—the first global climate accord that commits all signatories to climate action. The Agreement acts as an action plan for the international community to address climate change, but there is still work to be done.

The pledges and promises made in the COP21 document are not enough by themselves to save the planet from the worst impacts of climate change. In fact, they would lead to a world nearly twice as warm and climate-impacted as the aspirational 1.5 °C target laid out in the Paris text. Now that the foundation for action is in place, however, there is much that we can all do to ensure that the Paris Agreement lives up to its commitment to limiting temperature rise to 1.5 °C. The true assessment of the success or failure of COP21 will be years in the future, when we can look back and see what actions it inspired and how the world responded to its call to action.

Climate Generation's delegation of educators attend COP21 in Paris in 2015. (Photo courtesy of Climate Generation: A Will Steger Legacy)

Post COP21, climate action has ramped up at all scales and in all sectors. Leading up to and beyond COP21, the actions of states, cities, businesses, and individuals have been critical in showing the world that we are collectively up to the task of realizing a zero-carbon, 100% clean energy future.

However, the 2016 US elections were a wake-up call. In the election of Trump, climate action faced the daunting obstacle of financial interests' intent on maintaining their power and profits at the expense of our collective future. Yet despite this setback, the movement of people for justice, for climate action, and for an inclusive vision of society continues to elevate the promise of the Paris Agreement.

The resistance to the Trump Administration's intent to withdraw the United States from the Paris Agreement was, and continues to be, strong. People took to the streets, businesses stepped up their commitments, and local and state governments increased their clean energy pledges. The People's Climate March turned out hundreds of thousands, a powerful reminder that the climate movement is not going away; indeed it is stronger, more diverse, and more resilient than ever. US states have stepped up to announce "We Are Still In" despite the decisions of our federal government and businesses, cities, universities, tribes, and nonprofits have also signed on to work toward reducing carbon emissions and mitigating climate change.

Post COP21: Subsequent Action Toward 2020 and Beyond

COP24 in Katowice, Poland in 2018 saw the creation of the official rulebook for implementation of the Paris Agreement. The 156-page rulebook was written and agreed upon by more than 190 participating countries. These rules dictate how countries tackle climate change, beginning in 2020 and beyond. Every participating country, whether developed or developing, is expected to follow the same standard for measuring emissions and tracking climate policies. The rulebook also expects richer countries to spell out the financial support they will offer to assist poorer nations as they navigate the clean energy transition and build resilience against natural disasters. Now, country representatives will devise how they plan to ramp up their pledges to cut carbon emissions before the 2020 conference, COP26, in Chile.

The rulebook is more comprehensive than many expected to come out of the conference amidst the running theme of coal throughout the talks and tense political debates. The business community provided strong climate leadership and a presence at COP24 as a critical way to show the enduring US commitment to its Paris Agreement pledge. However, the voices of young people, indigenous groups, vulnerable communities, and people of civil society are saying it does not go far enough to connect the science the Intergovernmental Panel on Climate Change report communicated—that we have less than 12 years to cut carbon pollution in half to avoid the global temperature rise of 1.5 degrees Celsius—with the necessary requirement and accountability for action now.

Education for Action

Nicole Rom

We know we need bold climate leadership within a limited time frame, approximately 10 years to truly chart a different course for our country, our economy, and our planet. What is required is an overhaul of our energy systems, building and transportation infrastructure, how we produce our food, protect our health, and comprehensive education and training to prepare a new green workforce and just transition for resilient communities.

Established in 2006, Climate Generation: A Will Steger Legacy's story stems from founder, Will Steger, an eyewitness to climate change from a lifetime of polar expeditions. Climate Generation's mission is to empower individuals and their communities to engage in solutions to climate change. The organization is unabashedly hopeful and solutions-focused, recognizing a variety of solutions are required to chart a path toward a decarbonized, equitable, and resilient world.

Climate Generation provides educators, youth, policymakers, communities, and business leaders with the resources and opportunities to engage in solutions to climate change. Across these different audiences, Climate Generation programming aims to:

- Build climate literacy
- Develop powerful climate advocates
- Elevate leadership within an individual's sphere of influence.

Climate Generation empowers high school youth to take action for a just transition for a climate resilient future for all. (Photo courtesy of Climate Generation: A Will Steger Legacy)

Climate Generation programing is grounded in personal storytelling as a tool for action. Indeed, whereas Will Steger's powerful eyewitness perspective once brought alive climate impacts in faraway regions, now everyone is experiencing climate change. Climate Generation's work is also grounded in equity, recognizing that climate change disproportionately hurts the most marginalized people in society—including people of color, people from working class backgrounds, women, and LGBTQ people.

Climate Generation has trained and supported over 5000 educators annually with science-based educational curriculum and is nationally recognized as a "go-to" organization for climate change education. They have successfully passed nation-leading renewable energy legislation, aggressive carbon reduction goals, and energy efficiency improvements in Minnesota, where they are headquartered. For 10 years, Climate Generation has also led multi-sector delegations of youth, educators, policymakers, and business leaders to the international climate summits, bringing Minnesota and the Midwest to the international stage and translating the outcome of each summit to their audiences. Finally, their work with communities, youth leaders, and businesses has galvanized climate action in schools, workplaces, communities, and in public policy.

Climate Generation builds the comfort and confidence for its audiences to understand, talk about, and engage in solutions to climate change. While there is virtually unanimous scientific agreement about climate change, due to both the inherent complexity of the topic and the social controversies surrounding it, confusion and doubt often persist. If the nation is to address climate change, it must begin with a public that is climate-literate. Action begins with education.

Starting with our educational system is critical. Teaching and understanding climate change is a process involving scientific inquiry and educational pedagogy; it is not about politics or partisanship. Educators report lack of knowledge about climate change [3]. Broadening their understanding through professional development and curriculum resources strengthens their ability to teach the topic and answer colleagues, students, and parents who often do not know the facts of climate change.

Climate change education means being able to understand the basics of Earth's climate system, to know how to assess scientifically credible information about climate, to communicate about climate change in a meaningful way, and, most importantly, to be able to make informed and responsible decision regarding our actions that affect the climate [4]. For example, we should know the reason for the seasons, the basic dynamics of the greenhouse effect and the carbon cycle, and the differences between weather and climate.

Will Steger shares his eyewitness to climate change presentation with a classroom of young students. (Photo courtesy of Climate Generation: A Will Steger Legacy)

Standard curriculum and textbook cycles are often slow and subject to state and local review and debate, leaving them disconnected from new findings in climate science. For example, observed sea ice melt and changes in ice sheets are occurring faster than models had predicted. This disconnect quickly leads to outdated educational resources.

Climate change education must be based in peer-reviewed, consensus-based science. For this reason, Climate Generation curriculum materials are aligned to the Climate and Energy Literacy principles. "Climate Literacy: The Essential Principles of Climate Science" and the Energy Literacy Principles are a product of the U.S. Global Change Research Program and were compiled by an interagency group, led by NOAA. Climate Generation offers curriculum resources for Grades 3–12 which can be downloaded free from their website. Curriculum topics include: climate change basics; climate change communication; climate change impacts on the polar regions, Minnesota, and the National Parks; international climate policy; and the basics of energy, energy efficiency, and renewables. Their resources are tied to online materials that include videos, interactive games, and blogs. Each resource is aligned to state and/or national standards in science, as well as language arts, STEM, geography, and mathematics and support the Next Generation Science Standards. Climate Generation curriculum resources are free to download and available online: https://www.climategen.org/take-action/teach-climate-change/curriculum/

Climate Generation recognizes the need for quality education materials and educator support focused on climate change for a number of reasons. First, climate change is currently not included in most education curriculum. Teachers cite a lack of comfort, lack of time and, in some cases, opposition from parents, administrators, and even students (BOSE 2011) [1]. The recently developed Next Generation Science Standards (2012) explicitly include climate change and present an opportunity to ensure integration of climate change within science classrooms. Almost 20 states around the country have adopted or used these standards to develop their own. Science educators need to have access to quality, science-based materials and professional development as climate change becomes a core subject to include in the science classroom.

Educators build their comfort and confidence to teach interdisciplinary climate change topics in their educational setting. (Photo courtesy of Climate Generation: A Will Steger Legacy)

Second, there is a movement at a policy level to make it difficult for teachers to include climate change in their classroom, despite an overwhelming consensus among scientists of the reality of climate change and of humans as the main driver (NCSE website) [2]. The recent *Mixed Messages* report indicated that many teachers are unaware of the 97% consensus among climate scientists that anthropogenic climate change is happening [3]. In addition, the ongoing mailing of climate misinformation from the Heartland Institute, to as many as 200,000 science teachers around the country indicates an all-out, direct assault on climate change education that must be addressed.

The Paris Agreement reached at the COP21 talks offered an opportunity and roadmap for climate change education. It provided the foundation for a climate-resilient future, while elevating the importance of preparing today's students to implement the policies and develop the innovations needed to realize that future. We must prepare educators to teach about climate change science and solutions, and develop students with the twenty-first century skills to mitigate and adapt to the impacts of climate change.

Climate Generation is working to mainstream climate change education across the country in all subjects, preparing students for a future built on innovation and green STEM (Science, Technology, Engineering, and Math) career opportunities. They recognize the complexity of climate change and the importance of interdisciplinary climate change education. Climate Generation's curriculum resources are aligned to science, social studies, and language arts standards and encourage student thinking beyond science alone. Their professional development opportunities support teaching across the curriculum and have been attended by teachers of all disciplines over the last 14 years. Their professional development opportunities prepare educators to integrate climate change into their educational setting using best practices, while also providing the space for educators to learn deeply about the most current, relevant, and pressing issue their students face. They ultimately increase the climate literacy of educators and their students.

While education is critical, action is also needed in our energy system as we decarbonize and electrify everything on renewable energy sources. Climate change, an environmental and moral issue, is also a unifying issue. It affects all of us; therefore, the solution requires all of us. Individual action leads to collective action. But individual action alone will not solve the problem. We need to demand that our elected officials act to create solutions to climate change. State and local initiatives are proving that answers exist. To reinforce and expand these efforts, we need federal action that triggers solutions on a national scale. Currently, US businesses, universities, states, cities, and nonprofits are stepping up in the absence of national action.

The effects of climate change are pervasive. We cannot delay in slowing and reversing this trend. Our health, economy, national security, and the environment demand it. We have a responsibility to prepare and empower people of all ages to tackle climate change.

References

1. Board on Science Education. A framework for K-12 science education: practices, cross- cutting concepts, and core ideas: National Academies Press; 2011.
2. National Center for Science Education: Climate. Available at: http://ncse.com/climate. [Accessed May 2012].
3. National Center for Science Education: Mixed Messages Report, 2016: https://ncse.ngo/files/MixedMessagesReport.pdf
4. United States Global Change Research Program. Essential principles of climate science: a guide for individuals and communities; 2009.

Chapter 5
Arctic Connections to Global Warming and Health

M. Torre Jorgenson and Janet C. Jorgenson

Global warming is having increasing consequences on ecosystems and public health [1], especially in the Arctic where climate warming is occurring at double the rate of the rest of the Earth and is causing rapid biophysical changes [2]. Observations from field monitoring, indigenous ecological knowledge, and remote sensing are providing overwhelming evidence that large environmental and social changes are already being affected by changes in climate, sea ice, snow, glaciers, permafrost, fires, vegetation growth, marine mammal behavior, and human infrastructure and social systems. Because the Arctic is strongly connected to the global climate system, the changes are creating positive feedbacks to the global climate system that are accelerating global warming.

The past and projected future changes present an enormous challenge to society in how to adapt and mitigate environmental and health effects, such as expanding oil development, mining, shipping, commercial fishing, damage to village infrastructure, access to subsistence foods, and increasing exposure to contaminants and diseases. The Arctic biome, then, provides important early warning signs regarding the consequences of global warming on ecosystem and societal health. In this chapter, we examine the importance of the Arctic within a global context by highlighting prominent observations on environmental and societal changes, comparing projections of future changes under differing scenarios of greenhouse gas emissions and identifying feedbacks to the global climate system, and discussing how environmental and health policy is mitigating, or failing to mitigate, the effects of a changing Arctic. Because the Arctic encompasses nine nations and has important connections to the global

M. T. Jorgenson (✉)
Alaska Ecoscience, Fairbanks, AK, USA
e-mail: ecoscience@alaska.net

J. C. Jorgenson
U.S. Fish and Wildlife Service, Retired Arctic National Wildlife Refuge, Fairbanks, AK, USA
e-mail: Janet_Jorgenson@fws.gov

K. E. Pinkerton, W. N. Rom (eds.), *Climate Change and Global Public Health*,
Respiratory Medicine, https://doi.org/10.1007/978-3-030-54746-2_5

climate and economy, there have been numerous international and regional syntheses of Arctic conditions that provide a foundation for our chapter [2–7].

Rapid Environmental and Societal Change in the Arctic

Ecosystems Under Stress

Nearly all aspects of Arctic ecosystems are under stress from the rapid warming. Mean annual air temperatures in the Northern Hemisphere have increased by almost 2 °C since 1980, about 2.5 times faster than that for the whole Earth, with winters warming considerably faster than summers [2]. Recent records show that during 2011–2015 the biome was warmer than at any time since instrumental records began in around 1900. While climates have undergone large fluctuations associated with the coming and going of ice ages over the last three million years, the rate of warming is the greatest since the end of the Last Glacial Maximum around 12 thousand years ago and is approaching the warmest period over the last 130 thousand years. A unique aspect of the Arctic is that it has a large portion of the Earth's cryosphere, land and water that is frozen for most of the year in the form of sea and lake ice, glaciers, snow, and permafrost. Because water changes state from ice to water at 0 °C, the region is especially vulnerable to small changes in climate and there is a growing body of evidence that Arctic ecosystems are reaching a tipping point.

Sea ice, as measured by its annual minimum extent, has declined from 8 million km^2 in 1979 to 4.1 million km^2 in 2016 [2]. Sea-ice thickness in the central Arctic Ocean declined from an average of 3.6 m in 1975 to 1.3 m in 2012. This loss of summer ice is having profound effects on mammals, birds, and fish that depend on the ice for feeding, breeding, and movement [8]. Particularly notable are the impacts to polar bears, which have recently been listed as a threatened species due to habitat changes, walrus and seals that depend on the ice as a birthing and feeding platform, and seabirds that nest on land but feed along the highly productive margins of the sea ice. In addition to the sea ice loss, the ice is becoming younger and thinner, which contributes to the formation of melt ponds that absorb more solar energy, and to less resistant ice that makes shipping more feasible.

Mountain glaciers, ice caps, and the Greenland ice sheet have been declining faster since 2000 than in the previous decade [2]. All Arctic regions lost land ice mass between 2003 and 2014, averaging 413 ± 40.6 Gt/y. Ice losses were greatest from Greenland (64% of the Arctic total), followed by Arctic Canada (14%) and Alaska (12%). Siberian glaciers have been decreasing in area in response to summer temperature increases and a relatively small snowfall increase. The ice loss, mostly from Greenland, is equivalent to a 1.1 ± 0.1 mm rise in eustatic sea level per year, representing two-thirds of the global land ice contribution to sea level rise.

Snow cover across the Arctic has been decreasing in its annual duration by 2–4 days per decade since the 1980s, based on satellite monitoring and ground monitoring stations [2]. For most regions the snow-free season is 1–2 weeks longer than in the 1980s. There is substantial variability in its distribution over time, however, because it is affected by warming temperatures, increased moisture availability, changing atmospheric circulation, changing vegetation, increased frequency of winter thaws, and rain-on-snow events. A shorter snow-cover season has been linked to declining access to country foods for some northern communities and to reduced periods for off-road vehicle traffic associated with oil exploration and development.

Permafrost occurs under the land in ~24% of the Northern Hemisphere and is fundamental to ecological processes by supporting the ground surface, facilitating wetland development by impeding subsurface drainage, and storing vast amounts of soil organic carbon in the frozen soils. Long-term borehole monitoring has shown that permafrost temperatures have risen by up to 2 °C since ~1980, and the southern limit of permafrost has moved northward in Russia and Canada [2]. How permafrost thaws depend on the type and amount of ground ice present in the soil, with most of the ice occurring as segregated ice in the top 1–3 m of the upper permafrost [9]. Permafrost also can contain various forms of massive ice bodies, including wedge-shaped ice bodies 2–4 m wide that extend 3–4 m deep below the surface; injection ice that forms thick horizontal sheets of pure ice beneath the surface or in large ice-cored mounds called pingos; and buried glacial ice that has persisted since the last ice age. Thermokarst, the collapse of the ground surface after thaw, is widespread across the Arctic due to the high ice contents of permafrost soils that make the surface especially vulnerable to climate change and disturbance from human activity [9]. In particular, there has been an abrupt increase is the degradation of ice wedges due to recent warming [10, 11]. When permafrost thaws, nearly all aspects of the ecosystems change in response to deepening active layers (zone of summer thaw), changes in surface water, thawing soil conditions, and shifts in plant species as soils either drain downward in upland areas or become flooded in lowland areas.

Ecosystem patterns and processes are causing profound shifts across the Arctic related to forest and shrubland expansion, fires, thermokarst, lake drainage, coastal erosion and flooding, and river channel dynamics [8, 12, 13]. While forest expansion into Arctic and alpine areas has been slow, shrubs have been found to increase in stature and progressively expand and infill patchy landscapes [14]. Fire, which has important ecological impacts on vegetation succession, wildlife use, permafrost stability, and carbon cycling, has long been the dominant factor in boreal forest dynamics and is become more prevalent in the Arctic. Fires in boreal forests of Siberian Russia tend to be roughly ten times more numerous, three times larger on average, and three time more frequent than in Canada [15]. Comparisons with other long-term fire records in Alaska reveal that the tundra biome can sustain a wide range of burning, with fire return intervals ranging from as low as 30 years to more than 5000 years [16]. Lakes are abundant in the Arctic due to the thawing and collapsing of permafrost, but these lakes also are sensitive to being lost by lateral drainage through gullies eroding through permafrost or by river tapping [13]. Finally, the loss of sea ice has led to larger wave fetch and energy that has caused increased erosion of the Arctic coastlines [17], and led to loss of critical habitat for marine mammals (Fig. 5.1). Adequately assessing the ecological responses to climate change, however, will require large improvements in our understanding of the ecological complexity of the Arctic [18].

Fig. 5.1 Photos of distressed polar bears feeding on human-supplied food onshore at the Inupiat village of Kaktovik, Alaska, (FWS photo) and walruses aggregating on the beach near Pt Lay, Alaska, due to loss of sea ice. (NOAA photo)

Perhaps the most vulnerable region in the Arctic to climate and coastal changes is Alaska's Yukon-Kuskokwim Delta, where a number of environmental and societal factors combine to synergistically exacerbate the impacts of climate change [19]. Declining sea ice in the Bering Sea, rising sea level, and subsiding land due to degrading permafrost plateaus interact on the extremely flat coastal wetlands to increase the magnitude, extent, and inland extent of flooding and salinization. The region has globally important coastal wetlands that are critical breeding and brood rearing habitat to many waterfowl that migrate there from around the world. There are ~25,000 Yup'ik indigenous people residing in 17 villages that depend on the subsistence resources in the region, including freshwater, berries, waterfowl, birds, and sea mammals. Finally, many of the villages are threatened by coastal erosion, collapsing infrastructure from degrading permafrost, and shifting food resources that affect livelihoods and community health.

Future Projections and Feedbacks to the Global Climate System

To project future global warming, numerous global circulation models have been developed to project temperature changes under a range of representative concentration pathways (RCPs) from potential greenhouse gas emissions [4]. Two commonly referenced pathways include the high scenario (RCP8.5) that envisions a future where annual greenhouse gas emissions increase significantly throughout the twenty-first century before leveling off by 2100, and a low scenario (RCP4.5) that envisions substantial emission reductions that result in emissions 85% lower than in RCP8.5 [6]. The impacts of the projected warming in Arctic, however, not only will have dire consequences for the Arctic but extend globally due to feedbacks that amplify the warming beyond the direct effects of anthropogenic greenhouse

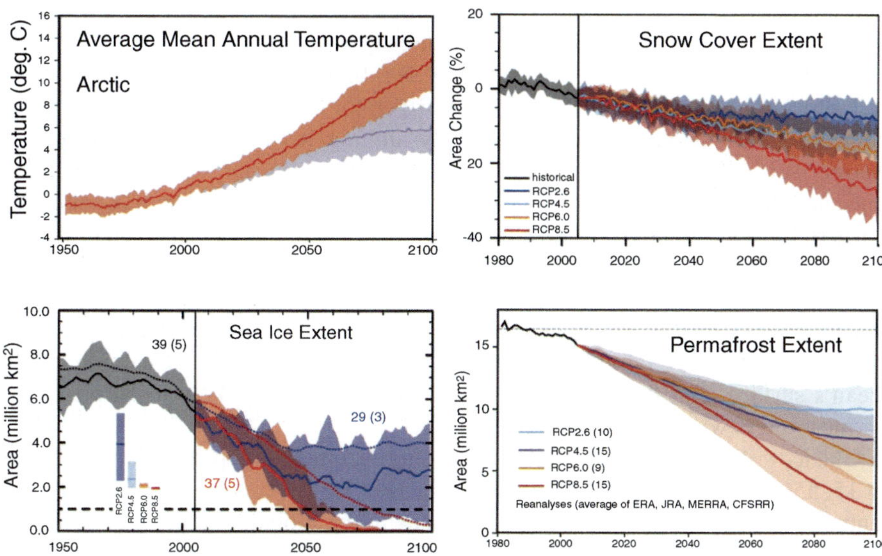

Fig. 5.2 Projected changes in winter air temperatures in the Arctic, and snow cover, sea ice, and permafrost in the Northern Hemisphere. (Adapted from [4])

emission. These feedbacks include the role of sea ice and snow in cooling the planet, of cold Arctic Ocean waters in driving the north Atlantic conveyor belt and ocean circulation, of melting glaciers contributing to sea-level rise, and of thawing of permafrost soils that release additional greenhouse gasses. The changes underway appear to even be affecting weather patterns in lower latitudes, through effects of the Arctic Oscillation and jet stream patterns [2].

Projections of future air temperatures in the Arctic, based on the mean results of 30 models run under RCP-forcing scenarios [5], indicate winter temperatures averaged over the Arctic will warm by 4 °C under RCP4.5 and by 12 °C under RCP8.5 by the end of the century (Fig. 5.2). This Arctic amplification of temperatures, relative to global projections, is due primarily to surface albedo effects of reduced snow and ice, warming of the newly ice-free Arctic Ocean, and the vertical structure of temperature and water vapor in the atmosphere at high latitudes that allows less energy to be radiated out to space [2].

The ice-albedo feedbacks play a fundamental role in affecting global climate because of the highly reflective white surface of snow, sea ice, and glaciers [4]. Decreases in these surface manifestations of the cryosphere increase the amount of solar energy absorbed, leading to more warming and more ice loss. The two most important factors in this feedback are changes in seasonal snow cover and sea ice loss. Snow cover during spring in the Northern Hemisphere is projected to decrease by 7% for RCP2.6 and by 25% for RCP8.5 by the end of the twenty-first century [4]. Sea ice covering the Arctic Ocean during summer is projected to be nearly all gone within this century, likely within the next 30 to 40 years [2].

Rapid sea ice loss and thinning has numerous global implications [2]. The reduction in maximum winter sea ice extent is already affecting ocean conditions along the southern margin, such as in the Bering Sea. In addition to the ice-albedo feedback process, the Arctic acts as a global refrigerator by drawing warm ocean water from the south, cooling it, and ultimately sinking it toward the ocean bottom. Warmer surface water moves in to replace the sinking water, creating ocean currents, and thus has a major influence on global climate. For example, it accounts for northern Europe's relatively mild climate compared with that of Canadian provinces at the same latitude and it keeps the tropics cooler than they would be otherwise. More open water is expected in winter, affecting temperatures and the exchange of moisture between the atmosphere and ocean, leading to more extreme weather locally and at lower latitudes. Melting sea ice contributes to freshening of the ocean's surface, and the thinner single-season ice is more likely to be pushed to warmer waters where it will melt.

Glacier volume across the globe, excluding Antarctica, is projected to decrease by 15–55% for RCP2.6, and by 35–85% for RCP8.5 by the end of the twenty-first century [4]. This projected ice loss feeds back to the global system by contributing to sea level rise, increasing freshwater input to the ocean, and affecting ocean circulation. Global mean sea level rise for 2081–2100 will likely be 0.26–0.55 m for RCP2.6, 0.32–0.63 m for RCP4.5, and 0.45–0.82 m for RCP8.5, with glaciers accounting for 15–35% [4]. There remains large uncertainty, however, in the collapse of marine-based sectors of the Antarctic ice sheet, which could cause sea level to rise substantially above current projections. Freshwater from the melting Greenland ice sheet into the Labrador Sea has increased by 50% in less than 20 years and may influence the North Atlantic thermohaline circulation, storm severity, and the Jet Stream [2].

Permafrost extent in the Northern Hemisphere is projected to decrease by 15–87% for RCP4.5 and 30–99% for RCP8.5 by the end of the twenty-first century [2, 20]. The large uncertainty in the projections reflects the sensitivity of permafrost temperatures to a wide range of surface conditions, such as air temperatures, energy balance, snow, soil conditions, and surface water. The thawing of permafrost could lead to large greenhouse gas emissions to the atmosphere because of the enormous stocks of organic carbon long sequestered in frozen soil in Northern Hemisphere estimated to be 1330–1580 Pg carbon, almost twice the amount contained in the atmosphere and an order of magnitude greater than the amount in plant biomass and litter [21]. Microbial decomposition of organic carbon in the soils under well-drained conditions generally leads to release of carbon dioxide, and under flooded conditions to methane, which is ~30 times more potent than carbon dioxide as a heat-trapping gas. Under current warming trends 5% to 15% of the soil organic carbon stored in circumpolar permafrost soils is considered vulnerable to release into the atmosphere by 2100 [22]. Modeled carbon emissions projected under various warming scenarios translate into a range of 0.13–0.27 °C additional global warming by 2100 and up to 0.42 °C by 2300, but currently remain one of the least constrained biospheric feedbacks to climate [4, 21, 23]. An even larger uncertainty is the fate of large stocks of methane hydrates, a frozen and highly concentrated form of methane that is widespread in sediments of marine continental margins and permafrost areas [24]. While there has long been concern about catastrophic releases from the dissociation of gas hydrates that could exacerbate greenhouse warming, more observational data and improved modeling are needed to better characterize the interactions between climate warming and hydrate emissions.

As the environmental conditions in the Arctic change, so do the ecosystems adapted to the cold and icy conditions [8]. The changes will lead to both winners and losers. At the bottom of the marine food chain, primary production by phytoplankton in the Arctic increased by 20% between 1998 and 2009 (and the increase has been as much as 70% in the Kara Sea and 135% in the Siberian sectors of the eastern Arctic Ocean). On land, a multi-decadal satellite record indicates that the Arctic had been becoming increasingly green, caused in part by the growth of deciduous shrubs, but has recently showed a remarkable, but poorly understood browning since 2011 [25]. For terrestrial vegetation, winners will likely be deciduous shrubs and losers will be mosses and lichens [26]. Wildlife losers include walrus and polar bear populations that have tended to decline because of reductions in sea ice [8]. Caribou and wild reindeer herd across the Arctic tundra have declined by nearly 50% over the last two decades [7]. Ocean acidification due to increased carbon dioxide uptake in warmer seas also can harm some marine life and the fisheries associated with them [27]. Others adapt: some fish stocks have moved, and flourished, as a result of warmer waters, such as cod stocks in the Barents Sea and off the coast of Greenland that have become more productive and moved further north.

Environmental and Health Policy for a Changing Land

The changing Arctic will severely challenge the ability of society and ecosystems to adapt to changing physical and ecological conditions, but also provides opportunities to exploit resources that become more readily accessible. Already there is substantial damage to community infrastructure from coastal and river erosion, permafrost degradation, and fires. The effects of climate change also are having large impacts on food security and community health. The warming Arctic, however, provides new opportunities for shipping across the Arctic Ocean, fishing, oil and mining development, as well as military expansion into the Arctic region. These changes are creating difficult and long-term challenges in how government policies can best balance economic development, societal needs, and environmental protection. Below we highlight some of the most pertinent issues by summarizing how communities can adapt to the new Arctic, identifying new frontiers that are opening up for development, discussing proposed oil development in Arctic National Wildlife Refuge as a case history of how government policy is attempting to address competing interests, and end by evaluating the threats to food security and community health.

Adapting to a New Arctic

There are large contrasts between North America, Scandinavia, and Russia in terms of native culture, economies, and history of development that strongly influence how societies are adapting to the rapid changes in the Arctic. In northern Canada and Alaska, as well as the Far East of Russia, northern settlements are primarily inhabited by Inuit and Yup'ik populations totaling ~165,000 people [28]. The majority of Inuit reside in small, remote, coastal communities, with economies composed of waged employment and subsistence resource harvesting. In contrast, settlements in northern Russia are more closely tied with industrial and military development, although there are still sizable populations of indigenous peoples including the Nenets, Khanty, Evenk, and Saami. Overall, there were about four million people living in the Arctic in 2013, and indigenous people make up about 10% of the population [5]. The Arctic population has declined 1.4% from 2000 to 2010, primarily due to collapse of economies in northern Russia, while populations in Alaska and Canada have increased at 13%, nearly the same rate as global population growth. Most of the increase in North America has been due to increased local birth rates, while the large change in Russia is mostly due to people leaving the depressed economy.

The changing physical and biological environment of the Arctic creates numerous types of vulnerabilities, but also a range of opportunities for adaptation [29]. The opportunities include improved opportunities for agriculture and biofuels from warming temperatures, increased access to marine and river transport from reduced ice, more hydroelectric potential, learning and innovation from increased cultural diversity, and substitution of locally produced food and fuel for expensive imports. Below, we selectively highlight one of the more critical adaptation challenges, that of dealing with coastal erosion.

The effects of sea ice retreat, sea level rise, increased wave erosion, coastal flooding, and permafrost collapse already are all too real for many villages along the Arctic coastline. These synergistic effects have led to the emerging concept of "Usteq," which embodies the Yup'ik word for eroding and collapsing permafrost land. In Alaska, 31 rural Alaska communities face significant damage to infrastructure, and the need for intensive erosion protection, or even relocation [30]. A major factor contributing to this vulnerability is that over the past century many rural Alaska villages developed with little thought about terrain conditions and economic sustainability associated with the long-term costs for energy, transportation, water and sewer, freight delivery, air access, and other community infrastructure. Recently, at three of these villages, Shishmaref, Kivalina, Shaktoolik, intensive bank stabilization structures have been constructed, while at Newtok the erosion and flooding is so severe that the village is being relocated (Fig. 5.3). At the individual level, many villagers have opted to simply abandon the land for better opportunities in regional cities.

Governments in Alaska at the village, regional, state, and federal levels have responded to this climate-driven crisis by assessing the impacts and threats through

Fig. 5.3 Aerial views of Newtok, AK, where the tidal river is eroding the bank at an average rate of ~20 m/yr., and of Shishmaref, AK, where a massive revetment was constructed to prevent severe shoreline erosion associated with the loss of sea ice. (Photos from Alaska Shorezone Project, NOAA)

technical investigations, developing plans to mitigate the problems, and by constructing protective structures or relocating villages. Technical assessments have been conducted by village councils, such as the Newtok Village Council [31] as a requirement for federal funding. State agencies, such as the Division of Community and Regional Affairs, collect and coordinate community information. At the Federal level, agencies include: the U.S. Army Corps of Engineers, which does erosion and flooding assessment and geotechnical investigations [32]; the Denali Commission, which supports a variety of rural programs, including village infrastructure protection [33]; and the General Accounting Office [30]. Overall, there are 46 federal programs within ten Departments that offer some kind of technical or grant assistance to Alaskan villages. While this provides numerous pathways for assistance, it is also an enormous barrier in terms of competing jurisdictions and multi-agency coordination. More importantly, agencies often avoid responsibility because of the huge costs involved; for just the four villages mentioned above, the Denali Commission estimated that the unmet needs for infrastructure protection exceed more than $200 million. In an estimate by USCOE, the cost for moving the 372 people of Newtok to the new village site of Mertarvik, alone, could be $130 million. Recently, relocation of the airport at Kaktovik in northern Alaska, which was subject to frequent flooding from storms at sea, cost $50 million. Overall, costs of infrastructure damage from climate change across the Arctic and boreal regions of Alaska are projected to total $4.2 billion for the RCP4.5 scenario and $5.5 billion for RCP8.5 scenario [34].

Of great importance for adaptation to the changing Arctic is the increasing trend of robust participation of local residents, and indigenous peoples in particular, in Arctic decision-making and continued innovation in governance [29, 35]. In the example above, village and regional governments have been instrumental in planning the adaptation strategies. In a broader context, governance has been expanded through international organizations, such as the Inuit Circumarctic Council, Russian Association of Indigenous Peoples of the North, and the Saami Council. Many of the organizations represent their people as Permanent Participants on the Arctic

Council, a high-level intergovernmental forum that addresses issues faced by Arctic nations and their indigenous people. While there has been increasing self-determination and improved indigenous representation in regional, national, and international governments and institutions, the demands on local representatives are stretching human capabilities and fiscal resources to the limit. The emergence of Arctic identities and a sense of Indigenous and, more broadly, Northern identity is becoming an asset.

New Frontiers for Development

The remoteness and harsh environment of the Arctic has long kept it on the frontiers of development, yet there have been rapid changes over the last several decades. One of the first large-scale developments in the region was military construction of the distant early warning (DEW) stations along the US and Canadian Arctic coastline in the 1950s. Oil exploration began in the 1940s and oil has been produced onshore since the 1970s. Even offshore drilling began as early as the 1970s. The large-scale development of the Arctic associated with oil development, as well as mining, shipping, fisheries, and military expansion, present unique and rapidly evolving risks to ecosystem and human health.

Oil and gas development in the Arctic boomed in the 1970s with the discovery of massive oil reserves at Prudhoe Bay in northern Alaska in 1968. But it wasn't until the 2000s that oil and gas development took off in Norway, with production from the Snøhvit field starting in 2006, and in Russia, with gas production in Yurkharovskoye field starting in 2003 and oil shipping from the Varandey terminal on the Pechora Sea coast through Arctic waters starting in 2008. More recently, large-scale developments in Russia have increased dramatically with gas production from the giant Bovanenkovskoye field on the Yamal peninsula starting in 2012, and the Prirazlomnaya offshore platform starting in 2014 [36]. Overall, the Arctic contributed one tenth of the world's total oil production and one-quarter of the world's gas production in 2007, of which 80% of the oil and 99% of the gas was extracted in the Russian Arctic [2]. The US Geological Survey (2008) estimates that the Arctic holds 13% of the world's undiscovered oil and 30% of undiscovered natural gas, and that 84% of those resources lie offshore. These huge reserves have spurred intensive offshore drilling in the United States and Canadian Beaufort Sea, the United States and Russian Chukchi Sea, and Greenland over the last decade. Climate change and loss of sea ice are easing drilling and shipping, but are also impeding access through reductions in snow cover needed for seismic exploration and cold temperatures needed for ice-road access to remote sites. While oil development provides great economic promise, oil spills from drill sites, pipelines, and ships remain a significant threat to ecosystems and human health. The spill of 40 million liters of crude oil from the grounding of the Exxon Valdez in 1989 and the leaking 15–350 million liters of oil from corroding pipelines in the Pechora region of Russia in 1994 stand out as the largest disasters, yet smaller spills remain chronic and frequent.

Mining in the Arctic has a long history and recently mining companies have increased their investment in the region, offering long-term economic development with a large local workforce [37, 38]. Currently, Russia has 25 mines, including the large Norilsk Nickel mine started in 1935 with the world's largest smelting complex. Alaska has one of the largest lead-zinc mines in the world, the Red Dog mine. Canada has seen large recent expansion of diamond mining, and the large Mary River iron ore mine of Baffin Island started up in 2013. In Scandinavia, there are several mining prospects, with the world's largest underground mine at Kirkenes, Norway. Geological prospects in the Arctic have not been fully explored and more world class prospects could be discovered. Development, however, remains challenging due to remoteness, lack of infrastructure, and high costs. Mining and associated processing can have large environmental and health effects. At the Norilsk nickel smelter, four million tons of cadmium, copper, lead, nickel, arsenic, selenium and zinc are released into the air every year, resulting in one of the ten most polluted cities in the world and a zone of dead vegetation extending 50 km from the smelter. At the Red Dog Mine, which produces 10% of the world's zinc supplies, the lead-zinc concentrate is trucked by road to a loading dock on the Chukchi Sea and elevated lead and zinc concentrations have been found in the tundra as far as 1 km from the road.

Shipping in the Arctic Ocean is still mostly regional traffic associated with military activities, oil development, and fishing, but with the projected decline in Arctic sea ice the opening of shorter trade routes across the Arctic Ocean will have global economic implications [39, 40]. The reduction in summer sea ice will likely lead to increased transit shipping that uses the Arctic Ocean as a shortcut between Pacific and Atlantic ports, resulting in large cost savings due to reduced fuel consumption. This will also reduce global shipping emissions. European routes to Asia could become 10 days faster via the Arctic than alternatives by midcentury, and 13 days faster by late century, while North American routes become 4 days faster [40]. The shipping season is projected to reach 4–8 months under the RCP8.5 scenario, double that of RCP2.6, with ice-strengthened vessels possibly able to transit the ocean 10–12 months a year by late century. Significant threats from increased shipping to the Arctic marine environment include the release of oil through accidental or illegal discharge, ship strikes on marine mammals, spread of alien species, disruption of marine mammal activities, and noise [39].

Arctic fisheries amount to only ~5% of the overall global catch [38], but are important to regional economies and to local subsistence use. Commercial fisheries are a dominant portion of the economies of Greenland (90%) and Iceland (33%), but account for relatively small export earnings for Norway (6%), the United States (<1%), and Russia (<1%). Commercial fishing has boomed in recent years as ship voyages increased from 30 in 2005 to 221 in 2010, and the Greenlandic shrimp catch increased 50% in the early 2000s. While climate warming may be responsible for an increase in phytoplankton that supports the food chain, the sustainability of Arctic fisheries is of concern. Prodded by a letter from over 2000 scientists and the Utqiaġvik Declaration from the Inuit Circumpolar Council, an international agreement was reached in 2018 by nine nations and the European Union to halt unregulated fishing in the central Arctic Ocean. The agreement represents the first proactive ecosystem-based approach undertaken to conserve resources in the Arctic Ocean. The interests of northern fishing communities are often opposed to oil development and large-scale shipping, as people fear the consequence of oil spills.

Threats of Oil and Gas Development in the Arctic Refuge

The long-running controversy over oil and gas development in the Arctic National Wildlife Refuge (ANWR) in northern Alaska has become a global symbol of the struggle between environmental protection and the energy requirements of a rapidly growing global population. Even during the establishment of the Alaska National Interest Lands Act of 1980, an inability to decisively determine the future of environmental policy in ANWR led to a compromise for the 1002 Area (named for Section 1002 of the Act), where most of ANWR became wilderness while the 1002 Area on the North Slope remained open to the possibility of oil exploration (Fig. 5.4). The struggle for lasting protection for the whole wildlife refuge has become even more intense due to the continued expansion of oil and gas developments across northern Alaska, the impacts of climate change, the global imperative to reduce emissions from fossil fuel burning, and the need for oil revenues by local and state governments in the face of declining revenues and increasing obligations.

Since the discovery of enormous oil reserves in northern Alaska in 1968, oil development quickly expanded from Prudhoe Bay and Kuparuk, two of the largest

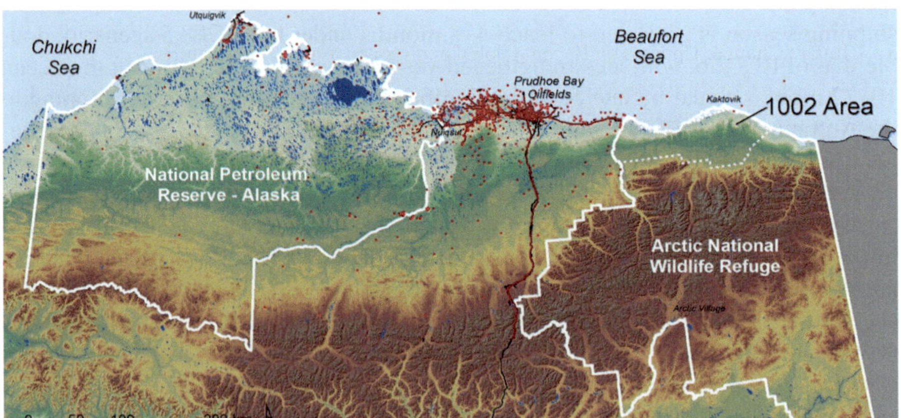

Fig. 5.4 Map of oil and gas development in Arctic Alaska relative to federal lands in the National Petroleum Reserve – Alaska and the 1002 Area of ANWR. Development includes roads (black lines), pipelines (red lines), and exploration well sites (red dots)

oilfields in the United States, to more than 20 oilfield units by 2018, extending all the way from the border of the ANWR to the National Petroleum Reserve – Alaska and toward the foothills (Fig. 5.4). This development has provided tremendous profits for the industry and income to the State of Alaska that has been used to fund government, eliminate income taxes, establish the Permanent Fund saving account, and provide an annual payout to state residents. Regional and local village governments also have benefitted greatly by taxing oilfield properties and using the funds to expand infrastructure and social services for the predominantly Iñupiat residents. But the oil development has had extensive direct and indirect impacts associated with seismic exploration, exploratory well sites, ice roads, gravel roads and pads, gravel mines, water impoundments, surface water use, dust from roads, permafrost degradation, oil spills, contaminants from drilling waste, air pollution, displacement of caribou calving, and attraction of predators that affect bird populations [41, 42]. In addition, A Health Impact Assessment by [43] identified social and health concerns related to oil development, including: dietary changes that increase high blood pressure and diabetes; rising rates of substance abuse, domestic abuse and suicide; exacerbation of asthma; and exposure to organic pollutants.

Soon after establishment of the 1002 Area in 1980, seismic exploration was conducted during the winters of 1984–1985 and the first exploratory well (Chevron KIC #1) was drilled on land inholdings owned by Kaktovik Iñupiat Corporation during winters of 1985–1986. Off-road vehicle traffic associated with the seismic exploration created ~4000 km of trails. Long-term monitoring of the trail damage showed that 79% of the trails had some level of disturbance in 1985 and disturbance persisted in 5% of the trails in 2009. Long-term damage mainly occurred where mechanical disruption of the insulating blanket of vegetation and organic soil caused thawing of permafrost and permanent track depressions [44]. While vegetation at most study plots recovered to pre-disturbance levels, some plots showed persistent shifts in plant species associated with permafrost degradation and increased wetting. At the KIC well, which was drilled from a temporary timber pad, vegetation remained much altered in 2018. Thermokarst has been extensive and led to ponding and exposure of contaminants from buried drilling wastes, requiring expensive remediation over the past decades.

With the passage of tax legislation in 2017 that required oil and gas leasing in the 1002 Area, the battle over the future of the Arctic's preeminent ecological reserve began anew. The initial skirmish focused on the Environmental Impact Statement for the leasing program under the National Environmental Protection Act. Environmental and indigenous groups have used this process to identify the numerous, cumulative, and persistent impacts that are likely to occur from oil development, and inadequacies in the EIS that are likely to be the basis of litigation to delay or halt the development. The potential consequences of oil development would likely be severe, given the: sensitivity of the ecosystems and permafrost to disturbance: the importance of the area to calving caribou (Fig. 5.5), denning polar bears, and migrating birds; and the dependence on the Gwich'in peoples on the Porcupine Caribou Herd for subsistence and maintaining their culture. Together with the projected impacts of climate change, the urgent need to greatly reduce global

Fig. 5.5 Caribou aggregation during the calving season on the coastal plain of the 1002 Area of the Arctic National Wildlife Refuge. (FWS photo)

dependence on fossil fuels, and the fragility of Arctic ecosystems, the need for new oil and gas development here, and elsewhere throughout the Arctic should be avoided.

Food Security

In the Arctic, where household livelihoods and community food systems are tightly connected to weather and the land, changing climate and socioeconomic conditions are significantly altering the ability of Arctic indigenous communities to achieve food security with locally available food resources [28, 45]. The challenge of obtaining sufficient food is exacerbated by high rates of poverty due to the lack of sufficient wage employment, the high cost of living in the North, changing knowledge systems and food sharing practices, population growth, and wildlife management practices that are typically beyond local control. Livelihoods traditionally centered on the harvest of wild, "country foods," are transitioning to a cash economy, with increasing reliance on industrially produced, store-bought foods that are prohibitively expensive for local incomes. In Nunavut, Canada, nearly 70% of the Inuit preschoolers have been found to reside in food-insecure households and in Fennoscandia reindeer herding that is important to the traditional Sami food culture is threatened by climate change [46]. While commercially available foods help with nutrition, the availability and quality of these foods depends on an uncertain global food system and the ability to pay for them. More importantly, imported foods often do not fulfill many of the roles that country foods have played in these communities and cultures. For example, in northern Alaska, the hunting of bowhead whales is of fundamental importance to the cultural traditions, societal relationships, and nutrition of the Iñupiat, as is caribou to the Gwich'in.

Conventional approaches to measuring food security often fail to adequately assess the transition in the nature and availability of foods [45]. Thus, there is an urgent need to better understand the evolving role of traditional activities that produce country foods in northern societies, and their evolving interdependencies with the industrial sector. More broadly, food security is a matter of human health that depends on both shifts in local resources and the influence of global pressures, and will require international efforts to help the Inuit and other circumpolar indigenous groups adapt to the changing Arctic [28].

Community Health

The effects of climate change on Arctic ecological and social systems are having large impacts on community health. Over the last six decades, indigenous peoples have seen an unprecedented transformation of their way of life, including diminution of land rights, resettlement, relocation of children to residential schools, industrial development, and changing food sources. These social changes, in addition to more recent environmental changes, are affecting diets (discussed above), rates of substance abuse and suicide; and accidental deaths from changing ice and storm conditions. Further risks are caused by exposure to organic pollutants from long range transport, and to contaminated water from damaged infrastructure (fuel storage, landfills, sewage lagoons).

While cancer and heart disease are the leading causes of death in the Arctic, particularly for people over 45, suicide and unintentional injury are the leading cause of death for people less than 45, especially for native people [47, 48]. Greenland and Chukotka, Russia have the highest suicide rates in the world at nearly 80 per 100,000 each year, about six times higher than the overall rates for the United States and Canada. In Alaska, the annual suicide rate of native people has hovered around 40 per 100,000 people over the last decade, twice the statewide average [48]. For Canada, the suicide rate in Nunavut is 10 times the rest of Canada. The crisis is particularly manifest in young people that are struggling to adapt to the social, political, economic, and environmental changes that characterize rapid modernization. In response to this crisis, numerous efforts have been undertaken, such as the National Inuit Suicide Prevention Strategy, that include early intervention and prevention programs that are critically important in reducing the risk and occurrence of suicide.

Unintentional injuries caused by poisoning, drowning, off-road vehicles, and hypothermia, also are unusually common for Arctic residents, and was the leading cause of death in Alaska during 2012–2015 for adults 25–44 year old, at an astounding annual rate of 117 per 100,000 people. It accounts for nearly a quarter of all years of potential life lost from premature death [48]. Poisoning from alcohol, illegal drugs, and prescription drugs was the most common cause of death, followed by drowning. Accidental drowning has been identified as a particular concern from unpredictable transportation routes on sea and river ice, and from rougher seas, associated with climate change.

Contaminants are still being transported to and recycled within the Arctic environment despite global action through the Stockholm Convention to reduce the production and use of POPs [5]. Acerbating this risk to humans are changes in the structure and dynamics of the Arctic food webs that could affect contaminant levels in Arctic subsistence species that are important to the traditional diet. The main source of contaminant exposure is the consumption of traditional foods of marine origin, such as whales, seals, polar bears, and some fish species. Based on international collaborative efforts to monitor human health across nine Arctic nations (Fig. 5.6), there are now substantial data extending back as far as the 1980s for evaluating the evolving health status of Arctic populations, especially indigenous peoples [5]. To assess exposure to contaminants in the Arctic environment, AMAP has compiled a vast amount of data on contaminant levels in human tissues, especially in hair and blood, and in some studies even human milk. Exposure levels vary

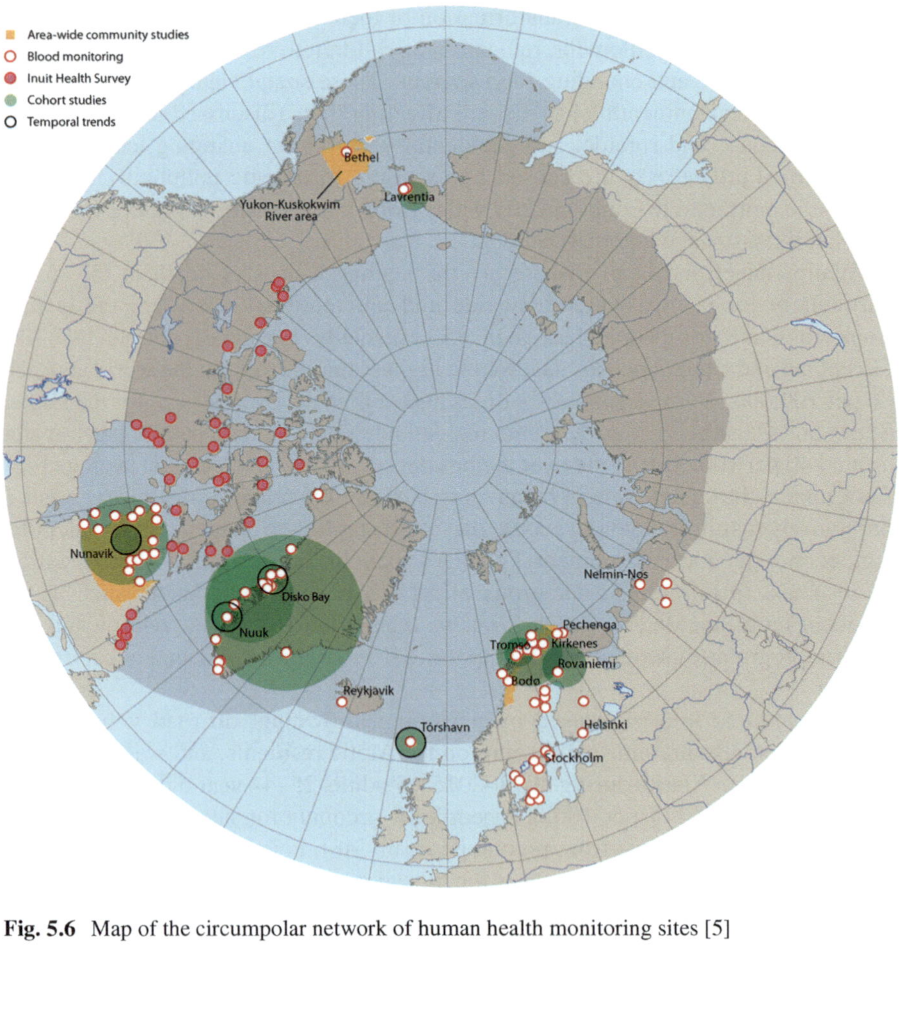

Fig. 5.6 Map of the circumpolar network of human health monitoring sites [5]

in different regions of the Arctic, which can be largely explained by variations in contaminant levels in the traditional diet. Results indicate concentrations for most POPs in Arctic biota, particularly PCBs and DDT, are declining, contrary to their earlier increasing trends. Despite elevated levels, certain high-concern contaminants (e.g., p,p'-DDT, PCB153, HCB, Hg) evident in monitoring of people in Eurasia and North America are continuing to decrease relative to previous years. Dietary precautions, however, are still advised for pregnant women. The decline in measured body burdens, after considering the population demographics, suggests international efforts to manage the risks of long-range transport of contaminants may be having a positive effect. The potential implications for human health, however, highlight the clear need to continue biomonitoring of contaminants.

An emerging issue that is unique to the Arctic is the threat of the spread of old diseases from corpses exposed by thawing permafrost. Permafrost soils are an ideal environment for preserving bacteria for very long periods of time [49], and bacteria and viruses have been successfully revived from 30,000-year-old permafrost in Alaska and Siberia [50, 51]. As climate change has been accelerating the thawing of permafrost soils, there have been rare instances of human and animal bodies being exposed that contain dormant viruses and bacteria. On the Yamal Peninsula in Russia in 2016, a young boy died and at least 20 people were hospitalized after being infected by anthrax, presumably from reindeer that died of anthrax over 75 years ago and were buried, with the carcasses subsequently exposed by thawing during a summer heat wave. During that summer, more than 2000 reindeer became infected, which then led to the small number of human cases. Frequent outbreaks of anthrax caused the death of 1.5 million reindeer in Siberian Russia between 1897 and 1925, with the reindeer usually buried in shallow permafrost soils [51]. Overall, anthrax among people or reindeer has been reported in more than 200 Yakutia settlements that are located near the reindeer burial grounds. In 1997, the body of a victim of the 1918 Spanish flu was exhumed from permafrost in the graveyard at Brevig Mission, western Alaska, and tissue from her lungs were used to sequence gene segments from RNA fragments, providing some concern that viable viruses from the pandemic influenza might be able to persist [52]. Victims of smallpox and the bubonic plague are also buried in cemeteries in Siberia and may become vectors after exposure from permafrost thawing [51]

Addressing future environmental and health policies for a region, where many of the current drivers of biophysical and socioeconomic changes are projected to continue or intensify, will require strong resolve and innovative approaches for adapting to the changing Arctic. An important fundamental step is to recognize that the northern people and their institutions are integral components of ecological systems [29]. Adaptation will require diverse strategies, ranging from the development and strengthening of surveillance and early warning systems for both ecosystems and community health, community empowerment and education on risks posed, to the promotion of sustainable development [28]. Northern peoples, however, have already shown a strong capacity to adapt to the enormous changes of the past century, and thus are well equipped to face the large challenges posed by climate change.

Conclusion

The Arctic in strongly linked to the global climate system through the roles of sea ice in cooling the planet, of cold Arctic Ocean waters affecting ocean circulation, of melting glaciers contributing to sea-level rise, and of thawing of permafrost soils that release greenhouse gasses from decomposing peat. With the Arctic warming at 2.5 times the rate of the rest of the Earth, biophysical responses in the Arctic are creating positive feedbacks to the global climate system that are accelerating global warming. The Arctic also is biologically connected to the rest of the world through annual migrations of whales, fish, and birds, airborne movement of contaminants, and human demographic responses to economic booms and crashes. With air temperatures projected to rise 4 to 12 °C during the next century, future changes in sea ice, snow, glaciers, permafrost, soil carbon, vegetation, fish, and wildlife are projected to be severe. This presents an enormous challenge to society and governments as to how to adapt and mitigate the environmental and health effects, such as expanding oil development, mining, shipping, commercial fishing, damage to village infrastructure from coastal erosion and collapsing permafrost, unpredictable access to subsistence foods, and increasing exposures to contaminants and diseases. The Arctic biome, then, provides important early warning signs regarding the consequences of global warming on ecosystem and societal health.

References

1. Pinkerton KE, Rom WN, editors. Global climate change and public health. New York: Springer; 2014.
2. AMAP. Snow, water, ice and permafrost in the Arctic (SWIPA) 2017. Oslo: Arctic Monitoring and Assessment Programme; 2017.
3. ACIA. Arctic climate impact assessment: Cambridge University Press; 2005.
4. IPCC. In: Stocker TF, Qin D, Plattner G-K, Tignor M, Allen SK, Boschung J, Nauels A, Xia Y, Bex V, Midgley PM, editors. Climate change 2013: the physical science basis. Contribution of Working Group I to the Fifth Assessment Report of the Intergovernmental Panel on Climate Change. Cambridge, UK and New York: Cambridge University Press; 2013.
5. AMAP. Human health in the Arctic. Oslo: Arctic Monitoring and Assessment Programme; 2015.
6. USGCRP. Climate science special report: fourth National Climate Assessment, vol. I. Washington, DC: U.S. Global Change Research Program; 2017. https://doi.org/10.7930/J0J964J6.
7. Osborne E, Richter-Menge J, Jefferies MO. Arctic report card 2018: NOAA; 2018. https://arctic.noaa.gov/Report-Card/Report-Card-2018
8. Meltofte H, Barry T, Berteaux D, Bültmann H, Christiansen JS, Cook JA, Dahlberg A, Daniëls FJ, Ehrich D, Fjeldså J, Fridriksson F, Ganter B. Arctic biodiversity assessment. Status and trends in Arctic biodiversity. Akureyri: Conservation of Arctic Flora and Fauna (CAFF); 2013.
9. Jorgenson MT. Thermokarst terrains. In: Schroeder J, Giardino R, Harbor J, editors. Treatise on geomorphology, Glacial and periglacial geomorphology, vol. 8. San Diego: Academic Press; 2013. p. 313–24.
10. Jorgenson MT, Shur YL, Pullman ER. Abrupt increase in permafrost degradation in Arctic Alaska. Geophys Res Lett. 2006;33:L02503. https://doi.org/10.1029/2005GL024960.

11. Liljedahl AK, Boike J, Daanen RP, Fedorov AN, Frost GV, Grosse G, Hinzman LD, Iijma Y, Jorgenson JC, Matveyeva N, Necsoiu M, Raynolds MK, Romanovsky VE, Schulla J, Tape KD, Walker DA, Wilson CJ, Yabuki H, Zona D. Pan-Arctic ice-wedge degradation in warming permafrost and its influence on tundra hydrology. Nat Geosci. 2016;9:312–8. https://doi.org/10.1038/ngeo2674.
12. Jorgenson MT, Marcot BG, Swanson DK, Jorgenson JC, DeGange AR. Projected changes in diverse ecosystems from climate warming and biophysical drivers in Northwest Alaska. Clim Chang. 2015;130:131–44. https://doi.org/10.1007/s10584-014-1302-1.
13. Jorgenson MT, Grosse G. Remote sensing of landscape change in permafrost regions. Permafr Periglac Process. 2016;27(4):324–38.
14. Myers-Smith IH, Forbes BC, Wilmking M, Hallinger M, Lantz TC, et al. Shrub expansion in tundra ecosystems: dynamics, impacts and research priorities. Environ Res Lett. 2011;6(045509)
15. de Groot WJ, Cantin AS, Flannigan MD, Soja AJ, Gowman LM, Newbery A. A comparison of Canadian and Russian boreal forest fire regimes. For Ecol Manag. 2013;294:23–34.
16. Higuera PE, Chipman ML, Barnes JL, Urban MA, Hu FS. Variability of tundra fire regimes in Arctic Alaska: millennial-scale patterns and ecological implications. Ecol Appl. 2011;21(8):3211–26.
17. Lantuit H, Overduin PP, Wetterich S. Recent progress regarding permafrost coasts. Permafr Periglac Process. 2013;24:120–30.
18. Post E, Forchhammer MC, Bret-Harte MS, Callaghan TV, Christensen TR, Elberling B, Fox AD, Gilg O, Hik DS, Høye TT, Ims RA, Jeppesen E, Klein DR, Madsen J, McGuire MA, Rysgaard S, Schindler DE, Stirling I, Tamstorf MP, Tyler NJC, van der Wal R, Welker J, Wookey PA, Schmidt NM, Aastrup P. Ecological dynamics across the arctic associated with recent climate change. Science. 2009;325:1355–8.
19. Jorgenson MT, Ely C, Terenzi J. Monitoring and predicting the effects of climate change on coastal ecosystems on the Yukon-Kuskokwim Delta. In: Reynolds JH, Wiggins HV, editors. Shared science needs: report from the Western Alaska landscape conservation cooperative science workshop, vol appendix 5. Anchorage: Western Alaska Landscape Conservation Cooperative; 2012. p. 130–7.
20. Slater AG, Lawrence DM. Diagnosing present and future permafrost from climate models. J Clim. 2013;26(15):5608–23.
21. Schuur EAG, McGuire AD, Schädel C, Grosse G, Harden JW, Hayes DJ, Hugelius G, Koven CD, Kuhry P, Lawrence DM, Natali SM, Olefeldt D, Romanovsky VE, Schaefer K, Turetsky MR, Treat CC, Vonk JE. Climate change and the permafrost carbon feedback. Nature. 2015;520(7546):171–9. https://doi.org/10.1038/nature14338.
22. Schuur EAG, McGuire AD, Romanovsky V, Schädel C, Mack M. Chapter 11: Arctic and boreal carbon. In: Cavallaro N, Shrestha G, Birdsey R, et al., editors. Second state of the carbon cycle report (SOCCR2): a sustained assessment report. Washington, DC: U.S. Global Change Research Program, 2018. p. 428–68. https://doi.org/10.7930/SOCCR2.2018.
23. McGuire AD, Lawrence DM, Koven C, Clein JS, Burke E, Chen G, Jafarov E, MacDougall AH, Marchenko S, Nicolsky D, Peng S. Dependence of the evolution of carbon dynamics in the northern permafrost region on the trajectory of climate change. Proc Natl Acad Sci. 2018;115(15):3882–7.
24. Ruppel CD, Kessler JD. The interaction of climate change and methane hydrates. Rev Geophys. 2017;55(1):126–68.
25. Phoenix GK, Bjerke JW. Arctic browning: extreme events and trends reversing arctic greening. Glob Chang Biol. 2016;22(9):2960–2.
26. Elmendorf SC, Henry GHR, Hollister RD, Bjork RG, Boulanger-Lapointe N, et al. Plot-scale evidence of tundra vegetation change and links to recent summer warming. Nat Clim Chang. 2012;2:453–7.
27. AMAP. AMAP assessment 2013: Arctic Ocean acidification. Oslo: Arctic Monitoring and Assessment Programme (AMAP); 2013.

28. Ford JD, McDowell G, Pearce T. The adaptation challenge in the Arctic. Nat Clim Chang. 2015;5(12):1046.
29. Chapin FS, Hoel M, Carpenter SR, Lubchenco J, Walker B, Callaghan TV, Folke C, Levin SA, Mäler KG, Nilsson C, Barrett S. Building resilience and adaptation to manage Arctic change. Ambio. 2006;35(4):198–203.
30. GAO. Alaska native villages: limited progress has been made on relocating villages threatened by flooding and erosion, vol. GAO-09-551. Washington, DC: Government Accounting Office; 2009.
31. Newtok Village Council. Newtok Village Tribal Hazard Mitigation Plan Update. 2015.
32. Engineers USACo. Alaska baseline Erosion assessment: study findings and technical report. Anchorage. 2009.
33. Denali Commission. Village infrastructure protection program: partnering to build community resilience, vol. VIP Fact Sheet 2019. Anchorage. 2019.
34. Melvin AM, Larsen P, Boehlert B, Neumann JE, Chinowsky P, Espinet X, Martinich J, Baumann MS, Rennels L, Bothner A, Nicolsky DJ. Climate change damages to Alaska public infrastructure and the economics of proactive adaptation. PNAS. 2017;114(2):E122–31. https://doi.org/10.1073/pnas.1611056113.
35. Larsen JN, Fondahl G. Arctic human development report: regional processes and global linkages, vol. II. Copenhagen: Nordic Council of Ministers; 2015. https://doi.org/10.6027/TN2014-567.
36. Kontorovich AE. Oil and gas of the Russian Arctic: history of development in the 20th century, resources, and strategy for the 21st century science first hand 41 (N2). 2015.
37. Bandopadhyay S, Nelson MG. Mining in the Arctic. Leiden: A.A. Balkema; 1992.
38. Emmerson C, Lahn G. Arctic opening: opportunity and risk in the high north. London: Lloyd's and Chatham House; 2012.
39. Arctic Council. Arctic marine shipping assessment 2009 report. 2009.
40. Melia N, Haines K, Hawkins E. Sea ice decline and 21st century trans-Arctic shipping routes. Geophys Res Lett. 2016;43(18):9720–8.
41. National Research Council (NRC). Cumulative environmental effects of oil and gas activities on Alaska's north slope. Washington, DC: National Academies Press; 2003.
42. Raynolds MK, Walker DA, Ambrosius KJ, Brown J, Everett KR, Kanevskiy M, Kofinas GP, Romanovsky VE, Shur Y, Webber PJ. Cumulative geoecological effects of 62 years of infrastructure and climate change in ice-rich permafrost landscapes, Prudhoe Bay oilfield, Alaska. Glob Chang Biol. 2014;20(4):1211–24. https://doi.org/10.1111/gcb.12500.
43. Wernham A. Inupiat health and proposed Alaskan oil development: results of the first integrated health impact assessment/environmental impact statement for proposed oil development on Alaska's north slope. Eco Health. 2007;4(4):500–13.
44. Jorgenson JC, Ver Hoef J, Jorgenson MT. Long-term recovery patterns of arctic tundra after winter seismic exploration. Ecol Appl. 2010;20(1):205–21.
45. Loring PA, Gerlach SC. Food, culture, and human health in Alaska: an integrative health approach to food security. Environ Sci Pol. 2009;12(4):466–78.
46. Nilsson LM, Destouni G, Berner J, Dudarev AA, Mulvad G, Odland JØ, Parkinson A, Tikhonov C, Rautio A, Evengård B. A call for urgent monitoring of food and water security based on relevant indicators for the Arctic. Ambio. 2013;42(7):816–22.
47. Young TK, Revich B, Soininen L. Suicide in circumpolar regions: an introduction and overview. Int J Circumpolar Health. 2015;74(1):27349.
48. Alaska Native Epidemiology Center. Alaska native health status report: second edition. Alaska native tribal health consortium, Anchorage. 2017.
49. Jansson JK, Taş N. The microbial ecology of permafrost. Nat Rev Microbiol. 2014;12(6):414.
50. Legendre M, Bartoli J, Shmakova L, Jeudy S, Labadie K, Adrait A, Lescot M, Poirot O, Bertaux L, Bruley C, Couté Y. Thirty-thousand-year-old distant relative of giant icosahedral DNA viruses with a pandoravirus morphology. Proc Natl Acad Sci. 2014;111(11):4274–9.
51. Revich BA, Podolnaya MA. Thawing of permafrost may disturb historic cattle burial grounds in East Siberia. Glob Health Action. 2011;4(1):8482.
52. Taubenberger JK, Kash JC. Insights on influenza pathogenesis from the grave. Virus Res. 2011;162(1–2):2–7.

Chapter 6
Assessing the Health Risks of Climate Change

Kristie L. Ebi

Introduction

Climate change is altering everyday weather patterns, including changing average temperature and precipitation; increasing the frequency, intensity, duration, and spatial extent of some extreme weather and climate events; and altering sea level [12, 13]. Climate-sensitive health outcomes include injuries, illnesses, and deaths associated with extreme events, such as floods, droughts, and heat waves, or with changes in air quality; increases in the geographic range, seasonality, and/or intensity of transmission of infectious diseases, such as diarrheal disease and vector-borne diseases (e.g., dengue and malaria); undernutrition; and health consequences associated with diffuse, delayed, and/or cascading effects of climate change or the actions taken to prepare for and address risks (e.g., occupational impacts, undernutrition, conflict, migration, and mental stress) [7, 25].

Policy- and decision-makers, public health and health-care agencies and institutions, and the general public want to understand to what extent these changes could affect their health and that of their families, today and in the future. Providing the answer is more complex than for more traditional health risks. This chapter discusses some challenges with using traditional risk assessment approaches to estimate the health risks of climate change. It then presents a conceptual approach for thinking about assessing health risks, followed by a discussion of the process used by the Intergovernmental Panel on Climate Change to conduct their assessments.

K. L. Ebi (✉)
University of Washington, Seattle, WA, USA
e-mail: krisebi@uw.edu

K. E. Pinkerton, W. N. Rom (eds.), *Climate Change and Global Public Health*, Respiratory Medicine, https://doi.org/10.1007/978-3-030-54746-2_6

Traditional Risk Assessment Approaches Are Inappropriate for Estimating the Health Risks of Climate Change

Public health has a long history of determining whether an agent presents a risk to health, where risk is defined as probability times consequence. Methods and tools to assess whether an agent could harm human health range from the Bradford-Hill criteria [3] to the International Agency for Research on Cancer [14] to toxicological risk assessments for environmental stressors [18]. At their simplest, these approaches ask whether this is sufficient information to determine if an agent is a hazard to health and, if so, to determine exposure-response relationships and to characterize the extent of human exposures. These are combined, often including safety factors designed to protect those most vulnerable to adverse impacts, to produce a quantitative or qualitative statement about the probability and degree of harm to the exposed populations [19]. This information is intended to be used by policymakers in formulating standards for "safe" exposures.

Climate change challenges the assumptions underlying these traditional approaches to assessing risk, including the following [1]:

- An exposure to a specific agent causes one or a limited number of adverse health outcome.
- The health outcome associated with an exposure is distinctive.
- There is an unexposed control group against which to compare; and.
- The causal association is direct.

The basic assumption of traditional risks assessment is that a defined exposure to a specific agent causes an adverse health outcome in exposed populations, with specific groups at particular risk. The reality is that climate change affects health through relatively direct and indirect pathways. High ambient temperatures increase morbidity and mortality from, for example, cardiovascular, respiratory, and kidney diseases, by directly affecting human physiology. High temperatures also can reduce crop yields, leading to increases in undernutrition; this pathway is indirect. Further, loss of access to critical resources resulting in food and water insecurity can lead to migration, with its attendant consequences for human health and well-being. As these simple examples show, there is a wide range of exposures related to climate change that can result in injuries, illnesses, and deaths.

A further complexity is that multiple weather variables may be associated with a single health outcome, with the variables varying geographically. The geographic distributions and seasonal variations of many infectious diseases indicate the potential importance to disease patterns of weather and seasonal-to-interannual climate variability [20]. Temperature, precipitation, and humidity can affect vector survival, reproduction, development, and biting rates, as well as pathogen reproduction and development, thus affecting the timing and intensity of outbreaks. Further, the weather variables of importance can vary geographically. For example, Thomson [26] found a geographically complex association between malaria incidence and the timing of the onset and retreat of seasonal rains in Nigeria, with rainfall onset related to the El Niño Southern Oscillation and the Northern Annular Mode, and retreat related to the North Atlantic Oscillation and the East Pacific or West Pacific circulation index.

Traditional risk assessment assumes the health outcome associated with an exposure is distinctive and the association between immediate cause (e.g., cigarette smoking) and health impact (e.g., lung cancer) is direct. However, the number of health outcomes affected by changes in weather patterns and sea-level rise because of climate change covers most current health concerns [7]. Any health outcome sensitive to weather or climate could be affected by climate change. Further, an exposure-response relationship may not be applicable across all temporal and spatial scales. A few examples illustrate the challenge. Heat waves kill unnecessarily; the extent to which a heat wave is a risk depends on the population's acclimatization to hot weather, the number of previous heat waves that season, the proportion of the population with increased sensitivity (e.g., older adults, the prevalence of diabetes, the proportion using certain drugs, etc.), mortality rates the previous winter, the effectiveness of the local early warning system, etc. ([15]; e.g., [23]). In some regions of Africa, malaria follows the rains, in others it follows drought [21]. The same magnitude typhoon hitting Japan will have very different consequences from one hitting the Philippines [27]. An evaluation of the flood risk in Sri Lanka depends on the question being asked; storm surges affect coastal regions fairly infrequently, with large consequences when they do [27]. Inland areas have more frequent and less intense flooding events that affect more communities and their livelihoods. Increased investment in risk reduction activities, from strengthening housing to moving buildings at particular risk to early warning systems, could reduce vulnerability over time, so the consequences of a heavy precipitation event would change over time.

The range of climate-sensitive health outcomes is increasing. For example, there is growing evidence that high ambient temperature can trigger adverse birth outcomes, such as preterm birth, low birth weight, and stillbirth [31]. Another example is that mental health outcomes can be associated with changing weather patterns, including post-traumatic stress disorder, anxiety, and depression [8].

There are no unexposed, control groups against which to compare. We are all exposed to climate change, although the degree of exposure to specific changes in weather variables varies across spatial and temporal scales. Flooding, droughts, and wildfires affect some populations, while sea-level rise and storm surges affect others. Further, the association between weather factors and health outcomes may not be static across time. A changing climate can result in key weather variables crossing thresholds that result in large changes in the geographic range or incidence of a health outcome. One example is the 2004 outbreak of *Vibrio parahaemolyticus* in Alaska, the leading cause of seafood-associated gastroenteritis in the United States; outbreaks are typically associated with the consumption of raw oysters gathered from warm-water estuaries [17]. The consumption of raw oysters was the only significant predictor of illness. All oysters associated with the outbreak were harvested when mean daily water temperatures exceeded 15.0 °C (the theorized threshold for the risk of *V. parahaemolyticus* illness). Between 1997 and 2004, mean water temperatures in July and August at the implicated oyster farm increased 0.21 °C per year. 2004 was the first year during which mean daily water temperatures did not drop below 15.0 °C during the outbreak. The outbreak extended by 1000 km the northernmost documented source of oysters that caused illness due to *V. parahaemolyticus*.

The assumption of a direct, causal association between climate change and health outcomes is rarely satisfied. Climate-sensitive health outcomes often have many, interrelated causes, of which weather is only one factor; feedback mechanisms may be important. The causal chain between exposure to a pathogen and disease is complex: exposure is necessary but not sufficient to cause disease. Therefore, weather and climate may not be the primary drivers of a health outcome, which is not to discount their importance, but to acknowledge the importance of studying health outcomes using systems-based approaches that include the social, economic, and political factors influencing disease risk.

Determining the Extent to Which Climate Change Has Affected Health

Analyzing the extent to which climate change has altered the burden of climate-sensitive health outcomes is complex because climate change refers to changes in weather patterns over decades or longer. A formal statistical method, detection and attribution, can be used to determine how recent climate change affected morbidity and mortality. This method determines whether the occurrence of adverse health outcomes has changed relative to a baseline, and for those outcomes where there has been a change, then determines the extent to which the change could be attributed to climate change [6]. Case studies for heat waves, Lyme disease in Canada, and *Vibrio* emergence in northern Europe highlight evidence that a portion of the changes in rates and geographic distribution of these health outcomes can be attributed to recent climate change. This evidence is useful to inform risk management and to inform communication about the health risks of climate change.

Analyzing relationships between climate change and health outcomes requires decade-long data sets; such data sets are available for meteorological data, but are rare in the health sector. Analyzing these data requires selecting a baseline for comparison because there is no natural baseline in a changing climate. Climate analyses use baselines of 20–30 years to avoid annual variability in weather variables affecting the analyses.

Risk Management as an Appropriate Framework for Assessing the Health Risks of Climate Variability and Change

Risk management is a more appropriate framework for assessing the health risks of climate change because it recognizes and addresses the challenges outlined to using a traditional risk assessment. A risk-based framing explicitly recognizes the future will differ in many ways from the present, with uncertainties as to the timing and magnitude of change.

In this framework, the magnitude and extent of the health risks of climate change is a function of the interactions between the hazards of a changing climate; the exposure(s) to climate change-related alterations in weather patterns (and the implications of associated changes, such as changing crop yields) of importance for health; and the vulnerabilities of the exposed human and natural systems (e.g., changing crop yields can have differential consequences depending on the availability of other food sources, etc.). Hazards are changes in the mean and variability of temperature, precipitation, and other weather variables associated with climate-sensitive health outcomes.

Understanding differences in vulnerability is necessary for estimating the possible health risks of climate change, and for designing effective and efficient adaptation options to prepare for and manage risks [5]. A large number of factors determine vulnerability, including poverty (although all poor people are not equally at risk), demographics (although not all population groups are equally vulnerable to each outcome), wealth and income distribution, status of the public health infrastructure, access to medical care, behavioral factors, individual physiological factors, and a wide range of social and cultural factors [4]. Vulnerability can be much more important than climate change in determining impacts over the short-term [11]. Policies to address these vulnerabilities may have commonalities across regions and sectors, but need to be tailored to specific circumstances. Further, policies need to balance competing demands, such as water needs across agriculture, other economic sectors and tourism, urban areas, recreational use, health concerns, and others.

Vulnerability, Capacity, and Adaptation Assessments to Estimate Local to Regional Health Risks of Climate Change

Vulnerability, capacity, and adaptation assessments (V&As) are used from local to national scales to assess the possible health risks of climate change within the context of adaptive risk management [2, 28]. V&As are a process and instrument to establish partnerships and obtain information for understanding and addressing climate change-related risks. They can also provide the knowledge needed to realize potentially large health co-benefits from well-designed adaptation and greenhouse gas mitigation policies.

The ultimate goal of a V&A is to provide evidence of current, emerging, and projected health risks of climate change in a particular region; to identify populations at particular risk; and to recommend policies and measures that, if implemented, over short- to longer-time scales can effectively manage the risks. These policies and measures include modifications to sectoral plans, such as vector-borne disease control programs, to explicitly incorporate the challenges and opportunities of climate change (e.g., adaptation) and greenhouse gas reductions to reduce the challenges faced by health systems later in the century (e.g., mitigation). These assessments are participatory in nature, engaging a range of stakeholders from within and outside health systems, such as meteorological services, ministries responsible for water and agriculture, NGOs involved in adaptation, and others.

For low- and middle-income countries, the Health component of National Adaptation Plans (HNAPs) [29] and the Operational Framework for Building Climate Resilient Health Systems [30] are complementary tools designed by WHO to apply information collected from a V&A to define strategic goals and plans for building health resilience to climate change. HNAPs, as sector-specific adaptation plans, highlight national health adaptation goals to be achieved over a specific time frame and with available resources.

Key functions of V&As include the following [28]:

- Improving evidence and understanding of the current associations between weather/climate and health outcomes, including the populations most vulnerable to these risks
- Providing health and emergency management officials, stakeholders, and the public with information on the magnitude and pattern of current and projected future health risks associated with climate variability and change, and identifying vulnerabilities in the health system itself
- Identifying opportunities to incorporate climate change challenges and opportunities into existing policies and programs designed to manage health risks associated with weather and climate, and to develop new programs where necessary to prevent and reduce the severity of future risks
- Serving as a baseline analysis against which future changes in risks and in climate change-related policies and programs can be monitored
- Facilitating collaborations with sectors, such as water and infrastructure, to promote activities to improve population health in a changing climate
- Strengthening the case for investment in health protection

Assessments by the Intergovernmental Panel on Climate Change (IPCC)

The Intergovernmental Panel on Climate Change (IPCC) conducts periodic assessments of the scientific, technical, and socioeconomic information relevant for understanding anthropogenic climate change, its potential impacts, and options for mitigation and adaptation. These assessments are based on expert judgment evaluation of the literature (peer- and non-peer-reviewed) combined with the collective experience and judgment of a group of individuals chosen because of their diverse and relevant expertise [22]. The chapters produced are thoroughly reviewed by the worldwide scientific community to ensure the assessments reflect the literature base.

The United Nations Environment Programme and the World Meteorological Organization established the IPCC as a unique collaboration between the scientific community and policymakers, with governments (through their Focal Points) providing guidance and input at several stages during the process to the scientists conducting an assessment. IPCC reports are mandated to be comprehensive, objective, and balanced [10]. Additional requirements are to describe different scientific, technical, and socioeconomic views on a subject, and for an assessment to be policy-relevant and policy-neutral. The assessments aim to inform national governments about the most up-to-date scientific thinking and to highlight possible policy options to address current and projected risks, without promoting one set of options over another. The members of the IPCC are the world governments.

When the governments decide to initiate an assessment, the first step is for governments to select individuals to lead the three Working Groups (WGs) in the IPCC: WGI assesses the science of climate change; WGII assesses impacts, adaptation, and vulnerability; and WGIII assesses mitigation options. It is the IPCC Panel that decides whether to prepare a report, including its scope, outline, and work plan, in consultation with the respective WG. Policymakers and other users of IPCC Reports may be consulted to identify key policy-relevant issues. Once an outline is agreed upon, Governments and IPCC Observer Organizations are requested to nominate experts to be coordinating lead authors (CLAs), lead authors (LAs), and review editors (REs). Author teams were constructed with attention to scientific qualifications, the needed range of institutional and disciplinary perspective, and adequate regional and gender balance, while also involving the next generation of climate scientists.

As required by the IPCC Principles and Procedures, there are two reviews of a report, the Expert Review (First Order Draft) and the Government and Expert Review (Second Order Draft). These reviews involve hundreds of reviewers who submit thousands of comments. There were more than 42,000 review comments for the five chapters in the IPCC Special Report on Warming of 1.5 °C (https://www.ipcc.ch/sr15/). A requirement of the IPCC process is that authors must provide written responses to all comments submitted during these review periods; a considerable task. Review Editors, a unique feature of IPCC reports, are involved in the process starting with the First Order Draft review, representing the reviewers and ensuring that each comment is considered and appropriately addressed.

The last step in the process for a WG contribution to an assessment cycle is for the Summary for Policymakers (SPM) to be approved line-by-line in a Working Group session. Every sentence in a Summary for Policymakers is discussed and agreed (by consensus) between the authors who drafted the SPM and governments. Authors and governments want to ensure the SPM is not only an accurate assessment of the state of knowledge, but also that it communicates key findings clearly in understandable language to policymakers. When an SPM is approved, the governments then accept the underlying report [10]. This close and ongoing dialogue at the science-policy interface ensures an assessment achieves its mandate and requirements; it is a unique feature of the IPCC process. This process was intensively reviewed and endorsed with some modifications by the Inter Academy Council [9] .

Science is only one input into decision-making [24]; policymakers also take into consideration social and cultural values and perspectives, practical issues (from technological to political), and other factors when developing and implementing a policy. Policies need to be specific to a national (or sub-national) context, including level of development, current and projected vulnerabilities, current and projected climate variability and change, and many other factors. For example, policies to enhance food security in a changing climate will depend on a wide range of issues, such as causes of food insecurity in the region of interest, crops grown, water availability, transport, trade policies, and others.

IPCC CLAs and LAs are not only experts in their field, they also willingly donate considerable time and intellect to an IPCC assessment. CLAs can expect to commit approximately six to nine months of full-time activity between appointment and the approval session, and LAs can expect to commit approximately four to six months of full-time activity. The time committed is voluntary; the IPCC does not support the time scientists spend working on an assessment. A WG Technical Support Unit provides some support for aspects of report development, but not for reviewing literature and writing text. The IPCC has a Trust Fund that covers travel and per diem to lead author meetings for authors and review editors from developing countries and countries with economies in transition. Developed country governments are expected to cover travel and per diem for their authors and review editors.

No chapter team has experts for every issue that will be covered. Many chapters will cite roughly 1000 references. For example, the human health chapter typically includes topics such as the current burden of climate-sensitive health outcomes; vulnerability of children and older adults; projected changes in undernutrition, infectious diseases, emerging zoonotic diseases, morbidity and mortality due to extreme weather events; experience with adaptation; costs of action and of inaction on climate change; and co-benefits of mitigation policies. Therefore, expertise is drawn from authors on other chapters and from the wider scientific community through selection of Contributing Authors (CAs). CAs are selected as needed to write about a specific topic or contribute a case study to illustrate a particular point.

In reporting the key conclusions from their chapter, authors describe the certainty in those findings using calibrated uncertainty language [16]. This language aims to facilitate clear communication of the degree of certainty in assessment findings, including findings that span a range of possible outcomes. It also aims to avoid descriptions of uncertainties using casual terms that may imply different meanings to different disciplines and/or in different languages. The degree of certainty in key findings is described using two metrics [16]:

- Confidence in the validity of a finding, based on the type, amount, quality, and consistency of evidence (e.g., mechanistic understanding, theory, data, models, expert judgment) and the degree of agreement. Confidence is expressed qualitatively.
  - Evidence and agreement are each categorized on a three-point scale (for nine combinations). The author team's evaluation of evidence and agreement is the basis for the level of confidence assigned to each key finding. The description of the author team's evaluation of evidence and agreement is called a traceable account. Each key finding is presented in a chapter's Executive Summary, including reference to the chapter section containing the traceable account for the finding.

  - When there is sufficient evidence and agreement, these can be synthesized into one metric to describe a qualitative level of confidence, where level of confidence is designated as very low, low, medium, high, and very high.

- Quantified measures of uncertainty in a finding, such as a probabilistic estimate of a specific occurrence or range of outcomes. Probabilistic information may originate from statistical or modeling analyses, expert elicitation of views, or other quantitative information.

Discussion

Climate change presents a wide range of risks to human health that vary spatially and temporally. Traditional risk assessment approaches are ill-suited to understanding the complex interactions leading to adverse health impacts when exposure is one of many factors affecting the health burden. Focusing instead on risk management, based on vulnerability and adaptation assessments, provides the policy-relevant information needed to effectively manage changing health risks over time. Expert judgment processes, such as that used by the IPCC, are important national approaches to understanding risks and how they could change with changes in climate, development, and other factors. As the literature base on the health impacts of climate change expands, meta-analytic and other techniques may be possible to increase the robustness of key findings.

Whatever the approach used to assess the health risks of climate change, the goal should be to provide information relevant for developing strategies, policies, and measures to protect the most vulnerable, today and in the future.

References

1. Bernard SM, Ebi KL. Comments on the process and product of the health impacts assessment component of the United States National Assessment of the potential consequences of climate variability and change. Environ Health Perspect. 2001;109(Suppl 2):177–84.
2. Berry P, Enright PM, Shumake-Guillemot J, Villalobos Prats E, Campbell-Lendrem D. Assessing health vulnerabilities and adaptation to climate change: a review of international progress. Int J Environ Res Public Health. 2018;15:2626. https://doi.org/10.3390/ijerph15122626.
3. Bradford-Hill A. The environment and disease: association or causation? Proc Royal Soc Med. 1965;58:295–300.
4. Cardona O-D, van Aalst MK, Birkmann J, Fordham M, McGregor G, Perez R, Pulwarty RS, ELF S, Sinh BT. Determinants of risk: exposure and vulnerability. In: Field CB, Barros V, Stocker TF, Qin D, Dokken DJ, Ebi KL, Mastrandrea MD, Mach KJ, Plattner G-K, Allen SK, Tignor M, Midgley PM, editors. Managing the risks of extreme events and disasters to advance climate change adaptation. A special report of working groups I and II of the intergovernmental panel on climate change. Cambridge, UK, and New York: Cambridge University Press; 2012. p. 65–108.

5. Ebi KL. Public health responses to the risks of climate variability and change in the United States. J Occup Environ Med. 2009;51:4–12.
6. Ebi KL, Ogden NH, Semenza JC, Woodward A. Detecting and attributing health burdens to climate change. Environ Health Perspect. 2017;125(8). https://ehp.niehs.nih.gov/doi/full/10.1289/EHP1509.
7. Haines A, Ebi K. The imperative for climate action to protect health. N Engl J Med. 2019;380:263–73.
8. Hayes K, Poland B. Addressing mental health in a changing climate: incorporating mental health indicators into climate change and health vulnerability and adaptation assessments. Int J Environ Res Public Health. 2018;15:1806.
9. Inter Academy Council. Climate change assessments, review of the processes and procedures of the IPCC. Inter Academy Council, Amsterdam. 2010. http://reviewipcc.interacademycouncil.net/; Accessed 14 July 2019.
10. IPCC. Decisions taken with respect to the review of IPCC processes and procedures: procedures. IPCC 34th Session, 18–19 Nov 2011, Kampala. 2011. http://ipcc.ch/pdf/ipcc-principles/appendix_a_decision.pdf; Accessed 14 July 2019.
11. IPCC. Summary for Policymakers. In: Field CB, Barros V, Stocker TF, Qin D, Dokken DJ, Ebi KL, Mastrandrea MD, Mach KJ, Plattner G-K, Allen SK, Tignor M, Midgley PM, editors. Managing the risks of extreme events and disasters to advance climate change adaptation. A special report of working groups I and II of the intergovernmental panel on climate change. Cambridge, UK, and New York: Cambridge University Press; 2012. p. 1–19.
12. IPCC. In: Stocker TF, Qin D, Plattner G-K, Tignor M, Allen SK, Boschung J, Nauels A, Xia Y, Bex V, Midgley PM, editors. Climate change 2013: the physical science basis. Contribution of working group I to the fifth assessment report of the intergovernmental panel on climate change. Cambridge, UK, and New York: Cambridge University Press; 2013. 1535 pp.
13. IPCC. Summary for Policymakers. In: Masson-Delmotte V, Zhai P, Pörtner H.-O, Roberts D, Skea J, Shukla PR, Pirani A, Moufouma-Okia W, Péan C, Pidcock R, Connors S, Matthews JBR, Chen Y, Zhou X, Gomis MI, Lonnoy E, Maycock, Tignor M, Waterfield T, editors. Global warming of 1.5 °C. An IPCC Special report on the impacts of global warming of 1.5 °C above pre-industrial levels and related global greenhouse gas emission pathways, in the context of strengthening the global response to the threat of climate change, sustainable development, and efforts to eradicate poverty. Geneva: World Meteorological Organization. 2018. 32 pp.
14. International Agency for Research on Cancer (IARC). Preamble. IARC monographs on the evaluation of carcinogenic risks to humans. World Health Organization, IARC, Lyon; 2006. Available at http://monographs.iarc.fr/ENG/Preamble/index.php. Accessed 14 July 2019.
15. Kovats RS, Hajat S. Heat stress and public health: a critical review. Annu Rev Public Health. 2008;29:41–55.
16. Mastrandrea MD, Field CB, Stocker TF, Edenhofer O, Ebi KL, Frame DJ, Held H, Kriegler E, Mach KJ, Matschoss PR, Plattner G-K, Yohe GW, Zwiers FW. Guidance note for lead authors of the IPCC fifth assessment report on consistent treatment of uncertainties. Intergovernmental Panel on Climate Change (IPCC). 2010. http://www.ipcc-wg2.gov/meetings/CGCs/index.html#UR; Accessed 14 July 2019.
17. McLaughlin JB, DePaola A, Bopp CA, Martinek KA, Napolilli NP, Allison CG, Murray SL, Thompson EC, Bird MM, Middaugh JP. Outbreak of *Vibrio parahaemolyticus* gastroenteritis associated with Alaskan oysters. N Engl J Med. 2005;353:1463–70.
18. National Research Council. Risk assessment in the Federal Government: managing the process. Washington, DC: National Academy Press; 1983.
19. National Research Council Committee on Risk Assessment of Hazardous Air Pollutants. Science and judgment in risk assessment. Washington, DC: National Academy Press; 1994.
20. National Research Council. Under the weather: climate, ecosystems, and infectious diseases. Washington, DC: National Academy Press; 2001.
21. Parham PE, Michael E. Modelling climate change and malaria transmission. Adv Exp Med Biol. 2010;673:184–99.

22. Parry M, Carter T. Climate Impact and Adaptation Assessment: a guide to the IPCC approach. London: Earthscan Publications; 1998.
23. Rocklov J, Forsberg B, Meister K. Winter mortality modifies the heat-mortality association the following summer. Eur Respir J. 2009;33:245–51.
24. Scheraga J, Ebi K, Moreno AR, Furlow J. From science to policy: developing responses to climate change. In: McMichael AJ, Campbell-Lendrum D, Corvalan CF, Ebi KL, Githeko A, Scheraga JD, Woodward A, editors. Climate change and human health: risks and responses. Geneva: WHO/WMO/UNEP; 2003.
25. Smith KR, Woodward A, Campbell-Lendrum D, Chadee DD, Honda Y, Liu Q, Olwoch JM, Revich B, Sauerborn R. Human health: impacts, adaptation, and co-benefits. In: Field CB, Barros VR, Dokken DJ, Mach KJ, Mastrandrea MD, Bilir TE, editors. Climate change 2014: impacts, adaptation, and vulnerability. Part A: global and sectoral aspects. Contribution of working group II to the Fifth assessment report of the intergovernmental panel on climate change. Cambridge, UK and New York: Cambridge University Press; 2014. p. 709–54.
26. Thomson AJ. Climate indices, rainfall onset and retreat, and malaria in Nigeria. J Vector Borne Diseases. 2010;47:193–203.
27. United Nations International Strategy on Disaster Reduction Secretariat. Global assessment report on disaster risk reduction. 2009. http://www.preventionweb.net/english/hyogo/gar/report/index.php?id=9413; Accessed 14 July 2019.
28. World Health Organization and Pan American Health Organization. Protecting Health from Climate Change: Vulnerability and Adaptation Assessment. World Health Organization; 2013. Available online: http://www.who.int/iris/handle/10665/104200; Accessed on 14 July 2019.
29. World Health Organization. WHO guidance to protect health from climate change through health adaptation planning. Geneva; 2014. Available online: http://apps.who.int/iris/bitstream/handle/10665/137383/9789241508001_eng.pdf?sequence=1&isAllowed=y; Accessed on 14 July 2019.
30. World Health Organization. Operational framework for building climate resilient health systems. World Health Organization; 2015. Available online: http://apps.who.int/iris/bitstream/10665/189951/1/9789241565073_eng.pdf?ua=1; Accessed on 14 July 2019.
31. Zhang Y, Yu C, Wang L. Temperature exposure during pregnancy and birth outcomes: an updated systematic review of epidemiological evidence. Environ Pollut. 2017;225:700–12.

Chapter 7
Heat Waves and Rising Temperatures: Human Health Impacts and the Determinants of Vulnerability

Helene G. Margolis

Introduction

Globally, heat waves account for dramatic increases in mortality and morbidity; however, there is increasing awareness that day-to-day increases in temperature contribute to a significant risk of heat-related morbidity and mortality (HRMM) that over one or more warm seasons may exceed the public health burden of heat waves. Climate change has already and will continue to increase both average ambient temperatures and the frequency and intensity of excursions above those averages (i.e., heat waves or extreme heat events) and will thereby lead directly and indirectly to amplification of the risk of HRMM. This chapter provides a brief synopsis of our current knowledge about thermoregulation, thermotolerance, and the pathophysiology of heat stroke, and the multiple determinants of health and illness that influence the risk of HRMM and that collectively define vulnerability. A particular focus is on two vulnerable populations, older adults and children. An Environmental Health Multiple-Determinants Model of Vulnerability is presented as a conceptual framework to integrate that knowledge, with the intent of providing a tool that can facilitate compilation and translation of the information to interventions and adaptation strategies relevant at the individual level and/or subpopulation and population levels and at one or more geopolitical scales in developing and/or developed nations. Three overarching strategies for HRMM risk reduction are discussed, including Extreme Heat Event and Warm Season Heat Preparedness and Response Action Plans, Promote Good Health and Access to Quality Healthcare (reduces risk and increases resiliency), and Reduce/Manage Potential Exposure(s) (individual, community) to Ambient Heat and Other Physical Environmental Stressors. A key focus of this chapter is on integration and translation of knowledge.

H. G. Margolis (✉)
Department of Internal Medicine, General Medicine Division, Center for Health Services Research in Primary Care, Women's Center for Health, University of California, Davis, Davis, CA, USA
e-mail: hgmargolis@ucdavis.edu

K. E. Pinkerton, W. N. Rom (eds.), *Climate Change and Global Public Health*, Respiratory Medicine, https://doi.org/10.1007/978-3-030-54746-2_7

Over evolutionary time scales, humans have evolved to tolerate ambient heat across a fairly wide range of environmental conditions; that ability is enabled by behavioral and complex biological/physiological thermoregulatory adaptations that serve to maintain an average core body temperature within a narrow life-sustaining range around 37 °C (98.6 °F) [1] regardless where they live or where their ancestors evolved [2]. Under past and present climatic conditions, human populations around the globe have been and continue to be exposed to periods of *extreme* high temperatures that pose a risk of adverse health impacts, which include, but are not limited to, a suite of mild-to-severe conditions within the rubric of "heat-related illness (HRI)," and acute exacerbations of prevalent chronic diseases [3, 4], as well as death that may or may not be attributed as a direct or indirect consequence of heat exposure or a combination of heat and comorbidity. Climate change has already and will continue to increase both average ambient temperatures and the frequency and intensity of excursions above those averages [5] and will thereby lead directly and indirectly to amplification of the risk of heat-related morbidity and mortality (HRMM) [6]. (Key terms used in this chapter are defined in Table 7.1).

Table 7.1 Glossary of terms

Heat-related morbidity and mortality (HRMM) : This term is used here to reflect the full-spectrum of causes of illness or death, including heat-related illness (HRI; a clinically defined spectrum of conditions associated with excessive heat stress). The abbreviation HRI is used when explicitly referring to one or more conditions within the spectrum of heat-related illnesses
Heat wave (extreme heat event): There is no universally accepted definition of "heat wave"; however, commonly applied criteria include the occurrence of temperatures, or a temperature plus humidity metric (e.g., heat index or humidex) above a threshold level that persists over 2 or 3 consecutive days. The term extreme (or excessive) heat event (EHE) is generally used synonymously with "heat wave"; for the purposes of this chapter, the term is used to represent any extreme excursion above usual average temperature conditions that may pose a health risk, regardless of whether it meets criteria for designation as a heat wave
Vulnerability: the definition applied in this chapter (see text) has a public health orientation and differs from the definition used by the IPCC (Climate Change 2007: Synthesis Report), which states: *Vulnerability is the degree to which a system is susceptible to, and unable to cope with, adverse effects of climate change, including climate variability and extremes. Vulnerability is a function of the character, magnitude, and rate of climate change and variation to which a system is exposed, its sensitivity, and its adaptive capacity*
Heat stress: *heat stress* is defined as the total heat load on the body from metabolic heat production plus external environmental factors; and *Heat Strain* is the total physiological stresses resulting from heat stress. An alternate common *heat stress* definition combines heat load and its consequences: *Heat Stress is any combination of work, airflow, humidity, air temperature, thermal radiation, or internal body condition that strains the body as it tries to regulate its temperature. When the strain to regulate body temperature exceeds the body's capability to adjust, heat stress has become excessive* (US navy definition)
Heat acclimatization: The terms *heat acclimatization* and *heat acclimation* are often used interchangeably; however, acclimatization refers to adaptations that develop as a result of challenges in the natural environment (e.g., physical training in a hot country), and acclimation refers to similar adaptations acquired from experimental exposure to artificial conditions
Climate change mitigation strategies (CCMS) : Actions to limit further climate change by reducing the production of greenhouse gases (GHG)
Climate change adaptation strategies (CCAS): actions to lessen the adverse impacts by preparing for inevitable changes in climate and climate variability

Modern societies, especially politically and economically stable nations, have social systems that include mechanisms designed to protect the stability of the society by reducing the health risks and/or increase the resilience of the overall population during natural disasters, including heat waves or more generally "extreme heat events" (EHE) (see Table 7.1). Thus one would expect, at least in developed nations, for there to be sufficient experience and knowledge, guidance, policies, and infrastructure to adequately protect the population's health during EHE. This expectation was proven wrong in 2003, when an intense and extended heat wave and exceptionally hot summer in Europe claimed about 70,000 lives [7])—with about 15,000 deaths occurring in France alone [8]. Extreme heat exposure remains the leading cause of weather-related deaths in the United States [9]. Although the death toll paled in comparison to the 2003 European heat wave, the summer 2006 California heat wave, which affected most of the State and was of unprecedented intensity (with both extreme high daytime maximum temperatures and high nighttime minimum temperatures) and duration (about 17 days) [10, 11], had a very significant public health burden. That event is estimated to have resulted in over 600 excess deaths [12, 13] and about 1200 excess hospitalizations and 16,000 excess emergency department contacts for a variety of causes [14]. The economic cost of the health impacts (mortality and morbidity) of that event has been estimated to have been $5.4 billion [15].

Importantly, although less dramatic than a heat wave-related sudden upsurge in deaths and illnesses, there are significant health risks associated with day-to-day excursions in temperature above local warm season means that might not meet a definition of "extreme" heat and that might not be perceived by the overall population and specific at-risk subpopulations as hazardous [16–21]. In a meta-regression analysis using published results from multiple cities around the world, it was estimated that in nearly half of those locations, the risk of all-cause (all-age) mortality increased by one to three percent (1–3%) per 1 °C increase above the city-specific threshold (i.e., the temperature at which the mortality/morbidity indicator is lowest or the temperature where there is a sharp increase in a nonlinear exposure-response function) with the effect estimate (i.e., slope of linear-response function) varying by different city-specific characteristics and a general trend for the thresholds to be higher in locations closer to the equator [17]. Geographic patterns in effects have been reported in a number of studies, for example, heat-related mortality in the United States tends to be greater in communities in cooler climates than in warmer climates; the smaller effect in warmer areas has been attributed to adaptation through physiological, behavioral, technological means [22–27].

A few studies have evaluated the added heat wave effect above the overall warm season increase in mortality. For example, in a meta-analysis of seven California counties, the July 2006 heat wave was associated with a 9% (95% CI: 1.6, 16.3) increase in all-cause daily mortality per 10 °F (5.6 °C approximately) change in apparent temperature or about threefold the effect estimated over the entire warm season (May–September) or July only in 1999–2005 [13]. That magnitude of added heat wave effect is consistent with those observed for some European cities [16]. Over one or multiple warm seasons and over large geographic areas with exposed populations, the increased risks associated with non-extreme temperatures, reflected in increases in numbers of deaths and emergency department visits or hospitalizations, are a major contributor to the cumulative public health and healthcare burden of ambient heat, potentially greater than heat wave periods (which are relatively rare) [16, 17, 28].

Organizations charged with protecting public health during natural disasters are becoming more aware of the potential for health effects (mortality or morbidity) to occur not only during EHE but also at less-than-extreme temperatures common over a warm season. However, most if not all of those organizations continue to use extreme heat alert systems and HRMM risk-reduction strategies that are formulated from an "emergency response" perspective and involve implementation of public health protection protocols that are triggered by forecasted or observed temperatures (or other biometeorological measures) that meet criteria for "extreme" heat conditions. Furthermore, to date, those criteria are always based on exposure-response functions derived from mortality studies, in part because there are overall and for specific locations far fewer studies of ambient heat impacts on morbidity than on mortality. Given that even under current climatologic conditions, ambient heat continues to lead to significant morbidity and mortality, despite the fact that HRI is potentially preventable [3, 9, 29] as is most of the excess HRMM observed in epidemiologic studies makes it clear that improved approaches for prevention of HRMM need to be developed and implemented in the near term. It will be essential to augment the emergency response approach and add a broad suite of strategies that aim to diminish individual and population risk under the full range of ambient heat conditions, not just extremes. To that end, it is necessary to identify the populations, subpopulations, and individuals at elevated risk and to define and understand the independent and joint influence of determinants that contribute to greater (or diminished) *vulnerability* (see Table 7.1 and next section).

Furthermore, while epidemiologic observations and research conducted at the population level is critically important and has been invaluable in guiding current strategies for reducing HRMM, the existent burden of HRMM and the amplified challenges to public health posed by climate change and other global changes, such as migration to urban areas or increased prevalence of chronic diseases, that are adversely affecting population health and resilience make it essential that the science upon which risk-reduction strategies are based is broadened. Major advances in our understanding of the pathophysiology of HRI and how it may be related to underlying health status, in particular the role of the immune system (innate and adaptive) and systemic inflammation and oxidative stress [1, 3, 30–32], can provide critical insights to which individuals and populations are most susceptible to HRMM and can guide identification of efficacious and cost-effective interventions.

This chapter provides a brief synopsis of our current knowledge about the multiple determinants of health and illness that influence the risk of HRMM and that collectively define vulnerability. A conceptual framework to integrate that knowledge is presented, with the intent of providing a tool that can facilitate compilation and translation of the information to interventions and adaptation strategies relevant at the individual level and/or subpopulation and population levels and at one or more geopolitical scales in developing and/or developed nations. The scope of this chapter does not allow a comprehensive exposition of the determinants of risk for all vulnerable populations; however, recent advances in knowledge about thermoregulation and risk factors in older adults and children are briefly discussed. Strategies for HRMM prevention are identified.

Vulnerable Populations: Multiple Determinants of Ambient Heat Health Impacts

Populations and Subgroups at Elevated Risk: Insights from Epidemiology

Identification of vulnerable populations for the purposes of developing public health approaches to prevention of HRI and HRMM is primarily based on epidemiologic studies that utilize routinely collected administrative data (death certificates, hospital admissions, and emergency department contacts). A number of mortality and morbidity studies (case-control, case cross-over, time-series, and case-series) have evaluated the impacts of ambient heat on specific subpopulations defined by diagnosis group (i.e., to identify cases for specific-cause analyses), age, sex, race/ethnicity, or activity if the data are available (e.g., occupational workers, athletes) and/or evaluated the influence of population-specific or location-specific factors, such as socioeconomic indicators or co-exposure to air pollution either as potential confounders or as effect modifiers. Direct comparison of individual epidemiologic study results is challenging due to differences in study populations, locations, and designs, in particular the use of different temperature indicators and/or different definitions of a heat wave, and whether potential confounding or modifying factors have been considered [18, 33]. Importantly, the commonly used epidemiologic data and study designs preclude detailed examination of individual-level factors, such as obesity or comorbidity and treatment, or location-time-activity patterns that can modify exposures and that may account for the enhanced risk observed at the population level; thus, clear attribution of the elevated risk to just biological susceptibility or another factor is not possible. (For reviews of the epidemiologic literature on temperature effects on all-cause or specific causes of mortality, see Hajat and colleagues [17] and Odame et al. [149], or morbidity see Ye et al. [18], and Bunker et al. [120] for mortality and morbidity effects in elderly populations [120], and see Xu and colleagues for heat wave effects on children's health [160]. In addition, see Smith and colleagues (2012) for a discussion of heat wave definitions [33]).

Among the different studies, there is heterogeneity in the results for some key factors, that is, whether there is an effect or association and the direction and magnitude of the association, with some of the differences likely a function of whether the study is examining mortality or morbidity and the specific diagnoses being examined [17, 18, 34]. Age, specifically older adults (usually defined as ≥60 years of age) and the very young (infants, children <5 years of age), is among the strongest and most consistent predictors of elevated risk for HRMM [14, 17, 18, 34]. There are mixed results for sex, with some studies indicating no influence, and others suggesting women or men are at greater risk (often dependent on the health outcome) [17, 18, 34]. Predisposing chronic diseases (e.g., psychiatric illness and neurological disorders, pulmonary diseases, chronic kidney disease (CKD), obesity, diabetes and cardiovascular diseases) are also consistently implicated in elevated risk for HRMM [17, 18, 34, 35]. Dehydration (and hypohydration) in general, and dehydration associated with medications (neurological and non-neurological) that impair thermoregulation or thirst regulation have also consistently been associated with elevated risk of mortality and morbidity [43, 159].

Other factors prognostic of increased risk of HRMM include: being confined to bed, not leaving home daily, and being unable to care for oneself [36]; various general indicators of being socially isolated (e.g., living alone, presence of or frequency of social contacts, or linguistically isolated) [36–40]; and persons who are socioeconomically disadvantaged [36–40]. Interestingly, some studies have indicated the higher risk associated with socioeconomic factors exists for American but not European cities [41], although in France during the 2003 heat wave, for older adults income was associated with greater risk of mortality [42]. Factors associated with lower risk include air conditioning (as indicated by air conditioning saturation in a community or evidence of functional/used home air conditioning), visiting cool environments, and increasing social contacts [36, 44].

Importantly, there is mounting epidemiologic evidence that chronic diseases not only predispose individuals to elevated risk of HRMM but heat stress is associated with onset of some chronic diseases or disease manifestations. Psychiatric illnesses and CKD are prime examples of heat stress-associated health outcomes of emergent concern.

In the case of psychiatric illnesses, recent studies have observed higher temperatures, especially in conjunction with other environmental and socioeconomic stressors, to be positively associated with violence towards others and towards self [117, 146, 152, 153, 155]. Those relatively short-term exposure-response relations cannot be viewed in isolation of evidence pointing to climate change's longer-term consequences on mental health in the general population and those already struggling with mental health disorders [116, 126, 157, 158, 162]. For example, slow-moving disasters, such as drought, may affect mental health over many years [158], and individuals who experience a natural disaster, such as a hurricane or wildfire, may have symptoms, such as those associated with Post-traumatic Stress Disorder (PTSD), long after the disaster-related acute stress exposure [114, 122, 125, 129, 154, 156, 161]. These observations are consistent with studies of chronic stress that indicate a potentially diminished psychological and physiological ability to cope with subsequent exposures to stress, especially when the initial psychological stress exposure occurs in early childhood [119, 123, 132]. With climate change-associated extreme events occurring more frequently and being superimposed on the residual psychological dysfunction of prior acute and longer-term stress-inducing events, it will be increasingly important to quantify, understand and address mental health impacts. The epidemiologic associations between heat stress and mental health or maladaptive behavior, and the physiologic basis of those associations as well as the role of contextual risk factors, such as socioeconomic factors and social cohesion, require further study.

Since 2002, an emergent epidemic of rapidly progressing CKD unexplained by traditional risk factors (hypertension and diabetes) has been reported among outdoor laborers, especially sugarcane and other agricultural workers in multiple regions of Central America and southwestern Mexico [121, 124, 128, 130, 141, 150]. The disease occurs most commonly among young and middle-aged men and has a high fatality rate [124]. This aggressive form of CKD is alternatively referred to as Mesoamerican nephropathy (MeN) reflecting the region in which it was initially identified, or chronic kidney disease of unknown origin (CKDu) as the etiology remains unknown and the pathophysiology has not been fully elucidated. [124] Epidemiologic evidence most strongly points to repeated exposures to heat stress under conditions of intense physical activity accompanied by dehydration as a likely cause [147]; other hypothesized causes (or cofactors) include, but are not limited to, exposure to agrochemicals, heavy metals or metalloids, or infections [124]. There are also reports of a similar form of CKD among agricultural workers in Sri Lanka [136], and India [148], suggesting there is a far greater global burden of this disease than is currently recognized. It is important to note that there is a complex interplay among common chronic diseases, including CKD, regardless of etiology, which elevates risk of heat stress and HRMM [143].

Environmental Health Multiple-Determinants Model of Vulnerability

Multiple (or Multi-) Determinants Models (MDM) are increasingly being used (qualitatively and quantitatively) to evaluate complex multifactorial chronic disease processes and incorporate consideration of a broad range of risk factors, especially host factors and social determinants of health. This approach is consistent with a paradigm shift by major public health organizations (e.g., WHO, US NIH, and CDC) from a model that just focuses on the determinants of health and disease at the individual level to a holistic model that considers the individual and populations within the context of their physical, societal/cultural, and economic environments across the lifespan [45, 46]. As is the case for complex diseases, complex environmental problems require a holistic approach. Figure 7.1 presents a schematic of the *Environmental Health Multiple-Determinants Model of Vulnerability (MDMv)*, which is proposed here as a conceptual framework to evaluate the local-to-global health impacts of climate change in general, and for the purposes of this chapter ambient heat in particular. Importantly, this model is consistent with and applicable to *precision (or personalized) medicine* research and practice. Vulnerability factors and their relative importance may differ at the individual and population levels and at different geographical scales or geopolitical domains, and there can be cross-scale interactions among factors. Furthermore, the presence and importance of a given factor or factors can change over time, affecting one or more scales differently. Table 7.2 lists observed and putative determinants of vulnerability for HRI and HRMM; selected factors are discussed further above.

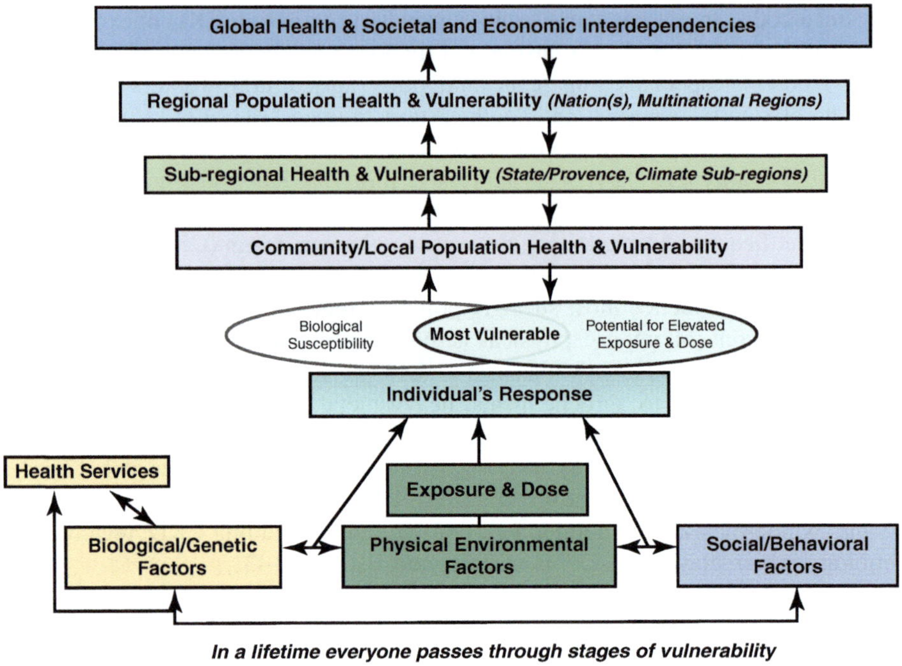

Fig. 7.1 Schematic of the *Environmental Health Multiple-Determinants Model of Vulnerability.* The premise of the model is that, as for most public health issues, there are disparities in how and the extent to which physical environmental factors (e.g., heat, air pollution, water quality/access) impact different populations and subgroups. Further, the health impacts of environmental factors on populations begins with impacts on individuals, and in a lifetime everyone passes through stages and degrees of vulnerability, with potential lifetime cumulative influences (positive and negative) affecting risk. Vulnerability is greatest among individuals (or subpopulations) who are most biologically susceptible and who have the largest exposure to one or more environmental hazards (depicted by Venn diagram). Vulnerability for development and severity of heat stress/heat strain and subsequent risk of illness or death (whether considering individual risk or population risk), is a function of complex interrelationships among *biologic factors*, including those that confer innate biologic sensitivity and/or resilience to an environmental insult (e.g., sex, race/ethnicity, oxidative stress, nutritional status, comorbidities and related treatments, and genetics/epigenetics); *physical environment and exposure characteristics* (e.g., physical/chemical nature of the exposure, duration and dose, coincident environmental stressors (such as water and/or food scarcity, air pollution)); and the *social, behavioral, and economic factors* that may influence (or be associated with) both biologic response and exposure (e.g., access to healthcare, social isolation, location-time-activity patterns, disparate neighborhood exposure levels)

Biological Adaptations to Heat Stress and Susceptibility and Pathophysiology of Heat Illness

To facilitate the understanding of the potential source of biological susceptibility, this section provides an overview of the normal physiologic responses involved in maintenance of thermal homeostasis (thermoregulation and acclimatization) and

Table 7.2 Determinants of heat-related morbidity and mortality

Susceptibility:
Biological/Physiological/Clinical Factors
Age (< 5 years, teens, ≥ 60 years)
Sex/Gender
Race/Ethnicity
Genetics/Epigenetics
Health Status
  Dehydration/hypohydration
  Nutrition
  Physical Fitness
  Obesity/Overweight
  Oxidative Stress & Inflammation
Communicable Diseases:
  Water- & food borne diseases (Diarrheal)
  Influenza & other acute viral infections
Chronic Diseases:
  CVD, respiratory (asthma, COPD),
  diabetes, renal insufficiency,
  immunologic disorders,
  neurologic disorders, mental illness
Medications & Pharmacologic Agents

Health Care Services:
Access, adequacy, quality of care
  Clinical management of chronic disease

Physical Environmental Factors
Temperature, Humidity
↑ Long-term Average Temperature
↑ Freq. Hot Days/Nights
↑ Freq. Heat Waves/Extreme Heat Events (EHE)
↑ Intensity, duration, geographic extent EHE
↓ Freq. Cold Days/Nights
Sub-regional/local scale influences on meteorology
  Topography
  Coastal (e.g., cloud cover) & sea surface temp.
  Land surface characteristics
    (e.g., soil moisture, irrigation, vegetation)
  Built environment (e.g., impervious surfaces)
Coincident Challenges
  Air Pollution (additive, synergistic): ozone, MVE
  Water Quantity/Quality
  ↑ Heavy Rainfall Events
    (without ↑ in total annual precipitation)
  ↓ Snowfall & Snow pack
  ↓ Mountain Glaciers
  ↑ Drought *(Areas, Freq. & Duration)*
  ↓ Soil Moisture *(Met. Feedbacks)*
  ↑ Extreme High Sea Level (Storm surges)

Factors that Modify Exposures
Location-Time-Activity Patterns
  Age-related differences: Children, older adults
  Time(s) outdoors & indoors
Physical or cognitive development or impairment
Built Environment – Outdoor & Indoor Factors
  Impervious surfaces
  Community Design (Trees/vegetation, land-use)
  Building Age, type, condition, heating/
    cooling systems (presence/usage), indoor air
    quality & ventilation

Social/Cultural/Behavioral/Economic Factors
Demographic
Age
Gender/Sex
Race/Ethnicity
Education
Economic
Built Environment
Location-Time-Activity Patterns
Building Age, type, condition, heating/
  cooling systems (presence/usage)
Time-outdoors (work, leisure activities)
Community-level factors (e.g., design,
  assets such as parks)
Disparities (and consequences of
  disparities) in quality of indoor
  environments: residences, schools, work
Social/cultural influences
  (clothing, climate-influenced behaviors)
Lifestyle Factors
Physical activity (daily-living activities;
  exercise (recreational))
Water/diet/nutrition: quality, quantity,
  subsistence cultures' food sources
Psychosocial support
Living conditions (e.g., alone & isolated,
  crowding)
Psychosocial stress
Community infrastructure

cellular adaptations (thermotolerance), and the pathophysiological consequences when the body's heat load exceeds its cooling capacity. It is beyond the scope of this chapter to provide detailed information on the prevention, diagnosis, and treatment of HRI (or of other heat-related morbidity) in the general or vulnerable populations; in addition to authoritative medical texts, that information is available from other sources, including for the general population [47], and for older adults [48–50], infants and children [39, 51–53], athletes [39, 54], the occupationally exposed [55–58], persons with alcohol, drug, and mental health disorders [59], and those taking medications (neurologic and non-neurological) [43].

Thermoregulation, Acclimatization, and Thermotolerance

Thermoregulation is a collective of mechanisms, behavioral and physiological, by which humans (and other homeotherms) maintain thermal homeostasis, and avoid development of, or minimize the adverse consequences of *heat stress* (see Table 7.1). *Behavioral Thermoregulation* ultimately aims to reduce exposure by modifying the microclimate (e.g., through clothes, buildings (residence and work), air conditioning) and by modifying location-time-activity patterns. The focus here is on *Physiological Thermoregulation*, which involves integrated biological processes that serve to balance the body's heat gain (from internal heat generated via mechanical work [i.e., physical activity] and basal metabolic processes, and/or gained from environmental heat exposure) and heat dissipation to the environment so as to

$$S = M_{(b+w)} \pm K \pm C \pm R - E$$

Where S = net heat storage (in tissues)

M = Metabolic heat production (basal metabolism (b) + mechanical work (w))

K = Conduction

C = Convection

R = Radiation

E = Evaporation

Fig. 7.2 Heat Balance Equation. There is continuous heat exchange between the body and the environment that can be described and quantified by the Heat Balance Equation. Storage (S) of heat is a function of metabolic heat (M) produced by basal metabolic processes (*b*) and heat generated by physical activity (i.e., mechanical work (*w*) of which only a portion of the energy generated is expended on the work itself), the gain or loss of heat through conduction (K), convection (C), and radiation (R), and heat dissipation through evaporation (E).

maintain the core body temperature (T_c). The T_c is the operating temperature of vital organs in the head or trunk and must be maintained in a narrow range 35–40 °C (95–104 ° F) with an usual target temperature of 37 °C (98.6 ° F) at rest [1, 2, 4]. For healthy subjects at rest, there can be between- and within-subject variation of T_c of up to about 1 °C due to a number of factors, for example, diurnal fluctuations, menstrual cycle phase, acclimatization to heat, exercise-related fitness level, and age-related differences [2, 4, 60]. For most healthy (unclothed) humans at rest, ambient temperatures of 24–29 °C (75.2–84.2 ° F) are thermoneutral, that is, there is no heat transfer between the body and the environment and basal metabolic processes generate sufficient heat to maintain T_c at the target temperature [2, 4]. The summertime ambient temperature range for thermal comfort (i.e., when an individual expresses satisfaction with their thermal environment) is 23–27 °C (73.4–80.6 ° F) [61].

Heat balance (i.e., where heat gain equals heat dissipation) requires the continuous transfer of energy, most of which is in the form of heat, across tissues within the body, and between the body and the environment; the transfer of heat follows basic laws of thermodynamics and has been well characterized and quantified in terms of the heat balance equation [4, 61]. A simple form of the equation is shown Fig. 7.2.

For an in-depth discussion of the quantitative aspects of heat balance, see [4]. The flow of heat is from warmer to cooler media. Within the body, the tissues store the heat, with tissue average temperatures and capacity to store and transfer heat varying by tissue type. For example, adipose tissue (i.e., fat) has lower heat capacity [62, 63], and its conductivity is about one-third that of muscle, with the rate of heat flow substantially slower (14 kcal/h for fat and 40 kcal/h for muscle) [4]. Convective heat transfer is involved in the flow of heat via the blood from working muscles to the core and from the core to the surface tissues [2, 4]. Conductive heat transfer occurs between tissues that are in direct contact, with the net heat flow from the core to the surface [2, 4]. Heat exchange between the body and the environment is primarily through radiation, convection, and evaporation (most important for dissipation of heat in warm environments) with all three processes occurring at the skin, but only convection and evaporation occurring in the respiratory tract (i.e., air is usually cooler and dryer than exhaled air) [4]. Notably, for a person at rest, radiation (in the form of infrared rays) is the primary pathway by which the body loses heat to the environment; however, the temperature gradient between the skin and the environment influences whether there is heat loss or gain via radiation. Heat gain from solar radiant energy or from solid objects, such as paved surfaces, can be a significant contributor to heat stress. Conduction usually plays a negligible role in body-to-environment heat transfer; however, it has an important role in treatment of extreme hyperthermia if the patient is immersed in a cool water bath (or shower) to facilitate rapid cooling (with careful monitoring of patient T_c to prevent overcooling) [64]. Clothing can significantly affect heat gain and heat loss (by impeding evaporation and heat transfer) and can be a major contributor to uncompensable heat stress, for example, in occupational workers wearing heavy impermeable clothing [4, 62].

Within a 1 °C rise in blood temperature, afferent heat receptors in the body core and skin transmit signals to the central nervous system's (CNS) primary thermoregulatory centers in the preoptic and anterior hypothalamus, where thermodetectors sensitive to increases in their own temperature trigger an efferent response. That response includes a suite of physiologic processes that ensure adequate energy and oxygen, while increasing flow of the heated blood from the core and working muscles to the surface of the body from where the heat can be dissipated to the environment, primarily by an increase in sweating (rate and the number of eccrine sweat glands activated) [4, 65]. (Temperature receptors in other CNS sites (e.g., medulla) also play a role, and there are thermal receptors outside the CNS (e.g., in heart, and pulmonary vessels), the role of which is not known [4].) Blood flow to the skin is the result of active sympathetic cutaneous vasodilatation. Increased heart rate, cardiac output, and minute ventilation rate facilitate the shift in blood to the body surface [3, 65]. Efficiency of cooling by evaporation of sweat depends on the air velocity and the water vapor pressure gradient between the skin and the air surrounding the body. The greater the water saturation of air the less cooling can occur. For the thermoregulatory response to be sustained, there must be adequate water intake and electrolyte supplementation to offset the losses [3, 4, 65].

Heat Acclimatization and Thermotolerance

Repeated exposure to either passive-heat or exercise-heat stress with attendant increases in T_c leads to physiological adaptations, referred to as *heat acclimatization* (see Table 7.1) that enhance perception of thermal comfort, increase work/athletic performance, and ultimately mitigate risk of heat-related morbidity [1, 66, 67]. There are various definitions of *Thermotolerance* (*aka* thermal or heat tolerance) in the literature; however, as defined by Moseley [67] it is "a cellular adaptation caused by a single severe but nonlethal heat exposure that allows the organism to survive a subsequent and otherwise lethal heat stress." Thermotolerance is associated with the presence (and upregulated gene expression) of families of heat shock proteins (HSP) , which protect cells and tissues from initial damage and accelerate repair if damage occurs as a result of heat stress, as well as a variety of other insults [1, 67]. The HSP have different cellular locations and functions that include binding to and processing of denatured proteins, management of protein fragments, maintenance of structural proteins, and chaperone of proteins across cell membranes [1, 67]. Acclimatization and thermotolerance are usually considered separately; however, there is evidence they are related through a shared dependence on the Heat Shock Response [67, 68] or more broadly the Stress Response [67, 69]. In that context, acclimatization can be viewed as a whole organism adaptation, of which thermotolerance—a cellular adaptation—is one part. After exposure to repeated heat-exercise stress, there is a reduction in gastrointestinal barrier permeability (discussed further in section on HRI pathophysiology), and there is an increase in cytoprotective HSP70 along with a decrease in plasma levels of tumor necrosis factor-alpha (TNF-α) and the pro-inflammatory interleukins (IL) IL6 and IL10, leading to lower levels of cellular and systemic markers of heat strain [68]. It should be noted that the complex array of cytokines involved in the systemic inflammatory response syndrome (SIRS) have both a role in promoting and resolving the SIRS [31].

Most of the information on acclimatization in humans is derived from sports physiology or military medicine research on acclimation among young healthy study subjects, usually males, that examined the immediate and/or adaptive physiologic responses from short-term exposures to heat-exercise stress under experimental (i.e., controlled) conditions. Short-term acclimation and acclimatization reflect similar physiologic adaptations that develop (or decay in the absence of heat-stress exposure) over a period of days to weeks [66]. There are very few published studies of long-term acclimatization (or habituation), which occurs over a period of years and reflects both the short-term physiologic adaptations and other usually poorly characterized physiological, behavioral, and technological adaptations by populations and individuals. There is also little published research on acclimation/acclimatization in the general population or vulnerable subgroups, such as the elderly, children, or those with chronic medical conditions.

When acclimatized, an individual's metabolic rate and T_c are lower at rest, accompanied by a lower heart rate, and under conditions of heat stress, there is an increase in stroke volume and blood/plasma volume, a reduced loss of electrolytes in sweat/urine, and increased thermal tolerance (i.e., cellular stress protein adaptations) [1, 65, 66, 68]. Among the physiologic adjustments that underlie those changes are a lower T_c threshold required for sweating to be initiated and the sweat rate is greater per degree rise in T_c, which enhances evaporative heat loss and the ability to lower skin and core temperatures [4, 66]. Also, skin vasodilatation and core-to-skin heat transfer is initiated at lower T_c and skin blood flow is higher for a given T_c [66]. The physiologic systems involved in acclimatization adapt at different rates, with changes in heart rate and plasma volume occurring first, then the reduction in resting T_c, and finally changes to sweat and sweat rate [66].

The rate of induction of heat acclimatization is exponential with 75% of the adaptations occurring within about the first 4–6 days of heat-exercise stress exposure and almost complete adaptation present after about 7–10 days [66]. One recommended protocol to achieve acclimatization is a single-daily exposure of about 100 min, with a work rate sufficient to increase T_c to 38.5 °C (101.3 ° F) [66]. Moseley [67] has noted that passive heat exposure-induced hyperthermia is usually associated with only partial acclimatization. Once heat acclimatized, unless there is repeated heat-exercise or passive heat exposure(s), there is a decay in acclimatization that can occur in as little as a week, with the decline in the different physiologic systems' adaptations occurring in reverse order of induction [66]. Depending on the interval without exposure to heat stress, re-acclimation is more rapid than initial acclimation. There is far less research on the time course of acclimatization decay and re-acclimatization or the determinants of those rates. One rule of thumb has been that for every 2 days without heat stress exposure, there is 1 day of acclimatization lost; however, more recent research suggests that decay occurs more slowly and that at least for healthy young adults they can safely return to work or athletic competition after as long as a month away from heat stress conditions [66].

Adaptations associated with thermotolerance, that is, the HSP response, are evident within several hours of heat stress exposure (messenger RNA levels peak within the first hour) and increase for several days [1, 67]; however, the duration of the adaptations is only for 2–7 days (in contrast to acclimatization, which is indefinite as long as a person has periodic mild elevations in T_c) [67]. After the initial exposure, HSP synthesis is a function of the intensity, duration, and cumulative effects of subsequent heat-stress exposures [1]. Importantly, although passive heat exposure and physical exercise can independently trigger HSP synthesis, there is a greater HSP response when those two stressors are combined as compared to either one alone [1].

It is important to emphasize that, although there is a paucity of data for the general or vulnerable populations, it is known that the time required to acclimatize or to see significant decay in acclimatization and to re-acclimatization can vary substantially depending on an individual's age, health status (especially by physical fitness, obesity (adiposity), or cardiopulmonary diseases), and the type of exposure (i.e., passive heat or heat-exercise exposure).

Heat Stress-Related Morbidity and Pathophysiology of Severe Heat-Related Illness

Any individual, regardless of age, sex, or health status, can develop heat stress if engaged in intense physical activity and/or exposed to environmental heat (dry or humid), especially if they are not acclimatized. If heat stress exceeds the physiologic capacity to cool and T_c rises, then a range of heat-related symptoms and conditions can develop. The medical conditions that result from heat stress/heat strain and fall within the formal classification of *Heat-Related Illness* (HRI) represent a spectrum that starts with relatively mild and easily treated illness (heat cramps, heat edema, and heat syncope) and progresses in severity to heat exhaustion and then to heat stroke, an extreme medical emergency. While the mild conditions may not be

Table 7.3 Heat-related illness: heat cramps, heat edema, heat syncope, and heat exhaustion[a]

Heat cramps: severe painful cramping of muscles in the legs or abdomen are the hallmarks of heat cramps, which result from electrolyte disturbance, most notably when plasma sodium levels fall significantly below normal. heat cramps are commonly caused by exertion, with profuse sweating, and often occur during cool down after activity has stopped. stopping intense activity and consumption of drinks with electrolytes (e.g., some sports drinks) to replenish fluid volume and electrolytes is usually sufficient treatment
Heat edema: swelling in the legs due to accumulation of fluids in the tissues; results from prolonged dilatation of the small arteries in the legs, especially after prolonged standing or sitting still in the heat. Treatment is to increase circulation (venous return) by alternating between elevating the legs and gently moving them
Heat syncope: sudden loss of consciousness (fainting), usually preceded by light-headedness or weakness, can result from orthostatic hypotension related to peripheral blood pooling. Loss of consciousness can be prevented by sitting or lying down at the initial signs of illness (dizziness, weakness)
Heat exhaustion: extreme depletion of blood plasma volume, which may be coincident with low plasma levels of electrolytes, as well as peripheral blood pooling, can lead to heat exhaustion. Core temperature may be in the normal range or slightly elevated, but less than 40 °C. symptoms can include generalized malaise, weakness, nausea, vomiting, headache, tachycardia, and hypotension. Although there can also be mild disorientation, the absence of clear neurologic complications distinguishes heat exhaustion from heat stroke
If heat exhaustion is suspected, the recommended course of action is to immediately move the affected individual to a cool environment and give them fluids supplemented with electrolytes. It may be necessary to actively cool the person by loosening clothing, increasing air flow across the skin, for example with a fan while misting or wiping them down with cool water, or placing ice packs on their extremities. Massage of extremities to mitigate vasoconstriction associated with use of cold water or ice is usually recommended

[a]Heat stroke is the most extreme form of HRI and is discussed in main text

life threatening, to prevent progression to more serious HRI, they should be treated appropriately and taken as warning signs to immediately remove an affected individual from the exposure situation. Table 7.3 provides an overview of the milder forms of HRI; the focus below is on the most severe condition—heat stroke.

Heat Stroke

Heat exhaustion may be the early stage of heat stroke [54], and within a 24-h period, if untreated, it can progress to heat stroke; thus, to prevent heat stroke and improve patient outcome, treatment should begin at the first signs of heat exhaustion [137]. Heat exhaustion does not necessarily present with definitive symptoms, therefore it is frequently misdiagnosed, commonly as acute viral infection, leading to delayed treatment. Importantly, acute viral or bacterial infections coincident with heat stress are implicated in increased risk of heat stroke [31], as well as sudden infant death syndrome (SIDS) in infants who were also more heavily wrapped in clothing [70]. Heat stroke is typically divided into two types: "Exertional Heat Stroke" as the name implies involves strenuous physical activity usually under high temperature conditions to which the person was not acclimatized and usually affects healthy older teens and young adults, such as athletes, occupational workers, and soldiers. "Classic heat stroke," by definition, does not involve exertion and usually affects biologically susceptible individuals, such as infants and young children, the elderly, persons with chronic illness and/or taking medications (prescribed or over-the-counter), as well as persons with alcohol or drug dependencies and with mental illness or neurologic conditions [43, 59]. It is imperative that measures be taken to prevent and/or aggressively treat heat stroke, which, even if treated, can have a crude mortality rate as high as 50%, and a large proportion of heat stroke survivors suffer permanent neurologic damage [3, 71]. Among 58 survivors of near-fatal classic heat stroke that occurred during the 1995 Chicago heat wave, 33% had substantial functional impairment at discharge from the hospital and had not improved at 1-year follow-up [71]. The sequelae of heat stroke-related multiorgan system dysfunction/failure (discussed below) can persist months or years after the initial treatment, thereby increasing the risk of mortality over the long term [31].

For both types of heat stroke, the clinical definition is when a person's body core temperature rises above 40 °C (104 ° F) and there are CNS neurologic complications (e.g., initially headache, dizziness, and weakness followed by hallucinations, combative behavior, coma, and seizures) [3, 31]. The more quickly the patient receives treatment to bring down their T_c to 39 °C (102 ° F) or below (ideally within 30 min of presentation [71]), and supportive therapies such as replacement of blood volume and electrolytes are administered, the less likely are severe complications and the better the prognosis [3, 31]. Although the clinical criteria and overall treatment of both types of heat stroke are essentially the same, a number of differences in patient characteristics, including signs and symptoms have been noted [65] that reflect the population subgroups commonly affected and that may require medical interventions specific to their unique physiology and medical status. For example, in classic heat stroke, sweating is usually absent, respiratory alkalosis is a dominant feature, coagulopathies (i.e., disseminated intravascular coagulation (DIC)) is mild, and if present rhabdomyolysis is rarely severe, whereas in exertional heat stroke sweating is often present, respiratory alkalosis is mild, DIC is marked, and rhabdomyolysis is severe [65].

Heat-Related Illness: Pathophysiology

Over the past 2 decades, research has led to critical insights to the pathophysiology of heat stroke [3, 31, 65]; based on that information, Bouchama and Knochel [3] proposed that heat stroke be defined as *a form of hyperthermia associated with a systemic inflammatory response leading to a syndrome of multiorgan dysfunction in which encephalopathy predominates* [3]. It has long been known that heat stroke is associated with an overload of the thermoregulatory response, including reduced capacity to increase cardiac output due to water and electrolyte depletion, cardiovascular disease, or medications or alcohol and illicit drugs, which affect cardiovascular, respiratory, or neurologic function [3, 43, 59]. As the T_c rises above 40 °C (104 ° F), there is tissue injury, with the extent of injury a function of the level and duration of heating [3]; the acute injury triggers the acute phase response (APR). It is now recognized that an upregulated APR and oxidative stress (likely both a precipitant and a downstream consequence of the APR) and possible altered expression of cytoprotective HSP are central to the pathophysiology of heat stroke [1, 3]. The cytotoxic effects of heat and the APR-associated inflammatory and coagulation responses of the affected individual contribute to the multiorgan injury [31]. As noted above, as part of the normal thermoregulatory process in response to hyperthermia, that is, increased T_c, the circulation of blood is shifted to the skin and working muscles and away from vital organs, including the gastrointestinal tract; this can lead to ischemia of the gut and intestinal hyperpermeability. An emerging body of evidence, primarily from animal models, indicates that endotoxemia resulting from intestinal hyperpermeability and leakage into the circulation may contribute to the progression from heat stress to heat stroke [1, 3, 31, 65].

Within the scope of this chapter, it is not possible to review the literature on this critical line of investigation linking heat stroke and the heat-stress response, oxidative stress and systemic inflammation, and the complex interplay between the innate and adaptive immune systems' responses (for an overview see Leon and Helwig [31], or for additional information see [138, 139, 144, 145]). However, it is important to note that over the past two decades a robust body of evidence has linked systemic and/or organ-/tissue-specific inflammation and oxidative stress pathways to: aging [72]; to the pathophysiology underlying a number of chronic diseases and related conditions (e.g., atherosclerosis and cardiovascular disease) [73, 74], chronic respiratory disease (e.g., asthma and chronic obstructive pulmonary disease (COPD)) [73, 75, 76], diabetes and obesity [77, 78]; and as potential mechanisms by which ambient air pollution increases the risk of acute exacerbations of those chronic diseases/conditions and/or contributes to their development and severity [79–84]. Furthermore, oxidative stress may impair the protective heat shock response [30], potentially reducing thermotolerance and increasing risk and severity of heat stroke. The implication of these observations is that individuals with chronic health conditions/diseases who already have high levels of oxidative stress and chronic inflammation are at elevated risk of HRI [31], and that this is an important underlying mechanism that contributes to the excess acute cardiovascular, respiratory, and diabetes cases associated with ambient heat. This will be an important area of further delineation and research, as it also opens the door to many more clinical and public health intervention options.

Vulnerable Populations: Determinants of Thermoregulatory Capacity

The strongest and most consistent observations in epidemiologic studies have been an elevated risk for HRMM among older adults, children, and people with chronic diseases regardless of age. There are physiologic attributes specific to older adults and children that affect thermoregulation (described below); however, recent literature suggests age per se is not in and of itself necessarily the major driver of risk, but rather it is the common (often interrelated) correlates of age specific to these age groups that contribute greater risk. Some of these factors are shared determinants of risk (SDR), that is, factors that impact these and other population subgroups.

Older Adults

Under resting thermoneutral conditions, older men and women have been reported to have lower T_c than younger adults; however, after accounting for factors, such as nutrition, comorbidity, and medication effects, the differences in T_c related to age essentially disappear [49]. The number of sweat glands and sweat gland function, in particular the amount of sweat produced per gland, diminishes with aging [49]. Sweating rate of older adults has been reported to be diminished under passive heat exposure; this appears to be a function of maximal oxygen uptake (VO_{2max}; a measure of aerobic capacity) rather than chronological age [49]. Chronological age-related reductions in skin blood flow do occur (attributed to reduced superficial microvasculature), accompanied by lower cardiac outputs and less redistribution of blood flow from the splanchnic and renal circulations [48, 49], with some yet to be understood sex differences in the central cardiovascular changes observed under conditions of heat stress [49]. Overall with age there is potential for greater heat gain and a diminished capacity for heat dissipation, especially via evaporation, as a result of the changes to sweating capacity and cardiovascular function noted above and an increase in body mass (and associated increase in adiposity). The greater the body mass, the more heat is generated for a given activity level [48], and the smaller the surface area to body mass ratio, so cooling capacity is diminished. In addition to adipose tissue acting as insulation and impeding heat exchange, there are less heat-activated sweat glands found in skin covering adipose tissue [48]. Importantly, with aging, peripheral and central thermosensor neurons are less sensitive and respond less effectively to temperature changes, with the result that the elderly have a decreased perception of heat along with less effective heat dissipation mechanisms [48], which together has important implications for HRMM risk and prevention.

A number of chronic medical conditions disproportionately affect older adults and predispose them to heat illness.(48) Cardiovascular disease is the most important, with direct effects on thermoregulatory mechanisms and capacity, for example, heart failure and myocardial infarction affect cardiac output and potentially cutaneous vasodilatation. Atherosclerosis, hypertension, and type II diabetes mellitus reduce vascular compliance and can directly affect thermoregulatory capacity [48]. Chronic respiratory diseases, such as COPD and asthma, can impair thermoregulatory capacity (due to diminished ability to provide sufficient oxygen to support increased energy demands) and contribute to hypoxemia that amplifies tissue damage and the risk and severity of heat stroke. Reduced fluid and electrolyte retention and dehydration are associated with aging-related renal insufficiency and with diabetes (type II diabetes mellitus, diabetes insipidus)-related renal damage and impaired renal function. Hypohydration and dehydration are common among older adults, who in addition to changes in renal function also experience a decreased sense of thirst, or to manage bladder control problems they (or their caregivers) may limit their fluid intake [50]. Obesity and/or lower lean body mass are common among the elderly, and as described above can directly affect thermoregulation and risk of HRI. And as noted above, cardiovascular and respiratory diseases, diabetes, and obesity/overweight are associated with elevated oxidative stress and chronic inflammation, which can contribute to pathophysiology and risk of heat stroke. Hyperthyroidism (via increased metabolic heat production or hyperpyrexia), and extensive skin damage or disease, can also directly affect thermoregulatory mechanisms [48]. Neurologic and psychiatric disorders that disproportionately affect older adults may directly impact CNS thermoregulatory centers and efferent responses and/or contribute to behaviors (e.g., wearing excess clothing or not removing themselves from excessive heat exposure) and social conditions (e.g., being socially isolated) that increase the risk of HRI [48]. A point of concern for the elderly, and an area that has not received much consideration in the context of direct or indirect influence on HRI, is nutritional deficiencies, such as inadequate intake of antioxidant-rich foods. Many of the above conditions occur concurrently, with complex physiologic and clinical interrelationships, including treatment and disease management that further complicates delineating a clear path to HRI risk prevention strategies. For example, recommendations to increase fluid intake to prevent hypohydration/dehydration may be contraindicated for a person with heart failure or with renal failure on hemodialysis, thus fluid intake and signs and symptoms must be carefully monitored for such patients. Medications may play a critical role in altering risk for HRI [43, 59]. While the literature focuses on increased risks of HRI and HRMM associated with commonly prescribed or over-the-counter medications, there may also be protective effects afforded by medications, such as anti-inflammatory agents.

Infants and Children

A number of studies point to increased risk of HRMM among children, especially those less than 5 years of age [14, 41, 85] and adolescent athletes [86]. Heat stroke is the third leading cause of death among high school athletes in the United States [86].

Most of the information on heat stress and HRI in children is in the context of exercise and physical activity, which by default focuses on school-aged children (e.g., ≥5 years of age). Despite the epidemiological evidence pointing to infants and very young children being at especially high risk, there is a paucity of literature that discusses thermoregulation or risk factors (other than extreme exposures, such as being left in a car) for HRI in this age group, especially infants. There is a rich literature on hypothermia in neonates and on SIDS. From birth through age 3 months, an infant's metabolic rate increases, the ratio of body mass to surface area increases, and at 3 months there is a thicker layer of subcutaneous fat which together shifts thermal balance towards heat conservation [87]. Some research on SIDS has pointed to a combination of ambient heat and concurrent viral infection in conjunction with excess covering (e.g., blankets or clothing), especially of the head where 40% of heat production and 85% of heat loss occurs in an infant in bed (elevated head/brain temperature could affect thermoregulation and respiratory control); the risk of SIDS was greater in infants older than 2–3 months as compared to those younger [70, 87]. It was suggested that an increase in metabolic rate associated with viral infection in the older infants reflected an acute phase response, which would not be as well developed in some younger infants [70].

There are physiological differences between children and adults, including morphologic, metabolic, cardiovascular, and sweating capacity that traditionally have been viewed as conferring less thermotolerance and greater risk of heat stress and HRI among children [52, 88]. Children (past early infancy) have a higher body surface to mass ratio which can increase heat gain from the environment (when ambient temperature is greater than skin temperature), and depending on the water vapor pressure of the air (or humidity) evaporative cooling by sweating may not be sufficient to compensate for that gain. Younger children are less metabolically efficient when walking or running such that their oxygen consumption and heat production are greater than those of adults engaged in a similar level of activity, thus potentially increasing heat strain. (This is less of a factor for non-weight-bearing exercises such as cycling or rowing [52]). When children are exercising in heat, heat convection to the body surface (and cooling) may be compromised (relative to similar heat loads in an adult) as a result of the combined cardiac output demands of working muscles and of moving blood to their larger body surface area. Under similar conditions of ambient heat children have a higher skin blood flow (and peripheral vasodilatation), which compromises venous return and in turn cardiac output and potentially thermoregulation and/or exercise performance. The greatest difference between children and adults is their sweating rates (absolute, relative to body surface, and per gland), and there are apparent sex differences, with lower sweat rates more pronounced in boys compared to men, than in girls compared to women [52]. Children also take longer to acclimatize than do adults [53].

Based on recent research, it has been suggested that due to compensatory mechanisms children's thermoregulatory capacity may be more similar to adults than traditionally accepted, at least under less extreme environmental conditions [52, 89]. This position has been adopted in the 2011 revised American Academy of Pediatrics Council on Sports Medicine and Fitness and Council on School Health Policy statement-Climatic Heat Stress and Exercising Children and Adolescents [89]. A number of risk factors for exertional HRI (heat exhaustion and heat stroke) other than age-specific differences in thermoregulation were identified, including: current or recent illnesses that alter hydration status or thermoregulation (e.g., gastrointestinal illness and/or fever); chronic clinical conditions (diabetes insipidus, type II diabetes mellitus, obesity, juvenile hyperthyroidism (Graves disease), and cystic fibrosis); medications (e.g., dopamine-reuptake inhibitor to treat attention deficit/hyperactivity disorder or enhance performance, or diuretics); any other acute or chronic medical condition or an injury that affects water-electrolyte balance, thermoregulation or exercise-heat tolerance; and lastly Sickle cell trait, which can contribute to risk and severity or complications of HRI [89].

Chronic respiratory diseases (allergic airways diseases and asthma), and obesity and associated with it type II diabetes mellitus have reached epidemic proportions among children, especially in developed nations. (In developing nations obesity is also epidemic; however, there are complex interrelationships between malnutrition in children and obesity in adults [90]). The pathways by which these conditions can amplify risk of HRI or HRMM in children are for the most part the same as noted above for the general population and older adults and will not be revisited here. However, in the context of climate change and the projected increases in ground level ozone (a potent oxidant), it is also important to note that children are especially vulnerable for developing chronic respiratory disease. They are biologically more susceptible due to their developing respiratory tracts and immune system, and they have potential for greater exposures and doses of air pollution as their breathing rates relative to body size are greater than adults, and they spend more time outdoors. In a cohort of children in southern California, participation in three or more team sports (an indicator of intense physical activity outdoors) in communities with high ozone was associated with a threefold higher risk of developing new onset asthma, as compared with children playing no sports. No effect of sports was observed in low ozone communities [91]. In another study of children with asthma, anti-inflammatory medication was observed to modify (diminish) the effect of air pollution on asthma symptoms [92]. There is also accumulating evidence that dietary intake of antioxidants (e.g., vitamin C), and specific genetic polymorphisms that are associated with antioxidant capacity, independently and/or jointly can modify the effects of ozone on children's lung function and growth [93, 94].

Determinants of Thermoregulatory Capacity: Additional Population Subgroups

Sex/Gender

Epidemiologic studies have yielded heterogeneous results when sex/gender is considered as a risk factor for HRI or HRMM. Most past research on thermoregulation has been in young healthy men and has not explicitly examined thermoregulation in women or sex-related differences in men and women. A review by Kaciuba-Uscilko and Grucza [60] concluded that *despite a smaller sweating response to heat load in women than in men, there are no substantial sex differences in the effectiveness of thermoregulation, except those that resulted from differences in body size and composition and physical working capacity*. They noted there were sex-hormone-related fluctuations in body temperature and some thermoregulatory processes during the menstrual cycle and in menopause; however, the mechanisms by which sex hormones affect thermoregulation require further study. To the extent there is differential distribution of predisposing chronic conditions/diseases or that lifestyle factors and location-time-activity patterns differ among men/boys and women/girls, the impacts of ambient heat and risk of HRMM would be expected to differ.

Race/Ethnicity

A review of temperature regulation and ethnicity by Lambert et al. [95] provides insights to variation in physiological traits across human populations that developed over the long term as a function of different climatic conditions. They noted the evidence suggests the differences reflect phenotypic rather than genotypic variation [95]. As in the case of sex-related differences in risk, differential distribution of predisposing chronic conditions/diseases across race/ethnicities also would affect the impacts of ambient heat. For a review of the racial and socioeconomic disparities in heat-related health effects and possible mechanisms see Gronlund [131]. Disentangling the complex relations between physiological and morphological characteristics (and potentially the underlying genetics) that affect thermoregulatory capacity in warm/hot climatic conditions, from the social, behavioral, economic, and environmental determinants of health that affect overall health (resiliency) and risk of HRI and/or HRMM poses significant challenges. There are both challenges and research opportunities afforded by the increasing ethnic diversity of many nations resulting from modern migrations facilitated by population mobility.

Genetics/Epigenetics

Research on genetic polymorphisms and epigenetic processes that modulate (increase/diminish) susceptibility to physiological heat stress, oxidative stress, and/or the heat shock response associated with environmental challenges (e.g., heat, air pollution, toxins) or specific diseases/conditions and subsequent risk and severity of heat illness are areas of intense investigation [96, 97, 127, 133, 134]. This research offers future promise of identifying the most at-risk individuals and subpopulations to target interventions for prevention. It may also provide more definitive insights to a biological basis for observed variation in risk of HRMM among different race/ethnic groups or between females and males.

Global Environmental and Societal Challenges Affecting Population Vulnerability

Global warming, in addition to increasing land surface average temperatures and frequency of EHE that are of greater intensity and duration [5], will also lead to other concurrent environmental changes, such as increased occurrence of droughts and extreme precipitation events, to sea level rise and higher storm surges, and to higher levels of air pollution, most notably ozone [6], the independent and joint effects of which will significantly affect the ability of ecosystems and human populations to cope with changes in temperatures. From a global health perspective, the most important coincident challenge will be hydrological system perturbations and downstream consequences on water and food security, and energy production and distribution (e.g., due to infrastructure damage), which have direct and indirect impacts on individuals', populations', and societal adaptive capacity. Of critical importance is that not only will there be coincident challenges to health within a given region, there is mounting scientific evidence that synoptic climatic processes are leading to coupled extreme weather events in distant regions. For example, EHE and extended droughts in Russia have been climatically tied to extreme precipitation events in Pakistan [98]. Among the effects these extreme weather events have locally are impacts on water availability and quality, and on crop production. A related concern is there is high confidence that many semiarid areas (e.g., the Mediterranean Basin, western United States, southern Africa, and northeastern Brazil) will experience decreased water resources [6]; many of these areas are among the most productive agricultural regions globally. Thus, not only is water and food security impacted within each affected region, the overall capacity for the international community to provide aide to any one region is diminished due to multiple regions being affected and potentially needing aide at the same time.

While global warming discussions usually note average *global* increases in temperature (land and ocean), at the local and subregional scales (e.g., subcommunity, community), there exist large variations in land surface temperatures—averages and excursions above averages (variability), and with climate change the degree to which temperature will increase in a given location will also vary and not always predictably. For example, climate models predict that year-round average temperatures throughout California will keep increasing with warming more pronounced in the summer than in the winter season, and depending on the general circulation model (GCM) and greenhouse gas (GHG) emissions scenario, the summer (July–September) increases range from 1.5 to 6 °C (2.7–10.8 °F) [99]. Also predicted is greater warming in inland areas, as compared with coastal locations (within ~50 km of the coast) with the increase as much as 4 °C (7.2 °F) higher in the interior land areas as compared to the coast [99]. As elsewhere, the frequency, intensity, duration, and geographic extent of EHE are predicted to increase in California; a trend already evident in the past decade along with the emergence of EHE characterized by higher humidity and higher minimum (overnight) temperatures [10]. Urbanization/suburbanization accounts for areas with the largest increases; however, there are also many rural areas that have experienced substantial temperature increases [99, 100]. That noted, the urban heat island effect can contribute to ambient temperatures being more than 10 °C higher than neighboring rural areas. Among the factors that contribute to this phenomenon is greater heat generation from local sources, such as vehicles and other machinery; dark surfaces with low albedo (i.e., reflectivity) that absorb and reradiate heat; low vegetation density and commensurate reduction in capacity to cool through evapotranspiration; and layout and design of buildings and other structures (e.g., urban canyons, height) that result in heat retention [101, 102]. Interestingly, independent of climate zone, metropolitan population size or rate of metropolitan population growth, over the last half century the rate of increase in the annual number of EHE was reported to be greater in metropolitan regions characterized by greater urban sprawl compared with more compact metropolitan regions

[101]. The primary mechanism attributed to this observation was the rate of deforestation in more sprawling areas and the associated loss of regional vegetative land cover [101].

Human populations are not just facing unprecedented environmental changes but also global societal and demographic shifts. Key among the societal changes is the migration from rural communities to densely populated urban locations where, in addition to higher temperatures, there are other challenges to health [103]. In developing nations, migrants tend to be poor and frequently end up in "irregular settlements" where there is little or no heath protective infrastructure, such as sewer systems and reliable potable water sources [29, 104]. In these settlements, as well as many other urban and rural communities in developing nations, water- and food-borne diseases, especially diarrheal diseases among infants and children under 5 years of age, remain a leading cause of illness and premature preventable deaths, despite the eradication and improved management of many communicable diseases that have been achieved globally [105]. Even in developed nations, populations that are economically disadvantaged (and/or medically underserved) or displaced (e.g., due to natural disasters) are also at elevated risk of communicable diseases, as was seen in the aftermath of Hurricane Katrina in the United States [106]. Diarrheal and other communicable diseases, including intercurrent infections, can predispose affected individuals to heat stress and HRMM [3, 71, 89]. Wherever populations reside, work, or recreate, insufficient access to potable water increases the risk of hypohydration and dehydration and, in turn, to increased risk of heat stress and HRMM in general and HRI in particular.

Strategies to Reduce Vulnerability and Incidence of Heat-Related Morbidity/Mortality

As noted at the beginning of this chapter, the existent and projected large public health and healthcare burden associated with ambient heat requires that the emergency response approach to EHE be augmented with strategies that reduce individual and population risk of HRMM over the full range of ambient heat conditions. Effective policies and interventions require knowledge, not assumptions about who is at risk, the drivers of that risk, and where and when those determinants of risk are greatest, as well as the efficacy of risk-reduction strategies. Within the framework of an Environmental Health Multiple-Determinants Model of Vulnerability (Fig. 7.1; Table 7.2) that incorporates knowledge from different disciplines, it is possible to identify the factors that independently or jointly confer increased (or diminished) risk of HRMM within the general population and within or across specific subpopulations already identified as *vulnerable*. In addition to developing/implementing evidence-based *Extreme Heat Event and Warm Season Heat Preparedness and Response Action Plans*, two other overarching and interrelated strategies are self-evident: *Promote Good Health & Access to Quality Healthcare (reduces risk and increases resiliency)* and *Reduce/Manage Potential Exposure(s)* (*individual, community*) *to Ambient Heat and Other Physical Environmental Stressors*. To be efficacious and resource-efficient, all three strategies require a coordinated "top-down" and "bottom-up" approach involving governments, non-governmental organizations, communities, and strong partnerships with diverse stakeholders (e.g., public health officials, healthcare and social service providers, educators, athletic coaches, and other private sector participants, such as faith-based organizations). The translation of those broad strategies to specific actions is where careful integrative considerations of the multiple determinants of risk becomes most critical, and the implementation is most challenging, especially in light of climate change-related environmental shifts. The discussion below primarily focuses on examples of translation and integration in the context of the two overarching strategies and heat-health action plans.

Promote Good Health and Access to Quality Healthcare

The above overview of normal thermoregulatory processes, pathophysiology of severe HRI (heat stroke), and the characteristics of older adults and children that affect their risk for HRMM highlighted key points of knowledge. Most notably, the recurrent theme for both age groups (with special considerations for infants) and applicable to other age groups is that individuals (females and males) who are more physically fit, have greater percent lean body mass, are adequately hydrated, and are not afflicted with a chronic disease (especially cardiovascular, respiratory, neurological, renal, or diabetes), and do not have an acute intercurrent infection, are less biologically susceptible to HRI and HRMM. That is because they have the physiological reserves to experience moderate-to-extreme heat stress and heat strain and still maintain thermal homeostasis, with less cell and tissue damage, and low risk of acute cardiopulmonary events or other complications of heat strain. In addition, physiological acclimatization can further reduce susceptibility and enhance resilience to heat stress/heat strain. Although far from being fully elucidated in the context of the sequelae from heat stress to heat exhaustion and heat stroke, a biological mechanism that unifies these observations in the healthy heat acclimatized phenotype is a lower level of oxidative stress and less chronic low-grade inflammation and potentially modulation of the acute phase response and stress response (e.g., downregulation of pro-inflammatory cytokines and upregulation of HSP response) that together confer greater thermotolerance. Beyond thermotolerance, there may be important co-benefits of enhancing the HSP response. HSP have the potential to alter obesity-induced insulin resistance (via preventing inflammatory disruption of insulin signaling), and lower HSP expression has been observed in human diabetes patients [78]; thus maintenance of HSP expression may be a pathway by which insulin resistance and diabetes are or could be improved with exercise [78] (and potentially exercise-heat acclimatization protocols).

Thus, the broadest recommendation to diminish HRMM across an entire population over the long term, with near-term benefits, is to invest in and capitalize on public health programs and interventions that aim to improve health and prevent/manage common chronic diseases, especially through improved nutrition and increased physical activity, as well as prevent/manage communicable diseases with specific consideration of the impacts (e.g., via dehydration, fever) on risk of HRMM. Integral to achieving that overall aim is to ensure access to healthcare (especially preventive medicine), and ensure clinicians and other healthcare service providers or points of patient contact (e.g., pharmacists) are informed about the HRMM risk factors relevant to their patients and measures that can be taken to manage that risk. This approach can contribute significantly to reducing the pressures on the public health infrastructure created by the global demographic trend towards older populations, and the global increase in prevalence of chronic diseases and obesity, as well as climate change.

Reduce/Manage Potential Exposures to Ambient Heat and Other Physical Environmental Stressors

Achieving "good health" and reducing HRMM, especially as the climate changes, will require concurrently addressing physical environmental stressors. In addition to advocating for and investment in pollution prevention programs at all geopolitical scales, specific actions need to be developed/implemented to reduce potential exposures (to heat, chemical and/or infectious agents) experienced by populations and individuals at the local scale. For example, when making the recommendation to increase physical activity (e.g., to manage weight), assuming the majority of the population does not have options to exercise in indoor locations (with healthful environmental conditions), there also has to be guidance on minimizing exposure to ambient air pollution, which can vary substantially temporally (e.g., diurnally and seasonally) and spatially at the local scale (e.g., neighborhood-to-neighborhood, proximity to a roadway), as well as provide advice to avoid the hottest time of the day (which usually is also coincident with the highest ozone levels). If the individual has compromised health, even if an apparently relatively benign condition, such as being overweight (but not obese and with no other health problems), or if they are taking medications that predispose them to heat stress/heat strain, they need to be alerted to their potentially heightened susceptibility to heat strain and risk of HRI or HRMM. Warnings to acclimatize before engaging in outdoor physical activities need to be accompanied by specific guidance on how to acclimatize. Such guidance is available for athletes (e.g., see Bergeron [89] and Pryor [151]); however, few if any of the documents that recommend acclimatization specifically address the issue of co-exposure to air pollution or aeroallergens. Currently there is little or no published quantitative information that specifically outlines or provides the basis for acclimatization protocols (that consider both exercise-heat exposure and passive heat exposure) for the general healthy population or subgroups defined by age and/or specific health conditions. This is an area of investigation that should be a priority.

Access to an air-conditioned cooler environment has consistently been associated with lower risk of HRMM over usual summertime and extreme heat conditions [22, 27, 71, 107, 108, 115, 118, 131, 140, 142]. And during EHE, recommendations to use air conditioning or move to an air-conditioned location, including public access cooling centers, have become a cornerstone of HRMM prevention strategies. There are however a number of potential pitfalls to this strategy. Even in developed nations, the energy generation and distribution infrastructure may not be able to support energy demands during EHE of long duration and large geographic extent, especially if there is increased penetration of AC into homes and businesses. During the 2006 California heat wave that also affected other western states (that can share energy resources with California), there were near failures of the power supply, with some areas experiencing brownouts. If there are coincident extreme weather events, such as hurricanes or storm surges, the energy infrastructure, including power plants, is at risk. In consideration of climate change and the need to reduce GHG emissions, unless sufficient (truly non-polluting) "green energy" is available, reliance on air conditioning may be counterproductive for health in the near and longer term. Public gathering places, such as older schools or workplaces, and eldercare residential facilities often do not have air conditioning, even in developed nations. Many populations (e.g., in irregular settlements) or individuals within populations (e.g., urban or rural poor in older residences) do not have nor is it feasible for them to have and/or use an air conditioner. A related concern is that the recommendation to avoid heat exposure by going indoors is not universally protective due to highly variable indoor heat and air quality conditions. Furthermore, by avoiding any heat exposure, the opportunity for acclimatization is diminished.

With respect to recommendations to minimize heat exposure, a critical caution regarding the use of fans is warranted. It is not recommended to use fans to prevent an individual from becoming overheated under certain climatic conditions of high humidity (greater than about 33% relative humidity) and high temperatures (i.e., temperature is ≥32.3 °C (90 ° F)); when temperatures are above 37.8 °C (100 ° F), fans may actually contribute to heat stress and subsequent illness (37). However, the use of a fan in conjunction with wetting down the skin of a person showing signs of heat stress or illness can facilitate evaporation and the cooling process.

Clearly, completely abandoning air conditioning as a solution for HRMM prevention is not recommended or feasible. However, more sustainable strategies that focus on reducing heat exposure by modifying the built environment to minimize heat gain (inside buildings and outside) and maximize heat loss and transfer from inhabited areas can reduce the need for air conditioning. Increasingly national and provincial municipal governments are developing/implementing sustainability policies and plans that include improved community design and land-use planning (e.g., increase green space, and rerouting of traffic to decrease vehicle miles traveled), retrofitting existing buildings (e.g., with green roofs, energy efficient windows), and replacing pavement with pervious surfaces. In addition to reducing temperatures (and potentially air pollution exposures), many of these strategies also promote increased physical activity and positively enhance the psychosocial environment and livability of a neighborhood and community and ultimately improve overall health [109, 135].

Extreme Heat Event and Warm Season Heat Preparedness and Response Action Plans

Formal EHE emergency response plans developed and implemented by government organizations at the national, regional, and local levels can significantly reduce HRMM. Comprehensive guidelines and considerations for designing and implementing heat-health action plans focused on emergency response to EHE have been developed by the WHO (Europe) [110]; the guidelines include principles and core elements (summarized in Table 7.4) of a potentially optimum system to prevent EHE-related HRMM that can be adapted to different geopolitical scales and infrastructures. Rather than reiterating recommendations contained in that document, the focus here will be on some of the issues related to enhancing HRMM risk-reduction plans to improve their efficacy during EHE, as well as potentially extending their application to an entire warm season.

Table 7.4 Principles and core elements of heat-health action plans as delineated by the World Health Organization[aa]

Principles
• Use existing systems and link to general emergency response arrangements
• Adopt a long-term approach
• Be broad (i.e., emergency response requires multiagency and multi-sector participation)
• Communicate effectively
• Ensure that responses to heat waves do not exacerbate the problem of climate change
• Evaluate (a key public health principle—evaluate efficacy of an intervention or strategy)
Core elements for implementation of a heat-health action plan
• Establish agreement on a lead organization
• Accurate and timely alert systems (i.e., heat-health warning systems to trigger weather-related warnings, determine the threshold for action, and communicate risks)
• A heat-related health information plan (what to communicate, to whom, and when)
• A reduction in indoor heat exposure (medium- and short-term strategies)
• Particular care for vulnerable population groups
• Preparedness of the health and social care system
• Long-term urban planning
• Real-time surveillance and evaluation

[a]World Health Organization: Europe. Heat-health action plans: guidance. 2008. Copenhagen, Denmark. http://www.euro.who.int/__data/assets/pdf_file/0006/95919/E91347.pdf

The particular issues were identified after the 2006 California heat wave, when the State's Contingency Plan for Excessive Heat Emergencies was reviewed by officials and scientists from public health and emergency response organizations and the US National Oceanic and Atmospheric Administration's National Weather Service (NWS) with the aims to improve heat alert system(s) and emergency preparedness and response, including medical resource planning, and the public health messages and interventions especially those targeted to vulnerable populations. Key gaps in information and limitations in prior studies upon which those systems are based were identified. Among the major issues raised during the evaluation was the need for local scale (i.e., subcommunity, such as neighborhood or US census tract) environmental and population data, and two key questions regarding criteria for issuance of heat alerts, including: (1) *Should the definition of a heat wave and heat alert criteria be based on morbidity rather than mortality-response studies as currently done*? (2) *Should the temperature indicator thresholds be lowered to account for the HRMM that occurs during less than extreme conditions*? Subsequent considerations highlighted issues related to risk communication and engaging the public. A discussion of these issues follows.

Local-scale population and environmental information (in urban, suburban, and rural areas) is required to identify high-risk locations and vulnerable populations and individuals, as well as establish mechanisms to contact those individuals in order for local government agencies (and nongovernmental organizations (NGOs)) to target public health and individual clinical or exposure mitigation interventions and allocate resources to *prevent* HRMM. An example of why local-scale information across the urban-to-rural gradient (i.e., not just urban areas) is necessary lies in the fact that while only 6% of California's population lives in areas designated as rural, the rural populations tend to be older, with about 20% of Californians ≥65 years of age living in a rural area [111, 112]. The older adults residing in rural areas tend to be less healthy, with higher rates of overweight/obesity, physical inactivity and food insecurity, and less access to medical resources, than older adults living in suburban areas; for a number of measures, rural older adults are more similar to their urban counterparts than to those in suburban areas [111]. Prior epidemiologic evidence of spatial heterogeneity in HRMM indicates that exposure-response relations derived from one community may not be applicable in another location [27], which combined with differential distribution of vulnerable populations reinforces the need for location-specific data at the finest spatial resolution possible. Community vulnerability mapping, facilitated by the use of geographic information systems (GIS) and advances in geospatial analysis, including methods of protecting confidentiality of individuals [28] is an important tool to identify at-risk populations, determinants of risk, and evaluate efficacy of interventions through ongoing surveillance.

The need for local-scale information partly informs the answer to the first question. (*Should the definition of a heat wave and heat alert criteria be based on morbidity rather than mortality-response studies*?) In general, administrative morbidity data (e.g., emergency department contacts, hospitalizations) are less readily available (especially for research) and there can be wide variation in quality and content.

However, when they are available, the benefits are that there are many more observations representing a broader cross-section of the population, and heat-related morbidity outcomes occur more frequently than deaths, providing significantly larger sample sizes, which usually provides greater spatial coverage and density at finer spatial resolution (e.g., patient residence Zip Code [postal code]). These attributes facilitate evaluation of HRMM risk and vulnerability factors at a fine spatial scale and the provision of local information. There are also good reasons for reliance on mortality as an endpoint. Vital statistics death data are almost always available and are collected with some degree of consistency, their use generally generates less concern with issues of confidentiality, and there are long records across many years lending them to time-series analyses and application of similar heat-mortality modeling strategies in diverse locations. However, use of mortality data has the implicit assumption that deaths represent the most extreme endpoint of a fixed chain of events, that is, people are exposed to heat, get sick, and then die, and those deaths can (always) be used as a marker of a relevant population exposure and of a predictable risk. Evidence suggests this is not necessarily the case, as mortality may strike quickly prior to the notice of emergency responders and affects elderly, socially isolated, and nonmobile populations [113, 114]. Thus, to the extent the spatial distribution of vulnerable subgroups more likely to die does not track with subgroups who are more likely to contact an emergency department, mortality-based analyses, and heat alert criteria derived from those analyses from one location would not necessarily provide the best information to reduce risk of morbidity or mortality in another location.

An analysis of hospitalizations and emergency department visits (ED) for all-causes and selected causes during the 2006 California heat wave revealed an intriguing and important observation related to spatial variation in different health outcomes [14]. In that analysis, the State was divided into six geographic regions, based approximately on climate zones, each comprising multiple counties. Risk ratios (RR) that compared rates during the heat wave and during a referent period (each period = 17 days) in the same summer were computed. Unexpectedly, while the highest risk of HRI ED visits (RR = 23.1, 95% CI: 15.1, 37.1) occurred in the usually cooler region of central coast counties (including San Francisco), there were too few hospitalizations to calculate a risk estimate (due to small cell sizes and required data suppression) for that region (and two other regions). In contrast, in the Central Valley (a much warmer region), the HRI ED-visit risk was substantially lower, but risk of hospitalization for HRI was very high (RR = 17.1, 95% CI: 9.8, 36.3). That observation is of particular interest because when the ~140 coroner-reported deaths attributed to hyperthermia (126 of the cases were classic heat stroke) during the heat wave were evaluated, the majority occurred in the Central Valley, which is a more rural agricultural region and an area with many socioeconomic-driven health disparities [35]. Taken together, the findings indicate the importance of examining/comparing different measures of health impacts— ED, hospitalizations, and deaths—for which the spatial heterogeneity may reflect a variety of determinants of risk that could influence/inform intervention and adaptation strategies. Thus, when possible, heat alert criteria would ideally reflect the composite information.

With respect to the second question, there are practical reasons for continuing to use extreme temperature thresholds (usually the 95th or 99th percentile of daily maximum temperature or temperature-humidity index) to trigger emergency response protocols and to develop supplemental strategies to diminish the health risks associated with usual warm season elevated temperatures. The primary reason being in many locations lower thresholds would be met repeatedly (if not almost continuously), especially during the hottest months. For example, in a Zip Code-level analysis of emergency department visits in California in the warm seasons (May–September) of 2005–2008, significant increases in patients diagnosed with electrolyte imbalance were observed when deviation of the daily maximum temperature from the Zip Code-specific seasonal mean daily maximum temperature was +6 °C (about the 88th percentile for most locations) [28]. Thus, redefining the threshold criteria for issuance of heat alerts based on this relatively low threshold would not likely be the optimum strategy to reduce public health risk. Not only is it impractical and a resource burden to keep the emergency response and public health infrastructures for EHE risk mitigation in a near-constant state of activation, the communities and populations would likely become desensitized to public health messages about the potential health risks of heat exposure and not take requisite precautions even when a severe EHE is forecast.

There must be a careful balance between informing and overwhelming (and desensitizing) the public with information on risk and prevention of HRMM across the full range of ambient heat exposures. This becomes even more of an issue when trying to share information about joint hazards (e.g., heat and air pollution), while also trying to promote health-protective measures such as exercise. Thus, one of the most critical elements of any heat-health action plan, whether aimed at just EHE or also considering less-than-extreme temperatures, is an evidence-based well-designed communication and education-outreach plan (e.g., the heat-related health information plan suggested by WHO). An essential part of the plan is ensuring the public health messages and recommended actions are correct and that they are effective, and if they are not effective, the reasons and how to remedy the deficits. A prime example of an action that could be effective, but is not always is the recommendation, usually targeted to older adults or those with chronic health conditions, to use home air conditioning or go to an air-conditioned location, such as a "cooling center." Experience in California and elsewhere indicates cooling centers are often underutilized, including by older adults, which has led some municipalities to consider not opening centers to save the expense of their operation. Among the recognized ancillary actions required to increase use of centers (cooling or for other emergencies) is to identify persons needing transportation to the center and then provide that service. In addition, emergency plans must consider care of companion animals as many people will not evacuate if they have to leave their pets behind.

It is well established that public health messaging can be a powerful tool for health promotion and protection, and obtaining such information from multiple sources (top-down (e.g., government issued health warnings) and bottom-up (e.g., healthcare provider)) can enhance the public's awareness and adoption of health-protective measures (to improve overall health or in emergencies). However, the implications of the observations about perception of individual risk among vulnerable populations strongly point to the need for innovative approaches and testing the efficacy of those approaches, as well as additional research. That said, the reasons vulnerable populations may not take health-protective measures (even when they are aware of a heat alert and heard public health warnings), such as using a home air conditioner, are complex and may reflect their knowledge, attitudes, and beliefs about the level of personal risk related to their age or chronic illness [115]. For example, as noted by Richard et al. [115], many older adults do not see themselves as old or at risk, and the individuals who believe limitations in their lives are related to aging are less likely to adopt preventive or adaptive behaviors. Socioeconomic deterrents to air conditioning use may be less of a factor than perception of risk [115]. In addition, the source of information about their vulnerability, including from their physicians, may not influence their perception of risk or adoption of protective measures [115]. Direct one-on-one contact and provision of education and assistance is one solution when individuals cannot due to mental or physical limitations, or who do not of their own accord, take preventive measures.

In general, and to enhance the efficacy of direct contacts, there is an urgent need to engage and educate a wider range of stakeholders, especially social service and healthcare providers, and persons in direct contact with vulnerable populations than are currently knowledgeable and proactive about reducing risk of HRMM among the populations with which they interact. In addition to older adults, the chronically ill and socially isolated, this is especially important for reducing risk of HRMM among infants and children. Children's physical and emotional development and their location-time-activity patterns clearly can contribute to differences in ambient heat exposures, exercise-related heat loads, and ultimately to risk of heat stress and HRI. Infants do not have the motor skills to remove blankets or remove themselves from hot environments [51], young children may continue to play outside even when overheating (past their thermal comfort zone) and often do not know/or sense the need to drink fluids [86], and young athletes may push themselves well past thermal comfort levels that are signaling heat stress and illness onset [86]. It thus becomes imperative that adults (parents and other caregivers, teachers, sports coaches, and observers) be cognizant of the risks and remedies and ensure all precautions and necessary actions be implemented to guarantee the safety of children. Specific guidance for each group needs to be built into the heat plan communication and education element.

A key to reducing HRMM is to have a full heat-health action plan with all the elements outlined (Table 7.4); if the requisite resources (including data on where vulnerable individuals/populations reside and the optimum mode for directly contacting them) are not available at the outset, then the plan should include specific contingencies to fill resource gaps, and timelines and steps to build the infrastructure. Unfortunately, even in developed nations EHE emergency response plans are often not available or of inconsistent quality, as was found to be the case in a survey of selected municipal heat wave response plans from cities in the United States that had a history of or were at risk for heat-related mortality [113]. Adding elements to plans to address HRMM that occur at less-than-extreme temperatures will add a layer of complexity; however, with climate-change-related rising temperatures and increased variability superimposed on the existent risks, this is an essential task. Regardless of the apparent completeness of the plan, once developed it will need to be regularly evaluated for its efficacy and updated to reflect lessons learned.

Conclusion

The rapid convergence of all of the climatologic and anthropologic changes in the present and over the very near term (next 2 or 3 decades) and throughout the twenty-first century exceed the current adaptive capacity of many if not most human social systems around the globe to cope with rising temperatures and increasing frequency and magnitudes of EHE. At all levels—from global to local—there needs to be proactive development of a broad range of strategies to reduce the societal, public health, and healthcare burden of HRMM, especially through primary and secondary prevention of chronic and communicable diseases. This will require an integrated multidisciplinary approach to evaluate and define the problem, including the determinants of individual and population vulnerability for HRMM, and develop the solutions in consideration of those vulnerabilities reflecting both morbidity and mortality. The conceptual framework of the Environmental Health Multiple-Determinants Model of Vulnerability provides a tool that allows quantitative and qualitative consideration of factors that independently or jointly confer increased (or diminished) risk of HRMM and identification of strategies to reduce that risk, including those that might not be evident when the problem is viewed less holistically. Furthermore, it fosters multidimensional thinking when developing/applying solutions, including revealing opportunities to integrate climate change mitigation and adaptation strategies that can realize co-benefits for public health and environmental welfare, and/or identify potential adverse unintended consequences of strategies.

Fortunately, through strategic development and implementation of "top-down" and "bottom-up" HRMM risk mitigation policies and actions that are coordinated with and leverage existing global, regional, national, and local public health and healthcare services programs targeting the root causes of poor health, as well as programs aimed at pollution (including GHG) and exposure prevention, significant progress can be made towards reducing HRMM efficaciously and cost effectively. The global interconnectedness of economies and of the health and welfare of populations creates an imperative for nations to work together to prevent and/or respond to all of those challenges.

References

1. Sawka MN, et al. Human adaptations to heat and cold stress. In: RTO HFM symposium on "Blowing hot and cold: protecting against climatic extremes". Dresden, Germany: RTO-MP; 2001. p. KN4-1–15.
2. Hanna JM, Brown DE. Human heat tolerance: an anthropological perspective. Annu Rev Anthropol. 1983;12:259–84.

3. Bouchama A, Knochel JP. Heat stroke. N Engl J Med. 2002;346(25):1978–88.
4. Wenger CB. Human adaptation to hot environments. In: Pandolf KB, Burr RE, editors. Medical aspects of harsh environments. Washington, DC: Office of The Surgeon General Department of the Army, United States of America; 2002. p. 51–86.
5. Meehl G, Tebaldi C. More intense, more frequent, and longer lasting heat waves in the 21st century. Science. 2004;305:994–7.
6. IPCC. Intergovernmental panel on climate change, climate change 2007: synthesis report. Contribution of Working Group I to the Fourth Assessment Report of the Intergovernmental Panel on Climate Change. 2007. p. 103.
7. Robine JM, et al. Death toll exceeded 70,000 in Europe during the summer of 2003. C R Biol. 2008;331(2):171–8.
8. Fouillet A, et al. Excess mortality related to the august 2003 heat wave in France. Int Arch Occup Environ Health. 2006;80(1):16–24.
9. Luber G, McGeehin M. Climate change and extreme heat events. Am J Prev Med. 2008;35(5):429–35.
10. Gershunov A, Cayan D, Iacobellis S. The great 2006 California heat wave: signal of an increasing trend. J Clim. 2009;22:6181–203.
11. Kozlowski DR, Edwards LM. An analysis and summary of the July 2006 record-breaking heat wave across the state of California. NOAA Western Regional Tech Attach, No. 07-05 (Feb 27, 2007). Salt Lake City: NOAA; 2007.
12. Hoshiko S, et al. A simple method for estimating excess mortality due to heat waves, as applied to the 2006 California heat wave. Int J Public Health. 2010;55(2):133–7.
13. Ostro BD, et al. Estimating the mortality effect of the July 2006 California heat wave. Environ Res. 2009;109(5):614–9.
14. Knowlton K, et al. The 2006 California heat wave: impacts on hospitalizations and emergency department visits. Environ Health Perspect. 2009;117(1):61–7.
15. Knowlton K, et al. Six climate change-related events in the United States accounted for about $14 billion in lost lives and health costs. Health Aff (Millwood). 2011;30(11):2167–76.
16. Hajat S, et al. Impact of high temperatures on mortality: is there an added heat wave effect? Epidemiology. 2006;17(6):632–8.
17. Hajat S, Kosatky T. Heat-related mortality: a review and exploration of heterogeneity. J Epidemiol Community Health. 2010;64(9):753–60.
18. Ye X, et al. Ambient temperature and morbidity: a review of epidemiological evidence. Environ Health Perspect. 2012;120(1):19–28.
19. Zanobetti A, et al. Summer temperature variability and long-term survival among elderly people with chronic disease. Proc Natl Acad Sci U S A. 2012;109(17):6608–13.
20. McMichael AJ, et al. International study of temperature, heat and urban mortality: the 'ISOTHURM' project. Int J Epidemiol. 2008;37(5):1121–31.
21. McMichael AJ, Woodruff RE, Hales S. Climate change and human health: present and future risks. Lancet. 2006;367(9513):859–69.
22. Chestnut LG, et al. Analysis of differences in hot-weather-related mortality across 44 US metropolitan areas. Environ Sci Technol. 1998;1:59–70.
23. Haines A, et al. Climate change and human health: impacts, vulnerability and public health. Public Health. 2006;120(7):585–96.
24. Haines A, Patz JA. Health effects of climate change. JAMA. 2004;291(1):99–103.
25. Kalkstein LS, Greene JS. An evaluation of climate/mortality relationships in large U.S. cities and the possible impacts of a climate change. Environ Health Perspect. 1997;105(1):84–93.
26. Ebi KL, et al. Weather changes associated with hospitalizations for cardiovascular diseases and stroke in California, 1983–1998. Int J Biometeorol. 2004;49(1):48–58.
27. Anderson BG, Bell ML. Weather-related mortality: how heat, cold, and heat waves affect mortality in the United States. Epidemiology. 2009;20(2):205–13.
28. Roberts EM, et al. Personal communication. 2012.
29. Durand-Lasserve A, Clerc V. Regularization and integration of irregular settlements: lessons from experience (working paper no. 6): Urban Management and Land; 1996.
30. Adachi M, et al. Oxidative stress impairs the heat stress response and delays unfolded protein recovery. PLoS One. 2009;4(11):e7719.

31. Leon LR, Helwig BG. Heat stroke: role of the systemic inflammatory response. J Appl Physiol. 2010;109(6):1980–8.
32. Horowitz M, Robinson SD. Heat shock proteins and the heat shock response during hyper-thermia and its modulation by altered physiological conditions. Prog Brain Res. 2007;162:433–46.
33. Smith T, Zaitchik B, Gohlke J. Heat waves in the United States: definitions, patterns and trends. Clim Chang. 2013;118:811–25.
34. Basu R. High ambient temperature and mortality: a review of epidemiologic studies from 2001 to 2008. Environ Health. 2009;8:40.
35. Trent RB, et al. Review of July 2006 heat wave related fatalities in California. Sacramento: California Department of Public Health; 2008. http://www.cdph.ca.gov/HealthInfo/injviosaf/Documents/HeatPlanAssessment-EPIC.pdf.
36. Bouchama A, et al. Prognostic factors in heat wave related deaths: a meta-analysis. Arch Intern Med. 2007;167(20):2170–6.
37. CDC. Heat-related deaths—Los Angeles County, California, 1999–2000, and United States, 1979–1998. MMWR Morb Mortal Wkly Rep. 2001;50(29):623–6.
38. Medina-Ramon M, et al. Extreme temperatures and mortality: assessing effect modification by personal characteristics and specific cause of death in a multi-city case-only analysis. Environ Health Perspect. 2006;114(9):1331–6. http://dx.doi.org/. https://doi.org/10.1289/ehp.9074.
39. Naughton GA, Carlson JS. Reducing the risk of heat-related decrements to physical activity in young people. J Sci Med Sport. 2008;11(1):58–65.
40. Naughton MP, et al. Heat-related mortality during a 1999 heat wave in Chicago. Am J Prev Med. 2002;22(4):221–7.
41. Kovats RS, Hajat S. Heat stress and public health: a critical review. Annu Rev Public Health. 2008;29:41–55.
42. Belmin J, et al. Level of dependency: a simple marker associated with mortality during the 2003 heat wave among French dependent elderly people living in the community or in institutions. Age Ageing. 2007;36(3):298–303.
43. Stollberger C, Lutz W, Finsterer J. Heat-related side-effects of neurological and non-neurological medication may increase heat wave fatalities. Eur J Neurol. 2009;16(7):879–82.
44. Fouillet A, et al. Has the impact of heat waves on mortality changed in France since the European heat wave of summer 2003? A study of the 2006 heat wave. Int J Epidemiol. 2008;37(2):309–17.
45. Gehlert S, et al. Targeting health disparities: a model linking upstream determinants to downstream interventions: knowing about the interactions of societal factors and disease can enable targeted interventions to reduce health disparities. Health Aff (Millwood). 2008;27(2):339–49.
46. Liburd LC, Sniezek JE. Changing times: new possibilities for community health and Well-being. Prev Chronic Dis. 2007. http://www.cdc.gov/pcd/issues/2007/jul/07_0048.htm. Accessed July 2007.
47. Blum LN, Bresolin LB, Williams MA. From the AMA Council on scientific affairs heat-related illness during extreme weather emergencies. JAMA. 1998;279(19):1514.
48. Allen AJ, Segal-Gidan F. Heat-related illness in the elderly. Clin Geriatr. 2007;15(7):37–45.
49. Kenney WL, Munce TA. Invited review: aging and human temperature regulation. J Appl Physiol. 2003;95(6):2598–603.
50. Worfolk JB. Heat waves: their impact on the health of elders. Geriatr Nurs. 2000;21(2):70–7.
51. Jardine DS. Heat illness and heat stroke. Pediatr Rev. 2007;28(7):249–58.
52. Falk B, Dotan R. Children's thermoregulation during exercise in the heat: a revisit. Appl Physiol Nutr Metab. 2008;33(2):420–7.
53. Bytomski JR, Squire DL. Heat illness in children. Curr Sports Med Rep. 2003;2(6):320–4.
54. Howe AS, Boden BP. Heat-related illness in athletes. Am J Sports Med. 2007;35(8):1384–95.
55. Bates G, Gazey C, Cena K. Factors affecting heat illness when working in conditions of thermal stress. J Hum Ergol (Tokyo). 1996;25(1):13–20.

56. Jay O, Kenny GP. Heat exposure in the Canadian workplace. Am J Ind Med. 2010;53(8):842–53.
57. Rodahl K. Occupational health conditions in extreme environments. Ann Occup Hyg. 2003;47(3):241–52.
58. Parsons KC. International standards for the assessment of the risk of thermal strain on clothed workers in hot environments. Ann Occup Hyg. 1999;43(5):297–308.
59. Cusack L, de Crespigny C, Athanasos P. Heat waves and their impact on people with alcohol, drug and mental health conditions: a discussion paper on clinical practice considerations. J Adv Nurs. 2011;67(4):915–22.
60. Kaciuba-Uscilko H, Grucza R. Gender differences in thermoregulation. Curr Opin Clin Nutr Metab Care. 2001;4(6):533–6.
61. Epstein Y, Moran DS. Thermal comfort and the heat stress indices. Ind Health. 2006;44(3):388–98.
62. McLellan TM. The importance of aerobic fitness in determining tolerance to uncompensable heat stress. Comp Biochem Physiol A Mol Integr Physiol. 2001;128(4):691–700.
63. Selkirk GA, McLellan TM. Influence of aerobic fitness and body fatness on tolerance to uncompensable heat stress. J Appl Physiol. 2001;91(5):2055–63.
64. Gagnon D, et al. Cold-water immersion and the treatment of hyperthermia: using 38.6 degrees C as a safe rectal temperature cooling limit. J Athl Train. 2010;45(5):439–44.
65. Hales JRS, Hubbard RW, Gaffin SL. Limitation of heat tolerance. In: Fregly MJ, Blatteis CM, editors. Handbook of physiology, environmental physiology, Suppl 14. New York: Wiley; 1996.
66. Weller AS, et al. Quantification of the decay and re-induction of heat acclimation in dry-heat following 12 and 26 days without exposure to heat stress. Eur J Appl Physiol. 2007;102(1):57–66.
67. Moseley PL. Heat shock proteins and heat adaptation of the whole organism. J Appl Physiol. 1997;83(5):1413–7.
68. Kuennen M, et al. Thermotolerance and heat acclimation may share a common mechanism in humans. Am J Physiol Regul Integr Comp Physiol. 2011;301(2):R524–33.
69. Moseley P. Stress proteins and the immune response. Immunopharmacology. 2000;48(3):299–302.
70. Fleming PJ, et al. Thermal balance and metabolic rate during upper respiratory tract infection in infants. Arch Dis Child. 1994;70(3):187–91.
71. Dematte JE, et al. Near-fatal heat stroke during the 1995 heat wave in Chicago. Ann Intern Med. 1998;129(3):173–81.
72. Finkel T, Holbrook NJ. Oxidants, oxidative stress and the biology of ageing. Nature. 2000;408(6809):239–47.
73. Lee IT, Yang CM. Role of NADPH oxidase/ROS in pro-inflammatory mediators-induced airway and pulmonary diseases. Biochem Pharmacol. 2012;84(5):581–90.
74. Libby P. Inflammation in atherosclerosis. Nature. 2002;420(6917):868–74.
75. Agarwal SK, et al. Airflow obstruction, lung function, and risk of incident heart failure: the atherosclerosis risk in communities (ARIC) study. Eur J Heart Fail. 2012;14(4):414–22.
76. Park HS, Kim SR, Lee YC. Impact of oxidative stress on lung diseases. Respirology. 2009;14(1):27–38.
77. Emanuela F, et al. Inflammation as a link between obesity and metabolic syndrome. J Nutr Metab. 2012;2012:476380.
78. Whitham M, Febbraio MA. HSP and diabetes. In: Asea AAA, Pederson BK, editors. Heat shock proteins and whole body physiology. New York: Springer; 2010. p. 3–18.
79. Plummer LE, Smiley-Jewell S, Pinkerton KE. Impact of air pollution on lung inflammation and the roll of toll-like receptors. Int J Infereron Cytokine Mediator Res. 2012;4: 43–57.
80. Breton CV, et al. Genetic variation in the glutathione synthesis pathway, air pollution, and children's lung function growth. Am J Respir Crit Care Med. 2011;183(2):243–8.
81. Islam T, et al. Glutathione-S-transferase (GST) P1, GSTM1, exercise, ozone and asthma incidence in school children. Thorax. 2009;64(3):197–202.

82. Gilliland FD, et al. A theoretical basis for investigating ambient air pollution and children's respiratory health. Environ Health Perspect. 1999;107(Suppl 3):403–7.
83. Franchini M, et al. Air pollution, vascular disease and thrombosis: linking clinical data and pathogenic mechanisms. J Thromb Haemost. 2012;10(12):2438–51.
84. Miller MR, Shaw CA, Langrish JP. From particles to patients: oxidative stress and the cardiovascular effects of air pollution. Futur Cardiol. 2012;8(4):577–602.
85. Basu R, Ostro BD. A multicounty analysis identifying the populations vulnerable to mortality associated with high ambient temperature in California. Am J Epidemiol. 2008;168(6):632–7.
86. Coris EE, Ramirez AM, Van Durme DJ. Heat illness in athletes: the dangerous combination of heat, humidity and exercise. Sports Med. 2004;34(1):9–16.
87. Fleming PJ, Azaz Y, Wigfield R. Development of thermoregulation in infancy: possible implications for SIDS. J Clin Pathol. 1992;45(11 Suppl):17–9.
88. Falk B. Effects of thermal stress during rest and exercise in the paediatric population. Sports Med. 1998;25(4):221–40.
89. Bergeron MF, Devore C, Rice SG. Policy statement—climatic heat stress and exercising children and adolescents. Pediatrics. 2011;128(3):e741–7.
90. Prentice AM. The emerging epidemic of obesity in developing countries. Int J Epidemiol. 2006;35(1):93–9.
91. McConnell R, et al. Asthma in exercising children exposed to ozone: a cohort study. Lancet. 2002;359(9304):386–91.
92. Delfino RJ, et al. Association of asthma symptoms with peak particulate air pollution and effect modification by anti-inflammatory medication use. Environ Health Perspect. 2002;110(10):A607–17.
93. Moreno-Macias H, et al. Ozone exposure, vitamin C intake, and genetic susceptibility of asthmatic children in Mexico City: a cohort study. Respir Res. 2013;14(1):14.
94. Romieu I, et al. Genetic polymorphism of GSTM1 and antioxidant supplementation influence lung function in relation to ozone exposure in asthmatic children in Mexico City. Thorax. 2004;59(1):8–10.
95. Lambert MI, Mann T, Dugas JP. Ethnicity and temperature regulation. Med Sport Sci. 2008;53:104–20.
96. Horowitz M. Heat acclimation and cross-tolerance against novel stressors: genomic-physiological linkage. Prog Brain Res. 2007;162:373–92.
97. Horowitz M, Kodesh E. Molecular signals that shape the integrative responses of the heat-acclimated phenotype. Med Sci Sports Exerc. 2010;42(12):2164–72.
98. Lau WKM, Kim K-M. The 2010 Pakistan flood and Russian heat wave: teleconnections of hydrometeorological extremes. J Hydrometeorol. 2012;13:392–403.
99. Cayan D, et al. Climate scenarios for California. Sacramento: California Energy Commission, California Climate Change Center; 2006. p. 52.
100. CDPH. Heat-related illness and mortality: information for the public health network in California. In: public health impacts of climate change in California: community vulnerability assessments and adaptation strategies. Sacramento: California Department of Public Health, Public Health Institute; 2007. http://www.ehib.org/papers/Heat_Vulnerability_2007.pdf
101. Stone B, Hess JJ, Frumkin H. Urban form and extreme heat events: are sprawling cities more vulnerable to climate change than compact cities? Environ Health Perspect. 2010;118(10):1425–8.
102. Oke T. The energetic basis of the urban heat island. Q J R Meteorol Soc. 1982;108(455):1–24.
103. United Nations. World urbanization prospects: the 2007 revision. New York: United Nations; 2008.
104. UNHabitat. Cities and climate change global report on human settlements 2011. United Nations Human Settlements Programme. London, UK/Washington, DC: Earthscan Ltd/ Earthscan LLC; 2011.
105. WHO. World health statistics, vol. 2012. Geneva: World Health Organization; 2012.
106. CDC. Infectious disease and dermatologic conditions in evacuees and rescue workers after hurricane Katrina—multiple states, august–September, 2005. MMWR Morb Mortal Wkly Rep. 2005;54(38):961–4.

107. Keatinge WR. Death in heat waves. BMJ. 2003;327(7414):512–3.
108. Semenza JC, et al. Excess hospital admissions during the July 1995 heat wave in Chicago. Am J Prev Med. 1999;16(4):269–77.
109. Dannenberg AL, Frumkin H, Jackson R. Making healthy places: designing and building for health, well-being, and sustainability. Washington, DC: Island Press; 2011.
110. WHO. Heat-health action plans: guidance. Europe. Copenhagen: World Health Organization; 2008.
111. Durazo EM, et al. The health status and unique health challenges of rural older adults in California. Los Angeles: University of California, Los Angeles (UCLA) Center for Health Policy Research; 2011.
112. CHHSA, Strategic plan for an aging California population. Getting California ready for the "baby boomers". Sacramento: California Health and Human Services Agency, Editor; 2003. p. 258.
113. Bernard SM, McGeehin MA. Municipal heat wave response plans. Am J Public Health. 2004;94(9):1520–2.
114. Diaz J, Linares C, Tobias A. A critical comment on heat wave response plans. Eur J Pub Health. 2006;16(6):600.
115. Richard L, Kosatsky T, Renouf A. Correlates of hot day air-conditioning use among middle-aged and older adults with chronic heart and lung diseases: the role of health beliefs and cues to action. Health Educ Res. 2011;26(1):77–88.
116. Aitsi-Selmi A, Murray V. Protecting the health and Well-being of populations from disasters: health and health Care in the Sendai Framework for disaster risk reduction 2015–2030. Prehosp Disaster Med. 2016;31(1):74–8.
117. Basu R, Gavin L, Pearson D, Ebisu K, Malig B. Examining the association between apparent temperature and mental health-related emergency room visits in California. Am J Epidemiol. 2018;187(4):726–35.
118. Benmarhnia T, Deguen S, Kaufman JS, Smargiassi A. Review article: vulnerability to heat-related mortality: a systematic review, meta-analysis, and meta-regression analysis. Epidemiology. 2015;26(6):781–93.
119. Bucci M, Marques SS, Oh D, Harris NB. Toxic stress in children and adolescents. Adv Pediatr Infect Dis. 2016;63(1):403–28.
120. Bunker A, Wildenhain J, Vandenbergh A, Henschke N, Rocklov J, Hajat S, Sauerborn R. Effects of air temperature on climate-sensitive mortality and morbidity outcomes in the elderly; a systematic review and meta-analysis of epidemiological evidence. EBioMedicine. 2016;6:258–68.
121. Butler-Dawson J, Krisher L, Yoder H, Dally M, Sorensen C, Johnson RJ, Asensio C, Cruz A, Johnson EC, Carlton EJ, Tenney L, Asturias EJ, Newman LS. Evaluation of heat stress and cumulative incidence of acute kidney injury in sugarcane workers in Guatemala. Int Arch Occup Environ Health. 2019;
122. Cherry KE, Sampson L, Nezat PF, Cacamo A, Marks LD, Galea S. Long-term psychological outcomes in older adults after disaster: relationships to religiosity and social support. Aging Ment Health. 2015;19(5):430–43.
123. Chung EK, Siegel BS, Garg A, Conroy K, Gross RS, Long DA, Lewis G, Osman CJ, Jo Messito M, Wade R Jr, Shonna Yin H, Cox J, Fierman AH. Screening for social determinants of health among children and families living in poverty: a guide for clinicians. Curr Probl Pediatr Adolesc Health Care. 2016;46(5):135–53.
124. Correa-Rotter R, Garcia-Trabanino R. Mesoamerican nephropathy. Semin Nephrol. 2019;39(3):263–71.
125. Dai W, Wang J, Kaminga AC, Chen L, Tan H, Lai Z, Deng J, Liu A. Predictors of recovery from post-traumatic stress disorder after the dongting lake flood in China: a 13–14 year follow-up study. BMC Psychiatry. 2016;16(1):382.
126. Dodgen D, Donato D, Kelly N, La Greca A, Morganstein J, Reser J, Ruzek J, Schweitzer S, Shimamoto MM, Thigpen Tart K, Ursano R. Ch. 8: mental health and Well-being. The impacts of climate change on human health in the United States: a scientific assessment. Washington, DC: U. S. G. C. R. Program; 2016. p. 217–46.

127. Ely BR, Lovering AT, Horowitz M, Minson CT. Heat acclimation and cross tolerance to hypoxia: bridging the gap between cellular and systemic responses. Temperature (Austin). 2014;1(2):107–14.
128. Gallo-Ruiz L, Sennett CM, Sanchez-Delgado M, Garcia-Urbina A, Gamez-Altamirano T, Basra K, Laws RL, Amador JJ, Lopez-Pilarte D, Tripodis Y, Brooks DR, McClean MD, Kupferman J, Friedman D, Aragon A, Gonzalez-Quiroz M, Scammell MK. Prevalence and risk factors for CKD among brickmaking Workers in La Paz Centro, Nicaragua. Am J Kidney Dis. 2019;
129. Gibbs L, Waters E, Bryant RA, Pattison P, Lusher D, Harms L, Richardson J, MacDougall C, Block K, Snowdon E, Gallagher HC, Sinnott V, Ireton G, Forbes D. Beyond bushfires: community, resilience and recovery - a longitudinal mixed method study of the medium to long term impacts of bushfires on mental health and social connectedness. BMC Public Health. 2013;13:1036.
130. Gonzalez-Quiroz M, Pearce N, Caplin B, Nitsch D. What do epidemiological studies tell us about chronic kidney disease of undetermined cause in Meso-America? A systematic review and meta-analysis. Clin Kidney J. 2018;11(4):496–506.
131. Gronlund CJ. Racial and socioeconomic disparities in heat-related health effects and their mechanisms: a review. Curr Epidemiol Rep. 2014;1(3):165–73.
132. Hornor G. Resilience. J Pediatr Health Care. 2017;31(3):384–90.
133. Horowitz M. Epigenetics and cytoprotection with heat acclimation. J Appl Physiol (1985). 2016;120(6):702–10.
134. Horowitz M. Heat acclimation-mediated cross-tolerance: origins in within-life epigenetics? Front Physiol. 2017;8:548.
135. James P, Hart JE, Banay RF, Laden F. Exposure to greenness and mortality in a Nationwide prospective cohort study of women. Environ Health Perspect. 2016;124(9):1344–52.
136. Jayasumana C. Chronic interstitial nephritis in agricultural communities (CINAC) in Sri Lanka. Semin Nephrol. 2019;39(3):278–83.
137. Kenny GP, Wilson TE, Flouris AD, Fujii N. Heat exhaustion. Handb Clin Neurol. 2018;157:505–29.
138. King MA, Leon LR, Morse DA, Clanton TL. Unique cytokine and chemokine responses to exertional heat stroke in mice. J Appl Physiol. (1985). 2017;122(2):296–306.
139. King MA, Leon LR, Mustico DL, Haines JM, Clanton TL. Biomarkers of multiorgan injury in a preclinical model of exertional heat stroke. J Appl Physiol. (1985). 2015;118(10):1207–20.
140. Kravchenko J, Abernethy AP, Fawzy M, Lyerly HK. Minimization of heatwave morbidity and mortality. Am J Prev Med. 2013;44(3):274–82.
141. Kupferman J, Ramirez-Rubio O, Amador JJ, Lopez-Pilarte D, Wilker EH, Laws RL, Sennett C, Robles NV, Lau JL, Salinas AJ, Kaufman JS, Weiner DE, Scammell MK, McClean MD, Brooks DR, Friedman DJ. Acute kidney injury in sugarcane workers at risk for Mesoamerican nephropathy. Am J Kidney Dis. 2018;72(4):475–82.
142. Lane K, Wheeler K, Charles-Guzman K, Ahmed M, Blum M, Gregory K, Graber N, Clark N, Matte T. Extreme heat awareness and protective behaviors in New York City. J Urban Health. 2014;91(3):403–14.
143. Lee WS, Kim WS, Lim YH, Hong YC. High temperatures and kidney disease morbidity: a systematic review and meta-analysis. J Prev Med Public Health. 2019;52(1):1–13.
144. Leon LR. Common mechanisms for the adaptive responses to exercise and heat stress. J Appl Physiol (1985). 2016;120(6):662–3.
145. Leon LR, Bouchama A. Heat stroke. Compr Physiol. 2015;5(2):611–47.
146. Levy BS, Sidel VW, Patz JA. Climate change and collective violence. Annu Rev Public Health. 2017;38:241–57.
147. Madero M, Garcia-Arroyo FE, Sanchez-Lozada LG. Pathophysiologic insight into MesoAmerican nephropathy. Curr Opin Nephrol Hypertens. 2017;26(4):296–302.
148. O'Callaghan-Gordo C, Shivashankar R, Anand S, Ghosh S, Glaser J, Gupta R, Jakobsson K, Kondal D, Krishnan A, Mohan S, Mohan V, Nitsch D, Prabhakaran AP, Tandon N, Narayan

KMV, Pearce N, Caplin B, Prabhakaran D. Prevalence of and risk factors for chronic kidney disease of unknown aetiology in India: secondary data analysis of three population-based cross-sectional studies. BMJ Open. 2019;9(3):e023353.
149. Odame EA, Li Y, Zheng S, Vaidyanathan A, Silver K. Assessing heat-related mortality risks among rural populations: a systematic review and meta-analysis of epidemiological evidence. Int J Environ Res Public Health. 2018;15(8):1597–611.
150. Ordunez P, Nieto FJ, Martinez R, Soliz P, Giraldo GP, Mott SA, Hoy WE. Chronic kidney disease mortality trends in selected Central America countries, 1997-2013: clues to an epidemic of chronic interstitial nephritis of agricultural communities. J Epidemiol Community Health. 2018;72(4):280–6.
151. Pryor JL, Johnson EC, Roberts WO, Pryor RR. Application of evidence-based recommendations for heat acclimation: individual and team sport perspectives. Temperature (Austin). 2019;6(1):37–49.
152. Schinasi LH, Hamra GB. A time series analysis of associations between daily temperature and crime events in Philadelphia, Pennsylvania. J Urban Health. 2017;94(6):892–900.
153. Schmeltz MT, Gamble JL. Risk characterization of hospitalizations for mental illness and/or behavioral disorders with concurrent heat-related illness. PLoS One. 2017;12(10):e0186509.
154. Schwartz RM, Gillezeau CN, Liu B, Lieberman-Cribbin W, Taioli E. Longitudinal impact of hurricane Sandy exposure on mental health symptoms. Int J Environ Res Public Health. 2017;14(9):957–68.
155. Thompson R, Hornigold R, Page L, Waite T. Associations between high ambient temperatures and heat waves with mental health outcomes: a systematic review. Public Health. 2018;161:171–91.
156. Thordardottir EB, Valdimarsdottir UA, Hansdottir I, Hauksdottir A, Dyregrov A, Shipherd JC, Elklit A, Resnick H, Gudmundsdottir B. Sixteen-year follow-up of childhood avalanche survivors. Eur J Psychotraumatol. 2016;7:30995.
157. Trombley J, Chalupka S, Anderko L. Climate change and mental health. Am J Nurs. 2017;117(4):44–52.
158. Vins H, Bell J, Saha S, Hess JJ. The mental health outcomes of drought: a systematic review and causal process diagram. Int J Environ Res Public Health. 2015;12(10):13251–75.
159. Westaway K, Frank O, Husband A, McClure A, Shute R, Edwards S, Curtis J, Rowett D. Medicines can affect thermoregulation and accentuate the risk of dehydration and heat-related illness during hot weather. J Clin Pharm Ther. 2015;40(4):363–7.
160. Xu Z, Sheffield PE, Su H, Wang X, Bi Y, Tong S. The impact of heat waves on children's health: a systematic review. Int J Biometeorol. 2014;58(2):239–47.
161. Zahran S, Peek L, Snodgrass JG, Weiler S, Hempel L. Economics of disaster risk, social vulnerability, and mental health resilience. Risk Anal. 2011;31(7):1107–19.
162. Ziegler C, Morelli V, Fawibe O. Climate change and underserved communities. Prim Care. 2017;44(1):171–84.

Chapter 8
Climate, Air Quality, and Allergy: Emerging Methods for Detecting Linkages

Patrick L. Kinney, Perry E. Sheffield, and Kate R. Weinberger

Introduction

Climate factors like temperature, wind, and precipitation play important roles in determining patterns and concentrations of air pollution over multiple scales in time and space [1, 2]. These may operate through changes in air pollution emissions, transport, dilution, chemical transformation, and eventual deposition of air pollutants, especially for secondary pollutants like ozone (O_3). Naturally occurring air contaminants of relevance to human health, including airborne pollens, also may be influenced by climate. Thus, there are a range of air contaminants, both anthropogenic and natural, for which climate change impacts are of potential importance.

O_3 is a serious health concern and is formed in the lower atmosphere by reactions involving precursor air pollutants in the presence of sunlight. The key precursor pollutants for O_3 formation are nitrogen oxides (emitted mainly by burning of fuels) and volatile organic compounds (VOCs) (emitted both by burning of fuels and by evaporation from vegetation and stored fuels). Because O_3 formation increases with greater sunlight and higher temperatures, it reaches unhealthy levels primarily during the warm half of the year. It has been firmly established that breathing O_3 can cause inflammation in the deep lung as well as short-term, reversible decreases in

P. L. Kinney
Department of Environmental Health, Boston University School of Public Health, Boston, MA, USA

P. E. Sheffield
Environmental Medicine & Public Health, Icahn School of Medicine at Mount Sinai, New York, NY, USA
e-mail: perry.sheffield@mssm.edu

K. R. Weinberger (✉)
School of Population and Public Health, University of British Columbia, Vancouver, BC, Canada
e-mail: kate.weinberger@ubc.ca

K. E. Pinkerton, W. N. Rom (eds.), *Climate Change and Global Public Health*, Respiratory Medicine, https://doi.org/10.1007/978-3-030-54746-2_8

lung function. In addition, epidemiologic studies have demonstrated that O_3 can increase the risk of asthma-related hospital visits and premature mortality [3–6].

Airborne allergens are substances present in the air that stimulate an allergic response in sensitized individuals upon inhalation [7]. Outdoor pollens are one important class of airborne allergens. Pollens are released by plants at specific times of the year that depend to varying degrees on temperature, sunlight, and moisture [8]. Thus, airborne pollen concentrations are sensitive to climate variability and change.

The influence of climate on air quality is substantial and well established [1, 2], giving rise to the expectation that changes in climate are likely to alter patterns of air pollution concentrations. Higher temperatures hasten the chemical reactions that lead to O_3 formation. Higher temperatures, and perhaps elevated carbon dioxide (CO_2) concentrations, also lead to increased emissions of O_3-relevant VOC precursors by vegetation [9]. Weather patterns influence the movement and dispersion of all pollutants in the atmosphere through the action of winds, vertical mixing, and rainfall. Air pollution episodes can occur with atmospheric conditions that limit both vertical and horizontal dispersion. Emissions from power plants increase substantially during heat waves when air conditioning use peaks. Finally, the production and distribution of airborne allergens, such as pollens and molds, are highly influenced by weather phenomena and also have been shown to be sensitive to atmospheric CO_2 levels [10]. For example, the timing of phenological events, such as flowering and pollen release, are closely linked with temperature [8].

Human-induced climate change is likely to alter the distributions over both time and space of all of the meteorologic factors discussed above, which could in turn lead to changes in air contaminants. One concern is that multiple interacting exposures could be affected simultaneously by climate change, leading to enhanced adverse health impacts. For example, the severe heat wave in 2003 in France was associated with elevated levels of ozone [11].

Research into the potential effects of climate change on air quality and human health is challenging, due in part to the highly interdisciplinary nature of the underlying science. Expertise is needed across a range of disciplines that have not often been linked in the past, including but not limited to climate data acquisition and processing, climate modeling, air quality modeling, exposure assessment, epidemiology, and clinical science. After teams are formed, they need to learn to communicate effectively so that the research can proceed productively, which can take considerable time and effort. An important technical challenge is the need to take the broad-scale predictions generated by global climate models and make them relevant and meaningful for impact assessments on fine geographic scales. Outputs from global models typically are resolved on a scale of hundreds of kilometers. Development and integration of research on the human dimensions of global environmental change require downscaling these projections to the regional metropolitan scale (tens of miles/kilometers or finer). Data at these finer scales facilitate planning for mitigation and adaptation strategies.

In the remainder of this chapter, we present two case studies investigating health impacts of climate change in New York City (NYC), one involving mortality effects of heat and air quality projected into the future and the other examining the influence of pollen on allergic responses in the present day, to illustrate and illuminate the challenges and potentials for climate, air quality, and health research. We also highlight recent developments in interdisciplinary research projecting the future health impacts of climate-driven changes in the pollen season.

Case Study 1: Climate, O_3, and Heat in the NYC Metropolitan Region

Episodes of heat and/or O_3 are current risk factors for adverse health effects in many urban areas around the world, and NYC is no exception. The NYC metropolitan region has often been out of attainment with the O_3 air quality standard, and heat waves are on the increase, a trend that is likely to continue for several decades. Future O_3 concentrations and resulting health effects will depend both on precursor emissions and on climate conditions. Here, we focus on the climate effect, holding O_3 precursor emissions constant.

The New York Climate and Health Project (or NYCHP) was designed to project future health impacts of climate-related changes in temperatures and ground-level O_3 concentrations [12]. We compared acute summertime heat- and O_3-related mortality from the past (using data from the 1990s) to several future decades (modeled for the 2020s, 2050s, and 2080s). We used a three-part methodology to assess these health impacts. First, we estimated coefficients describing mortality effects of temperature and O_3 using historical (1990–1999) death, weather, and air quality data for the study area. Next, we developed an integrated modeling system to project future environmental conditions under two scenarios of climate change, including modules for global climate, regional climate, and regional air quality. Third, the exposure-response coefficients were combined with the projections of future temperature and O_3 to estimate mortality in future decades under a changing climate. This is a good example of a project requiring multidisciplinary expertise, including climate and air quality modelers, public health scientists, and others.

Epidemiologic Analysis of Historical Data

Mortality data were obtained from the US National Center for Health Statistics (NCHS) for 1990–1999. Daily death counts were computed for each of 31 counties comprising the NYC metropolitan area for all internal causes (ICD-9 codes 0–799.9 for 1990–1998 and ICD-10 codes A00-R99 for 1999), excluding accidental causes and those among nonresidents. Air quality data were obtained for all O_3 monitoring stations within the study area. Of 39 stations that reported summer season data, those with fewer than 80% non-missing days were removed from further analyses, leaving 16 stations. Daily mean temperature (T_{ave}) (°F) data were obtained from the

National Oceanic and Atmospheric Administration (NOAA) National Climatic Data Center (NCDC) data inventory. Stations within the study area with at least 80% non-missing T_{ave} data included 16 meteorological stations (not the same as those where O_3 monitoring took place).

Coefficients representing the effects of O_3 and temperature on daily mortality were estimated using a Poisson generalized additive regression model with log daily death counts as the outcome variable, applied to the 31-county area as a whole. Analysis was restricted to the period between June 1 and August 30 for the years 1990–1999, to be consistent with the future projections (see below). Based on prior studies, we used T_{ave} at lag 0 and the 2-day average of the 1-hour daily maximum O_3 from lags 0 and 1. O_3 was treated as a linear term in the model, whereas temperature was modeled as a 3° polynomial in order to capture nonlinear effects at high temperatures. We examined possible confounding effects of particulate matter with aerodynamic diameter less than 10 μm (PM_{10}) on the relationship between O_3 and mortality, and found no evidence of such confounding in our dataset, consistent with previous work [6, 13, 14].

Future Projections of Temperature and O_3

As described previously [15, 16], we used the Goddard Institute for Space Studies (GISS) coupled global ocean/atmosphere model, driven by two different greenhouse gas scenarios of the fourth assessment report of the Intergovernmental Panel on Climate Change (IPCC): A2 and B2. The A2 scenario assumes relatively high and the B2 relatively low emissions of greenhouse gases over the twenty-first century. The GISS global model was linked via initial and boundary conditions to the Penn State-NCAR Mesoscale Model 5 (MM5) regional climate model. MM5 was run on two nested domains of 108 and 36 km over the United States. To simulate O_3 concentrations, the Community Mesoscale Air Quality (CMAQ) model was run at 36 km and took its initial conditions from the GISS-MM5 simulations [16]. For CMAQ the 1996 US Environmental Protection Agency (EPA) National Emission Trends (NET) database was processed by the Sparse Matrix Operator Kernel Emissions Modeling System (SMOKE) . The simulation periods were June–August 1993–1997, June–August 2023–2027, June–August 2053–2057, and June–August 2083–2087. In the work presented here, O_3 precursor emissions were held constant in the baseline and future simulations in order to isolate the climate effect. The MM5 model simulated T_{ave} and the CMAQ model simulated 1-hour daily maximum O_3 concentrations across the model domain in summers for these four future decades. Gridded temperatures and O_3 concentrations were interpolated to county geographic centroids using inverse distance weighting. Because of model biases for temperature, the modeled temperatures were converted to anomalies (i.e., monthly difference between future decadal estimate and baseline estimate), and these were used to adjust observed temperatures obtained from NCDC from the baseline period

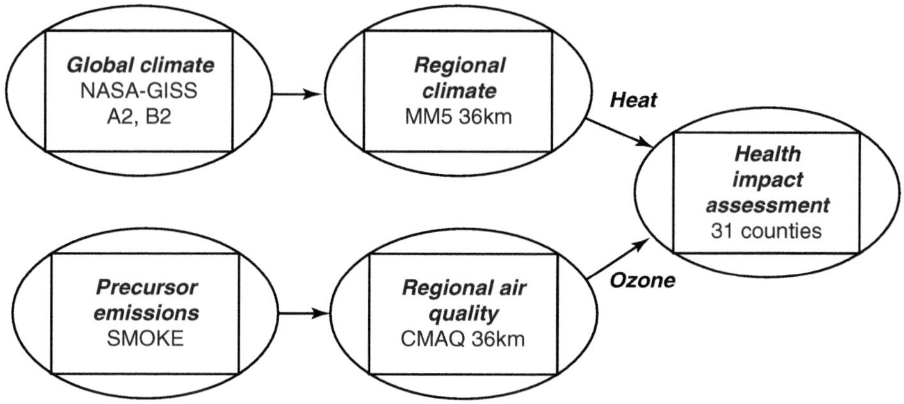

Fig. 8.1 Goddard Institute for Space Studies (GISS) coupled global ocean/atmosphere model

(1990s) to future decades. Further details are given in Knowlton et al. 2004 [17]. The modeling system is shown schematically in Fig. 8.1.

Health Impact Assessment

The daily model simulations of temperature and O_3 on a 36 km grid were combined with the exposure-response functions developed above to compute mortality risks in the baseline and future time periods. In order to isolate the impacts of climate change on future regional mortality, we held population constant at the Census 2000 county totals. We also held mortality rates constant at county-specific mean 1990s reference rates for the same reason. Preliminary analysis of the mortality and temperature data suggested that days with mean temperatures below 63.6 °F were not associated with excess mortality; thus, we only estimated mortality above this threshold temperature.

Results

Statistically significant coefficients of both temperature and O_3 on mortality were observed in the epidemiologic analysis of data from the 1990s, with results as follows:

$$\text{heat-related mortality} = (\text{Population}/100{,}000) * (\text{County daily mortality rate}) \\ * \left[\exp\left((\text{Temp} * 0.29193) + (\text{Temp}\,{}^\wedge 2 * -0.00434) + (\text{Temp}\,{}^\wedge 3 * 0.00002152)\right) - 1\right]$$

$$O_3\text{-related mortality} = (\text{Population}/100{,}000) * (\text{County daily mortality rate}) * [\exp(\max O_3 * 0.00045738) - 1]$$

Total temperature- or O_3-related deaths in the June–August period, averaged over each decade, were computed and compared to those in the 1990s. Figure 8.2 shows the regional distribution of percentage changes in heat-related mortality by the 2050s under the A2 greenhouse gas emissions scenario, and Fig. 8.3 shows the O_3 effects for the same conditions. While highly populated counties showed greater absolute numbers of heat-related deaths, higher percentage increases occurred in non-urban counties on the perimeter of the study area. For O_3, higher concentrations by the 2050s spread beyond the urban core into non-urban counties along the SW–NE prevailing wind directional axis, resulting in larger percentage increases in mortality in those locations.

Table 8.1 and Fig. 8.4 show the projected evolution over time of heat vs. O_3 mortality impacts under the A2 scenario. In the 1990s, summer O_3-related mortality was on par with heat-related mortality in a typical summer, but by mid-century this could change with heat-related mortality approximately doubling as compared to

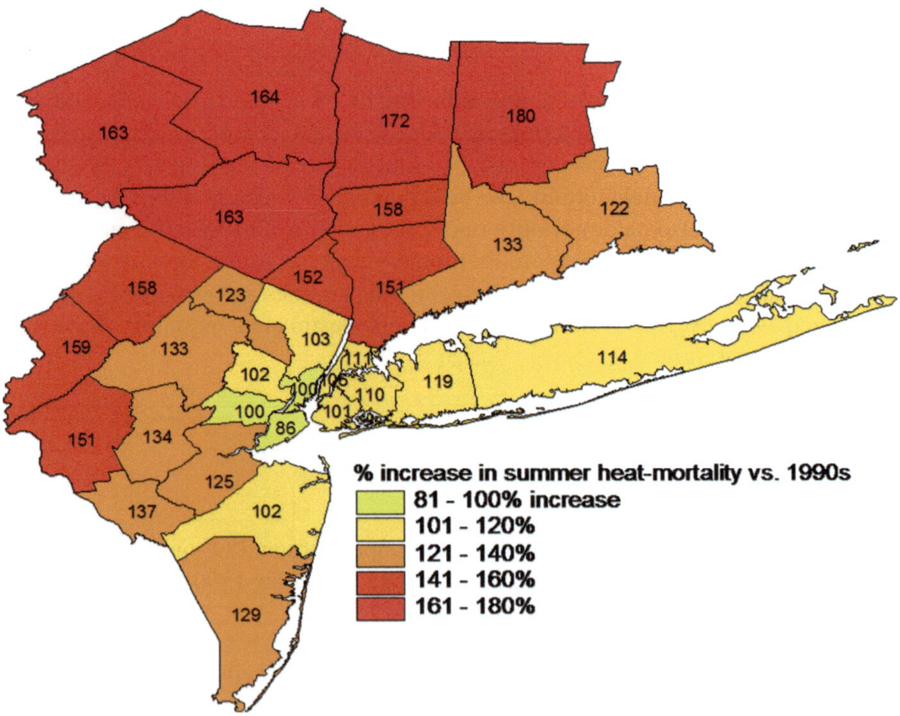

Fig. 8.2 The regional distribution of percentage changes in heat-related mortality by the 2050s under the A2 greenhouse gas emissions scenario

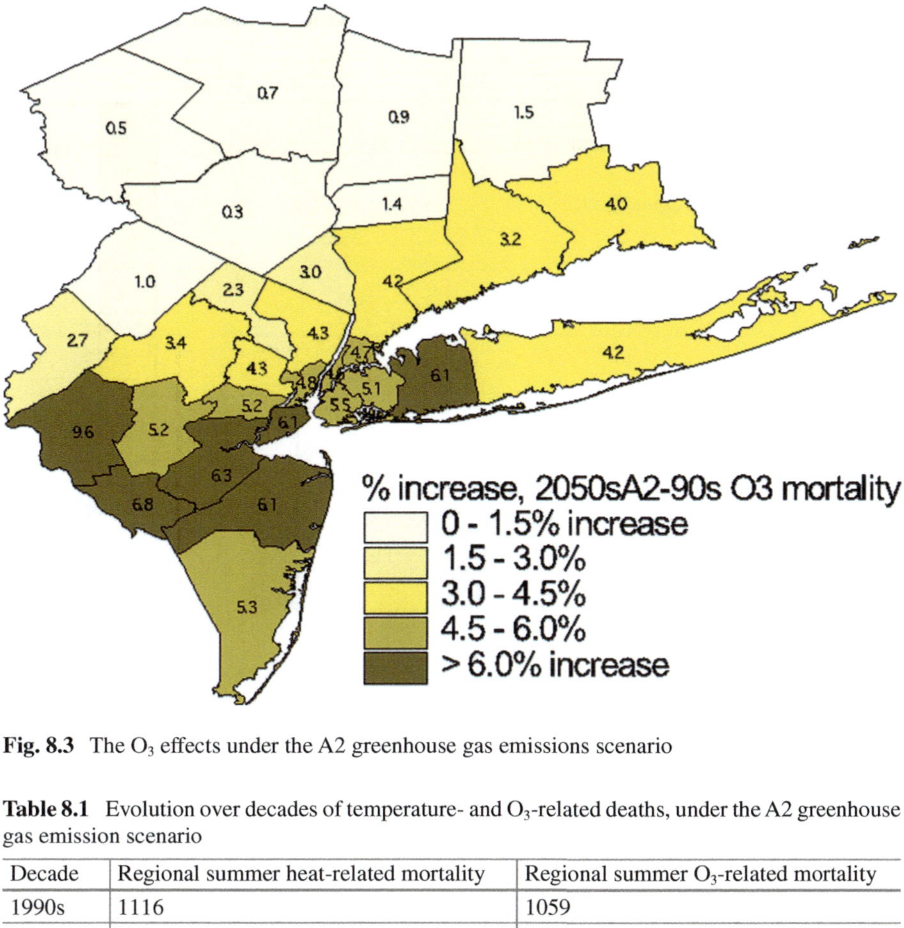

Fig. 8.3 The O_3 effects under the A2 greenhouse gas emissions scenario

Table 8.1 Evolution over decades of temperature- and O_3-related deaths, under the A2 greenhouse gas emission scenario

Decade	Regional summer heat-related mortality	Regional summer O_3-related mortality
1990s	1116	1059
2020s	1542 *38% increase* vs. *1990s*	1174 *11% increase* vs. *1990s*
2050s	2347 *110% increase* vs. *1990s*	1108 *5% increase* vs. *1990s*
2080s	5533 *396% increase* vs. *1990s*	1266 20% *increase* vs. *1990s*

the 1990s, but with O_3-mortality increasing by just 5%. By the 2080s, heat-related mortality could be over four times that from O_3.

Sensitivity analyses compared the 2050s B2 scenario mortality projections to that of the A2 scenario (Table 8.2) and found approximately 27% fewer heat-related deaths with the B2 scenario. These represent potential health benefits of more aggressive greenhouse gas regulatory schemes. While larger O_3-related mortality was projected for the New York metro region under the B2 scenario assumptions, different patterns across the eastern United States were seen; across the eastern United States as a whole, O_3 was projected to increase more under the 2050s A2 scenario than under the B2 scenario.

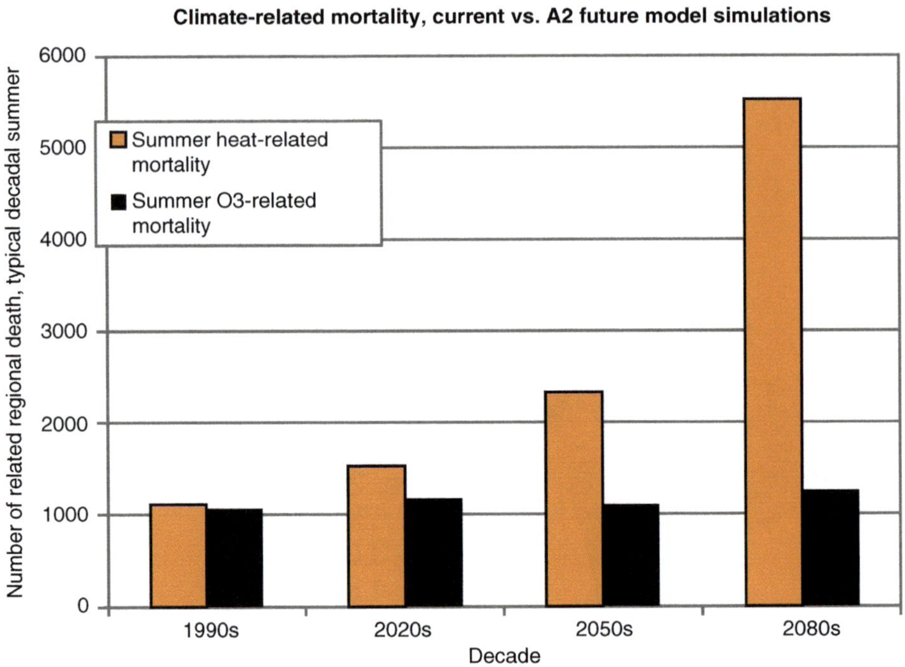

Fig. 8.4 The projected evolution over time of heat vs. O_3 mortality impacts under the A2 scenario

Table 8.2 Comparison of temperature- and O_3-related deaths in a typical summer for the 2050s vs. the 1990s, under two different greenhouse gas emission scenarios

	1990s	2050s B2 (lower CO_2 emissions)	2050s A2 (higher CO_2 emissions)
Heat-related mortality	1116	2013 *80% increase relative to 1990s*	2347 *110% increase relative to 1990s*
O_3-related mortality	1059	1139 *7.6% increase relative to 1990s*	1108 *4.6% increase relative to 1990s*

Discussion and Implications

This work illustrates an interdisciplinary study to develop local-scale projections of some possible health impacts of climate change in the NYC metropolitan region. In the United States, health policy decisions (emergency planning, hospital surveillance, etc.) are often made by county health departments, so climate impact projections are likely to be most meaningful if framed at the county level. Further, in the absence of federal regulations, greenhouse gas emission control policies often begin at the local level. If in the future the potential health impacts of climate change are monetized and become part of cost–benefit regulatory schemes, then risk assessments such as this could provide information useful not only to public health care infrastructure planning but also to regulators and legislative policymakers. An important limitation of this work is that we did not account for possible acclimatization to heat effects over multiple years as warming trends continue. This is an area in need of future research.

Case Study 2: Spring Pollen Peaks and Over-the-Counter Allergy Medication Sales

Introduction

Studies of the onset and duration of the pollen season have revealed substantial advances in the start date of the season that are consistent with recent warming trends [18–27]. In addition to earlier onset of the pollen season and possibly enhanced seasonal pollen loads in response to higher temperatures and resulting longer growing seasons, there is evidence that CO_2 rise itself may cause increases in pollen levels. For example, experimental studies have shown that elevated CO_2 concentrations stimulate greater vigor, pollen production, and allergen potency in ragweed [10, 28, 29]. In ragweed – arguably the most important pollen in the United States with up to 75% of hay fever sufferers sensitized [30] – significant differences in allergenic pollen protein were observed when comparing plants grown under historical CO_2 concentrations of 280 parts per million (ppm), recent past concentrations of 370 ppm, and potential future concentrations of 600 ppm [29]. Interestingly, significant differences in ragweed productivity were observed in outdoor plots situated in urban, suburban, and rural locales, where measurable gradients were observed in both CO_2 concentrations and temperatures. Cities are not only heat islands but also CO_2 islands and thus represent, to some extent, proxies for a future warmer, high CO_2 world [10]. With warming over the longer term, changing patterns of plant habitat and species density are likely, with gradual movement northward of cool-climate species, like maple, birch, and northern spruce [31].

Concentrations of various pollen taxa are associated with multiple measures of allergic and respiratory morbidity, including higher rates of allergic sensitization [32, 33], tendency toward increased asthma episodes [34], higher numbers of asthma-related emergency department (ED) visits [35–38] and hospital admissions [39, 40], and higher numbers of allergic rhinitis-related ED visits [41] and physician visits [42]. What remains unknown is whether, and to what extent, recent trends in pollen seasons may be linked with upward trends in allergic diseases like hay fever and asthma that have been seen in recent decades.

Allergic rhinitis, a type of allergic airway disease that is a risk factor for increased asthma severity [43], decreases the quality of life of a substantial proportion of the US population (10–30% of adults and up to 40% of children) and imposes large costs on our health care system [30, 44, 45]. Symptomatic relief of allergic rhinitis primarily involves ambulatory care and self-administration of medications. Thus, studies that look at more severe health outcomes like ED visits, hospitalizations, and physician visits likely only capture a small fraction of the population affected.

Thus, to examine whether pollen concentrations are temporally linked to allergic rhinitis symptoms more broadly, we analyzed the association of daily tree pollen peaks and over-the-counter (OTC) allergy medication sales over a 6-year period in the NYC metropolitan area [46].

Data and Methods

Airborne pollen was collected with a Burkard volumetric spore trap (Burkard Manufacturing Co., Rickmansworth, UK) located on the rooftop of Calder Hall at Fordham University's Louis Calder Biological Station in Armonk, New York. This station, which is located about 30 miles north of mid-town Manhattan, is the closest long-term, nearly continuous pollen record for the NYC region. Trained counters carried out microscopic analysis of daily pollen slides for 6 years from 2003 to 2008. From these slides, we computed daily concentrations of three genera of tree pollen: maple (*Acer* spp.), birch (*Betula* spp.), and oak (*Quercus* spp.). These subtypes were selected because they are clinically relevant aeroallergens in the United States [47] during the early season (March through May) and have well-established sensitization patterns in populations from the northeast region of the United States [48, 49]. The peak date – defined as the day on which the pollen concentration was highest – was identified for each of the three pollen types in each year.

Daily temperature data from LaGuardia International airport were downloaded from the NCDC. Data for $PM_{2.5}$ were obtained from US EPA's Air Quality System. The temporal variations of $PM_{2.5}$ across 21 sites were highly correlated ($r > 0.85$). Therefore, we computed the average of multiple sites, taking into consideration the difference in site-specific means and standard deviations [50].

Data on OTC pharmacy sales are reported electronically to the NYC Department of Health and Mental Hygiene on a daily basis from over 200 store locations, disproportionately in Manhattan but also from the other four NYC boroughs and nearby suburbs in New York State and New Jersey. The store locations in this database cover approximately 30% of retail pharmacies in NYC [51]. For this analysis, the following brand-name and generic products were classified as allergy medications: Alavert, Benadryl, Claritin, loratadine, Sudafed, and Tavist, as well as other oral and nasal spray medications that include the word "allergy" in their name. Eye drops and topical creams were not included.

We used an indicator variable (1 for peak dates; 0 otherwise) for the tree pollen peak dates between March and May each year for each genus. There was a total of 18 pollen peak dates over the 6-year study period. A regression model was used to estimate the impact of the tree pollen peak dates on the daily allergy medication sales, adjusting for potential confounding factors. We examined lags 0–6 days from the pollen peak dates (i.e., we compared today's allergy medication sales with today's tree pollen peak, today's allergy medication sales with yesterday's pollen peak, and so on). We first included individual lags of the tree pollen peak date indicator to determine the lag structure of associations and then included all of the 7-day lags to estimate the multi-day effects (i.e., an unconstrained distributed lag model). Covariates considered in the regression model included a day-of-week indicator variable, a year indicator variable, and air pollution and temperature variables to capture the effects of temperature on allergy symptoms [52] or on purchasing behaviors. Further details are given in Sheffield et al. 2011 [46].

Results and Discussion

Figure 8.5 shows time-series plots of OTC allergy medication sales for the entire city during the years 2003–2008, with tree pollen peak dates superimposed for maple, oak, and birch. Peaks dates for elm (*Ulmus* spp.) are also shown but were excluded from the analysis due to lack of a visual relationship with OTC sales. The tree pollen peak dates appear to coincide with sharp peaks in the spring medication sales. A general upward trend in sales across years can partially be explained by the number of stores reporting, which increased from 206 in 2003 to 231 in 2008.

Figure 8.6 shows the estimated impacts of tree pollen peaks when all the lagged peak date indicators were included simultaneously in the regression model. The largest statistically significant impact occurred at lag 2 day (28.7% [95%CI: 17.4, 41.2]), followed by lag 1 day. In the distributed lag model, the sum of the effects

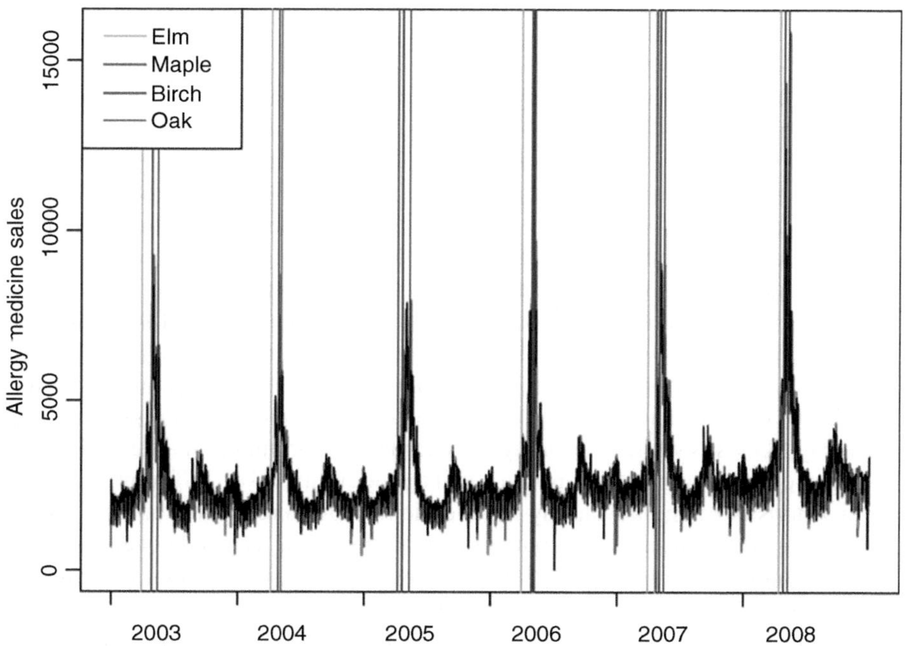

Fig. 8.5 Time-series plots of OTC allergy medication sales for New York City during the years 2003–2008, with tree pollen peak dates superimposed for maple, oak, birch, and elm

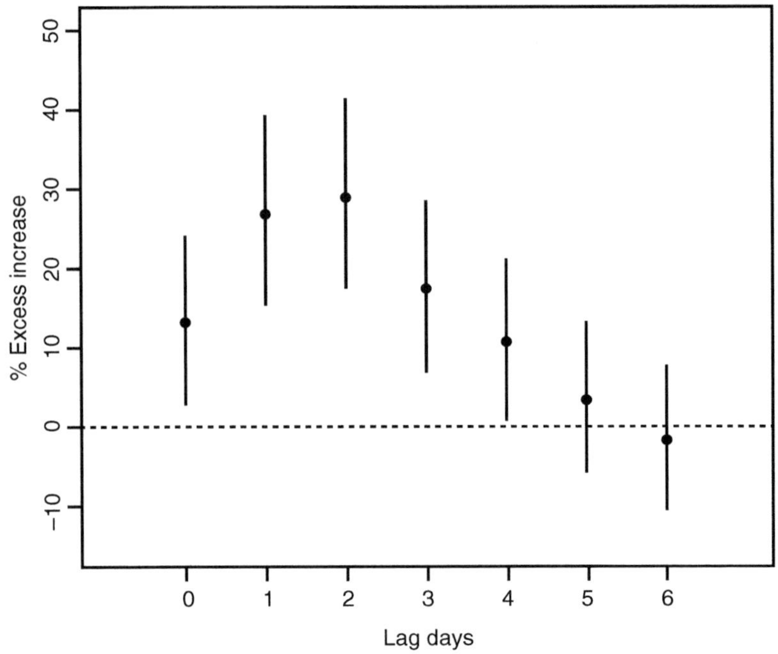

Fig. 8.6 The estimated impacts of tree pollen peaks when all the lagged peak date indicators were included simultaneously in the regression model

over the 7-day period was 141.1% (95%CI: 79.4, 224.1). These results were not substantially different in sensitivity analyses that tested alternative covariates and modeling methods.

These findings suggest that monitoring OTC medication sales may be a useful method of population surveillance for allergic illness and the impact of pollen. Our findings are generally but not entirely consistent with those of other studies examining the relation of ambient pollen to minor allergic illness. In an urban area in France, insurance claims were used to show that daily purchases of prescription allergy medications were associated with same-day concentrations of some tree pollens and grass pollen while controlling for weather and air pollution [53]. A study in Ottawa, Canada, found no effect of tree pollen on ED visits for conjunctivitis and rhinitis, but ragweed and fungal spore concentrations appeared to be associated with same-day ED visits while controlling for weather and air pollution. The exploration of lagged effects was not described in detail by the authors [41]. In Toronto, Canada, physician visits among the elderly for allergic rhinitis were associated with 10-day average ragweed concentrations but not with air pollution; they did not analyze pollen types other than ragweed [42]. One strength of our study is that it includes a fuller examination of lags than these previous studies.

An advantage of using OTC medications is that this health outcome reflects minor illness, as many will not seek health care nor have claims filed for prescription medications for allergic rhinitis. The observed associations support the use of genus-specific tree pollen season charts in clinical allergy practice, which were not being used in allergy clinics in New York at the time this research was conducted (personal correspondence, president of the NY Allergy Society, January 2010). However, limitations of this approach include the possibility that individuals may self-medicate using previously purchased OTC medications, that the single purchase of an OTC allergy medication could result in usage at multiple different times other than the day of purchase, that available in-home medications may vary within a calendar year, that pollen concentrations lower than the peak concentration may be related to OTC medication sales, and that pollen taxa other than the three genera included in this analysis may contribute to allergic rhinitis symptoms. Indeed, more recent work suggests that a larger set of spring pollen taxa – including sycamore (*Platanus* spp.), ash (*Fraxinus* spp.), and hickory (*Carya* spp.) – are associated with OTC allergy medication sales in NYC [54]. Thus, the analysis presented here likely underestimates the overall contribution of pollen to the use of OTC allergy medications. Furthermore, purchase of an OTC allergy medication does not describe frequency of use, severity of symptoms, or the number of individuals using a particular medication.

In contrast to the first case study presented in this chapter, this study did not directly address the role of climate factors in driving variations in pollen exposure and allergic health responses, either in the present day or projected into the future. While climate change is anticipated to alter the timing and severity of the pollen season as well as potentially the allergenicity of pollen grains, quantitative projections of the future burden of allergic disease attributable to climate change are relatively rare. Recent work [55] suggests that the magnitude of the impact of climate change on pollen-associated allergic disease could be large. In this proof-of-concept study, the research team combined output from global climate models with previously published, quantitative estimates of (1) the relationship between climate variables (temperature, precipitation) and the pollen season and (2) the relationship between daily pollen concentrations and emergency department (ED) visits for asthma. Using this information, the authors estimated the change in asthma ED visits attributable to changes in the length of the oak pollen season through the year 2090 compared to the present day in the northeastern, southeastern, and midwestern United States. The authors estimated that by the year 2090, the climate-driven changes in oak pollen season length could lead to a 10% increase in asthma ED visits across the study regions under a high greenhouse gas emissions scenario. Estimates were smaller for more proximal years (e.g., 2050) and for a more moderate emissions scenario.

Notably, the study described above was limited to a single pollen type (i.e., oak). Indeed, a key challenge in this line of work is the need to make projections for multiple allergenic pollen types with differing relationships to both climate variables and health outcomes. Other challenges include the sparseness of daily pollen-monitoring stations (at least in the United States), the paucity of epidemiologic studies relating concentrations of specific pollen types to health outcomes at the population level, and difficulty disentangling the health consequences of different pollen types that are highly correlated with each other in time. Efforts to address these analytic and data availability challenges would contribute to work in this field. Additional directions for future research could include assembling interdisciplinary teams to simultaneously model climate-driven changes in the pollen season and health outcomes, incorporating projected changes in the geographic distribution of pollen-producing species, and further exploring potentially synergistic relationships between air pollutants and pollen, both in the present-day and projected into the future.

Summary

The two case studies reviewed above demonstrate some of the methods that have recently been applied to study climate interactions with human health, mediated by temperature, air pollution, and/or airborne pollen. These examples demonstrate some of the characteristic features of emerging research in this area, including the formation of interdisciplinary teams, merging of health and climate data, the role of geographical downscaling, and the tools of health impact assessment. These approaches and others will be needed to further examine linkages between climate variations and human health impacts across a range of disease outcomes.

Acknowledgments This work was supported by the US Environmental Protection Agency (US EPA) under Science to Achieve Results (STAR) grant R828733. Additional support was provided by the National Institute of Environmental Health Sciences Center grant ES09089 and from the National Aeronautics and Space Administration/Goddard Institute for Space Studies Climate Impacts Group. Although the research described in this chapter has been funded wholly or in part by the US EPA, it has not been subjected to the agency's required peer and policy review and therefore does not necessarily reflect the views of the agency and no official endorsement should be inferred.

References

1. Kinney PL. Climate change, air quality, and human health. Am J Prev Med. 2008;35(5):459–67.
2. Jacob DJ, Winner DA. Effect of climate change on air quality. Atmos Environ. 2009;43(1):51–63.
3. Peel JL, Tolbert PE, Klein M, Metzger KB, Flanders WD, Todd K, et al. Ambient air pollution and respiratory emergency department visits. Epidemiology. 2005;16(2):164–74.
4. Peel JL, Metzger KB, Klein M, Flanders WD, Mulholland JA, Tolbert PE. Ambient air pollution and cardiovascular emergency department visits in potentially sensitive groups. Am J Epidemiol. 2007;165(6):625–33.
5. Kinney PL, Ozkaynak H. Associations of daily mortality and air pollution in Los Angeles County. Environ Res. 1991;54(2):99–120.
6. Levy JI, Chemerynski SM, Sarnat JA. Ozone exposure and mortality: an empiric bayes metaregression analysis. Epidemiology. 2005;16(4):458–68.
7. Nielsen GD, Hansen JS, Lund RM, Bergqvist M, Larsen ST, Clausen SK, et al. IgE-mediated asthma and rhinitis I: a role of allergen exposure? Pharmacol Toxicol. 2002;90(5):231–42.

8. Cecchi L, D'Amato G, Ayres JG, Galan C, Forastiere F, Forsberg B, et al. Projections of the effects of climate change on allergic asthma: the contribution of aerobiology. Allergy. 2010;65(9):1073–81.
9. Hogrefe C, Leung LR, Mickley L, Hunt SW, Winner D. Considering climate change in U.S. air quality management. J Air Waste Manag Assoc. 2005;October:35–40.
10. Ziska LH, Gebhard DE, Frenz DA, Faulkner S, Singer BD, Straka JG. Cities as harbingers of climate change: common ragweed, urbanization, and public health. J Allergy Clin Immunol. 2003;111(2):290–5.
11. Filleul L, Cassadou S, Medina S, Fabres P, Lefranc A, Eilstein D, et al. The relation between temperature, ozone, and mortality in nine French cities during the heat wave of 2003. Environ Health Perspect. 2006;114(9):1344–7.
12. Knowlton K, Hogrefe C, Lynn B, Rosenzweig C, Rosenthal J, Kinney PL. Impacts of heat and ozone on mortality risk in the New York City metropolitan region under a changing climate. In: Thomson MC, Garcia-Herrera R, Beniston M, editors. Seasonal forecasts, climate change and human health. Dordrecht: Springer; 2008. p. 143–60.
13. Ito K, De Leon SF, Lippmann M. Associations between ozone and daily mortality: analysis and meta-analysis. Epidemiology. 2005;16(4):446–57.
14. Bell ML, Dominici F, Samet JM. A meta-analysis of time-series studies of ozone and mortality with comparison to the national morbidity, mortality, and air pollution study. Epidemiology. 2005;16(4):436–45.
15. Hogrefe C, Lynn B, Civerolo K, Ku J-Y, Rosenthal J, Rosenzweig C, et al. Simulating changes in regional air pollution over the eastern United States due to changes in global and regional climate and emissions. J Geophys Res. 2004;109(D22301):1–13.
16. Hogrefe C, Biswas J, Lynn B, Civerolo K, Ku J-Y, Rosenthal J, et al. Simulating regional-scale ozone climatology over the eastern United States: model evaluation results. Atmos Environ. 2004;38(17):2627–38.
17. Knowlton K, Rosenthal JE, Hogrefe C, Lynn B, Gaffin S, Goldberg R, et al. Assessing ozone-related health impacts under a changing climate. Environ Health Perspect. 2004;112(15):1557–63.
18. Root TL, Price JT, Hall KR, Schneider SH, Rosenzweig C, Pounds JA. Fingerprints of global warming on wild animals and plants. Nature. 2003;421(6918):57–60.
19. Beggs PJ, Bambrick HJ. Is the global rise of asthma an early impact of anthropogenic climate change? Environ Health Perspect. 2005;113(8):915–9.
20. Beggs PJ. Impacts of climate change on aeroallergens: past and future. Clin Exp Allergy. 2004;34(10):1507–13.
21. Clot B. Trends in airborne pollen: an overview of 21 years of data in Neuchâtel (Switzerland). Aerobiologia. 2003;19(3/4):227–34.
22. Emberlin J, Detandt M, Gehrig R, Jaeger S, Nolard N, Rantio-Lehtimaki A. Responses in the start of Betula (birch) pollen seasons to recent changes in spring temperatures across Europe. Int J Biometeorol. 2002;46(4):159–70.
23. Galan C, Garcia-Mozo H, Vazquez L, Ruiz L, de la Guardia CD, Trigo MM. Heat requirement for the onset of the Olea europaea L. pollen season in several sites in Andalusia and the effect of the expected future climate change. Int J Biometeorol. 2005;49(3):184–8.
24. Rasmussen A. The effects of climate change on the birch pollen season in Denmark. Aerobiologia. 2002;18(3):253–65.
25. Teranishi H, Kenda Y, Katoh T, Kasuya M, Oura E, Taira H. Possible role of climate change in the pollen scatter of Japanese cedar Cryptomeria japonica in Japan. Clim Res. 2000;14(1):65–70.
26. van Vliet AJH, Overeem A, De Groot RS, Jacobs AFG, Spieksma FTM. The influence of temperature and climate change on the timing of pollen release in the Netherlands. Int J Climatol. 2002;22(14):1757–67.

27. Huynen M, Menne B, Behrendt H, Bertollini R, Bonini S, Brandao R, et al. Phenology and human health: allergic disorders. Report of a WHO meeting, Rome, Italy; 2003. Available: http://www.euro.who.int/__data/assets/pdf_file/0013/240232/e79129.pdf
28. Ziska LH, Caulfield FA. Rising CO2 and pollen production of common ragweed (Ambrosia artemisiifolia), a known allergy-inducing species: implications for public health. Aust J Entomol. 2000;27:893–8.
29. Singer BD, Ziska LH, Frenz DA, Gebhard DE, Straka JG. Research note : increasing Amb a 1 content in common ragweed (Ambrosia artemisiifolia) pollen as a function of rising atmospheric CO 2 concentration. Funct Plant Biol. 2005;32(7):667–70.
30. American Academy of Allergy Asthma and Immunology (AAAAI). Allergy statistics; 2000. Available: http://www.aaaai.org/about-the-aaaai/newsroom/allergy-statistics.aspx
31. Rosenzweig C, Casassa G, Karoly DJ, Imeson A, Liu C, Menzel A, et al. Assessment of observed changes and responses in natural and managed systems. In: Parry ML, Canziani OF, Palutikof JP, van der Linden PJ, editors. Climate change 2007: impacts, adaptation and vulnerability contribution of working group II to the fourth assessment report of the intergovernmental panel on climate change. Cambridge, UK/New York: Cambridge University Press; 2007. p. 79–131.
32. Bjorksten F, Suoniemi I. Time and intensity of first pollen contacts and risk of subsequent pollen allergies. Acta Med Scand. 1981;209(4):299–303.
33. Porsbjerg C, Linstow ML, Nepper-christensen SC, Rasmussen A, Korsgaard J, Nolte H, et al. Allergen sensitization and allergen exposure in Greenlander Inuit residing in Denmark and Greenland. Respir Med. 2002;96(9):736–44.
34. Delfino RJ, Zeiger RS, Seltzer JM, Street DH, McLaren CE. Association of asthma symptoms with peak particulate air pollution and effect modification by anti-inflammatory medication use. Environ Health Perspect. 2002;110(10):A607–17.
35. Babin SM, Burkom HS, Holtry RS, Tabernero NR, Stokes LD, Davies-Cole JO, et al. Pediatric patient asthma-related emergency department visits and admissions in Washington, DC, from 2001–2004, and associations with air quality, socio-economic status and age group. Environ Health. 2007;6:9.
36. Lierl MB, Hornung RW. Relationship of outdoor air quality to pediatric asthma exacerbations. Ann Allergy Asthma Immunol. 2003;90(1):28–33.
37. Wang HC, Yousef E. Air quality and pediatric asthma-related emergencies. J Asthma. 2007;44(10):839–41.
38. Zhong W, Levin L, Reponen T, Hershey GK, Adhikari A, Shukla R, et al. Analysis of short-term influences of ambient aeroallergens on pediatric asthma hospital visits. Sci Total Environ. 2006;370(2–3):330–6.
39. Dales RE, Cakmak S, Judek S, Coates F. Tree pollen and hospitalization for asthma in urban Canada. Int Arch Allergy Immunol. 2008;146(3):241–7.
40. Im W, Schneider D. Effect of weed pollen on children's hospital admissions for asthma during the fall season. Arch Environ Occup Health. 2005;60(5):257–65.
41. Cakmak S, Dales RE, Burnett RT, Judek S, Coates F, Brook JR. Effect of airborne allergens on emergency visits by children for conjunctivitis and rhinitis. Lancet. 2002;359(9310):947–8.
42. Villeneuve PJ, Doiron MS, Stieb D, Dales R, Burnett RT, Dugandzic R. Is outdoor air pollution associated with physician visits for allergic rhinitis among the elderly in Toronto, Canada? Allergy. 2006;61(6):750–8.
43. Thomas M. Allergic rhinitis: evidence for impact on asthma. BMC Pulm Med. 2006;6(Suppl 1):S4.
44. Soni A. Allergic rhinitis: trends in use and expenditures, 2000 and 2005. 2008. Available: http://meps.ahrq.gov/mepsweb/data_files/publications/st204/stat204.pdf
45. Nielsen-Wolter Kluwer Health. Consumer behavior and managed care impact of the Zyrtec Rx-to-OTC switch; 2008. Available: http://download.lww.com/wolterskluwer_vitalstream_com/PermaLink/Allergy-whitepaperFinal.pdf

46. Sheffield PE, Weinberger KR, Ito K, Matte TD, Mathes RW, Robinson GS, et al. The association of tree pollen concentration peaks and allergy medication sales in New York City: 2003–2008. ISRN Allergy. 2011;2011:537194.
47. U.S. EPA. A review of the impact of climate variability and change on aeroallergens and their associated effects. Washington, DC; 2008. Available: https://cfpub.epa.gov/ncea/risk/recordisplay.cfm?deid=190306
48. Lin RY, Clauss AE, Bennett ES. Hypersensitivity to common tree pollens in New York City patients. Allergy Asthma Proc. 2002;23(4):253–8.
49. White JF, Levin L, Villareal M, Murphy K, Biagini R, Wellinghoff L, et al. Lack of correlation between regional pollen counts and percutaneous reactivity to tree pollen extracts in patients with seasonal allergic rhinitis. Ann Allergy Asthma Immunol. 2005;94(2):240–6.
50. Zanobetti A, Schwartz J, Dockery DW. Airborne particles are a risk factor for hospital admissions for heart and lung disease. Environ Health Perspect. 2000;108(11):1071–7.
51. Das D, Metzger K, Heffernan R, Balter S, Weiss D, Mostashari F. Monitoring over-the-counter medication sales for early detection of disease outbreaks – New York City. MMWR Suppl. 2005;54:41–6.
52. Cruz AA, Togias A. Upper airways reactions to cold air. Curr Allergy Asthma Rep. 2008;8(2):111–7.
53. Fuhrman C, Sarter H, Thibaudon M, Delmas MC, Zeghnoun A, Lecadet J, et al. Short-term effect of pollen exposure on antiallergic drug consumption. Ann Allergy Asthma Immunol. 2007;99(3):225–31.
54. Ito K, Weinberger KR, Robinson GS, Sheffield PE, Lall R, Mathes R, et al. The associations between daily spring pollen counts, over-the-counter allergy medication sales, and asthma syndrome emergency department visits in New York City, 2002–2012. Environ Health. 2015;14:71.
55. Anenberg SC, Weinberger KR, Roman H, Neumann JE, Crimmins A, Fann N, et al. Impacts of oak pollen on allergic asthma in the United States and potential influence of future climate change. GeoHealth. 2017;1:80–92.

Chapter 9
The Human Health Co-benefits of Air Quality Improvements Associated with Climate Change Mitigation

George D. Thurston and Michelle L. Bell

Overview

Fossil fuel combustion processes that generate greenhouse gases (GHG) also emit and/or cause the creation of other harmful air pollutants. Thus, while policies designed to avert the course of climate change would eventually result in direct human health benefits from lessened global temperature profile changes and associated impacts, they would also bring much more immediate ancillary human health co-benefits from the associated reduced ground-level air pollution [1–6]. Multiple measures aimed at reducing GHG emissions, notably the reduced use of fossil fuels, such as coal, can also improve local air quality, most notably particulate matter (PM) and ozone (O_3) air pollution. Further, whereas the benefits from climate change mitigation would materialize far in the future, these co-benefits, or ancillary benefits, would provide much more immediate "return on investment" in climate change mitigation. Thus, as detailed below, the near-term human health co-benefits of climate mitigation (e.g., fossil fuel emission reductions) may provide the most economically compelling justification for immediate action toward climate change mitigation. Here, we focus on the health impacts of PM and ozone, two key air pollutants that have substantial impacts on human health and are also likely to be affected by policies aimed at reducing GHG emissions.

G. D. Thurston (✉)
NYU School of Medicine, New York University, New York, NY, USA
e-mail: george.thurston@nyu.edu

M. L. Bell
School of Forestry and Environmental Studies, Yale University, New Haven, CT, USA
e-mail: michelle.bell@yale.edu

K. E. Pinkerton, W. N. Rom (eds.), *Climate Change and Global Public Health*, Respiratory Medicine, https://doi.org/10.1007/978-3-030-54746-2_9

Health Effects of Particulate Matter

Tropospheric aerosols that affect climate change also have significant human health implications. A wealth of scientific literature clearly links particulate matter with numerous adverse health effects. Indeed, a US Environmental Protection Agency (USEPA) assessment of human health effect benefits of the Clean Air Act attributed nearly 90% of the estimated monetary valuation of the human health effect benefits from the Act during 1990–2010 to reductions in PM [7, 8].

Short-Term Exposure Effects of PM

Acute (short-term) exposure to particulate air pollution has been found to be associated with increases in the rates of daily asthma attacks, hospital admissions, and mortality. PM exposure has been associated with increased risk of respiratory hospital admissions and mortality in the United States, and in other cities throughout the world, including national multi-city studies [9–17].

In addition to lung damage, recent epidemiological and toxicological studies of PM air pollution have shown adverse effects on the heart, including an increased risk of heart attacks. For example, when PM stresses the lung (e.g., by inducing edema), it places extra burden on the heart, which can induce fatal complications for persons with cardiac problems. Indeed, Peters and colleagues [18] found that elevated concentrations of fine particles (PM $\leq$ 2.5 μm in aerodynamic diameter, i.e., $PM_{2.5}$) in the air could elevate the risk of myocardial infarctions (MIs) within a few hours and extending 1 day after $PM_{2.5}$ exposure.

Epidemiologic research conducted in the United States and elsewhere has indicated that acute exposure to PM air pollution is associated with increased risk of mortality. For example, a national multi-city time-series statistical analysis of mortality and PM $\leq$ 10 μm in aerodynamic diameter (PM_{10}) air pollution in 90 US cities indicated that an increase of 10 μg/m^3 in daily PM_{10} was associated with an increase of approximately 0.3% in the daily risk of death [19]. This result of a 0.3% change in the daily mortality rate is tied to the increment of pollution; in other words, a pollution increase larger than 10 μg/m^3 would be associated with an even larger increase in the risk of mortality. Further, such added risks apply to the entire population and accumulate on every day of exposure until they account for many deaths from air pollution globally each year.

Long-Term Exposure Effects of PM

In addition to the health effects associated with acute exposure to PM pollution, long-term chronic exposure to particles is also associated with an increased lifetime risk of death and has been estimated to take years from the life expectancy of people living in the most polluted cities, relative to those living in cleaner cities. The first study to show the association between fine particulate matter mass ($PM_{2.5}$) was a cross-sectional study that compared metropolitan area death rates in high and low air pollution exposure cities, after adjusting for potentially confounding factors in the populations, such as age, income, education, and race [20]. These results have since been confirmed by cohort studies that followed large groups of individuals in various cities over time that are able to control for potential confounding factors on an individual level. For example, in the Six-Cities Study, which was a key basis for the setting of the USEPA's original health-based regulation for a $PM_{2.5}$ annual standard in 1997, Dockery and colleagues analyzed survival probabilities among 8111 adults living in six cities in the central and eastern portions of the United States during the 1970s and 80s [21]. The cities were as follows: Portage, WI (P); Topeka, KS (T); a section of St. Louis, MO (L); Steubenville, OH (S); Watertown, MA (M); and Kingston-Harriman, TN (K). Air quality was averaged over the period of study in order to study long-term (chronic) effects. It was found that, even after adjusting for potentially confounding factors such as age, sex, race, smoking, etc. at the individual level, the long-term risk of death increased with fine particle exposure level.

It has also been shown that long-term exposure to combustion-related fine particulate air pollution is an important environmental risk factor for cardiopulmonary and lung cancer mortality (see Fig. 9.1). Indeed, this study indicates that the increase in the risk of lung cancer from long-term exposure to $PM_{2.5}$ was of roughly the same size as the increase in lung cancer risk of a nonsmoker who breathes passive smoke while living with a smoker, or about a 20% increase in lung cancer risk [22].

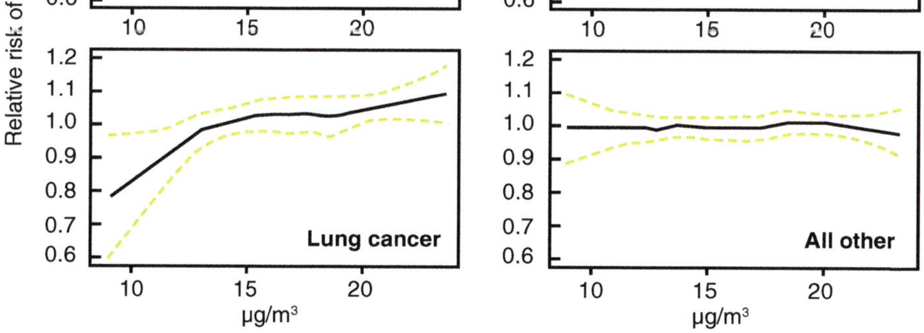

Fig. 9.1 The cardiac, lung, and cancer mortality risks of long-term fine PM exposure increase monotonically with exposure (*RR* = relative risk, relative to the mean exposure). (Adapted From: Pope [22])

Other studies indicating health risk from chronic exposure to PM include a multi-city US study finding that a 10 μg/m^3 increase in yearly $PM_{2.5}$ is associated with approximately a 11–21% increase in mortality [23]. A systematic review of research on long-term PM exposure found that, collectively, the studies indicate a 15–21% increase in mortality per 10 μg/m^3 $PM_{2.5}$ [24]. Since that time, there have been analyses in other large and well-characterized cohorts, such as the NIH-AARP cohort, finding consistent results [25].

Health Effects of PM Constituents

Particulate matter is a complex mixture of a wide array of chemical constituents, and its chemical composition varies seasonally and regionally [26]. For example, some particles may have a larger contribution of sulfate, whereas others may have more nitrate. The chemical structures of such ambient particles depend in large measure on their respective sources [27]. While most past studies have investigated the effects of the PM *mass* concentration on human health effects, newer studies have begun to evaluate the mortality impacts of PM by specific constituents or sources, including two key *aerosol constituents* that can affect climate change: sulfates and elemental black carbon (BC) soot. However, different types of particles have very different climate implications, with sulfates having a climate-cooling forcing, while elemental BC soot is a climate-warming constituent [28].

With regard to acute health effects of PM components, it has been found that coal-burning-related sulfate-containing aerosols were among those most associated with increases in daily mortality [29]. In addition, Bell and colleagues [30] found that communities with higher $PM_{2.5}$ content of nickel (Ni), vanadium (V), and elemental carbon (EC) and/or their related sources yielded higher risks of hospitalizations associated with short-term exposure to $PM_{2.5}$. Lall and colleagues [31] similarly found that EC of traffic origins was associated with higher risk of cardiovascular disease (CVD) hospital admissions in New York than $PM_{2.5}$ mass in general. In a study of mortality in New York, Ito and colleagues [32] have reported that coal combustion–related components (e.g., selenium (Se) and sulfur) were associated with CVD mortality in summer, whereas the traffic-related EC showed associations with CVD mortality throughout the year. Zhou and colleagues [33] investigated the $PM_{2.5}$ components and gaseous pollutants associated with mortality in Detroit, MI, and Seattle, WA, similarly finding that CVD and respiratory mortality were most associated with warm season secondary aerosols (e.g., sulfates) and traffic markers (e.g., EC) in Detroit, while in Seattle, the component species most closely associated with mortality included those for cold season traffic and other combustion sources, such as residual oil and wood burning. In addition, diesel traffic–derived EC has been implicated as a factor in increased risk of acute asthma morbidity [34]. A systematic review found evidence that BC and EC $PM_{2.5}$ are associated with cardiovascular endpoints, but not sufficient evidence to distinguish between these components or between them and $PM_{2.5}$ total mass [35]. Overall, these studies of $PM_{2.5}$ components and constituents largely indicate that both EC and sulfates (and their associated sources, including diesel traffic and coal burning) were among the most explanatory of the acute adverse health effects of $PM_{2.5}$.

With regard to the long-term effects of PM air pollution, Ozkaynak and Thurston [19] conducted the first source apportionment of $PM_{2.5}$-mortality effects, finding that sulfate-related particles, largely from coal burning, were most associated with the mortality impacts of long-term exposure to $PM_{2.5}$. Ostro and colleagues [36] examined daily data from 2000 to 2003 on mortality and $PM_{2.5}$ mass and components, including elemental and organic carbon (EC and OC), nitrates, sulfates, and various metals. The authors examined associations of $PM_{2.5}$ and its constituents with daily counts of several mortality categories: all-cause, cardiovascular, respiratory, and mortality age > 65 years, finding the strongest associations between mortality and sulfates and several metals. Ostro and colleagues [37] also used data from a prospective cohort of active and former female public school professionals to develop estimates of long-term exposures to $PM_{2.5}$ and several of its constituents, including EC, OC, sulfates, nitrates, iron (Fe), potassium (K), silicon (Si), and zinc (Zn), finding increased risks of all-cause and cardiopulmonary mortality from exposure to constituents derived from combustion of fossil fuel (including diesel), as well as those of crustal origin. In addition, Smith and colleagues [38] undertook a meta-analyses of existing time-series studies as well as an analysis of a cohort of 352,000 people in 66 US cities during 18 years of follow-up of the ACS cohort, finding total mortality effects from long-term exposure to both the elemental BC and sulfate components of $PM_{2.5}$ aerosols. More recently, Thurston *and colleagues* partitioned the earlier Pope et al.'s $PM_{2.5}$-mortality associations (as shown in Fig. 9.1)[21] into results instead based on risks associated with contributions to $PM_{2.5}$ by source category, finding that $PM_{2.5}$ from fossil fuel combustion sources, especially from coal combustion, was most strongly associated with increased ischemic mortality risk in the ACS cohort (Fig. 9.2) [39].

Recent evidence indicates that long-term air pollution exposure, especially from traffic-related air pollution (TRAP), can induce induction of new-onset asthma in children. The Southern California Children's Health Study (CHS) found an especially increased risk of childhood new-onset asthma from TRAP at home residence [40]. Carlsten and colleagues found $PM_{2.5}$ to be the TRAP pollution component most associated with new-onset childhood asthma in a susceptible population [41].

Health Effects of Tropospheric Ozone

Tropospheric ozone is a highly reactive pollutant that is common in the urban environment, as it largely results from emissions from fossil fuel combustion. O_3 is a secondary pollutant, meaning it is not directly emitted into the air by fossil fuel combustion sources, but rather is formed through complex nonlinear reactions from O_3 precursors, volatile organic compounds (VOCs) and nitrogen oxides (NO_x), in the presence of sunlight. Sources of VOCs and NO_x include transportation, industry,

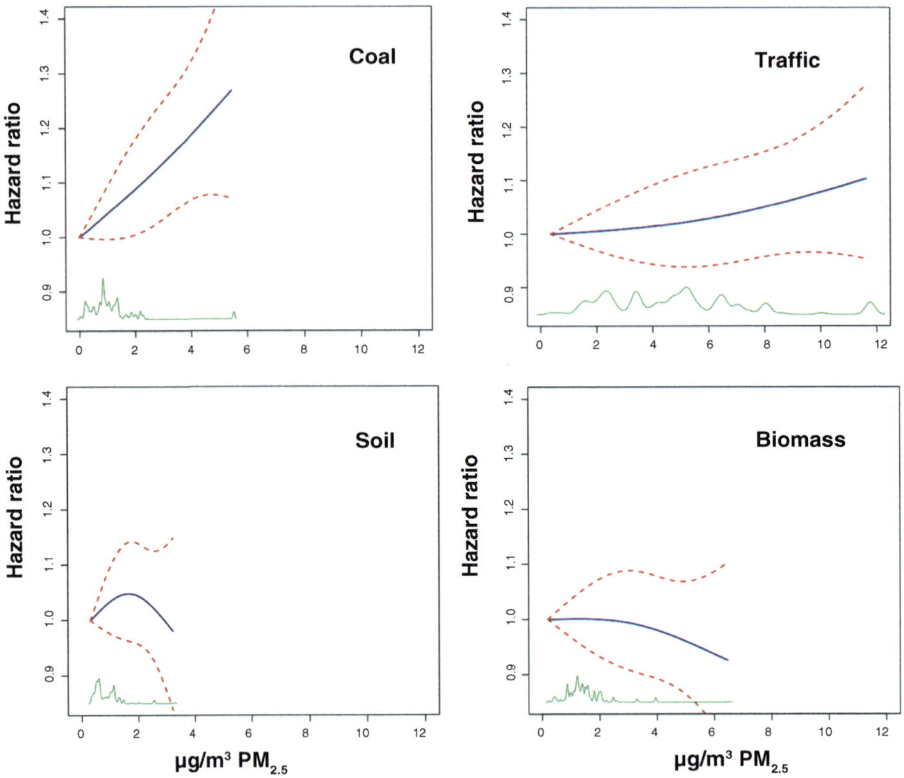

Fig. 9.2 Concentration-response curve (solid lines) and 95% confidence intervals (dashed lines) for source-specific $PM_{2.5}$ mass in the US American Cancer Society (ACS) cohort. (Thurston et al. [39])

and power plants. Both VOCs and NO_x have natural sources, of which vegetative emissions are key contributors of VOCs. Levels of O_3 are especially of concern in urban environments, and in fact, in the United States, more persons live in areas that exceed the health-based regulations for O_3 than for any other criteria pollutant. Ozone is a growing problem in developing regions of the world, with expanding transportation networks and industry. Thus, this pollutant is not only a global warming pollutant, but also has significant global health impacts [42].

Tropospheric ozone, which occurs at the surface layer of the Earth, should be distinguished from stratospheric ozone, which is present higher (10–50 km) in Earth's atmosphere. While the same chemical form, tropospheric ozone is harmful as it is present in the breathing layer and can be inhaled, whereas stratospheric ozone provides a protective layer, the "ozone layer," against ultraviolet (UV) radiation. This radiation is associated with increased risk of adverse health outcomes, including skin cancer and cataracts. Substances that deplete stratospheric ozone (e.g., chlorofluorocarbons, hydrochlorofluorocarbons, halons) are regulated in order to prevent and mitigate what is commonly referred to as the "ozone hole" in the stratosphere.

These chemicals had been used in refrigeration and air conditioning, such as for appliances and motor vehicles. Many substances that deplete the ozone layer also contribute to climate change, with global warming potential that can greatly exceed that of CO_2. Some tropospheric ozone is the result of intrusion of stratospheric ozone; however, this occurrence is a minor contributor to tropospheric ozone. Thus, tropospheric O_3 exposure is of predominant human health effects concern, except to those in high-altitude aircraft, who can potentially receive more substantial stratospheric O_3 exposures [43].

Short-Term Exposure Health Effects of O_3

The scientific evidence for the respiratory morbidity effects from acute exposure to O_3 is well documented. Animal toxicological studies have indicated that chronic O_3 exposure caused structural changes in the respiratory tract, and simulated seasonal exposure studies in animals have also suggested that such exposures might have cumulative impacts, providing evidence of a biological foundation for the associations observed in population-based studies [42]. Epidemiologic studies have also observed that reduced lung function growth in children is associated with exposure to O_3 [44–47] and that O_3 exposure can affect lung exposure in the elderly [48]. Based on evidence from animal toxicological studies, short-term and sub-chronic exposures to O_3 can cause morphological changes in the respiratory systems of a number of species, including primates.

Following chronic O_3 exposure, structural changes have been observed in the centriacinar region, the region typically affected in most chronic airway diseases of the human lung. In addition, a substantial number of human exposure studies have been published that have provided important information on lung inflammation and epithelial permeability. Mudway and Kelly [49], for example, examined O_3-induced inflammatory responses and epithelial permeability with a meta-analysis of 21 controlled human exposure studies, finding that polymorphonuclear neutrophils (PMN) influx in healthy subjects is associated with total O_3 dose product of O_3 concentration, exposure duration, and minute ventilation. Overall, animal toxicological studies indicate that short-term and sub-chronic exposures to O_3 can cause morphological changes in the respiratory systems, particularly in the centriacinar region of the lung [42]. Thus, there is strong supportive evidence from both acute epidemiological studies and toxicological studies of respiratory morbidity that ozone exposure can have serious respiratory morbidity health effects.

O_3 exposure has also been found to be associated with short-term increases in the risk of mortality as a result of acute exposures. Indeed, robust associations have been identified between various measures of daily O_3 concentrations and increased risk of mortality [42]. In addition, multiple studies have conducted meta-analyses of O_3-mortality associations [50–53]. Combined O_3 excess mortality risk estimates from the meta-analyses by Bell et al. [54], Ito et al. [55], and Levy and colleagues [56] were also all very consistent. Associations have also been observed in other study designs, including a multi-city time series of 95 US urban cities over a 14-year period, finding a 0.52% (95% interval 0.27, 0.77%) increase in mortality risk for a 10 ppb increase in daily ozone over the previous week [14].

Adverse O_3 effects have also been observed for cardio-respiratory mortality. The Air Pollution and Health: A European Approach (APHEA2) project examined ozone and mortality for 23 European cities with at least 3 years of data [57]. A 10 μg/m^3 increase in the 1-hour max ozone was associated with a 0.33% (0.17, 0.52%) increase in mortality risk, with associations also observed for cardiovascular and respiratory deaths. A case-crossover study of 14 US cities found a 0.23% (0.01, 0.44%) increase in mortality risk for a 10 ppb increase in daily maximum ozone levels, with matching on days of similar temperature [58].

A systematic review of studies on O_3 and health found strong evidence of higher associations based on age with higher effects for older persons, unemployment, or lower occupational status; limited/suggestive evidence for higher associations for women; and weak evidence of higher effects for racial/ethnic minorities [59].

There is evidence that the association between ozone and mortality persists at low concentrations. A study of 98 US urban communities with 14 years of data used several modeling approaches to investigate the shape of the exposure–response curve [60]. The first method assumed that any level of ozone could potentially be associated with mortality risk; this is the traditionally applied time-series approach. The second method examined the subset of data below specified values of 5–60 ppb, at 5 ppb increments, for daily ozone. A threshold model was fit to assume no association for ozone levels below a specified threshold value and a traditional shape for higher ozone levels. The final approach used a nonlinear function of ozone levels to allow a flexible relationship between ozone and mortality. None of the alternative models found evidence of a threshold at policy-relevant concentrations. The study found that associations were significant at levels nearing natural background concentrations and levels below the USEPA's National Ambient Air Quality Standard at the time of the study.

Several studies have examined whether associations between short-term exposure to ozone and risk of mortality are confounded by airborne particles, which have demonstrated links with mortality, as discussed above. The most common approach, to include a variable for particulate matter (e.g., $PM_{2.5}$ concentration) in the model, was found to result in little change to ozone effect estimates [50, 52, 55, 57, 58]. Figure 9.3 shows the relationship between short-term exposure to ozone and risk of mortality with and without adjustment for PM_{10} [14]. Other approaches to exploring confounding have also provided evidence for the hypothesis that the ozone-mortality association is not confounded by particulate matter, including at low levels of O_3 [61].

Some segments of the population may face a disproportionate burden from ozone pollution. Communities with higher unemployment had higher effect estimates for short-term ozone and mortality for 98 US urban communities [62]. A higher proportion of Black/African-American residents was also associated with higher effect

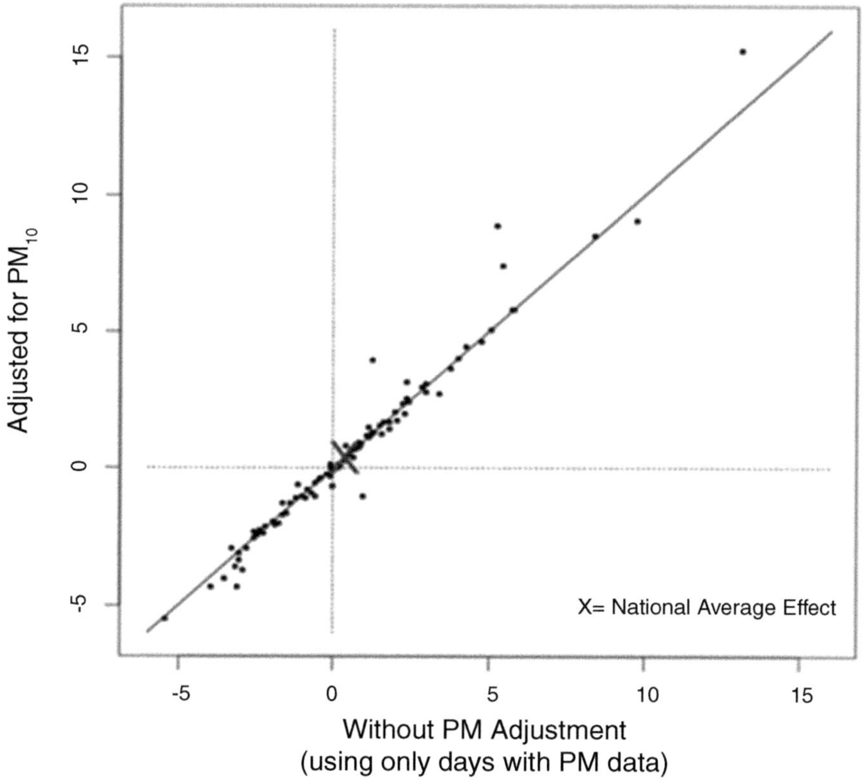

Fig. 9.3 Percentage change in risk of mortality for 10 ppb increase in ozone, with and without adjustment PM_{10} [14]. *Note: Each dot represents a community-specific estimate. The X represents the national average.* (Adapted from Bell et al. [14])

estimates. These findings may relate to differences in baseline healthcare status, access to health care, or exposure patterns. However, the impact of population characteristics on ozone effect estimates is not fully understood. Findings on socioeconomic status and effect modification of short-term ozone associations are not consistent across the few studies that have investigated this issue. In Mexico City, socioeconomic status did not demonstrate clear patterns for ozone and mortality associations [63]

Overall, there is substantial and growing body of evidence on acute adverse effects of O_3, and it can be concluded that robust associations have been identified between various measures of daily O_3 concentrations and increased risk of mortality. Further, the scientific evidence covers a variety of study designs and locations, and studies have consistently demonstrated an acute mortality effect of ozone that is not confounded by particulate matter.

Long-Term Exposure Mortality Effects of O_3

A limited number of epidemiologic studies have assessed the relationship between long-term exposure to O_3 and mortality. The US EPA's 2013 O_3 Integrated Science Assessment (ISA) concluded that sufficient evidence exists to suggest a causal relationship between chronic O_3 exposure and increased risk for mortality in humans [42], and more recent evidence is consistent with a relationship between long-term ozone exposure and an increased risk of mortality.

In the Harvard Six Cities Study, adjusted mortality rate ratios were examined in relation to long-term mean O_3 concentrations in six cities: Topeka, KS; St. Louis, MO; Portage, WI; Harriman, TN; Steubenville, OH; and Watertown, MA [21]. Mortality rate ratios were adjusted for age, sex, smoking, education, and body mass index. Mean O_3 concentrations from 1977 to 1985 ranged from 19.7 ppb in Watertown to 28.0 ppb in Portage. Long-term mean O_3 concentrations were not found to be associated with mortality across these six cities. However, the authors noted: "The small differences in ozone levels among the (six) cities limited the power of the study to detect associations between mortality and ozone levels." In addition, while total and cardio-pulmonary mortality were considered in this study, respiratory mortality, which may have been more directly affected, was not specifically considered.

In a subsequent large prospective cohort study of approximately 500,000 US adults, Pope and colleagues examined the effects of long-term exposure to air pollutants on mortality in the American Cancer Society (ACS) Cancer Prevention Study II [22]. While no consistently significant positive associations were observed between O_3 and mortality, the mortality risk estimates were larger when analyses considered more accurate exposure metrics, rising when the entire period was considered compared to analysis using just the start of the study period, and becoming marginally significant when the exposure estimates were restricted to the summer months (July to September), especially when considering cardiopulmonary deaths (HR = 1.09, 95%ile confidence interval [CI] = 0.99–1.19 per 60 ppb mean daily 1-hour maximum).

In an extended follow-up analysis of the ACS cohort [64], ozone effects were tested for associations with cardio-pulmonary deaths subdivided into respiratory and cardiovascular, separately, as opposed to combined in the earlier work. This analysis utilized the ACS cohort with data from 1977 through 2000 (mean O_3 concentration ranged from 33.3 to 104.0 ppb). In two-pollutant models, $PM_{2.5}$ was associated with the risk of death from cardiovascular causes, whereas ozone was associated with the risk of death from respiratory causes. As shown in Fig. 9.4, exposure to O_3 was positively associated with risk of death from respiratory causes. The relative risk of death from respiratory causes = 1.040 (95% CI = 1.010–1.067) was found to be associated with an increment in mean ozone-season (April 1 to September 30) daily maximum O_3 concentration of 10 ppb. The association of ozone with risk of death

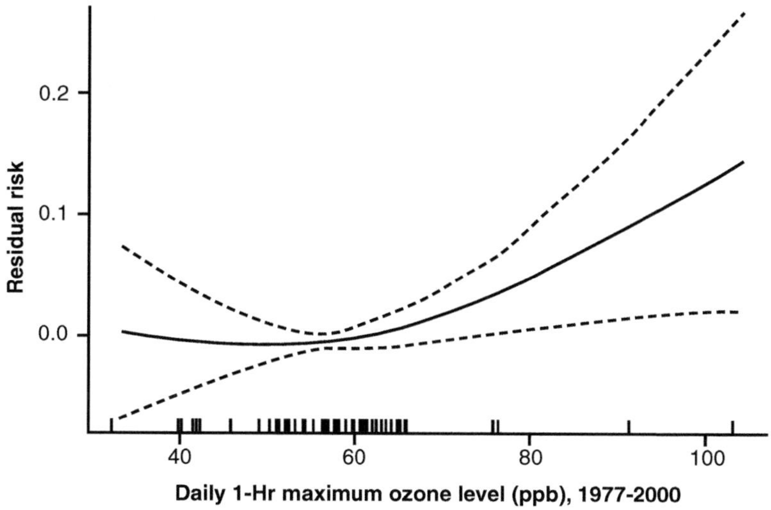

Fig. 9.4 Relationship between respiratory mortality risk and long-term O_3 in the ACS cohort

from respiratory causes was insensitive to adjustment for confounders and to the type of statistical model. Overall, this analysis strongly suggests that, while long-term exposure to $PM_{2.5}$ increases the risk of cardiac death, long-term exposure to O_3 is specifically associated with an increased risk of respiratory death.

More recently, an analysis was conducted of ozone effects on mortality in the NIH-AARP cohort [65]. Long-term annual average exposure to O_3 was significantly associated with deaths (per 10 ppb annual average 8-hour daily maximum) from cardiovascular disease (HR = 1.03; 95% CI: 1.01–1.06), ischemic heart disease (HR = 1.06; 95% CI: 1.02–1.09), respiratory disease (HR = 1.04; 95% CI: 1.00–1.09), and chronic obstructive pulmonary disease (HR = 1.09; 95% CI: 1.03–1.15) in single-pollutant models. The results were robust to alternative models and adjustment for co-pollutants (fine particulate matter and nitrogen dioxide), although some evidence of confounding by temperature was observed. Interestingly, significantly elevated respiratory disease mortality risk associated with long-term O_3 exposure was found among those living in locations with high temperature (p-interaction < 0.05), suggesting that climate change may increase the effects of O_3 beyond those found today.

At this time, the United States and most other nations have only a short-term exposure air quality standard for ozone (e.g., maximum allowable 8-hour average). The literature now indicates there is a need for a long-term (e.g., annual or ozone season average concentration) air quality standard to more effectively protect public health from the effects of cumulative ozone exposures on our health.

Ancillary Health Benefits of Climate Change Mitigation

Framework of Climate Mitigation Co-benefits Assessment

Figure 9.5 displays the relationships among the health consequences of climate change and air quality policies and the general framework of how these responses can be assessed. Air quality policies are routinely evaluated in terms of the estimated health outcomes avoided and their economic impact [7, 8]. However, an assessment of the health impacts of GHG strategies often considers only consequences in the far future (i.e., left side of Fig. 9.5), without integration of the short-term benefits of related policies [66]. Well-informed public health and environmental strategies require full consideration of consequences, including co-benefits and potential ancillary harms.

A broad array of tools to evaluate the health-related ancillary costs and benefits of climate change is currently available, and some examples are provided in italics in Fig. 9.5. As described in detail by Bell and colleagues, the general structure for most assessments involves three key steps: (1) estimating changes in air pollutant concentrations, comparing levels in response to GHG mitigation to concentrations under a baseline "business-as-usual" scenario; (2) estimating the adverse health impacts avoided from reduced air pollution; and (3) for some studies, estimating the monetary benefit from these averted health consequences, often with comparison to the cost of the climate change mitigation measure [67].

The first step in such a co-benefit analysis is often the development of emissions scenarios and information regarding how emissions translate into pollutant concentrations, such as with air quality modeling systems. The second step employs concentration-response functions from existing epidemiological studies on ambient air pollution and health. The third stage utilizes a variety of techniques to translate health benefits into monetary terms. Potential additional steps include sensitivity analysis, such as applying multiple climate change scenarios or concentration-response functions for health effects.

Studies of Health and Air Pollution Benefits and Costs of Climate Change Mitigation

A variety of studies have been conducted to estimate the health and air pollution ancillary benefits and costs from GHG reduction, with a wide range of methods and study areas. Energy scenarios, emission inventories, and global change and regional air quality modeling systems have been linked to estimate the short-term incremental changes in public health and the environment that could result from various GHG mitigation policies [68, 69].

There are now numerous analyses indicating substantial health co-benefits from reductions in PM pollution that can be induced by GHG mitigation measures that

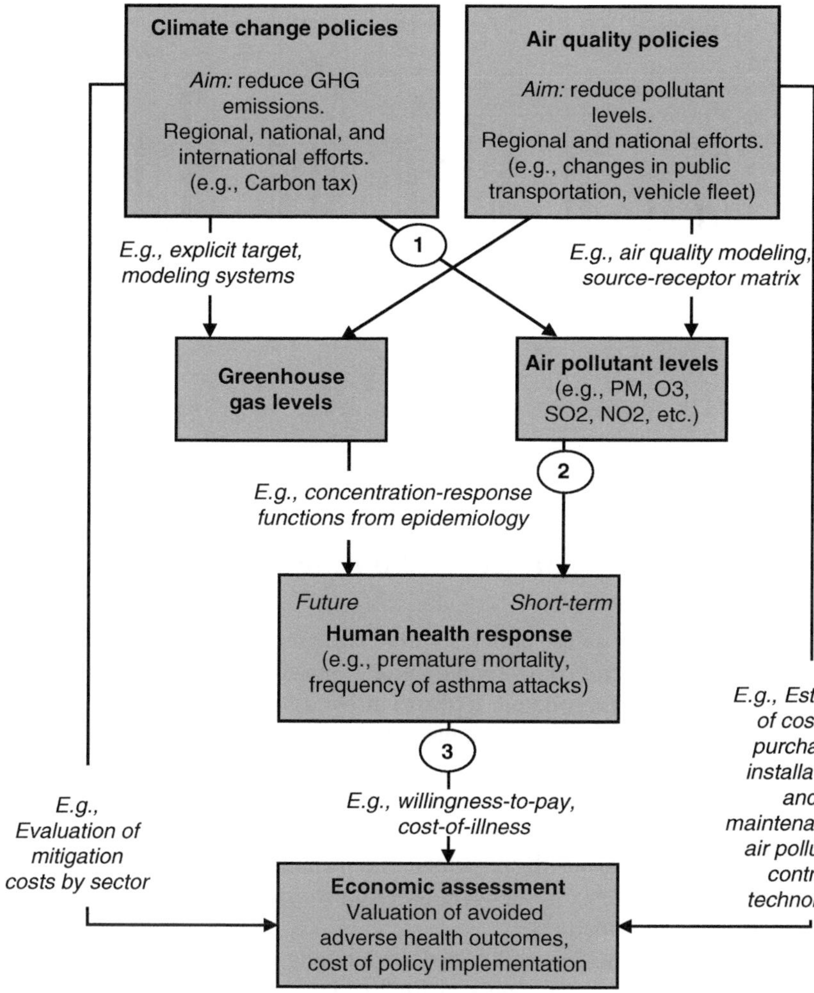

Fig. 9.5 Framework of air pollution co-benefit estimation [67]

involve reductions in fossil fuel combustion emissions. As shown in Fig. 9.6, a study of New York and three Latin American cities identified significant health benefits from reducing GHG, including about 64,000 cases of avoided premature mortality over a 20-year period [70]. Countrywide assessments of GHG mitigation policies on public health have been produced for Canada [71] and selected energy sectors in China [72, 73], under differing baseline assumptions. A synthesis of research on co-benefits and climate change policies in China concluded that China's Clean Development Mechanism potentially could save 3000–40,000 lives annually through co-benefits of improved air pollution [74]. Several of the earliest studies were made of the links between regional air pollution and climate policy in Europe [75–77].

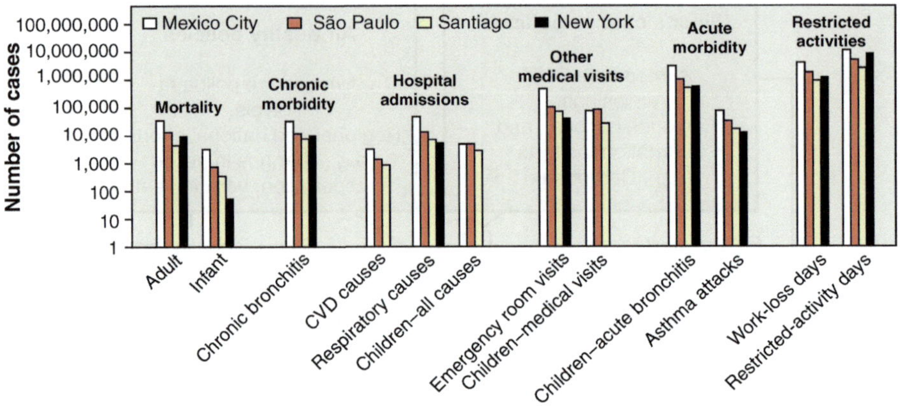

Fig. 9.6 Estimated potential human health benefits from reductions in air pollution associated with implementing GHG mitigation measures in four cities (2001–2020) [70]

Monetary Valuations of Mitigation Co-benefits

To help decision-makers assess policies with a wide array of health consequences, outcomes are often converted into comparable formats. One used approach is to convert health outcomes into economic terms to allow direct comparison of costs and benefits. There are several common approaches for economic valuation of averted health consequences (step 3 of Fig. 9.5): cost of illness (COI), human capital, willingness to Pay (WTP) methods, and quality-adjusted life year (QALY) approaches. The COI method totals medical and other out-of-pocket expenditures and has been used for acute and chronic health endpoints. For instance, separate models of cancer progression and respiratory disease were used to estimate medical costs from these diseases over one's lifetime [78].

Early attempts to value mortality risk reductions applied the human capital approach, which estimates the "value of life" as lost productivity. This method is generally recognized as problematic and not based on modern welfare economics, where preferences for reducing death risks are not captured. Another limitation is incorporation of racial- or gender-based discrimination in wages. This method assigns value based solely on income, without regard to social value, so unpaid positions, such as homemaker, and lower paid positions, such as social worker, receive lower values. Because data are often available for superior alternatives, this approach is rarely used in health benefit studies today.

WTP, alternatively, generates estimates of preferences for improved health that meet the theoretical requirements of neoclassical welfare economics, by aiming to measure the monetary amount persons would willingly sacrifice to avoid negative health outcomes. Complications arise in analysis and interpretation because changes in environmental quality or health often will themselves change the real income (utility) distribution of society. A valuation procedure that sums individual

WTP does not fully capture individual preferences about changes in income distribution. Another complication is that the value of avoided health risk may differ by the type of health event and age. The QALY approach attempts to account for the quality of life lost by adjusting for time "lost" from disease or death, but these estimates may be very insensitive to different severities and types of acute morbidity [79].

Estimating the ancillary public health consequences of GHG policies is a challenging task, drawing upon expertise in economics, emission inventories, air pollution modeling, and public health. However, most assessments to date have focused more heavily on one aspect of the framework (i.e., a portion of Fig. 9.6), such as estimation of changes in air pollutant concentrations, health response, or economic analysis.

The global co-benefits regarding mortality reductions caused by reduced PM air pollution that would be achieved by going ahead with global climate CO_2 emissions reductions have recently been estimated [80, 81]. As presented by West and colleagues, climate mitigation measures will result in the greatest reductions in PM mortality in those places that implement the most CO_2 reductions (Fig. 9.7a), and the financial valuation of the mortality benefits will outweigh the costs of the climate mitigation measures in those localities that do so (Fig. 9.7b). As such, the health co-benefits, and their financial valuations, make a strong case for moving ahead with climate mitigation measures that would at the same time reduce emissions of PM from fossil fuel sources, such as from coal combustion.

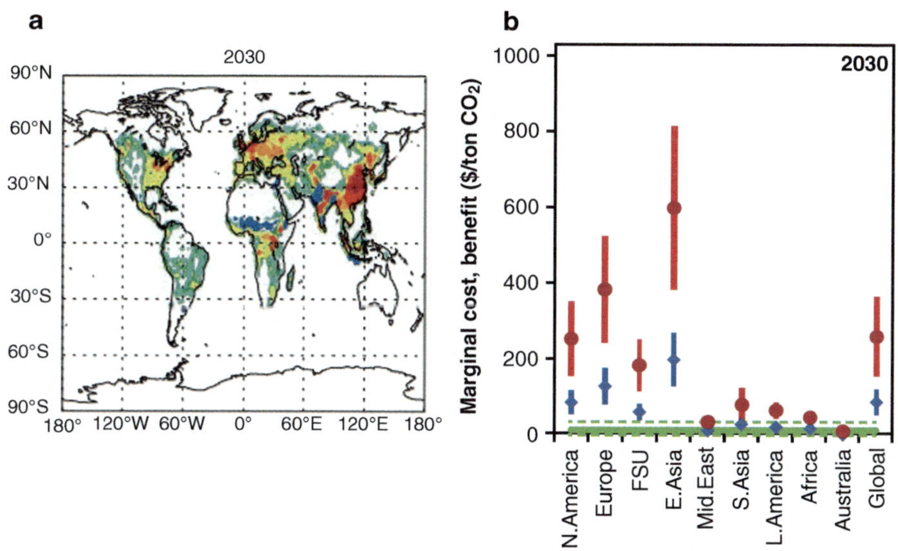

Fig. 9.7 (**a**) The avoided premature mortality from $PM_{2.5}$ in 2030 (deaths per year per 1000 km^2) due to climate change mitigation measures and (**b**) he range of associated marginal benefits ($/ton CO_2) by region [80] (Red: high estimate, blue: low estimate), relative to the range of expected mitigation cost (green lines)

Results from current ancillary benefits studies may be underestimated due to unquantified benefits, as only a subset of the health consequences from air pollution have adequate exposure–response relationships [82–85]. A USEPA evaluation of the Clean Air Interstate Rule (CAIR) noted numerous unquantified health impacts, such as chronic respiratory damage for O_3, loss of pulmonary function for PM, and lung irritation for NO_x [8]. The nature of unquantified effects is continually evolving. Some pollution and health relationships considered unquantifiable by USEPA in the past have since been identified, such as PM air pollution's association with lung cancer [22, 86], which is now recognized by the International Agency for Research on Cancer (IARC). Furthermore, some endpoints may be included in one analysis, but regarded as too uncertain for another, perhaps due to a different study location or differences in researchers' judgment. One approach to addressing health endpoints with uncertain concentration-response functions is to include these effects qualitatively in the discussion of unquantified benefits. Another is to incorporate these effects within a sensitivity analysis.

Valuations of mortality risk reductions associated with environmental policies are usually the largest category of benefits, both among health responses and compared to other attributes. For instance, a USEPA analysis of the Clean Air Act estimated a value of $100 billion annually for reduced premature mortality out of $120 billion in total benefits, compared to costs of approximately $20 billion [8]. European and Canadian studies similarly found that mortality risk dominates analysis of pollution reductions [87, 88]. Next to mortality, reductions in the probability of developing a chronic respiratory disease have been estimated to have the highest monetary value, recognizing that values for other types of diseases are sparse. Recent evidence indicates that reduced incidence of childhood asthma should be added to the human health and financial benefits of cleaner air [89, 90]. Thus, the health benefits can only consider what evidence there is available "under the streetlamp" of past research, while the countervailing costs of cleanup are much easier to completely compile, so such analyses of the health benefits vs. costs of cleaner air will unfortunately always be conservative.

The Stern Review addressed a wide range of global benefits and costs associated with climate change, including air pollution co-benefits [91]. Citing a study by the European Environmental Agency, the review notes that limiting the global mean temperature increase to 2 °C would lead to annual savings in the implementation of existing European air pollution control measures of €10 billion and additional avoided annual health costs of €16–46 billion. Even larger co-benefits are estimated in developing countries, including via the substitution of modern fuels for biomass. The Stern Review also recognizes some of the trade-offs between climate change objectives and local air quality gains. For instance, switching from petrol to diesel reduces carbon dioxide (CO_2) emissions, but increases PM_{10} and NO_x emissions. Other GHG-mitigating actions present fewer environmental trade-offs (e.g., reductions in aircraft weight can decrease CO_2 emissions and simultaneously improve local air quality).

Aerosol health benefits have not been fully incorporated into past cost–benefit modeling as to how much the world should optimally mitigate climate change. By addressing this, Scrovronick and colleagues have recently found that, when both aerosol co-benefits and co-harms are taken fully into account, optimal climate policy results in immediate net benefits globally, in contradiction to findings from prior cost–benefit models that omitted these effects [92]. They estimate that global health benefits from climate policy could reach trillions of dollars in valuation annually, but these benefits will vary as a function of the air quality policies that nations adopt independently of climate change. The authors conclude that, depending on how society values better health, economically optimal levels of mitigation can be designed to be consistent with a target of 2 °C global average change or lower.

Overall, though still a work in progress, the present techniques available for the analyses of the ancillary public health costs and benefits are adequate and appropriate for implementation by those comparing the relative merits and overall value of various GHG mitigation policies. Estimates of considerable benefits that remain after a variety of sensitivity analyses can alleviate some concerns regarding limitations of individual methods or assumptions. Because of their large health impacts per amount of energy produced, fossil fuel (and especially coal) combustion air pollution mitigation strategies should be considered as a key factor in the choice of GHG policies, and these health benefits should be noted as a potentially major local incentive for programs to reduce GHG emissions.

Implications

The anthropogenic contribution to the climate change pollutants is largely caused by the same activities that cause most human health effects of air pollution. This indicates that, if a city, state, or nation acts to reduce the combustion of fossil fuels and the air pollution caused by them, it will reap not only the climate change benefits but also the localized health benefits associated with that air pollution reduction. Thus, substantial near-term air pollution–associated health benefits of climate control measures can go to the cities and countries that act most vigorously to control their combustion emissions of greenhouse gases. These local and near-term health "co-benefits" of reductions in the air pollution from fossil fuel combustion should be considered in the overall analysis, including economic consequences, for climate change mitigation measures.

Competing Interests The authors declare that they have no competing interests.

Authors' Contributions Both authors made substantial contributions to the conception and design of this paper, were involved in drafting and revising of the manuscript, and have approved the final version.

References

1. Cifuentes L, Borja-Aburto VH, Gouveia N, Thurston GD, Davis DL. Climate change: hidden health benefits of greenhouse gas mitigation. Science. 2001;293(5533):1257–9.
2. Thurston GD. Air pollution, human health, climate change and you. Thorax. 2007;62:748–9.
3. Walsh MP. Ancillary benefits for climate change mitigation and air pollution control in the world's motor vehicle fleets. Annu Rev Public Health. 2008;29:1–9.
4. Smith KR, Haigler E. Co-benefits of climate mitigation and health protection in energy systems: scoping methods. Annu Rev Public Health. 2008;29:11–25.
5. Watts N, Adger WN, Agnolucci P, Blackstock J, Byass P, Cai W, Chaytor S, Colbourn T, Collins M, Cooper A, Cox PM, Depledge J, Drummond P, Ekins P, Galaz V, Grace D, Graham H, Grubb M, Haines A, Hamilton I, Hunter A, Jiang X, Li M, Kelman I, Liang L, Lott M, Lowe R, Luo Y, Mace G, Maslin M, Nilsson M, Oreszczyn T, Pye S, Quinn T, Svensdotter M, Venevsky S, Warner K, Xu B, Yang J, Yin Y, Yu C, Zhang Q, Gong P, Montgomery H, Costello A. Health and climate change: policy responses to protect public health. Lancet. 2015;386(10006):1861–914. https://doi.org/10.1016/S0140-6736(15)60854-6. Epub 2015 Jun 25.
6. Rice MB, Thurston GD, Balmes JR, Pinkerton KE. Climate change: a global threat to cardiopulmonary health. Am J Respir Crit Care Med. 2014;189(5):512–9.
7. U.S. Environmental Protection Agency (USEPA). The benefits and costs of the clean air act 1970 to 1990. Washington, DC; 1997.
8. U.S. Environmental Protection Agency (USEPA). The benefits and costs of the clean air act 1990 to 2010. EPA-410-R-99-001. Washington, DC; 1999.
9. Thurston GD, Ito K, Hayes CG, Bates DV, Lippmann M. Respiratory hospital admissions and summertime haze air pollution in Toronto, Ontario: consideration of the role of acid aerosols. Environ Res. 1994;65(2):271–90.
10. Thurston GD, Ito K, Kinney PL, Lippmann M. A multi-year study of air pollution and respiratory hospital admissions in three New York State metropolitan areas: results for 1988 and 1989 summers. J Expo Anal Environ Epidemiol. 1992;2(4):429–50.
11. Schwartz J, Kocnig J, Slater D, et al. Particulate air pollution and hospital emergency visits for asthma in Seattle. Am Rev Respir Dis. 1993;147:826–31.
12. Dominici F, Peng RD, Bell ML, Pham L, McDermott A, Zeger SL, Samet JM. Fine particulate air pollution and hospital admission for cardiovascular and respiratory diseases. JAMA. 2006;295(10):1127–34.
13. Peng RD, Samoli E, Pham L, Dominici F, Touloumi G, Ramsay T, Burnett RT, Krewski D, Le Tertre A, Cohen A, Atkinson RW, Anderson HR, Katsouyanni K, Samet JM. Acute effects of ambient ozone on mortality in Europe and North America: results from the APHENA study. Air Qual Atmos Health. 2013;6(2):445–53.
14. Bell ML, McDermott A, Zeger SL, Samet JM, Dominici F. Ozone and short-term mortality in 95 US urban communities, 1987-2000. J Am Med Assoc. 2004;292:2372–8.
15. Cai J, Peng C, Yu S, Pei Y, Liu N, Wu Y, Fu Y, Cheng J. Association between $PM_{2.5}$ exposure and all-cause, non-accidental, accidental, different respiratory diseases, sex, and age mortality in Shenzhen, China. Int J Environ Res Public Health. 2019;16(3) https://doi.org/10.3390/ijerph16030401.
16. Li M, Wu Y, Tian YH, Cao YY, Song J, Huang Z, Wang XW, Hu YH. Association between $PM_{2.5}$ and daily hospital admissions for heart failure: a time-series analysis in Beijing. Int J Environ Res Public Health. 2018;15(10) https://doi.org/10.3390/ijerph15102217.
17. Sarkodie SA, Strezov V, Jiang Y, Evans T. Proximate determinants of particulate matter ($PM_{2.5}$) emission, mortality and life expectancy in Europe, Central Asia, Australia, Canada, and the US. Sci Total Environ. 2019;683:489–97.

18. Peters A, Dockery DW, Muller JE, Mittleman MA. Increased particulate air pollution and the triggering of myocardial infarction. Circulation. 2001;103(23):2810–5.
19. Dominici F, McDermott A, Daniels M, Zeger SL, Samet JM. Revised analyses of the national morbidity, mortality, and air pollution study: mortality among residents of 90 cities. J Toxicol Environ Health A. 2005;68:1071–92.
20. Ozkaynak H, Thurston GD. Associations between 1980 U.S. mortality rates and alternative measures of airborne particle concentration. Risk Anal. 1987;7:449–60.
21. Dockery D, Pope CA 3rd, Xu X, Spengler JD, Ware JH, Fay ME, Ferris BG Jr, Speizer FE. An association between air pollution and mortality in six U.S. cities. N Engl J Med. 1993;329(24):1753–9.
22. Pope CA III, Burnett RT, Thun MJ, Calle EE, Krewski D, Ito K, Thurston GD. Lung cancer, cardiopulmonary mortality and long-term exposure to fine particulate air pollution. JAMA. 2002;287(9):1132–41.
23. Eftim SE, Samet JM, Janes H, McDermott A, Dominici F. Fine particulate matter and mortality: a comparison of the six cities and American Cancer Society cohorts with a Medicare cohort. Epidemiology. 2008;19(2):209–16.
24. Chen H, Goldberg MS, Villeneuve PJ. A systematic review of the relation between long-term exposure to ambient air pollution and chronic diseases. Rev Environ Health. 2008;23(4):243–97.
25. Thurston GD, Ahn J, Cromar K, Shao Y, Reynolds H, Jerrett M, Lim C, Shanley R, Park Y, Hayes RB. Ambient particulate matter air pollution exposure and mortality in the NIH-AARP diet and health cohort. Environ Health Perspect. 2016;124(4):484–90. https://doi.org/10.1289/ehp.1509676. Epub 2015 Sep 15.
26. Bell ML, Dominici F, Ebisu K, Zeger SL, Samet JM. Spatial and temporal variation in $PM_{2.5}$ chemical composition in the United States for health effects studies. Environ Health Perspect. 2007;115(7):989–95.
27. Spengler JD, Thurston GD. Mass and elemental composition of fine and coarse particles in six U.S. cities. J Air Pollut Control Assoc. 1983;33:1162–71.
28. National Academy of Science (NAS). Radiative forcing of climate change: expanding the concept and addressing uncertainties. Washington DC: National Research Council. Committee on Radiative Forcing Effects on Climate, Climate Research Committee, National Academy Press; 2005.
29. Thurston GD, Ito K, Mar T, Christensen WF, Eatough DJ, Henry RC, Kim E, Laden F, Lall R, Larson TV, Liu H, Neas L, Pinto J, Stölzel M, Suh H, Hopke PK. Workgroup report: workshop on source apportionment of particulate matter health effects – intercomparison of results and implications. Environ Health Perspect. 2005;113(12):1768–74.
30. Bell ML, Ebisu K, Peng RD, Samet JM, Dominici F. Hospital admissions and chemical composition of fine particle air pollution. Am J Respir Crit Care Med. 2009;179(12):1115–20. Epub 2009 Mar 19.
31. Lall R, Ito K, Thurston G. Distributed lag analyses of daily hospital admissions and source-apportioned fine particle air pollution. Environ Health Perspect. 2011;119(4):455–60. PMC3080925.
32. Ito K, Mathes R, Ross Z, Nádas A, Thurston GD, Matte T. Fine particulate matter constituents associated with cardiovascular hospitalizations and mortality in New York City. Environ Health Perspect. 2011;119(4):467–73. [Epub ahead of print].
33. Zhou J, Ito K, Lall R, Lippmann M, Thurston GD. Time-series analysis of mortality effects of fine particulate matter components in Detroit and Seattle. Environ Health Perspect. 2011;119(4):461–6. [Epub ahead of print] Online 30 Dec 2010.
34. Spira-Cohen A, Chen LC, Kendall M, Lall R, Thurston GD. Personal exposures to traffic-related air pollution and acute respiratory health among Bronx school children with asthma. Environ Health Perspect. 2011;119(4):559–65. PMCID: PMC3080941.

35. Kirrane EF, Luben TJ, Benson A, Owens EO, Sacks JD, Dutton SJ, Madden M, Nichols JL. A systematic review of cardiovascular responses associated with ambient black carbon and fine particulate matter. Environ Int. 2019;127:305–16.
36. Ostro B, Feng WY, Broadwin R, Green S, Lipsett M. The effects of components of fine particulate air pollution on mortality in California: results from CALFINE. Environ Health Perspect. 2007;115(1):13–9.
37. Ostro B, Lipsett M, Reynolds P, et al. Long-term exposure to constituents of fine particulate air pollution and mortality: results from the California teachers study. Environ Health Perspect. 2010;118(3):363–9. Epub 2009 Oct 26.
38. Smith KR, Jerrett M, Anderson HR, et al. Public health benefits of strategies to reduce greenhouse-gas emissions: health implications of short-lived greenhouse pollutants. Lancet. 2009;374(9707):2091–103. Epub 2009 Nov 26.
39. Thurston GD, Burnett RT, Turner MC, Shi Y, Krewski D, Lall R, Ito K, Jerrett M, Gapstur SM, Diver WR, Pope CA. Ischemic heart disease mortality and long-term exposure to source-related components of U.S. fine particle air pollution. Environ Health Perspect. 2016;124(6):785–94. https://doi.org/10.1289/ehp.1509777. Epub 2015 Dec 2.
40. McConnell R, Islam T, Shankardass K, Jerrett M, Lurmann F, Gilliland F, Gauderman J, Avol E, Künzli N, Yao L, Peters J, Berhane K. Childhood incident asthma and traffic-related air pollution at home and school. Environ Health Perspect. 2010;118(7):1021–6.
41. Carlsten C, Dybuncio A, Becker A, Chang-Yeung M, Brauer M. Traffic-related air pollution and incident asthma in high-risk birth cohort. Occup Environ Med. 2011;68:291–5.
42. U.S. EPA. Integrated Science Assessment (ISA) of ozone and related photochemical oxidants (final report, Feb 2013). U.S. Environmental Protection Agency, Washington, DC, EPA/600/R-10/076F, 2013.
43. National Academy of Science (NAS). The airliner cabin environment: air quality and safety. Washington, DC: National Research Council. Committee on Airliner Cabin Air Quality, National Academy Press; 1986.
44. Gauderman WJ, McConnell R, Gilliland F, London S, Thomas D, Avol E, Vora H, Berhane K, Rappaport EB, Lurmann F, Margolis HG, Peters J. Association between air pollution and lung function growth in southern California children. Am J Respir Crit Care Med. 2000;162(4 Pt 1):1383–90.
45. Gauderman WJ, Gilliland GF, Vora H, Avol E, Stram D, McConnell R, Thomas D, Lurmann F, Margolis HG, Rappaport EB, Berhane K, Peters JM. Association between air pollution and lung function growth in southern California children: results from a second cohort. Am J Respir Crit Care Med. 2002;166(1):76–84.
46. Rojas-Martinez R, Perez-Padilla R, Olaiz-Fernandez G, Mendoza-Alvarado L, Moreno-Macias H, Fortoul T, McDonnell W, Loomis D, Romieu I. Lung function growth in children with long-term exposure to air pollutants in Mexico City. Am J Respir Crit Care Med. 2007;176:377–84.
47. Hwang BF, Chen YH, Lin YT, Wu XT, Leo Lee Y. Relationship between exposure to fine particulates and ozone and reduced lung function in children. Environ Res. 2015;137:382–90.
48. Zhang J, Sun H, Chen Q, Gu J, Ding Z, Xu Y. Effects of individual ozone exposure on lung function in the elderly: a cross-sectional study in China. Environ Sci Pollut Res Int. 2019;26(12):11690–5.
49. Mudway IS, Kelly FJ. An investigation of inhaled ozone dose and the magnitude of airway inflammation in healthy adults. Am J Respir Crit Care Med. 2004;169:1089–95.
50. Thurston GD, Ito K. Epidemiological studies of acute ozone exposures and mortality. J Expo Anal Environ Epidemiol. 2001;11:286–94.
51. Levy JI, Carrothers TJ, Tuomisto JT, Hammitt JK, Evans JS. Assessing the public health benefits of reduced ozone concentrations. Environ Health Perspect. 2001;109:1215–26.
52. Stieb DM, Judek S, Burnett RT. Meta-analysis of time-series studies of air pollution and mortality: effects of gases and particles and the influence of cause of death, age, and season. J Air Waste Manage Assoc. 2002;52:470–84.

53. World Health Organization (WHO). Meta-analysis of time-series studies and panel studies of particulate matter (PM) and ozone (O3): report of a WHO task group. Copenhagen: WHO Regional Office for Europe; document no. EUR/04/5042688. 2004. Available: http://www.euro.who.int/document/E82792.pdf [18 November, 2004].
54. Bell ML, Dominici F, Samet JM. A meta-analysis of time-series studies of ozone and mortality with comparison to the national morbidity, mortality, and air pollution study. Epidemiology. 2005;16:436–45.
55. Ito K, De Leon SF, Lippmann M. Associations between ozone and daily mortality, analysis and meta-analysis. Epidemiology. 2005;16:446–57.
56. Levy JI, Chemerynski SM, Sarnat JA. Ozone exposure and mortality: an empiric bayes metaregression analysis. Epidemiology. 2005;16(4):458–68.
57. Gryparis A, Forsberg B, Katsouyanni K, Analitis A, Touloumi G, Schwartz J, Samoli E, Medina S, Anderson HR, Niciu EM, Wichmann HE, Kriz B, Kosnik M, Skorkovsky J, Vonk JM, Dörtbudak Z. Acute effects of ozone on mortality from the "air pollution and health: a European approach" project. Am J Resp Crit Care Med. 2004;170:1080–7.
58. Schwartz J. How sensitive is the association between ozone and daily deaths to control for temperature? Am J Resp Crit Care Med. 2005;171:627–31.
59. Bell ML, Zanobetti A, Dominici F. Who is more affected by ozone pollution? A systematic review and meta-analysis. Am J Epidemiol. 2014;180(1):15–28.
60. Bell ML, Peng RD, Dominici F. The exposure-response curve for ozone and risk of mortality and the adequacy of current ozone regulations. Environ Health Perspect. 2006;114(4): 532–6.
61. Bell ML, Kim JY, Dominici F. Potential confounding of particulate matter on the short-term association between ozone and mortality in multisite time-series studies. Environ Health Perspect. 2007;115:1591–5.
62. Bell ML, Dominici F. Effect modification by community characteristics on the short-term effects of ozone exposure and mortality in 98 U.S. communities. Am J Epidemiol. 2008;167:986–97.
63. O'Neill MS, Loomis D, Borja-Aburto VH. Ozone, area social conditions, and mortality in Mexico City. Environ Res. 2004;94:234–42.
64. Jerrett M, Burnett RT, Pope CA III, Ito K, Thurston G, Krewski D, Shi Y, Calle J, Thun M. The contribution of long-term ozone exposure to mortality. N Engl J Med. 2009;360:1085–95.
65. Lim CC, Hayes RB, Ahn J, Shao Y, Silverman DT, Jones RR, Garcia C, Bell ML, Thurston GD. Long-term exposure to ozone and cause-specific mortality risk in the U.S. Am J Respir Crit Care Med. 2019; https://doi.org/10.1164/rccm.201806-1161OC. [Epub ahead of print].
66. Ebi K, Mills DM, Smith JB, Grambsch A. Climate change and human health impacts in the United States: an update on the results of the U.S. National Assessment. Environ Health Perspect. 2006;114:1318–24.
67. Bell ML, Davis DL, Cifuentes LA, Krupnick AJ, Morgenstern RD, Thurston GD. Ancillary human health benefits of improved air quality resulting from climate change mitigation. Environ Health. 2008;7:41. Review
68. Burtraw D, Toman M. The benefits of reduced air pollutants in the U.S. from greenhouse gas mitigation policies. Washington, DC: Resources for the Future; 1997. Discussion Paper 98-01-REV.
69. McCarthy JJ, Canziani OF, Leary NA, Dokken DJ, White KS, editors. Climate change 2001: impacts, adaptation & vulnerability. Contribution of working group II to the third assessment report of the intergovernmental panel on climate change. Cambridge, UK: Cambridge University Press; 2001.
70. Cifuentes L, Borja-Aburto VH, Gouveia N, Thurston GD, Davis DL. Assessing the health benefits of urban air pollution reductions associated with climate change mitigation (2000–2020): Santiago, São Paulo, Mexico City, and New York City. Environ Health Perspect. 2000;2001(109):S419–25.

71. Last J, Trouton K, Pengelly D. Taking our breath away: the health effects of air pollution and climate change. Vancouver: David Suzuki Foundation; 1988.
72. Wang X, Smith KR. Near-term health benefits of greenhouse gas reductions: a proposed assessment method and application in two energy sectors of China. Geneva: World Health Organization. WHO/SDE/PHE/99.1; 1999.
73. Cao J, Ho MS, Jorgenson DW. "Co-benefits" of greenhouse gas Mitigation policies in China. Washington, DC: Resources for the Future; 2008.
74. Vennemo H, Aunan K, Jinghua F, et al. Domestic environmental benefits of China's energy-related CDM potential. Clim Chang. 2006;75:215–39.
75. Alcamo J, Mayerhofer P, Gaurdans R, et al. An integrated assessment of regional air pollution and climate change in Europe: findings of the IAR-CLIM project. Environ Sci Pol. 2002;5:257–72.
76. van Harmelen T, Bakker J, de Vries B, van Vuuren D, den Elzen J, Mayerhofer P. Long-term reductions in costs of controlling regional air pollution in Europe due to climate policy. Environ Sci Pol. 2002;5:349–65.
77. Working Group on Public Health and Fossil Fuel Combustion. Short-term improvements in public health from global-climate policies on fossil-fuel combustion: an interim report. Lancet. 1997;350:1341–9.
78. Hartunian NS, Smart CN, Thompson MS. The incidence and economic costs of major health impairments: a comparative analysis of Cancer, motor vehicle injuries, coronary heart disease, and stroke. Lexington: Lexington Books; 1981.
79. Miller W, Robinson LA, Lawrence R, editors. Valuing health for regulatory cost-effectiveness analysis. Institute of Medicine (IOM) committee to evaluate measures of health benefits for environmental, health, and safety regulation. Washington, DC: National Academies Press; 2006. http://www.nap.edu/catalog.php?record_id=11534.
80. West JJ, Smith SJ, Silva RA, et al. Co-benefits of global greenhouse gas Mitigation for future air quality and human health. Nat Clim Chang. 2013;3(10):885–9.
81. Thurston GD. Mitigation policy: health co-benefits. Nat Clim Chang. 2013;3:863–4.
82. Committee on Estimating the Health-Risk-Reduction Benefits of Proposed Air Pollution Regulations, National Research Council. Estimating the public health benefits of proposed air pollution regulations. Washington, DC: National Academies Press; 2002.
83. U.S. Environmental Protection Agency (USEPA). Regulatory impact analysis for the final clean air interstate rule. EPA-452/R-05-002. Washington, DC; 2005.
84. Voorhees AS, Sakai R, Araki S, Sato H, Otsu A. Benefits analysis of nitrogen dioxide control programmes: a case-study of Chiyoda-ku, Tokyo. J Environ Plann Manage. 2001;44:149–65.
85. Voorhees SA. Benefits analysis of particulate matter control programs – a case study of Tokyo. J Risk Res. 2005;8:331–29.
86. Filleul L, et al. Twenty five year mortality and air pollution: results from the French PAARC survey. Occup Environ Med. 2005;62:453–60.
87. Miller W, Robinson LA, Lawrence RS, editors. Air quality valuation model documentation, for Health Canada. Boulder: Stratus Consulting; 1999.
88. Bickel P, Friedrich RE. ExternE, externalities of energy, methodology 2005 update. Luxemburg: European Commission; 2005.
89. Khreis H, Kelly C, Tate J, Parslow R, Lucas K, Nieuwenhuijsen M. Exposure to traffic-related air pollution and risk of development of childhood asthma: a systematic review and meta-analysis. Environ Int. 2017;100:1–31.
90. Garcia E, Berhane KT, Islam T, McConnell R, Urman R, Chen Z, Gilliland FD. Association of changes in air quality with incident asthma in children in California, 1993–2014. JAMA. 2019;321(19):1906–15.
91. Stern N. The economics of climate change: the Stern review. New York, NY: Cambridge University Press; 2007.
92. Scovronick N, Budolfson M, Dennig F, Errickson F, Fleurbaey M, Peng W, Socolow RH, Spears D, Wagner F. The impact of human health co-benefits on evaluations of global climate policy. Nat Commun. 2019;10(1):2095.

Chapter 10
Asthma, Hay Fever, Pollen, and Climate Change

Anthony Szema, Jonathan Li, Ashlee Pagnotta, Malvika Singh, and Jo' Ale White

Climate change, if present, is associated with atmospheric warming—so-called "global warming"—as well as volatility in weather patterns, leading to more severe winters at a given latitude (since cold air typically further north in latitude is pushed south) and hotter summer months, when the Earth is closer to the Sun [1]. Hot

A. Szema (✉)
Three Village Allergy & Asthma, PLLC, South Setauket, NY, USA

Northwell Health International Center for Excellence in Deployment Health and Medical Geosciences, Manhasset, NY, USA

Department of Occupational Medicine, Epidemiology, and Prevention at Zucker School of Medicine at Hofstra/Northwell, Hempstead, NY, USA

Donald and Barbara Zucker School of Medicine at Hofstra/Northwell, Hempstead, NY, USA

Columbia University Global Psychiatric Epidemiology Group CDC NIOSH U01 OH011308-01 9/11, Trauma and Toxicity in Childhood: Longitudinal Health and Behavioral Outcomes, New York, NY, USA

Faculty Committee on Health Professions (Pre-Med Committee) Stony Brook University, Stony Brook, NY, USA

RDS2 Solutions, Inc., Stony Brook, NY, USA

Division of Pulmonary and Critical Care, Northwell Health, Manhasset, NY, USA

Division of Allergy/Immunology, Northwell Health, Manhasset, NY, USA

Department of Occupational Medicine, Epidemiology and Prevention, Donald and Zucker School of Medicine at Hofstra/Northwell, Hempstead, NY, USA

J. Li
Three Village Allergy & Asthma, PLLC, South Setauket, NY, USA

Sidney Kimmel Medical College at Thomas Jefferson University, Philadelphia, PA, USA

A. Pagnotta · M. Singh · J. A. White
Three Village Allergy & Asthma, PLLC, South Setauket, NY, USA

Stony Brook University, Stony Brook, NY, USA

K. E. Pinkerton, W. N. Rom (eds.), *Climate Change and Global Public Health*, Respiratory Medicine, https://doi.org/10.1007/978-3-030-54746-2_10

weather generates more pollen from plants [2]. Cold weather is associated with asthma emergency room visits in New York City [3]. More pollen causes more disease, not only allergically induced but also non-allergic, since ragweed, for example, produces more reactive oxygen species (ROS), so it may produce inflammation, leading to upper and lower respiratory tract symptoms, even in those persons without allergic asthma, rhinitis, or conjunctivitis [4]. Clean Air Act enforcement may reduce sources of anthropogenic heat [5].

Does climate change have two potentially deleterious effects on human health: (1) prolonged and more severe pollen seasons, leading to (2) worsened asthma and allergies? If true, these downstream consequences may pose significant risks in terms of patient care costs, lost time from work, morbidity, and, possibly, mortality.

1. *For the first question, supporting the possible concept of prolonged and more severe pollen seasons from hot weather, the duration of ragweed pollen season has been increasing as a function of latitude in North America*—associated with delay in first frost by 27 days and lengthening of the frost-free period at latitudes above 44°N since 1995 [2]. In Turkey, daily mean temperature and levels of sunshine are associated with more severe pollen counts [6].

P. J. Beggs has reviewed work of other investigators, who have noted an association between *increases in carbon dioxide (CO_2) concentration in ambient air and increases in pollen*, even irrespective of temperature. Ziska and Caulfield determined that ragweed pollen (Ambrosia *artemisiifolia* L.) production increased from preindustrial times to the present. Wayne identified a twofold increase in atmospheric CO_2 concentration led to a significant increase in ragweed pollen production.

Ziska noted a CO_2 temperature gradient between rural and urban areas such that the higher CO_2 concentration and air temperature of the urban area resulted in ragweed in air at higher concentrations. In another study, Ambrosia taxa actually decreased, while concentrations of Juniper tree pollen (Juniperus), Quercus, Carya, and *Betula* (birch tree pollen) increased. Speiksma studied *Betula* pollen in five European cities from 1961 to 1193 and found slightly rising trends over this time. Teranishi found that over a 15-year period from 1983 to 1998, Japanese cedar pollen (*Cryptomeria japonica*) significantly correlated between total pollen count in a year and temperature in July the previous year [7].

Not only do increased pollen counts provoke allergic disease, but also the *potency or allergenicity* of pollen is concerning. Birch pollen grown at two temperatures differing by 1.1 °C yielded significantly stronger allergenicity in pollen from trees grown at higher temperatures. Hjelmroos found that heterogeneity of antigenic proteins was more diverse in pollen from the south side of trees, supporting the concept that higher temperature from the south side of trees may modulate this phenomenon.

Longer grass pollen seasons with earlier start dates have been associated with increases in cumulative temperatures over 5.5 ° C during winter–early spring (January March). Emberlin showed that start dates of the birch pollen season advance 6 days over 10 years, for birch pollen, based on changes in spring temperatures in four out of six sites in Europe.

In Italy, from 1981 to 2000, temperature warming was associated with an earlier initiation of the pollen season. In particular, a plant family called Urticaceae had prolongation of its pollen season—critically important, since this is clinically significant in that region. A World Health Organization report concluded that an earlier start and peak of the pollen season is more pronounced in species that start flowering earlier in the year—and the duration of the season is extended in some summer and late flowering species.

In North America, earlier start dates for juniper trees (*Juniperus*) and related taxa *Ulmus* and *Morus* have been studied. Actually, an earlier start time was associated with increasing winter temperatures. Other studies for the Japanese cedar (*C. japonica*) have noted the first date of the pollen season advanced from 1983 to 1998, from mid-March to late February, according to the mean February temperature. Ziska found that higher CO_2 concentrations and air temperature of the urban area resulted in earlier ragweed seasons, compared to rural areas.

Heat may change plant and pollen distribution at a given latitude. Predictions of extending the northern limit of birch by several hundred kilometers and increasing the altitudinal tree line have also been modeled with contraction of the distribution in the south. *Plantago lanceollata*, a common allergen producer, benefited from more abundance after experimental studies of climate, soil, fertility, and disturbance, though other species declined or became extinct.

A recent report found that the duration of the ragweed (*Ambrosia* spp.) pollen season has been increasing in recent decades as a function of latitude in North America. These latitudinal effects leading to increasing season length were associated with a delay in the first frost of the fall season and lengthening of the frost-free period. A significant increase in the length of the ragweed pollen season was found between 13 and 27 days at latitudes above 44°N since 1995.

These data support the Intergovernmental Panel on Climate Change Projections, which notes enhanced warming is a function of latitude. Greater exposure times to seasonal allergens may therefore occur with subsequent effects on human health. For example, 10% of the U.S. population is estimated to be ragweed sensitive. As an explanation for the increased prevalence of allergic disease worldwide over the past 30 years, ragweed is an importnat factors, and climate change is a plausible etiologic agent.

Shoot growth, water use efficiency, and phenological phases (leaf unfolding, needle flush, flowering) are other plant attributes potentially affected by warming. Increased CO_2 concentration near perennial ryegrass seedlings leads to increased shoot growth and increased biomass. In addition, Lindroth showed that carbon-to-nitrogen ratios, or C:N, as well as starch concentrations and condensed tannin, of paper birch significantly increased in response to increased CO_2. Other variables that increase include below-ground mass, carbon, nitrogen, hexose sugar, gas exchange properties, water use efficiency, and total mass.

The growing season can start earlier with warmth. In Europe, *B. pubescens* and *Quercus robur* have such phenological phases: (1) leaf unfolding, (2) needle flush, and (3) flowering spring events. These advanced by 6.3 days, while autumn events were delayed by 4.5 days, resulting in a longer growing season lengthening by

10.8 days since the 1960s. This has been called the “anthropogenic greenhouse effect.” Other studies have shown that elevated CO_2 concentration decreased seed weight, increased germination percentage and rate, and increased seedling size for the progeny of *P. lanceolata*. Both higher CO_2 concentration and air temperature of the urban area led to ragweed plants, which grew quicker and generated more above-ground biomass than rural areas.

While this chapter has a focus on pollen, it is possible that other sources of aeroallergens, such as cockroach (*Blattidae*), may proliferate in the presence of higher temperatures, since they would be able to survive passage between buildings, thwarting extermination efforts [7].

2. *Our research group determined that for the latter question—cold weather and asthma—atmospheric temperature indeed impacts New York City asthma emergency room visits.* Temperature data were recorded in Central Park from 2000 to 2007. Particulate matter (PM2.5) air pollution data—the concentration of 2.5 μm-sized particles per cubic meter of air—were collected from the Bronx, Queens, and Manhattan from 2006 to 2009. Pollen counts were measured in Brooklyn in 2008. We obtained New York City asthma emergency room visit data from nyc.gov from 2000 to 2007. Relations among these data were determined based on correlation coefficients. There was a reverse relationship between asthma indicators and temperature; that is, *extremely low temperature was associated with higher asthma discharge rates in the Bronx* [3].
3. *Does more pollen cause more disease?*

Prolonged pollen seasons may increase the duration of human exposure to aeroallergens and may increase the risk of allergic sensitization. In persons with allergic disease, a longer pollen season may increase the duration of allergy symptoms. Higher concentrations of atmospheric pollen may also increase the severity of allergic symptoms [2].

Ragweed (known as *Artemisia* species) pollen represents a major cause of allergy in Central Europe. Variations in the pollen season, the influence of climate variables, and the prevalence of pollinosis were analyzed in Ponzan, western Poland, between 1995 and 2004. The Artemisia species pollen season grew longer due to a clear advance in the starting day and only a slightly earlier endpoint; the peak day also came slightly earlier. *Temperature was directly correlated with daily Artemisia species pollen levels; relative humidity was inversely correlated.* Figure 10.1 shows that ragweed pollen counts increase in season and are predicted to increase over time, supporting the possibility that more pollen will cause more disease in the future.

Twelve percent of patients had a positive skin prick test reaction to Artemisia species. Their symptoms were rhinitis and conjunctivitis (15%), atopic dermatitis (15%), chronic urticaria (14.3%), bronchial asthma (2.4%), and facial and disseminated dermatitis (1.3%). Chronic urticaria, though present in this series, likely was unrelated to seasonal pollen. Elevated specific IgE concentrations were detected

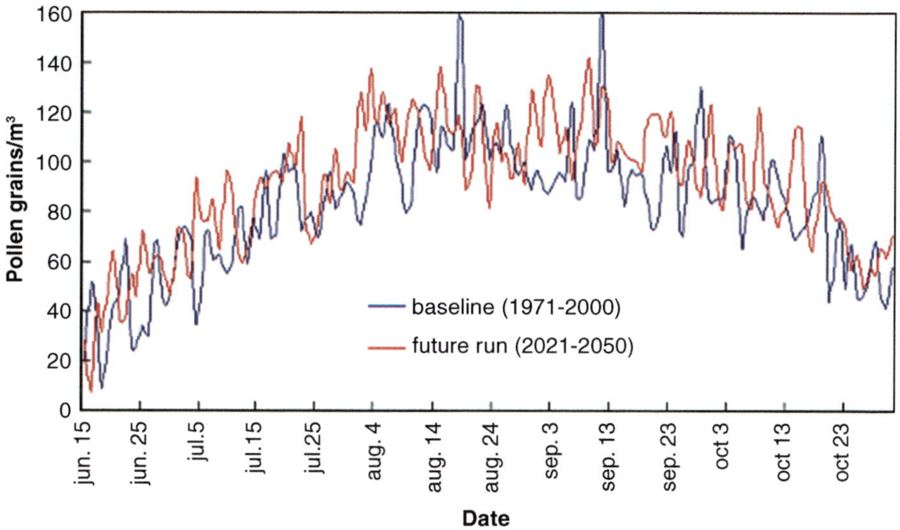

Fig. 10.1 Ragweed pollen counts over time in Europe. Ragweed pollen counts increase in season and are predicted to rise in Europe over time. The hotter summer months in August/September are associated with more release of pollen. From the Climate Change and Variability: Impact on Central and Eastern Europe website coordinated by the Max Planck Institute of Meteorology. http://www.clavier-eu.org/?q=node/880. (Used with permission from Dr. Pálvölgyi et al. [81] CLAVIER)

in the sera of 10.1% patients. Pollen season intensity was also found to be highly influenced by rainfall in the previous weeks. Trends toward earlier season starts and longer duration, possibly caused by climate change, may have had an impact on this allergic Polish population [4].

Another study relates geo-climate effects on asthma and allergic diseases in adults in Turkey (PARFAIT study). Evaluation of 25,843 questionnaires from parents of 25,843 primary schoolchildren in 14 cities indicated that mean annual temperature was significantly associated with the prevalence of asthma and wheezing in both genders. Eczema and temperature were associated in female subjects. Asthma in women was associated with mean annual humidity in the air. Annual number of days with snow was associated with wheezing [8].

In Japan, cypress and cedar plantations account for one-fourth of the population suffering from hay fever in the spring. Kouji Murayama, quoted in *Nature* (vol 43, April 28, 2005, p. 1059), points to global warming as linking summer temperatures to the amount of pollen produced the following spring and that these data already provide the basis for pollen forecasts. Tokyo's average yearly temperature has increased by 3 °C since 1890 and is predicted to rise up to 3.5 °C by the end of the century. If this is indeed the case, then it is possible that the number of hay fever sufferers will rise by 40% by the year 2050. Thus, global warming has the potential to magnify an already entrenched, important health problem in Japan.

Global warming may be additive with higher levels of industrial carbon dioxide and diesel exhaust. However, even economic factors may intensify the problem, since unmaintained cedar and cypress plantations allow trees to mature to their prime pollen-producing age. A solution would be to replace these pollen-producing trees with pollen-free cypress and cedar, an approach that may take decades to implement.

Pollen types are temporally related seasonally. In the northeast United States, tree pollen sheds in the spring, grass pollen is released during summer, and weed typically is disbursed in late summer (classically taught as August 15, especially with ragweed). Figure 10.2 shows seasonal distribution of pollen in the United States.

These large pollen grains, about 5 μm in size or larger, are deemed too small to be respirable and rather deposit in the ocular conjunctiva to cause allergic conjunctivitis—watery, itchy, red eyes are sequellae. These pollen grains also contact the nasal mucosa and trigger allergic rhinitis or hay fever via an IgE-mediated mechanism, in those allergically sensitized.

For ragweed pollen, even in those not allergically sensitized, reactive oxygen species are produced to incite inflammation. Runny, itchy nose; postnasal drip; repetitive sneezing; stuffiness/congestion; and dry cough are cardinal symptoms.

Physical exam signs related to histamine release include allergic shiners (dark, puffy eyes from histamine release), Dennie-Morgan Lines (lines below the eyelid from histamine release), nasal crease (from rubbing the nose in an upward fashion leading to bent cartilage in the nose), and the nasal salute (rubbing nose vertically in an inferior to superior direction) (Fig. 10.3). Since pollen grains are too large to be respirable, they do not directly reach the bronchi. However, pollen-induced asthma does occur and manifests late in the season and after it ends.

For grass, in particular, the English have noted "thunderstorm asthma" when respirable particles become airborne during gusts of wind. The reason for the lack of immediate asthma symptoms may be the location of allergens in pollen. Important allergens are on the outside of the cell membrane called the exine. They are actually not produced by the pollen cell itself but are "stuccoed" onto the exine by other cells of the male flower. Considerable amount of allergens remain behind for weeks after pollen is shed. Also, allergens extracted from pollen by raindrops may lead to airborne dust particles after drying. So, asthma symptoms may begin after hay fever symptoms and persist longer [9].

Figure 10.4 describes the cascade of pollen inhalation leading to allergic asthma. Pollen is inhaled and the protein antigens in pollen are engulfed by antigen-presenting cells, such as the macrophage, which degrades protein into peptides. The peptides are shuttled to the macrophage surface and presented to an activated T cell in the context of major histocompatibility complex type II. The T cell, when activated, engages a B cell via CD40 ligand (also known as CD154) interactions with CD40, using accessory molecules CD80 on the T cell and CD86 on the B cell.

Depending on the cytokine environment near these cells, for example, if IL-4 is present, then the B cell will differentiate into a plasma cell and class switch from IgM to IgE in order to make immunoglobulin molecules IgE, the allergic antibody. These IgE allergic antibodies bind to IgE receptors in mast cells, and when two IgE molecules are in close proximity, they dimerize and engage the mast cell to release its content of pro-inflammatory pre-formed mediators, such as histamine,

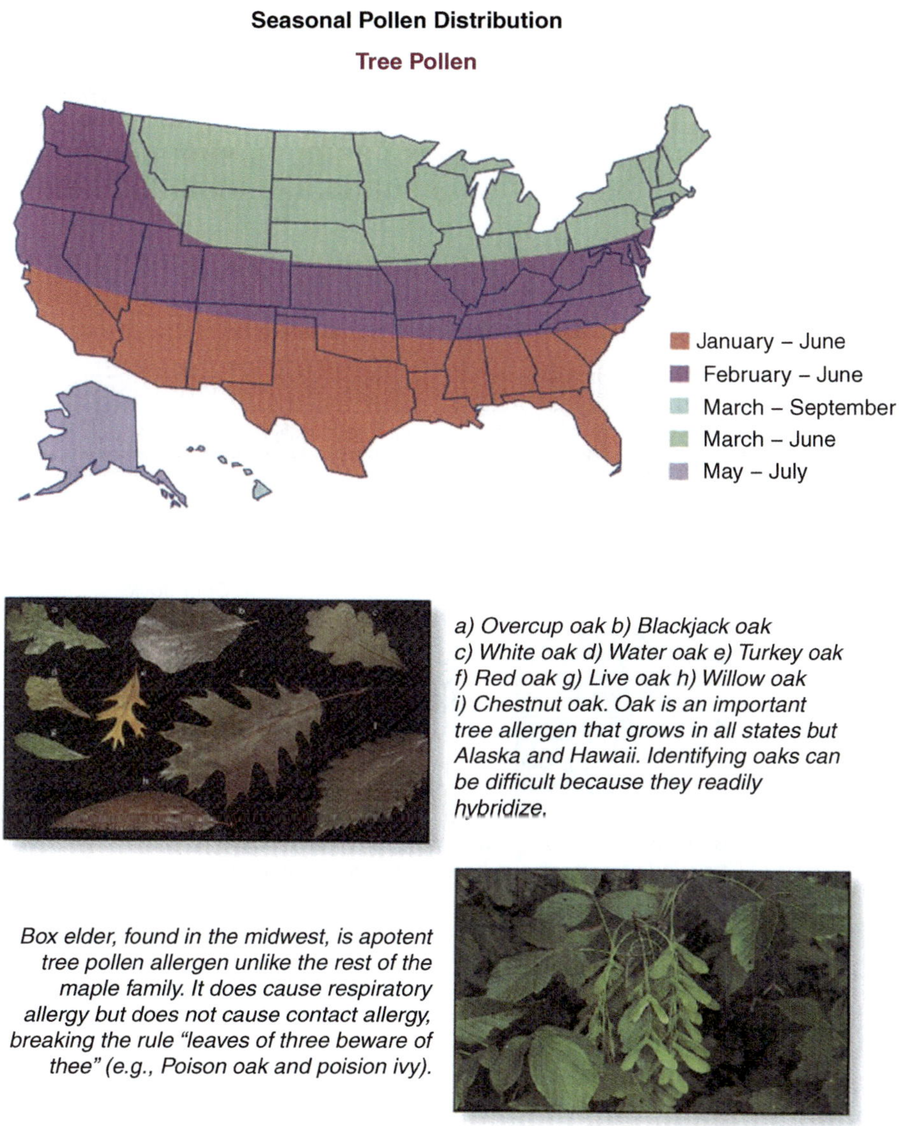

Fig. 10.2 Distribution of pollen in the United States. Seasonal pollen distribution graphics from Jelks [82]. (Used with permission from Mary L. Jelks, MD, FAAAAI)

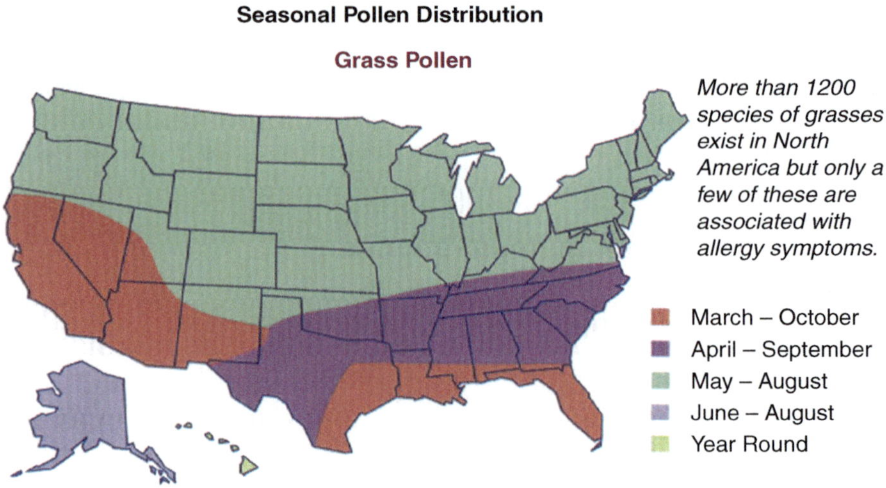

Three examples of grasses that cause allergy:

Timothy Grass and Timothy grass bloom

Sour dock

Bermuda grass can grow in the open or in beauity traveled area as shown here in cracks of side walk.

Fig. 10.2 (continued)

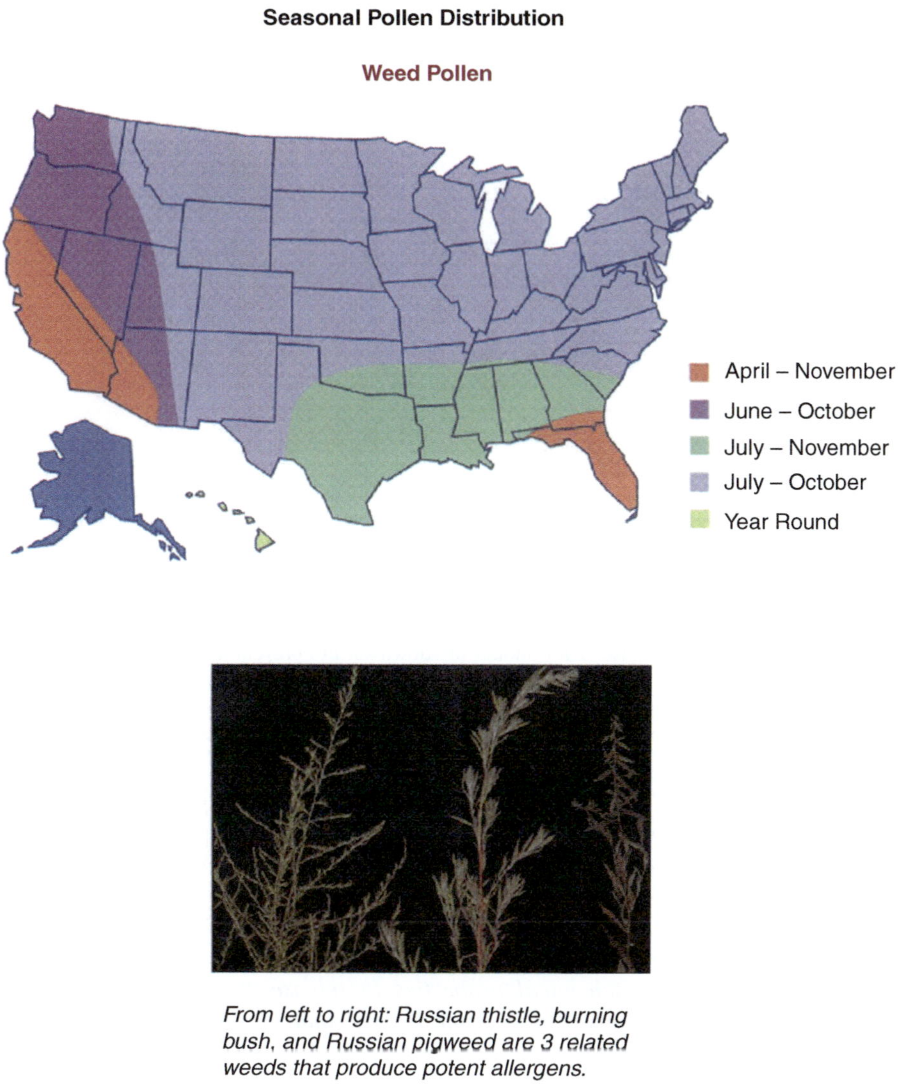

Fig. 10.2 (continued)

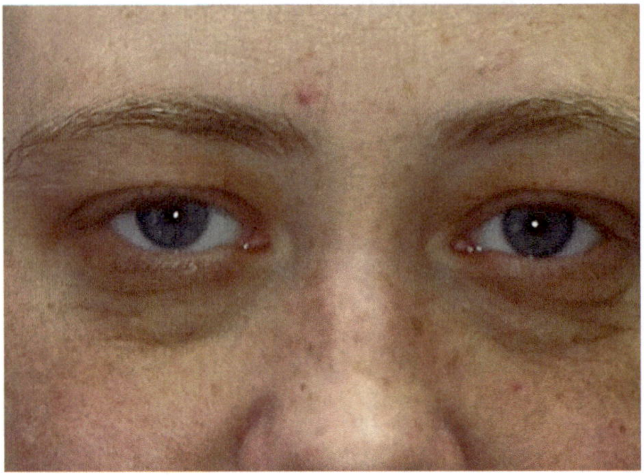

Fig. 10.3 Cardinal physical examination signs of atopic disease. Allergic shiners in a 34-year-old pregnant woman. The swollen dark eyelids are from histamine release. The patient appears tired despite many hours of sleep. The physical appearance makes her appear older than her stated age. Dennie-Morgan lines are horizontal lines across eyelids. A nasal crease is the horizontal band across the bridge of the nose. The nasal crease is caused by upward rubbing of the nose in itchy patients. The physical act of upward rubbing the nose is called the "allergic salute," which is responsible for the nasal crease. They should rub downward to prevent this permanent sign. This patient was admitted with throat closure and uvular swelling after inhaling hyacinth pollen at her house at the onset of the spring vernal equinox. Photo of patient, who gave permission, was taken by medical assistant Shauna McCleary from the Stony Brook Allergy & Asthma practice

which causes clinical airway constriction and gastrointestinal symptoms and vascular inflammation. Generation of pro-inflammatory cytokines, such as IL-5, will recruit the eosinophil allergic cell to release its hydrogen peroxide, which damages airway epithelium.

Is there an additional effect of non-allergenic air pollution acting in concert with aeroallergens?

Ragweed has physicochemical properties to release reactive oxygen species (ROS) to cause disease; a two-hit hypothesis may come into play for those allergic to ragweed. For those not sensitized, ROS may play a role solely; for those with allergies, ROS plus IgE-mediated disease would enhance the inflammation [10].

Changes in production, dispersion, and allergen content of pollen and spores, both region- and species specific, may have been influenced by *urban air pollutants interacting directly with pollen* [11].

While the incidence of allergy and asthma appears to be increasing worldwide, residents of urban areas more frequently experience these conditions than rural dwellers. Outdoor air pollution concentrations result from intense energy consumption and exhaust emissions from automobiles.

Urban air pollution is a serious public health hazard. Laboratory studies have confirmed epidemiologic evidence that air pollution adversely affects lung function in asthmatics. Damage to airway mucous membranes and impaired mucociliary

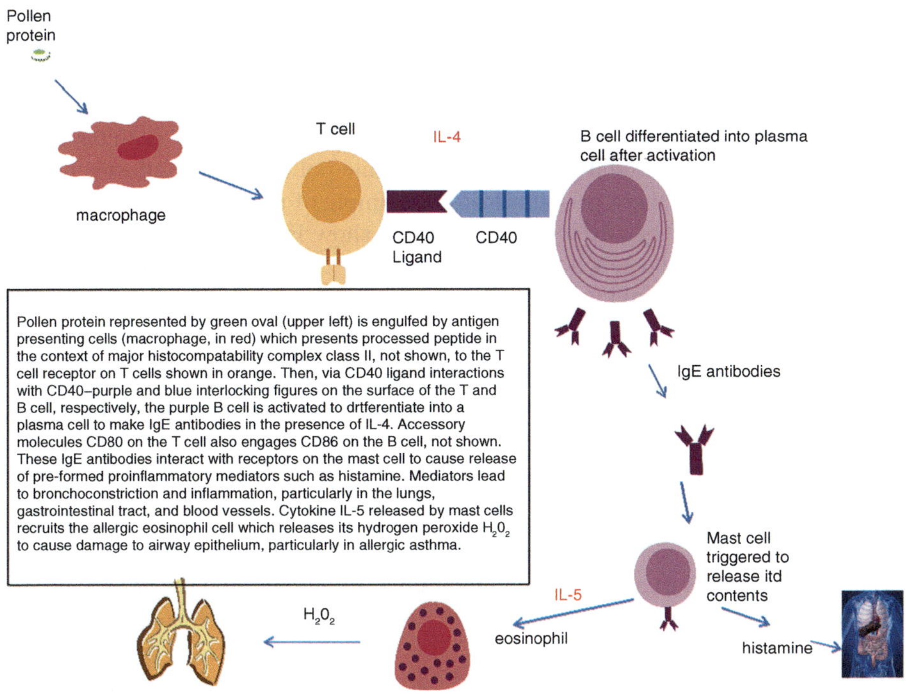

Fig. 10.4 The allergic cascade simplified: from pollen inhalation to disease

clearance caused by air pollution may facilitate access of inhaled allergens to the immune cells in the airway, thus *promoting sensitization of the airway*. Consequently, *a more severe allergic antibody (immunoglobulin IgE-mediated) response to aero-allergens and airway inflammation could account for increasing prevalence of allergic respiratory diseases in polluted urban areas.*

The most abundant components of urban air pollution entail high levels of vehicle traffic with airborne particulate matter called PM10 and PM2.5, nitrogen dioxide, and ozone [5]. Diesel exhaust is particularly troublesome, since it increases the production of allergic IgE antibodies [12]. Ozone levels have been modeled to track asthma emergency room visits and are predicted to be associated with increased pediatric emergency room visits for asthma for the next decade. Changing levels of ozone could lead to a 7.3% increase in asthma-related emergency room visits by children, ages 0–17.

This asthma and ozone study, led by Perry Sheffield, MD at Mount Sinai School of Medicine, used regional and atmospheric chemistry models. Regional climate and air quality information was linked to New York State Department of Health records of pediatric, asthma-related emergency room visits in 14 counties that are part of the New York City metropolitan area. They simulated ozone levels for June through August for five consecutive years in the 2020s and compared them with 1990s levels. They then determined a median increase of 7.3% in ozone-related asthma emergency department visits, with increases ranging from 5.2% to 10.2% per county [13].

If the earth's temperature is increasing—from fossil fuel combustion and greenhouse gas emissions from energy supply, transport, industry, and agriculture—then climate change altering the concentration and distribution of air pollutants, and interfering with the seasonal presence of allergenic pollens in the atmosphere, will significantly prolong these periods [14]. An example of melting of even glacial ice over time is seen in Figs. 10.5 and 10.6, which show Hubbard Glacier, Alaska, in 1986, the last year ice reached "the gap" to land. In 2011, the gap was wide and ice floes were melting, shrinking the size of the glacier.

4. *The Clean Air Act gives Americans the opportunity to attenuate anthropogenic climate change, like industrial air pollution, thereby alleviating a man-made scourge of heat-induced increased aeroallergen concentrations.* I testified before Congress about the need to fund the EPA and the Clean Air Act, and in this chapter, based on a letter published in the January 2012 issue of the *Journal of Occupational and Environmental Medicine*, I re-affirm my position [5, 15].

As an update to our original chapter, the following section focuses on climate change with respect to hurricanes and their effects on aspects on respiratory health, such as asthma and allergy to mold.

Fig. 10.5 Hubbard Glacier Alaska circa 1986. In 1986, Hubbard Glacier, Alaska, squeezed the passage between Russell Fiord (background) and Disenchantment Bay (foreground) in this photo taken the last time Hubbard "galloped" and closed the passage. (This photo was downloaded from the US Forest Service public website. http://www.fs.fed.us/r10/tongass/forest_facts/photogallery/hubbard_photos.html)

Fig. 10.6 Hubbard Glacier Alaska, July 2011. Hubbard Glacier photo taken by Dr. Anthony Szema, July 2011, aboard the MS Westerdam, Holland America Line. Note the melting ice floes. There is a gap between the glacier and land to the right

Climate Change, Hurricanes, and Respiratory Health

In 2018, the Intergovernmental Panel on Climate Change (IPCC) released a special report summarizing the global impact of continued climate change. It estimated that current global temperatures rose approximately 1.0 °C (range of 0.8–1.2 °C) since pre-industrial times[1] and predicted that if global warming continues along the current trajectory, it will reach an increase of 1.5–2.0 °C [16]. This rise in temperature leads to air holding more moisture, which in turn causes an increase of allergen concentration [17]. These increases in global temperatures will impact human health through a myriad of mechanisms. For example, regions suffering from water scarcity are projected to experience even greater severity, leading to widespread famine, health effects of dehydration and heat exposure, large-scale migration, and political unrest. Risk of exposure to vector-borne diseases (i.e., malaria, tick-borne illness, Zika, dengue fever) will increase as warming temperatures allow greater geographical distribution of disease-harboring organisms. Coastal regions will be impacted by

[1] Pre-industrial: The multi-century period prior to the onset of large-scale industrial activity around 1750. The reference period 1850–1900 is used to approximate pre-industrial global mean surface temperature [16].

rising sea levels and increasing severity of weather catastrophes [16, 18]. This water exposure promotes the growth of mold, which may impact asthma and allergies in susceptible patients [19–22].

Hurricanes are one such weather catastrophe predicted to worsen with increasing climate change. In the North Atlantic seas alone, the World Climate Research Program estimates that the number of extreme hurricanes will increase annually by over 50% for every 1 °C increase in surface temperature [23]. Some models predict that this degree of temperature change will spur a two- to sevenfold increase in the number of Hurricane Katrina–sized events [24]. The amount of precipitation generated by each hurricane has also increased as a result of climate change [25]. While some studies suggest that no appreciable trend has been observed in hurricane landfall frequency or intensity from 1900 to 2017 [26], multiple other reports suggest that the intensity, frequency, and duration of hurricanes affecting North America have been increasing since the early 1980s [27–29]. Figure 10.7 depicts the long-term trend in Atlantic tropical cyclone formation. These changes are projected to worsen due to human activity related to carbon dioxide production, vehicle emissions, and particulate matter air pollution [30]. In addition, present-day hurricanes have caused increasingly more damage due to population growth and city wealth

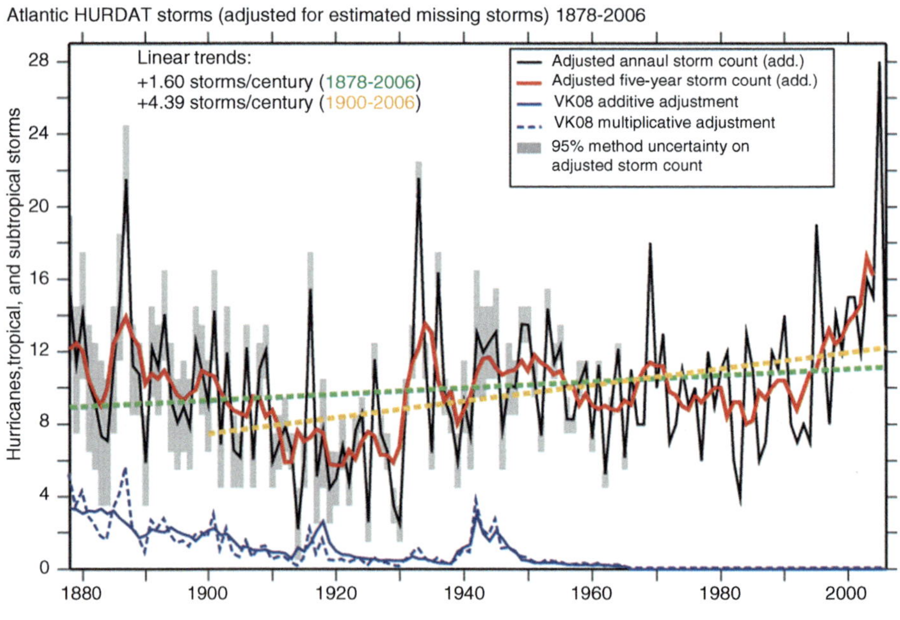

Fig. 10.7 Prior to the advent of satellite tracking of tropical cyclones, these were only opportunistically observed while at sea. As a result, the number of tropical cyclones was systematically underestimated. This chart is an attempt to look at the long-term trend in Atlantic tropical cyclone formation while adjusting for lack of data on storms that never struck land. (*This image is in the public domain because it contains materials that originally came from the U.S. National Oceanic and Atmospheric Administration, taken or made as part of an employee's official duties* via https://upload.wikimedia.org/wikipedia/commons/1/1b/Adjusted_Tropical_Storm_Count.png)

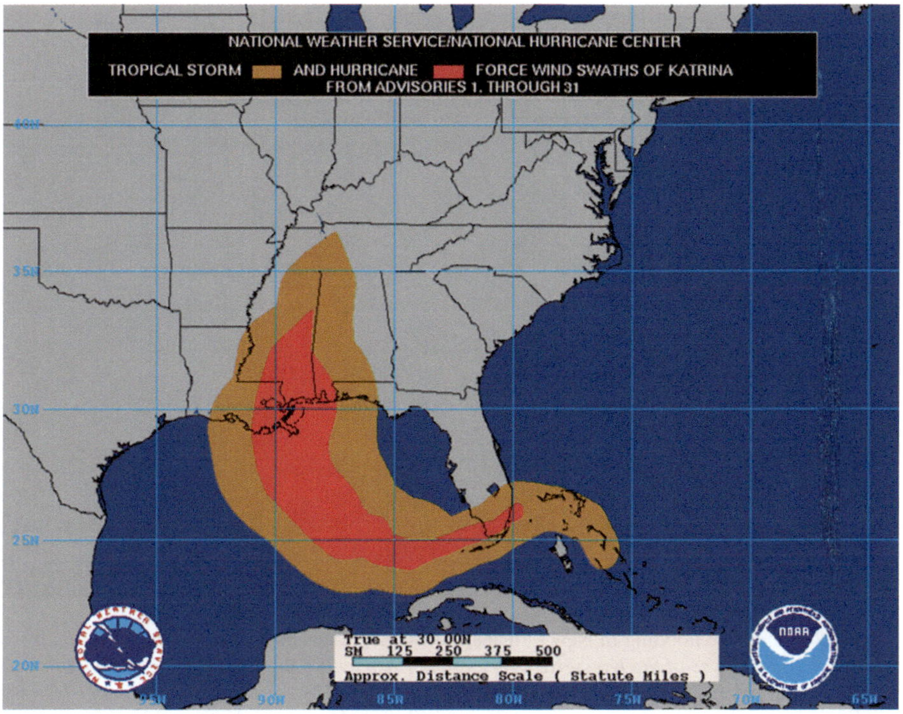

Fig. 10.8 The path of Hurricane Katrina, abstracted from satellite imaging data. (*This image is in the public domain because it contains materials that originally came from the U.S. National Oceanic and Atmospheric Administration, taken or made as part of an employee's official duties*— http://www.nhc.noaa.gov/archive/2005/graphics/AT12/31.AL1205S.GIF)

[26]. Below, we will review the impact of three major North American hurricanes from the last decade and their impact on human respiratory health, such as asthma and allergies.

Hurricane Katrina (2005)

Hurricane Katrina was the deadliest hurricane to hit the United States since 1928 [31]. This Category 5[2] hurricane made landfall in August 2005 and affected southern Florida, eastern Texas, and, most notably, New Orleans, Louisiana [31]. Figure 10.8 illustrates the path of Hurricane Katrina and where landfall was first made. Katrina caused catastrophic damage with an approximate death toll of 1200 people (1000 in Louisiana) and an estimated $75 billion in damage in New Orleans alone [31–33]. Intense wind speeds, storm surge, levee failures, and heavy rainfall

[2] 157 mph or higher (Saffir-Simpson Wind Scale).

Fig. 10.9 Extensive flooding in New Orleans, Louisiana, after Hurricane Katrina. (*This image is a work of an employee of the Executive Office of the President of the United States, taken or made as part of that person's official duties. As a work of the US federal government, the image is in the public domain.*—https://commons.wikimedia.org/wiki/File:Hurricane_Katrina_Flooding.jpg)

resulted in catastrophic flooding (depicted in Fig. 10.9), which subsequently caused hazardous environmental exposures, including indoor mold, chemicals, and infectious pathogens.

Mold

Following Katrina, the resulting damp, humid environment fostered widespread fungal growth in damaged homes. Indoor mold exposure increases the risk for upper respiratory tract symptoms, such as throat irritation, coughing, wheezing, nasal congestion, and irritation of the eyes and skin [17]. In October 2005, the Centers for Disease Control (CDC) conducted a thorough investigation of 112 homes. They observed visible mold in 44% of homes; 16% had heavy mold growth and 28% had light growth. A visual assessment of water damage in the homes determined the intensity of mold growth. When these data were extrapolated to the entire affected New Orleans area, they estimated 194,000 homes experienced visible mold growth and 70,000 homes had heavy mold growth [34]. Across multiple studies, the predominant fungal species identified in flood-affected homes included: *Aspergillus, Penicillium, Trichoderma,* and *Paecilomyces* [34, 35]. Exposure to these fungi may cause a variety of symptoms (Table 10.1) and can increase the likelihood of

Table 10.1 [45–49] Lists fungi that were found in the homes affected by Hurricane Katrina, according to the CDC [34].

Fungi	Description	Symptoms
Aspergillus	Genus comprises several hundred species. These fungi can cause diseases, including localized infections, allergic responses, and allergic bronchopulmonary aspergillosis (ABPA)	Wheezing, coughing, chest pain, and fever
Penicillium	Responsible for decomposing organic matter and inducing rot in food. Some *Penicillium* species are common indoor aeroallergens	Weight loss, diarrhea, dyspnea, swollen lymph nodes, abdominal pain and fever
Trichoderma	Present in soils, this *ascomycete* can be found in various environments. This genus is known to have symbiotic relationships with plants, protecting them from toxic chemicals. Several species are reported to be human pathogens	Exacerbation of asthma, sinusitis, urticaria, blocked nose, rhinitis, otitis, hoarseness, ache in joints, myalgia, and tiredness
Paecilomyces	Heat-resistant fungi can deteriorate foods, paper, and grain. They produce mycotoxins. This genus can be found in acidic environments and is often recovered from the air and soil	Itching, sweating, fever, fatigue, and dyspnea
Mucor	Mainly soilborn but can be found in decaying organic matter, compost, or rotten wood. This genus can cause a fungal infection known as *Mucormycosis,* particularly in diabetics	Fever, coughing, chest pain, dyspnea, nausea and vomiting

experiencing asthmatic and allergic symptoms [36]. Air sampling studies inside flood-affected homes have detected mold, endotoxin, volatile organic compounds (VOCs), and (1 → 3)-β-D glucan.[3] [35, 38, 39] Mold and endotoxin levels may surpass those found in commercially mold-rich environments (wastewater and agriculture) [35]. These levels decreased significantly after cleaning and personal protective equipment (PPE) use, which significantly decreased inhalational exposure [35]. Thus, cleanup workers and returning residents were at high risk for exposure if not properly protected. In *Drosophila*, VOCs generated from post-Katrina fungal isolates of *Aspergillus*, *Mucor*, *Penicillium*, and *Trichoderma* are embryotoxic and cytotoxic, causing developmental defects and cell death [38]. Exposure to microbial VOCs during childhood are associated with an increased risk for asthma [40]. Post-Katrina mold components have also been studied in relation to particulate size, specifically (1 → 3)-β-D glucan and endotoxin. Small (<1.8 μm) particulates from flood-water-affected materials were associated with (1 → 3)-β-D glucan and particulates >1.8 μm were associated with endotoxin [39]. Particulates less than 2.5 μm are able to penetrate deep into the distal airways. Exposure to β-D glucan is associated with fungal sensitization, airway hyperresponsiveness, asthma severity, and resistance to steroid therapy [41–43]. Cleanup efforts have been shown to decrease levels of detectable mold; however, mycotoxins persist beyond 2 years post cleanup and do not correlate with mold levels. Therefore, post-hurricane cleanup efforts

[3] A fungal cell wall component with known toxic and inflammatory effects [37].

warrant the use of PPE long after the initial event, and continued air filtration is necessary even after mold has been eliminated [44].

After Hurricane Katrina, a number of population studies were conducted to assess the clinical impact of widespread mold growth. Rusiecki et al. studied mold exposure among 2834 U.S. Coast Guard personnel. They found that 20.1% ($n = 595$) reported mold exposure. Of those with mold exposure, 90% reported mild respiratory problems and 10% reported severe respiratory symptoms. Sinus infection occurred among 12% of responders, and a positive odds ratio of 10.39 was found between mold exposure and sinus infection. Mold exposure was the main reason responders sought medical treatment [50]. This study highlights a preventable cause of morbidity among rescue workers. PPE is essential for these personnel. The Head-off Environmental Asthma in Louisiana (HEAL) study, started in 2007, was conducted to characterize post-Katrina exposures to mold and allergens in New Orleans children with asthma. Among 182 children studied, over half (62%) were living in water-damaged homes. Of the antigens detected, *Alternaria* was detected in almost all homes (98%), with more than half (58%) having concentrations >10 μg/g. Other allergens (mouse, cockroach, dustmite, (1 → 3)-β-D glucan, endotoxin) detected in this study were low in comparison to similar studies [29]. Despite cleanup efforts, mold has been a persistent problem in homes of asthmatic children, even in homes not damaged by water. The long-term impact of these exposures on atopy and asthma morbidity are unknown. Finally, a unique case of concurrent *Aspergillus fumigatus* and mucor infection occurred in a cardiac transplant patient following Katrina. It was suspected that the post-Katrina environment and cleaning efforts contributed to this dual invasive mold infection [51]. Practitioners should be aware of the increased risk for invasive fungal infection among immunocompromised patients following hurricanes due to elevated microbial load and airborne distribution.

Volatile Organic Compounds

In the aftermath of Hurricane Katrina and Rita, FEMA[4] supplied over 100,000 temporary housing units (THUs). The USEPA reported that the building material within THUs emitted high levels of aldehyde and other VOCs at levels exceeding the NIOSH-recommended exposure limit by a factor of 10 or more. Both formaldehyde and VOCs are known to cause acute respiratory irritation, and chronic exposure increases the risk of developing a chronic respiratory disease. The release of formaldehyde in THUs increased significantly with both temperature and relative humidity [36]. Due to the hot, humid environment of post-Katrina Louisiana, exposure to these respiratory irritants was increased, and many citizens displaced from their homes were at high risk for exposure.

[4] Federal Emergency Management Agency.

Infectious Disease

There was a notable increase in the number of respiratory infections that occurred following Hurricane Katrina. The Louisiana Department of Health and New Orleans Public Health Response Team surveyed New Orleans hospitals for conditions with infectious disease epidemic potential. In total, 869 cases were identified, with 188 cases (21.6%) linked to respiratory infection within 2 weeks of Katrina making landfall [43]. In Mississippi, respiratory infections ($n = 37$) were the main reason shelters called an infectious disease hotline [44]. Uniquely, there were four cases of *Coccidioidomycosis* following Katrina in New Orleans, which is a non-endemic region for this fungi. Two patients were immunocompromised (HIV); two were immunocompetent. These cases highlight the impact hurricanes can have on the geographic distribution and impact of infectious organisms [45].

Particulate Matter

Particulate matter (PM) describes air pollution that is respirable with the capacity to penetrate deep in the lungs (PM10) and even cross the blood-lung barrier (PM2.5), causing systemic effects. There are two classes of PM monitored by the U.S. Environmental Protection Agency (USEPA) designated by size. PM10 describes particulates less than 10 μm in diameter, and PM2.5 describes particulates less than 2.5 μm in diameter [52]. PM2.5 spiked days prior to and after Hurricane Katrina hit Baton Rouge, Louisiana. There were transiently high, unsafe levels[5] of chromium, nickel, and manganese detected in the air. Grab samples from the ground contained polycyclic aromatic hydrocarbons compounds (PAH), endotoxins, and the aforementioned metals. Six years after collection, Bourgeois and Owens exposed A549 cells[6] to PM2.5. They found PM2.5 to be cytotoxic. Exposed cells exhibited signs of elevated oxidative stress, apoptosis, necrosis, and a pro-inflammatory cytokine profile of IL-6, IL-8, and TNF-α. These endotoxins were not cytotoxic on their own. The investigators concluded that the transition metals were responsible for the cytotoxicity of Katrina-related PM2.5 [53]. Airborne PM10 levels recorded in the New Orleans area post-Katrina were variable, ranging from 70 to 688 $\mu g/m^3$ in Lakeview (residential areas) [54]. Acute exposure to elevated PM10 and PM2.5 are known to exacerbate respiratory conditions. Furthermore, evidence is emerging that PM2.5 exacerbates non-respiratory conditions due to absorption into the systemic circulation [52]. Practitioners should be aware of the impact hurricanes have on PM levels

[5]Compared to IRIS (Integrated Risk Information System), OSHA (Occupational Safety and Health Administration), NIOSH (National Institute for Occupational Safety and Health), and ACGIH (American Conference of Governmental Industrial Hygienists) exposure limits.

[6]Adenocarcinomic human alveolar basal epithelial cells.

and the inflammatory nature of the PM. They should monitor high-risk patients for exacerbations of chronic respiratory disease due to elevated PM environments.

Floodwater

Hurricanes cause a transient, storm-associated rise in sea level. This phenomenon is known as a storm surge. Hurricane Katrina caused a massive storm surge along the Gulf Coast that ranged from 8 ft. to 16 ft. and flooded approximately 80% of New Orleans [31]. Analysis of Katrina-related floodwaters has yielded high numbers of *Aeromonas* species, pathogenic *Vibrio* species, and other coliform bacteria, likely from the mixing of sewage water with storm surge and rain water [55]. Analysis of the floodwater sediment detected elevated aldrin (an organochlorine insecticide), arsenic, lead, and semi-volatile organic compounds (SVOCs) above USEPA levels [54, 55]. Tak et al. reported on floodwater-related symptoms among 525 firefighters after Katrina. Of those exposed, 38% ($n = 201$) reported one or more new-onset respiratory symptoms, such as sinus congestion (28% [$n = 145$]), throat irritation (17% [$n = 92$]), and cough (24% [124]). Floodwater contact with skin, nose, mouth, and/or eyes was associated with an increased rate of new-onset upper respiratory symptoms (PR[7] = 1.9; 95%CI, 1.1–3.1) compared to those who were not exposed to floodwater [56]. Hurricane Katrina floodwater contained pathogenic bacteria and toxins. Exposure among rescue workers caused a variety of symptoms, including upper respiratory symptoms. Physical contact with hurricane water should be minimized through the use of PPE.

Floodwaters also deposited large amounts of sediment in homes. Smaller-sized sediment particles (clay and silt) were found to selectively deposit inside flooded homes [57]. The Krumbein phi scale categorizes sediment based on size. It defines silt as 3.9–62.5 μm and clay as 0.98–3.9 μm [58]. Mean aerodynamic diameter of these sediment deposits ranged from 3 to 5 μm [54]. Due to their size, when aerosolized during cleanup, the sediment is small enough to be inhaled into the distal airways. In mice, these particulates are shown to cause neutrophilic pulmonary inflammation, airway resistance, hyperresponsiveness to methacholine challenge, oxidative stress, and expression to pro-inflammatory cytokines TNF-α and IL-6 [54].

Large Cohort Studies

Few studies report the collective incidence of respiratory symptoms in the context of Hurricane Katrina's exposome. Rath et al. analyzed self-reported respiratory symptoms (upper and lower) among children and adolescents ($n = 1243$) following

[7] Prevalence ratio.

Hurricane Katrina who had sundry deleterious environmental exposures (roof/glass/storm damage, mold, dust, flood damage, smoke/fumes, and chemicals). Within the cohort, 17.2% had asthma and 12% had allergies prior to Katrina. After Katrina, there was a significant increase in respiratory symptoms. Although some children missed doctor's visits and ran out of medications, it was not enough to account for the significant increase in respiratory symptoms. Thus, this increase was attributed to Katrina-related environmental exposures [59].

In 2012, Rando et al. studied post-Katrina symptomatology among New Orleans restoration workers. They found that symptoms included transient fever/cough (29%), sinus symptoms (48%), pneumonia (3.7%), and new-onset asthma (4.5%). Prevalence rate ratios for post-Katrina sinus symptoms (PRR = 1.3; CI: 1.1, 1.7) and fever and cough (PRR = 1.7; CI: 1.3, 2.4) were significantly elevated overall for those who did restoration work, and prevalence increased with restoration work hours. Prevalence rate ratios with restoration work were also elevated for new-onset asthma (PRR = 2.2; CI: 0.8, 6.2) and pneumonia (PRR = 1.3; CI: 0.5, 3.2) but were not statistically significant. Overall, lung function was slightly decreased but was not significantly different between those with and without restoration work exposure. Post-Katrina restoration work was associated with moderate adverse effects on respiratory health, including sinusitis and toxic pneumonitis [24]. In 2014, Rando et al. reviewed post-Katrina exposures among New Orleans restoration workers. They found high average levels of exposure to PM2.5, PM10, endotoxin, and (1 → 3, 1 → 6)-β-D-glucan. Exposure levels decreased rapidly in the first year following Katrina and continued to decrease until 2012, when levels stabilized. They concluded that their review supported published reports of respiratory illness among restoration workers following Hurricane Katrina [60].

PTSD and Asthma/Smoking Behavior

Natural disasters are psychologically traumatizing due to loss of homes, loved ones, and separation of families. There was an increased incidence of psychiatric disease (PTSD, suicidality) following Hurricane Katrina [61]. Arcaya et al. reported an association between post-traumatic stress disorder (PTSD) and post-Katrina asthma exacerbations. They surveyed students at New Orleans colleges located in low-income areas, who were also parents ($n = 1000$) ages 18–34 years, with at least one child under the age of 19 years. They found a positive association between PTSD avoidance symptoms and post-hurricane asthma episodes [62]. This association between PTSD and asthma is well documented among post-9/11 rescue workers and suspected to be due to changes in gene expression from chronic stress [63–65]. Aside from stress-induced physiologic changes, PTSD avoidance symptoms were associated with unhealthy coping mechanisms, such as cigarette smoking, which may also exacerbate asthma as first-, second-, and third-hand smoke [66]. Alexander et al. surveyed residents with a smoking history from the New Orleans area who were directly affected by the hurricane ($n = 1003$) and the Memphis area who were indirectly impacted ($n = 1000$). They found an increased probability of smoking relapse following Hurricane Katrina that was directly related to the number of hurricane-related events that took place in each area. Post-disaster healthcare should be conducted with a multidisciplinary approach. Psychiatric disease has a significant impact on asthma severity, and it is an essential component of asthma management following a severe natural disaster.

Accessibility to Healthcare Facilities and Medicine

Access to healthcare is significantly impacted by natural disasters. Providers and patients are displaced, and healthcare facilities are destroyed. Rudowitz et al. reported medication loss and prescription refill burdens of relief teams following extreme weather occurrences, including Hurricane Katrina. Patients were unable to reach their pharmacies to refill medications and medical aids, such as insulin pens. A significant problem faced by relief teams was that patients were unable to recall their medical history, the names of their medications, and their medication dosages. In some instances, self-harm was inflicted by psychiatric patients directly as a result of not having necessary prescriptions [67].

For Katrina evacuees sheltered at the Lowry Air Force Base in Denver, Colorado, the local health department conducted a survey on new evacuees. They found that 60.2% had at least one family member who needed prescription medications. Of those who required medications, 43% went without their medication [68]. These data support thorough evacuation preparation for natural disaster events and the immediate awareness of medication accessibility. Individuals may survive a natural disaster, such as Hurricane Katrina, yet the aftermath can be more dangerous if one is exposed to subsequent environmental dangers and lack proper medications and treatments to manage his/her conditions.

Hurricane Sandy (2012)

Late October 2012, Hurricane Sandy became the largest Atlantic hurricane ever recorded. It was 1100 miles wide and made landfall in Atlantic City, New Jersey, as a post-tropical cyclone (Figs. 10.10 and 10.11). Flooding occurred along the east coast of the United States as well as the Caribbean region and Canada [69]. There were an estimated minimum of 142 direct deaths and $50 billion in preliminary damage [70]. The majority of deaths due to this disaster occurred from falling trees and drowning. Elderly were especially vulnerable, as many deaths resulted from power loss. In New York City, approximately half deaths were of individuals 65 years or older. Many homes were destroyed and many lives were lost by this "Superstorm" [69]. Kim et al. reported a 6% increase in 1-month and 7% increase in 3-month all-cause mortality. They found that excesses in death due to noninfectious respiratory

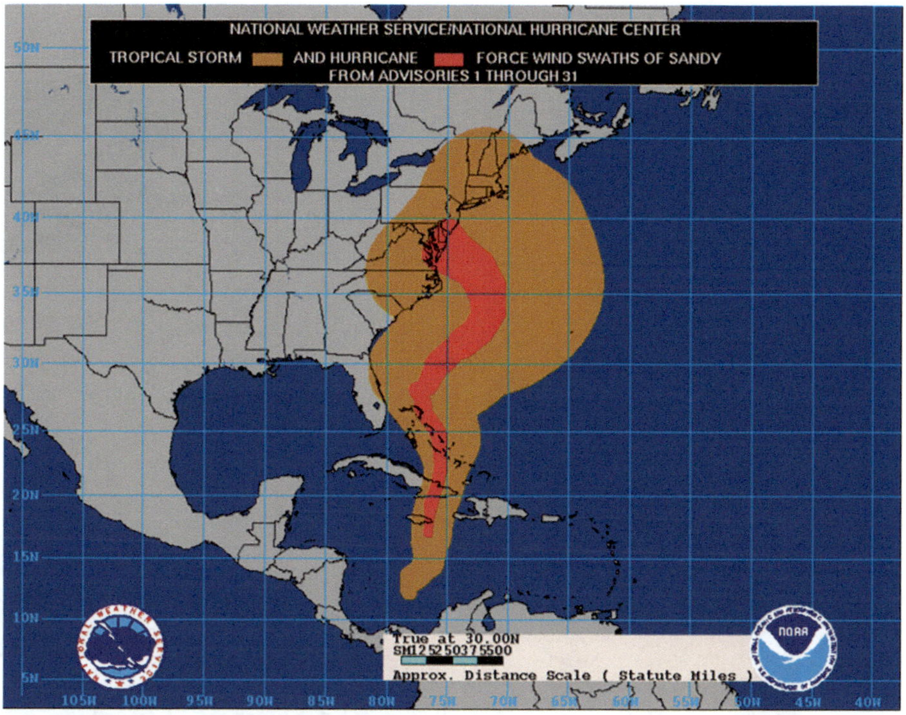

Fig. 10.10 The path of Hurricane Sandy, abstracted from satellite imaging data. (*This image is in the public domain because it contains materials that originally came from the U.S. National Oceanic and Atmospheric Administration, taken or made as part of an employee's official duties*—https://upload.wikimedia.org/wikipedia/commons/d/d7/Hurricane_Sandy_cumulative_wind_history_211847P_sm.gif)

and infectious disease occurred further out from the hurricane event, suggesting a longer disease process or latency period prior to the onset of disease [71].

Hurricane Sandy affected air quality in New York City. In the days following landfall at the sites where the hurricane hit the most (including Lower Manhattan), there was an increase in fine particulate matter that exceeded the USEPA's recommended 24-hour standard. The storm surge following Hurricane Sandy caused flooding that ranged from 2 to 9 ft. above ground level [51]. Similar to Hurricane Katrina, sediment deposition from floodwaters created an inhalational exposure for workers and returning residents [72]. Destruction of city infrastructure and loss of power by Hurricane Sandy led to increased gasoline exposure from the use of generators. Kim et al. report that in comparison to the 4 years prior to Sandy, there was an 18- to 283-fold increase in gasoline exposure that occurred after the hurricane. Most exposures occurred in men (83%). The most common cohort was men over the age of 20 years (91.9%). Although 61.5% of the exposed were asymptotic, some experienced moderate-to-toxic effects manifesting mainly as gastrointestinal or pulmonary symptoms [73].

Fig. 10.11 Satellite image of Hurricane Sandy approaching the New Jersey coastline the morning of October 29, 2012. (*This image is in the public domain because it contains materials that originally came from the U.S. National Oceanic and Atmospheric Administration, taken or made as part of an employee's official duties* – https://upload.wikimedia.org/wikipedia/commons/9/90/Hurricane_Sandy_morning_October_29_2012.jpg)

Due to the extensive flooding and water damage, mold exposure was also a problem in affected homes and buildings. Saporta and Hurst tested sensitivity to mold allergens via intradermal skin testing to 18 molds after Hurricane Sandy and Hurricane Irene. They found that the number of patients allergic to one mold increased by 30% and the number allergic to all 18 molds increased by 17% [74]. Another study was conducted on asthmatic children (n = 58), between the ages of 5 and 16, living in Sandy-affected homes in New York City. These children had dust collected from their homes. The dust was analyzed by quantitative polymerase chain reaction (PCR) for 36 various fungi and compared to the dust of houses not affected by the hurricane. Dust from houses damaged by the hurricane had *Acremonium strictum, Aspergillus fumigatus, Aspergillus niger, Aspergillus penicillioides, Cladosporium cladosporioides, Epicoccum nigrum, Mucor amphibiorum,*

Penicillium purpogenum, and *Scopulariopsis brevicauli.* The most abundant species of fungi included *Alternaria alternata* (36%), *Candida albicans* (22%), *Aureobasidium pullulans* (19%), *Aspergillus fumigatus* (17%), *Helminthosporium halodes* (17%), *and Mucor racemosus (17%).* Approximately 50% of the children tested had measurable IgE to at least one of the fungi tested in the study [75]. Therefore, sensitization to mold increased after Hurricane Sandy. Hurricane Sandy was also temporally associated with a polymicrobial fungal outbreak (non-candidal) among patients at one institution's burn ICU, suggesting that severe weather events may also influence nosocomial infection risk and rate [76].

Gargano et al. surveyed 3835 individuals with previous exposure to the World Trade Center disaster. They assessed for exposures during reconstruction, to mold/damp environments, and other respiratory irritants, to the incidence of lower respiratory symptoms (LRS.) The symptoms studied were wheezing, persistent cough, or shortness of breath reported on ≥1 day of the 30 days preceding survey completion. Approximately one-third of the individuals tested reported LRS after the hurricane. A dose–response relationship was observed between the number of different hurricane exposures and incidence of LRS [77].

Hurricane Maria (2017)

In 2017, there were numerous natural disasters throughout the United States, Central America, and the Caribbean—Hurricane Harvey, Irma, Maria, José, and Katia. Hurricane Maria damaged the island of Puerto Rico on September 20, 2017. This was a Category 4 hurricane, which impacted multiple Caribbean islands, but damaged Puerto Rico the most (Fig. 10.12). Wind speeds reached as high as 155 miles per hour (mph), with over 1 ft. of rainfall on most of the island, eliminating over 80% of power on the island [78]. Below are some pictures of the effects that Hurricane Maria had on the land and homes (Figs. 10.13, 10.14, 10.15, and 10.16).

After Hurricane Maria struck, there was an increase in accumulations of certain allergens, including pollen and molds, which have led to increases in respiratory diseases and allergic disorders [17, 79]. Why such a tremendous increase in allergens after the hurricane? The volume and range of aeroallergens are dependent on the temperature, humidity, and rainfall in a given area (which would have caused the increase in allergen concentration after the storm). The average annual temperature in Puerto Rico and other Caribbean islands have increased approximately 0.5 °C in the past few years, and for every degree (Celsius) that the temperature increases, the air can hold 7% more water. This, in turn, leads to the air gaining more humidity, which can lead to increased allergen concentrations [17]. San Juan, Puerto Rico, has had more days of extreme heat than in recent years. These warmer days can lead to exacerbations asthma [79].

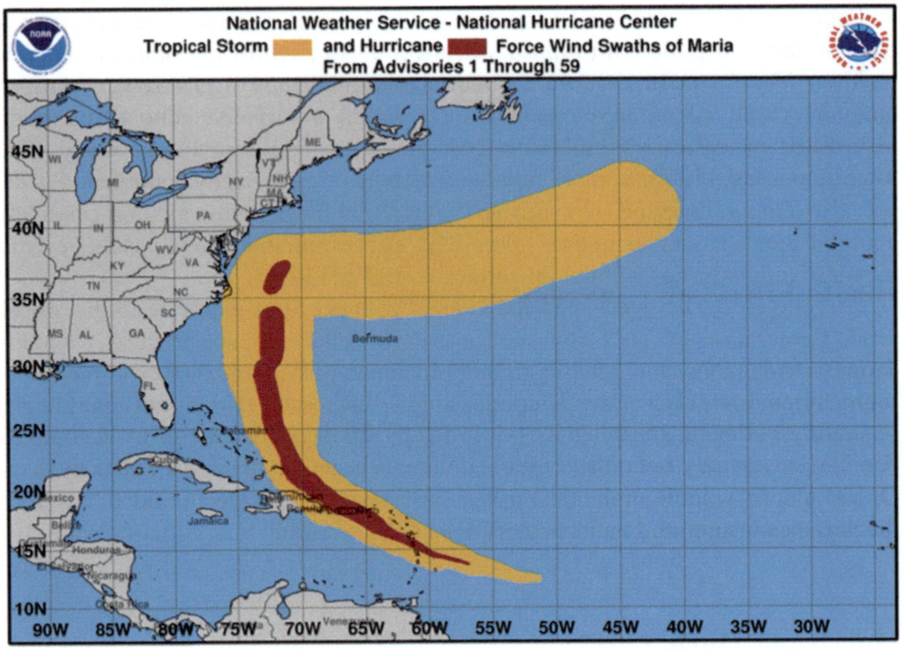

Fig. 10.12 The path of Hurricane Maria, abstracted from satellite imaging data. (*This image is in the public domain because it contains materials that originally came from the US National Oceanic and Atmospheric Administration, taken or made as part of an employee's official duties* – https://upload.wikimedia.org/wikipedia/commons/7/74/2017_Hurricane_Maria_wind_history.png)

Medical literature regarding Hurricane Maria is less robust vs. Hurricane Katrina. However, we know mold-infested domiciles, in general, were not adequately remediated post-Hurricane Maria. The photographs above show wreckage that was prevalent throughout the island. This detritus was not cleaned up for months after the storm because roads were inaccessible and power was down. Many families and businesses also used generators, since greater than half the island lost power. These fuel-powered generators expelled fumes harmful to those who already had respiratory diseases [80]. Recent press reports indicate doctors in Puerto Rico are concerned by the amount of asthma cases that have risen since Hurricane Maria. In May 2018, mold aeroallergen concentrations were the highest ever recorded in Puerto Rico [80].

In summary, climate change can induce catastrophic hurricane events, which have immediate and long-lasting respiratory/allergic health effects. Public policy should address not only preventive measures to avert future disasters but also still must remediate the aftermath of the aforementioned "once-in-a-lifetime weather extremes".

Fig. 10.13 Visual mold on the exterior of a post-hurricane residence in Utuado, Puerto Rico. (Permission granted by photographer Tyler Turner)

Fig. 10.14 Damaged home removed from its foundation in Utuado, Puerto Rico. (Permission granted by photographer Tyler Turner)

Fig. 10.15 Mold inside a bedroom from a post-hurricane home in Toa Baja, Puerto Rico, located north of San Juan, on August 3, 2018. (Permission granted by photographer Yamela Cando)

Fig. 10.16 A moldy bathroom in the same home as image 3. (Permission granted by photographer Yamela Cando)

Bibliography

1. Petoukhov V, Semenov VA. A link between reduced Barents-Kara sea ice and cold winter extremes over northern continents. J Geophys Res. 2010;115 https://doi.org/10.1029/2009JD013568.
2. Ziska L, Knowlton K, Rogers C, et al. Recent warming by latitude associated with increased length of ragweed pollen season in Central North America. Proc Natl Acad Sci. 2011;108:4248–51. https://doi.org/10.1073/pnas.1014107108.
3. Chen Y, Forsyth E, Pan K, et al. Atmospheric temperature & pollen counts impact New York City asthma ER visits. J Allergy Clin Immunol. 2010;125:AB208. https://doi.org/10.1016/j.jaci.2009.12.813.
4. Stach A, García-Mozo H, Prieto-Baena JC, et al. Prevalence of Artemisia species pollinosis in western Poland: impact of climate change on aerobiological trends, 1995–2004. J Investig Allergol Clin Immunol. 2007;17:39–47.
5. Ziska LH. Rising atmospheric carbon dioxide and plant biology: the overlooked paradigm. DNA Cell Biol. 2008;27:165–72. https://doi.org/10.1089/dna.2007.0726.
6. Kizilpinar I, Civelek E, Tuncer A, et al. Pollen counts and their relationship to meteorological factors in Ankara, Turkey during 2005–2008. Int J Biometeorol. 2011;55:623–31. https://doi.org/10.1007/s00484-010-0363-8.
7. Beggs PJ. Impacts of climate change on aeroallergens: past and future. Clin Htmlent Glyphamp Asciiamp Exp Allergy. 2004;34:1507–13. https://doi.org/10.1111/j.1365-2222.2004.02061.x.
8. Metintas S, Kurt E, PARFAIT Study Group. Geo-climate effects on asthma and allergic diseases in adults in Turkey: results of PARFAIT study. Int J Environ Health Res. 2010;20:189–99. https://doi.org/10.1080/09603120903456828.
9. Peden DB, Bush RK. Advances in environmental and occupational respiratory disease in 2010. J Allergy Clin Immunol. 2011;127:696–700. https://doi.org/10.1016/j.jaci.2011.01.030.
10. Dharajiya NG, Bacsi A, Boldogh I, Sur S. Pollen NAD(P)H oxidases and their contribution to allergic inflammation. Immunol Allergy Clin N Am. 2007;27:45–63. https://doi.org/10.1016/j.iac.2006.11.007.
11. Cecchi L, D'Amato G, Ayres JG, et al. Projections of the effects of climate change on allergic asthma: the contribution of aerobiology: climate change and aerobiology. Allergy. 2010; https://doi.org/10.1111/j.1398-9995.2010.02423.x.
12. Diaz-Sanchez D, Dotson AR, Takenaka H, Saxon A. Diesel exhaust particles induce local IgE production in vivo and alter the pattern of IgE messenger RNA isoforms. J Clin Invest. 1994;94:1417–25. https://doi.org/10.1172/JCI117478.
13. Sheffield PE, Knowlton K, Carr JL, Kinney PL. Modeling of regional climate change effects on ground-level ozone and childhood asthma. Am J Prev Med. 2011;41:251–7. https://doi.org/10.1016/j.amepre.2011.04.017.
14. D'Amato G, Cecchi L, D'Amato M, Liccardi G. Urban air pollution and climate change as environmental risk factors of respiratory allergy: an update. J Investig Allergol Clin Immunol. 2010;20:95–102; quiz following 102
15. Szema AM. Climate change, allergies, and asthma. J Occup Environ Med. 2011;53:1353–4. https://doi.org/10.1097/JOM.0b013e318237a00d.
16. Masson-Delmotte V, Zhai P, Portner H, et al. IPCC, 2018: summary for policymakers. In: global warming of 1.5°C. an IPCC special report on the impacts of global warming of 1.5°C above pre-industrial levels and related global greenhouse gas emission pathways, in the context of strengthening the global response to the threat of climate change, sustainable development, and efforts to eradicate poverty. Geneva. Switzerland: World Meterological Organization; 2018.
17. Institute of Medicine (U.S.). Damp indoor spaces and health. Washington, DC: National Academies Press; 2004.
18. Crimmins A, Balbus J, Gamble J, et al. The impacts of climate change on human health in the United States: a scientific assessment. Washington, DC: U.S. Global Change Research Program; 2016.

19. Jacob B, Ritz B, Gehring U, et al. Indoor exposure to molds and allergic sensitization. Environ Health Perspect. 2002;110:647–53. https://doi.org/10.1289/ehp.02110647.
20. Blatter J, Forno E, Brehm J, et al. Fungal exposure, atopy, and asthma exacerbations in puerto rican children. Ann Am Thorac Soc. 2014;11:925–32. https://doi.org/10.1513/AnnalsATS.201402-077OC.
21. Karvonen AM, Hyvarinen A, Korppi M, et al. Moisture damage and asthma: a birth cohort study. Pediatrics. 2015;135:e598–606. https://doi.org/10.1542/peds.2014-1239.
22. Sharpe RA, Thornton CR, Tyrrell J, et al. Variable risk of atopic disease due to indoor fungal exposure in NHANES 2005–2006. Clin Exp Allergy. 2015;45:1566–78. https://doi.org/10.1111/cea.12549.
23. Bender MA, Knutson TR, Tuleya RE, et al. Modeled impact of anthropogenic warming on the frequency of intense Atlantic hurricanes. Science. 2010;327:454. https://doi.org/10.1126/science.1180568.
24. Grinsted A, Moore JC, Jevrejeva S. Projected Atlantic hurricane surge threat from rising temperatures. Proc Natl Acad Sci U S A. 2013;110:5369–73. https://doi.org/10.1073/pnas.1209980110.
25. Risser MD, Wehner MF. Attributable human-induced changes in the likelihood and magnitude of the observed extreme precipitation during hurricane Harvey. Geophys Res Lett. 2017;44:12, 457–12, 464 https://doi.org/10.1002/2017GL075888.
26. Klotzbach P, Bowen S, Pielke R Jr, Bell M. Continental U.S. hurricane landfall frequency and associated damage: observations and future risks. Am Meteor Soc. 2018;99(7):1359–76.
27. Melillo JM, Richmond T (T. C.), Yohe GW. Climate change impacts in the United States: the third national climate assessment. U.S. Global Change Research Program. 2014;841 https://doi.org/10.7930/J0Z31WJ2.
28. Webster PJ, Holland GJ, Curry JA, Chang H-R. Changes in tropical cyclone number, duration, and intensity in a warming environment. Science. 2005;309:1844. https://doi.org/10.1126/science.1116448.
29. Elsner JB, Kossin JP, Jagger TH. The increasing intensity of the strongest tropical cyclones. Nature. 2008;455:92–5. https://doi.org/10.1038/nature07234.
30. U.S. Global Change Research Program, Wuebbles DJ, Fahey DW, et al. Climate science special report: fourth national climate assessment, vol. I: U.S. Global Change Research Program; 2017.
31. Knabb R, Rhome J, Brown D. Tropical cyclone report hurricane Katrina 23–30 August 2005: National Hurricane Center; 2005.
32. Brunkard J, Namulanda G, Ratard R. Hurricane Katrina deaths, Louisiana, 2005. Disaster Med Public Health Prep. 2008;2:215–23. https://doi.org/10.1097/DMP.0b013e31818aaf55.
33. Facts + Statistics: Hurricanes. Insurance Information Institute.
34. Riggs MA, Rao CY, Brown CM, et al. Resident cleanup activities, characteristics of flood-damaged homes and airborne microbial concentrations in New Orleans, Louisiana, October 2005. Environ Res. 2008;106:401–9. https://doi.org/10.1016/j.envres.2007.11.004.
35. Chew GL, Wilson J, Rabito FA, et al. Mold and endotoxin levels in the aftermath of hurricane Katrina: a pilot project of homes in New Orleans undergoing renovation. Environ Health Perspect. 2006;114:1883–9. https://doi.org/10.1289/ehp.9258.
36. Baxi SN, Portnoy JM, Larenas-Linnemann D, et al. Exposure and health effects of fungi on humans. J Allergy Clin Immunol Pract. 2016;4:396–404. https://doi.org/10.1016/j.jaip.2016.01.008.
37. Williams DL. Overview of (1 → 3)-beta-D-glucan immunobiology. Mediat Inflamm. 1997;6:247–50. https://doi.org/10.1080/09629359791550.
38. Inamdar AA, Bennett JW. Volatile organic compounds from fungi isolated after hurricane katrina induce developmental defects and apoptosis in a *Drosophila melanogaster* model: volatile organic compounds from fungal isolates induce developmental defects and apoptosis. Environ Toxicol. 2015;30:614–20. https://doi.org/10.1002/tox.21933.

39. Adhikari A, Jung J, Reponen T, et al. Aerosolization of fungi, (1→3)-β-d glucan, and endotoxin from flood-affected materials collected in New Orleans homes. Environ Res. 2009;109:215–24. https://doi.org/10.1016/j.envres.2008.12.010.
40. Choi H, Schmidbauer N, Bornehag C-G. Volatile organic compounds of possible microbial origin and their risks on childhood asthma and allergies within damp homes. Environ Int. 2017;98:143–51. https://doi.org/10.1016/j.envint.2016.10.028.
41. Zhang Z, Biagini Myers JM, Brandt EB, et al. β-Glucan exacerbates allergic asthma independent of fungal sensitization and promotes steroid-resistant TH2/TH17 responses. J Allergy Clin Immunol. 2017;139:54–65.e8. https://doi.org/10.1016/j.jaci.2016.02.031.
42. Oluwole O, Rennie DC, Senthilselvan A, et al. The association between endotoxin and beta-(1 → 3)-D-glucan in house dust with asthma severity among schoolchildren. Respir Med. 2018;138:38–46. https://doi.org/10.1016/j.rmed.2018.03.015.
43. Maheswaran D, Zeng Y, Chan-Yeung M, et al. Exposure to Beta-(1,3)-D-glucan in house dust at age 7–10 is associated with airway hyperresponsiveness and atopic asthma by age 11–14. PLoS One. 2014;9:e98878. https://doi.org/10.1371/journal.pone.0098878.
44. Bloom E, Grimsley LF, Pehrson C, et al. Molds and mycotoxins in dust from water-damaged homes in New Orleans after hurricane Katrina. Indoor Air. 2009;19:153–8. https://doi.org/10.1111/j.1600-0668.2008.00574.x.
45. Schuster A, Schmoll M. Biology and biotechnology of Trichoderma. Appl Microbiol Biotechnol. 2010;87:787–99. https://doi.org/10.1007/s00253-010-2632-1.
46. Peltola J, Andersson MA, Haahtela T, et al. Toxic-metabolite-producing bacteria and fungus in an indoor environment. Appl Environ Microbiol. 2001;67:3269–74. https://doi.org/10.1128/AEM.67.7.3269-3274.2001.
47. Moreira DC, Oliveira MME, Borba CM. Human pathogenic Paecilomyces from food. Microorganisms. 2018;6 https://doi.org/10.3390/microorganisms6030064.
48. Steiner B, Aquino VR, Paz AA, et al. Paecilomyces variotii as an emergent pathogenic agent of pneumonia. Case Rep Infect Dis. 2013;2013:273848. https://doi.org/10.1155/2013/273848.
49. Petrikkos G, Skiada A, Lortholary O, et al. Epidemiology and clinical manifestations of mucormycosis. Clin Infect Dis Off Publ Infect Dis Soc Am. 2012;54(Suppl 1):S23–34. https://doi.org/10.1093/cid/cir866.
50. Rusiecki JA, Thomas DL, Chen L, et al. Disaster-related exposures and health effects among US coast guard responders to hurricanes Katrina and Rita: a cross-sectional study. J Occup Environ Med. 2014;56:820–33. https://doi.org/10.1097/JOM.0000000000000188.
51. Webb BJ, Blair JE, Kusne S, et al. Concurrent pulmonary Aspergillus fumigatus and Mucor infection in a cardiac transplant recipient: a case report. Transplant Proc. 2013;45:792–7. https://doi.org/10.1016/j.transproceed.2012.03.056.
52. Kim K-H, Kabir E, Kabir S. A review on the human health impact of airborne particulate matter. Environ Int. 2015;74:136–43. https://doi.org/10.1016/j.envint.2014.10.005.
53. Bourgeois B, Owens JW. The influence of hurricanes Katrina and Rita on the inflammatory cytokine response and protein expression in A549 cells exposed to PM $_{2.5}$ collected in the Baton Rouge–Port Allen industrial corridor of Southeastern Louisiana in 2005. Toxicol Mech Methods. 2014;24:220–42. https://doi.org/10.3109/15376516.2014.881945.
54. Wang K, You D, Balakrishna S, et al. Sediment from hurricane katrina: potential to produce pulmonary dysfunction in mice. Int J Clin Exp Med. 2008;1:130–44.
55. Presley SM, Rainwater TR, Austin GP, et al. Assessment of pathogens and toxicants in New Orleans, LA following hurricane Katrina. Environ Sci Technol. 2006;40:468–74.
56. Tak S, Bernard BP, Driscoll RJ, Dowell CH. Floodwater exposure and the related health symptoms among firefighters in New Orleans, Louisiana 2005. Am J Ind Med. 2007;50:377–82. https://doi.org/10.1002/ajim.20459.
57. Ashley NA, Valsaraj KT, Thibodeaux LJ. Elevated in-home sediment contaminant concentrations – the consequence of a particle settling-winnowing process from hurricane Katrina floodwaters. Chemosphere. 2008;70:833–40. https://doi.org/10.1016/j.chemosphere.2007.07.010.

58. Kennish MJ. Encyclopedia of estuaries. Dordrecht: Springer Science+Business Media; 2016.
59. Rath B, Young E, Harris A, et al. Adverse respiratory symptoms and environmental exposures among children and adolescents following hurricane Katrina. Public Health Rep. 2011;126:853–60.
60. Rando RJ, Kwon C-W, Lefante JJ. Exposures to thoracic particulate matter, endotoxin, and glucan during post-hurricane Katrina restoration work, New Orleans 2005–2012. J Occup Environ Hyg. 2014;11:9–18. https://doi.org/10.1080/15459624.2013.839879.
61. Kessler RC, Galea S, Gruber MJ, et al. Trends in mental illness and suicidality after hurricane Katrina. Mol Psychiatry. 2008;13:374–84. https://doi.org/10.1038/sj.mp.4002119.
62. Arcaya M, Lowe S, Rhodes J, et al. Association of PTSD symptoms with asthma attacks among hurricane Katrina survivors. J Trauma Stress. 2014;27:725–9.
63. Rosenberg SL, Miller GE, Brehm JM, Celedón JC. Stress and asthma: novel insights on genetic, epigenetic, and immunologic mechanisms. J Allergy Clin Immunol. 2014;134:1009–15. https://doi.org/10.1016/j.jaci.2014.07.005.
64. Mindlis I, Morales-Raveendran E, Goodman E, et al. Post-traumatic stress disorder dimensions and asthma morbidity in world trade center rescue and recovery workers. J Asthma Off J Assoc Care Asthma. 2017;54:723–31. https://doi.org/10.1080/02770903.2016.1263650.
65. de la Hoz RE, Jeon Y, Miller GE, et al. Post-traumatic stress disorder, bronchodilator response, and incident asthma in world trade center rescue and recovery workers. Am J Respir Crit Care Med. 2016;194:1383–91. https://doi.org/10.1164/rccm.201605-1067OC.
66. Baschnagel J, Coffey S, Schumacher J, et al. Relationship between PTSD symptomatology and nicotine dependence severity in crime victims. Addict Behav. 2008;33:1441–7.
67. Rudowitz R, Rowland D, Shartzer A. Health care in New Orleans before and after hurricane Katrina. Health Aff. 2006;25:w393–406.
68. Dippold L, Patnaik J, Vogt R, et al. Illness surveillance and rapid needs assessment among hurricane Katrina evacuees – Colorado, September 1–23, 2005. MMWR. 2006;55:244–7.
69. Purdy E. Hurricane Sandy. Salem Press Encyclopedia.
70. Blake E, Kimberlain T, Berg R, et al. Tropical cyclone report: hurricane Sandy (AL182012) 22–29 October 2012: National Hurricane Center; 2013.
71. Kim S, Kulkarni PA, Rajan M, et al. Hurricane Sandy (New Jersey): mortality rates in the following month and quarter. Am J Public Health. 2017;107:1304–7. https://doi.org/10.2105/AJPH.2017.303826.
72. Manuel J. The long road to recovery: environmental health impacts of hurricane Sandy. Environ Health Perspect. 2013;121:a152–9. https://doi.org/10.1289/ehp.121-a152.
73. Kim HK, Takematsu M, Biary R, et al. Epidemic gasoline exposures following hurricane Sandy. Prehosp Disaster Med. 2013;28:586–91. https://doi.org/10.1017/S1049023X13009023.
74. Saporta D, Hurst D. Increased sensitization to mold allergens measured by intradermal skin testing following hurricanes. J Environ Public Health. 2017;2017:5.
75. Divjan A, Acosta LM, Sobek E, et al. IgE antibodies to fungi among asthmatic children living in homes damaged by hurricane Sandy in New York City. J Allergy Clin Immunol. 2016;137:AB180. https://doi.org/10.1016/j.jaci.2015.12.723.
76. Sood G, Vaidya D, Dam L, et al. A polymicrobial fungal outbreak in a regional burn center after hurricane Sandy. Am J Infect Control. 2018;46:1047–50. https://doi.org/10.1016/j.ajic.2018.01.011.
77. Gargano LM, Locke S, Jordan HT, Brackbill RM. Lower respiratory symptoms associated with environmental and reconstruction exposures after hurricane Sandy. Disaster Med Public Health Prep. 2018;12:697–702. https://doi.org/10.1017/dmp.2017.140.
78. Irfan U. One of the clearest signs of climate change in hurricanes Maria, Irma, and Harvey was the rain: Vox Media; 2017.
79. Palmer J. Surviving climate change: lessons from hurricane Maria: Cable News Network; 2018.

80. Associated Press. Puerto Rico faces a spike in asthma cases following hurricane Maria. NBCUniversal News Group; 2018.
81. Pálvölgyi T, Szabó É, Makra L. Ragweed impact case study (in: evaluation and assessment of various impacts in the framework of CLAVIER project – climate change and variability: impact on Central and Eastern Europe, www.clavier-eu.org). Scientific report. Budapest: Env-in-Cent; 2009.
82. Jelks M. Allergy plants that cause sneezing and wheezing. Tampa: World Wide Printing; 1989. p. 34–5.

Chapter 11
California and Climate Changes: An Update

Rupa Basu

Introduction

According to California's Fourth Climate Assessment Report [1], temperatures have been warming in California because of climate change, with current annual temperatures one to two degrees Fahrenheit (°F) above what was observed during the first half of the century. These temperature increases have been linked with increased deaths from various causes and cardio-respiratory diseases and injuries, among other public health impacts. This chapter focuses on epidemiologic studies of temperature and adverse health outcomes in California. With the large, demographically diverse population distributed throughout the state as well as geographic variations in temperature exposure captured by a vast monitoring network, health outcomes, including mortality and morbidity, such as hospital and emergency room visits and adverse birth outcomes, can be studied.

Furthermore, California is unique since temperature and humidity tend to be relatively mild in areas where most of the population resides, while pollutant levels are generally high with distinct sources and patterns of exposure. Furthermore, people spend more time outdoors throughout the year, lending them the potential for more exposure to heat, air pollution, smoke, as well as vector-borne diseases. Air conditioning (AC) use is not a surrogate for socioeconomic status, as it may be in other parts of the country. Many homes in coastal areas don't have AC installed because predominantly cool temperatures have minimized the need for them, although coastal homes tend to be more expensive and consist of wealthier populations. Thus, people living in coastal areas may be more impacted by a heat wave, since many do not have AC in their homes and are not acclimatized to high ambient temperatures.

R. Basu (✉)
Office of Environmental Health Hazard Assessment, Air and Climate Epidemiology Section, Oakland, CA, USA
e-mail: Rupa.Basu@oehha.ca.gov

K. E. Pinkerton, W. N. Rom (eds.), *Climate Change and Global Public Health*, Respiratory Medicine, https://doi.org/10.1007/978-3-030-54746-2_11

Summary of Epidemiologic Studies of Temperature and Mortality

Temperature and Mortality Studies in the United States Including California

Previous studies of heat waves or elevated temperature and mortality have been documented worldwide and summarized in epidemiologic review articles [2–5]. Exposure to heat has often been defined by apparent temperature (also known as heat index or humidex), a combination of temperature and relative humidity, as a continuous variable or as extreme temperatures above or below a specific threshold.

Few investigators examining temperature and mortality in the United States have included cities or counties in California as a part of their analyses. Among the first studies of temperature and cardio-respiratory mortality was conducted by Basu et al. [6] using National Morbidity and Mortality Air Pollution Study (NMMAPS) data from 20 metropolitan areas in the United States. The investigators reported a positive association between temperature and mortality during the summer (June–August) for all regions and mostly null or negative associations during all other seasons. The southwest region, consisting of Phoenix, AZ; San Diego, CA; Santa Ana, CA; Los Angeles, CA; and San Bernardino, CA, had the highest regional association with an odds ratio of 1.15 (95% confidence interval (CI): 1.07, 1.24) per 10 °F increase in mean daily temperature, adjusted for dew point temperature to account for humidity. In this study, the time-stratified case-crossover approach using logistic regression models and the time-series analysis using Poisson regression models produced virtually identical results. Since this study was based on 1 year of data in 1992, more studies of multiple areas over a longer time period are warranted.

Other investigators expanded the NMMAPS data to include 95 US cities from 1987 to 2000 [7, 8]. Synergistic effects between ozone and temperature on cardiovascular mortality was observed during the summer months in most regions, including Southern California [8]. Barnett [7] included 107 cities in this analysis to compare findings between the summers of 1987 and 2000. He reported an elevated risk in 1987 for temperature and cardiovascular mortality that was not observed in 2000. Similar to the Basu et al. [6] study, regional analyses showed that southern California had among the greatest associations in 1987, but also had the largest decline in 2000. The author attributes the diminished effect partially to the increased availability of AC. However, racial/ethnic disparities exist for access to AC in the US, and not everyone is able to afford using it [9]. In another study using data provided by a housing survey [10], usage of ACs significantly reduced the associations between temperature and hospital admissions for several outcomes, but still remained elevated even after controlling for socioeconomic status [11]. Furthermore, prolonged and widespread AC use can lead to power brownouts and blackouts. Thus, AC is not a viable solution to mitigate heat-related health impacts entirely, particularly for the vulnerable populations that are most affected.

In another case-crossover study of temperature and mortality in 50 US cities using data from 1989 to 2000, investigators explored extreme temperatures, using various cut-off values for temperature [12]. In their analysis of over 6 million observations, mortality was found to increase with extreme heat (5.74%, 95% CI: 3.38, 8.15). Although no estimates were provided for California specifically, Los Angeles and San Diego were included in the overall analysis. The largest associations were generally observed in cities with milder summers, less AC, and higher population density. In another case-only study using the same data, Medina-Ramon et al. [13] found that older subjects, diabetics, Blacks, and those dying outside a hospital were more susceptible to extreme heat.

Temperature and Mortality Studies in California

In the first epidemiologic study of temperature and mortality focusing on California, temperature and mortality data from nine counties in California were analyzed, including Contra Costa, Fresno, Kern, Los Angeles, Orange, Riverside, Sacramento, San Diego, and Santa Clara, which comprise approximately 65% of the State's population and include northern and southern California as well as inland and coastal regions [14]. To focus on heat effects, data were limited to the warm season from May through September 1999–2003. County-specific estimates were obtained followed by an overall combined estimate using the random effects model in meta-analyses [15]. Same-day lag was found to have the best data fit and the highest risk estimates for the 248,019 deaths included in this study, demonstrating the acute effect of temperature on mortality and the importance of immediate public health responses to prevent heat-related deaths and illnesses. Each 10 °F increase corresponded to a 2.3% increase in mortality (95% CI: 1.0, 3.6) in the time-stratified case-crossover analysis (Fig. 11.1) for all nine counties combined, with similar results produced from the time-series analysis. No criteria air pollutant examined, including ozone, fine particles, carbon monoxide, and nitrogen dioxide, was found to be a significant confounder or effect modifier. Regional differences between coastal and inland areas were observed, and thus, region-specific policies are necessary. An association between background levels of apparent temperature and mortality in California was observed without focusing on extremes in apparent temperature or heat waves. The findings from this study are comparable to temperature and mortality in other regions in the US using the same methods [16].

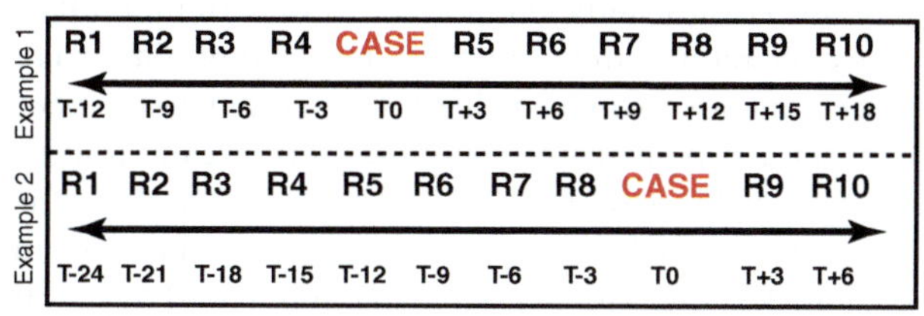

Fig. 11.1 CASE: case period; R1–R10: referent periods 1–10 every third day in the same month and year; T0: time that case occurred (death date); T − 24…T + 18: time that referent periods occurred

Vulnerable Subgroups

In another case-crossover study examining temperature and mortality in California, several vulnerable subgroups were identified [17]. Identifying susceptible populations in local regions is essential for preventing mortality and morbidity from ambient heat exposure. A total of 231,676 non-accidental deaths were included to evaluate several disease categories. Each 10 °F increase in mean daily apparent temperature corresponded to a 2.6% (95% CI: 1.3, 3.9) increase in cardiovascular disease mortality, with elevated risk especially found for ischemic heart disease. Acute myocardial infarction and congestive heart failure also had greater associations, although respiratory disease mortality did not. High risks were also found for persons at least 65 years of age (2.2%, 95% CI: 0.04, 4.0), infants 1 year of age and under (4.9%, 95% CI: −1.8, 11.6), and Black non-Hispanic racial/ethnic group (4.9%, 95% CI: 2.0, 7.9). No differences were found by gender or education level. Thus, persons at risk for cardiovascular disease, the elderly, infants, and Blacks, among others, should be targeted to prevent mortality associated with high apparent temperature.

Because of the findings from this study, a case-crossover study was conducted to focus on apparent temperature and mortality specifically for infants under 1 year of age [18]. Although very little previous research had focused on infants, they are unable to control their core body temperature via thermoregulation efficiently like the elderly and other vulnerable populations. Thus, infants are potentially vulnerable to the impact of high temperatures. From May through October 1999–2011, 12,356 infants who died from all causes and specifically from congenital malformations, sudden infant death syndrome, abnormal gestation duration, respiratory causes, and circulatory causes were identified. Although not statistically significant, the percentage change in odds was 4.4% (95% CI: −0.3, 9.2) for all-cause mortality per 10 °F increase for average of same-day and previous 3 days apparent temperature. These deaths as well as those caused by short gestation duration were highest for Black infants, while White infants had elevated risk for deaths from respiratory causes. Differences were also found for neonates (infants aged 28 days and under) and post-neonates (infants above 28 days and under 1 year) and coastal and inland regions. As with the original temperature and mortality study, these associations remained even after considering possible confounding by criteria air pollutants.

Mortality Displacement

In a time-series study, the potential effect of mortality displacement in the relationship between apparent temperature and mortality was explored [19]. Mortality displacement, also known as harvesting, refers to the phenomenon in which a specific exposure, such as temperature, impacts already frail individuals whose deaths may have been brought forward only by a few days. Significant associations were observed for the same-day (4.3% per 10 °F increase in apparent temperature, 95% CI: 3.4, 5.2) continuing up to a maximum of 4 days following apparent temperature exposure for non-accidental mortality. Similar patterns of risk were found for mortality from cardiovascular diseases, respiratory diseases, and among children 0–18 years and those 50+ years. Since no significantly negative effects were observed in the following single or cumulative days, evidence of mortality displacement was not found. Thus, apparent temperature and mortality in California is a valid public health concern, with the most significant impacts occurring on the same day and up to 3 days following exposure, and appears to have a broad impact on the general population and disease- and age-specific subgroups.

Heat Waves and Mortality

The estimates provided in the previous section discussing studies conducted in California were based on background apparent temperature, including both heat wave and non–heat wave periods. Thus, they do not capture the worst case scenario, as would be observed during heat wave periods. Ostro et al. [20] investigated the July 2006 heat wave in California from July 14 to August 1, 2006. County coroners reported that high ambient temperatures caused 142 deaths. However, heat wave deaths are likely to be underreported due to a lack of a clear case definition and the multi-factorial nature of mortality [2]. Furthermore, no systematic definition for heat-related deaths currently exists in the United States or California specifically, so they are often only reported during a heat wave when no other cause of death can be identified. Daily data were collected for mortality, relative humidity, ambient temperature, and ozone in seven California counties impacted by the July 2006 heat wave. The combined meta-analytic results suggested a 9% (95% CI: 1.6, 16.3) increase in daily mortality per 10 °F change in apparent temperature, which is more than three times greater than the estimate for the warm season and corresponds to 450–600 deaths, which is approximately three to four times greater than the coroner's estimates. Evidence from the July 2006 heat wave suggests that elevated nighttime temperatures and higher levels of relative humidity were primary factors for increased deaths. Heat waves in California have generally been more influenced by increased nighttime temperatures (Fig. 11.2).

Summary of Epidemiologic Studies of Temperature and Morbidity

Temperature and Hospitalizations/Emergency Room Visits

Using the same nine counties as the mortality analyses, apparent temperature and hospitalizations from various causes were evaluated [21]. The study population consisted of 597,735 individuals who had unscheduled hospital visits with selected diagnoses and lived within 10 km of a temperature monitor in an effort to refine

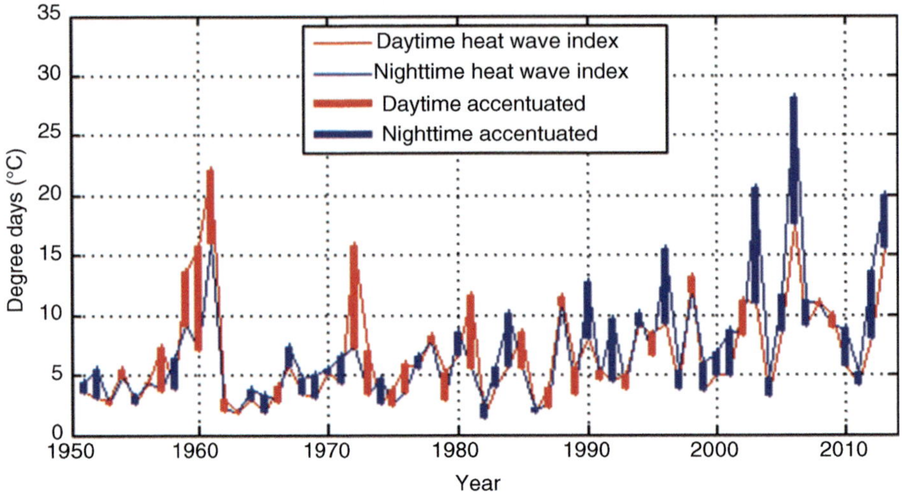

Fig. 11.2 Evidence for increasing nighttime temperatures in California. (Source: California-Nevada Climate Applications Program [CNAP], 2015)

exposure assessment from using entire county to characterize apparent temperature exposure. A 10 °F increase in mean apparent temperature was associated with a 3.5% (95% CI: 1.5, 5.6) increase in several disease-specific outcomes, such as ischemic stroke, all respiratory diseases (2.0%, 95% CI: 0.7, 3.2), pneumonia (3.7%, 95% CI: 1.7, 3.7), dehydration (10.8%, 95% CI: 8.3, 13.6), diabetes (3.1%, 95% CI: 0.4, 5.9), and acute renal failure (7.4%, 95% CI: 4.0, 10.9). There was little evidence for confounding by either fine particles or ozone.

In a case-crossover study of over 1.2 million emergency room visits in 16 climate zones [22] in California, the study population also consisted of cases who resided within 10 km of a temperature monitor, but this time in the same climate zone rather than the same county (see Fig. 11.3) [23]. Climate zones were developed by the California Energy Commission based on weather and energy use patterns and may be more relevant for capturing temperature exposure for populations. Significant positive associations were observed for same-day apparent temperature and ischemic heart disease (per 10 °F: 1.7%; 95% CI: 0.2, 3.3), ischemic stroke (2.8%; 95% CI: 0.9, 4.7), cardiac dysrhythmia (2.8%; 95% CI: 0.9, 4.9), hypotension (12.7%; 95% CI: 8.3, 17.4), diabetes (4.3%; 95% CI: 2.8, 5.9), intestinal infection (6.1%; 95% CI: 3.3, 9.0), dehydration (25.6%; 95% CI: 21.9, 29.4), acute renal failure (15.9%; 95% CI: 12.7, 19.3), and heat illness (393.3%; 95% CI: 331.2, 464.5). Statistically significant negative associations were found for aneurysm, hemorrhagic stroke, and hypertension. These estimates all remained relatively unchanged after adjusting for criteria air pollutants, with the exception of pneumonia and all respiratory diseases, which were confounded by nitrogen dioxide and carbon monoxide. Risks often varied by age or racial/ethnic group. Again, prevention strategies for morbidity during heat exposure require an immediate response, and should

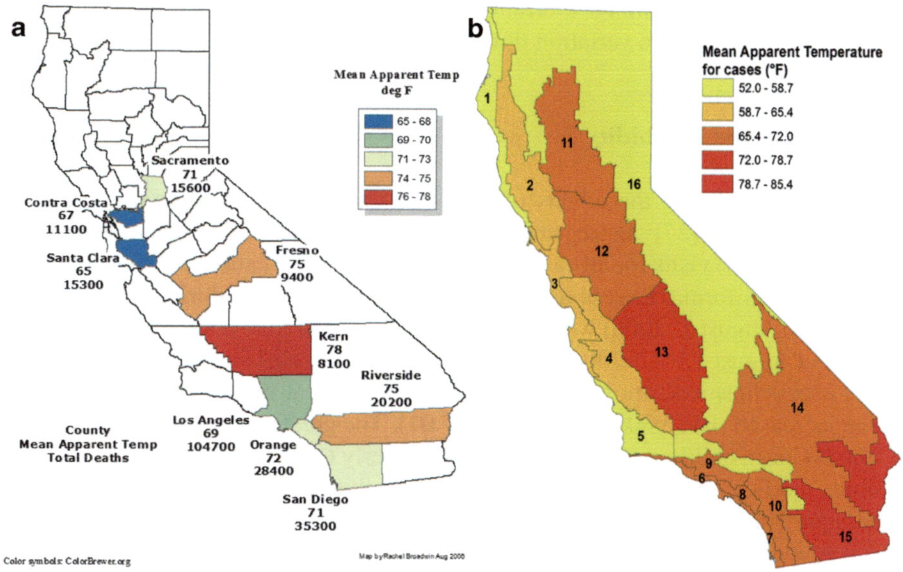

Fig. 11.3 Refining exposure assessment for apparent temperature from using (**a**) 9 California counties to (**b**) 16 climate zones as designated by the California Energy Commission (apparent temperature ranges are shown for Basu et al. [14] mortality study and Basu et al. [26] mental health study, respectively)

consider those who are at greatest risk for cardiovascular disease as well as the elderly, children, and minority race/ethnic groups. Since both positive and negative associations were observed, it is important to consider cause-specific disease subgroups rather than relying on an outcome such as "all cardiovascular diseases," which would underestimate the positive associations.

Recent evidence has suggested that violence, aggression, and risky behaviors increase during hotter temperatures, proposing that heat may impact neurological outcomes [24]. Certain medications, such as beta-blockers to treat heart disease or anti-depressants, may contribute to increased risk as well as suppression of neurotransmitters that correspond with thermoregulation (see Biological Mechanisms below) [25]. Thus, a study of temperature and mental health–related outcomes, including 219,942 emergency room visits of neurotic and psychotic symptoms as well as self-injury/suicide, and intentionavl injury/homicide, was conducted in California [26]. Apparent temperatures during both the warm and cold seasons were associated with increases in the risk of emergency room visits for all mental health outcomes studied. During the warm season, a 10 °F increase in same-day mean apparent temperature was associated with 4.8% (95% CI: 3.6, 6.0), 5.8% (4.5, 7.1), and 7.9% (7.3, 8.4) increases in the risk of emergency room visits for mental health disorders, self-injury/suicide, and intentional injury/homicide, respectively. Hispanics, Whites, persons aged 6–18 years, and females were at greatest risk for most outcomes, although other variations were observed between race/ethnic group, age, and sex. These findings warrant further study in other locations, particularly areas with more climate variation throughout the year.

Heat Waves and Morbidity

In a study examining the effects of the 2006 California heat wave on morbidity, Knowlton et al. [27] aggregated county-level hospitalizations and emergency department (ED) visits for all causes and for some specific causes for six geographic regions of California. Excess morbidity and rate ratios (RRs) during the heat wave (July 15 to August 1, 2006) were calculated and compared to a referent period (July 8–14 and August 12–22, 2006). During the heat wave, 16,166 excess ED visits and 1182 excess hospitalizations occurred. ED visits for heat-related causes were found to be increased (RR 6.30, 95% CI: 5.67, 7.01). The greatest risks were found in the Central Coast, children (0–4 years), and the elderly (≥65 years of age). Acute renal failure, cardiovascular diseases, diabetes, electrolyte imbalance, and nephritis also had significantly increased risk. Some regions with relatively mild temperatures were found to be at increased risk, suggesting the influential roles of population acclimatization and biological adaption.

In a more recent study linking temperature and hospitalizations from May through October 1999–2009 from cardiovascular disease, respiratory disease, dehydration, acute renal failure, heat illness, and mental health, there were 11,000 excess hospitalizations that were due to extreme heat [28]. However, the majority of the 19 regional heat waves that were identified were not accompanied by a heat advisory or warning from the National Weather Service. Heat waves were defined as at least two consecutive days where mean apparent temperature exceeded the 95th percentile in the region. Using the same data [29], higher temperatures and heat waves were compared with hospitalizations to evaluate the "added" effect of a heat wave. Admissions for acute renal failure, appendicitis, dehydration, ischemic stroke, mental health, noninfectious enteritis, and primary diabetes were significantly increased with higher temperatures. Additional heat wave associations were observed for acute renal failure and dehydration. Higher temperatures also had statistically significant decreases in hypertension admissions, respiratory admissions, and respiratory diseases with secondary diagnoses of diabetes. These findings show that both temperature and heat wave exposures can exert effects independently.

Temperature and Adverse Birth Outcomes

In the first large-scale study of temperature and preterm delivery (20 to <37 gestational weeks) in the United States, Basu et al. [30] examined approximately 60,000 births spanning 16 counties in California from May through September 1999–2006. Apparent temperature was significantly associated with preterm birth for all mothers, regardless of maternal race/ethnic group, age, education, or infant sex. Per

10 °F increase in weekly average apparent temperature, an 8.6% (6.0, 11.3) increase in preterm delivery was found. Greater associations were observed for younger mothers, African Americans, and Asians, which may be markers of lower socioeconomic status. These associations were found to be independent of criteria air pollutants. Similar results were found when considering spontaneous preterm delivery during the warm season in a study using 14,466 electronic medical records of pregnant women from the Kaiser Permanente Division of Research in Northern California [31]. Mothers who were younger, Black, Hispanic, or underweight; smoked or consumed alcohol during pregnancy; or had preexisting or gestational hypertension, diabetes, or pre-eclampsia were found to be at greatest risk [32]. Several national and international studies have confirmed positive associations between heat and preterm delivery since the initial study in California was conducted in 2010.

Other adverse birth outcomes, such as stillbirth and low birth weight (LBW), have relatively little data to support the hypothesis that pregnant women and their developing fetuses are vulnerable populations from other adverse birth outcomes. In a recent case-crossover study, 8510 fetal deaths (≥20 weeks' gestation) were examined to estimate the association with mean apparent temperature in May through October 1999–2009. A 10.4% change (4.4, 16.8) in odds of stillbirth for every 10 °F increase in apparent temperature (cumulative average of lags 2–6 days) was found. Risk varied by maternal race/ethnicity and was greater for younger mothers, less-educated mothers, and male fetuses. The highest risks were observed during gestational weeks 20–25 and 31–33. No associations were found during the cold season (November–April), and the observed associations were independent of air pollutants. In another recent study, 43,629 full-term LBW infants and 2,032,601 normal-weight infants in California were evaluated from 1999 to 2013. For full gestation, a 13.0% change (4.1, 22.7) per 10 °F increase in apparent temperature was observed above 55 °F, with the greatest association for third trimester exposure above 60 °F. As in the previous stillbirth study, mothers who delivered male infants or gave birth during the warm season conferred the greatest risks. In contrast, mothers who were older were more vulnerable for term LBW, as were Black mothers. Maternal factors, such as age, race/ethnicity, and education, are often indicators for socioeconomic status.

Using data from 2002 through 2008 of 12 US sites, including Los Angeles, CA, investigators reported findings on preterm delivery for hot and cold extremes above the 90th percentile and below the 10th percentile, respectively, particularly during gestational weeks 34 and 36–38 [33]. In the same case-crossover study, an estimated 12–16% increase was found during the warm season of May through October and 4–5% decrease during the cold season of November through April were reported. Extreme temperatures also affected stillbirth from cold and hot temperatures during the full pregnancy but not the first or second trimesters using the same data [34], with the week prior to delivery during the warm season having a 6% (95% CI: 3–9%) increase in odds of stillbirth. The same data were used to analyze small for gestational age and term LBW infants with a slightly smaller population of 220,572 singleton births [35]. Again, both cold and heat extremes (this time defined as below the 5th percentile and above the 95th percentile, respectively) increased risk, while third trimester (RR: 1.31, 95% CI: 1.15–1.50) and whole pregnancy (2.49, 2.20–2.83) risks were the greatest for term LBW during the warm season, as was found in the study of California. First trimester, second trimester, and whole pregnancy during the cold season, however, were associated with increased term LBW risk. No consistent evidence was found between temperature and small for gestational age.

Future Projections for Temperature and Mortality

Using various climate models, greater increases in summer temperatures compared to winter temperatures are predicted. Based on the higher A1 emission scenarios, heat waves and extreme heat in Los Angeles are expected to be six to eight times more frequent, with heat-related excess mortality increasing five to seven times by the year 2100 [36]. The projections were slightly lower for the lower B2 emission scenarios. Under a more recent higher emission RCP 8.5 scenario, a net increase in annual temperature-related deaths per million people by 2086–2095 is projected, with about 627 (95% CI: 239–11,018) deaths per million in Los Angeles specifically [37]. Other investigators also predicted a significant increase in heat events with longer duration and greater frequency over the twenty-first century, particularly for coastal areas of California [38]. By the 2090s, annual mortality could rise to a total of 4684–8757 deaths per year in California depending upon the scenario used from the General Circulation Model. The elderly over 65 years and urban centers are likely to face the greatest impact. A similar prediction was made in another study, with the central estimate of annual mortality ranging from 2100 to 4300 for the year 2025 and from 6700 to 11,300 for 2050 [39]. A 10% and 20% increase in AC use would generate reductions of 16% and 33% in the years 2025 and 2050, respectively. National US estimates, including California-specific studies of annual incidence of heat-related mortality, were found to be 3700–3800 from all causes, 3500 from cardiovascular disease, and 21,000–27,000 from non-accidental deaths from May through September 2048–2052 relative to 1999–2003 using the A1 emissions scenario [40].

Biological Mechanisms

Since heat-related mortality and morbidity have multiple etiologies, a clear biologic mechanism or cause is unknown. Susceptible individuals such as the elderly, infants, children, or those with lower socioeconomic status may not be able to thermoregulate efficiently or have the financial means to mitigate heat exposure. When body temperatures rise, the body generally shifts blood flow from the vital organs to the skin's surface in an effort to cool down [41]. Thus, thermoregulation may be inadequate when too much blood is diverted from the vital organs, especially the heart [42]. Increased blood viscosity, elevated cholesterol levels associated with higher temperatures, and a higher sweating threshold have also been reported in vulnerable subgroups [43]. However, heat exposure decreases blood pressure, and therefore, both positive and negative associations are often observed for cause-specific cardiovascular diseases (Fig. 11.4). A possible mechanism for heat exposure and adverse birth outcomes may be increased dehydration, which decreases uterine blood flow and increases pituitary secretion of antidiuretic hormone and oxytocin to induce labor [44].

To attempt to understand the biological mechanisms involved with temperature and cardiovascular disease outcomes, a study examined biomarkers for inflammation (hs-C reactive protein), hemostasis (fibrinogen and factor VII, both of which aid with coagulation; tPA protein that is involved with blood clot resorption and PAI-1 counteracts it), and lipids (high-density lipoprotein [HDL]/aka "good cholesterol;" low-density lipoproteins [LDL]/aka "bad cholesterol;" total cholesterol; triglycerides) using data from the Study of Women's Health Across the Nation cohort from six sites, including two in California [45]. More significant ($p < 0.10$) negative associations were found during the warm season for various lag times, specifically for hs-CRP, fibrinogen, tissue plasminogen activator antigen (tPA-ag), tissue

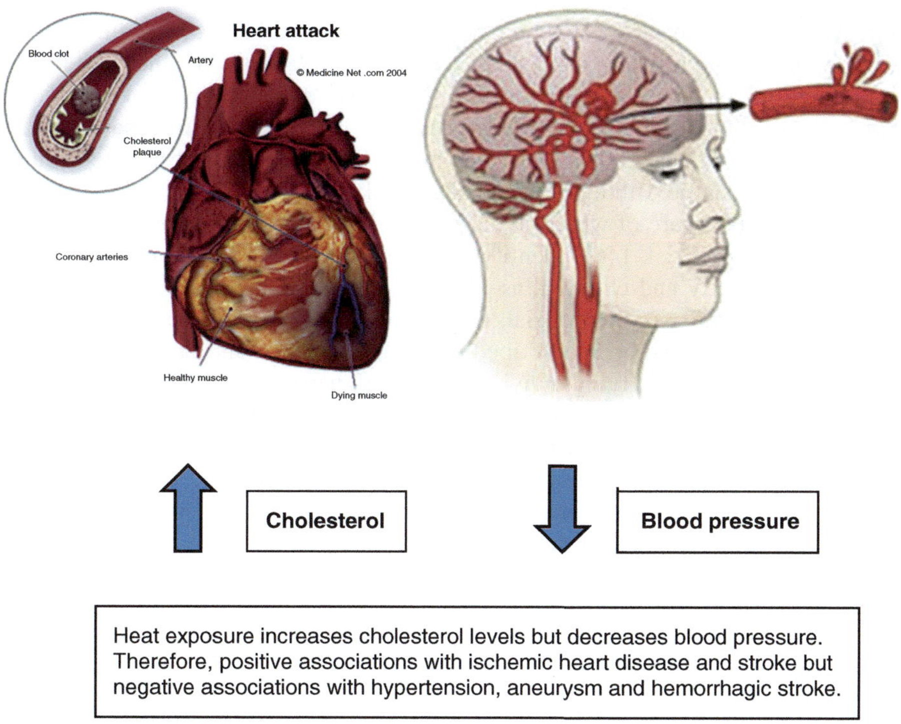

Fig. 11.4 Illustration of mechanisms of heat exposure on heart attacks and strokes

plasminogen activator antigen (PAI-1), Factor VIIc, high-density lipoprotein, and total cholesterol. During the cold season, significant negative associations for fibrinogen and HDL but significant positive associations for tPA-ag, PAI-1, and triglycerides were observed during various lag times. Thus, inflammation, hemostatis, and lipid fluctuations are all possible biological mechanisms for increased cardiovascular disease following ambient heat exposure. With the exception of ozone, air pollutants did not confound these associations.

Conclusions

Public health impacts of climate change in California are expected to be broad, including direct impacts from increased temperature and extreme weather events. Most epidemiologic studies of temperature and mortality or morbidity have been conducted in the past two decades. Prior to that, research was focused on case reports following heat waves, rather than using background apparent temperature as a measure of exposure. However, this topic warrants more research focusing on the southwest and California specifically for interventions for heat exposure as well as studies of wildfires and droughts.

Several research questions remain regarding the relationship between temperature, heat waves, and subsequent human mortality and morbidity. More information from public health research is needed to provide the National Weather Service the best measure of heat warning that is predictive of mortality and morbidity. Recommendations should be developed based on the characteristics that comprise the most effective heat-warning systems in the United States and how to develop such systems locally. Although individuals may know about heat-warning systems, they may not be aware of what actions need to be taken or perceive themselves as being at increased risk [46]. Identifying co-morbidities in vulnerable subgroups such as the elderly and children, as well as communicating precautionary efforts that can be administered, is crucial. Expansion of personal heat exposure assessment studies, using methods described previously by Basu and Samet [47], would be informative for identifying individual high-risk personal and housing characteristics as well as for understanding the biological mechanism between heat exposure and associated morbidity and mortality. Since heat waves are expected to occur more frequently with longer duration, the focus of epidemiologic studies should be on the higher end of temperature exposure or temperature extremes, as they will continue to have the greatest public health impact with increasing greenhouse gas emissions predicted in the future [48].

Disclaimer The opinions expressed in this chapter are solely those of the author and do not represent the policy or position of the State of California or the California Environmental Protection Agency.

References

1. California's Fourth Climate Assessment Report Sacramento, CA: Governor's Office of Planning and Research, California Natural Resources Energy, California Energy Commission. 2018; http://www.climateassessment.ca.gov/
2. Basu R, Samet JM. Relation between elevated ambient temperature and mortality: a review of the epidemiologic evidence. Epidemiol Rev. 2002;24(2):190–202.
3. Basu R. High ambient temperature and mortality: a review of epidemiologic studies from 2001 to 2008. Environ Health. 2009;8(40) https://doi.org/10.1186/476-069X-8-40.
4. Benmarhnia T, Deguen S, Kaufman JS, Smargiassi A. Vulnerability to heat-related mortality: a systematic review, meta-analysis, and meta-regression analysis. Epidemiology. 2015;26(6):781–93.
5. Song X, Wang S, Hu Y, Yue M, Zhang T, Liu Y, Tian J, Shang K. Impact of ambient temperature on morbidity and mortality: an overview of reviews. Sci Total Environ. 2017;586:241–54.
6. Basu R, Dominici F, Samet JM. Temperature and mortality among the elderly in the United States: a comparison of epidemiologic methods. Epidemiology. 2005;16(1):58–66.
7. Barnett AG. Temperature and cardiovascular deaths in the US elderly: changes over time. Epidemiology. 2007;18(3):369–72.
8. Ren C, Williams GM, Morawska L, Mengersen K, Tong S. Ozone modifies associations between temperature and cardiovascular mortality: analysis of the NMMAPS data. Occup Environ Med. 2008;65(4):255–60.
9. O'Neill MS, Zanobetti A, Schwartz J. Disparities by race in heat-related mortality in four US cities: the role of air conditioning prevalence. J Urban Health. 2005;82(2):191–7.
10. Commission CE. California statewide residential appliance saturation study: final report. Sacramento, California: California Energy Commission. 2008; Report No: 400-04-009.
11. Ostro B, Rauch S, Green R, Malig B, Basu R. The effects of temperature and use of air conditioning on hospitalizations. Am J Epidemiol. 2010;172(9):1053–61.
12. Medina-Ramon M, Schwartz J. Temperature, temperature extremes, and mortality: a study of acclimatization and effect modification in 50 United States cities. Occup Environ Med. 2007;64(12):827–33.
13. Medina-Ramón M, Zanobetti A, Cavanagh DP, Schwartz J. Extreme temperatures and mortality: assessing effect modification by personal characteristics and specific cause of death in a multi-city case-only analysis. Environ Health Perspect. 2006;114(9):1331–6.
14. Basu R, Feng WY, Ostro BD. Characterizing temperature and mortality in nine California counties. Epidemiology. 2008;19(1):138–45.
15. DerSimonian R, Laird N. Meta-analysis in clinical trials. Control Clin Trials. 1986;7(3):177–88.
16. Zanobetti A, Schwartz J. Temperature and mortality in nine US cities. Epidemiology. 2008;19(4):563–70.
17. Basu R, Ostro BD. A multicounty analysis identifying the populations vulnerable to mortality associated with high ambient temperature in California. Am J Epidemiol. 2008;168(6):632–7.
18. Basu R, Pearson D, Sie L, Broadwin R. A case-crossover study of temperature and infant mortality in California. Paediatr Perinat Epidemiol. 2015;29(5):407–15.
19. Basu R, Malig B. High ambient temperature and mortality in California: exploring the roles of age, disease, and mortality displacement. Environ Res. 2011;111(8):1286–92.
20. Ostro BD, Roth LA, Green RS, Basu R. Estimating the mortality effect of the July 2006 California heat wave. Environ Res. 2009;109(5):614–9.
21. Green RS, Basu R, Malig B, Broadwin R, Kim JJ, Ostro B. The effect of temperature on hospital admissions in nine California counties. Int J Public Health. 2010;55(2):113–21.
22. California Energy Commission (CEC). California statewide residential appliance saturation study: final report. 400-04-009; 2004.
23. Basu R, Pearson D, Malig B, Broadwin R, Green RS. The effect of elevated ambient temperature on emergency room visits among vulnerable subgroups in California. Epidemiology. 2012;23(6):813–20.

24. Page LA, Hajat S, Kovats RS. Relationship between daily suicide counts and temperature in England and Wales. Br J Psychiatry J Ment Sci. 2007;191:106–12.
25. Martin-Latry K, Goumy MP, Latry P, Gabinski C, Begaud B, Faure I, et al. Psychotropic drugs use and risk of heat-related hospitalisation. Eur Psychiatry. 2007;22(6):335–8.
26. Basu R, Gavin L, Pearson D, Ebisu K, Malig B. Examining the association between apparent temperature and mental health-related emergency room visits in California. Am J Epidemiol. 2018;187(4):726–35.
27. Knowlton K, Rothkin-Ellman M, King G, Margolis HG, Smith D, Solomon G, et al. The 2006 California heat wave: impacts on hospitalizations and emergency department visits. Environ Health Perspect. 2009;117(1):61–7.
28. Guirguis K, Gershunov A, Tardy A, Basu R. The impact of recent heat waves on human health in California. J App Met Clim. 2014;53(1):3–19.
29. Sherbakov A, Malig B, Guirguis K, Gershunov A, Basu R. A time-series analysis of heat waves and hospital visits in California, 1999–2009. Environ Res. 2017;160:83–90.
30. Basu R, Malig B, Ostro B. High ambient temperature and the risk of preterm delivery. Am J Epidemiol. 2010;172(10):1108–17.
31. Avalos LA, Chen H, Li DK, Basu R. The impact of high apparent temperature on spontaneous preterm delivery: a case-crossover study. Environ Health. 2017;16(1):5. https://doi.org/10.1186/s12940-017-0209-5.
32. Basu R, Chen H, Li DK, Avalos LA. The impact of maternal factors on the association between temperature and preterm delivery. Environ Res. 2017;154:109–14. https://doi.org/10.1016/j.envres.2016.12.017.
33. Ha S, Liu D, Zhu Y, Kim SS, Sherman S, Mendola P. Ambient temperature and early delivery of singleton pregnancies. Environ Health Perspect. 2017;125(3):453–9.
34. Ha S, Liu D, Zhu Y, Soo Kim S, Sherman S, Grantz KL, et al. Ambient temperature and stillbirth: a multi-center retrospective cohort study. Environ Health Perspect. 2017;125(6):067011.
35. Ha S, Zhu Y, Liu D, Sherman S, Mendola P. Ambient temperature and air quality in relation to small for gestational age and term low birthweight. Environ Res. 2017;155:394–400.
36. Hayhoe K, Cayan D, Field CB, Frumhoff PC, Maurer EP, Miller NL, et al. Emissions pathways, climate change, and impacts on California. Proc Natl Acad Sci U S A. 2004;101(34):12422–7.
37. Weinberger KR, Haykin L, Eliot MN, Schwartz JD, Gasparrini A, Wellenius GA. Projected temperature-related deaths in the ten large US metropolitan areas under different climate change scenarios. Environ Int. 2017;107:196–204.
38. Sheridan S, Kalkstein L. A spatial synoptic classification approach to projected heat vulnerability in California under future climate changes scenarios. Prepared for the California Air Resources Board and California Environmental Protection Agency; 2011:1–144.
39. Ostro B, Rauch S, Green S. Quantifying the health impacts of future changes in California. Environ Res. 2011;111(8):1258–64.
40. Voorhees AS, Fann N, Fulcher C, Dolwick P, Hubbell B, Bierwagen B, et al. Climate change-related temperature impacts on warm season heat mortality: a proof-of-concept methodology using BenMAP. Environ Sci Technol. 2011;45(4):1450–7.
41. Charles L, Schauf DM, Moffett S. Human physiology: foundations and frontiers. 2nd ed. Oxford: Wm. C. Brown Publishers; 1998. p. 851.
42. Bouchama A, Knochel JP. Heat stroke. N Engl J Med. 2002;346(25):1978–88.
43. Astrand P-O, Rodahl K, Dahl HA, Stromme SB. Textbook of work physiology: physiological bases of exercise. 4th ed. Windsor, Ontario: McGraw-Hill; 2003. p. 649.
44. Stan C, Boulvain M, Hirsbrunner-Amagbaly P, Pfister R. Hydration for treatment of preterm labour. Cochrane Database Syst Rev. 2002;2:CD003096.
45. Basu R, May Wu X, Malig BJ, Broadwin R, Gold EB, Qi L, et al. Estimating the associations of apparent temperature and inflammatory, hemostatic, and lipid markers in a cohort of midlife women. Environ Res. 2016;152:322–7.

46. Kalkstein AJ, Sheridan SC. The social impacts of the heat-health watch/warning system in Phoenix, Arizona: assessing the perceived risk and response of the public. Int J Biometeorol. 2007;52(1):43–55.
47. Basu R, Samet JM. An exposure assessment study of ambient heat exposure in an elderly population in Baltimore, Maryland. Environ Health Perspect. 2002;110(12):1219–24.
48. Intergovernmental Panel on Climate Change (IPCC). Climate change 2014: impacts, adaptation, and vulnerability. Part A: global and sectoral aspects. Contribution of working group II to the fifth assessment report of the intergovernmental panel on climate change. [Field, C.B., V.R. Barros, D.J. Dokken, K.J. Mach, M.D. Mastrandrea, T.E. Bilir, M. Chatterjee, K.L. Ebi, Y.O. Estrada, R.C. Genova, B. Girma, E.S. Kissel, A.N. Levy, S. MacCracken, P.R. Mastrandrea, and L.L. White (eds.)]. Cambridge/New York: Cambridge University Press; 2014.

Chapter 12
Vector-Borne Diseases in a Changing Climate and World

Yesim Tozan, Ora Lee H. Branch, and Joacim Rocklöv

Introduction

The past few decades have marked significant reductions in many important infectious diseases at a global scale owing to large-scale concerted prevention and control efforts, significant advances in medical care and treatment, improved access to water and sanitation services, and overall development [1–3]. However, vector-borne diseases that are transmitted by mosquitoes, ticks, and other insect vectors continue to take a heavy toll on populations around the world, particularly in developing countries in tropical and subtropical regions [2]. Every year, more than 1 billion people are infected and more than 700,000 die from vector-borne diseases, including malaria, dengue, schistosomiasis, lymphatic filariasis, onchocerciasis, Chagas disease, and leishmaniasis [4]. The World Health Organization (WHO) estimates that these prominent vector-borne diseases together account for about 17% of the global burden of all infectious diseases [4]. In addition, other vector-borne diseases take a heavy toll on a population due to their targeting pregnant women and infants. For example, Zika virus is associated with congenital morbidity and

Y. Tozan (✉)
Global Health and Environmental Public Health Sciences Program, College of Global Public Health, New York University, New York, NY, USA
e-mail: tozan@nyu.edu

O. L. H. Branch
Department of Research, College of Health & Human Services, Concordia University, Portland, OR, USA
e-mail: obranch@cu-portland.edu

J. Rocklöv
Department of Public Health and Clinical Medicine, Section of Sustainable Health, Umeå University, Umeå, Sweden
e-mail: joacim.rocklov@umu.se

K. E. Pinkerton, W. N. Rom (eds.), *Climate Change and Global Public Health*, Respiratory Medicine, https://doi.org/10.1007/978-3-030-54746-2_12

mortality even in symptomless mothers and fathers, which is a particularly insidious public health and reproductive population threat [5].

Vector-borne diseases are long known to be highly sensitive to the changes in weather and climate. The research community has extensively studied these diseases to understand the effects of climate and climate variability on their transmission dynamics [6–8]. Malaria, one of the most prominent vector-borne diseases and the primary focus of large-scale control and elimination efforts, has been decreasing in incidence globally [1], while persisting or increasing in certain locations [9, 10]. The incidence of arboviral diseases, such as dengue and chikungunya, have been increasing steadily to the great concern of the global health community over the past few decades [11, 12]. These arboviruses are now responsible for explosive and periodic epidemics in endemic populations [11, 13–15] and have already caused outbreaks in immunologically naïve populations in previously disease-free areas where competent mosquito vectors are readily present [16, 17]. Although climatic factors have been shown to influence the transmission dynamics of these arboviruses, a combination of non-climatic factors, such as unplanned urbanization, land use changes, and increasing human mobility at local and global scales, is likely to have complex effects on vector breeding habitats and vector–host–pathogen interactions in a warming climate and have contributed to their geographic spread [16]. These non-climatic factors are now better understood as drivers of arboviral disease transmission and outbreaks specifically [18–23] and vector-borne disease transmission more generally [6, 24]. In a globalized world, vector-borne disease risks are further mediated by factors such as emergence and spread of drug and insecticide resistance, for instance, in the case of malaria [25, 26], and fluctuating resources for disease prevention and control efforts due to constantly shifting public health priorities, as experienced during the 2016 Zika virus epidemic [27].

Against this backdrop, the current scientific evidence indicates a dramatic expansion in the geographic range of mosquito-borne diseases in the coming decades in response to rising global temperatures and more variable weather [28]. Much research has centered on the use of statistical and mathematical models to assess future changes in transmission risks of most prominent vector-borne diseases, such as malaria [6] and dengue [8, 29]. Most of these works considered the climatic changes predicted by the Global Climate Models (GMCs) at regional and global scales according to a range of greenhouse gas emission trajectories during the twenty-first century [30]. The trajectories represent a range of greenhouse gas emission scenarios for the world, known as the representative concentration pathways (RCPs), and are developed by the scientific community at the request of the Intergovernmental Panel on Climate Change (IPCC) [31]. One of the important challenges for vector-borne disease control is, however, to consider the dynamic and complex interplay of the aforementioned climatic and non-climatic factors that affect population exposure to the changes in transmission risks. Ultimately, the risk of vector-borne disease in a population is determined by that population's vulnerability, which is a measure of the capacity available to adapt and respond to the changes in the environmental suitability for mosquito vectors, pathogen replication, and disease transmission [28]. In this chapter, we review the current status and challenges in understanding vector-borne diseases dynamics in a rapidly changing climate and environment. We focus on mosquito-borne diseases due to their current high burden and widespread global distribution and highlight current knowledge base and knowledge gaps on this topic to stimulate future research in this field.

Mosquito-Borne Diseases: Drivers, Dynamics, and Risk

Climatic Drivers of Mosquito-Borne Diseases

Development rates of parasites and arboviruses that cause disease and of mosquito vectors that transmit these pathogens are sensitive to even small changes in temperature and precipitation. Understanding how the environmental suitability for mosquito vectors, pathogen replication, and disease transmission changes in response to short-term climate variability and longer-term climate change is key to our understanding of the consequences for human exposure [28].

Mosquito vector abundance and distribution are strongly influenced by climatic conditions. In the immature stages of their life cycle, mosquito vectors depend on fresh and clean water for oviposition and breeding. Air and water temperature govern the development rate of mosquitoes from larvae to pupae. Overall, a warmer environment leads to faster development. Recent studies have also shown that temperature during larval development has a direct effect on adult mosquito size [32]. Adult mosquito size can affect a number of epidemiologically significant traits, such as longevity, length of gonotrophic cycle (time between two consecutive blood meals), blood meal size, biting rate, immunocompetence, and infection intensity, all of which in turn are known to affect mosquito survival and pathogen development within vectors [32]. In the adult stages, mosquito vectors are affected by temperature predominantly for their survival. The relationship between vector survival and temperature has been studied extensively in experimental conditions. Overall, mosquito mortality increases rapidly with decreasing temperature [33]. Humid environments, on the other hand, are shown to favor vector survival at the same temperature and precipitation levels. Small increases in temperature lead to faster blood meal digestion, shorter gonotrophic cycle, and increased biting rates [34], resulting in a higher capacity for mosquito vectors to transmit the pathogen among humans, known as vectorial capacity.

The replication of parasites and arboviruses within vectors, which is another important determinant of disease transmission, is long known to be temperature sensitive [35]. Specifically, the duration of pathogen development within vectors, known as the extrinsic incubation period, is dependent on the temperature surrounding vectors. There is a threshold temperature below which a pathogen will not continue to develop in the vector (e.g., 15.4 °C for malaria parasite and 19 °C for dengue virus [36]). On the flip side, some pathogens stop developing above certain temperatures (e.g., 34.4 °C for malaria parasite and 31.7 °C for dengue virus [36, 37]). Any increase in temperature within these thresholds typically accelerates pathogen development within vectors and is expected to increase vectorial capacity. The extrinsic incubation period gets significantly prolonged at temperatures below the threshold value. Therefore, at decreasing temperatures, vectors may not outlive the extrinsic incubation period and are less likely to transmit pathogens to hosts. Above the threshold temperature, the development rates of both vectors and pathogens increase; however, such warming may also be associated with a drier climate characterized by reduced rainfall where vectors are less likely to survive and breed, and these conditions may negatively affect transmission dynamics [38].

In reality, mosquito vectors do not only experience a constant mean temperature, but are also exposed to fluctuating temperatures throughout their life cycle. Short-term variability in temperature, as well as extreme temperatures, may affect the ability of mosquito vectors to effectively transmit pathogens. For instance, diurnal temperature ranges (DTRs) have been found to be more important than changes in average temperature for transmission of malaria and dengue [39–42]. One study on malaria showed that daily temperature fluctuations can significantly alter the extrinsic incubation period of the malaria parasite and hence transmission rates [40]. The study findings indicated that DTRs around >21 °C slowed parasite development compared with constant mean temperatures, whereas DTRs around <21 °C sped up parasite development. On the other hand, extreme hot temperatures may increase mosquito mortality and decrease vectorial capacity and transmissions risk [43]. Further, a recent study on dengue examined the combined effect of temperature and DTR on the epidemic potential of this disease worldwide over a 200-year period (1901–2099) using historical and predicted climate data under the high greenhouse gas emission scenario (i.e., RCP8.5) [8]. The study found small increases in the dengue epidemic potential over the past 100 years while predicting larger increases in temperature by the end of this century in Northern Hemisphere regions (Fig. 12.1). Further, the study reported an overall increasing trend in the epidemic potential in temperate regions over time. These findings indicate that short-term temperature fluctuations and extreme temperatures need to be considered when investigating the impact of longer-term climatic changes on transmission dynamics of mosquito-borne diseases.

Similarly, precipitation extremes, whether associated with heavy rainfall or drought, have been found to be more important than changes in average precipitation [44]. Although heavy rainfall may temporarily reduce the risk of transmission by flushing out larvae and pupae from breeding sites, residual water pools create optimum breeding grounds for vectors. Extreme events, such as flooding and drying of riverbeds, may have a greater impact on the life cycle of mosquito vectors and the incubation of parasites and arboviruses, and strong associations were found between precipitation anomalies and mosquito-borne diseases [45–47]. According to the 2018 report of the Lancet Countdown, changes in extreme precipitation and droughts have already been observed and vary regionally, with the most significant increases occurring in South America and southeast Asia, highlighting the varying impact of climate change in different parts of the world [28].

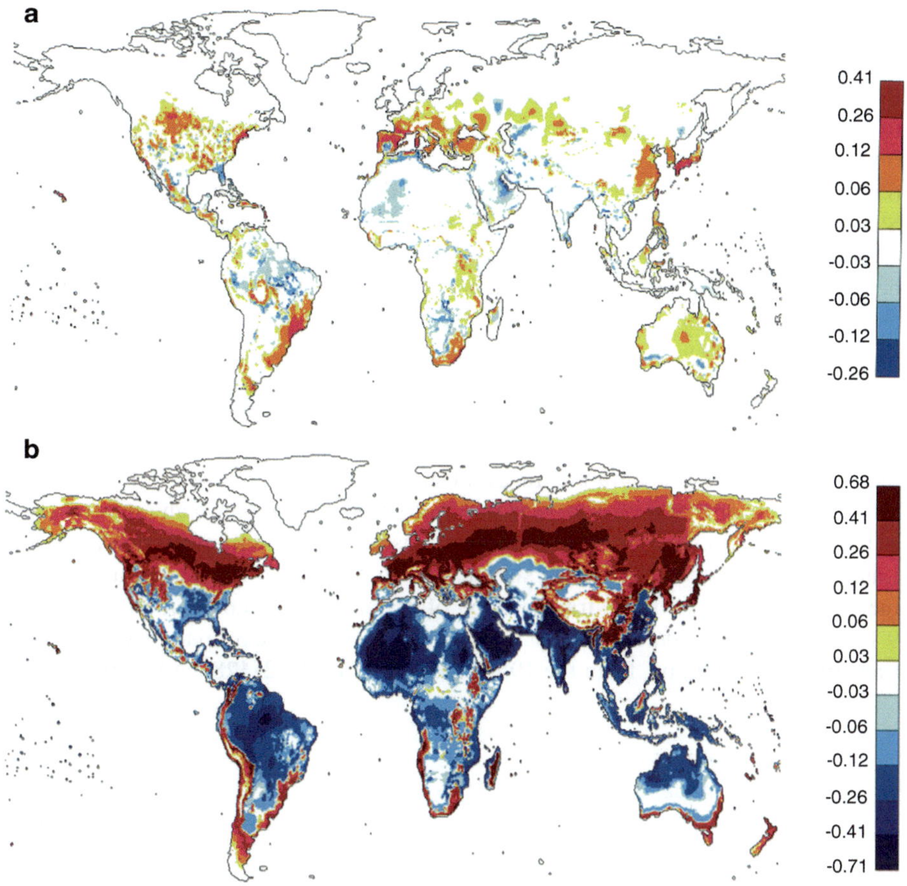

Fig. 12.1 Trend of global dengue epidemic potential (rVc) for the highest three consecutive months of the year. Differences in averaged rVc based on 30-year averages of temperature and DTR. (**a**) Differences between 1980–2009 and 1901–1930. (**b**) Differences between 2070–2099 and 1980–2009. The mean value of rVc was averaged from five global climate models under RCP8.5. The color bar describes the values of the rVc [8]

A growing number of studies have focused on investigating the associations between climatic factors and mosquito-borne disease risks across space and time. The objective is twofold. Identifying these spatiotemporal associations may improve our understanding of the impact of climatic variability on disease risk. Second, an improved understanding of the spatiotemporal distribution of disease risk may inform planning and implementation of effective interventions for mosquito-borne diseases. These studies have reported significant lag associations between temperature and rainfall and disease risk and have also revealed spatial heterogeneity in these associations [43, 48–51]. Of particular importance have been studies on the relationship of El Nino–Southern Oscillation to mosquito-borne diseases [17, 52].

A recent study on dengue in Sri Lanka found a strong association between the Oceanic Nino Index to weather patterns in the country and to dengue risk at a lag time of 6 months in a highly endemic district [53]. Furthermore, the study showed that increasing weekly mean temperature (above 29.8 °C) and increasing cumulative rainfall (above 50 mm) were associated with increased dengue risk, at 4 and 6 weeks of lag, respectively. This type of information can provide sufficient lead time to mount a timely response to abnormal disease events, including outbreaks. Shorter lags can, however, be expected in warmer climates where development rates of both vectors and parasites are faster. The study concluded that the considerable heterogeneity observed in dengue risk across the district at the same levels of temperature and rainfall could be due to the differences in population movements, human behavior, land use, and the effectiveness of dengue control interventions.

The aforementioned relationships highlight the complexities inherent in predicting the impact of changes in local climate and weather conditions on vector–pathogen–host systems in a rapidly changing climate. Despite this complexity, climatic factors constrain the geographic range of mosquito-borne diseases [54–58], determine their seasonal and year-to-year variability [59, 60], and have an important role in the longer-term shifts in their geographic distribution and transmission [61]. The Lancet Countdown's 2018 analysis showed that in 2016 the vectorial capacity increased by 27.6% for the transmission of malaria in highlands of Africa and by 9.1% for *Aedes aegypti* and 11.1% for *Aedes albopictus*, the primary vectors for the transmission of dengue, from the 1950s baseline (Fig. 12.2) [28]. According to this

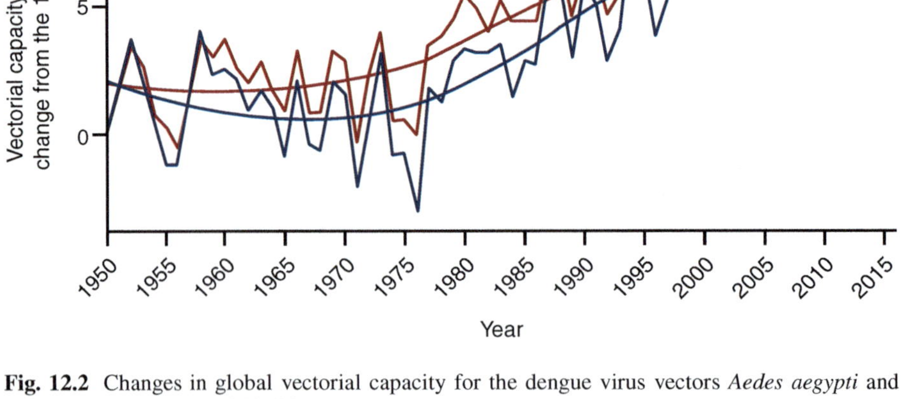

Fig. 12.2 Changes in global vectorial capacity for the dengue virus vectors *Aedes aegypti* and *Aedes albopictus* since 1950 [28]

analysis, the rise in global vectorial capacity for the transmission of dengue will continue given the projected increase in greenhouse gas emissions in the coming decades, which will be exacerbated by human mobility and globalization through their effects on the geographic spread of dengue virus and its competent mosquito vectors.

Non-climatic Drivers of Disease and Their Interactions with Climatic Drivers

In addition to climatic factors, there are a number of important non-climatic drivers of mosquito-borne diseases, such as changes in land use and human activity, human mobility, socioeconomic factors, and public health capacity. A comprehensive analysis of the impacts of climate variability and change should also consider the effects of non-climatic factors to determine the changes in exposure and vulnerability to mosquito-borne diseases in any given population. Changes in natural environments, such as unplanned urbanization and intensive agricultural activities with irrigation, combined with human mobility, can affect the distribution of mosquito vectors and pathogens, for instance, by increasing the environmental suitability for the proliferation of vectors as well as the density of susceptible populations.

The growth in air travel has accelerated introductions of pathogens to previously disease-free areas where competent vectors are already present and active. A recent study assessed the risk for chikungunya virus importation into France and Italy using air passenger journeys from international areas with active transmission as well as the risk of onward transmission by quantifying human mobility patterns using geocoded Twitter data during the 2017 outbreak [20]. The derived risk maps combining vectorial capacity and human mobility estimates had a good sensitivity in identifying at-risk areas for autochthonous chikungunya transmission during August–October 2017 (Fig. 12.3), with implications for targeting surveillance and outbreak response activities. Another recent study in Indonesia demonstrated the role of human mobility in the intra-urban spread of dengue by weighting local incidence data with geo-tagged Twitter data as a proxy for mobility patterns across 45 neighborhoods in Yogyakarta city during August 2016–June 2018 [19]. The study quantified the level of exposure to dengue virus in any given neighborhood by developing a new dynamic mobility-weighted incidence index, which was found to be a better predictor of dengue risk in a neighborhood than the recent transmission patterns in that neighborhood, or just the mobility patterns between neighborhoods.

The interactions between climatic, socioeconomic, and other factors are complex and dynamic. For example, in the case of dengue, non-climatic factors such as poor housing quality, limited access to safe water and sanitation, and poor waste management are likely to exacerbate the effects of climate change in crowded and highly

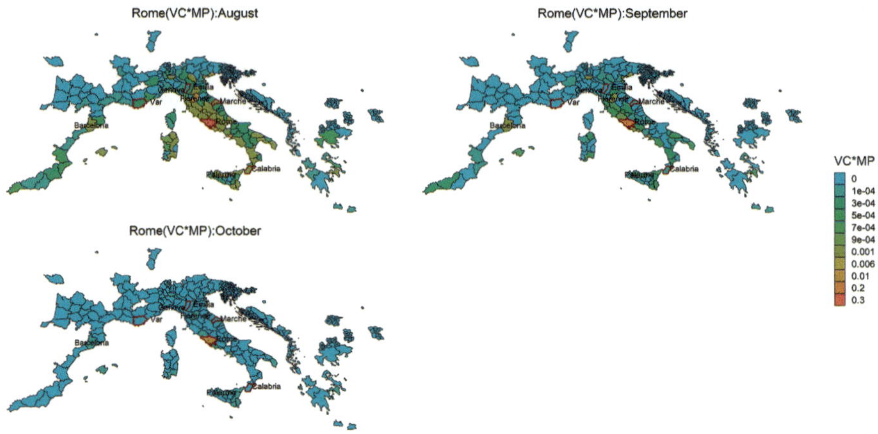

Fig. 12.3 Estimated areas of risk for chikungunya spread from the outbreak areas of Anzio and Rome in the Lazio region, Italy, based on combined VC and MP estimates, August–October 2017. Heavy outlines indicate the outbreak areas. MP, mobility proximity; VC, vectorial capacity [20]

connected urban settings. Long-term changes in climatic conditions and seasons can also affect mosquito vectors, human activity, and land use, which can further affect the spatial-temporal distribution and incidence of mosquito-borne diseases. Effective interventions and policies are key to respond to increasing risks from mosquito-borne diseases. These include awareness of mosquito-borne diseases and their impacts among key stakeholders, effective disease and vector surveillance systems, evidence-based prevention and control interventions, and increased support for research and development to understand current and future distribution of these diseases and their vectors. Socioeconomic conditions may, however, limit the capacity of public health systems to respond to changes, which in turn can increase the risks associated with any given level of climate change. For example, a statistical model, which combined projected changes in per capita gross domestic product with climate change, predicted that an additional 210 million people will be at risk of malaria by 2050 [62].

Human migration and other non-climate factors would arise from climate change and vector-borne diseases. For example, some populations might become denser after (or even in anticipation of) some geographic areas being non-desirable due to flooding or temperature changes.

As can be seen from the above discussion, combinations of these non-climatic drivers tend to increase vector abundance and/or host-vector interactions compared with the effects of each driver individually, with few combinations reducing the risks from mosquito-borne diseases [63]. The complexity of these interactions implies that the effects of climatic change on transmission risk will vary markedly by disease and geographic location in the face of non-climatic drivers of mosquito-borne diseases [64].

Predicting Climate Change Impacts on Mosquito-Borne Diseases: Current Status and Challenges

Scenario-based analysis is an integral part of climate change research and assessment. Last year marked the 30th anniversary of the IPCC, which has led the work on the development of the RCPs and, more recently, the Shared Socioeconomic Pathways (SSPs). The RCPs describe alternative scenarios of future radiative forcing relative to pre-industrial levels on the basis of a range of future global emission of greenhouse gases, which can then be expressed as an increase in the global mean surface temperature for the end of the twenty-first century [31]. While these climate change scenarios are not based on any socioeconomic narratives, emissions from short-term gases and land use changes are also incorporated [65]. Both GCM and regional climate models (RCM) use the RCPs as a basis for future projections of climate change over the course of this century.

Most studies have examined the changes in the range and intensity of mosquito-borne disease transmission according to these climate projections, employing an ensemble of climate models and using a diversity of modeling approaches ranging from statistical to mechanistic to hybrid [30, 66]. Climate projections over time can be made at global and regional scales, and this has allowed assessments of climate change impacts on disease transmission risk across different spatial and temporal scales [67]. Typically, assessments have used either the conservative low-emission scenario of RCP2.6 (radiative forcing of 2.6 W/m^2) or the drastic high-emission scenario of RCP8.5 (radiative forcing of 8.5 W/m^2), providing decision-makers with a range of worst-case and best-case scenarios. The increase in the global mean surface temperature for these two distinct scenarios is expected to range between 0.3–1.7 °C for RCP2.6 and 2.6–4.8 °C for RCP8.5 [68]. A recent review focusing on recent advances in modeling climate change impacts on mosquito-borne diseases has concluded that the use of different climate models and emission scenarios in future risk assessments has substantially improved with the availability of significantly greater funding for interdisciplinary research over the past decade [66]. The same review, however, highlighted that thorough validation of disease models is a continuing challenge due to a lack of field and laboratory data and that major uncertainties related to disease models, different climate models, and various emission scenarios should be clearly communicated to end users.

Over the past few years, the climate change research community has developed the SSPs for use within the scenario framework to represent different mitigation and adaptation challenges to climate change [63]. The SSPs comprise a set of five different socioeconomic development trajectories, describing a range of plausible futures under different demographic and economic development projections in which both the challenges to mitigation and adaptation are characterized as either high or low [69]. The SSPs are developed as reference pathways in the sense that these scenarios did not include any assumptions on climate change, its impacts, and climate policy responses, providing a starting point for developing integrated scenarios of the future [70]. A key aim of these integrated scenarios is to facilitate research to characterize the range of fundamental uncertainties in mitigation and adaptation efforts to achieve a given climate change target [69]. Future work must continue to combine the SSPs with the RCPs in earth system models for an integrated analysis of future climate impacts, vulnerabilities, adaptation, and mitigation [65]. The inclusion of socioeconomic development trajectories with climate change projections may allow for a more realistic assessment of future changes in the transmission risks of mosquito-borne diseases and support decision-maker needs from national to global levels [71].

Linking Climate and Climate Change Research to Health Policy and Programming: Current Initiatives

Research output on climate-driven risks on human health has increased significantly in recent years [72]. However, current research fails to match the demands of policymakers to enhance climate resilience in the health sector [64]. To increase the relevance to health programming, it is important to apply current research outputs to develop surveillance and response systems to anticipate, prevent, prepare for, and manage climate-related risks today [64]. In 2008, the member states represented on the World Health Assembly adopted a new resolution on health protection from climate change, harnessing a much higher level of commitment and engagement from the health sector [73]. The resolution called for close cooperation between the WHO, relevant organizations within and outside the United Nations, funding agencies, and member states to develop capacity to assess climate-driven risks on health and to implement effective response measures, by fostering interdisciplinary research and pilot projects in this area [64, 72].

Over the past decade, experience in strengthening the climate resilience of health systems has accumulated significantly through pilot projects, which aims to add resilience measures to the six building blocks (leadership and governance, health workforce, health information systems, essential medical products and technologies, service delivery) common to all health systems (Fig. 12.4) [74]. More specifically, there is now growing interest in climate data and information products to improve disease surveillance and response [75]. A central challenge to robust analyses of climate risks has been the very limited access to quality-controlled climate and weather data [76]. If available and accessible, climate data and information products can be used to answer the specific questions of the health community. To that end, the World Meteorological Organization has proposed a Global Framework on Climate Services (GFCS) to provide end users with policy relevant climate information and has been working closely with WHO to support the connection to health-policy makers. The GFCS approach has been piloted in a number of African countries, including Tanzania, Malawi, and Ethiopia, for malaria control and nutritional and disaster risks [77]. There are several other broad policy frameworks (e.g., the Paris Agreement) and global mechanisms (e.g., the United Nations Framework

Convention on Climate Change, the Climate for Development in Africa Initiative) that call for action to address the impact of climate change on health [78].

On the applied research front, a prominent example is the multi-country multi-year research initiative in sub-Saharan Africa implemented by the Special Programme for Research and Training in Tropical Diseases (TDR), with support from the International Development Research Center (IDRC). The TDR IDRC research initiative aims to understand the impact of climate change on population vulnerability to vector-borne diseases, including malaria, schistosomiasis, Rift Valley fever, and African trypanosomiasis. Taking a holistic approach, the various projects investigate the changing context of the environmental, social, economic, and climate conditions and their impact on vector-borne disease transmission and burden to get a better understanding of the adaptation needs.

The focus is on the most vulnerable populations, with the aim of developing decision support tools and strategies for adaptation to climate change, in line with the National Adaptation Plans for Climate Change [78].

There are a number of pressing needs at present. First, vector-borne disease control programs that integrate management of the risks brought by climate change should be strengthened [64, 76]. Political support and financial investments at national levels are key to scaling up innovative interventions and programs that address climate risks. There is also a need to better facilitate collaborations between Ministries of Health and other line ministries to ensure integration of health and

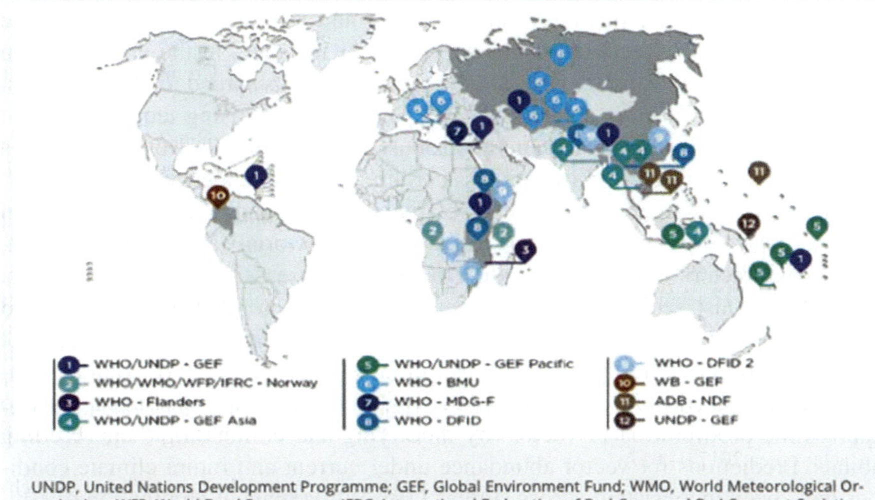

UNDP, United Nations Development Programme; GEF, Global Environment Fund; WMO, World Meteorological Organization; WFP, World Food Programme; IFRC, International Federation of Red Cross and Red Crescent Societies; BMU, Federal Minister for the Environment, Nature Conservation and Nuclear Safety (Germany); MDG-F, Millennium Development Goals Achievement Fund; DFID, Department for International Development (United Kingdom); WB, World Bank; ADB, Asian Development Bank; NDF, Nordic development Fund

Fig. 12.4 Completed, ongoing, or approved projects on health adaptation to climate change, 2008 to the present [74]

climate considerations in government-wide strategies while empowering public health practitioners, researchers, and communities [79, 80].

Future Directions for Research

Admittedly, studying vector-borne diseases is complex. Transmission results from dynamic and complex interactions between humans, vectors, and pathogens. These interactions are mediated by a multitude of climatic and environmental factors operating at multiple geographic and temporal scales and are further impacted by human activities and development. Numerous studies have examined the changes in the range and transmission risk of mosquito-borne diseases under current and future climatic conditions. Collectively, the results suggest that climate change, particularly rising temperatures, may cause the future range of disease vectors to expand from its present boundaries. Future studies are needed to ascertain the dependence of vectorial capacity parameters on temperature and also their sensitivity to diurnal temperature variations and further improve our currently limited understanding of these relationships. Such understanding is particularly important for *Ae. albopictus,* which is a competent vector for arboviruses. This vector has expanded its geographic range drastically over the past decade, including temperate areas in Europe, and is associated with dengue and chikungunya outbreaks in areas previously free of disease [20]. In relationship to temperature, studies that specifically incorporate urban heat islands are also needed. As a matter of fact, urban heat islands can be several degrees warmer than surrounding areas, and this increase in temperature may have profound effects on vector survival. For instance, in the case of urban dwelling of *Ae. aegypti*, urban heat islands can facilitate its spread to fringe areas in the United States, Europe, and China [16]. The effects of rising temperatures on vector population dynamics are generally robust across studies; however, the effects of rainfall fluctuations and shifts are less certain. Besides temperature and rainfall, there are other climatic factors, such as relative humidity and wind, that are known to affect vector development and survival and hence vectorial capacity. Future studies on these factors are also warranted to improve disease transmission models. Studies should also seek to better estimate and include vector densities that are field tested in the estimation of vectorial capacity. Vector density estimates should not only depend on climatic factors, such as rainfall, water temperature, and air temperature, but also consider land use, land cover, and local health system capacity to suppress the proliferation of vectors by modifying and eliminating their breeding habitats. Predictions for vector abundance under current and future climate conditions have been inconsistent across studies [16]. While manual elimination of containers that serve as breeding habitats may negatively affect the proliferation of *Ae. aegypti*, water storage practices because of drier conditions may favor it. It is important to bear in mind that disease models incorporating few biological and environmental factors may produce spurious estimates of vector abundance. Several of these factors are also likely to be affected by local socioeconomic and environmental conditions as well as vector control interventions in place.

One of the most challenging aspects of vector-borne disease prevention and control is the interdisciplinary nature of disease transmission and its drivers. Collaborations between researchers in physical science, epidemiology, and social science to better understand disease transmission dynamics have advanced considerably in recent years. These interdisciplinary collaborations have been encouraged by funding opportunities supported by the United States and European funding agencies and have substantially improved integrated assessments of disease transmission, predictability, and prevention. For the most part, these collaborations have involved diverse experts bringing their traditional analytical tools and study designs to the problem of vector-borne diseases, with minimal feedback across disciplines that limit more effective integration of techniques. For example, the epidemiological triangle of disease causation (agent–host–environment) often characterizes disease risk as discrete events between agents and hosts. Environment, often a distant third wheel, is usually considered as the place where agent–host interactions occur. This is where a large disconnect lies between epidemiology and land/climate scientists. In epidemiology, environment is a discrete space (e.g., community, political/administrative boundary, etc.) and is statistically modeled as a predictive variable of infection. However, land/climate scientists recognize that environmental characteristics are derived from modeled products (e.g., satellite imagery), and their use in epidemiology as input variables is actually continuous in space and time. The severing of a continuous ecological biome to examine discrete events can result in ecological fallacies or at least spurious relationships between environment and disease outcomes.

Overall, the application of satellite imagery to vector-borne diseases represents a unique opportunity. While multiple sources of information from satellite imagery and sensors are of interest to vector-borne disease risk monitoring and prediction, these satellite imagery sources and sensors are not designed with any specific consideration for what measurement characteristics would be most useful for vector-borne disease research or surveillance. Similarly, climate models are rarely optimized for vector-borne disease applications in their resolution, in periods of analysis, or even in the process simulations and model outputs. Of course, some limitations in these physical science techniques are difficult to overcome—high-resolution satellite-derived soil moisture measurements are expensive and sometimes impossible to obtain, and climate models are computationally intensive and are plagued by possibly irreducible uncertainties for both seasonal prediction and future climate change projections. Recognizing this, epidemiologists might need to alter the study designs and/or the surveillance networks to take full advantage of model results and satellite observations that are available.

In small research teams, there are opportunities for epidemiologists and land/climate scientists to collaborate during a given epidemic or during an eradication/elimination campaign. Epidemiologists might develop a model to test the effectiveness of focal screening for infection and treating individuals living near any given location [81]. However, in a resource-limited setting, there might not be enough public health resources to justify enacting this active surveillance and treatment public health campaign. Into the model, land/climate scientists might add the predicted climate and land use patterns. Satellite imagery and climate measures might be combined with epidemiologic surveillance and treatment protocols. The combined model would be a valuable tool for deciding upon public health strategies and priorities. Moreover, by modeling these factors together, we would increase our understanding of both climatic and non-climatic factors that impact vector-borne pathogens and the diseases they cause.

This suggests that collaborations that currently occur primarily at the scale of small research teams need to be moved upstream into satellite mission design, climate model development, and planning for health monitoring systems, so that the interdisciplinary nature of vector-borne disease problems is recognized in the design of the required research tools as well as in their application.

Acknowledgements This work was supported by two research grants from the Swedish Research Council Formas (grants no. 2018-01754 and 2017-01300).

Contributions YT wrote the initial draft. All authors critically reviewed and made extensive contributions to the final draft.

References

1. Bhatt S, Weiss DJ, Cameron E, Bisanzio D, Mappin B, Dalrymple U, et al. The effect of malaria control on Plasmodium falciparum in Africa between 2000 and 2015. Nature. 2015;526:207. https://doi.org/10.1038/nature15535. https://www.nature.com/articles/nature15535#supplementary-information
2. Dye C. After 2015: infectious diseases in a new era of health and development. Philos Trans R Soc Lond B Biol Sci. 369(1645):20130426. https://doi.org/10.1098/rstb.2013.0426. PubMed PMID: 24821913.
3. Weaver SC, Charlier C, Vasilakis N, Lecuit M. Zika, chikungunya, and other emerging vector-borne viral diseases. Annu Rev Med. 2018;69:395–408. https://doi.org/10.1146/annurev-med-050715-105122. Epub 2017/08/28. PubMed PMID: 28846489.
4. Organization WH. Vector-borne diseases 2016 [cited 2019 May 4]. Available from: http://www.who.int/en/news-room/factsheets/detail/vector-borne-diseases
5. Yun SI, Lee YM. Zika virus: An emerging flavivirus. J Microbiol (Seoul, Korea). 2017;55(3):204–19. https://doi.org/10.1007/s12275-017-7063-6. Epub 2017/03/01. PubMed PMID: 28243937.
6. Caminade C, Kovats S, Rocklov J, Tompkins AM, Morse AP, Colón-González FJ, et al. Impact of climate change on global malaria distribution. Proc Natl Acad Sci U S A. 2014;111(9):3286–91. https://doi.org/10.1073/pnas.1302089111. Epub 2014/02/03. PubMed PMID: 24596427.
7. Rocklöv J, Quam MB, Sudre B, German M, Kraemer MUG, Brady O, et al. Assessing seasonal risks for the introduction and mosquito-borne spread of Zika virus in Europe. EBioMedicine. 2016;9:250–6. https://doi.org/10.1016/j.ebiom.2016.06.009. PubMed PMID: 27344225.
8. Liu-Helmersson J, Stenlund H, Wilder-Smith A, Rocklöv J. Vectorial capacity of *Aedes aegypti*: effects of temperature and implications for global dengue epidemic potential. PLoS One. 2014;9(3):e89783. https://doi.org/10.1371/journal.pone.0089783.
9. Dhiman S. Are malaria elimination efforts on right track? An analysis of gains achieved and challenges ahead. Infect Dis Poverty. 2019;8(1):14. https://doi.org/10.1186/s40249-019-0524-x. PubMed PMID: 30760324.

10. Dhiman RC, Sarkar S. El Niño southern oscillation as an early warning tool for malaria outbreaks in India. Malar J. 2017;16(1):122. https://doi.org/10.1186/s12936-017-1779-y.
11. Stanaway JD, Shepard DS, Undurraga EA, Halasa YA, Coffeng LE, Brady OJ, et al. The global burden of dengue: an analysis from the global burden of disease study 2013. Lancet Infect Dis. 2016;16(6):712–23. https://doi.org/10.1016/S1473-3099(16)00026-8. Epub 2016/02/10. PubMed PMID: 26874619.
12. GBD DALYs and Hale Collaborators, CJL M, Barber RM, Foreman KJ, Abbasoglu Ozgoren A, Abd-Allah F, et al. Global, regional, and national disability-adjusted life years (DALYs) for 306 diseases and injuries and healthy life expectancy (HALE) for 188 countries, 1990–2013: quantifying the epidemiological transition. Lancet. 2015;386(10009):2145–91. https://doi.org/10.1016/S0140-6736(15)61340-X. Epub 2015/08/28. PubMed PMID: 26321261.
13. Smith KF, Goldberg M, Rosenthal S, Carlson L, Chen J, Chen C, et al. Global rise in human infectious disease outbreaks. J R Soc Interface. 2014;11(101):20140950. https://doi.org/10.1098/rsif.2014.0950. PubMed PMID: 25401184.
14. Khan K, Bogoch I, Brownstein JS, Miniota J, Nicolucci A, Hu W, et al. Assessing the origin of and potential for international spread of chikungunya virus from the Caribbean. PLoS Curr. 2014;6 https://doi.org/10.1371/currents.outbreaks.2134a0a7bf37fd8d388181539fea2da5. PubMed PMID: 24944846.
15. Bhatt S, Gething PW, Brady OJ, Messina JP, Farlow AW, Moyes CL, et al. The global distribution and burden of dengue. Nature. 2013;496(7446):504–7. https://doi.org/10.1038/nature12060. Epub 2013/04/07. PubMed PMID: 23563266.
16. Joacim R, Yesim T, Aditya R, Maquines OS, Bertrand S, Jon G, et al. Using big data to monitor the introduction and spread of chikungunya, Europe, 2017. Emerg Infect Dis J. 2019;25(6):1041. https://doi.org/10.3201/eid2506.180138.
17. Anyamba A, Chretien J-P, Britch SC, Soebiyanto RP, Small JL, Jepsen R, et al. Global disease outbreaks associated with the 2015–2016 El Niño Event. Sci Rep. 2019;9(1):1930. https://doi.org/10.1038/s41598-018-38034-z. PubMed PMID: 30760757.
18. Gubler DJ. Dengue, urbanization and globalization: the unholy trinity of the 21(st) century. Trop Med Health. 2011;39(4 Suppl):3–11. https://doi.org/10.2149/tmh.2011-S05. Epub 2012/04/14. PubMed PMID: 22500131; PubMed Central PMCID: PMCPMC3317603.
19. Ramadona AL, Tozan Y, Lazuardi L, Rocklov J. A combination of incidence data and mobility proxies from social media predicts the intra-urban spread of dengue in Yogyakarta, Indonesia. PLoS Negl Trop Dis. 2019;13(4):e0007298. https://doi.org/10.1371/journal.pntd.0007298. Epub 2019/04/16. PubMed PMID: 30986218; PubMed Central PMCID: PMCPMC6483276.
20. Rocklov J, Tozan Y, Ramadona A, Sewe MO, Sudre B, Garrido J, et al. Using big data to monitor the introduction and spread of chikungunya, Europe, 2017. Emerg Infect Dis. 2019;25(6):1041–9. https://doi.org/10.3201/eid2506.180138. Epub 2019/05/21. PubMed PMID: 31107221.
21. Stewart-Ibarra AM, Lowe R. Climate and non-climate drivers of dengue epidemics in southern coastal ecuador. Am J Trop Med Hyg. 2013;88(5):971–81. https://doi.org/10.4269/ajtmh.12-0478. Epub 2013/03/13. PubMed PMID: 23478584; PubMed Central PMCID: PMCPMC3752767.
22. Struchiner CJ, Rocklov J, Wilder-Smith A, Massad E. Increasing dengue incidence in Singapore over the past 40 years: population growth, climate and mobility. PLoS One. 2015;10(8):e0136286. https://doi.org/10.1371/journal.pone.0136286. Epub 2015/09/01. PubMed PMID: 26322517; PubMed Central PMCID: PMCPMC4554991.
23. Wilder-Smith A, Gubler DJ. Geographic expansion of dengue: the impact of international travel. Med Clin North Am. 2008;92(6):1377–90, x. https://doi.org/10.1016/j.mcna.2008.07.002. Epub 2008/12/09. PubMed PMID: 19061757.
24. Ermert V, Fink AH, Morse AP, Paeth H. The impact of regional climate change on malaria risk due to greenhouse forcing and land-use changes in tropical Africa. Environ Health Perspect. 2012;120(1):77–84. https://doi.org/10.1289/ehp.1103681. Epub 2011/09/07. PubMed PMID: 21900078.

25. Amato R, Pearson RD, Almagro-Garcia J, Amaratunga C, Lim P, Suon S, et al. Origins of the current outbreak of multidrug resistant malaria in Southeast Asia: a retrospective genetic study. bioRxiv. 2017;208371 https://doi.org/10.1101/208371.
26. Churcher TS, Lissenden N, Griffin JT, Worrall E, Ranson H. The impact of pyrethroid resistance on the efficacy and effectiveness of bednets for malaria control in Africa. eLife. 2016;5:e16090. https://doi.org/10.7554/eLife.16090. PubMed PMID: 27547988.
27. Wilder-Smith A, Gubler DJ, Weaver SC, Monath TP, Heymann DL, Scott TW. Epidemic arboviral diseases: priorities for research and public health. Lancet Infect Dis. 2017;17(3):e101–e6. https://doi.org/10.1016/S1473-3099(16)30518-7.
28. Watts N, Amann M, Arnell N, Ayeb-Karlsson S, Belesova K, Berry H, et al. The 2018 report of the <em>lancet</em> countdown on health and climate change: shaping the health of nations for centuries to come. Lancet. 2018;392(10163):2479–514. https://doi.org/10.1016/S0140-6736(18)32594-7.
29. Ryan SJ, Carlson CJ, Mordecai EA, Johnson LR. Global expansion and redistribution of Aedes-borne virus transmission risk with climate change. PLoS Negl Trop Dis. 2019;13(3):e0007213. https://doi.org/10.1371/journal.pntd.0007213.
30. Tjaden NB, Caminade C, Beierkuhnlein C, Thomas SM. Mosquito-borne diseases: advances in modelling climate-change impacts. Trends Parasitol. 2018;34(3):227–45. https://doi.org/10.1016/j.pt.2017.11.006.
31. van Vuuren DP, Edmonds J, Kainuma M, Riahi K, Thomson A, Hibbard K, et al. The representative concentration pathways: an overview. Clim Chang. 2011;109(1):5. https://doi.org/10.1007/s10584-011-0148-z.
32. Christiansen-Jucht C, Parham PE, Saddler A, Koella JC, Basáñez M-G. Temperature during larval development and adult maintenance influences the survival of Anopheles gambiae s.s. Parasit Vectors. 2014;7:489. https://doi.org/10.1186/s13071-014-0489-3. PubMed PMID: 25367091.
33. Alto BW, Bettinardi D. Temperature and dengue virus infection in mosquitoes: independent effects on the immature and adult stages. Am J Trop Med Hyg. 2013;88(3):497–505. https://doi.org/10.4269/ajtmh.12-0421. Epub 2013/02/06. PubMed PMID: 23382163; PubMed Central PMCID: PMCPMC3592531.
34. Beck-Johnson LM, Nelson WA, Paaijmans KP, Read AF, Thomas MB, Bjørnstad ON. The effect of temperature on Anopheles mosquito population dynamics and the potential for malaria transmission. PLoS One. 2013;8(11):e79276-e. https://doi.org/10.1371/journal.pone.0079276. PubMed PMID: 24244467.
35. DAVIS NC. The effect of various temperatures in modifying the extrinsic incubation period of the yellow fever virus in AËDES AEGYPTI*. Am J Epidemiol. 1932;16(1):163–76. https://doi.org/10.1093/oxfordjournals.aje.a117853.
36. Haider N, Kirkeby C, Kristensen B, Kjaer LJ, Sorensen JH, Bodker R. Microclimatic temperatures increase the potential for vector-borne disease transmission in the Scandinavian climate. Sci Rep. 2017;7(1):8175. https://doi.org/10.1038/s41598-017-08514-9. Epub 2017/08/16. PubMed PMID: 28811576; PubMed Central PMCID: PMCPMC5557972.
37. Bangs MJ, Pudiantari R, Gionar YR. Persistence of dengue virus RNA in dried *Aedes aegypti* (Diptera: Culicidae) exposed to natural tropical conditions. J Med Entomol. 2007;44(1):163–7. https://doi.org/10.1603/0022-2585(2007)44[163:podvri]2.0.co;2. Epub 2007/02/14. PubMed PMID: 17294936.
38. Thomson MC, Muñoz ÁG, Cousin R, Shumake-Guillemot J. Climate drivers of vector-borne diseases in Africa and their relevance to control programmes. Infect Dis Poverty. 2018;7(1):81. https://doi.org/10.1186/s40249-018-0460-1. PubMed PMID: 30092816.
39. Paaijmans KP, Read AF, Thomas MB. Understanding the link between malaria risk and climate. Proc Natl Acad Sci. 2009;106(33):13844–9. https://doi.org/10.1073/pnas.0903423106.
40. Paaijmans KP, Blanford S, Bell AS, Blanford JI, Read AF, Thomas MB. Influence of climate on malaria transmission depends on daily temperature variation. Proc Natl Acad Sci U S A. 2010;107(34):15135–9. https://doi.org/10.1073/pnas.1006422107. Epub 2010/08/09. PubMed PMID: 20696913.

41. Lambrechts L, Paaijmans KP, Fansiri T, Carrington LB, Kramer LD, Thomas MB, et al. Impact of daily temperature fluctuations on dengue virus transmission by *Aedes aegypti*. Proc Natl Acad Sci. 2011;108(18):7460–5. https://doi.org/10.1073/pnas.1101377108.
42. Blanford JI, Blanford S, Crane RG, Mann ME, Paaijmans KP, Schreiber KV, et al. Implications of temperature variation for malaria parasite development across Africa. Sci Rep. 2013;3:1300. https://doi.org/10.1038/srep01300. https://www.nature.com/articles/srep01300#supplementary-information
43. Hii YL, Rocklöv J, Ng N, Tang CS, Pang FY, Sauerborn R. Climate variability and increase in intensity and magnitude of dengue incidence in Singapore. Glob Health Action. 2009;2 https://doi.org/10.3402/gha.v2i0.2036. PubMed PMID: 20052380.
44. Linthicum KJAA, Britch SC, Small JL, Tucker CJ. Climate teleconnections, weather extremes, and vector-borne disease outbreaks. Washington, DC: National Academies Press (US); 2016. Available from: https://www.ncbi.nlm.nih.gov/books/NBK390440/
45. Beck-Johnson LM, Nelson WA, Paaijmans KP, Read AF, Thomas MB, Bjørnstad ON. The effect of temperature on Anopheles mosquito population dynamics and the potential for malaria transmission. PLoS One. 2013;8(11):e79276. https://doi.org/10.1371/journal.pone.0079276.
46. Anyamba A, Linthicum KJ, Small JL, Collins KM, Tucker CJ, Pak EW, et al. Climate Teleconnections and recent patterns of human and animal disease outbreaks. PLoS Negl Trop Dis. 2012;6(1):e1465. https://doi.org/10.1371/journal.pntd.0001465.
47. Chretien JP, Anyamba A, Bedno SA, Breiman RF, Sang R, Sergon K, et al. Drought-associated chikungunya emergence along coastal East Africa. Am J Trop Med Hyg. 2007;76(3):405–7. Epub 2007/03/16. PubMed PMID: 17360859.
48. Sewe M, Rocklov J, Williamson J, Hamel M, Nyaguara A, Odhiambo F, et al. The association of weather variability and under five malaria mortality in KEMRI/CDC HDSS in Western Kenya 2003 to 2008: a time series analysis. Int J Environ Res Public Health. 2015;12(2):1983–97. https://doi.org/10.3390/ijerph120201983. Epub 2015/02/13. PubMed PMID: 25674784; PubMed Central PMCID: PMCPMC4344705.
49. Sewe MO, Ahlm C, Rocklov J. Remotely sensed environmental conditions and malaria mortality in three malaria endemic regions in Western Kenya. PLoS One. 2016;11(4):e0154204. https://doi.org/10.1371/journal.pone.0154204. Epub 2016/04/27. PubMed PMID: 27115874; PubMed Central PMCID: PMCPMC4845989.
50. Ramadona AL, Lazuardi L, Hii YL, Holmner A, Kusnanto H, Rocklov J. Prediction of dengue outbreaks based on disease surveillance and meteorological data. PLoS One. 2016;11(3):e0152688. https://doi.org/10.1371/journal.pone.0152688. Epub 2016/04/01. PubMed PMID: 27031524; PubMed Central PMCID: PMCPMC4816319.
51. Chen MJ, Lin CY, Wu YT, Wu PC, Lung SC, Su HJ. Effects of extreme precipitation to the distribution of infectious diseases in Taiwan, 1994–2008. PLoS One. 2012;7(6):e34651. https://doi.org/10.1371/journal.pone.0034651. Epub 2012/06/28. PubMed PMID: 22737206; PubMed Central PMCID: PMCPMC3380951.
52. Johansson MA, Cummings DA, Glass GE. Multiyear climate variability and dengue – El Nino southern oscillation, weather, and dengue incidence in Puerto Rico, Mexico, and Thailand: a longitudinal data analysis. PLoS Med. 2009;6(11):e1000168. https://doi.org/10.1371/journal.pmed.1000168. Epub 2009/11/18. PubMed PMID: 19918363; PubMed Central PMCID: PMCPMC2771282.
53. Liyanage P, Tissera H, Sewe M, Quam M, Amarasinghe A, Palihawadana P, et al. A spatial hierarchical analysis of the temporal influences of the El Nino-southern oscillation and weather on dengue in Kalutara District, Sri Lanka. Int J Environ Res Public Health. 2016;13(11) https://doi.org/10.3390/ijerph13111087. Epub 2016/11/10. PubMed PMID: 27827943; PubMed Central PMCID: PMCPMC5129297.
54. Brady OJ, Golding N, Pigott DM, Kraemer MU, Messina JP, Reiner RC Jr, et al. Global temperature constraints on *Aedes aegypti* and Ae. albopictus persistence and competence for dengue virus transmission. Parasit Vectors. 2014;7:338. https://doi.org/10.1186/1756-3305-7-338. Epub 2014/07/24. PubMed PMID: 25052008; PubMed Central PMCID: PMCPMC4148136.

55. Paz S. Climate change impacts on West Nile virus transmission in a global context. Philos Trans R Soc Lond B Biol Sci. 2015;370(1665) https://doi.org/10.1098/rstb.2013.0561. Epub 2015/02/18. PubMed PMID: 25688020; PubMed Central PMCID: PMCPMC4342965.
56. Campbell LP, Luther C, Moo-Llanes D, Ramsey JM, Danis-Lozano R, Peterson AT. Climate change influences on global distributions of dengue and chikungunya virus vectors. Philos Trans R Soc Lond B Biol Sci. 2015;370(1665) https://doi.org/10.1098/rstb.2014.0135. Epub 2015/02/18. PubMed PMID: 25688023; PubMed Central PMCID: PMCPMC4342968.
57. Gething PW, Van Boeckel TP, Smith DL, Guerra CA, Patil AP, Snow RW, et al. Modelling the global constraints of temperature on transmission of Plasmodium falciparum and P. vivax. Parasit Vectors. 2011;4:92. https://doi.org/10.1186/1756-3305-4-92. Epub 2011/05/28. PubMed PMID: 21615906; PubMed Central PMCID: PMCPMC3115897.
58. Siraj AS, Santos-Vega M, Bouma MJ, Yadeta D, Ruiz Carrascal D, Pascual M. Altitudinal changes in malaria incidence in highlands of Ethiopia and Colombia. Science (New York, NY). 2014;343(6175):1154–8. https://doi.org/10.1126/science.1244325. Epub 2014/03/08. PubMed PMID: 24604201.
59. Servadio JL, Rosenthal SR, Carlson L, Bauer C. Climate patterns and mosquito-borne disease outbreaks in South and Southeast Asia. J Infect Public Health. 2018;11(4):566–71. https://doi.org/10.1016/j.jiph.2017.12.006. Epub 2017/12/25. PubMed PMID: 29274851.
60. Watts DM, Burke DS, Harrison BA, Whitmire RE, Nisalak A. Effect of temperature on the vector efficiency of *Aedes aegypti* for dengue 2 virus. Am J Trop Med Hyg. 1987;36(1):143–52. https://doi.org/10.4269/ajtmh.1987.36.143. Epub 1987/01/01. PubMed PMID: 3812879.
61. Liu-Helmersson J, Quam M, Wilder-Smith A, Stenlund H, Ebi K, Massad E, et al. Climate change and Aedes vectors: 21st century projections for dengue transmission in Europe. EBioMedicine. 2016;7:267–77. https://doi.org/10.1016/j.ebiom.2016.03.046. Epub 2016/06/21. PubMed PMID: 27322480; PubMed Central PMCID: PMCPMC4909611.
62. Béguin A, Hales S, Rocklöv J, Åström C, Louis VR, Sauerborn R. The opposing effects of climate change and socio-economic development on the global distribution of malaria. Glob Environ Chang. 2011;21(4):1209–14. https://doi.org/10.1016/j.gloenvcha.2011.06.001.
63. Sutherst RW. Global change and human vulnerability to vector-borne diseases. Clin Microbiol Rev. 2004;17(1):136–73. https://doi.org/10.1128/cmr.17.1.136-173.2004. Epub 2004/01/17. PubMed PMID: 14726459; PubMed Central PMCID: PMCPMC321469.
64. Campbell-Lendrum D, Manga L, Bagayoko M, Sommerfeld J. Climate change and vector-borne diseases: what are the implications for public health research and policy? Philos Trans R Soc Lond Ser B Biol Sci. 2015;370(1665):20130552. https://doi.org/10.1098/rstb.2013.0552. PubMed PMID: 25688013.
65. Riahi K, van Vuuren DP, Kriegler E, Edmonds J, O'Neill BC, Fujimori S, et al. The shared socioeconomic pathways and their energy, land use, and greenhouse gas emissions implications: an overview. Glob Environ Chang. 2017;42:153–68. https://doi.org/10.1016/j.gloenvcha.2016.05.009.
66. Tjaden NB, Suk JE, Fischer D, Thomas SM, Beierkuhnlein C, Semenza JC. Modelling the effects of global climate change on Chikungunya transmission in the 21(st) century. Sci Rep. 2017;7(1):3813. https://doi.org/10.1038/s41598-017-03566-3. Epub 2017/06/21. PubMed PMID: 28630444; PubMed Central PMCID: PMCPMC5476675.
67. Capinha C, Rocha J, Sousa CA. Macroclimate determines the global range limit of *Aedes aegypti*. EcoHealth. 2014;11(3):420–8. https://doi.org/10.1007/s10393-014-0918-y. Epub 2014/03/20. PubMed PMID: 24643859.
68. Collins M, Knutti R. Long-term climate change: projections, commitments and irreversibility. In: Stocker TF, editor. Climate change 2013: the physical science basis contribution of working group I to the fifth assessment report of the intergovernmental panel on climate change. Cambridge: Cambridge University Press; 2013. p. 1029–136.
69. O'Neill BC, Kriegler E, Riahi K, Ebi KL, Hallegatte S, Carter TR, et al. A new scenario framework for climate change research: the concept of shared socioeconomic pathways. Clim Chang. 2014;122(3):387–400. https://doi.org/10.1007/s10584-013-0905-2.

70. Kriegler E, O'Neill BC, Hallegatte S, Kram T, Lempert RJ, Moss RH, et al. The need for and use of socio-economic scenarios for climate change analysis: A new approach based on shared socio-economic pathways. Glob Environ Chang. 2012;22(4):807–22. https://doi.org/10.1016/j.gloenvcha.2012.05.005.
71. Rocklöv J, Tozan Y. Climate change and the rising infectiousness of dengue. Emerg Top Life Sci. 2019; https://doi.org/10.1042/etls20180123.
72. Hosking J, Campbell-Lendrum D. How well does climate change and human health research match the demands of policymakers? A scoping review. Environ Health Perspect. 2012;120(8):1076–82. https://doi.org/10.1289/ehp.1104093. Epub 2012/04/13. PubMed PMID: 22504669.
73. WHO. New World Health Assembly resolution on climate change and health 2008 [May 4, 2019]. Available from: https://www.who.int/globalchange/climate/EB_CChealth_resolution/en/
74. WHO. COP24 special report: health and climate change. Geneva: World Health Organization; 2018. Available from: http://www.who.int/iris/handle/10665/276405
75. Jancloes M, Thomson M, Costa MM, Hewitt C, Corvalan C, Dinku T, et al. Climate services to improve public health. Int J Environ Res Public Health. 2014;11(5):4555–9. https://doi.org/10.3390/ijerph110504555. PubMed PMID: 24776719.
76. Thomson MC. Emerging infectious diseases, vector-borne diseases, and climate change. In: Freedman B, editor. Global environmental change handbook of global environmental pollution, vol. 1. Dordrecht: Springer; 2014.
77. WHO. Climate change and health projects Geneva. Geneva: World Health Organization; 2019. [May 26, 2019]. Available from: www.who.int/globalchange/projects
78. Ramirez B. Diseases T-IRIoVB, Climate C. Support for research towards understanding the population health vulnerabilities to vector-borne diseases: increasing resilience under climate change conditions in Africa. Infect Dis Poverty. 2017;6(1):164. https://doi.org/10.1186/s40249-017-0378-z. PubMed PMID: 29228976.
79. Watts N, Adger WN, Agnolucci P, Blackstock J, Byass P, Cai W, et al. Health and climate change: policy responses to protect public health. Lancet. 2015;386(10006):1861–914. https://doi.org/10.1016/S0140-6736(15)60854-6.
80. Ebi KL, Semenza JC. Community-based adaptation to the health impacts of climate change. Am J Prev Med. 2008;35(5):501–7. https://doi.org/10.1016/j.amepre.2008.08.018. Epub 2008/10/22. PubMed PMID: 18929976.
81. Rosas-Aguirre A, Erhart A, Llanos-Cuentas A, Branch O, Berkvens D, Abatih E, et al. Modelling the potential of focal screening and treatment as elimination strategy for Plasmodium falciparum malaria in the Peruvian Amazon Region. Parasit Vectors. 2015;8:261. https://doi.org/10.1186/s13071-015-0868-4. PubMed PMID: 25948081.

Chapter 13
Dengue Fever and Climate Change

Lauren Cromar and Kevin Cromar

Introduction to Dengue Fever and Climate Change

Dengue fever is a viral tropical and subtropical mosquito-borne disease of special concern to public health in the context of a changing climate. Several factors have made this once localized disease rise to importance on the world stage during the latter half of the nineteenth century and climate change is expected to further its spread and intensity. Once isolated to a few areas in the tropics, dengue fever and its vectors have shown themselves to be highly adaptable to a wide variety of global environments and dengue fever is now the most rapidly spreading mosquito-borne disease in the world [157]. With a rapid spread fueled by globalization, transmission has increased dramatically in range and intensity and it is now found in 128 countries (Fig. 13.1). Today, an estimated 96 million clinically manifested cases occur annually out of an estimated 390 million total infections [12]. Population growth, unplanned and uncontrolled urbanization, and increased travel paired with ineffective vector control, disease surveillance, and inadequate public health infrastructure have been cited as drivers in the recent escalation of cases.

A growing public health concern exists not only due to the increased magnitude of incidence, but also to the escalating severity of its complications. Dengue hemorrhagic fever (DHF), which was recently reclassified as "severe dengue" by the World Health Organization, is a more serious and deadly form of the disease that has also become more prominent in recent decades. Currently, an estimated 500,000 people are hospitalized due to DHF each year with a case fatality rate of about 2.5% [156]. Most deaths occur in children, and DHF is now a leading cause of death in

L. Cromar
Tahoe Consulting, LLC, North Salt Lake, UT, USA

K. Cromar (✉)
Marron Institute of Urban Management, New York University, New York, NY, USA
e-mail: kevin.cromar@nyu.edu

K. E. Pinkerton, W. N. Rom (eds.), *Climate Change and Global Public Health*, Respiratory Medicine, https://doi.org/10.1007/978-3-030-54746-2_13

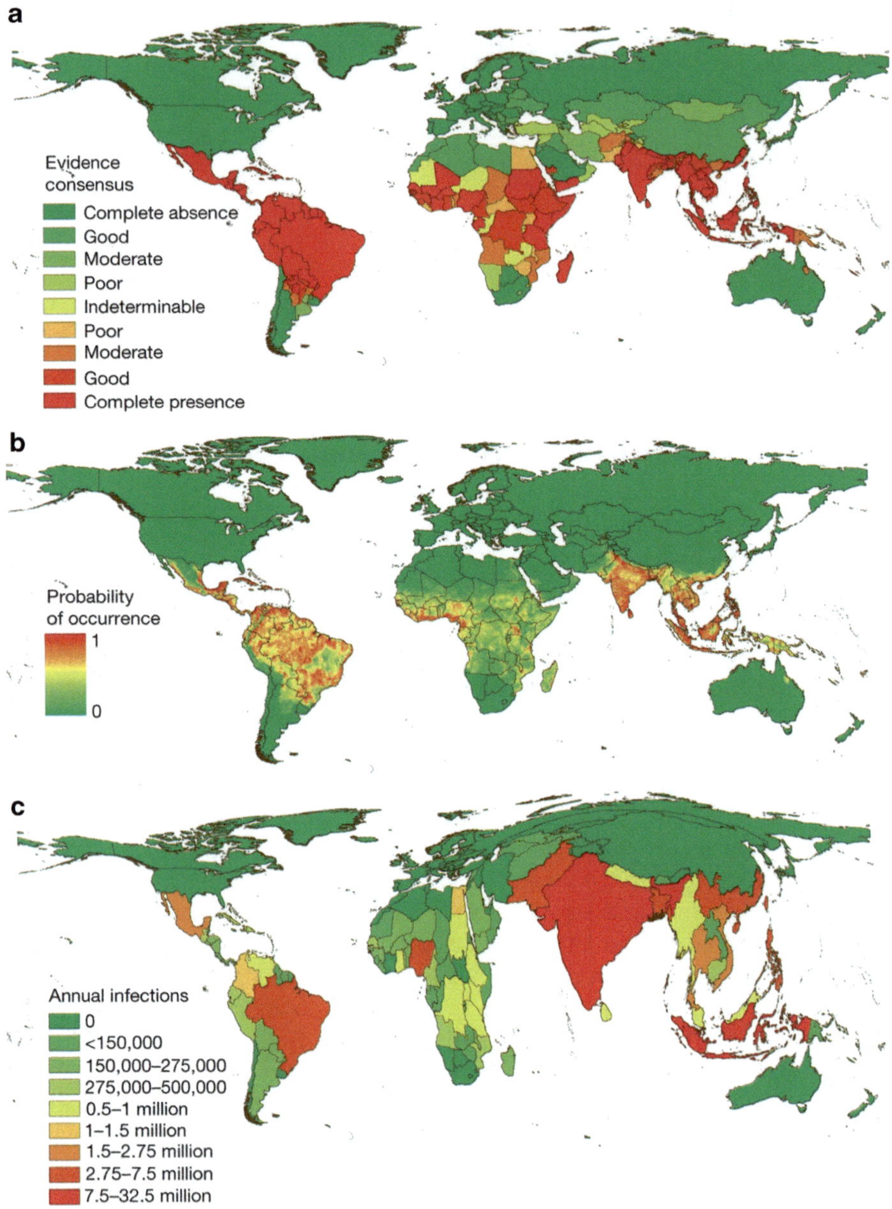

Fig. 13.1 Mapping the global evidence consensus, modeled risk, and estimated burden of dengue infections, 2010. **a**. Dengue infections have been recorded in over 128 counties, though inequalities in accurate diagnoses and reporting yield a skewed picture of actual distribution. **b**. Modeled probability of dengue occurrence finds that climatic and urbanization variables largely predict risk. c. Cartogram of estimated annual infections which are calculated to be around 390 million a year with 96 million infections being clinically apparent. [12]

children in several countries in Southeast Asia and in Latin America where the virus has an established history of hyperendemicity [121]. In areas such as South and Central America, where hyperendemicity is a new and growing occurrence, DHF rates have increased dramatically and experts fear that a continuing escalation of DHF incidence and mortality is inevitable.

The fact that dengue fever is a vector-borne disease makes it extremely sensitive to climatic variation. While many disease transmission routes have been linked to climate, the strongest relationships have been found in those spread through a mosquito vector [106]. Many observational studies have confirmed that dengue fever rates are often tightly correlated to climate conditions. This has been found for seasonal variation as well as for inter-annual departures from climatic norms. To better understand and predict dengue incidence, scientists have sought to define the relationships between climatic factors and the virus, its vectors, and the risk of transmission. Laboratory experiments have yielded strong mathematical relationships between climatic variables and many stages of the dengue transmission cycle. Modeling based on these relationships has been used to predict spatial and temporal variations in vector density and dengue risk in local areas as well as globally.

This framework of data and predictive tools has become especially informative in the context of global climate change along with the development of global climate modeling. A changing climate is predicted to expand the range of suitable habitat for dengue's mosquito vectors. Within that geographic range, greater portions of the world's population are predicted to live within a climate conducive to dengue epidemics. In addition to the modeled direct effects of a changing climate on epidemic potential, climate change is predicted to be significantly detrimental to societal and public health services stability in many of the same geographic areas where the population is already at risk of dengue or is expected to be under climate change scenarios.

Many of the same societal factors that have contributed to the recent global escalation in dengue fever incidence are likely to play an even larger role in the future. Changes in climate are expected to fuel breakdowns in ecological and economic systems, increase natural disasters, add to water insecurity, accelerate urban migrations, and result in widespread uncontrolled urbanization characterized by inadequate infrastructure, health services, and vector control. A more dengue-favorable climate combined with these climate-affected societal factors and projected population growth will result in the majority of the world's population living in areas at risk of dengue epidemics.

Dengue Disease

The dengue virus is a member of the family *Flaviviridae* along with West Nile virus and yellow fever virus. The dengue virus has four distinct serotypes, DENV-1, DENV-2, DENV-3, and DENV-4. Genetic differences between serotypes result in variations in the transmissibility and severity of the disease [140].

Infection from one of the dengue serotypes can cause a range of disease severity from asymptomatic cases to severe and even fatal infections. The most common manifestation of infection is dengue fever. Classical symptoms include a high fever with an abrupt onset accompanied by severe pain in the muscles and joints (thus earning it the name of "breakbone fever"), severe headaches, pain behind the eye, and a rash. Severe dengue (a newer classification including dengue hemorrhagic fever and dengue shock syndrome) is defined by the addition of shock or respiratory distress due to plasma leakage, clinically important bleeding, or severe organ impairment [117]. Unfortunately, there is no specific drug treatment for dengue fever, but supportive patient care can significantly improve disease progression and lower case fatality rates.

While the adaptive immune response from an initial infection proves to be protective from subsequent infections by the same serotype, it is not protective from infections from any of the three remaining serotypes. In fact, after a brief period of cross protection, antibodies from a previous infection of a different serotype are believed to be major factors in the development of DHF, the more severe form of the disease [64]. This occurs through a phenomenon known as antibody-dependent enhancement. Several other factors have been identified in triggering the development of DHF including varying degrees of virulence of the infecting strain and differences in the susceptibility of the patient due to age, immune status, race, and other genetic factors.

Vaccine development against dengue virus infection has proved challenging. As dengue fever can be caused by four separate serotypes of the virus, an effective vaccine must immunize against all four serotypes to be effective. Reinforcing this necessity is the fact that if a vaccine fails to provide immunity against one of the serotypes, an immunized individual is put at risk of developing a more severe disease via antibody-dependent enhancement. The requirement that four dengue serotype vaccines must be developed and combined in a single vaccine to preclude the development of DHF presents one of the largest challenges in vaccine development. Despite the challenges, several vaccine candidates have been developed and are at various stages of development, evaluation, and use. In 2016 a live recombinant tetravalent dengue vaccine developed by Sanofi Pasteur was made commercially available in eleven countries after passing stage III clinical trials. While the vaccine has been shown to be efficacious against severe dengue infection in those who have previously been infected with dengue virus, a greater risk of hospitalization and severe dengue were found in vaccinated individuals who had never been infected before [1]. Because of this, the World Health Organization recommends that the vaccine only be used in highly endemic areas and only in those individuals known to have been infected with dengue before [155]. As the greatest hope in reducing future dengue fever incidence lies in vaccination, it is hoped that future vaccine development efforts may prove successful.

Vectors and Transmission of Dengue

Dengue is transmitted by two mosquito species: primarily by *Aedes aegypti*, and secondarily by *Aedes albopictus*. *Ae. aegypti* are distributed around the globe in many areas throughout the tropics and subtropics and often invade farther north and south during the warmer summertime months [69, 71, 86]. Their current global distribution is at its widest historical point due in part to increased globalization, and international trade and travel [61].

Ae. aegypti exhibit a very high dengue virus infection rate, making them a very competent vector. It is this species that is primarily responsible for the high levels of endemic dengue fever in so many countries and for most of the explosive outbreaks that occur. *Ae. aegypti* are not vectors for dengue virus alone, but also carry the chikungunya virus, and the yellow fever virus which, despite the existence of an effective vaccine, still causes approximately 200,000 illnesses and 30,000 deaths per year. *Ae. aegypti* are also the primary vector for the Zika virus which made headlines in 2015 and 2016 as unprecedented outbreaks spread through much of the Americas, leaving a wake of miscarriages and infants born with microcephaly. Many of the same public health concerns relating to dengue fever and climate change are applicable to these other *Aedes*-borne viruses as well.

Ae. aegypti have several characteristics and tendencies which make them especially adapted to being an effective disease vector. They are almost entirely adapted to urban life, preferring to breed in or around homes in artificial household or yard water containers [94]. Common breeding sites include water storage drums, discarded automobile tires, vases, buckets, flower pots with saucers for water collection, and general trash (such as plastic containers) which can collect rainwater [63].

Adult mosquitoes have a tendency to live within homes and buildings, often taking refuge in the rafters and on the walls at night and feeding on the human inhabitants during the daytime hours [35]. Females are strongly anthropophilic, vastly preferring to feed on humans than on non-human mammals [135]. They are also easily interrupted while feeding and tend to have multiple feedings per completion of each gonotrophic cycle, thus allowing for disease transmission to multiple individuals [161].

Despite continued efforts to control *Ae. aegypti* around the globe, increasing dengue epidemics bear witness to their shortcomings and, at times, outright failures. Without an efficacious vaccine, preventative measures to combat dengue transmission are largely dependent on the control of vector populations. In the mid-twentieth century DDT was successfully used to eradicate *Ae. aegypti* from 19 different countries, but resistance to DDT and other pesticides has since developed, limiting their efficacy and requiring strategic use. Eliminating breeding habitats in and around homes is the primary method of vector control. This presents difficulties, however, as *Ae. aegypti* require very little water in which to lay their eggs and they utilize a wide range of breeding sites in urban and semi-urban habitats. Successful vector control campaigns require coordinated public health systems and consistent and substantial community involvement which are difficult to sustain. The lack of large-scale control of *Ae. aegypti* has been stated as being one of the most conspicuous failures in the public health sector. Trends in urban poverty and the expansion of slums will likely exacerbate efforts to control dengue rates as vector populations thrive in such areas and effective control is largely unattainable in such communities due to a multitude of factors [62].

A novel control method involving the infection of *Aedes* populations with *Wolbachia* bacteria has been developed and is being tested in many locations around the globe by the World Mosquito Program. In laboratory experiments, infection by *Wolbachia* in *Ae. aegypti* has resulted in a lengthened viral extrinsic incubation period (the time elapsed between viral infection of the mosquito and its becoming able to transmit the virus) and thus a decreased dengue virus transmission potential [26]. An initial trial in Townsville, Northern Australia, provided promising results, and it is hoped that this approach may be able to help ameliorate the global burden of dengue [113].

Aedes (*Stegomyia*) *albopictus*, a species commonly referred to as the "Asian tiger mosquito" are considered to be a secondary vector for the dengue virus and other arboviruses. Like *Ae. aegypti*, *Ae. albopictus* have experienced a wide expansion in global range but have done so in more recent history—with most of their continuing spread occurring over the last half century. The range of *Ae. albopictus* includes more northern and southern extremes than that of *Ae. aegypti* [15]. *Ae. albopictus* are highly invasive and ecologically adaptable [119]. They utilizes a wider range of habitats than *Ae. aegypti* and are more likely to be found in rural and suburban areas though they can also be found in urban areas as well. While *Ae. aegypti* are the primary vector responsible for most human dengue infections, *Ae. albopictus* have been shown to be the vector responsible for several dengue epidemics in locations scattered around the globe and play a larger role in dengue epidemiology in more temperate areas [13].

Ae. albopictus are less competent dengue vectors for several reasons. They are not as domesticated as *Ae. aegypti* and are less likely to be found indoors. The fact that they largely utilize natural breeding habitats and are less likely to become established in well-populated urban areas naturally results in their smaller role in precipitating outbreaks [60]. They are opportunistic feeders and do not display the marked preference for feeding on humans as do *Ae. aegypti* [119]. There is also evidence that *Ae. albopictus* have a lower oral receptivity and infection rate for the dengue virus [102]. Because of these factors, most scientific studies on dengue focus on the role of *Ae. aegypti* in dengue transmission.

Climatic Effects on Entomological and Viral Parameters in the Dengue Transmission Cycle

Investigations into the relationship between climate and dengue fever reveal that climatic factors strongly influence many of the biological and mechanical processes which drive dengue transmission. Many of the entomological variables which directly affect the severity of dengue epidemics are highly correlated with factors such as temperature, humidity, and rainfall. Not only do climatic factors primarily determine the range, density, and vector efficiency of *Ae. aegypti*, but they are also major factors in determining the rates of dengue virus multiplication and transmission (Fig. 13.2).

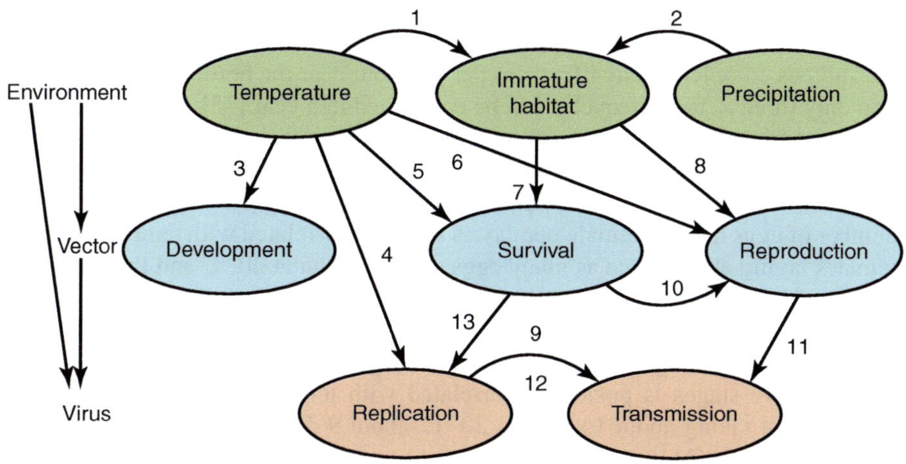

Fig. 13.2 Climatic variables impact the dengue virus transmission cycle in numerous ways through direct influence on vector and virus ecology. "Numbers identify relationships between variables. Habitat availability for mosquito larvae is influenced by temperature through evaporation and transpiration (1) and incoming precipitation (2). Temperature is a major regulator of mosquito development (3), viral replication within infected mosquitoes (4), mosquito survival (5), and the reproductive behavior of mosquitoes (6). Habitat availability is required for immature mosquito survival (7) and reproduction of adult mosquitoes (8). Faster mosquito development and increased survival will accelerate mosquito re production (9 and 10). Increased mosquito reproduction enhances the likelihood of transmission by increasing the number of blood feedings (11), whereas faster viral replication increases transmission by shortening the extrinsic incubation period (12). Last, increased survival of the adult mosquito increases the amount of viral replication (13)" [106]

Temperature

Of all the climatic factors that affect the lifecycle of *Ae. aegypti*, and consequently the transmission of dengue fever, the importance of temperature is perhaps the most conspicuous. The global range of *Ae. aegypti* is limited latitudinally and by altitude by decreasing temperatures. Beyond limiting *Ae. aegypti* distribution, temperature is an important factor in determining the population size within that range [45]. Field data has demonstrated a strong link between temperature and *Ae. aegypti* density over space and time, and it is understood that populations are generally favored with increasing temperatures.

The *Ae. aegypti* lifecycle displays several minimum and maximum temperature survival thresholds. Long-term exposure to temperatures under 10 °C or over 40°Cis generally lethal to eggs [51]. Upon hatching, larval and pupae survival is generally highest between 16 and 36 °C, dropping off steeply at lower and higher temperatures [30]. Adult activity and survival is limited outside of the range of 15–36°C [160]. In practice, however, *Ae. aegypti* populations have been found to survive despite extreme temperatures by taking refuge in or around buildings or breeding in large water storage tanks [35].

The time required to complete a lifecycle, a determinant in the size of the vector population, is also highly temperature-dependent. In most of their range, *Ae. aegypti* mosquitoes complete many lifecycles per year, but near the border of their distribution only three to four lifecycles can be completed in a year [35]. This difference is explained by temperature impacts on multiple stages throughout the lifecycle. For example, as temperatures decrease from 20 °C, females display a large delay in the time between blood meals and oviposition (the laying of eggs). Oviposition rate, the number of eggs laid per female per day, is strongly correlated with temperature with females laying about twice as many eggs at 25 °C than at 20 °C and three times as many at 31 °C [35, 160]. The incubation time before eclosion (hatching) is brief at high temperatures, taking only 2 days at 31 °C, and becoming longer as temperatures decreases, taking 20 days at 16 °C [51]. Upon eclosion, development through the immature stages is positively correlated with temperature with total time to development taking about 15 days at 20 °C, about 9 days at 25 °C, and about 6 or 7 days above 30 °C [160] (Fig. 13.3).

Relative Humidity

In addition to temperature impacts, the effects of relative humidity on the *Ae. aegypti* lifecycle are also significant. Relative humidity can play a role in evaporative loss of water from smaller containers serving as habitats for immature forms, but the most notable effects are limited to events within the adult and egg stages. Laboratory experiments with varying humidity levels tend to show graduated effects over a wide range of relative humidity levels as opposed to having defined threshold effects.

Under low-humidity conditions, females significantly delay oviposition and lay fewer eggs (e.g., an average of 11 eggs in 34% relative humidity compared to 31 eggs in 84% relative humidity over a 19-day period) [23]. Eggs remain viable for 2 months at 42% humidity, but for twice as long at 88% humidity [141].

The impact on adult survival is the most influential aspect of relative humidity in affecting dengue virus transmission rates. Because the latent period of the dengue virus within the mosquito can be fairly long, the lifespan of the mosquito is a critical factor to whether or not an infected mosquito may be able to transmit the virus before her death. The relationship between humidity and mosquito mortality is near linear with large differences existing even between higher levels of humidity [23, 88].

Precipitation

Water is a vital factor in the lifecycle of *Ae. aegypti* mosquitoes and it follows that rainfall is a driving force in population dynamics. *Ae. aegypti* eggs need standing water to hatch, and the pupal and larval stages are entirely aquatic and depend on

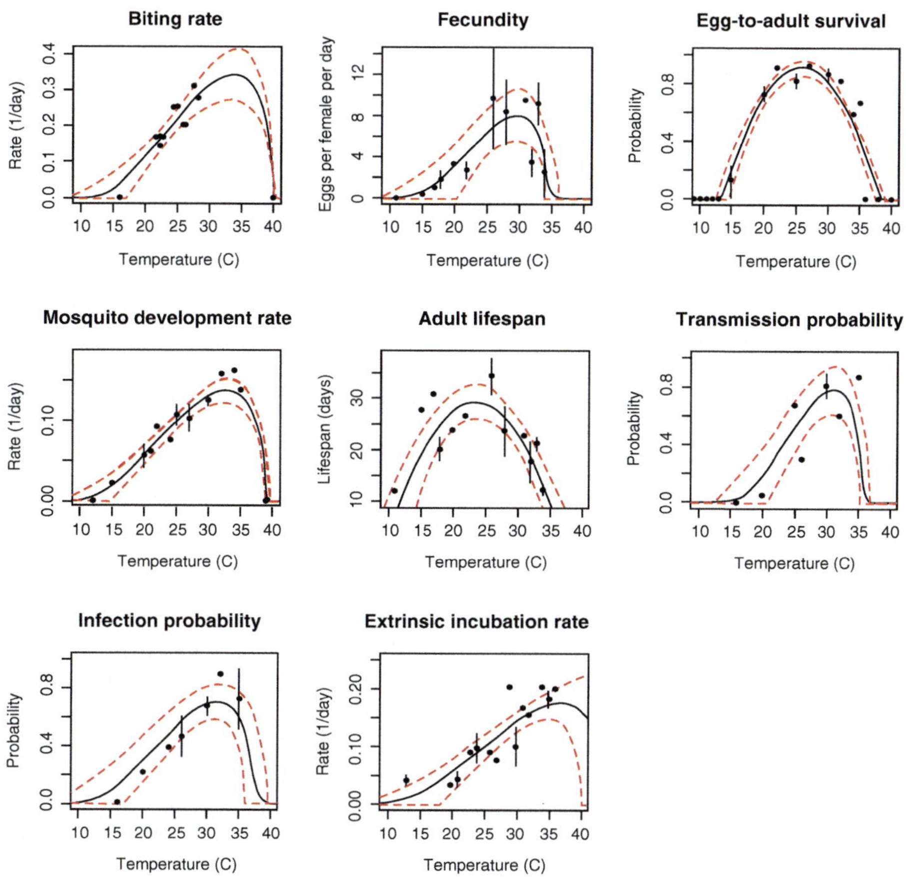

Fig. 13.3 A meta-analysis of numerous studies illustrates the thermal response of *Aedes aegypti* mosquito and dengue virus traits across environmental temperature ranges [103]

standing water in which to grow and develop. However, precipitation does not lend itself to the same experimentally derived quantitative relationships as temperature and humidity even though evidence of the importance of precipitation is found in many observational studies analyzing mosquito population dynamics in areas around the globe. Additionally, this relationship is often confounded by the fact that many of the containers that mosquitoes breed in are human-filled (such as those used for water storage) and can even be inversely related to rainfall in some locations.

Climatic Influence on Biting Behavior and Extrinsic Incubation Period

The role of biting rates in the epidemiology of dengue fever is critical. An increase in the biting rate not only increases the probability of a mosquito's becoming infected with the dengue virus, but also her ability to transmit it [137]. Whereas *Ae. aegypti* population size is linearly related to its vectorial capacity (the rate of future infections arising from each primary infection), biting rates are exponentially proportional to vectorial capacity. Thus, even small increases in the biting rate have the potential to result in relatively large increases in the incidence of dengue fever.

Higher temperatures result in a more rapid immature *Ae. aegypti* development time which in turn results in smaller adult mosquitoes with lower energy reserves. As this occurs, biting rates increase as females need to feed sooner and require more than one blood meal in order to complete a reproductive cycle [40, 50, 57, 94]. Experiments show a pattern of increasing biting rates with temperature until an optimal temperature around 30–35 °C is reached [93]. Humidity levels have also been found to promote increases in the general activity levels of the *Ae. aegypti* mosquito and positively correlate with increased biting rates [35].

Replication of the virus itself is also highly affected by temperature. The length of the extrinsic incubation time, or the period after a mosquito feeds on an infectious host until it is able to transmit the virus, is a crucial factor in transmission risk. When female mosquitoes bite an infected host, most are not capable of passing on the virus due to the fact that they often die before the extrinsic incubation time has completed [69]. More rapid viral replication makes this less likely, a process that has been shown to be highly temperature-dependent. For example, in mosquitoes infected with DEN-2, the extrinsic incubation time was found to be 7 days at 35°Cand 12 days at 30 °C. No viral transmission was found at 26 °C [150]. A similar study for all dengue virus serotypes found the mean extrinsic incubation time to be 6.5 days at 30 °C and 15 days at 25 °C [29].

Increased biting rates and a shorter extrinsic incubation time due to favorable climatic conditions may account for seasonal epidemics where a seasonally dynamic *Ae. aegypti* population has not been found and for epidemics in areas where *Ae. aegypti* density is below levels considered to be protective [114, 136, 139].

Daily Temperature Variation Effects on Vectorial Capacity

Dengue research has also looked beyond mean temperatures to examine the effects of the diurnal temperature range (DTR) which is the difference between the daily high and low temperature. It was found that at mean temperatures over 18 °C, increasing diurnal temperature range results in significantly fewer mosquitoes surviving long enough to become infected with the dengue virus [83]. Experiments on dengue viral replication in infected *Ae. aegypti* found that a large DTR resulted in a shorter extrinsic incubation time at low temperatures. At 20 °C the extrinsic incubation time was found to be 29.6 days, but when an 18 °C DTR was added, the extrinsic incubation time shortened to 18.9 days. Additionally, over twice the percentage of mosquitoes developed a disseminated infection than those at a constant 20 °C [24]. Researchers theorize that seasonal differences in daily temperature fluctuations may explain seasonal differences in dengue transmission rates in areas where the mean temperature does not vary significantly through the year. Laboratory experiments testing mosquito survival and infection rates in these modeled climatic conditions support this hypothesis [25].

Climate-Based, Mechanistic Dengue Transmission Models

As many facets of the growth, survival, reproduction, and behavior of *Ae. aegypti* and replication of the dengue virus have demonstrated clear relationships with climatic variables, several attempts have been made to mathematically model *Ae. aegypti* populations and the risk of dengue transmission based on a compilation of temperature-related factors. Modeling based on several entomological temperature-dependent variables calculates the optimal temperature for the maximum growth of *Ae. aegypti* populations to be 29.2 °C with an ideal range of 27–30 °C [159, 160]. More recent modeling, which also included viral replication and biting rates, found that transmission risk from *Ae. aegypti* exists between 17.8 and 34.6 °C, peaking at 29.1 °C, while *Ae. albopictus* was shown to be a more efficient at lower temperatures, peaking at 26.4 °C with a range of 16.2–31.6 °C (see Fig. 13.2) [103]. Separate modeling that incorporates daily temperature ranges along with mean temperatures postulates that *Ae. aegypti*/DENV vectorial capacity peaks at 29.3 °C, but that this is only true under conditions with "no to small" daily temperature fluctuations. For areas with larger daily temperature fluctuations, lower mean temperatures are more conducive to vectorial capacity [92]. This is also true in colder areas where vectorial capacity is increased when there is an increase in the daily temperature range.

Demonstrating the importance of temperature on dengue transmission, one study calculated temperature-dependent transmission thresholds in terms of the number of pupae per person required to maintain the dengue transmission cycle. These thresholds illustrate how hotter climates can make effective control efforts very difficult. For example, Bangkok, Thailand, has an observed 1.69 pupae per person on average [56, 142]. With an average summer temperature of 29.2 °C, the protective threshold is calculated at 0.29 pupae per person (assuming 33% seroprevalence of previous dengue infection), a value which would require control efforts to decrease *Ae. aegypti* prevalence by 83%. Mayaguez, Puerto Rico, on the other hand, has a similar average of pupae per person, 1.73, but, with a temperature of 26.6 °C, would need only a 40% effective control effort in order to bring pupae count down to a protective threshold of 1.05 per person [55].

Lessons Learned from Observational Studies

Observational studies that analyze the relationship between climatic variations and dengue fever incidence in locations around the globe yield further evidence of the relationship between climate and dengue in real-world settings. Despite the multiplicity of confounding factors, a wealth of strong observational evidence links dengue to climate and can help apprise scientists and public health officials of the effects of projected changes in climate on future dengue risk. Dengue and climate have been linked by a multitude of studies across both temporal and spatial scales utilizing various tools and methods of analysis. Analyses which test the correlation of dengue rates and climatic conditions in a specific location over time are most common.

The most common climatic factor associated with dengue incidence and/or *Ae. aegypti* population size in observational studies is rainfall. Rainfall is highly seasonal in many areas that experience dengue fever. In many such areas, *Ae. aegypti* populations and dengue infections become very low or nonexistent during the dry periods of the year when vector breeding becomes inhibited but increase sharply with the onset of the rainy season. While *Ae. aegypti* and dengue incidence are often strongly linked to the timing of rainfall, there is less evidence of a strong link to the magnitude of rainfall. In fact, an overabundance of rainfall may decrease the *Ae. aegypti* population by washing larvae from breeding containers [33, 77].

One caveat in the importance of rainfall is the fact that dry periods can increase household storage of water creating ideal breeding sites in close proximity to humans. Incidentally, water storage containers are one of the most productive vector breeding sites [85, 130]. Thus the vector cycle and dengue fever epidemics may also be induced by a lack of rainfall in areas where the water supply is unreliable or not easily accessible. Epidemics are also commonly linked to periods of moderate to severe drought. In such times, water storage becomes widespread and the emptying and cleaning of water containers is avoided [32, 42, 59, 122].

While the relationship between dengue and rainfall is clear in many locations, a strong correlation with temperature is often not found in observational studies. Some have cited these findings as evidence of a weak link between the two. However, these negative findings often occur in many dengue endemic countries that have little intra-annual temperature variation and rarely cool to levels which would inhibit mosquito activity. In such situations, temperature is sufficiently high during the period of highest rainfall, but a lack of rainfall inhibits the reproduction of the vector during the period of the year when temperatures are the highest.

Conversely, in areas where temperatures seasonally cycle to a level nonconductive to mosquito and viral activities, temperature is likewise found to be highly correlated with dengue incidence [17, 34, 73]. Additionally, when the effects of rainfall are taken out of the equation, such as in areas where there is sufficient rainfall year-round for vector breeding or where breeding sites are human-filled, a strong association with temperature is often revealed [7, 79, 101].

Further evidence of the importance of temperature can be found in spatially based studies comparing larger geographic regions where a strong association with annual temperature values is found more often than with rainfall levels. One meta-analysis of 33 observational studies linking dengue to temperature found that the ratio of dengue risk to an increase in temperature varied by location, but minimum, mean, and maximum temperature were all strongly correlated to dengue risk [49]. Minimum temperature was critical in some areas in determining mosquito survival and development rate. Mean temperature was most strongly associated with dengue risk, with 22° being the lower limit at which dengue risk increases significantly after which increasing temperatures were strongly correlated with increasing risk. It was determined that 29° was the upper limit for ideal mean temperature after which the risk of dengue transmission began to decline.

In areas with seasonal climates, it has been established that seasonal variations in temperature and rainfall drive the timing of "dengue seasons" but an understanding of what drives the magnitude of seasonal epidemics year to year is more useful. Of particular use are studies which analyze deviations in climate factors from normal cyclic levels. In Puerto Rico, for example, it was found that while intra-annual fluctuations in dengue incidence were driven by rainfall, year-to-year differences were temperature driven [76, 77]. In Thailand, the timing of rises in dengue incidence was also linked to the timing of rainfall. The magnitude of the outbreak intensity, however, was likewise found to be driven by increases in mean temperatures [143]. Similar conclusions have been found in other locations [7].

The El Niño Southern Oscillation (ENSO) phenomenon provides added data on dengue and climate relationships and affords additional clues on the effects of long-term climate change on dengue incidence and human health. ENSO-related deviations in climate mark a change from normal seasonal patterns allowing researchers to analyze dengue incidence under differing climatic conditions within the same geographical area. Regional prolonged dry conditions, altered rainfall patterns, and increases or decreases in temperature have been linked to El Niño and La Niña years as have changes in dengue incidence.

One of the most dramatic examples of ENSO-related epidemics occurred in conjunction with the 1997–1998 El Niño. This El Niño event proved to be the strongest in recorded history and was linked to catastrophic weather and profound widespread health effects, including severe dengue epidemics. In Asia, many countries and urban areas saw the highest rates of dengue-related morbidity and mortality on record [27, 31, 108, 109]. Retrospective analyses linked many of these epidemics to the El Niño conditions. In Indonesia, for example, a severe epidemic was preceded by a 2-month delay of the rainy season and was accompanied by elevated temperatures. Analyses found that the high temperatures played a major role in precipitating the explosive outbreaks [10, 39]. The link between ENSO-related warmer temperatures and decreased precipitation with increased dengue rates in Indonesia has been confirmed by decades of weather and dengue data [58].

Populations at Risk: Present and Future

Global Circulation Models (GCM) allow researchers to marry currently established patterns of vector and disease distribution with projections of future climate scenarios. Due to inherent differences in modeling approaches, variable selection, and data inclusion, there are some dissimilarities in conclusions from various models [99, 128].

Because of the highly climate-dependent nature of dengue's vectors, GCMs are also useful in postulating future spread of climatic suitability for mosquito vector populations. Both *Ae. aegypti* and *Ae. albopictus* have been shown to be highly adaptable to new areas as their global range has expanded dramatically over the past century. One study using environmental niche modeling as well as modeling based on historic data projects that human-mediated factors will further the expansion of their range until they reach the limit of current climatic suitability. It is predicted that *Ae. aegypti* will do so around 2020 and that *Ae. albopictus* will reach the limits of their current suitable range between 2030 and 2050 [82].

Significantly greater expansion is predicted when climate change is factored in. Advanced by increasingly hospitable environmental conditions, the highly adaptable *Ae. albopictus* are predicted to spread to 20 additional countries by 2080 ranging across a total of 197 nations. In the same time period, *Ae. aegypti* are predicted to mostly concentrate within its tropical and subtropical range while spreading to a few new temperate areas adding in three countries by 2018 making a total 159 countries that comprise its future territory. By 2050, it was estimated that 49% of the world's population will be living in an area supporting vector populations and therefore at risk of dengue transmission. Researchers warn that human populations, already at an increasing risk from greater urbanization and interconnectivity, will face greater risks of dengue infection under unmitigated climate change.

A similar study likewise projected a relatively small increase in range for *Ae. aegypti* but warned of the large increase in human exposure due to population growth in at-risk areas. They projected that the geographical range of *Ae. aegypti* would expand by 8% (moderate emissions pathway) to 13% (high emissions pathway) by 2061–2080 due to climate change. When climate change and population growth projections were factored in, however, the number of people at risk were calculated to increase by 2.2–2.5 billion (59–65%) for a low population growth pathway and 4.8–5.1 billion (127–134%) for a high population growth pathway [100].

Along with the mosquito vector population's distribution, their efficiency at transmitting the virus is also climate-dependent and can be forecasted with GCMs. The greatest increases in epidemic potential will be during the summer months at the borders of current dengue distribution where the climate already supports *Ae. aegypti* populations but does not support endemics due to climate-limited factors such as extended extrinsic incubation periods. Under different climate change scenarios, the projected potential period of transmission is expanded in subtropical climates, and in some tropical areas, the transmission period is extended to support endemicity year-round [75] (Fig. 13.4).

More recent findings on the impact of diurnal temperature range on *Ae. aegypti*'s vectorial capacity have also been paired with GCMs giving alternate insights into the future change in risk distribution [92]. Most mean temperature-based models project an intensification of dengue risk in already endemic areas due to warming global temperatures. However, when diurnal temperature ranges are also factored in, many of the tropical and subtropical regions in which dengue is endemic actually show a reduction in *Ae. aegypti* vectorial capacity by 2070–2090. It is important to note, however, that the calculated relative vectorial capacity is still sufficiently high to result in sustained transmission in these regions and that relative vectorial capacity assumes a 1:1 human to mosquito ratio which varies greatly in the real world.

The main effect of adding diurnal temperature ranges into future projections is that it results in an increase of epidemic potential in temperate areas by a much larger degree than an increase in mean temperature alone. The dramatic increase in epidemic potential spreads through large areas in the Northern Hemisphere as well as parts of the Southern Hemisphere. However, these calculations are based on laboratory data from *Ae. aegypti*, more efficient vectors than *Ae. albopictus* which are much more likely to be found in more extreme temperate regions [91].

Non-climatic Risk Factors

Because of the multiplicity of factors which contribute to dengue transmission within a community, climatic factors can only account for a portion of transmission-supporting conditions and cannot explain all the variations in dengue rates. Viral

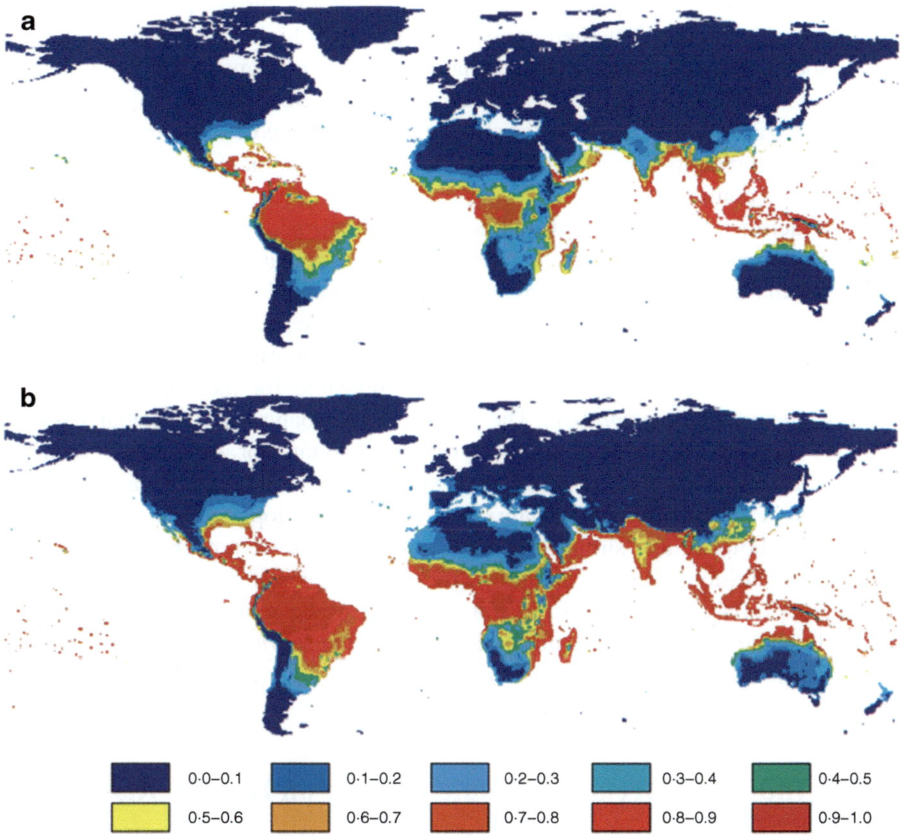

Fig. 13.4 Estimated population at risk of dengue infection in 1990 (**a**) and in 2085 (**b**). A number of studies using differing methodologies have used GCMs to project the future distribution of dengue infection risk. While the projections from these studies often disagree in the details, they all anticipate an expansion of transmission risk to new areas and an intensification of existing risk, particularly along the borders of its current range. In this study, current dengue fever transmission was modeled using a logistical regression model on the basis of long-term vapor pressure (a measure which is also related to temperature and rainfall) with 89% accuracy. Pairing their findings with GCM modelling, it was projected that in 2085, 50–60% of the global population will be at risk of dengue transmission as opposed to 35% who would be at risk in the absence of climate change [66]

introduction, immunological susceptibility, human population density, contact between humans and mosquitoes, housing quality, availability of breeding containers, and mosquito control programs are among the list of factors that also shape transmission risk.

Whether on a global or local scale, most climate-based models do not incorporate the multiplicity of factors which modulate the actual risk of dengue despite climatic suitability. Demographic, societal, and public health factors play a pivotal role in contributing to dengue transmission in some areas and eliminating it in others, even if a vector population is present. The availability of reliable piped water, which eliminates the need for water storage, has been cited as one of the largest protective factors [134]. Factors which limit contact between vectors and hosts such as well-sealed homes with air conditioning or screens, wide-spread automobile use, and a population which spends most of its time indoors further prohibit transmission in wealthier areas. The hallmarks of an area with high risk of dengue include lower socioeconomic status, high population density, low-quality housing, lack of waste-removal services, lack of health-services, and poor vector control. In endemic countries, transmission intensity is also largely influenced by the immunological state of the population, a factor which will likely play an increasingly large role as dengue rates continue to rise.

Two factors which will greatly affect the distribution and magnitude of dengue incidence across the globe are urbanization and population growth. The global urban population is projected to grow by 2.5 billion people by 2050 with a disproportional amount of that growth projected to occur in less-developed nations which are already at high risk of dengue [118]. Dengue is primarily an urban disease, and urbanization is one of the driving factors responsible for the continuing global establishment of *Ae. aegypti* [61]. The conditions in urban areas, particularly poorly planned urban areas, frequently result in ideal habitats for *Ae. aegypti* with the accompanying overcrowding and human density providing the means for high transmission rates [80]. Rapid urbanization in underdeveloped regions often results in informal housing and slums, the conditions of which have been blamed for the epidemic conditions in many countries [62]. Such communities often lack health services, a reliable and accessible water supply, waste removal services, surface water drainage systems, and a multitude of additional services which cause them to become prime habitats for *Ae. aegypti* [147].

Other Indirect Effects of Climate Change on Dengue Risk

In addition to directly affecting vector populations and viral replication rates, climate change also has a tremendous potential to increase the risk of dengue indirectly [67]. Dengue is strongly influenced by socio-economic factors, and major economic sectors such as agriculture and fishing in many at-risk countries are highly vulnerable to variations in climate, leaving their economies vulnerable [37]. Population displacement and increased rural-to-urban migrations are likely in many areas.

Thus, one projected result of climate change is that urban communities will experience an even greater influx of people putting further strain on public health, infrastructure systems, and water resources often already stressed by current population growth and urbanization trends. Additionally, climate change-related alterations in rainfall patterns, surface water availability, and sea level–related intrusion of saltwater into water tables can further exacerbate water shortages resulting in increased neccesity of water storage, thus providing habitats for vector populations.

Refugee conditions, which also have been linked to dengue outbreaks, may also be climatically induced by increased ethnic conflict aggravated by climate-related economic failures. Any climatic event that acts to destabilize communities, displace populations, breakdown infrastructure, limit public health and services, or cause a lapse in coordinated vector control efforts has the potential to increase the risk of dengue.

Regional Implications

Asia

Dengue has a long history in the Asian region. Rapid urbanization following WWII led to epidemic conditions in Southeast Asia and the first major epidemics of DHF. A disproportionate number of dengue cases occur in the Southeast and South Central regions of Asia. Asia currently bears 70% of the current global dengue fever burden with approximately 67 million annual apparent infections [12]. Travel between areas and the co-circulation of multiple dengue virus serotypes have resulted in a state of hyperendemicity. Thus, DHF incidence has since risen dramatically and has been a major cause of hospitalizations and death in children since the mid-1970s [63].

Looking forward, Southeast and South Asian countries, where large populations coexist with the highest levels of climatic dengue risk, face multiple obstacles. Much of the population currently lives in rural areas, but the region has been experiencing a massive shift toward urbanization. In Southeast Asia, urbanization increased from 15.6% to 49% since 1950 and is projected to be 66% by 2050. Rural-to-urban migrations paired with a likewise rapid growth in population size will greatly increase the number of people who will be living in conditions where dengue thrives. In India alone, a nation which already shoulders 34% of the global dengue burden (see Fig. 13.1), the urban population is projected to grow by over 400 million by the year 2050 [12]. Contradicting low official reporting statistics, it was recently found that an average of 60% of children from eight sites around the country tested positive for a past dengue infection, many for more than one dengue virus serotype. In one site in Mumbai, 80% of children tested positive [14]. These findings are on par with similar studies done in other highly endemic countries of Southeast Asia. Despite the high disease burden and heightened public health concern toward dengue, local governments have largely failed to mount or sustain successful control efforts. Additionally, forecasting warns that many Asian countries where dengue is endemic are among those most at risk for the destabilizing effects of climate change [38]. Heat stress, extreme precipitation events, inland and coastal flooding, drought, and water scarcity are predicted with very high confidence. These pose risks to urban areas, many of which already lack essential infrastructure and services or are located in vulnerable areas.

Modeling of climatic suitability to dengue transmission confirms that many parts of Southeast Asia are at extremely high levels of risk [66, 129]. Asia has already experienced warming and an increase in temperature extremes over the past century which have been linked to increases in dengue transmission in retrospective observational studies. GCM-based modeling predicts an increase of the epidemic potential within these areas, and a spread of high dengue transmission risk to many parts of South Central Asia where warming is predicted to be even greater than the global mean [66]. While much of India is currently modeled to be at a lower transmission risk level, future predictions indicate it as having the largest expansion of high transmission risk to new areas. The inclusion of diurnal temperature ranges in modeling, however, results in somewhat lowered current and future projected risks for much of South and Southeast Asia but projects a large geographical spread of moderate and high risk toward the Asian northern latitudes should vector populations also spread [92].

Endemic regions serve as a reservoir for the importation of the dengue virus by travelers to surrounding non-endemic areas that host vector species. For example, despite Japan's temperate climate, the first dengue outbreak in 70 years occurred in Tokyo during the warm summer months in 2014 where it was found that *Ae. albopictus* was the vector responsible [124]. Higher transmission rates within endemic countries coupled with increasing regional interconnectivity and an expansion of suitable climate can be anticipated to result in more widespread epidemics within the Asian region.

Australia and Oceania

In Australia, dengue is currently limited to the northern part of Queensland where outbreaks occur when the dengue virus is imported by travelers from endemic Southeast Asian countries. While dengue is not endemic to Australia, *Ae. aegypti* are established in parts of Queensland and outbreaks have occurred with increasing frequency and magnitude over the past few decades. In the past, *Ae. aegypti* have ranged far south along the eastern regions and modeling confirms the area remains climatic suitable. GCM-based modeling suggests that climate change will result in a spread of increased risk of dengue in the country, particularly down the more populated eastern coastal regions [98, 154]. Conversely, it has been noted that epidemic potential may be reduced in southern Queensland due to evaporative loss of breeding sites and a decline in mosquito survivorship due to higher temperatures [153].

A large focus has been on how human adaptation to climate change will increase the risk of dengue. Drier conditions and water shortages in southeast Australia have resulted in the installation of many governments-subsidized and ad hoc water storage tanks in many cities and towns which are ideal breeding grounds for *Ae. aegypti* and may play a role in allowing a broader range for *Ae. aegypti* populations and a spread of dengue risk [11]. Actual dengue risk will be moderated by housing, cultural, socioeconomic, and other factors which already limit transmission.

Since 1970, dengue epidemics have spread across many Pacific islands with all four viral serotypes now in circulation [68]. While historical arbovirus outbreaks in the region have been somewhat sporadic, concurrent outbreaks of dengue, chikungunya, and Zika viruses have become a regular phenomenon since 2010 [21, 131]. The total population of the Pacific island nations and territories is small, totaling around eight million people scattered among thousands of islands. It is estimated that the countries of Oceania contribute about 180,000, or 0.2%, of global apparent dengue infections, but the incidence levels during outbreaks can be high [12]. During a 2012–2013 dengue outbreak in New Caledonia where dengue is now endemic, 10,987 cases were confirmed in a population of just 259,000 [131]. Largely due to their population size, Papua New Guinea, Fiji, and Samoa are predicted to experience the highest numbers of annual infections [12]. While there have been some improvements in reporting in the Pacific region, it is estimated that most infections go unrecognized or unreported.

Ae. aegypti were widely spread throughout the Pacific by human commercial travel during the nineteenth and twentieth centuries and act as the primary arbovirus vector in the region [21]. They are established on most islands, Fotuna and a few isolated islands being the only exception. *Ae. albopictus* were first introduced to Fiji in 1989 and have since spread to a number of islands. In Hawaii, where dengue has recently reemerged, both species have been separately implicated in outbreaks. In 2001, 122 cases of dengue were confirmed on three of the six islands after over a half a century of its absence [43]. It was determined that *Ae. albopictus* were responsible for the outbreak. More recently, a larger outbreak occurred during the 2015–2016 ENSO event during which 264 cases co-localized with *Ae. aegypti* populations were confirmed [87].

A changing climate has been sighted as a possible contributor to increasing dengue incidence in the Pacific, but much of the rise can be confidently attributed to increasing international travel as evidenced by concurrent or subsequent epidemics often occurring in islands separated by great distances [68, 152]. Rapid urbanization accompanied by poor public health practices, inadequate infrastructure, poor waste management practices, and water storage practices also account for the increase of dengue and other diseases in many small island states [112].

Dengue rates in the Pacific have been linked with climatic fluctuations, particularly with ENSO events [68]. As ENSO events have increased in recent history, peaks in dengue incidence corresponding to a positive southern oscillation index (La Niña conditions) have been observed across many countries. During La Niña years, much of the Central Pacific region tends to experience warmer temperatures and increased precipitation. In New Caledonia, which is less effected by ENSO events than more equatorial Oceanic countries, separate modeling did not find associations with ENSO events, but did find a positive correlation between dengue rates and the rise in average temperature and humidity [41].

Temperature increases for small islands are projected to be generally less than the increase of the global mean due to the fact that the greatest warming is projected to be over large land masses. Humidity-based modeling suggests that climate change will result in an increase in dengue risk with the largest increases occurring in Hawaii, New Caledonia, Fiji, and Vanuatu [98]. Under a high emissions scenario, the equatorial Pacific is projected to experience significant increases in precipitation which, if the relationship between dengue and wetter La Niña conditions holds, could also add to an increase in dengue risk [74].

Islands have intrinsically heightened vulnerability and low adaptive capacity to weather events and climatic shifts [96, 112]. Many of the Pacific Islands are among the most vulnerable areas to the health impacts of climate change [97]. A rise in sea levels, increases in intensities of natural catastrophes such as hurricanes, cyclones, and storm surge, and ecological damage can result in the destabilization of economic, health services, and sanitation systems that are protective of the risk of dengue and many other pathogens. Islands also have limited capacity to manage excess precipitation or water shortages which can additionally result in an expansion of breeding opportunities to dengue vectors. Effective dengue risk management is dependent on coordinated public health programs and well-managed urban planning. Many Pacific island nations already struggle with high poverty rates and poor infrastructure and are highly vulnerable to the impacts of climate change on various economic sectors [132]. Human displacement, unplanned urbanization, and low socioeconomic indices are well documented to be associated with increased dengue risk. As a result, it is likely that the greatest climate change associated increase in dengue risk for many island nations will likely be from factors such as sea level changes, extreme weather events, and the societal effects of climate change rather than the direct effects of altered mean temperature and precipitation.

The Middle East and North Africa

Dengue has recently re-emerged in the Middle East and North Africa (MENA) region, causing sporadic yet increasingly common outbreaks after half a century of its absence [126]. Egypt recorded its first outbreak in a decade in 2015 [115]. In Sudan, which has a more diverse climate, repeated outbreaks have yielded high seroprevalence in various regions [46]. Yemen has been experiencing dengue outbreaks with increasing frequency since the year 2000 [36]. Dengue outbreaks have also become more common in Saudi Arabia since they were first recorded in 1994, particularly in the western regions. While seroprevalence studies from the region have generally shown low levels of infection, a recent seroprevalence study of blood donors in Saudi Arabia found evidence of a past dengue infection in 39% of individuals, and evidence of a recent infection in an additional 5% [8].

Global-based environmental modeling predicts low–to-no suitability for *Aedes* vector populations throughout much of the region, [78, 82, 86] but both *Ae. aegypti* and *Ae. albopictus* populations have been reported separately or jointly in Yemen,

Saudi Arabia, Oman, Palestine, Lebanon, Syria, Turkey, Egypt, Sudan, and Algeria [72]. To reconcile these differences, a recent MENA-specific empirical model based only on data points from similar climates was developed. In global-based modeling, temperature has been determined to be the most important predictor of vector presence followed by precipitation and vegetation indices [82]. In contrast, it was found that in the MENA region and similar climates, factors related to urbanization and population density along with precipitation play the largest roles in determining suitability. The resulting risk maps found numerous areas of suitable habitat coinciding with urban centers in coastal regions, along the Nile, and the other inland densely populated areas. More southern and eastern regions with higher precipitation levels were determined to beat broader risk for hosting vector populations. The risk maps validate the growing concern expressed by a small group of scientists and public health officials that *Ae. aegypti* may reinvade previous territory in the area or that both vector species may find new urban habitats due to human activities in the region.

Among the factors believed to be responsible for the recent re-emergence of dengue in the region are rapid urbanization, the development of slums and shanty towns, long-lasting conflicts, deteriorating public health services, large numbers of displaced populations, and increasing viral introduction through migrations and travel [5, 126]. A recent large dengue outbreak has been tied to the current civil war in Yemen which has displaced over two million people into camp shelters. Poor infrastructure and hygiene conditions have resulted in numerous mosquito breeding sites due to open water storage, discarded plastic containers, and poor surface water drainage in communities. In 2015, over 6777 dengue cases were reported including an extreme spike in cases from the region most impacted by the civil war [2].

Global modeling of the probability of dengue transmission also predicts low-to-no risk across the region, with predicted areas risk being limited to the southern regions of the Arabian Peninsula or coastal areas of the Mediterranean and Red Seas [12, 66, 92, 129]. The highly arid climate is not ideal for the establishment of vector populations in the natural environment, but limited field reports often find *Ae. aegypti* breeding in water storage containers which partly explains their strong spatial relationship with human settlements [72, 163]. In terms of future modeling, the region is predictably not anticipated to become more hospitable to vector populations in a warmer climate, though shifting patterns of precipitation may change the distribution of vectors and transmission risk [78, 82]. Humidity-based modeling projects intensification and expansion of high transmission risk in the southern region of the Arabian Peninsula and along much of the coastal regions of the Red Sea and the Persian Gulf [66].

Much of the MENA region experiences average surface temperatures in or past the upper limits of the ideal range for dengue virus transmission, but when diurnal temperature ranges are considered in conjunction with average temperatures, risk levels for sustained transmission persist across the entire region [92]. However, as temperatures rise over the twenty-first century, the central region of the Arabian Peninsula and part of Sudan are anticipated to fall below the epidemic potential threshold during the hottest months of the year. Outside of this region, moderate levels of transmission risk are projected year-round.

There is great diversity among countries in this region, but in many nations numerous environmental and social stressors result in an increased vulnerability to epidemics of communicable diseases. This is due to factors such as water scarcity, food insecurity, armed conflicts, protracted humanitarian crises, displaced populations, unplanned urbanization, and inadequate governmental and health systems equipped to monitor and respond to dynamic public health threats [18, 65]. It is likely that climate change will exacerbate many of these factors. Additionally, many countries host a large immigrant workforce from dengue-endemic countries and are experiencing increasing travel to and from dengue endemic countries in the form of religious pilgrimage [4, 5]. As dengue rates in other areas rise, an influx in viral introductions to countries in this region can be anticipated.

Africa

One of the largest challenges of forecasting the impact that climate change will have on the risk of dengue fever in African nations is due to the fact that the current burden of dengue in Africa is poorly defined. Diagnostic capacity is limited, and uncomplicated dengue cases can be mistaken for other febrile illnesses such as malaria, which are common in African nations. Dengue is one of the many illnesses for which systematic surveillance and reporting do not exist, with Cape Verde being the only nation to report dengue incidence to the World Health Organization [12]. It is becoming increasingly recognized, however, that dengue fever likely presents a significant but hidden health burden for much of the continent. A growing number of outbreak reports, serological surveys, and outbreaks of other arboviruses transmitted by the same vectors add evidence that true rates are overlooked.

Recent modeling suggests that 750,000 people or 63% of the African population live in areas at risk of dengue transmission [151]. In their study quantifying the global burden of dengue, Bhatt, et al. predicted that dengue is present in 46 African countries, 32 of which have over 50,000 infections a year. They believe that the burden of dengue in Africa is on par with that in the Americas with 15.7 million apparent infections and 48.4 million in apparent infections a year, about 16% of the global total [12].

The possibility of a significant burden of dengue in Africa is further evidenced by the endemicity of yellow fever and other viruses carried by *Ae. aegypti* and *Ae. albopictus* [86]. Yellow fever is endemic in 34 African countries, and while its incidence is also difficult to quantify, it causes an estimated 29,000–60,000 deaths per year despite the existence of an effective vaccine [158]. A number of mosquito species serve as vectors for yellow fever in its sylvatic transmission cycle, but *Ae. aegypti* serve as the primary vector for large epidemics in urban areas [151]. The chikungunya and Zika viruses, for which *Ae. agypti* and *Ae. albopictus* serve as vectors, have also caused outbreaks in many African countries [86].

While numerous seroprevalence studies quantifying the incidence of past dengue infections have been done in other continents, surprisingly few have been done in

Africa. Those that have been done show a wide variability in seroprevalence across areas, with low evidence of past infection being found particularly in rural areas. Most studies are of febrile patients and predictably identify high rates of malaria, but significant rates of past, if not acute, dengue infections are also often revealed. A study of febrile patients from two hospitals in northwest Ethiopia, where no reports of dengue existed before 2013, found 40% and 27.5% seroprevalence [52]. In one of the largest commercial centers in Sudan, a 47.6% seroprevalence was found with sleeping outdoors and living in low-income significantly increasing risk of infection [46]. In Nigeria, where the highest number of dengue infections are predicted to occur, different types of seroprevalence studies have again yielded a wide range of results, but a 2017 study found that 46% of febrile patients were recently infected with dengue [48, 81]. To better understand and quantify the current dengue burden in Africa, the Dengue Vaccine Initiative has initiated large multi-disciplinary population-based epidemiological studies in three African locations [90]. They believe that the results can be used to infer the larger African dengue burden and hope that they will be useful in future in developing future public health strategies.

While the evidence of dengue in Africa requires interpretation, its presence is widely recognized in the islands off its coast. Dengue is endemic year-round on the island of Réunion which has been the scene of unusually large epidemics. In nearby Seychelles, one epidemic involved over 80% of the population [20]. On the opposite coast of Africa, Cape Verde has been vigilant since a 2009 outbreak in which over 21,000 cases and four deaths were reported despite the fact that dengue had never previously been reported in the country.

Before their worldwide expansion, *Ae. aegypti* are believed to have originated from Africa (along with some of the arboviruses it carries such as chikungunya, yellow fever, Zika, and possibly dengue) [123]. A clear picture of their current distribution in Africa is difficult to make due to lack of reporting, but their presence has been noted in many countries and ecological modeling indicates that a wide distribution through much of sub-Saharan Africa is likely [78, 82]. At least two forms of *Ae. Aegypti* can be found in Africa: the domestic form that spread globally and the ancestral sylvan form that remained. While both can transmit the virus, they differ in behavior, ecology, distribution, and potentially in their competence as a vector, thus complicating risk projections [151]. *Ae. albopictus* are relatively new to the area, having first been reported in Nigeria in 1991. Their distribution is not predicted to be as broad as *Ae. aegypti*'s. Nevertheless, large areas of sub-Saharan Africa are modeled to be environmentally suitable and limited field reports of *Ae. albopictus* populations have identified them in a number of countries, particularly in West Africa [151]. While *Ae. aegypti* are believed to be the primary dengue vector in Africa, *Ae. albopictus* has also been recognized as a public health threat due to its role in dengue and chikungunya outbreaks in Gabon and as the sole vector in the large epidemics on Seychelles and Réunion [20, 120, 148].

Climate-based projections of dengue transmission risk forecast an extension of risk into more southern areas as temperatures rise. Temperature-based modeling shows that much of Africa is climatically conducive to dengue virus transmission, though humidity-based modeling restricts higher risk-levels to the more tropical regions in West and East Africa [66, 92]. While there is disagreement on the expansions and contractions of the geographical boundaries of risk, they agree on the fact that much of the arid regions of North Africa will remain less hospitable to vector populations or dengue transmission due to low moisture and high temperatures (see Fig. 13.4) [66, 78, 92, 100].

Scientific literature on the effects of climate change on dengue in Africa is generally limited to global-based studies while there is a more thorough African-specific discussion of malaria. Conclusions from malaria-based studies cannot be accurately generalized to dengue fever, however, due to differences in the ecology of the vector and infectious agents. Among other differences, malaria vector species are more likely to be found in rural areas where they primarily utilize natural breeding sites and are thus more sensitive to changes in land use and rainfall patterns. In contrast, the greatest opportunities for dengue transmission are more likely to be driven by urbanization in this historically rural continent. Climate change–based modeling does not generally project significant changes in the climatically suitable ranges of the dengue vectors. Within that range, the growth of urban environments can be anticipated to be selective for both *Ae. aegypti* and *Ae. albopictus* populations [3, 89, 110, 162].

By the year 2010, only 38.9% of Africa was urbanized (compared to 78.6% of Latin America) [146]. Low population density is a protective factor in dengue transmission and is offered as one explanation for historically low dengue reporting in Africa [129]. Projections of the highest African dengue incidence are centered around areas of high population density with the greatest number of infections estimated to occur in the urban regions of Nigeria (which is estimated to experience 4.2 million apparent infections a year) [12]. Urbanization in Africa is increasing and is projected to climb to 58.9% by the year 2050 [146]. Additionally, the fertility rates of its populations are the highest in the world. When combined, these two facts forecast a substantial addition to the urban population. The African population living in urban areas was estimated at 408 million in the year 2010, but by 2050, it is projected to grow to nearly 1.5 billion. Many of the regions which are projected to experience the greatest increases of population growth and urbanization in the coming decades are the same areas which are projected to become climatically at high risk for dengue transmission. Nigeria, which is located in the region of modeled prime climatic suitability for dengue transmission and has reported *Ae. aegypti* and *Ae. albopictus* populations, is on track to become the third most populated country in the world by 2050, more than doubling its current urban population by adding an additional 287 million people to its cities.

Looking forward, changes in climate-based vector distributions and vectorial capacity may contribute to an increase in dengue risk among African nations. More confidently, as areas become increasingly urbanized and densely populated, coordinated and consistent vector surveillance and control efforts will be required to compensate for the anticipated heightened risk of dengue transmission. Unfortunately, successful control efforts are difficult to sustain and require significant governmental involvement and public health resources. Endemic poverty, low access to improved sanitation and water resources, and poor infrastructure are common in many parts of Africa and will make dengue control even more difficult. Climate change is likely to exacerbate nearly every stressor which currently plagues African nations. The continent's low adaptive capacity makes it especially at risk to the changing climate, and future projections do not paint an encouraging picture of a continent with the capability of fighting the growing threat of dengue [111].

Europe

Until very recently, natural transmission of dengue had not occurred in European countries since the 1920s, although historically both *Ae. aegypti* and dengue were reported in a number of European countries [70]. Before their elimination, *Ae. aegypti* populations ranged along the Mediterranean shores around seaports and historical reports of their presence in several other countries exist, though evidence that some of these other populations were well-established is less strong. Dengue epidemics have occurred in Spain, the Canary Islands, and in many Mediterranean nations [127]. One of the worst epidemics occurred in refugee camps in Greece in 1927–1928 when an estimated 650,000 infections and 1000 deaths occurred. Since the mid-twentieth century, *Ae. aegypti* have largely disappeared from Europe due to increased hygiene, reliable and piped water supplies, and the use of insecticide [70].

Currently, *Ae. aegypti* are absent in the European area with populations only being found on its periphery on the Portuguese island of Madeira and on the eastern coast of the Black Sea [9]. After re-colonizing Madeira in 2004, they were responsible for a 2012 outbreak in which over 2000 cases occurred. In contrast, after being transported to Italy in used tires in the 1990s, *Ae. albopictus* have spread to much of southern Europe and are quickly expanding their range northward [9, 82]. Local dengue transmission has been limited but has been found in France, Croatia, and Spain, highlighting their ability to serve as vectors in European locations [47, 146]. The widespread establishment of *Ae. albopictus* raises concern due to the large numbers of travelers traveling to Europe who could potentially be infected with the dengue virus. In response to their invasion, a recent study found that in 2010, over 700,000 travelers from dengue-affected countries arrived at European airports located in areas with *Ae. albopictus* populations [146].

Climate-based modeling predicts that *Ae. albopictus* will continue their spread through Europe due to the presence of yet-unreached currently climatically suitable territory. Climate change is anticipated to further expand their range. One study predicts it to ultimately include wide areas of France and Germany by 2080, while portions of eastern Europe become less hospitable due to increased aridity [78, 82]. Another study projects a northward shift in *Ae. albopictus* populations with climatic hotspots in Portugal, the Southern United Kingdom, western Germany, the Benelux, Slovakia, Cyprus, Bulgaria, Macedonia, Hungary, and Turkey [22]. *Ae. aegypti* are modeled to be more constrained in their expansion with their range limited to isolated locations in southern European under even the most extreme climate change scenarios.

Modeling based on *Ae. aegypti* vectorial capacity confirms current European climatic suitability for dengue transmission during the summer months and posits that vectorial capacity has already been increased due to higher temperatures and wider diurnal temperature ranges [91]. When *Ae. albopictus* is specified as the vector, however, the intensity and range of epidemic potential are significantly decreased and limited to more southern regions. By the end of the century, a northward shift of higher epidemic potential is projected due to climate change as well as a prolonged transmission season with the highest risk levels found in the coastal areas of Southern Europe and in the Central Eastern regions, though none that could sustain year-round transmission [144]. Under a RPC2.6 scenario, the expansion in epidemic potential is more modest, but under RPC8.5, *Ae. albopictus* are anticipated to be able to sustain epidemics in regions as far north as Paris in the 2060s and Berlin in the 2070s. Were the highly competent vector *Ae. aegypti* present, high epidemic potentials would be anticipated through much of Europe in both conservative and high climate change scenarios, but these risks are contraindicated by vector-population-based modeling which does not project an accompanying expansion of the *Ae. aegypti* range. Though, as they have been found in a broader (yet still limited) European distribution in the past, public health officials are vigilant for signs of re-introduction.

Concern over possible future dengue transmission in the Mediterranean region is not unwarranted as several characteristics of the region add to the danger of transmission. Cities are generally densely populated. Windows are generally left open during the summer months as air conditioners are rarely used. In addition, contact with the vector can occur during activities and social gatherings which typically are held outdoors [127]. It should be noted, however, that the same factors which have inhibited outbreaks in the past decades can be expected to dampen the effects of calculated climate-based epidemic potentials and that it remains to be seen how much a changing climate will actually affect transmission in real-world setting which will depend on a multiplicity of factors.

Latin America

Of all the regions in the world, the emergence of dengue and DHF in recent decades has been the most dramatic in the Americas. Historically, outbreaks of dengue occurred sporadically in the Americas for hundreds of years. In 1947, a coordinated hemisphere-wide effort by the Pan American Health Organization to eradicate *Ae. aegypti* proved hugely successful. By the early 1960s, *Ae. aegypti* were eradicated from most of their previous territory with dengue infections largely disappearing as well. Unfortunately, control efforts soon lapsed and *Ae. aegypti* quickly disseminated to nearly every country in the Western Hemisphere. The increase in international travel and commerce spread both vector and virus which flourished in the conditions of rapidly urbanizing nations [107]. By the 1980s major epidemics began again.

Over the last four decades, dengue fever cases have increased many-fold. The American region reported approximately a million cases for all of the 1980s, but for the years 2014–2018, an average of 1.38 million cases were reported per year. DHF cases have also increased with approximately 13,400 in the 1980s to 172,000 in the decade of 2000–2009. While reporting to the Pan American Health Organization is considerably better than that for many parts of the world, dengue rates are still consistently underreported [145]. It is believed that over 13 million apparent dengue infections occur annually out of over 40 million in apparent infections, accounting for 14% of the global burden [12].

Serological testing confirms the wide-spread nature of dengue infections in the Americas. A number of studies from a variety of countries show evidence that often most, if not nearly all, tested populations have been previously infected [145]. In Recife, Brazil, a survey performed in three socioeconomically diverse areas found significantly differing levels of seropositivity according to the poverty levels of the neighborhoods [16]. In prosperous socioeconomic areas, the seroprevalence was 74.3%, in intermediate areas, the seroprevalence was 87.4%, and in deprived areas, the seroprevalence increased to 91.1% where 59% of children had already been infected before the age of 5. Increased international travel has increased viral introduction, and outbreaks in the Americas have now been caused by all four serotypes. As co-circulation of multiple serotypes continues to become more widespread, the potential for complications from serious dengue infection increases, though most of Latin America has successfully reduced case fatality rates to less than 1% through better patient management [116, 156].

The IPCC warns with high confidence that changes in climate patterns are already negatively affecting human health by promoting dengue fever and other health conditions and that they will continue to exacerbate future health risk. Warming trends are predicted to continue with larger than global mean increases of warming in most of Central and South America [95]. Changes in precipitation patterns are predicated, leading to increased precipitation in some areas, but notable reductions in others as well as an increased probability of regional droughts and water supply shortages. Dengue fever rates have been strongly associated with climatic variables in the region and with ENSO events [17, 53, 77, 84, 104, 133, 145, 149]. In particular, the ENSO event of 2014–2016 coincided with historically high transmission rates of dengue [6]. Temperatures were elevated to a record degree and drought conditions prevailed for much of Latin America. The region reported an unprecedented 2.4 million infections in 2015 and another 2.2 million the following year, over half of which occurred in Brazil [116].

Climate change–based modeling predicts an extension of at-risk areas into larger regions of Mexico, Brazil, Peru, and Ecuador and predicts lengthened transmission periods for many areas [66, 75]. Recent modeling based on multiple climatic variables predicts that the increases of dengue risk in northern regions of South America and the southern regions of Central America will be among the highest global increases [128]. The inclusion of diurnal temperature ranges into projections indicates that risk may be reduced in equatorial regions while still remaining above epidemic thresholds [92].

An increase in mean temperature (0.7–1 °C in Central America and most of South America) and changes in rainfall patterns have already been recorded in Latin America [145]. Whether a changing climate is to blame for the increase of dengue in the region has been a heavily debated topic. While climate is undeniably linked to dengue, it is not sufficient to explain the explosive rise in dengue rates. The *Ae. aegypti* eradication campaign coincided with a remarkable period of urbanization in Latin America in which it became the most urbanized region of the developing world. The 1980s, however, brought a serious debt crisis to the region, resulting in a deterioration of economic and social conditions, large inequalities, unemployment, poverty, and failures in health systems. These conditions, paired with the reestablishment of the vector and increased mobility of people and virus within and between countries, have resulted in circumstances prime for dengue transmission.

The United States

Locally acquired dengue is largely absent in the continental United States, but it has not always been so. Pandemics that stretched through the Caribbean and Gulf region also struck the southern states through the first half of the twentieth century with an epidemic occurring as far north as Philadelphia during an unusually hot summer in 1780 [44].

Ae. Aegypti can currently be found in a number of large pockets along the southern US border and the east coast [82]. Over the past several decades, *Ae. albopictus* have become established in much of the eastern half of the United States. Despite the presence of anthroponotic disease vectors, high standards of living typical for US residents have largely limited contact between mosquitoes and humans, essentially eliminating viruses with no animal reservoir, thus requiring constant transmission between humans and mosquitoes in order to remain in circulation. Reliable piped water, sanitation, air conditioning, window screens, and indoor lifestyles are among factors which limit the possibility of transmission [105].

In contrast, Mexican states bordering the United States have experienced repeated large epidemics of dengue. Beginning in 1980, after a long absence, small dengue outbreaks began to occur sporadically along the Texas-Mexico border. While these outbreaks highlight the potential for dengue transmission within the United States, they also underscore the conditions inherent to American society which are prohibitive to dengue. In a 2005 outbreak, a handful of people in Brownsville, Texas, were diagnosed with dengue fever, but over the border in Tamaulipas, Mexico, over 7000 cases were reported [125]. Abundant *Ae. aegypti* mosquitoes were found breeding in both cities. The use of air conditioning, which limits vector-human contact, and larger lot sizes, which result in less dense urban environments, were found to be protective factors during the outbreak.

Dengue is now a leading cause of acute febrile illness in travelers returning to the United States from the Caribbean, South America, and Asia. Viral introductions have resulted in subsequent small outbreaks of locally acquired dengue infections in

2009, 2010, and 2013 in Florida and again in Brownsville in 2013 [54]. During the pan-American Zika outbreak of 2016 the virus was introduced to American soil by travelers and *Ae. aegypti* served as the vector propagating local transmission in Florida and Texas resulting in 224 identified infections [28].

Climate-based modeling confirms that the current climate can support dengue transmission in the southernmost states during the summer months. Sustained transmission is limited by lower winter temperatures for all mainland US locations, however, precluding endemicity [19, 103]. GCM-based projections anticipate a small northern spread of *Ae. aegypti* populations which will primarily concentrate within their current range [82]. A much larger expanded distribution of *Ae. albopictus* is projected through eastern North America. The western half of the United States is projected to become less hospitable to vector populations as the climate becomes increasingly hot and arid [82, 92]. While mosquito season may be extended in many areas, the period of potential viral transmission will likewise grow but to a lesser degree and is still projected to be limited to southern areas during southern months, again precluding the chance of sustained year-round transmission. South Florida, an area with high tourism rates and thus higher risk of viral introduction, is projected to have the highest future levels of climate-based transmission risk [19]. Modeling that factors in diurnal temperature ranges projects a broader spread of moderate transmission risk to more northern temperate areas but is based on *Ae. aegypti* data and does not account for the fact that only the less efficient *Ae. albopictus* are anticipated to establish in northern areas [92, 103].

Whether the recent outbreaks of dengue in the United States mark the beginning of a larger scale reemergence has been a topic of debate. As both dengue incidence and international travel rise across the globe, so too will the number of viral introductions to the United States. The fact that dengue has been largely absent in the United States despite the presence of vector populations paired with the substantial disparity of rates across the border is the most convincing argument that dengue outbreaks will remain a rarity. The societal factors which have proved so effective against transmission are unlikely to change. It remains to be seen whether increased risk due to an altered climate will result in more frequent outbreaks of dengue in the United States.

References

1. Aguiar M, Stollenwerk N, Halstead SB. The impact of the newly licensed dengue vaccine in endemic countries. PLoS Negl Trop Dis. 2016;10(12):e0005179. https://doi.org/10.1371/journal.pntd.0005179.
2. Alghazali KA, Teoh BT, Loong SK, Sam SS, Che-Mat-Seri NA, Samsudin NI, et al. Dengue outbreak during ongoing civil war, Taiz, Yemen. Emerg Infect Dis. 2019;25(7) https://doi.org/10.3201/eid2507.180046.
3. Alirol E, Getaz L, Stoll B, Chappuis F, Loutan L. Urbanisation and infectious diseases in a globalised world. Lancet Infect Dis. 2011;11(2):131–41. https://doi.org/10.1016/s1473-3099(10)70223-1.

4. Altassan KK, Morin C, Shocket MS, Ebi K, Hess J. Dengue fever in Saudi Arabia: a review of environmental and population factors impacting emergence and spread. Travel Med Infect Dis. 2019; https://doi.org/10.1016/j.tmaid.2019.04.006.
5. Amarasinghe A, Letson GW. Dengue in the Middle East: a neglected, emerging disease of importance. Trans R Soc Trop Med Hyg. 2012;106(1):1–2. https://doi.org/10.1016/j.trstmh.2011.08.014.
6. Anyamba A, Chretien JP, Britch SC, Soebiyanto RP, Small JL, Jepsen R, et al. Global disease outbreaks associated with the 2015–2016 El Nino event. Sci Rep. 2019;9(1):1930. https://doi.org/10.1038/s41598-018-38034-z.
7. Arcari P, Tapper N, Pfueller S. Regional variability in relationships between climate and dengue/DHF in Indonesia. Singap J Trop Geogr. 2007;28(3):251–72. https://doi.org/10.1111/j.1467-9493.2007.00300.x.
8. Ashshi AM, Alghamdi S, El-Shemi AG, Almdani S, Refaat B, Mohamed AM, et al. Seroprevalence of asymptomatic dengue virus infection and its antibodies among healthy/eligible Saudi blood donors: findings from holy Makkah City. Virology (Auckl). 2017;8:1–5. https://doi.org/10.1177/1178122x17691261.
9. European Centre for Disease Prevention and Control. Mosquito maps. 2019. Retrieved from https://ecdc.europa.eu/en/disease-vectors/surveillance-and-disease-data/mosquito-maps
10. Bangs MJ, Larasati RP, Corwin AL, Wuryadi S. Climatic factors associated with epidemic dengue in Palembang, Indonesia: implications of short-term meteorological events on virus transmission. Southeast Asian J Trop Med Public Health. 2006;37(6):1103–16.
11. Beebe NW, Cooper RD, Mottram P, Sweeney AW. Australia's dengue risk driven by human adaptation to climate change. PLoS Negl Trop Dis. 2009;3(5):e429. https://doi.org/10.1371/journal.pntd.0000429.
12. Bhatt S, Gething PW, Brady OJ, Messina JP, Farlow AW, Moyes CL, et al. The global distribution and burden of dengue. Nature. 2013;496(7446):504–7. https://doi.org/10.1038/nature12060.
13. Bonizzoni M, Gasperi G, Chen X, James AA. The invasive mosquito species *Aedes albopictus*: current knowledge and future perspectives. Trends Parasitol. 2013;29(9):460–8. https://doi.org/10.1016/j.pt.2013.07.003.
14. Brady OJ, Gething PW, Bhatt S, Messina JP, Brownstein JS, Hoen AG, et al. Refining the global spatial limits of dengue virus transmission by evidence-based consensus. PLoS Negl Trop Dis. 2012;6(8):e1760. https://doi.org/10.1371/journal.pntd.0001760.
15. Brady OJ, Golding N, Pigott DM, Kraemer MUG, Messina JP, Reiner RC, et al. Global temperature constraints on Aedes aegypti and Ae. albopictus persistence and competence for dengue virus transmission. Parasites Vectors. 2014;7:338. https://doi.org/10.1186/1756-3305-7-338.
16. Braga C, Luna CF, Martelli CM, de Souza WV, Cordeiro MT, Alexander N, et al. Seroprevalence and risk factors for dengue infection in socio-economically distinct areas of Recife, Brazil. Acta Trop. 2010;113(3):234–40. https://doi.org/10.1016/j.actatropica.2009.10.021.
17. Brunkard JM, Cifuentes E, Rothenberg SJ. Assessing the roles of temperature, precipitation, and ENSO in dengue re-emergence on the Texas-Mexico border region. Salud Publica Mex. 2008;50(3):227–34. Retrieved from <Go to ISI>://WOS:000255699700006.
18. Buliva E, Elhakim M, Tran Minh NN, Elkholy A, Mala P, Abubakar A, Malik S. Emerging and reemerging diseases in the World Health Organization (WHO) Eastern Mediterranean region-progress, challenges, and WHO initiatives. Front Public Health. 2017;5:276. https://doi.org/10.3389/fpubh.2017.00276.
19. Butterworth MK, Morin CW, Comrie AC. An analysis of the potential impact of climate change on dengue transmission in the Southeastern United States. Environ Health Perspect. 2017;125(4):579–85. https://doi.org/10.1289/ehp218.
20. Calisher CH, Nuti M, Lazuick JS, Ferrari JD, Kappus KD. Dengue in the Seychelles. Bull World Health Organ. 1981;59(4):619–22.
21. Calvez E, Guillaumot L, Millet L, Marie J, Bossin H, Rama V, et al. Genetic diversity and phylogeny of Aedes aegypti, the Main Arbovirus vector in the Pacific. PLoS Negl Trop Dis. 2016;10(1):e0004374. https://doi.org/10.1371/journal.pntd.0004374.

22. Caminade C, Medlock JM, Ducheyne E, McIntyre KM, Leach S, Baylis M, Morse AP. Suitability of European climate for the Asian tiger mosquito Aedes albopictus: recent trends and future scenarios. J R Soc Interface. 2012;9(75):2708–17. https://doi.org/10.1098/rsif.2012.0138.
23. Canyon DV, Hii JLK, Muller R. Adaptation of Aedes aegypti (Diptera: Culicidae) oviposition behavior in response to humidity and diet. J Insect Physiol. 1999;45(10):959–64. https://doi.org/10.1016/s0022-1910(99)00085-2.
24. Carrington LB, Armijos MV, Lambrechts L, Scott TW. Fluctuations at a low mean temperature accelerate dengue virus transmission by Aedes aegypti. PLoS Negl Trop Dis. 2013b;7(4):e2190. https://doi.org/10.1371/journal.pntd.0002190.
25. Carrington LB, Seifert SN, Armijos MV, Lambrechts L, Scott TW. Reduction of Aedes aegypti vector competence for dengue virus under large temperature fluctuations. Am J Trop Med Hyg. 2013a;88(4):689–97. https://doi.org/10.4269/ajtmh.12-0488.
26. Carrington LB, Tran BCN, Le NTH, Luong TTH, Nguyen TT, Nguyen PT, et al. Field- and clinically derived estimates of Wolbachia-mediated blocking of dengue virus transmission potential in Aedes aegypti mosquitoes. Proc Natl Acad Sci U S A. 2018;115(2):361–6. https://doi.org/10.1073/pnas.1715788115.
27. Cazelles B, Chavez M, McMichael AJ, Hales S. Nonstationary influence of El Nino on the synchronous dengue epidemics in Thailand. PLoS Med. 2005;2(4):313–8. https://doi.org/10.1371/journal.pmed.0020106.
28. Centers for Disease Control and Prevention. Zika virus: 2016 case counts. April 24, 2019. Retrieved from https://www.cdc.gov/zika/reporting/2016-case-counts.html
29. Chan M, Johansson MA. The incubation periods of Dengue viruses. PLoS One. 2012;7(11):e50972. https://doi.org/10.1371/journal.pone.0050972.
30. Chang LH, Hsu EL, Teng HJ, Ho CM. Differential survival of Aedes aegypti and Aedes albopictus (Diptera: Culicidae) larvae exposed to low temperatures in Taiwan. J Med Entomol. 2007;44(2):205–10. https://doi.org/10.1603/0022-2585(2007)44[205:Dsoaaa]2.0.Co;2.
31. Chantha N, Guyant P, Hoyer S. Control of DHF outbreak in Cambodia. Dengue Bull. 1998;22:69–74.
32. Chareonviriyaphap T, Akratanakul P, Nettanomsak S, Huntamai S. Larval habitats and distribution patterns of Aedes aegypti (Linnaeus) and Aedes albopictus (Skuse), in Thailand. Southeast Asian J Trop Med Public Health. 2003;34(3):529–35.
33. Chien LC, Yu HL. Impact of meteorological factors on the spatiotemporal patterns of dengue fever incidence. Environ Int. 2014;73:46–56. https://doi.org/10.1016/j.envint.2014.06.018.
34. Chowell G, Sanchez F. Climate-based descriptive models of dengue fever: the 2002 epidemic in Colima, Mexico. J Environ Health. 2006;68(10):40–4. Retrieved from <Go to ISI>://WOS:000237816400004.
35. Christophers S. Aedes aegypti (L.) the yellow fever mosquito: its life history, bionomics and structure. London: Cambridge University Press; 1960.
36. Ciccozzi M, Lo Presti A, Cella E, Giovanetti M, Lai A, El-Sawaf G, et al. Phylogeny of dengue and Chikungunya viruses in Al Hudayda governorate, Yemen. Infect Genet Evol. 2014;27:395–401. https://doi.org/10.1016/j.meegid.2014.08.010.
37. Climate Change 2014: impacts, adaptation, and vulnerability. Part A: global and sectoral aspects. Contribution of working group II to the fifth assessment report of the intergovernmental panel on climate change. 2014. Retrieved from Cambridge, UK and New York.
38. Climate Change 2014: impacts, adaptation, and vulnerability. Part B: regional aspects. Contribution of working group II to the fifth assessment report of the intergovernmental panel on climate change. 2014. Retrieved from Cambridge, UK and New York.
39. Corwin AL, Larasati RP, Bangs MJ, Wuryadi S, Arjoso S, Sukri N, et al. Epidemic dengue transmission in southern Sumatra, Indonesia. Trans R Soc Trop Med Hyg. 2001;95(3):257–65. https://doi.org/10.1016/s0035-9203(01)90229-9.
40. Costero A, Edman JD, Clark GG, Kittayapong P, Scott TW. Survival of starved Aedes aegypti (Diptera: Culicidae) in Puerto Rico and Thailand. J Med Entomol. 1999;36(3):272–6. https://doi.org/10.1093/jmedent/36.3.272.

41. Descloux E, Mangeas M, Menkes CE, Lengaigne M, Leroy A, Tehei T, et al. Climate-based models for understanding and forecasting dengue epidemics. PLoS Negl Trop Dis. 2012;6(2):e1470. https://doi.org/10.1371/journal.pntd.0001470.
42. Eamchan P, Nisalak A, Foy HM, Chareonsook OA. Epidemiology and control of dengue virus infections in Thai villages in 1987. Am J Trop Med Hyg. 1989;41(1):95–101.
43. Effler PV, Pang L, Kitsutani P, Vorndam V, Nakata M, Ayers T, et al. Dengue fever, Hawaii, 2001–2002. Emerg Infect Dis. 2005;11(5):742–9. https://doi.org/10.3201/eid1105.041063.
44. Ehrenkranz NJ, Ventura AK, Cuadrado RR, Pond WL, Porter JE. Pandemic dengue in Caribbean countries and the southern United States – past, present and potential problems. N Engl J Med. 1971;285(26):1460–9. https://doi.org/10.1056/nejm197112232852606.
45. Eisen L, Monaghan AJ, Lozano-Fuentes S, Steinhoff DF, Hayden MH, Bieringer PE. The impact of temperature on the bionomics of Aedes (Stegomyia) aegypti, with special reference to the cool geographic range margins. J Med Entomol. 2014;51(3):496–516. https://doi.org/10.1603/me13214.
46. Eldigail MH, Adam GK, Babiker RA, Khalid F, Adam IA, Omer OH, et al. Prevalence of dengue fever virus antibodies and associated risk factors among residents of El-Gadarif state, Sudan. BMC Public Health. 2018;18(1):921. https://doi.org/10.1186/s12889-018-5853-3.
47. European Centre for Disease Prevention and Control. Local transmission of dengue fever in France and Spain – 2018 – 22 October 2018 [Press release].
48. Fagbami AH, Onoja AB. Dengue haemorrhagic fever: an emerging disease in Nigeria, West Africa. J Infect Public Health. 2018;11(6):757–62. https://doi.org/10.1016/j.jiph.2018.04.014.
49. Fan J, Wei W, Bai Z, Fan C, Li S, Liu Q, Yang K. A systematic review and meta-analysis of dengue risk with temperature change. Int J Environ Res Public Health. 2014;12(1):1–15. https://doi.org/10.3390/ijerph120100001.
50. Farjana T, Tuno N. Multiple blood feeding and host-seeking behavior in Aedes aegypti and Aedes albopictus (Diptera: Culicidae). J Med Entomol. 2013;50(4):838–46. https://doi.org/10.1603/me12146.
51. Farnesi LC, Martins AJ, Valle D, Rezende GL. Embryonic development of Aedes aegypti (Diptera: Culicidae): influence of different constant temperatures. Memorias Do Instituto Oswaldo Cruz. 2009;104(1):124–6. https://doi.org/10.1590/s0074-02762009000100020.
52. Ferede G, Tiruneh M, Abate E, Wondimeneh Y, Damtie D, Gadisa E, et al. A serologic study of dengue in Northwest Ethiopia: suggesting preventive and control measures. PLoS Negl Trop Dis. 2018;12(5):e0006430. https://doi.org/10.1371/journal.pntd.0006430.
53. Ferreira MC. Geographical distribution of the association between El Nino South Oscillation and dengue fever in the Americas: a continental analysis using geographical information system-based techniques. Geospat Health. 2014;9(1):141–51. https://doi.org/10.4081/gh.2014.12.
54. Florida Department of Health. Martin county dengue outbreak and serosurvey, Martin County, Florida. 2014. Retrieved from http://www.floridahealth.gov/diseases-and-conditions/dengue/_documents/mc-dengue-survey-summary-june2014.pdf
55. Focks DA, Brenner RJ, Hayes J, Daniels E. Transmission thresholds for dengue in terms of Aedes aegypti pupae per person with discussion of their utility in source reduction efforts. Am J Trop Med Hyg. 2000;62(1):11–8.
56. Focks DA, Haile DG, Daniels E, Mount GA. Dynamic life table model for Aedes aegypti (Diptera: Culicidae): analysis of the literature and model development. J Med Entomol. 1993a;30(6):1003–17. https://doi.org/10.1093/jmedent/30.6.1003.
57. Focks DA, Haile DG, Daniels E, Mount GA. Dynamic life table model for Aedes aegypti (diptera: Culicidae): simulation results and validation. J Med Entomol. 1993b;30(6):1018–28. https://doi.org/10.1093/jmedent/30.6.1018.
58. Gagnon AS, Bush ABG, Smoyer-Tomic KE. Dengue epidemics and the El Nino southern oscillation. Clim Chang. 2001;19:34–43.
59. Ghosh SN, Pavri KM, Singh KR, Sheikh BH, D'Lima LV, Mahadev PV, Ramachandra Rao T. Investigations on the outbreak of dengue fever in Ajmer City, Rajasthan State in 1969. Part I. epidemiological, clinical and virological study of the epidemic. Indian J Med Res. 1974;62(4):511–22.

60. Gratz NG. Critical review of the vector status of Aedes albopictus. Med Vet Entomol. 2004;18(3):215–27. https://doi.org/10.1111/j.0269-283X.2004.00513.x.
61. Gubler D. Dengue, urbanization and globalization: the unholy trinity of the 21st century. Trop Med Health. 2011;39:3–11.
62. Gubler DJ. The changing epidemiology of yellow fever and dengue, 1900 to 2003: full circle? Comp Immunol Microbiol Infect Dis. 2004;27(5):319–30. https://doi.org/10.1016/j.cimid.2004.03.013.
63. Gubler DJ. Dengue/dengue haemorrhagic fever: history and current status. Novartis Found Symp. 2006;277:3–16; discussion 16–22, 71–13, 251–253. https://doi.org/10.1002/0470058005.ch2.
64. Guzman MG, Alvarez M, Halstead SB. Secondary infection as a risk factor for dengue hemorrhagic fever/dengue shock syndrome: an historical perspective and role of antibody-dependent enhancement of infection. Arch Virol. 2013;158(7):1445–59. https://doi.org/10.1007/s00705-013-1645-3.
65. Habib RR, Zein KE, Ghanawi J. Climate change and health research in the eastern Mediterranean region. EcoHealth. 2010;7(2):156–75. https://doi.org/10.1007/s10393-010-0330-1.
66. Hales S, de Wet N, Maindonald J, Woodward A. Potential effect of population and climate changes on global distribution of dengue fever: an empirical model. Lancet. 2002;360(9336):830–4. https://doi.org/10.1016/s0140-6736(02)09964-6.
67. Hales S, Edwards S, Kovats R. Impacts on health of climate extremes. In: Climate change and human health. Geneva: World Health Organization; 2003.
68. Hales S, Weinstein P, Woodward A. Dengue fever epidemics in the South Pacific: driven by El Nino Southern Oscillation? Lancet. 1996;348(9042):1664–5. https://doi.org/10.1016/s0140-6736(05)65737-6.
69. Halstead SB. Dengue virus-mosquito interactions. Annu Rev Entomol. 2008;53:273–91. https://doi.org/10.1146/annurev.ento.53.103106.093326.
70. Holstein M. Dynamics of Aedes aegypti distribution, density and seasonal prevalence in the Mediterranean area. Bull World Health Organ. 1967;36(4):541–3.
71. Hopp M, Foley J. Global-scale relationships between climate and the dengue fever vector, Aedes aegypti. Clim Chang. 2001;48(2–3):441–63.
72. Humphrey JM, Cleton NB, Reusken CB, Glesby MJ, Koopmans MP, Abu-Raddad LJ. Dengue in the Middle East and North Africa: a systematic review. PLoS Negl Trop Dis. 2016;10(12):e0005194. https://doi.org/10.1371/journal.pntd.0005194.
73. Hurtado-Diaz M, Riojas-Rodriguez H, Rothenberg SJ, Gomez-Dantes H, Cifuentes E. Short communication: impact of climate variability on the incidence of dengue in Mexico. Trop Med Int Health. 2007;12(11):1327–37. Retrieved from <Go to ISI>://WOS:000251716900009.
74. IPCC. Climate change 2014: synthesis report. Contribution of working groups I, II and III to the fifth assessment report of the intergovernmental panel on climate change. Geneva: IPCC; 2014.
75. Jetten TH, Focks DA. Potential changes in the distribution of dengue transmission under climate warming. Am J Trop Med Hyg. 1997;57(3):285–97.
76. Johansson MA, Dominici F, Glass GE. Local and global effects of climate on dengue transmission in Puerto Rico. PLoS Negl Trop Dis. 2009;3(2) https://doi.org/10.1371/journal.pntd.0000382.
77. Jury MR. Climate influence on dengue epidemics in Puerto Rico. Int J Environ Health Res. 2008;18(5):323–34. https://doi.org/10.1080/09603120701849836.
78. Kamal M, Kenawy MA, Rady MH, Khaled AS, Samy AM. Mapping the global potential distributions of two arboviral vectors Aedes aegypti and Ae. albopictus under changing climate. PLoS One. 2018;13(12):e0210122. https://doi.org/10.1371/journal.pone.0210122.
79. Keating J. An investigation into the cyclical incidence of dengue fever (vol 53, pg 1587, 2001). Soc Sci Med. 2002;55(9):1691. https://doi.org/10.1016/s0277-9536(02)00197-1.
80. Knudsen AB, Slooff R. Vector-borne disease problems in rapid urbanization: new approaches to vector control. Bull World Health Organ. 1992;70(1):1–6.

81. Kolawole OM, Seriki AA, Irekeola AA, Bello KE, Adeyemi OO. Dengue virus and malaria concurrent infection among febrile subjects within Ilorin metropolis, Nigeria. J Med Virol. 2017;89(8):1347–53. https://doi.org/10.1002/jmv.24788.
82. Kraemer MUG, Reiner RC Jr, Brady OJ, Messina JP, Gilbert M, Pigott DM, et al. Past and future spread of the arbovirus vectors Aedes aegypti and Aedes albopictus. Nat Microbiol. 2019;4(5):854–63. https://doi.org/10.1038/s41564-019-0376-y.
83. Lambrechts L, Paaijmans KP, Fansiri T, Carrington LB, Kramer LD, Thomas MB, Scott TW. Impact of daily temperature fluctuations on dengue virus transmission by Aedes aegypti. Proc Natl Acad Sci U S A. 2011;108(18):7460–5. https://doi.org/10.1073/pnas.1101377108.
84. Laureano-Rosario AE, Garcia-Rejon JE, Gomez-Carro S, Farfan-Ale JA, Muller-Karger FE. Modelling dengue fever risk in the State of Yucatan, Mexico using regional-scale satellite-derived sea surface temperature. Acta Trop. 2017;172:50–7. https://doi.org/10.1016/j.actatropica.2017.04.017.
85. Lenhart AE, Castillo CE, Oviedo M, Villegas E. Use of the pupal/demographic-survey technique to identify the epidemiologically important types of containers producing Aedes aegypti (L.) in a dengue-endemic area of Venezuela. Ann Trop Med Parasitol. 2006;100:S53–9. https://doi.org/10.1179/136485906x105516.
86. Leta S, Beyene TJ, De Clercq EM, Amenu K, Kraemer MUG, Revie CW. Global risk mapping for major diseases transmitted by *Aedes aegypti* and *Aedes albopictus*. Int J Infect Dis. 2018;67:25–35. https://doi.org/10.1016/j.ijid.2017.11.026.
87. Lew RJ, Tsai WY, Wang WK. Dengue outbreaks in Hawai'i after WWII – a review of public health response and scientific literature. Hawaii J Med Public Health. 2018;77(12):315–8.
88. Lewis D. Observations on Aedes aegypti under controlled atmospheric conditions. Bull Entomol Res. 1933;24:363–72.
89. Li Y, Kamara F, Zhou G, Puthiyakunnon S, Li C, Liu Y, et al. Urbanization increases Aedes albopictus larval habitats and accelerates mosquito development and survivorship. PLoS Negl Trop Dis. 2014;8(11):e3301. https://doi.org/10.1371/journal.pntd.0003301.
90. Lim JK, Carabali M, Lee JS, Lee KS, Namkung S, Lim SK, et al. Evaluating dengue burden in Africa in passive fever surveillance and seroprevalence studies: protocol of field studies of the dengue vaccine initiative. BMJ Open. 2018;8(1):e017673. https://doi.org/10.1136/bmjopen-2017-017673.
91. Liu-Helmersson J, Quam M, Wilder-Smith A, Stenlund H, Ebi K, Massad E, Rocklov J. Climate change and aedes vectors: 21st century projections for dengue transmission in Europe. EBioMedicine. 2016;7:267–77. https://doi.org/10.1016/j.ebiom.2016.03.046.
92. Liu-Helmersson J, Stenlund H, Wilder-Smith A, Rocklov J. Vectorial capacity of Aedes aegypti: effects of temperature and implications for global dengue epidemic potential. PLoS One. 2014;9(3):e89783. https://doi.org/10.1371/journal.pone.0089783.
93. Lumsden WHR. Observations on the effect of microclimate on biting by Aedes-Aegypti (L) (Dipt Culicid). J Exp Biol. 1947;24(3–4):361–73. Retrieved from <Go to ISI>://WOS:A1947UF60700010.
94. Macdonald WW. Aedes aegypti in Malaya. II. Larval and adult biology. Ann Trop Med Parasitol. 1956;50(4):399–414. Retrieved from <Go to ISI>://MEDLINE:13395330.
95. Magrin GO, Marengo JA, Boulanger J-P, Buckeridge MS, Castellanos E, Poveda G, et al. Central and South America. In: Climate change 2014: impacts, adaptation, and vulnerability. Part B: regional aspects. Contribution of working group II to the fifth assessment report of the intergovernmental panel on climate change. Cambridge, UK/New York: Cambridge University Press; 2014.
96. Mavian C, Dulcey M, Munoz O, Salemi M, Vittor AY, Capua I. Islands as hotspots for emerging mosquito-borne viruses: a one-health perspective. Viruses. 2018;11(1) https://doi.org/10.3390/v11010011.
97. McIver L, Kim R, Woodward A, Hales S, Spickett J, Katscherian D, et al. Health impacts of climate change in Pacific Island countries: a regional Assessment of vulnerabilities and adaptation priorities. Environ Health Perspect. 2016;124(11):1707–14. https://doi.org/10.1289/ehp.1509756.

98. McMichael AJ, Woodruff R, Whetton P, Hennessy K, Nicholls N, Hales S, et al. Human health and climate change in Oceania: a risk assessment. Canberra: Commonwealth Department of Health and Ageing; 2002.
99. Messina JP, Brady OJ, Pigott DM, Golding N, Kraemer MU, Scott TW, et al. The many projected futures of dengue. Nat Rev Microbiol. 2015;13(4):230–9. https://doi.org/10.1038/nrmicro3430.
100. Monaghan AJ, Sampson KM, Steinhoff DF, Ernst KC, Ebi KL, Jones B, Hayden MH. The potential impacts of 21st century climatic and population changes on human exposure to the virus vector mosquito Aedes aegypti. Clim Chang. 2018;146(3–4):487–500.
101. Moore CG, Cline BL, Ruiztiben E, Lee D, Romneyjoseph H, Riveracorrea E. Aedes-aegypti in Puerto-Rico – environmental determinants of larval abundance and relation to dengue virus transmission. Am J Trop Med Hyg. 1978;27(6):1225–31. https://doi.org/10.4269/ajtmh.1978.27.1225.
102. Moore PR, Johnson PH, Smith GA, Ritchie SA, Van Den Hurk AF. Infection and dissemination of dengue virus type 2 in Aedes aegypti, Aedes albopictus, and Aedes scutellaris from the Torres Strait, Australia. J Am Mosq Control Assoc. 2007;23(4):383–8. https://doi.org/10.2987/5598.1.
103. Mordecai EA, Cohen JM, Evans MV, Gudapati P, Johnson LR, Lippi CA, et al. Detecting the impact of temperature on transmission of Zika, dengue, and chikungunya using mechanistic models. PLoS Negl Trop Dis. 2017;11(4):e0005568. https://doi.org/10.1371/journal.pntd.0005568.
104. Moreno-Banda GL, Riojas-Rodriguez H, Hurtado-Diaz M, Danis-Lozano R, Rothenberg SJ. Effects of climatic and social factors on dengue incidence in Mexican municipalities in the state of Veracruz. Salud Publica Mex. 2017;59(1):41–52. https://doi.org/10.21149/8414.
105. Moreno-Madrinan MJ, Turell M. History of mosquitoborne diseases in the United States and implications for new pathogens. Emerg Infect Dis. 2018;24(5):821–6. https://doi.org/10.3201/eid2405.171609.
106. Morin CW, Comrie AC, Ernst K. Climate and dengue transmission: evidence and implications. Environ Health Perspect. 2013;121(11–12):1264–72. https://doi.org/10.1289/ehp.1306556.
107. Musso D, Rodriguez-Morales AJ, Levi JE, Cao-Lormeau VM, Gubler DJ. Unexpected outbreaks of arbovirus infections: lessons learned from the Pacific and tropical America. Lancet Infect Dis. 2018;18(11):E355–61. https://doi.org/10.1016/s1473-3099(18)30269-x.
108. Muto R. Summary of dengue situation in WHO Western Pacific Region. Dengue Bull. 1998;22:12–9.
109. Nagao Y, Thavara U, Chitnumsup P, Tawatsin A, Chansang C, Campbell-Lendrum D. Climatic and social risk factors for Aedes infestation in rural Thailand. Tropical Med Int Health. 2003;8(7):650–9. https://doi.org/10.1046/j.1365-3156.2003.01075.x.
110. Ndenga BA, Mutuku FM, Ngugi HN, Mbakaya JO, Aswani P, Musunzaji PS, et al. Characteristics of Aedes aegypti adult mosquitoes in rural and urban areas of western and coastal Kenya. PLoS One. 2017;12(12):e0189971. https://doi.org/10.1371/journal.pone.0189971.
111. Niang I, Ruppel OC, Abdrabo MA, Essel A, Lennard C, Padgham J, Urquhart P. Africa. In: Barros VR, Field CB, Dokken DJ, Mastrandrea MD, Mach KJ, Bilir TE, Chatterjee M, Ebi KL, Estrada YO, Genova RC, Girma B, Kissel ES, Levy AN, MacCracken S, Mastrandrea PR, White LL, editors. Climate change 2014: impacts, adaptation, and vulnerability. Part B: regional aspects. Contribution of working group II to the fifth assessment report of the intergovernmental panel on climate change. Cambridge, UK/New York: Cambridge University Press; 2014.
112. Nurse LA, McLean RF, Agard J, Briguglio LP, Duvat-Magnan V, Pelesikoti N, et al. Small islands. In: Climate change 2014: impacts, adaptation, and vulnerability. Part B: regional aspects. Cambridge, UK/New York: Cambridge University Press; 2014.
113. O'Neill SL, Ryan PA, Turley AP, Wilson G, Retzki K, Iturbe-Ormaetxe I, et al. Scaled deployment of Wolbachia to protect the community from dengue and other Aedes transmitted arboviruses. Gates Open Res. 2018;2:36. https://doi.org/10.12688/gatesopenres.12844.2.

114. Ooi EE, Goh KT, Gubler DJ. Denque prevention and 35 years of vector control in Singapore. Emerg Infect Dis. 2006;12(6):887–93. https://doi.org/10.3201/10.3201/eid1206.051210.
115. World Health Organization. Dengue fever – Egypt. 2015. Retrieved from https://www.who.int/csr/don/12-november-2015-dengue/en/
116. Pan American Health Organization. Reported cases of dengue fever in the Americas. Retrieved from http://www.paho.org/data/index.php/en/mnu-topics/indicadores-dengue-en/dengue-nacional-en/252-dengue-pais-ano-en.html
117. Pan American Health Organization. Dengue: guidelines for patient care in the Region of the Americas. Washington, DC: Pan American Health Organization; 2016.
118. Patz JA, Martens WJ, Focks DA, Jetten TH. Dengue fever epidemic potential as projected by general circulation models of global climate change. Environ Health Perspect. 1998;106(3):147–53. https://doi.org/10.1289/ehp.98106147.
119. Paupy C, Delatte H, Bagny L, Corbel V, Fontenille D. Aedes albopictus, an arbovirus vector: from the darkness to the light. Microbes Infect. 2009;11(14–15):1177–85. https://doi.org/10.1016/j.micinf.2009.05.005.
120. Paupy C, Ollomo B, Kamgang B, Moutailler S, Rousset D, Demanou M, et al. Comparative role of Aedes albopictus and Aedes aegypti in the emergence of dengue and chikungunya in Central Africa. Vector Borne Zoonotic Dis. 2010;10(3):259–66. https://doi.org/10.1089/vbz.2009.0005.
121. Phillips ML. Dengue reborn: widespread resurgence of a resilient vector. Environ Health Perspect. 2008;116(9):A382–8. https://doi.org/10.1289/ehp.116-a382.
122. Pontes RJ, Freeman J, Oliveira-Lima JW, Hodgson JC, Spielman A. Vector densities that potentiate dengue outbreaks in a Brazilian city. Am J Trop Med Hyg. 2000;62(3):378–83.
123. Powell JR, Tabachnick WJ. History of domestication and spread of Aedes aegypti--a review. Mem Inst Oswaldo Cruz. 2013;108(Suppl 1):11–7. https://doi.org/10.1590/0074-0276130395.
124. Quam MB, Sessions O, Kamaraj US, Rocklov J, Wilder-Smith A. Dissecting Japan's dengue outbreak in 2014. Am J Trop Med Hyg. 2016;94(2):409–12. https://doi.org/10.4269/ajtmh.15-0468.
125. Ramos MM, Mohammed H, Zielinski-Gutierrez E, Hayden MH, Lopez JL, Fournier M, et al. Epidemic dengue and dengue hemorrhagic fever at the Texas-Mexico border: results of a household-based seroepidemiologic survey, December 2005. Am J Trop Med Hyg. 2008;78(3):364–9.
126. Rathor HR. The role of vectors in emerging and re-emerging diseases in the Eastern Mediterranean Region. Dengue Bull. 2000;24:103–9.
127. Reiter P. Climate change and mosquito-borne disease. Environ Health Perspect. 2001;109(Suppl 1):141–61. https://doi.org/10.1289/ehp.01109s1141.
128. Rogers DJ. Dengue: recent past and future threats. Philos Trans R Soc Lond Ser B Biol Sci. 2015;370(1665) https://doi.org/10.1098/rstb.2013.0562.
129. Rogers DJ, Wilson AJ, Hay SI, Graham AJ. The global distribution of yellow fever and dengue. Adv Parasitol. 2006;62:181–220. https://doi.org/10.1016/s0065-308x(05)62006-4.
130. Romero-Vivas CME, Arango-Padilla P, Falconar AKI. Pupal-productivity surveys to identify the key container habitats of Aedes aegypti (L.) in Barranquilla, the principal seaport of Colombia. Ann Trop Med Parasitol. 2006;100:S87–95. https://doi.org/10.1179/136485906x105543.
131. Roth A, Mercier A, Lepers C, Hoy D, Duituturaga S, Benyon E, et al. Concurrent outbreaks of dengue, chikungunya and Zika virus infections – an unprecedented epidemic wave of mosquito-borne viruses in the Pacific 2012–2014. Euro Surveill. 2014;19(41):20929.
132. Russell L. Poverty, climate change and health in pacific island countries. Issues to consider in discussion, debate and policy development. 2011. Retrieved from https://ses.library.usyd.edu.au//bitstream/2123/9202/1/lrpacificislands2011.pdf
133. Sanchez AL, Sanchez NE, Sanchez AMG. Climatic phenomenon and meteorological variables influencing the dengue fever incidence in Colombian South Pacific region: Modeling

study. Ann Trop Med Public Health. 2017;10(6):1489–95. https://doi.org/10.4103/atmph.Atmph_395_17.
134. Schmidt WP, Suzuki M, Thiem VD, White RG, Tsuzuki A, Yoshida LM, et al. Population density, water supply, and the risk of dengue fever in Vietnam: cohort study and spatial analysis. PLoS Med. 2011;8(8):e1001082. https://doi.org/10.1371/journal.pmed.1001082.
135. Scott TW, Chow E, Strickman D, Kittayapong P, Wirtz RA, Lorenz LH, Edman JD. Blood-feeding patterns of Aedes-aegypti (Diptera, Culicidae) collected in a rural Thai village. J Med Entomol. 1993;30(5):922–7. https://doi.org/10.1093/jmedent/30.5.922.
136. Scott TW, Morrison AC. Aedes aegypti density and the risk of dengue-virus transmission. In: Takken W, Scott TW, editors. Ecological aspects for application of genetically modified mosquitoes. Dordrecht: Kluwer Academic Publishers; 2003.
137. Scott TW, Takken W. Feeding strategies of anthropophilic mosquitoes result in increased risk of pathogen transmission. Trends Parasitol. 2012;28(3):114–21. https://doi.org/10.1016/j.pt.2012.01.001.
138. Semenza JC, Sudre B, Miniota J, Rossi M, Hu W, Kossowsky D, et al. International dispersal of dengue through air travel: importation risk for Europe. PLoS Negl Trop Dis. 2014;8(12):e3278. https://doi.org/10.1371/journal.pntd.0003278.
139. Sheppard PM, Macdonald WW, Tonn RJ, Grab B. The dynamics of and adult population of Aedes aegypti in relation to dengue haemorrhagic fever in Bangkok. J Anim Ecol. 1969;38:661–702.
140. Soo KM, Khalid B, Ching SM, Chee HY. Meta-analysis of dengue severity during infection by different dengue virus serotypes in primary and secondary infections. PLoS One. 2016;11(5):e0154760. https://doi.org/10.1371/journal.pone.0154760.
141. Sota T, Mogi M. Interspecific variation in desiccation survival-time of Aedes (Stegomyia) mosquito eggs is correlated with habitat and egg size. Oecologia. 1992;90(3):353–8. https://doi.org/10.1007/bf00317691.
142. Southwood TR, Murdie G, Yasuno M, Tonn RJ, Reader PM. Studies on the life budget of Aedes aegypti in Wat Samphaya, Bangkok, Thailand. Bull World Health Organ. 1972;46(2):211–26.
143. Thammapalo S, Chongsuwiwatwong V, McNeil D, Geater A. The climatic factors influencing the occurrence of dengue hemorrhagic fever in Thailand. Southeast Asian J Trop Med Public Health. 2005;36(1):191–6.
144. Thomas SM, Fischer D, Fleischmann S, Bittner T, Beierkuhnlein C. Risk assessment of dengue virus amplification in Europe based on spatio-temporal high resolution climate change projections. Erdkunde. 2011;65(2):137–50. https://doi.org/10.3112/erdkunde.2011.02.03.
145. Torres JR, Orduna TA, Pina-Pozas M, Vazquez-Vega D, Sarti E. Epidemiological characteristics of dengue disease in Latin America and in the Caribbean: a systematic review of the literature. J Trop Med. 2017;2017:8045435. https://doi.org/10.1155/2017/8045435.
146. United Nations Department of Economic and Social Affairs Population Division. (2018). World urbanization prospects: the 2018 revision. New York: United Nations.
147. United Nations Human Settlements Programme. (2003). The challenge of slums: global report on human settlements. Retrieved from.
148. Vazeille M, Dehecq JS, Failloux AB. Vectorial status of the Asian tiger mosquito Aedes albopictus of La Reunion Island for Zika virus. Med Vet Entomol. 2018;32(2):251–4. https://doi.org/10.1111/mve.12284.
149. Vincenti-Gonzalez MF, Tami A, Lizarazo EF, Grillet ME. ENSO-driven climate variability promotes periodic major outbreaks of dengue in Venezuela. Sci Rep. 2018;8:11. https://doi.org/10.1038/s41598-018-24003-z.
150. Watts DM, Burke DS, Harrison BA, Whitmire RE, Nisalak A. Effect of temperature on the vector efficiency of Aedes-aegypti for dengue-2 virus. Am J Trop Med Hyg. 1987;36(1):143–52. https://doi.org/10.4269/ajtmh.1987.36.143.
151. Weetman D, Kamgang B, Badolo A, Moyes CL, Shearer FM, Coulibaly M, et al. Aedes mosquitoes and Aedes-borne arboviruses in Africa: current and future threats. Int J Environ Res Public Health. 2018;15(2) https://doi.org/10.3390/ijerph15020220.

152. World Health Organization. Western Pacific regional action plan for dengue prevention and control. Manila: World Health Organization Regional Office for the Western Pacific; 2016.
153. Williams CR, Mincham G, Faddy H, Viennet E, Ritchie SA, Harley D. Projections of increased and decreased dengue incidence under climate change. Epidemiol Infect. 2016;144(14):3091–100. https://doi.org/10.1017/s095026881600162x.
154. Woodruff R, Hales S, Butler C, McMichael AJ. Climate change and health impacts in Australia: effects of dramatic CO_2 emission reductions. Canberra: Australian National University; 2005.
155. World Health Organization. Dengue vaccine: WHO position paper, September 2018 – recommendations. Vaccine. 2018; https://doi.org/10.1016/j.vaccine.2018.09.063.
156. World Health Organization. Dengue and severe dengue. 2019a. Retrieved from https://www.who.int/en/news-room/fact-sheets/detail/dengue-and-severe-dengue
157. World Health Organization. Vector-borne diseases. 2019b. Retrieved from https://www.who.int/mediacentre/factsheets/fs387/en/index2.html
158. World Health Organization. Yellow fever: fact sheet. 2019c, 7 May 2019. Retrieved from https://www.who.int/news-room/fact-sheets/detail/yellow-fever
159. Yang HM, Macoris ML, Galvani KC, Andrighetti MT, Wanderley DM. Assessing the effects of temperature on the population of Aedes aegypti, the vector of dengue. Epidemiol Infect. 2009a;137(8):1188–202. https://doi.org/10.1017/s0950268809002040.
160. Yang HM, Macoris MLG, Galvani KC, Andrighetti MTM, Wanderley DMV. Assessing the effects of temperature on dengue transmission. Epidemiol Infect. 2009b;137(8):1179–87. https://doi.org/10.1017/s0950268809002052.
161. Yasuno M, Tonn RJ. Study of biting habits of aedes-aegypti in Bangkok-Thailand. Bull World Health Organ. 1970;43(2):319. Retrieved from <Go to ISI>://WOS:A1970H927900012.
162. Zahouli JBZ, Koudou BG, Muller P, Malone D, Tano Y, Utzinger J. Urbanization is a main driver for the larval ecology of Aedes mosquitoes in arbovirus-endemic settings in south-eastern cote d'Ivoire. PLoS Negl Trop Dis. 2017;11(7):e0005751. https://doi.org/10.1371/journal.pntd.0005751.
163. Zayed A, Awash AA, Esmail MA, Al-Mohamadi HA, Al-Salwai M, Al-Jasari A, et al. Detection of chikungunya virus in Aedes aegypti during 2011 outbreak in Al Hodayda, Yemen. Acta Trop. 2012;123(1):62–6. https://doi.org/10.1016/j.actatropica.2012.03.004.

Chapter 14
Climate Variability and Change: Food, Water, and Societal Impacts

Jonathan A. Patz

Past Climate Trends and Future Projections

Long-term climate change can be observed as a signal against a background of natural climate variability. Since instrument records are available only for the recent past (a period of less than 150 years), previous climates must be deduced from paleoclimatic records such as tree rings, pollen series, faunal and floral abundances in deep-sea cores, isotope analyses of coral and ice cores, and diaries and other documentary evidence. Surface temperatures in the mid- to late twentieth century appear to have been higher than they were during any similar period in the last 600 years in most regions, and in at least some regions warmer than in any other century for several thousand years [1].

Temperature changes are accelerating rapidly. According to the IPCC, human activities have caused about 1 °C of warming above pre-industrial times and at the current rate of greenhouse gas emissions will reach 1.5 °C warming by as early as 2030 – about a decade from now [2].

Climate Change, Sea-Level Rise, and Extremes in Climate Variability

Changing temperatures are only part of the story. Hot temperatures evaporate soil moisture more quickly, thereby leading to severe droughts, while warmer air can hold more moisture and result in heavy precipitation events; such "hydrologic

J. A. Patz (✉)
UW-Madison School of Medicine and Public Health, Department of Population Health Science, Madison, WI, USA
e-mail: patz@wisc.edu

K. E. Pinkerton, W. N. Rom (eds.), *Climate Change and Global Public Health*, Respiratory Medicine, https://doi.org/10.1007/978-3-030-54746-2_14

extremes" (floods and droughts) accompany warming temperatures within future climate change scenarios; both extremes are a concern to global public health.

The insurance and reinsurance industry is also worried about climate change. In 2011, the United States experienced 14 weather-related disasters exceeding $1 billion each in damage costs, a new record. Weather disasters since 1996 have been nearly twice as numerous and costly compared to the period from 1980 to 95 [3]. Of course, this growth is in part a result of development in vulnerable (especially coastal) areas, but the trend in weather extremes is part of the story.

If Greenland's ice sheet melts, sea level would rise by 23 feet (7 m). Over the past century, Greenland has already has *lost around 9,000 billion tons* of ice, accounting for 25 mm of sea-level rise [4]. But Greenland is dwarfed by the Antarctic with an ice volume that, if melted, could raise sea level 185 feet (57 m). The Antarctic is now losing six times as much ice as it was four decades ago. Also western North American glaciers are losing ice four times faster since the early 2000s [5] (Fig. 14.1).

Sea Surface Temperatures and Hurricanes

Sea surface temperatures have steadily increased over the last century, and more sharply over the last 35 years. The highest average sea surface temperatures were recorded from 1995 to 2004 [6]. Warmer ocean surface temperatures affect wind velocities in hurricanes. Hurricanes form only in regions where sea surface temperatures are above 26 °C [7] (see Keim and Miller). Since the 1950s, overall hurricane activity in the North Atlantic has doubled and the Caribbean has experienced a five-fold increase [8]. Hurricane intensity may also be associated with warmer temperatures [9, 10]. As Hurricane Katrina demonstrated in 2005, such events have enormous significance for public health (Fig. 14.2).

Vulnerable Geographic Regions

Certain regions and populations are more vulnerable to the health impacts of climate change [11]: areas bordering regions with a high endemicity of climate-sensitive diseases such as malaria; areas with an observed association between disease epidemics and weather extremes as with El Niño-linked epidemics; areas at risk from the combined impacts of climate relevant to health, such as stress on food and water supplies or risk of coastal flooding; and areas at risk from concurrent environmental or socioeconomic stresses, for example, local stresses resulting from land-use practices or an impoverished or undeveloped health infrastructure, with little capacity to adapt.

Vulnerability can also vary by neighborhood. For example, Uejio et al. [12] found that the number of heat distress calls in Phoenix, Arizona, was higher where city blocks had more impervious surfaces (indicating asphalt or concrete roads and

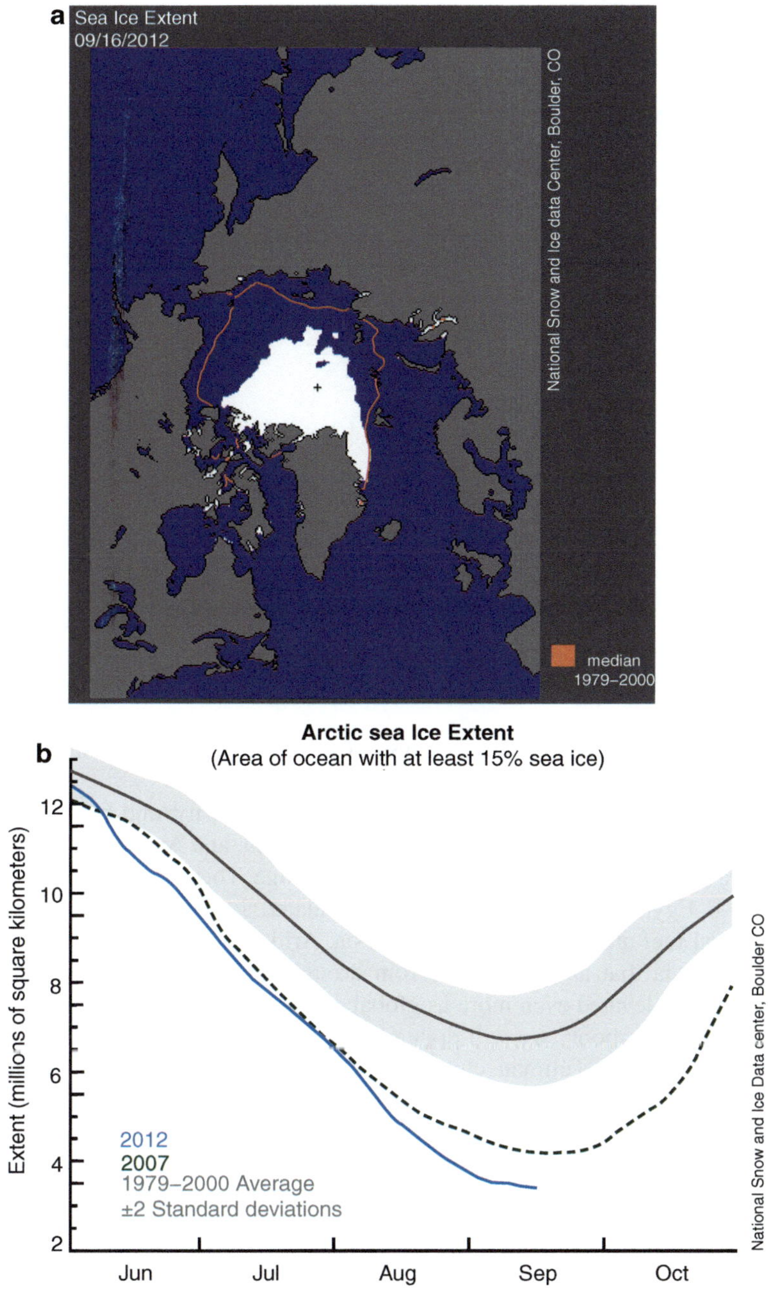

Fig. 14.1 Sea ice record melting

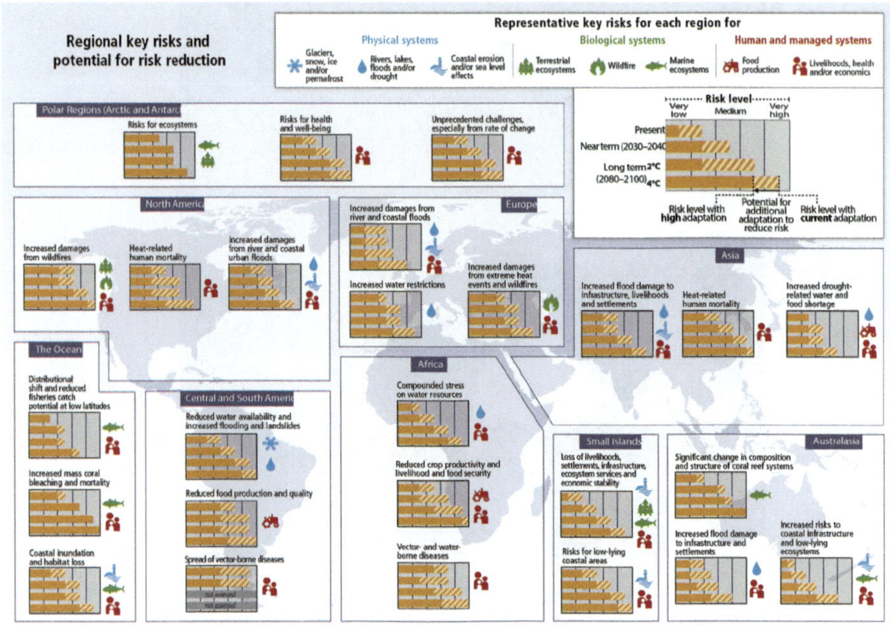

Fig. 14.2 Regional key risks and potential for risk reduction. (Source: IPCC [11])

buildings), and this was the primary cause of localized "urban heat island" intensification of temperatures.

Increases in floods and droughts, decreased food security, and biodiversity loss are special concerns for parts of Africa, Latin America, and Asia. Coastal and delta regions are at special risk even without climate change. These include coastal China, Bangladesh, Egypt, and especially densely populated, low-lying, small island states such as coral reef atolls throughout Polynesia. Arid regions such as eastern Africa and central Asia that already suffer from drought are likewise at increased risk. These risks are elevated even more as global climate warms [13].

Throughout this book, we will document many direct and indirect implications for human health due to climate change. This chapter focuses on threats to nutrition and safe water, risks from weather extremes and sea-level rise, and water- and food-borne infectious diseases. We then address public health responses to climate change and potential health "co-benefits" of greenhouse gas mitigation. Finally, the ethical dimensions of climate change and health are discussed.

Food Productivity and Malnutrition

It is no surprise that projected increases in frequency and intensity of climate extremes will have a major impact on crop and livestock production, as well as on the viability of fisheries [14–16]. Of course, the net effect on food production will vary from place to place. Changes will depend on several factors; their agents include temperature, precipitation, CO_2 levels (relating to the fertilization effect, for example), extreme climate variability, and sea-level rise. But indirect effects of climate-induced changes in soil quality, incidence of plant diseases, and increased weed, and insect populations could have just as large an effect on world food supplies. Higher heat and humidity will also increase food spoilage (discussed below). The last two decades have seen continuing deterioration of food production in Africa, caused in part by persistent drought. For some foods, nutritional quality (e.g., their protein content) will diminish as climate changes. Finally, the extent to which adaptive responses are available to farmers must be taken into account.

Food Production and Drought

Malnutrition remains one of the world's largest challenges to health. Eight hundred million people are currently undernourished [17]. Developing countries struggle with large and expanding populations and are particularly vulnerable to threats to food production. Projections forecast that drought-affected areas will increase, thereby exacerbating threats to agriculture, water supplies, energy production, and human health [18]. One-third of the world's population currently live in water-stressed countries, and that number is predicted to increase to five billion people by 2025.

In central Asia and southern Africa, stream flows are expected to fall, and this may affect the food supply. Mountain snow pack, glaciers, and small ice caps play a crucial role in freshwater availability at regional sites. Large losses from glaciers and reductions in snow cover over recent decades are likely to accelerate throughout the twenty-first century. This will reduce water availability and hydropower potential, and will change the seasonality of flows in regions supplied by melt water from major mountain ranges (e.g., the Hindu-Kush, the Himalayas, the Andes), where more than one-sixth of the world population resides [19]. Diarrhea and such diseases as scabies, conjunctivitis, and trachoma are associated with poor hygiene and can result from a breakdown in sanitation when water resources become depleted [20].

Despite significant agricultural technological advances, including irrigation, food production strongly depends on weather conditions. Most cultivars are growing close to their thermal optimum. Data from 23 global climate models show a high probability that the *average* growing season temperatures by the end of the century will exceed the *hottest* temperatures on record from 1900 to 2006 (Fig. 14.3) [21]. Lower yields are expected to occur at low latitudes due to heat stress, and crops will be subject to damage from flooding, erosion, and wildfires. The potential for global food production is projected to increase with increases in local average temperature over a range of 1–3 °C, but above this it is likely to decline [18]. Effects on global agricultural productivity will vary regionally; reductions will be especially acute in sub-Saharan Africa and South Asia [22]. Air temperatures above 30 °C (86 °F) generally reduce yields for rain-fed crops. Water stress exacerbates the effect of high temperatures because low soil moisture reduces evaporative cooling from the landscape [23].

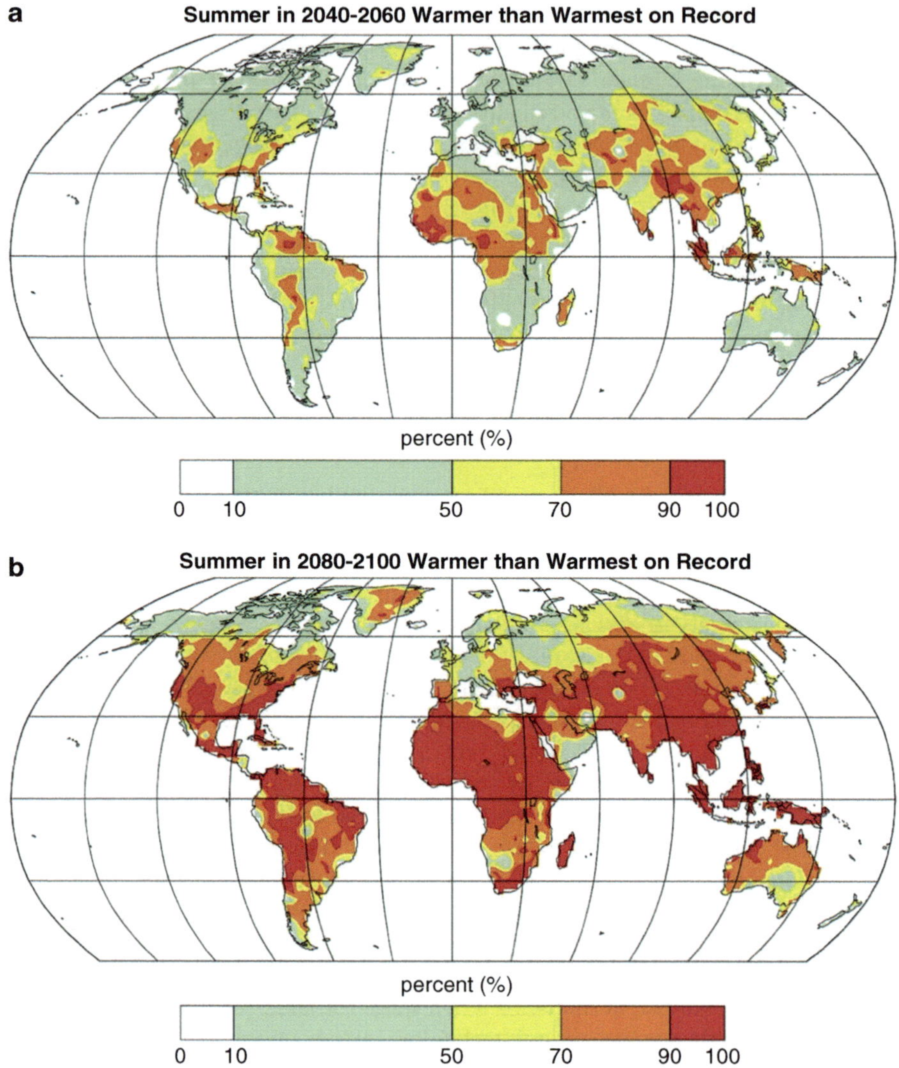

Fig. 14.3 Record summer temperatures and food supply. (Source: Battisi and Naylor [21])

According to one study [24], by the 2050s climate change would increase the risk of hunger from 34% currently to a level of 64–72%, unadjusted for potential adaptive interventions. Battisi and Naylor [21] found that reductions in regional productivity could destabilize food security to the extent that the number of people at risk for malnutrition could double by mid-century [21]. One study took the next step by estimating the human toll of such changes in worldwide malnutrition: Lloyd et al. [25] estimate that by 2030 climate change would lead to over 1.2 million malnutrition-related deaths. Considering the few crop monocultures on which global calories are derived from, crop diversity could help reduce such risks.

Crop yields are not the only concern. Nutritional value must be considered as well under a future climate regime. Some crops incorporate less nitrogen when CO_2 levels are elevated, resulting in lower protein content. Studies of barley, wheat, rice, potatoes, and soybeans show this reduced protein when crops are grown under high-CO_2 conditions. The magnitude of the effect varies with soil conditions, air quality, and other factors [26]. For populations that depend on these crops for their protein, the high-CO_2 effect could further threaten their nutritional status.

Nutritional quality of food may decrease from elevated carbon dioxide concentrations. For example, concentrations of iron and zinc fall in grains and legumes when exposed to higher carbon dioxide concentrations. Additionally, protein and amino acid concentrations drop in spring wheat (a major staple crop), but less nutritious non-structural carbohydrates and lipids significantly increases [23].

Of course, effects of climate change on malnutrition must be viewed in a broader context that takes into account other trends, such as climate effects on pollinators, as well as the greater portion of crops now used either to feed livestock or supply feedstock for biofuels. In addition, when climate change affects the prevalence of bacterial or parasitic infectious diseases –where nutrient absorption is limited – this indirect pathway will affect nutritional benefits [27].

Fisheries

The most abundant greenhouse gas, CO_2, when absorbed by the ocean, leads to acidification. Over the past 250 years, the uptake of anthropogenic carbon has reduced ocean pH by 0.1 units, a trend that is continuing. IPCC scenarios predict a drop in global surface ocean pH of between 0.14 and 0.35 units over the twenty-first century. While the effects of ocean acidification are not fully understood, this process may threaten marine shell-forming organisms (e.g., corals) and their dependent species [19]. Of course, climate change may also threaten fish populations through other mechanisms. The recent slowing of the North Atlantic Gulf Stream, for instance, may lower the abundance of plankton that support many fish larvae [28]. Declining larval populations will threaten the recovery of overexploited fish species.

Sea surface temperature change is the dominant driving force that shifts the geographical distribution of marine species. Warmer waters are oxygen-poor; when coupled with CO_2-induced ocean acidification, they pose substantial risks to marine ecosystems [29].

Coastal and island populations that rely on fish as their main source of protein could be threatened if global fisheries are further stressed. Worldwide, fish represent 16% of the animal protein consumed by people, with a higher proportion in some regions, for example, 26% in Asia. Climate change, together with such other pressures as overfishing, may have a seriously impact on this source of nutrition.

Finally, current understanding of ocean acidification direct effects on fisheries is limited but could be hugely impactful. What is known, however, is ocean acidity impacts on coral reefs – the ecosystem most critical in supporting many coastal tropical fisheries. One study estimates a 92% reduction in coral reef habitat by 2100 [30].

Extreme Weather Events and Health

Natural Disasters

Droughts, floods, and violent storms have claimed millions of lives during the past two decades, threatened many more millions of people, and caused billions of dollars in property damage. On average, such disasters have killed 123,000 people worldwide each year between 1972 and 1996 [31]. Africa suffers the highest rate of deaths related to disasters, although 80% of the people affected by natural disasters are in Asia. For every person killed in a natural disaster, one thousand people are estimated to be affected, either physically, mentally, or through loss of property or livelihood [32].

Mental health problems such as post-traumatic stress disorder (PTSD) can be pervasive after a disaster. Their persistence depends on how unexpected the event was, the intensity of the experience, the degree of personal and community disruption, and long-term exposure to the visual signs of the catastrophe [33]. PTSD symptoms have been found as high as 75% in refugee children and adolescents [34].

In poor countries, disasters can trigger large-scale dislocation of populations, often to jurisdictions ill prepared to receive them. Malnutrition and communicable diseases are prevalent in refugee populations. Displaced groups are also subjected to violence, sexual abuse, and mental illness. Generally, crude mortality rates in displaced populations may reach as high as 30 times the baseline with substantial mortality occurring in children under five [35]. Even in the United States, system failures were evident in the aftermath of hurricanes Katrina and Rita. Over 2000 Americans were killed during that hurricane season, more than double the average number lost to hurricanes in the United States [36]. The survivors of Katrina suffered twice the rate of mental illness after the disaster when compared to a similar New Orleans population prior to that hurricane [37].

Floods

Floods are the most frequently occurring type of natural disaster. Globally, the frequency of river flood events has been increasing, as well as economic losses, due to the expansion of population and property in flood plains [2]. In the United States, the heaviest 1% of rain events increased by 20% in the past century, while total precipitation increased by 7%. During the same period, there was a 50% increase in the frequency of days with precipitation over four inches in the upper Midwest [38].

Other regions, notably the South, have also seen strong increases in heavy downpours, with most of these coming in the warm season and almost all of the increase coming in the last few decades.

Populations are more vulnerable in floodplains and coastal zones. Degradation of the local environment can also contribute significantly to risk. Hurricane Mitch serves as one example; it was the most deadly hurricane to strike the Western Hemisphere in the last two centuries; the hurricane caused 11,000 deaths in Central America, with thousands of other people still recorded as missing. Many fatalities occurred from mudslides in deforested areas [39].

Wildfires

Hot temperatures combined with drought induce wildfires that threaten health both directly and through reduced air quality. Fire smoke carries a large amount of fine particulate matter that exacerbates cardiac and respiratory problems, such as asthma and chronic obstructive pulmonary disease (COPD). Drought-induced fires in Florida in 1998 were associated with increased hospital emergency room visits for asthma, bronchitis, and chest pain [40]. The incidence of extensive wildfires in the Western United States (counting those over 400 hectares) increased four-fold from the period 1970–1986 to 1987–2003 [41]. Higher springtime temperatures (0.87 °C warmer) that hasten spring snowmelt and result in a drop in soil moisture are considered driving factors that explain this increase in fires [41, 42]. Fire and climate change modeling for California has shown that the most severe effects of global climate change would occur in the Sierra foothills, where potentially catastrophic fires could increase by 143% in grassland and 121% in chaparral [43]. The same study showed that greater burn intensity would result from a predicted change in fuel moisture and wind speeds.

During the peak of the 2007 San Diego wildfires, emergency department visits for respiratory conditions increased by 34% and visits specific for asthma rose by 112% [44]. Outpatient visits for acute bronchitis remained elevated, 72% above the usual rate in the 5-day period following the peak fire period. Children under 4 years of age had a 136% increase in emergency department visits for asthma, and very young children aged 0–1 experienced a 243% increase.

Sea-Level Rise and Health

Thermal expansion of salt water alone (without adding glacial melt water) causes sea-level rise. One anticipated effect is an increase in flooding and coastal erosion in low-lying coastal areas. This will endanger large numbers of people; at present, thirteen of the world's twenty megacities are situated at sea level. Midrange estimates project a 40 cm sea-level rise by the 2080s. Under this scenario coastal regions at risk from storm surges will expand and the population at risk will increase from the current 75 to 200 million [13]. Greater sea-level rise would mean even more devastation. Nicholls and Leatherman (1995) showed that the extreme case of a one meter rise in sea level could inundate numerous low-lying areas, and impact 18.6 million people in China, 13 million in Bangladesh, 3.5 million in Egypt, and 3.3 million in Indonesia [45]. Countries similar to Egypt, Vietnam, and Bangladesh, as well as small island nations, are especially vulnerable, for several reasons. Coastal Egypt is already subsiding due to extensive groundwater withdrawal, and Vietnam and Bangladesh have heavily populated low-lying deltas along their coasts. In the United States, an estimated 20 million people will be affected by sea-level rise by the year 2030, either directly or indirectly by migration networks linking inland and coastal areas and their populations [46]. In addition, rising sea levels heighten storm surges and cause salination of coastal freshwater aquifers, and they disrupt stormwater drainage and sewage disposal. Armed conflict may be among the worst results emerging from forced population migrations [47].

Water- and Food-Borne Diseases

Waterborne diseases, from both freshwater and coastal marine waters, are likely to become a greater problem through climate change–related weather extremes. In freshwater systems, both water quantity and quality can be affected. In marine waters, changes in temperature and salinity will affect coastal ecosystems in ways that may increase the risk of certain diseases, and as cholera and food poisoning from toxic algal blooms.

Freshwater

Water quantity and quality play a large role in waterborne diseases, which are therefore particularly sensitive to changes in the hydrologic cycle.

The impact of climate change on water *quantity* is relatively straightforward. In some regions precipitation is expected to increase, whereas in others it is predicted to decline, even to the point of ongoing drought. Water shortages contribute to poor hygiene, and that in turn contributes to diarrheal disease, especially in poor countries. At the other extreme, flooding can contaminate drinking water across watersheds with runoff from sewage lines, containment lagoons (such as those used in livestock feeding operations), or nonpoint source pollution (such as agricultural fields).

Extreme weather events can affect water *quality* in more complex ways. Many community water systems are already overwhelmed by extreme rainfall events. Runoff can exceed the capacity of the sewer system or treatment plants, and these systems are designed to discharge the excess wastewater directly into surface water bodies [48, 49]. Urban watersheds receive more than 60% of their annual contaminant loads during storm events [50]. Turbidity also increases during storms, and studies have linked turbidity with illness in many communities [51, 52].

Disease outbreaks from most water-borne pathogens are distinctly seasonal, clustered in key watersheds, and associated with heavy precipitation [53]. In Walkerton, Ontario, in May 2000, heavy precipitation combined with failing infrastructure contaminated drinking water with *E. coli 0157:H7* and *Campylobacter jejuni*, resulting in 7 deaths and an estimated 2300 illnesses [54].

Intense rainfall can also contaminate recreational waters and increase the risk of human illness through higher bacterial counts [55]. This association is strongest at the beaches closest to rivers [56]. Enteric viruses are found at higher levels in both surface and ground water following heavy rainfall [57].

Cryptosporidiosis, one of the most prevalent diarrheal diseases in the world, is illustrative. Associated with domestic livestock, *cryptosporidium* is a protozoan that can contaminate drinking water during the periods of heavy precipitation. The oocyst is resistant to chlorine treatment. The 1993 cryptosporidiosis outbreak in Milwaukee, during which an estimated 403,000 people were exposed to contaminated water, followed unusually heavy spring rains and runoff from melting snow [58]. Similarly, studies of the Delaware River have shown that *Giardia* and *Cryptosporidium* oocyst counts correlate with the amount of rainfall [59]. In Walkerton, Ontario, in May 2000, heavy precipitation combined with failing infrastructure contaminated drinking water with *E. coli 0157:H7* and *Campylobacter jejuni*, resulting in 7 deaths and an estimated 2300 illnesses [54].

A nationwide analysis of waterborne disease outbreaks in the United States from 1948 to 1994 demonstrated a distinct seasonality, a spatial clustering in key watersheds, and an association with heavy precipitation; 67% of reported outbreaks were preceded by unusually rainy months (defined as rainfall in the upper 80th percentile based on a 50-year local baseline) [53]. A recent study from a pediatric hospital in Milwaukee found that admissions for acute gastrointestinal illness increased following rain [60]. Certain watersheds, by virtue of associated land-use patterns and the presence of human and animal feces, are at high risk of surface water contamination after heavy rains, and this seriously threatens the purity of drinking water.

Recreational waters are also contaminated by heavy rainfall. For example, extensive runoff leads to higher bacterial counts in rivers and at beaches in coastal areas and is strongest at the beaches closest to rivers [56]. This suggests that the public health risk of swimming at beaches increases with heavy rainfall, a predicted consequence of climate change.

Heavy rains can lead to flooding, which can also raise the risk of water-borne diseases such as *Cryptosporidium* and *Giardia*. In many communities where sewage and, infrastructure stormwater runoff are handled in a combined system, when heavy rainfall overwhelms storm drainage infrastructures, a combined sewer overflow (or CSO) event ensures. The highest levels of *E. coli* bacteria occur in surface waters in such cases (Fig. 14.4). Using 2.5 inches (6.4 cm) of daily precipitation as the threshold for initiating a CSO event, the frequency of these occurrences in Chicago is expected to rise by 50–120% by the end of this century [61]. This will

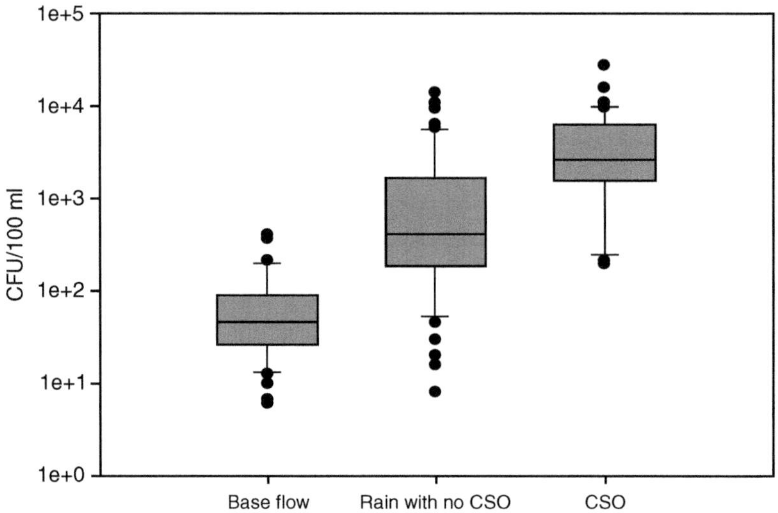

Fig. 14.4 Relationship between precipitation and *E coli* counts. (Source: McLellan)

pose increased risk to drinking and recreational water quality. The worldwide average for diarrheal diseases in the future is projected to rise 20% for the period 2040–69 and 29% for 2070–99 [62].

Marine Environments

The impact of climate change on the extent of sea ice melt is well documented [18]. Data since 1978 show average annual Arctic sea ice area has declined by 2.7% per decade (2.1–3.3), with decreases of 7.4% in summer (5.0–9.8) [18]. Atkinson et al. (2004) combined net sampling data on Antarctic krill from 1926 to 2003 to demonstrate the effect of the scope of sea ice on krill populations [63]. After controlling for populations of top-down predators and bottom-up resources, they found temporal links between summer krill density and the extent of winter sea ice the preceding year, perhaps related to larval overwintering. Krill have enormous effects on the entire Arctic ecosystem; they are one of the primary food sources for penguins, albatrosses, seals, and whales [63]. Subsequently, the humans who rely on these species for their food and livelihood are affected by krill prevalence.

Blooms of marine algae that can release toxins into the marine environment, including two groups, dinoflagellates and diatoms, are enhanced by warm water and elevated nitrogen levels. These harmful algal blooms – sometimes referred to as red tides – can cause acute paralytic, diarrheic, and amnesic poisoning in humans, as well as extensive die-offs of fish, shellfish, and marine mammals and birds that depend on the marine food web. Over the past, three decades the frequency and global distribution of harmful algal blooms appears to have increased, and more human intoxication from algal sources has occurred [64]. For example, during the 1987 El Niño, a bloom of *Gymnodinium breve,* previously confined to the Gulf of Mexico, extended northward after warm Gulf Stream water flowed far up the eastern United States coast. This resulted in human neurological poisonings from shellfish and in substantial fish kills [65]. Similarly that year, an outbreak of amnesic shellfish poisoning occurred on Prince Edward Island when warm eddies of the Gulf Stream neared the shore and heavy rains increased nutrient-rich runoff [66].

By the year 2100, a 4 °C increase in summer temperatures in combination with water column stratification would double the growth rates of several species of harmful algal blooms in the North Sea [67]. Biotoxins associated with warmer waters also include ciguatera (fish poisoning), which could extend its range to higher latitudes. An association has been found between ciguatera and sea surface temperature in some Pacific Islands [68].

Vibrio species are especially prolific in warm marine waters. Copepods (zooplankton), which feed on algae, can serve as reservoirs for *Vibrio cholerae* and other enteric pathogens. For example, in Bangladesh cholera follows seasonal warming of sea surface temperatures, which can enhance plankton blooms [69], and cholera cases fluctuate with temperature in coastal Africa as well [70, 71].

Vibrio species have expanded in northern Atlantic waters in association with warm water [72]. For example, in 2004 an outbreak of *V. parahaemolyticus* shellfish poisoning was reported from Prince William Sound in Alaska [73]. This pathogenic species of *Vibrio* had not previously been isolated from Alaskan shellfish due to the frigidity of Alaskan waters [73]. Water temperatures in the 2004 shellfish harvest remained above 15° C and mean water temperatures were significantly higher than the previous 6 years [73]. Northern expansion of *V. parahaemolyticus* has been documented in Europe [74], and a rising trend of *Vibrio* bacteria and shifts in plankton abundance have paralleled warming trends in the North Sea since 1987 [75]. Such evidence suggests the potential for warming sea surface temperatures to increase the geographic range of shellfish poisoning and *Vibrio* infections into temperate and even arctic zones.

The 1997 and 1998 El Niño event provided a natural experiment to examine temperature effects on diarrheal diseases, when winter temperatures in Lima, Peru, increased more than 5 °C above normal, and the daily hospital admission rates for diarrhea more than tripled the rates over the prior 5 years [76] (Fig. 14.5). Long-term studies of the El Niño Southern Oscillation, or ENSO, have confirmed this pattern. ENSO refers to natural year-to-year variations in sea surface temperatures, surface air pressure, rainfall, and atmospheric circulation across the equatorial Pacific Ocean. This cycle provides a model for observing climate-related changes in many ecosystems. ENSO has had an increasing role in explaining cholera outbreaks in recent years, perhaps because of concurrent climate change [77]. Overall there is growing evidence that climate change contributes to the risk of waterborne diseases in both marine and freshwater ecosystems.

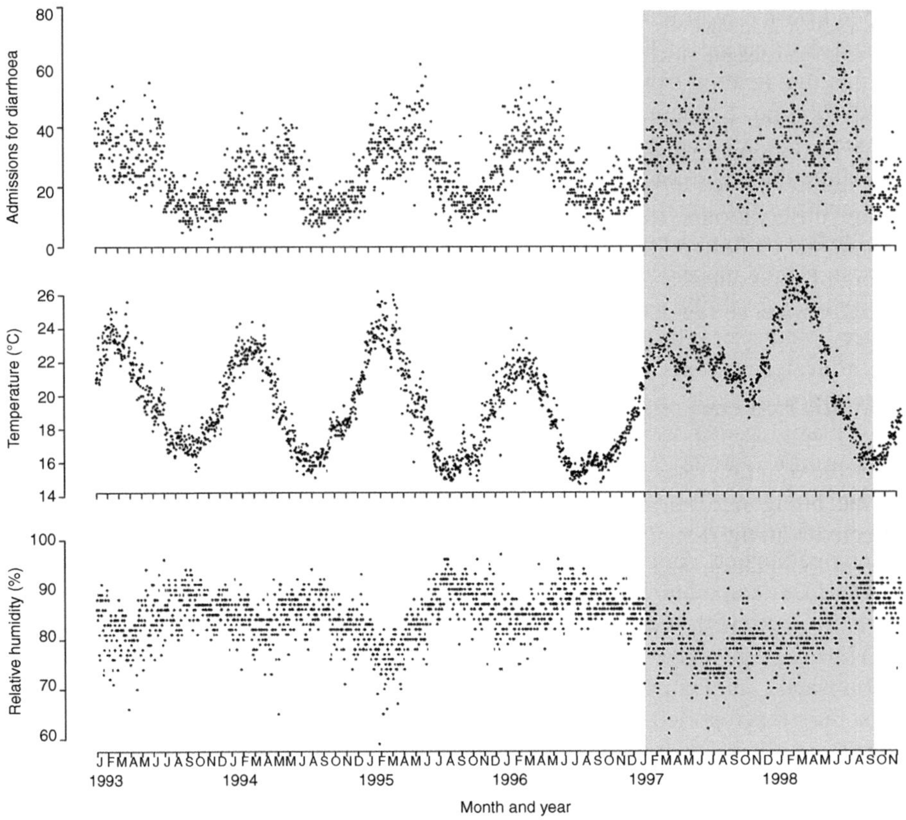

Fig. 14.5 Temperature and cholera relationship, Peru. (Source: Checkley et al. [76])

Food-Borne Diseases: Changes in temperature and/or humidity can alter the incidence of food-borne infectious diseases. In the United Kingdom, researchers have found a strong correlation between the incidence of food-borne disease and temperatures in the month preceding the illness [78], suggesting food poisoning or spoilage. Reported cases of food poisoning across Australia, Western and Central Europe, and Canada follow a near-linear relationship to each degree of increase in weekly temperature [18]. Temperature contributed to an estimated 30% of cases of salmonellosis in much of continental Europe, especially when they exceeded a threshold of 6 °C above average [79]. Monthly incidence of food poisoning in Britain is most strongly associated with temperatures during the previous 2–5 weeks [78]. Other food-borne agents, such as *campylobacter*, are also seasonal but not as strongly linked to temperature fluctuations. Food spoilage is temperature dependent since pest species, especially flies, rodents, and cockroaches, increase their contact with food as temperatures rise [80].

Land-Use Effects on Weather and Health

The center or core of most large cities is often much hotter than surrounding areas. The term "urban heat island" defines parts of a city that generate and/or retain heat as a result of roads, buildings, and industrial activities (Fig. 14.6). Black asphalt and other dark surfaces on roads, parking lots, and roofs have a low albedo (reflectivity); they absorb and retain heat, re-radiating it at night when the area would otherwise cool down. In addition, trees are relatively sparse in urban areas, so they provide less of the cooling effect associated with evapotranspiration. Global warming is expected to increase both heat and humidity, which will aggravate the effect of heat islands and increase heat stress on urban populations [81]. One study estimates the mean surface warming due to urban sprawl and land-use change to be 0.27 °C (0.49 °F) for the continental United States [82]. Urban areas may therefore face a compounded problem as they experience both global warming and localized warming from the heat island effect. Urban residents in developing countries may be especially vulnerable to morbidity and mortality during heat waves.

By the end of the twenty-first century, the number of heat wave days could double in Los Angeles [83] and quadruple in Chicago [84], if emissions are not reduced. A recent analysis of 21 United States cities found that the average number of deaths due to heat waves would more than double by 2050, even after controlling for acclimatization [85].

To understand the relationship between vector borne disease and climate, local landscapes need to be included in the analyses. For example, in the Amazon Basin, malaria incidence fluctuates with rainfall levels. Yet regional differences in the range of wetlands and surface water modify the effect of rainfall so much that in upland locations with sparse wetlands, malaria increases with rainfall, whereas in areas with abundant wetlands, it decreases [86]. In essence, climate effects must take into account local land cover data.

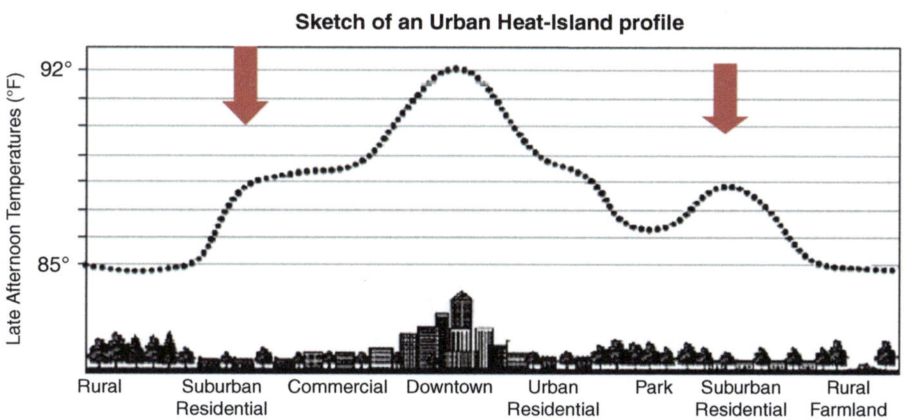

Fig. 14.6 Urban heat island effect. (Source: US EPA)

Ecosystems that preserve landscape integrity and biodiversity form the basis of many essential environmental services. However, global vegetation cover is changing far more rapidly than climate. Land cover is disrupted by such forces as deforestation, urban sprawl, industrial development, road construction, large water control projects – dams, canals, irrigation systems, and reservoirs – and climate change. Natural landscapes are being damaged or destroyed everywhere on a very large scale. A global pattern of landscape fragmentation has emerged.

The Public Health Response

Disentangling relationship between human health and climate change remains complex. The relationship is not always discernible, especially over short time spans. To understand and address such links requires systems thinking and consideration of many factors that range beyond health to such sectors as energy, transportation, agriculture, and development policy. Interdisciplinary collaboration is critical. A wide range of tools is needed, including innovative public health surveillance methods, geographically based data systems, classical and scenario-based risk assessment, and integrated modeling.

Mitigation and Adaptation

Two strategic approaches may be considered to address climate change. The first, mitigation, corresponds to primary prevention, and the second, adaptation, corresponds to secondary prevention.

Mitigation involves efforts to stabilize or reduce the production of greenhouse gases (and perhaps to sequester those gases that are produced). This goal can be achieved through policies and technologies that result in more efficient energy production and reduced energy demand. For example, sustainable energy sources such as wind and solar energy do not contribute to greenhouse gas emissions. Similarly, transportation policies that rely on walking, bicycling, mass transit, and fuel-efficient automobiles result in fewer greenhouse gas emissions than are produced by the current United States reliance on large automobiles with high fuel consumption for most transportation. Much energy use occurs in buildings, and green construction that emphasizes energy efficiency, together with electrical appliances that conserve energy, also plays a role in reducing greenhouse gas emissions (see Chap. 19). A final aspect of mitigation does not aim to reduce the production of greenhouse gases, but rather to accelerate their removal. Carbon dioxide sinks such as forests are effective in this regard, so land-use policies that preserve and expand forests are an important doctrine to mitigate global climate change.

Adaptation (or preparedness) refers to efforts to reduce the public health impact of climate change. For example, if we anticipate severe weather events such as hurricanes, then preparations by emergency management authorities and medical facilities can minimize morbidity and mortality. Similarly, public health surveillance systems can detect outbreaks of infectious diseases in vulnerable areas, a prerequisite to early control. Many of today's current challenges, such as deaths from heat waves, floods, and air pollution, will be exacerbated by climate change. Much preparedness can therefore be constructed from the analyses of the strengths and weaknesses of current prevention efforts, and a rethinking of potential thresholds that may change in the future. Examples are expected changes in the volume of stormwater runoff and the frequency of heat waves.

"Co-benefits" from Mitigating Climate Change

While the steps needed to address the evolution of climate may appear formidable, some of them – reducing greenhouse gas emissions, developing and deploying sustainable energy technologies, and/or adapting to climate change – yield multiple benefits [87]. This can make them especially attractive, cost-effective, and politically feasible. For example, urban tree planting helps reduce CO_2 levels, while at the same time it reduces the heat island effect and local energy demand, improves air quality, dampens noise levels, and provides an attractive venue for physical activities and social interaction [88]. Another example is the reduction of fossil fuel use in power plants. This is a principal strategy to reduce greenhouse gas emissions, and also a strategy to reduce air pollution [89]. A third example is sustainable community design [90]. Box 14.1 (below) shows the substantial health benefits gained by facilitating active transport (walking and bicycling) in and around urban settings.

In planning solutions such as sustainable communities, it is essential that the communities themselves be involved. Poor communities and communities of color bear a disproportionate vulnerability to many environmental health threats. These groups must be included when solutions are planned in order to preclude the possibility that the already large gap in access to healthy and desirable neighborhoods be widened. New areas must also be designed with cultural sensitivity and diversity in mind so that all people can be afforded a realistic opportunity to enjoy healthy new neighborhoods where environmental justice issues are considered at every level.

Box 14.1 Health Co-benefits of Greenhouse Gas Mitigation
With continued poor air quality in many regions of the world, coupled with accelerating trends in chronic diseases – many related to sedentary lifestyles afforded by the automobile – mitigating greenhouse gas emissions offers large and immediate health benefits [91]. These benefits will especially come from across the energy, transportation, and food sectors.

Energy

A recent study by Shindell et al. (2012) addressed tropospheric ozone and black carbon (BC) contribution to both degraded air quality and global warming [92]. The authors identified 14 best interventions targeting methane and BC emissions that reduce projected global mean warming ~0.5 °C by 2050. The resulting "co-benefit" was the avoidance of 0.7 to 4.7 million annual premature deaths from outdoor air pollution and increases annual crop yields by 30–135 million metric tons due to ozone reductions in 2030 and beyond. The valuation was dominated by health effects from reduced BC in the air. While this study was global in nature, the findings apply to any location with coal-fired power plants, the most substantial contributor to black carbon particulates. If 17% of US electricity came from solar, 1400 lives could be saved annually in Eastern US [93].

Transportation

Midwest Region Case study: Co-benefits of Alternative Transportation Futures from improving air Quality and Physical Fitness

The transportation sector produces one-third of US greenhouse gas emissions. Automobile exhaust contributes not only to GHGs but also contains precursors to fine particulate matter ($PM_{2.5}$) and ozone (O_3), posing public health risks. Adopting a low carbon transportation system with fewer automobiles, therefore, could have immediate health "co-benefits" via improved air quality. Grabow et al. (2012) modeled census tract-level mobile emissions for two comparative scenarios: current baseline versus a low carbon scenario where automobile trips shorter than five miles round-trip would be removed for the 11 largest metropolitan areas in the Midwestern United States. These relatively short car trips comprised approximately 20% of vehicle miles traveled for the region.

Across the upper Midwest study region of approximately 31.3 million people and 37,000 total square miles, mortality would decline by nearly 575 deaths per year from the benefit of improved air quality. Health benefits would also accrue in rural settings as well, with 25% air quality–related health benefits to populations outside metropolitan areas.

An active transport scenario was then added, with the assumption that 50% of the short trips (<5 miles) could by achieved by bicycle during the 4 months of most favorable weather conditions in the region. This theoretical maximum level of biking was selected because some locations in Europe have achieved this amount of bicycle commuting, and there already exists an observed trend of increasing bicycle share across all of the 11 Midwestern metropolitan areas [94]. This active transport scenario alone yielded savings of another 700 lives/

year and approximately \$3.8 billion/year from avoided mortality costs (95% CI: \$2.7, \$5.0 billion].

In summary, the estimated benefits of improved air quality and physical fitness from a green transportation scenario would be 1295 (95% CI: 912, 1636) lives saved and \$8 billion in avoided mortality and health care costs per year for the upper Midwest region alone. Nationally, there is already evidence that US cities with enhanced levels of active transport experience large health benefits; one study found that cities with the highest rates of commuting by bike or on foot have obesity and diabetes rates 20% and 23% lower, respectively, than cities with the lowest rates of active commuting [95].

Food Systems

According to the EAT-Lancet Commission on healthy diets from sustainable food systems, transforming our diets to sustainable food sources is required in order to both meet the UN Sustainable Development Goals (SDGs) and the Paris Climate Agreement (2019). Such transformational change will benefit both our health and the environment, but demand major shifts in our current dietary habits. By 2050, we will need to cut our consumption of red meat and sugar by half, and double our consumption of nuts, fruits, vegetables, and legumes.

From this transition, human health benefits would include: 10.8–11.6 million deaths, or a 19–24% reduction, annually. Additionally, shifting away from a meat-based diet potentially can reduce rates of deforestation, water consumption, nitrogen and phosphorus pollution, and CO_2 and nitrous oxide emissions [96].

Current food production contributes substantially to greenhouse gas emissions, including CO_2, methane, and nitrous oxide. Food production is the primary source of methane and nitrous oxide, with warming potentials 56 and 280 times that of CO_2, respectively [96]. Also, our food system demands on landcover conversion, especially from deforestation for cattle rangeland, wetland drainage, and soil tillage, further contribute to the problem.

Figure 14.7 from the Lancet EAT Commission Report shows the various contributions that terrestrial agriculture has made to climate change, and subsequently the necessary changes in agriculture required to stabilize climate below 2 °C warming. While production practices, such as improved efficiency in irrigation, cropping, fertilizer and manure management, and cutting food waste, will reduce emission from methane and nitrous oxide, increasing a plant-based diet will have the largest impact on greenhouse gas emission reductions.

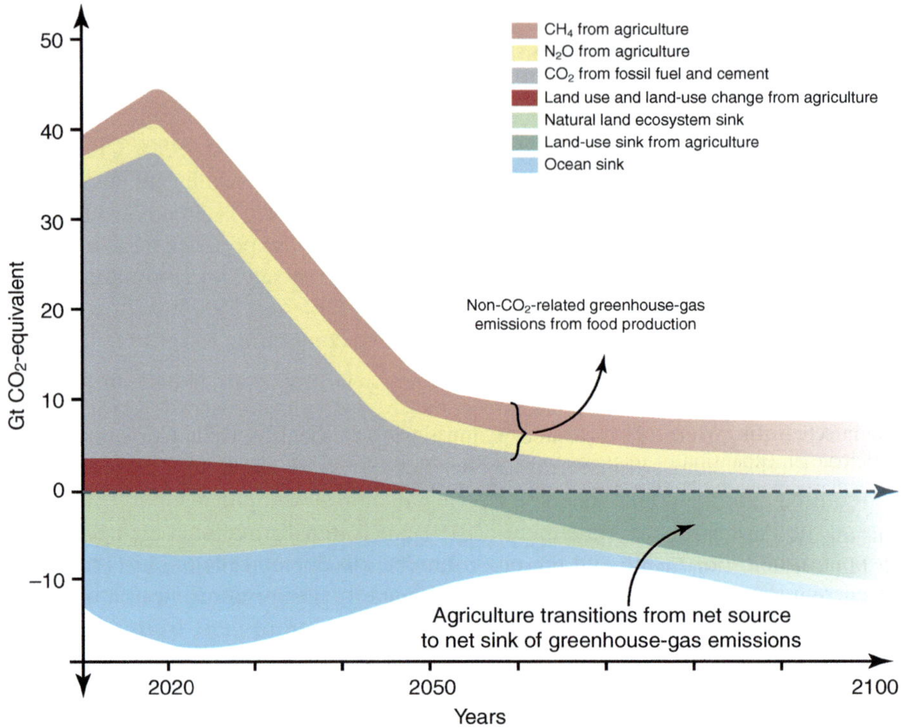

Fig. 14.7 Projections of global emissions to keep global warming to well below 2 °C, aiming for 1·5 °C Data are from Intergovernmental Panel on Climate Change AR5 (RCP2.6 data for nitrous oxide and methane) and Rockstrom et al. (for fossil-fuel emissions, land use, land-use change, and forestry, and biosphere carbon sinks). (Source: Willet et al. [96])

Side Effects and Unintended Consequences

When solutions are attempted through interventions that are too narrowly focused or lack involvement by the local community, steps taken to address climate change can have unintended consequences. A cautionary example is biofuel production, a rapidly growing industry driven by economic incentives and public policies. Worldwide biofuel production may quadruple within the next 15–20 years [97–99].

However, critics claim that large-scale production of biofuels diverts crops from use as food, thus creating scarcity and driving food prices higher [100, 101]. The extent of humanitarian food aid from the United States that is available for extremely impoverished countries is inversely correlated with commodity prices [102]. Demand for biofuels may also accelerate the conversion of forests to cropland. Paradoxically, this could increase carbon dioxide levels [103–106], and threaten biodiversity in sensitive areas [107]. It is quite surprising to learn that a full life cycle analysis for biofuels showed slightly higher particulate matter levels for corn-based ethanol compared to gasoline and cellulosic ethanol; growing corn for ethanol involves intense use of fertilizers and farm machinery, and may simply shift air pollution from urban locations toward rural areas [108]. Critics further claim that biofuel production is economically inefficient and relies heavily on subsidies [109]. Each of these claims is controversial; for example, some argue that food scarcity results from inadequate or inefficient distribution rather than from scarcity per se. In sum, the biofuel debate illustrates the potential to bring about unintended consequences that are especially harmful for vulnerable populations, and the need for careful analysis of each strategy proposed to address climate change [110].

Ethical Considerations

Climate change poses monumental ethical concerns in several ways. First, on a global scale, the nations that are responsible for most carbon emissions to date represent a small proportion of the world's population; they are relatively resilient to the effects of climate change. By contrast, the large population of the southern hemisphere – the poor countries – account for a relatively small cumulative share of carbon emissions, and present a very low per capita emission rate (although total emissions from developing nations are growing rapidly, and China surpassed the United States in 2006). The United States, with 5% of the global population, produces 25% of total greenhouse gas emissions. This discrepancy exemplifies the ethical implications posed by climate change on a global scale, shown graphically in Fig. 14.8. Poor populations in the developing world have little by way of industry, transportation, or intensive agriculture. They contribute only a fraction of the per capita greenhouse gases that the developed countries produce, and their capacity to protect themselves against the adverse consequences of emissions caused mostly by others is quite limited. Of course, if developing nations do not choose pathways that use more efficient energy technology, global climate change trends will intensify even as equity between rich and poor nations improves [47].

Within the United States, and within many other nations, a similar disparity exists. Poor and disadvantaged people will in many cases bear the brunt of climate change impacts, including those on health. This was graphically demonstrated in the aftermath of Hurricane Katrina, a disaster typical of those expected to increase with climate change. The poor populations of New Orleans and the nearby Gulf region were disproportionately likely to fail to evacuate, to suffer catastrophic disruption following the storm, and to be unable to recover [111–113].

Finally, an ethical issue arises with respect to intergenerational justice. Climate change holds the potential for enormous impacts on the health and well-being of future generations (Page 2007). Ethical and religious thinkers have pointed this out, and have argued that our generation owes a moral obligation to those who will follow to restore a sustainable climate.

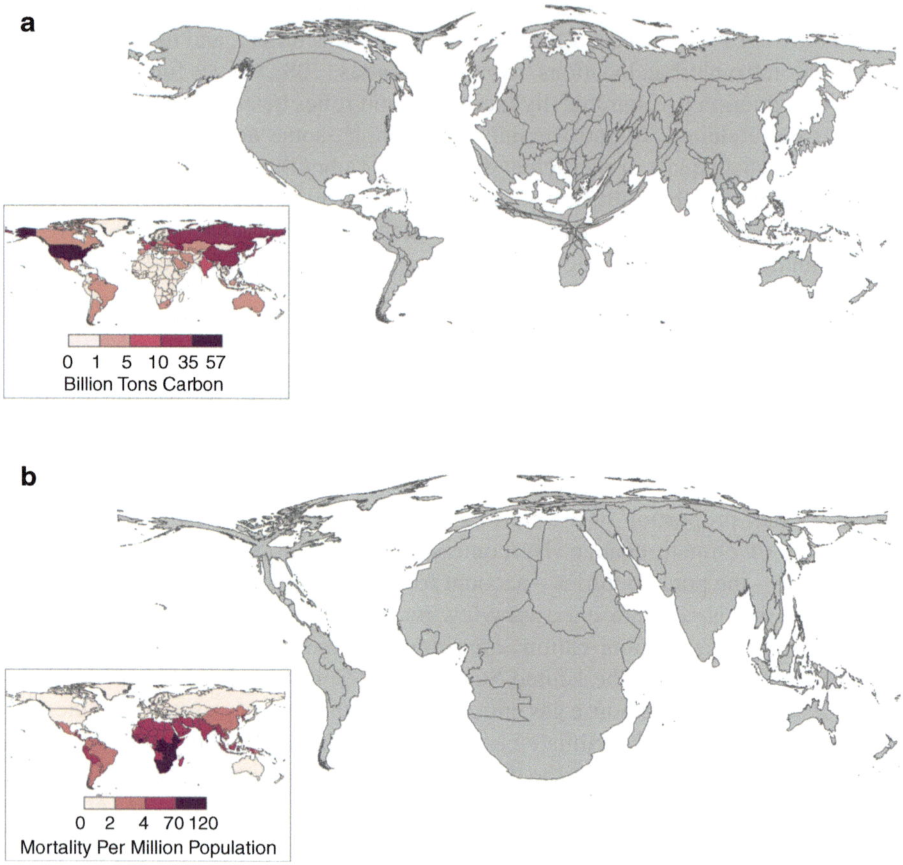

Fig. 14.8 Comparison of undepleted cumulative carbon dioxide (CO_2) emissions (by country) for 1950–2000 versus the regional distribution of four climate-sensitive health effects (malaria, malnutrition, diarrhea, and inland flood-related fatalities). (**a**) CO_2 emissions data source: Marland et al. [116]. (**b**) The Intergovernmental Panel on Climate Change (IPCC) "business as usual" greenhouse gas (GHG) emissions scenario, "IS92a" and the HadCM2 general circulation model (GCM) of the UK Hadley Centre were used to estimate climate changes relative to "baseline" 1961–1990 levels of GHGs and associated climate conditions. Existing quantitative studies of climate–health relationships were used to estimate relative changes in diarrhea, malaria, inland and coastal flooding, and malnutrition, for years from 2000 to 2030 (McMichael et al. 2004)

Financial Considerations

Every weather-related disaster has large economic costs as well as health cost. For example, estimates for Russia's 2010 heat wave are 55,000 deaths, 25% of annual crop failure, more than one million hectares of land destroyed by fire at, and economic losses at about $15 billion – or 1% gross domestic product (GDP) [114]. The most comprehensive economic analysis to emerge on climate change probably comes from the UK's Stern report.

According to the Stern Report [115] 5–6 °C warming would result in an average 5–10% loss in global GDP, with poor countries experiencing in excess of 10% loss of GDP. The report further describes analyses that include the full range of both impacts and possible outcomes, and under a Business as Usual scenario climate change would pose economic risks of between 5% and 20% per capita.

Summary

Climatologists now state with a high degree of certainty that global climate change is real, is advancing more rapidly than expected, and is caused by human activities, especially through fossil fuel combustion and deforestation. Environmental public health researchers, in assessing future projections for Earth's climate, have concluded that, on balance, adverse health outcomes will predominate under these changed conditions. The number of pathways through which climate change can affect the health of populations makes this environmental hazard one of the most perilous and intricate challenges that we face in this century. By contrast, the potential health co-benefits from departing from our current fossil fuel-based economy may offer some of the most beneficial health opportunities in over a century.

References

1. Nicholls N, et al. Observed climate variability and change. In: Houghton J, et al., editors. Climate change 1995: The science of climate change. Contribution of working group I to the second assessment report of the intergovernmental panel on climate change. New York: Cambridge University Press; 1996.
2. IPCC. Summary for policymakers. In: Masson-Delmotte V, Zhai P, Pörtner HO, Roberts D, Skea J, Shukla PR, Pirani A, Moufouma-Okia W, Péan C, Pidcock R, Connors S, Matthews JBR, Chen Y, Zhou X, Gomis MI, Lonnoy E, Maycock T, Tignor M, Waterfield T, editors. Global warming of 1.5°C. An IPCC Special Report on the impacts of global warming of 1.5°C above pre-industrial levels and related global greenhouse gas emission pathways, in the context of strengthening the global response to the threat of climate change, sustainable development, and efforts to eradicate poverty: Cambridge Press; 2018.
3. Tomanio J. Billion dollar weather. Nat Geo. Sept. 2012 issue: 53–54.
4. Bevis M, Harig C, Khan SA, Brown A, Simons FJ, Willis M, Fettweis X, van den Broeke MR, Madsen FB, Kendrick E, Caccamise DJ II, van Dam T, Knudsen P, Nylen T. Accelerating changes in ice mass within Greenland, and the ice sheet's sensitivity to atmospheric forcing. PNAS. 2019;116(6):1934–9; first published January 22, 2019. https://doi.org/10.1073/pnas.1806562116.
5. Rignot E, Mouginot J, Scheuchl B, van den Broeke M, van Wessem MJ, Morlighem M. Four decades of Antarctic ice sheet mass balance from 1979–2017. PNAS. 2019;116(4):1095–103; first published January 14, 2019. https://doi.org/10.1073/pnas.1812883116.
6. Gray W. Hurricanes: their formation, structure and likely role in the tropical circulation. In: Shaw DB, editor. Meteorology over the tropical oceans. London: Royal Meteorological Society; 1979.

7. Goldenberg SB, Landsea CW, Mestas-Nuñez AM, Gray WM. The recent increase in Atlantic hurricane activity: causes and implications. Science. 2001;293:474–9.
8. Webster PJ, Holland GJ, Curry JA, Chang H-R. Changes in tropical cyclone number, duration, and intensity in a warming environment. Science. 2005;309:1844–6.
9. Emanuel K, Sundararajan R, Williams J. Hurricanes and global warming: results from downscaling IPCC AR4 simulations. Bull Am Meteorol Soc. 2008;89:347–67.
10. Hess JJ, Malilay JN, Parkinson AJ. Climate change: the importance of place. Am J Prev Med. 2008;35:468–78.
11. IPCC. Climate change 2014: synthesis report. Contribution of working groups I, II and III to the fifth assessment report of the intergovernmental panel on climate change [Core writing team, R.K. Pachauri and L.A. Meyer (eds.)]. Geneva: IPCC; 2014. 151 pp.
12. Uejio CK, Wilhelmi OV, Golden JS, Mills DM, Gulino SP, Samenow JP. Intra-urban societal vulnerability to extreme heat: the role of heat exposure and the built environment, socioeconomics, and neighborhood stability. Health Place. 2011;17(2):498–507.
13. McCarthy J, et al., editors. Climate change 2001: Impacts, adaptation, and vulnerability. Contribution of working group II to the third assessment report of the intergovernmental panel on climate change. New York: Cambridge University Press; 2001.
14. Schimhuber J, Tubiello FN. Global food security under climate change. Proc Natl Acad Sci. 2007;104:19703–8.
15. Tubiello FN, Soussana J-F, Howden SM, Easterling W. Crop and pasture response to climate change. Proc Natl Acad Sci. 2007;104:19686–90.
16. Brown ME, Funk CC. Food security under climate change. Science. 2008;319:580–1.
17. World Health Organization. World health report 2002: reducing risks, promoting healthy life. Geneva: World Health Organization; 2002.
18. IPCC. Climate change 2007. Contribution of working groups I, II and III to the fourth assessment report of the intergovernmental panel on climate change [Core writing team, Pachauri, R.K and Reisinger, A. (eds.)]. Geneva: IPCC; 2007.
19. IPCC. Summary for policymakers. In: Climate change 2007: impacts, adaptation and vulnerability. Contribution of working group II to the fourth assessment report of the intergovernmental panel on climate change. Cambridge, UK: IPCC; 2007.
20. Patz JA. Public health risk assessment linked to climatic and ecological change. Hum Ecol Risk Assess. 2001;7(5):1317–27.
21. Battisi DS, Naylor RL. Historical warnings of future food insecurity with unprecedented seasonal heat. Science. 2009;323:240–4.
22. Cline WR. Global warming and agriculture impact estimates by country. Washington, DC: Center for Global Development and Peterson Institute for International Economics; 2007. 186 pp.
23. Myers SS, Smith MR, Guth S, Golden CD, Vaitla B, Mueller ND, Dangour AD, Huybers P. Climate change and global food systems: potential impacts on food security and undernutrition. Annu Rev Public Health. 2017;38:259–77.
24. Butt TA, McCarl BA, Angerer J, Dyke PT, Stuth JW. The economic and food security implications of climate change in Mali. Clim Chang. 2005;68(3):355–78.
25. Lloyd SJ, Kovats RS, Chalabi Z. Climate change, crop yields, and malnutrition: development of a model to quantify the impact of climate scenarios on child malnutrition. Environ Health Perspect. 2011;119(12):1817–23.
26. Taub D, Miller B. Effects of elevated CO_2 on the protein concentration of food crops: a meta-analysis. Glob Chang Biol. 2008;14:565–75.
27. Humphrey JH. Child undernutrition, tropical enteropathy, toilets, and handwashing. Lancet. 2009;374:1032–5.
28. Pauly D, Alder J. Marine fisheries systems. In: Hassan R, Scholes R, Ash N, editors. Ecosystems and human well-being: current state and trends, vol. 1. Washington, DC: Island Press; 2005.
29. Portner HO, Peck MA. Climate change effects on fishes and fisheries: towards a cause-and-effect understanding. J Fish Biol. 2010; https://doi.org/10.1111/j.1095-8649.2012.02783.

30. Speers AE, Besedin EY, Palardy JE, Moore C. Impacts of climate change and ocean acidification on coral reef fisheries: an integrated ecological-economic model. Ecol Econ. 2016;128:33–43.
31. Loretti A, Tegegn Y. Disasters in Africa: old and new hazards and growing vulnerability. World Health Stat Q. 1996;49:179–84.
32. International Federation of Red Cross and Red Crescent Societies. World disaster report 1997. New York: Oxford University Press; 1998.
33. Green BL. Assessing levels of psychological impairment following disaster: consideration of actual and methodological dimensions. J Nerv Ment Dis. 1982;170:544–8.
34. McCloskey LA, Southwick K. Psychosocial problems in refugee children exposed to war. Pediatrics. 1996;97:394.
35. Toole M, Waldman R. The public health aspects of complex emergencies and refugee situations. Annu Rev Public Health. 1997;18:283–312.
36. NOAA, Unified Synthesis Product. Global climate change impacts in the United States. World Health Organization, Geneva, Switzerland. 2009.
37. Kessler RC. Mental illness and suicidality after hurricane Katrina. Bull World Health Organ. 2006;84:930–93.
38. Kunkel KE, Easterling DR, Redmond K, Hubbard K. 2003: temporal variations of extreme precipitation events in the United States: 1895–2000. Geophys Res Lett. 2003;30:1900. https://doi.org/10.1029/2003GL018052.
39. National Climatic Data Center. Mitch: the deadliest atlantic hurricane since 1780; 1999. http://www.ncdc.noaa.gov/ol/reports/mitch/mitch.html
40. Centers for Disease Control and Prevention. Surveillance of morbidity during wildfires: 6 Central Florida, 1998. MMWR. 1999;48(4):78–9.
41. Westerling AL, Hidalgo HG, Cayan DR, Swetnam TW. Warming and earlier spring increase western U.S. forest wildfire activity. Science. 2006;313:940–3.
42. Running SW. Is global warming causing more, larger wildfires? Science. 2006;313:927–8.
43. Torn MS, Mills E, Fried J. Will climate change spark more wildfire damage? LBNL Report no. 42592. Lawrence: Lawrence Berkeley National Laboratory; 1998.
44. Hutchinson JA, Vargo J, Milet M, French NHF, Billmire M, Johnson J, et al. The San Diego 2007 wildfires and Medi-Cal emergency department presentations, inpatient hospitalizations, and outpatient visits: an observational study of smoke exposure periods and a bidirectional case-crossover analysis. PLoS Med. 2018;15(7):e1002601. https://doi.org/10.1371/journal.pmed.100260.
45. Nicholls R, Leatherman S. Global sea-level rise. In: Strzepek K, Smith J, editors. As climate changes: international impacts and implications. New York: Cambridge University Press; 1995. p. 92–123.
46. Curtis KJ, Schneider A. Understanding the demographic implication of climate change: estimates of localized population predictions under future scenarios of sea-level rise. Popul Environ. 2011; https://doi.org/10.1007/s11111-011-0136-2.
47. Patz JA, Kovats RS. Hotspots in climate change and human health. Br Med J. 2002;325:1094–8.
48. Perciasepe R. Combined sewer overflows: where are we four years after adoption of the CSO control policy? Report of the Environmental Protection Agency; 1998.
49. Rose JB, Simonds J. Kings county water quality assessment: assessment of public health impacts associated with pathogens and combined sewer overflows. Olympia: Washington State Department of Natural Resources; 1998.
50. Fisher GT, Katz BG. Urban stormwater runoff: selected background information and techniques for problem assessment with a Baltimore, Maryland, case study. 1988.
51. Morris RD, Naumova EN, Levin R, et al. Temporal variation in drinking water turbidity and diagnosed gastroenteritis in Milwaukee. Am J Public Health. 1996;86(2):237–9.
52. Schwartz J, Levin R, Hodge K. Drinking water turbidity and pediatric hospital use for gastrointestinal illness in Philadelphia. Epidemiology. 1997;8(6):615–20.
53. Curriero F, Patz JA, Rose J, Lele S. The association between extreme precipitation and waterborne disease outbreaks in the United States, 1948–1994. Am J Public Health. 2001;91:1194–9.

54. Hrudey SE, Payment P, Huck PM, et al. A fatal waterborne disease epidemic in Walkerton, Ontario: comparison with other waterborne outbreaks in the developed world. Water Sci Technol. 2003;47(3):7–14.
55. Schuster CJ, Ellis AG, Robertson WJ, et al. Infectious disease outbreaks related to drinking water in Canada, 1974–2001. Can J Public Health. 2005;96(4):254–8.
56. Dwight RH, Semenza JC, Baker DB, et al. Association of urban runoff with coastal water quality in Orange County. California. Water Environ Res. 2002;74(1):82–90.
57. Borchardt MA, Spencer SK, Kieke BA, Lambertini E, Loge FJ. Viruses in nondisinfected drinking water from municipal wells and community incidence of acute gastrointestinal illness. Environ Health Perspect. 2012;120:1272–9.
58. MacKenzie WR, et al. A massive outbreak in Milwaukee of cryptosporidium infection transmitted through the public water supply. N Engl J Med. 1994;331(3):161–7.
59. Atherholt TB, LeChevallier MW, Norton WD, Rosen JS. Effect of rainfall on Giardia and crypto. J Am Water Works Assoc. 1998;90(9):66–80.
60. Drayna P, McLellan SL, Simpson P, Li SH, Gorelick MH. Association between rainfall and pediatric emergency department visits for acute gastrointestinal illness. Environ Health Perspect. 2010;118:1439–43.
61. Patz J, Campbell-Lendrum D, Gibbs HK, Woodruff R. Health impact assessment of global climate change: expanding upon comparative risk assessment approaches for policy making. Annu Rev Publ Health. 2008;29:27–39.
62. Kolstad E, Johansson K. Uncertainties associated with quantifying climate change impacts on human health: a case study for diarrhea. Environ Health Perspect. 2011;119:299–305.
63. Atkinson A, Siegel V, Pakhomov E, Rothery P. Long-term decline in krill stock and increase in salps within the Southern Ocean. Nature. 2004;432:100–3. https://doi.org/10.1038/nature02950.1.
64. Van Dolah FM. Marine algal toxins: origins, health effects, and their increased occurrence. Environ Health Perspect. 2000;108(suppl. 1):133–41.
65. Tester PA, et al. An expatriate red tide bloom, transport, distribution, and persistence. Limnol Oceanogr. 1991;36:1053–61.
66. Hallegraeff GM. A review of harmful algal blooms and their apparent global increase. Phycologia. 1993;32(2):79–99.
67. Peperzak L. Future increase in harmful algal blooms in the North Sea due to climate change. Water Sci Technol. 2005;51(5):31–6.
68. Hales S, Weinstein P, Woodward A. Ciguatera fish poisoning, El Niño and Pacific Sea surface temperatures. Ecosyst Health. 1999;5:20–5.
69. Colwell RR. Global climate and infectious disease: the cholera paradigm. Science. 1996;274:2025–31.
70. Myers SS, Patz JA. Emerging threats to human health from global environmental change. Annu Rev Environ Res. 2009;34:223–52. https://doi.org/10.1146/annurev.environ.033108.102650.
71. IPCC, 2014: Climate Change 2014: Synthesis Report. Contribution of Working Groups I, II and III to the Fifth Assessment Report of the Intergovernmental Panel on Climate Change [Core Writing Team, R.K. Pachauri and L.A. Meyer (eds.)]. IPCC, Geneva, Switzerland, 151 pp.
72. Thompson JR, Randa MA, Marcelino LA, Tomita-Mitchell A, Lim E, Polz MF. Diversity and dynamics of a North Atlantic coastal *Vibrio* community. Appl Environ Microbiol. 2004;70(7):4103–10.
73. McLaughlin JB, DePaola A, Bopp CA, Martinek KA, Napolilli NP, Allison CG, Murray SL, Thompson EC, Bird MM, Middaugh JP. Outbreaks of *Vibrio parahaemolyticus* gastroenteritis associated with Alaskan oysters. N Engl J Med. 2005;353(14):1463–70.
74. Martinez-Urtaza J, Simental L, Velasco D, et al. Pandemic Vibrio parahaemolyticus 03:K6. Eur Emerg Infect Dis. 2005; https://doi.org/10.3201/eid1108.050322.
75. Vezzulli L, Brettar I, Pezzati E, et al. Long-term effects of ocean warming on the prokaryotic community: evidence from the vibrios. ISME J. 2012;6:21–30.
76. Checkley W, et al. Effects of El Nino and ambient temperature on hospital admissions for diarrhoeal diseases in Peruvian children. Lancet. 2000;355:442–50.

77. Rodo X, Pascual M, Fuchs G, Faruque AS. ENSO and cholera: A nonstationary link related to climate change? Proc Natl Acad Sci U S A. 2002;99(20):12901–6.
78. Bentham G, Langford IH. Climate change and the incidence of food poisoning in England and Wales. Int J Biometeorol. 1995;39:81–6.
79. Kovats RS, Edwards SJ, Hajat S, et al. The effect of temperature on food poisoning: a time-series analysis of salmonellosis in ten European countries. Epidemiol Infect. 2004;132(3):443–53.
80. Goulson D, Derwent LC, Hanley ME, Dunn DW, Abolins SR. Predicting calyptrate fly populations from the weather, and probable consequences of climate change. J Appl Ecol. 2005;42:795–804.
81. Kattenberg A, et al. Climate models: projections of future climate. In: Houghton J, et al., editors. Climate change 1995: the science of climate change. Contribution of working group I to the second assessment report of the intergovernmental panel on climate change. New York: Cambridge University Press; 1996.
82. Kalnay E, Cai M. Impact of urbanization and land-use change on climate. Nature. 2003;423:528–31.
83. Hayhoe K, et al. Emissions pathways, climate change, and impacts on California. PNAS. 2004;101(34):12422–7.
84. Vavrus S, Van Dorn J. Projected future temperature and precipitation extremes in Chicago. J Great Lakes Res. 2010;36:22–32.
85. Kalkstein LS, Greene JS, Mills DM, Perrin AD, Samenow JP, Cohen JC. Analog European heat waves for U.S. cities to analyze impacts on heat-related mortality. Bull Am Meteorol Soc. 2008;89(1):75–85.
86. Olson SH, Gangnon R, Eiguero E, Durieux L, Guegan JF, Foley JA, Patz JA. Surprises in the climate-malaria link in the Amazon. J Emerg Infect Dis. 2009;15(4):659–62.
87. Corfee-Morlot J, Agrawala S. The benefits of climate change policies: analytical and framework issues. Paris: Organisation for Economic Co-operation and Development; 2004.
88. Frumkin H, McMichael AJ. Climate change and public health: thinking, communicating, acting. Am J Prev Med. 2008;35:403–10.
89. Cifuentes L, Borja-Aburto VH, Gouveia N, Thurston G, Davis DL. Assessing the health benefits of urban air pollution reductions associated with climate change mitigation (2000–2020): Santiago, São Paulo, México City, and New York City. Environ Health Perspect. 2001;109(3S):419–25.
90. Wilkinson P, Smith KR, Beevers S, Tonne C, Oreszczyn T. Energy, energy efficiency, and the built environment. Lancet. 2007;370:1175–87.
91. Patz JA, Frumkin H, Holloway T, Vimont DJ, Haines A. Climate change: challenges and opportunities for Global Health. JAMA. 2014;312(15):1565–80.
92. Shindell D, Johan CI, JCI K, Vignati E, et al. Simultaneously mitigating near-term climate change and improving human health and food security. Science. 2012;335:183–9.
93. Abel D, Holloway T, Harkeya M, Rrushaja A, Brinkmanb G, Durana P, Janssenc M, Denholmb P. Potential air quality benefits from increased solar photovoltaic electricity generation in the eastern United States. Atmos Environ. 2018;175:65–74.
94. U.S. Census Bureau American community survey. 2008. American Factfinder. http://factfinder.census.gov. Accessed 1 Oct 2009.
95. Pucher J, Buehler R, Bassett DR, Dannenberg AL. Walking and cycling to health: a comparative analysis of city, state, and international data. Am J Public Health. 2010;100(10):1986–92. https://doi.org/10.2105/AJPH.2009.189324.
96. Willett W, Rockström J, Loken B, Springmann M, Lang T, Vermeulen S, et al. Food in the anthropocene: the EAT–Lancet Commission on healthy diets from sustainable food systems. Lancet. 2019;393:447–92. https://doi.org/10.1016/S0140-6736(18)31788-4.
97. International Energy Agency (2004), Biofuels for Transport: An International Perspective, OECD Publishing, Paris, https://doi.org/10.1787/9789264015135-en.
98. Himmel ME, Ding SY, Johnson DK, et al. Biomass recalcitrance: engineering plants and enzymes for biofuels production. Science. 2007;315(5813):804–7.
99. Fairless D. Biofuel: the little shrub that could—maybe. Nature. 2007;449(7163):652–5.

100. Food and Agriculture Organization. Biofuels: prospects, risks and opportunities. In: The state of food and agriculture. Rome: FAO; 2008.
101. United Nations-Energy. 2007.
102. Naylor RL, Liska AJ, Burke MB, et al. The ripple effect: biofuels, food security, and the environment. Environment. 2007;49(9):30–43.
103. Fearnside PM. Soybean cultivation as a threat to the environment in Brazil. Environ Conserv. 2001;28(1):23–38.
104. Morton DC, DeFries RS, Shimabukuro YE, et al. Cropland expansion changes deforestation dynamics in the southern Brazilian Amazon. Proc Natl Acad Sci U S A. 2006;103(39):14637–41.
105. Koh LP, Wilcove DS. Is oil palm agriculture really destroying tropical biodiversity? Conserv Lett. 2008;2(1):1–5.
106. Gibbs HK, Johnston M, Foley JA, et al. Carbon payback times for crop-based biofuel expansion in the tropics: the effects of changing yield and technology. Environ Res Lett. 3. https://doi.org/10.1088/1748-9326/3/3/034001.
107. Keeney D, Nanninga C. Biofuel and global biodiversity. Minneapolis: Institute for Agriculture and Trade Policy; 2008. Available: http://www.agobservatory.org/library.cfm?refid=102584
108. Hill J, et al. Climate change and health costs of air emissions from biofuels and gasoline. Proc Natl Acad Sci. 2009;106(6):2077–82.
109. Organization for Economic Cooperation and Development. Biofuel support policies: an economic assessment. Paris: OECD Publishing; 2008. Available: www.oecd.org/tad/bioenergy
110. Patz JA, Vavrus S, Uejio C, McClellan S. Climate change and waterborne disease risk in the Great Lakes region of the US. Am J Prev Med. 2008;35(5):451–8.
111. Dyson ME. Come hell or high water: hurricane Katrina and the color of disaster. New York: Basic Books; 2007.
112. Elliott JR, Pais J. Race, class, and hurricane Katrina: social differences in human responses to disaster. Soc Sci Res. 2006;35:295–321.
113. Pastor M, Bullard RD, Boyce JK, Fothergill A, Morello-Frosch R, Wright B. In the wake of the storm: environment, disaster and race after Katrina. New York: Russell Sage Foundation; 2006.
114. World Bank. Turn down the heat: why a 4°C warmer world must be avoided. Washington, DC: The World Bank; 2012.
115. Stern N. The economics of climate change: the Stern review. Cambridge/New York: Cambridge University Press; 2007.
116. Marland G, Boden TA, Andres RJ. Global, regional, and national fossil fuel CO_2 emissions. In: Trends: a compendium of data on global change. Oak Ridge: Carbon Dioxide Information Analysis Center, Oak Ridge National Laboratory, U.S. Department of Energy; 2007.
117. Grabow ML, Spak SN, Holloway T, Stone B, Mednick AC, Patz JA. Air quality and exercise-related health benefits from reduced car travel in the midwestern United States. Environ Health Persp. 2012;120:68–76. https://doi.org/10.1289/ehp.1103440.
118. McMichael AJ. Environmental and social influences on emerging infectious diseases: past, present and future. Philos Trans R Soc Lond B Biol Sci. 2004;359:1049–58.

Chapter 15
Hurricanes and Health: Vulnerability in an Age of Climate Change

Alexis A. Merdjanoff and Rachael Piltch-Loeb

Introduction

The prevalence and severity of hurricanes are increasing. Since Hurricane Katrina (2005), there have been a number of high-impact storms, including Superstorm Sandy (2012), Hurricane Harvey (2017), Hurricane Maria (2017), and Hurricane Michael (2018). The 2017 Atlantic hurricane season alone included a total of 17 named storms, all of varying intensity [1]. Accumulated cyclone energy (ACE) provides a summative measure of storm intensity. Higher wind speeds and/or longer-lived storms, both increase a storm's ACE value [2]. ACE was the highest on record for the month of September 2017 and seventh overall for the total hurricane season, based on data collected since 1851 [1]. For the United States, the 2017 hurricane season was the costliest season since records began in 1851 due to the trio of Category 4 hurricanes that made landfall including Harvey, Irma, and Maria [2]. Record rainfall occurred in Texas as a result of Hurricane Harvey, while record winds were recorded when Hurricane Irma slammed into Florida. The 2017 season was considered by many to be a "harbinger" of the future – indicating an increase in the future severity (although possibly decrease in frequency) over the remainder of the twenty-first century [1]. Sure enough, 2018 saw Hurricane Michael strike the Florida Panhandle, which was the first Category 5 hurricane to hit the continental United States since Hurricane Andrew in 1992.

A. A. Merdjanoff (✉)
School of Global Public Health, New York University, New York, NY, USA
e-mail: aam258@nyu.edu

R. Piltch-Loeb
T.H. Chan School of Public Health, Division of Policy Translation and Leadership Development, Harvard University, Boston, MA, USA
e-mail: piltch-loeb@hsph.harvard.edu

K. E. Pinkerton, W. N. Rom (eds.), *Climate Change and Global Public Health*, Respiratory Medicine, https://doi.org/10.1007/978-3-030-54746-2_15

There are several specific aspects of climate that contribute to the severity of a hurricane. These include factors such as vertical wind shear (the ways in which the winds change with height in the atmosphere); ocean water temperature (warm ocean water elevates maximum wind speeds); depth of warm ocean water (the deeper the warm temperatures the easier it is for the storm to maintain energy transfer as water churns); and atmospheric moisture-holding capacity (the ability to create and release moisture-increasing precipitation rates and total rainfall) [1]. While a single hurricane cannot be causally linked to climate change, climate change has increased the odds of extreme weather events occurring, as well as the intensity of those events. Global temperatures continue to rise and when combined with the sustained emission of fossil fuels have led to a greater amount of heat trapped within the lower atmosphere. This in turn leads to a greater energy flux within the atmosphere and results in more energetic and variable weather patterns [3]. Global warming modifies the environment, which increases the surface temperature of the water resulting in warm ocean waters and altered regional wind patterns. Altered wind patterns can result in extreme weather events and can also change precipitation patterns [4]. Climate change is also shifting the behavior of hurricanes, leading to unpredictability of flash flooding when storms make landfall, uncertain precipitation, and wind patterns [5].

Hurricanes are acute, collective traumas whose effects can last far beyond the meteorological event itself. There are several factors that contribute to the physical and mental health effects of a hurricane, including the severity, duration, surprise related to the event, and the context where the event takes place [4, 6]. Severity refers to how extreme the event conditions are, with more extreme conditions causing greater health impacts on exposed populations. Duration refers to how long the extreme conditions are experienced, with longer periods of exposure causing greater potential for damage to health. Surprise refers to how much notice there is before the event occurs, and in turn how prepared the population is to evacuate, get to shelter, and provide for pre-existing medical conditions. Health impacts from weather events are a product of local ability and local response. Context includes both the physical context where the storm hits such as the population density, geographic structure or isolation, and land use, as well as the population characteristics.

The following pages examine the physical and mental health effects of hurricanes within our changing climate. We explore a range of physical health effects, including mortality, morbidity, injuries, infectious disease, chronic disease, and mold-related illness. We also examine several mental health consequences of hurricane exposure, including post-traumatic stress disorder (PTSD) and risk factors for poor post-disaster mental health outcomes, as well as factors that protect against poor mental health. We use evidence from public health, sociology, psychology, and geography to highlight the health effects from the most recent, as well as most powerful, hurricanes to strike the continental United States, to demonstrate the wide-range of health consequences that can be caused by hurricane exposure. Our intent is to identify the wide-range of health consequences caused by direct and indirect hurricane exposure, and bring attention to areas that need further investigation as the intensity of hurricanes grows due to our changing climate.

Physical Health Effects of Hurricanes

Hurricanes can cause a variety of physical-health effects. Hurricanes directly and indirectly increase mortality and lead to long-term increases in morbidity [6]. When hurricanes make landfall, they cause both wind and water damage to the physical environment. Wind damage can topple trees and buildings, leaving streets full of debris, and cars and homes damaged. Flooding attributable to rising tides and rain water during the event also causes significant damage and can lead to both injuries and infectious disease concerns. Both wind and flooding can increase the risk of fatality.

Mortality

When a hurricane strikes, there are often immediate deaths as individuals are struck by objects moved by wind, trapped in unstable structures, or drown as a result of rising waters. In a review of death certificates from the 2 months after Katrina, drowning was the major cause of death, and people 75 years old and older were the most affected population cohort [7]. Differential vulnerability among low- and high-income countries and communities to flood exposure and its associated negative impacts is a key issue to address when discussing the flood's impact on human health [8]. Beyond the immediate direct effects of the storm, hurricanes knock down power lines, taking systems that provide electricity, heat, air conditioning, medical services, and clean water off the grid. The consequences of this means populations are exposed to the elements in new ways, and as a result mortality often increases, sometimes in immeasurable ways: injuries occur, infectious disease outbreaks arise, and chronic disease management becomes more difficult. This can lead to indirect mortality attributable to the hurricane. Following the 2017 hurricane season which impaled small islands in the Atlantic Ocean such as the US Virgin Islands and Puerto Rico, a contributor to increased mortality was the ongoing exposure of the storm-affected islands' tropical heat and humidity [1]. After Hurricane Maria (2017), excess mortality in Puerto Rico – which accounted for both direct and indirect deaths – was estimated to be as high as approximately 1000–4600 depending on the method used (i.e., death records vs. surveys) to estimate and attribute deaths [9, 10]. Many of these deaths occurred after the storm itself had passed, which demonstrates how indirect mortality can be even more deadly than mortality caused directly by a storm.

Injuries

Power lines often go down during an event. Most immediately, this can lead to increases in deaths and physical injuries caused by the movement of debris, destruction of shelter, and other changes to physical structures. In Hurricane Hugo (1989), almost 90% of over 2000 patients treated in emergency departments were hospitalized for injuries [11]. Nonfatal injuries together with the exacerbation of chronic illness are the leading causes of morbidity among affected residents and relief workers immediately following a hurricane. Most often, these initial impacts are due to flooding [11]. Following Katrina (2005), an active injury and illness surveillance system was established by the Centers for Disease Control and Prevention (CDC) along with the Louisiana Department of Health and Hospitals (LDHH) in functioning hospitals and medical clinics. The system recorded 7543 nonfatal injuries among residents and relief workers between September 8–October 14, 2005 [12]. The highest portion of these injuries were among middle-aged men who were likely most actively involved in clean-up and recovery efforts.

Infectious Diseases

Following an extreme meteorological event like a hurricane, there can be increases in several types of infectious diseases – meaning diseases that can be passed from person to person and from vector to human. However, the type and severity of these increases depend on geography, demography, economic conditions, the pre-existing infectious diseases in the affected population, and the cultural and political context. Hurricanes and other extreme weather events exacerbate underlying conditions that increase the prevalence of infectious diseases like poverty, hygiene, crowding, food insecurity, and vaccination or antibiotic access [3]. In high-income countries, the risk of post-flood disease outbreaks is low; however, it has been found to increase with the depth of flooding [8].

Rarely do hurricanes cause new infectious diseases; rather, they increase the transmission of infectious diseases and outbreaks because of the prolonged after-effects of the disaster. These after-effects change the ways in which humans and the environment interact, which shifts the patterns of disease transmission. Shifts in the physical environment attributable to population displacement, changes in air quality or humidity, and reduction in physical barriers like trees or ground cover can increase vector breeding sites and shift exposure to and proliferation of disease vectors like rodents or insects [13]. Flooding in particular can increase the breeding sites for mosquito populations, increasing the prevalence of endemic mosquito-borne illnesses such as Malaria, Dengue, West Nile Virus, or Zika, depending on the location of the event. Further, if the underlying population has low levels of immunity to vaccine-preventable diseases or insufficient vaccine coverage, there can be outbreaks.

Populations are often forced to relocate to sometimes unplanned and often overcrowded shelters in the aftermath of a hurricane. In these mass dwellings, there can be poor water and sanitation conditions, limited access to food, and poor hygiene practices. Because of the close proximity of individuals to each other, these areas can become a breeding ground for bacteria and make the spread of diseases more common. Consequences of flooding and movement of populations into overcrowded settings can cause an increase in disease outbreaks resulting from cross-contamination of water sources with fecal material and toxic chemicals. This often leads to an increase in diarrheal diseases especially in countries where pre-existing sanitation systems are limited. Even within the United States, various pathogens including norovirus, and toxigenic and non-toxigenic V. cholerae were confirmed among the populations displaced by Hurricane Allison (2001) [13]. Following Katrina, 27,000 evacuees sheltered at Houston's Astrodome and Reliant Park Complex. Transmission of communicable infectious diseases was an immediate public health concern to public health officials and responders, so a surveillance system was developed ad hoc. Fever, vomiting, diarrhea, sore throat, cough, runny nose, and rash emerged as the symptoms of primary concern. Cough and runny nose were the most frequently reported symptoms, with increases in both symptoms reported over time. A gastrointestinal outbreak of norovirus occurred [14]. In the days following Katrina, 4% of all evacuees in this shelter fell ill with norovirus-based gastroenteritis which was attributed to over-crowding, poor sanitation, and compromised health levels of evacuees [8]. In colder weather, acute respiratory infections (ARIs) risk may be increased due to overcrowding, poor ventilation and poor nutrition, and crowded shelters.

Chronic Diseases

In the aftermath of a hurricane, the burden of diseases attributable to chronic disease management can increase. Disaster-affected populations may carry a large and measurable burden of disabilities and chronic diseases, especially heart disease, cancer, stroke, diabetes, and chronic respiratory disorders [15]. Following Katrina, a limited needs assessment among individuals staying in evacuation centers, conducted in the field and reported to the Centers for Disease Control and Prevention (CDC), demonstrated that five of the top six conditions were all chronic diseases and that, other than injuries, the majority of medical and health visits were for medication refills, oral health issues, and other chronic health conditions [15]. Inability to maintain a stable medication uptake was the main barrier to continuity of care for chronic conditions during Katrina, with inadequate information and financial constraints as contributing factors [8].

Hurricanes often disrupt access to treatment or care for those with chronic conditions, making those conditions acute. Common reasons for disrupted treatment following a hurricane include problems accessing physicians or medications; new competing financial demands; transportation barriers; or competing demands on time. In a comparative study following Hurricane Katrina, researchers found nearly three-quarters of Katrina survivors had 1 or more chronic conditions in the year before the hurricane, and of these, 20.6% cut back or terminated their treatment because of the disaster [16]. Disruptions in treatment were higher among younger adults, those with housing needs, and those who were uninsured.

There are several specific chronic conditions that are exacerbated by hurricane exposure. Particular conditions are susceptible to these events – those which require access to medication, refrigeration for medication, chronic management, power for equipment such as dialysis, or ongoing treatment like chemotherapy [1]. Some of these examples include ischemic stroke survivors taking anticoagulants; people whose diabetes is controlled by insulin; heart attack survivors taking clot-preventing medications; people with severe lung disease receiving home oxygen therapy; people with hereditary blood disorders; and patients receiving hemodialysis for kidney failure. In a study following Superstorm Sandy that used Centers for Medicare & Medicaid Services claims data to explore the impact of the event on patients with end stage renal disease (ESRD), emergency department visits were nearly twice as high and hospitalizations were 50% higher for ESRD patients impacted by the storm compared to those living in areas not impacted by the storm, as well as patients living in the same area the year prior [17].

Mold-Related Illness

Flooding attributable to hurricanes creates the conditions for indoor mold growth and related adverse effects of exposure to indoor mold exposure [18]. Toxic diseases can then arise from exposure to mycotoxins produced by molds that grow following a hurricane [19]. Following Hurricane Katrina, studies found that homes with greater flood damage had higher levels of mold growth compared with homes with lower levels of flooding. Water intrusion due to roof damage was also associated with mold growth. CDC and the local health department in New Orleans conducted a surveillance assessment of mold growth in a sample of New Orleans homes to estimate population level prevalence of mold [20]. They saw visible mold growth was observed in 44% of homes; 19% had heavy mold growth, correlating with extent of water and wind damage.

Illnesses associated with fungal exposure depend on the route and magnitude of exposure and the immune status of the person exposed; however, the presence of mold in damp, indoor environments have been found to trigger upper respiratory tract symptoms, cough, and wheezing [21]. The effect of mold exposure may be influenced by the pre-existing conditions of those exposed [8]. Most commonly, those who are immunocompromised or those with a prior history of asthma or allergies are more likely to experience symptoms related to mold following a hurricane [18]. Exposure to mold following a hurricane will vary greatly depending on available housing stock – which means people may not need to live in a mold affected home during remediation efforts – whether or not individuals are exposed to mold during the clean-up process, and have pre-existing conditions. There is limited long-term follow-up on the impacts of mold following a hurricane though studies are ongoing.

Mental Health Effects of Hurricanes

Despite the recent increase of major hurricanes, our understanding of their effects on post-disaster mental health continues to evolve. The increased prevalence of mental illness following a disaster like a hurricane can stretch between 5% and 40%

[22]. This also involves a wide range of psychopathology, including symptoms such as anxiety, grief, and depression, as well as criterion-based conditions such as major depression disorder (MDD), generalized anxiety disorder (GAD), panic disorder (PD), and post-traumatic stress disorder (PTSD) [23]. These effects can unfold over the days, weeks, months, and years following an event. In the short-term, lack of basic necessities such as food, water, shelter, and medical care have been found to be strongly associated with post-traumatic stress (PTS) and general psychological distress (GPD) [24]. Alternatively, housing damage, financial loss, and family strain have been linked to long-term mental health challenges. It is important to note that even if survivors exhibit poor mental health symptoms most regain normal functioning over time [23, 25]. Additionally, a majority of survivors exhibit minimal or no mental health effects [26]. While there is a wide scope of mental health effects, research has shown that exposure and pre-existing inequalities both strongly predict post-disaster mental health outcomes. The following paragraphs will outline these relationships, as well as describe factors that can protect hurricane-exposed individuals from poor mental health outcomes.

Exposure and Post-disaster Mental Health

The relationship between hurricanes and mental health can be tied to intensity of exposure – it can essentially be considered a dose-response relationship [26]. Exposure can be measured at the individual-level, including the number of stressors and severity of exposure. Previous studies have found that as the number of stressors increases, so too do survivors' mental health symptoms [25, 27, 28]. Severity of exposure can be driven not only by direct exposure to a hurricane but also from secondary stressors – or mediators – that include family- or work-related stress [28]. Hurricane-related displacement, permanent relocation, and housing damage are risk factors for both PTSD and depression [25, 26]. Being injured and experiencing the death or injury of a family member or close friend are also related to poor mental health outcomes [26]. Secondary stressors such as income loss, job loss, and overall socioeconomic decline have been linked to long-term mental health effects, includ ing MDD [29]. The greater the loss and stressors related to the hurricane, the longer the adverse mental health consequences persist: symptoms lasted for more than a year after Hurricane Katrina hit and were most intense for those who experienced the greatest loss [30]. Oftentimes, and especially for those with greatest exposure, direct exposure and secondary stressors can interact: following Hurricane Katrina, there was a high prevalence of DSM-IV anxiety mood disorders that were associated with both ongoing stressors (i.e., poverty) and hurricane-related stressors [31].

Hurricanes can impact individuals both directly and indirectly but they also impact the communities in which they live. Because of this, it is important that when considering individual-level mental health effects that community exposure also be measured. Although there has been less focus on this type of ecological assessment, community exposure can be measured through conditions in the neighborhood or community – including places of worship or businesses – or by analyzing collective loss independent of personal loss [32]. Following Superstorm Sandy, Abramson et al. found that high community damage had detrimental effects on the mental health of New Jersey homeowners [33]. Individual-level exposure is considered to have a stronger effect on mental health although previous research suggests that individual exposure and community damage can interact [34].

PTSD is the most commonly studied post-disaster psychiatric disorder [25]. Numerous studies demonstrate that the risk of PTSD is associated with the severity of exposure [23]. Physical injury, threats to one's livelihood, and severe property destruction are predictive of high rates of PTSD. Residents of an affected area can be directly exposed or experience secondary stressors due to hurricane exposure. Survivors who are directly exposed are at an increased risk of PTSD [23]. Although PTSD is frequently studied, delayed onset of PTSD following hurricane exposure is common. For example, McLaughlin et al. found that an estimated 17% of residents from Katrina-affected areas had hurricane-related PTSD 7–19 months after the storm while 29% of the same residents exhibited symptoms 24–27 months after the event [35]. Findings such as these suggest that other measures of mental health stress are equally as important to estimate after a hurricane strikes. In addition to PTSD, longitudinal hurricane studies have relied upon measures of emotional distress or psychological well-being, which are strongly predictive of anxiety mood disorders and are linked to long-term recovery [36, 37]. Such findings suggest that it is important to expand the focus of mental health effects from hurricane exposure beyond PTSD.

Risk Factors for Post-Hurricane Mental Health Effects

Not all mental health effects are experienced equally. Hurricanes – and disasters, more generally – have the ability to reveal pre-existing inequalities, which means that particular individuals, groups, and communities can experience these mental effects more intensely than others. Women, children, racial and ethnic minorities, and lower socioeconomic status individuals and communities are often at the greatest risk of experiencing negative psychological health effects following a hurricane. Several studies have found that woman survivors are more likely to demonstrate post-disaster stress, depression, anxiety, or psychological disorders than men [38, 39]. More recently, in a study of gender differences in psychological reactions to Superstorm Sandy, women were more likely to report fear of future events than men [40]. Overall, gendered norms, the gendered division of labor, and gendered social structures place women at an elevated risk for poor mental health outcomes following a hurricane [41].

Children and adolescents tend to exhibit far-reaching cognitive and emotional changes ranging from depression, serious emotional disturbance, and suicidal ideation [42]. While these effects may not be fully realized until years after an event it is important to note that some children may be more vulnerable while others more resilient to hurricane exposure [42, 43]. Following Hurricane Katrina, children experienced disorientation; increases in stress and stress-related disorders; and behavioral problems [42]. Similarly, Abramson et al. found that children exposed to Hurricane Katrina were nearly five times as likely as a pre-Katrina cohort to exhibit serious emotional disturbance [44]. Children rely upon others for their recovery, including decisions that impact their household, neighborhood, and school, which makes them especially vulnerable to the effects of hurricane exposure.

In addition to gender, race and ethnicity are social identities that can shape psychological responses to hurricane exposure. Following Hurricane Andrew, Black (23%) and Latinx (38%) survivors exhibited higher rates of PTSD than Whites (15%), which suggests that structural racism and cultural-specific responses can shape psychological symptoms [45]. Similarly, in a study of the geography of post-disaster mental health after Superstorm Sandy, Hispanics living in Manhattan, non-Hispanic Black survivors living in Queens, and Asian survivors living in the Bronx exhibited higher levels of post-traumatic stress [46]. However, race and ethnicity are also tightly correlated with class and place. Katrina revealed that while a storm might strike a particular area, differences in socioeconomic status, race, and geography could mean that survivors had profoundly different experiences of the same event. These risk factors clearly do not operate in isolation. They can intersect, as demonstrated by a meta-analysis that showed African American women affected by Hurricane Katrina experienced emotional disturbances and stress, despite strengthened faith and cultural support [47]. Such studies provide evidence that there has been a theoretical and analytical shift to use an intersectional approach of race, class, gender and sexuality when exploring the relationship among social vulnerability, hurricane exposure and health impacts.

Social Support and Resilience as Protective Factors

While certain risk factors can increase vulnerability to the mental health effects of hurricanes, there are also protective factors that can buffer exposure to these traumatic events. Social support has been shown to both decrease exposure to disasters, including hurricanes, as well as the corresponding negative mental health effects [36]. Perceived social support has been consistently and positively associated with improved mental health following a disaster [48]. Following the 2004 Asian Tsunami, survivors emphasized the importance of extended social networks that helped their emotional well-being [49]. Alternatively, empirical evidence following Hurricane Katrina suggests that weak social networks, as well as a lack of social capital, can increase the likelihood of experiencing depression symptoms and PSD [50].

Another factor that often operates as a protective factor is prior trauma, including previous disaster exposure. Such findings support the stress inoculation hypothesis, which is a resilience concept that suggests those who have experienced previous life stress will be better prepared to handle future stress. Adults over the age of 55 who had experienced severe flooding experienced less mental health effects than less experienced adults [51].Similarly, Vanlandingham's work on the Vietnamese-American population living in New Orleans during Hurricane Katrina found they were more likely to recover than other racial or ethnic groups that were similarly exposed, which could be attributed to many living through the Vietnam War and migrating to the United States [52]. These findings suggest that, for some, prior trauma can build a reservoir of resilience that can be tapped when faced with disaster exposure.

Conclusion

Current Challenges and the Future of Hurricane-Health Research

The field of public health disaster science continues to grow, especially with the recent spate of powerful hurricanes. While our understandings of the link between hurricane exposure and health has expanded, there are also limitations with which we must contend. First, without pre-event data it is difficult to draw causal conclusions between hurricane exposure and health outcomes. The use of large population level data sets, such as Medicare or Medicaid data, is one way to identify changes in health status following a hurricane event. Research teams used this following Superstorm Sandy to identify effects on ESRD patients, and their approach can be replicated for other physical health conditions. However, these data are limited to those who are enrolled in these insurance programs. In the future, electronic health records maintained by hospital systems may be able to fill in for the rest of the population in a similar way. Gaining access to these data to understand physical impacts more rapidly and make real-time decisions to prevent long-term health consequences related to hurricanes must be a goal of public health practitioners.

The lack of pre-existing data also relates to environmental data. For example, to understand increases in mold following a hurricane and the potential effect of mold on changing physical health outcomes, there needs to be baseline data on mold in the home or outside area. Rarely would these data exist. The same challenge may apply to studies of changes in air quality or water toxicity levels. The underlying assumption is that hurricanes change these environmental exposures but to what extent is not wholly understood. Accurately measuring the changing environment would help researchers understand the way hurricane exposure directly changes the environment and human health. Second, hurricanes make for difficult data gathering environments. When an event occurs, the most immediate need is to get people out of harm's way. Pre-existing surveillance systems can be taken out by the storm. The immediate environmental impact means ephemeral data are lost because there is a lack of capacity to collect it during immediate response and relief efforts.

Specifically related to the study of post-disaster mental health, research tends to use validated screening scales rather than clinical interviews, which some argue are less precise [31]. Second, most post-hurricane studies are cross-sectional designs so there is limited research on long-term mental health effects [23]. Norris and colleagues describe the disaster literature as a "series of case studies," which makes it challenging to generalize [28]. This includes the difficulty of making comparisons across events, which is challenging because of sampling issues, preexisting inequalities, and differing levels of exposure. Because of this there is often a tendency to consider each disaster as unique due to the context within which it occurs. In more recent years, efforts continue to be made to generalize across events or at the very least, identify common denominators among similarly classified disasters. Researchers have also been focusing more on designing longitudinal studies that allow for such comparisons. These studies can shift understandings of not only the immediate or relatively short-term consequences of hurricanes but long-term impacts as well.

The effects of climate change are increasing the frequency and severity of hurricanes. As hurricanes become more common, the short and long-term health effects of these events will burden the population and healthcare system for generations to come. Improving the study of hurricane exposure and how it relates to physical and mental health can help the government and populations more effectively prepare for, respond to, and recover from such events.

References

1. Shultz JM, Kossin J, Shepherd JM, Ransdell J, Walshe R, Kelman I. Risks, health consequences, and response challenges for small-island-based populations: observations from the 2017 Atlantic hurricane season. Disaster Med Public Heal Prep. 2019;12(1):5–17.
2. Halverson JB. The costliest hurricane season in US history. Weatherwise. 2018;71(2):20–7.
3. McMichael AJ. Extreme weather events and infectious disease outbreaks. Virulence. 2015;6(6):543–7.
4. Hashim JH, Hashim Z. Climate change, extreme weather events, and human health implications in the Asia Pacific region. Asia Pacific J Public Heal. 2016;28(2_suppl):8S–14S.
5. Shultz JM, Kossin JP, Galea S. The need to integrate climate science into public health preparedness for hurricanes and tropical cyclones. JAMA. 2018;320(16):1637–8.
6. Woodward AJ, Samet JM. Climate change, hurricanes, and health. Am J Public Health. 2018;108(1):33–5. https://doi.org/10.2105/AJPH.2017.304197.
7. Brunkard J, Namulanda G, Ratard R. Hurricane Katrina deaths, Louisiana, 2005. Disaster Med Public Heal Prep. 2008;2(4):215–23.
8. Alderman K, Turner LR, Tong S. Floods and human health: a systematic review. Environ Int. 2012;47:37–47. https://doi.org/10.1016/j.envint.2012.06.003.
9. Santos-Lozada AR, Howard JT. Use of death counts from vital statistics to calculate excess deaths in Puerto Rico following hurricane Maria. JAMA. 2018;320(14):1491–3.
10. Kishore N, Marqués D, Mahmud A, et al. Mortality in Puerto Rico after hurricane Maria. N Engl J Med. 2018;379(2):162–70.
11. Diaz JH. The public health impact of hurricanes and major flooding. J Louisiana State Med Soc. 2004;156(3):145–50.
12. Sullivent EE III, West CA, Noe RS, Thomas KE, Wallace LJD, Leeb RT. Nonfatal injuries following hurricane Katrina – New Orleans, Louisiana, 2005. J Saf Res. 2006;37(2):213–7.

13. Kouadio IK, Aljunid S, Kamigaki T, Hammad K, Oshitani H. Infectious diseases following natural disasters: prevention and control measures. Expert Rev Anti-Infect Ther. 2012;10(1):95–104.
14. Murray KO, Kilborn C, desVignes-Kendrick M, et al. Emerging disease syndromic surveillance for hurricane Katrina evacuees seeking shelter in Houston's astrodome and Reliant Park complex. Public Heal Rep. 2009;124(3):364–71. http://www.jstor.org/stable/25682239
15. Mensah GA, Mokdad AH, Posner SF, et al. When chronic conditions become acute: prevention and control of chronic diseases and adverse health outcomes during natural disasters. Prev Chronic Dis. 2005;2(Special Issue):1–4.
16. Group THKCA, Kessler R. Hurricane Katrina's impact on the care of survivors with chronic medical conditions. J Gen Intern Med. 2007;22(9):1225–30. https://doi.org/10.1007/s11606-007-0294-1.
17. Kelman J, Finne K, Bogdanov A, et al. Dialysis care and death following hurricane Sandy. Am J Kidney Dis. 2015;65(1):109–15.
18. Barbeau DN, Grimsley LF, White LE, El-Dahr JM, Lichtveld M. Mold exposure and health effects following hurricanes Katrina and Rita. Annu Rev Public Health. 2010;31:165–78.
19. Meggs WJ. Epidemics of mold poisoning past and present. Toxicol Ind Health. 2009;25(9–10):571–6.
20. Centers for Disease Control and Prevention. Health concerns associated with mold in water-damaged homes after Hurricanes Katrina and Rita – New Orleans area, Louisiana, October 2005. MMWR Morb Mortal Wkly Rep. 2006;55(2):41.
21. Brandt M, Brown C, Burkhart J, et al. Mold prevention strategies and possible health effects in the aftermath of hurricanes and major floods. Morb Mortal Wkly Rep Recomm Rep. 2006;55(8):1–4.
22. Canino G, Bravo M, Rubio-stipec M, Woodbury M, Taylor P, Štipec MR. The impact of disaster on mental health: prospective and retrospective analyses. Int J Ment Health. 1994;19(1):51–69.
23. Neria Y, Nandi A, Galea S. Post-traumatic stress disorder following disasters: a systematic review. Psychol Med. 2008;38(4):467–80. https://doi.org/10.1017/S0033291707001353.
24. Chan CS, Rhodes JE. Measuring exposure in hurricane Katrina: A meta-analysis and an integrative data analysis. PLoS One. 2014;9(4) https://doi.org/10.1371/journal.pone.0092899.
25. Norris FH, Friedman MJ, Watson PJ, Byrne CM, Diaz E, Kaniasty K. 60,000 disaster victims speak: part I. An empirical review of the empirical literature, 1981–2001. Psychiatry. 2002;65(3):207–39. http://www.ncbi.nlm.nih.gov/entrez/query.fcgi?cmd=Retrieve&db=PubMed&dopt=Citation&list_uids=12405079
26. Neria Y, Shultz JM. Mental health effects of hurricane Sandy: characteristics, potential aftermath, and response. JAMA. 2012;308(24):2571–2. https://doi.org/10.1001/jama.2012.110700.
27. Briere J, Elliott D. Prevalence, characteristics, and long-term sequelae of natural disaster exposure in the general population. J Trauma Stress. 2000;13(4):661–79. http://www.johnbriere.com/JTSdisaster paper.pdf
28. Norris FH, Friedman MJ, Watson PJ. 60,000 disaster victims speak: part II. Summary and implications of the disaster mental health research. Psychiatry. 2002;65(3):240–60. http://www.ncbi.nlm.nih.gov/entrez/query.fcgi?cmd=Retrieve&db=PubMed&dopt=Citation&list_uids=12405080
29. Joseph NT, Myers HF. From the perspective of socioeconomic status decline. Health Psychol. 2015;33(2):139–46. https://doi.org/10.1037/a0031661.Conceptualizing.
30. Rhodes J, Chan C, Paxson C, Rouse CE, Waters M, Fussell E. The impact of hurricane Katrina on the mental and physical health of low-income parents in New Orleans. Am J Orthopsychiatry. 2010;80(2):237–47. https://doi.org/10.1111/j.1939-0025.2010.01027.x.
31. Galea S, Brewin CR, Gruber M, et al. Exposure to hurricane-related stressors and mental illness after hurricane Katrina. Arch Gen Psychiatry. 2007;64(12):1427–34.
32. Lindell MK, Prater CS. Assessing community impacts of natural disasters. Nat Hazards Rev. 2003;4(4):176–85. https://doi.org/10.1061/(ASCE)1527-6988(2003)4:4(176).
33. Abramson DM, Alst D Van, Merdjanoff A, et al. The hurricane Sandy person report: disaster exposure, health impacts, economic burden, and social well-being; 2015.

34. Phifer JF, Norris FH. Psychological symptoms in older adults following natural disaster: nature, timing, duration, and course. J Gerontol. 1989;44(6):S207–12. https://doi.org/10.1093/geronj/44.6.S207.
35. McLaughlin KA, Berglund P, Gruber MJ, Kessler RC, Sampson NA, Zaslavsky AM. Recovery from PTSD following hurricane Katrina. Depress Anxiety. 2011;28(6):439–46. https://doi.org/10.1002/da.20790.
36. Lowe SR, Chan CS, Rhodes JE. Pre-hurricane perceived social support protects against psychological distress: a longitudinal analysis of low-income mothers. J Consult Clin Psychol. 2010;78(4):551–60. https://doi.org/10.1037/a0018317.
37. Abramson DM, Stehling-Ariza T, Park YS, Walsh L, Culp D. Measuring individual disaster recovery: a socioecological framework. Disaster Med Public Heal Prep. 2010;4(Suppl 1):S46–54. https://doi.org/10.1001/dmp.2010.14.
38. Fothergill A. The neglect of gender in disaster work: an overview of the literature. Int J Mass Emerg Disasters. 1996;14(1):33–56.
39. Green BL. Identifying survivors at risk. In: Wilson JP, Raphael B, editors. International handbook of traumatic stress syndromes. Boston: Springer US; 1993. p. 135–44. https://doi.org/10.1007/978-1-4615-2820-3_11.
40. Hamama-Raz Y, Palgi Y, Shrira A, Goodwin R, Kaniasty K, Ben-Ezra M. Gender differences in psychological reactions to hurricane Sandy among New York metropolitan area residents. Psychiatry Q. 2015;86(2):285–96. https://doi.org/10.1007/s11126-014-9333-3.
41. Enarson E, Fothergill A, Peek L. Gender and disaster: foundations and new directions for research and practice. 2017:205–23. https://doi.org/10.1007/978-3-319-63254-4_11.
42. Peek L, Abramson DM, Cox RS, Fothergill A, Tobin J. Children and disasters. In: Rodriguez H, editor. Handbook of disaster research: Springer International Publishing AG; 2018. p. 243–62. https://doi.org/10.1177/0094306108037002227.
43. Fothergill A, Peek L. Children of Katrina. 1st ed. Austin: University of Texas Press; 2015.
44. Abramson DM, Park Y-S, Stehling-Ariza T, Redlener I. Children as bellwethers of recovery: dysfunctional systems and the effects of parents, households, and neighborhoods on serious emotional disturbance in children after hurricane Katrina. Disaster Med Public Heal Prep. 2010;4(S1):S17–27.
45. Perilla JL, Norris FH, Lavizzo EA. Ethnicity, culture, and disaster response: identifying and explaining ethnic differences in PTSD six months after hurricane Andrew. J Soc Clin Psychol. 2003;21(1):20–45. https://doi.org/10.1521/jscp.21.1.20.22404.
46. Gruebner O, Lowe SR, Sampson L, Galea S. The geography of post-disaster mental health: spatial patterning of psychological vulnerability and resilience factors in New York City after hurricane Sandy. Int J Health Geogr. 2015;14(1):1–13. https://doi.org/10.1186/s12942-015-0008-6.
47. Laditka SB, Murray LM, Laditka JN. In the eye of the storm: resilience and vulnerability among African American women in the wake of hurricane Katrina. Health Care Women Int. 2010;31(11):1013–27. https://doi.org/10.1080/07399332.2010.508294.
48. Bonanno GA, Brewin CR, Kaniasty K, La Greca AM, La Greca AM. Weighing the costs of disaster: consequences, risks, and resilience in individuals, families, and communities. Psychol Sci Public Interes Suppl. 2010;11(1):1–49. https://doi.org/10.1177/1529100610387086.
49. Ekanayake S, Prince M, Sumathipala A, Siribaddana S, Morgan C. "We lost all we had in a second": coping with grief and loss after a natural disaster. World Psychiatry. 2013;12(1):69–75. https://doi.org/10.1002/wps.20018.
50. Picou JS, Adeola FO. Social capital and the mental health impacts of hurricane Katrina: assessing long-term patterns of psychosocial distress. Int J Mass Emerg Disasters. 2014;32(1):121–56.
51. Norris FH, Murrell SA. Prior experience as a moderator of disaster impact on anxiety symptoms in older adults. Am J Community Psychol. 1988;16(5):665–83. https://doi.org/10.1007/BF00930020.
52. VanLandingham M. Weathering Katrina: culture and recovery among Vietnamese Americans. New York: Russell Sage Foundation; 2015.

Chapter 16
The Impact of Climate Change and Extreme Weather Conditions on Agricultural Health and Safety in California

Heather E. Riden, Emily Felt, and Kent E. Pinkerton

Introduction

Research indicates that the increasing frequency and severity of extreme weather events, which have been seen across the United States and in other countries, are related to climate change [1, 2]. The impacts of these events on agriculture are particularly important given its central role in food production, nutrition, employment, and the economy. Disruptions to agricultural production, changes in farm management and agricultural processes, uncertainty in agricultural markets, and exposure of workers, crops, and livestock to more extreme weather conditions are among the impacts of climate. Due to their outdoor work in all weather conditions, agricultural employers (i.e., farm owners, farm labor contractors) and farmworkers are exposed to increased occupational hazards in addition to the climate-related health risks faced by the general public. Farmworkers, or those hired to perform agricultural tasks, such as planting, irrigating, weeding, and harvesting, are at particular risk due to their socioeconomic status and vulnerability in the workplace. Unfortunately, there is a gap in existing literature regarding agricultural occupational health and safety and the hazards introduced or exacerbated by climate change and extreme weather.

This chapter describes the impact of extreme weather events on the health and safety of farmworkers with a focus on California. We use three extreme weather

H. E. Riden
Western Center for Agricultural Health and Safety and the Department of Public Health Sciences, University of California, Davis, Davis, CA, USA

E. Felt
Center for Health and the Environment, University of California, Davis, Davis, CA, USA

K. E. Pinkerton (✉)
Department of Pediatrics, School of Medicine, Department of Anatomy, Physiology and Cell Biology, School of Veterinary Medicine, Center for Health and the Environment, John Muir Institute of the Environment, University of California, Davis, Davis, CA, USA
e-mail: kepinkerton@ucdavis.edu

K. E. Pinkerton, W. N. Rom (eds.), *Climate Change and Global Public Health*, Respiratory Medicine, https://doi.org/10.1007/978-3-030-54746-2_16

conditions – drought, heat, and wildfires – to highlight this impact. We also explore the awareness and preparedness of agricultural employers and farmworkers using findings from our qualitative study in California [3]. The chapter then discusses existing policies and practices for protecting farmworkers from the impacts of extreme weather. Finally, recommendations for future research and translation of research findings are explored. Given the size and relevance of California's agricultural sector, its responses to extreme weather-related challenges have effects on consumers worldwide. In addition, thanks to its notable climate change adaptation and mitigation efforts, the state is positioned to develop occupational health and safety policies and practices that serve as examples for other agricultural states and regions.

Climate Change and Occupational Health and Safety

While the impacts of climate change on public health have received considerable academic and policy attention, impacts on occupational health and safety have been relatively overlooked, especially with respect to agricultural workers. Schulte and Chun conducted a systematic review of existing literature to identify the effects of climate change on worker health and safety [4]. They developed a framework with seven areas where an increase in prevalence, distribution, and severity of occupational illnesses and injuries may be observed as a result of climate change. These areas included (1) increased ambient temperature, (2) air pollution, (3) ultraviolet exposure, (4) extreme weather, (5) vector-borne diseases and expanded habitats, (6) industrial transitions and emerging industries, and (7) changes in the built environment [4]. In 2016, the framework was updated with new research findings and three additional considerations: (1) mental health effects of climate-related occupational hazards, (2) economic burden of climate-related occupational safety and health hazards, and (3) geoengineering and the potential for worker hazards [5]. The California Department of Public Health (CDPH) Climate Change and Health Equity Program identified outdoor workers and farmers as among the most vulnerable populations with respect the impacts of climate change. Notably, immigrants were also listed as one of the most vulnerable populations, and farmworkers in California are predominantly immigrants [6]. The effects of climate change on health and safety as described by Schulte and colleagues, and the vulnerable populations identified by the CDPH, highlight the need for a specific focus on the agricultural workforce. Increased ambient temperature, air pollution, extreme weather, vector-borne diseases, and mental health are all particularly relevant in the context of the agricultural industry. In this chapter we present a preliminary framework for climate change and agricultural health and safety (Fig. 16.1).

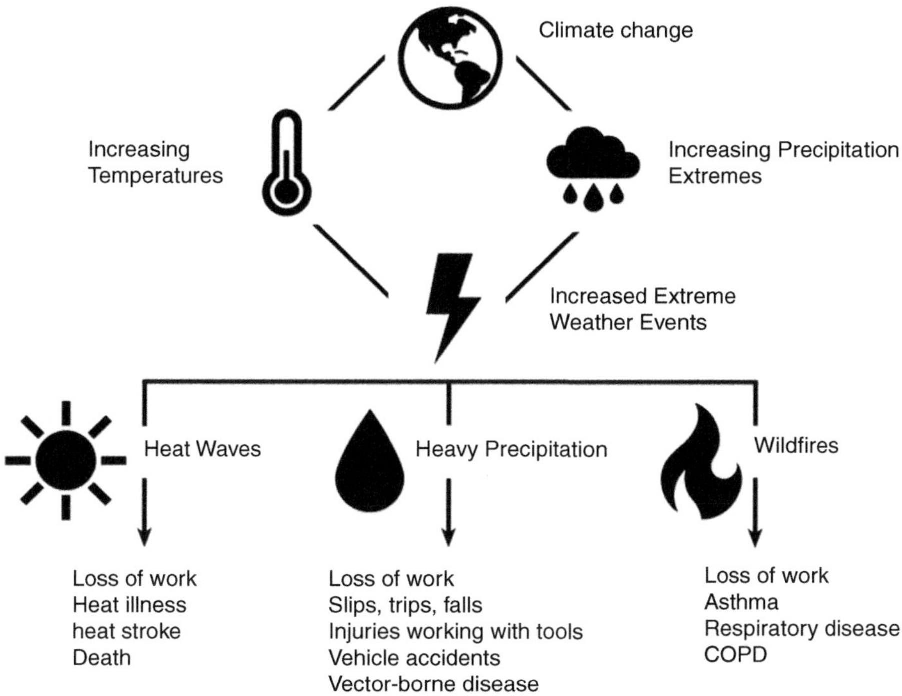

Fig. 16.1 Climate-related occupational hazards to agricultural workers

California Agriculture and Its Workforce

California produces one-third of the nation's vegetables and two-thirds of its fruits and nuts. In 2017, the state had over $20 billion in agricultural exports and registered $50 billion in agricultural cash receipts for over 400 commodities [7]. However, California's large agricultural sector relies on its Mediterranean climate, various micro-climates, and diverse range of marine and terrestrial ecosystems for water and agricultural inputs – all of which are experiencing changes due to climate change. The state has a mix of perennial crops (e.g., almonds, tree fruit, grapes) and annual crops (e.g., salad greens, tomatoes, melons) with varying labor demands with agricultural employment concentrated in the San Joaquin Valley, Central Coast, and Southern Coast, where many of the commodities produced are labor-intensive and hand-harvested. Weather factors into each stage in the growing process, from field preparation and planting to harvesting.

There has been a steady decrease in the size of the agricultural workforce of self-employed farmers, family farmers, and hired farmworkers since 1950. At the same time, the proportion of hired farmworkers in the overall agricultural workforce has increased [8]. California employs an estimated 800,000 farmworkers who work seasonally, rarely hold full-time equivalent jobs for an entire year, and earn an average annual income of less than $18,000 [9, 10]. An estimated 90% of California's farmworkers were born in Mexico, and approximately 60% are unauthorized to work in the United States [11]. In addition to the substantial Spanish-speaking farmworker population, California has a sizable number of workers who speak indigenous languages from Mexico (e.g., Mixteco, Zapotec, Trique), are recent immigrants, and are less likely to speak Spanish. Many farmworkers are paid piece-rate (i.e., by the number of crops picked vs. a set hourly rate), which can result in the pressure to continue working even while enduring physical risk in response to economic hardship [12, 13]. Federally, farmworkers are excluded from some labor law protections in the Fair Labor Standards Act and National Labor Relations Act, including overtime pay and collective bargaining. California has responded with state laws, including Assembly Bill 1066, which provides phased-in overtime pay for farmworkers. California farmworkers have had the right to organize since the mid-1970s, and only a few other states provide this protection.

Farmworker Occupational Hazards and Health Status

Farmers and farmworkers are at higher risk of fatal and non-fatal injuries than workers in most other sectors [14, 15]. In 2016, there were 417 farmer/farmworker deaths from work-related injuries recorded nationally [14]. Common agricultural occupational injuries include strains and sprains, falls, and musculoskeletal trauma. Farmworkers are also at risk of heat illness, hearing loss, pesticide exposure, and stress. The Mexican Immigration to California: Agricultural Safety and Acculturation (MICASA) study, which followed a cohort of immigrant farmworkers in California's Central Valley, found a 1-year cumulative agricultural injury incidence of 4.3%, with impacts from objects, falls, and cutting instruments as the most frequent causes of injuries [16]. While there has been a downward trend in occupational fatalities in California since 1999, agriculture had the highest fatality rate of all industries from 2013 to 2017, at 11 fatalities per 100,000 workers compared to the statewide average of 2.2 per 100,000 for all industries [17].

Farmworkers experience many of the same chronic diseases as the general population but have limited health care access. According to the U.S. Department of Labor's National Agricultural Worker Survey (NAWS), which provides information on the work history and health status of hired crop workers, less than a third of those in California have health insurance [18]. To add to this, few recent studies have systematically assessed farmworker health status. One such study, the California Agricultural Worker Health Survey (CAWHS), a statewide, cross-sectional survey conducted in 1999, found an obesity prevalence of 29.0% among male farmworkers and 38.0% among their female counterparts [19]. In contrast, obesity prevalence among Californians overall was 19.3% in 2001 and 27.0% in 2014 [20]. CAWHS also documented elevated cholesterol prevalence, as well as many previously undiagnosed health conditions in California farmworkers relative to the general population. Thus, promoting the health and safety of farmworkers in a context of increased hazards brought about by climate change is complicated by their existing higher risk of occupational injury, illness, and chronic disease, and exacerbated by insufficient health insurance coverage and access to health care [21].

Climate Change and Extreme Weather in California

Climate change is at least partially responsible for the increased incidence of extreme weather conditions [22]. Extreme weather refers to unpredictable, unexpected, and unusually severe weather as compared to the range of weather that has occurred in a particular area or region in the past. (Extreme weather is also referred to in the literature as *climate extremes* and *extreme climate events*.) Current predictions suggest the frequency of extreme weather events, including heat waves, cold waves, droughts, and intense precipitation and winds (e.g., thunderstorms, hurricanes, and tornadoes), will continue to increase over time. As the frequency and severity of these extreme events increase, so will wildfires and other downstream impacts of climate change [1].

Like other agricultural regions around the world, research suggests that California has experienced severe weather associated with climate change. Precipitation in California has become more unpredictable over time, with record-setting extremes for both precipitation and drought [23]. In January 2014, the governor of California declared a state of emergency after many consecutive years of drought with record temperature highs. The severe drought led to water shortages, groundwater overdraft, critically low streamflow levels, and increased wildfires. Researchers suggested a greater occurrence of drought years in California over the past two decades versus the past century, and indicated the state's relatively short rainy season, yielding low precipitation in the context of warmer temperatures, is more likely to result in future drought years [24]. Despite this, in late 2016 and early 2017, California experienced a season of extreme precipitation which resulted in severe flooding, overflow, and damage to the Oroville Dam, as well as the resultant evacuation of adjacent communities. While extreme weather events impact all members of society and occupational sectors, farmworkers and the agricultural industry are particularly vulnerable to these events.

Impact of Extreme Weather in Agriculture

Agriculture is directly dependent on climate and affected by the types of changes brought about by extreme weather. Direct impacts of extreme weather, such as flooding, drought, hurricanes, freezes, and extended heat waves, can result in the loss of crops and livestock. US agriculture sustained an estimated $700 billion in losses over the past 30 years (1980–2010), based on 90 extreme weather events [25]. Pathak and colleagues conducted a detailed review of climate change trends in California agriculture and found many negative impacts identifying potentially significant challenges to the industry in the future [26]. California is expected to experience increased variability in precipitation, snowpack, and extreme weather events such as heat waves, drought, and flooding. These changes will impact agricultural growing seasons, water availability, and pest life cycles. Climate impacts on crop yields are dependent on crop type, whether increased temperatures occur in summer vs. winter, and chill hours required for the crop [26]. While the economic impact of extreme weather on agricultural production should not be underestimated, and considerable literature examines short- and long-term impacts, corresponding impacts on the hired agricultural workforce are often overlooked. In addition to economic uncertainty experienced by agricultural employers and farmworkers, new occupational health and safety risks are likely to emerge from agricultural adaptations to climate change, such as the adoption of new crops or farm technology.

Perceptions of Risk: Impact of Extreme Weather on Occupational Health and Safety in California Agriculture

We are currently studying the experienced and anticipated impacts of extreme weather on occupational health and safety in agriculture through interviews and focus groups with agricultural employers and farmworkers, respectively. Our study focuses on extreme weather events, including heat, drought, rain, and wildfires, that are becoming more frequent throughout California. The research was approved by the University of California, Davis Institutional Review Board with interviews and focus groups conducted in 2018. The objective of the study is to identify workplace hazards and develop educational resources or other tools to assist agricultural employers and employees in adapting to an increase in extreme weather events. In the following sections, three impacts of extreme weather are examined: drought, heat, and wildfires. The ways in which these events exacerbate existing occupational health and safety risks of agricultural employers and farmworkers are discussed as reported in the preliminary results of our qualitative research as well as in the existing literature [3].

Case Study: Drought

A drought is defined as a prolonged period of below average precipitation that can be aggravated by hot temperatures. Climate change is likely to contribute to more frequent episodes of drought, which contribute to desertification, dust, air pollution, and airborne pesticide residue. These affect respiratory health among farmworkers and contribute to poor air quality for surrounding communities. Farmworker respiratory health is affected by dust from the soil, which may contain biological (e.g., microorganisms and mycotoxins) and chemical (e.g., gases and pesticides) allergens/immunogens. Farmworkers are exposed to dust while harvesting, pruning, or weeding crops; laying irrigation lines; dealing with livestock; or carrying out other tasks. A systematic review of respiratory health among farming populations found widespread prevalence of lung conditions such as asthma, chronic obstructive pulmonary disease (COPD), and decreased lung function [27]. More frequent droughts and the associated dust may exacerbate these conditions.

Coccidioidomycosis, commonly known as Valley Fever, is a respiratory disease that may be influenced by cycles of extreme precipitation and drought. Valley Fever is caused by a fungus that lives in the soil in dry climates, including portions of the California Central Valley and Central Coast as well as Arizona. While many people exposed to Valley Fever will not exhibit any symptoms, some individuals may experience pneumonia-like effects. In severe cases, Valley Fever may lead to death. Strategies to limit exposure of outdoor laborers (e.g., farmworkers) unable to cease work on windy days include the use of a National Institute of Occupational Safety and Health (NIOSH)-approved respirator and wetting the soil to reduce airborne dust.

During interviews carried out as a part of our research, agricultural employers discussed potential health and safety impacts of drought in the context of extreme heat and heat illness. Specifically related to drought, agricultural employers reported concerns about future access to water, feasibility of growing specific crops, and farming opportunities/capabilities. During focus groups, farmworkers reported an awareness of shorter rainy seasons and longer periods of drought. Most participants agreed that airborne dust was concerning and negatively impacting health, particularly during windy seasons or after drought. Dust was a serious concern for farmworkers because of the exposure to contaminants and pesticides that are mixed into the soil, inhaled, and absorbed through skin and eye contact or dermal-to-oral routes. Many farmworkers described the combination of breathing and working in dusty conditions, where pesticides and chemicals are prevalent, as a major health risk.

Photo credit: Joe Proudman for University of California, Davis

Case Study: Heat

A heat wave is defined as a period of at least five consecutive days in which the maximum daily temperature is more than 5 °C above the average maximum temperature [28]. Heat waves, which have a direct effect on agriculture, are predicted to increase in frequency and severity over the next century as a result of climate change [29]. Because average maximum temperatures are based on historical temperatures, the presence of heat waves contributes to increasing average maximum temperatures over time.

Outdoor workers have been identified as particularly vulnerable due to their increased exposure to heat and their physically demanding work [6, 30, 31]. The Centers for Disease Control and Prevention (CDC) reported that from 1992 to 2006, 423 workers in agricultural and non-agricultural industries died from heat exposure nationally; 16% of these deaths were of those engaged in crop production. The average annual heat-related fatality rate for these workers was 0.39 per 100,000 crop workers, compared to 0.02 per 100,000 for all US civilian workers [32].

Exposure to hot temperatures can lead to dehydration, nausea, exhaustion, stroke, and even death. The signs of these heat illnesses can be confused with other ailments and are not always easy to recognize. Farmworkers are at greater risk for heat illness than other outdoor workers because they wear extra clothing and personal protective equipment that make it difficult to stay cool.

As the climate warms and heat waves become more frequent and severe, the risk of heat exposure will disproportionately increase for outdoor versus indoor workers. Despite this risk, there is currently no federal standard, and only a few state standards (e.g., that in California), to protect farmworkers from heat exposure. The California Division of Occupational Safety and Health Administration (Cal/OSHA) requires all agricultural employers to have a heat illness prevention plan as described in the California Code of Regulations Section (§) 3395, Heat Illness Prevention [33]. The plan must include training on exposure to risk factors for heat illness, the importance of drinking water, and common signs and symptoms of heat illness. Employers must also provide at least one quart of cool fresh water per hour per worker as well as rest and shade whenever temperatures exceed 85 °F. Additional provisions specify guidelines on acclimatization and high-heat procedures.

In their study of heat illness among farmworkers in California, UC Davis researchers found that despite working on farms that were compliant with the Cal/OSHA heat-related illness prevention regulations, nearly 8% of workers were at risk of heat-related illness, nearly 12% were dehydrated at the end of the day, over 12% suffered reversible acute kidney injury over the work day, and 50% said they had at some point experienced a heat illness symptom while working [13, 34]. While the *Water. Rest. Shade.* message of the Cal/OSHA heat standard safety campaign is clear and widely recognized, barriers continue to exist related to adherence and the intended prevention of illness. For example, the piece-rate pay structure was found to significantly increase a workers' risk of heat illness symptoms in the California Heat Illness Prevention Study [12, 13]. Workers being paid a piece-rate are more likely to push themselves physically and not take needed breaks due to the financial incentive to harvest more during a fixed amount of time.

In our study, agricultural employers reported being very familiar with the California heat standard and had policies in place at their worksite to comply. Heat was the most cited challenge for managing employees in the field. Employers reported that heat waves and overall higher temperatures have an impact on both crops and workers – if extreme heat results in crop loss, workers are likely to seek other employment. Agricultural employers tended to place responsibility for self-care on farmworkers and deferred to the workers' choices in workplace clothing, as was found in other heat illness studies [12].

Farmworkers also expressed knowledge of the heat illness prevention standard and identified heat illness as their greatest occupational health hazard. They identified certain groups as being more susceptible to heat illness, including elderly workers or those who had pre-existing health conditions. Farmworkers described symptoms related to heat that they experienced first-hand or witnessed others experiencing while working; these symptoms included dizziness, nausea, vomiting, headaches, and fainting, among others. Most participants agreed that adverse effects from heat would be greatly reduced if employers took additional steps to both provide and promote measures for workers to avoid heat illness. For example, proximity to the bathrooms often played an important role in decisions regarding water consumption during high temperatures, particularly for females. Farmworkers reported that their ability to take precautions to avoid adverse health effects was highly dependent on supervisors and the work culture they created.

Case Study: Wildfires

Climate change will contribute to the increased frequency and severity of wildfires, which will impact agricultural employers and farmworkers. Exposure to wildfire smoke is strongly associated with mortality and respiratory morbidity including the exacerbation of asthma, COPD, bronchitis, and pneumonia [35–37]. While many negative health effects of acute wildfire smoke exposure in general have been identified, additional research is needed to identify the *long-term impact* of wildfire smoke exposure on human health [37]. Wildfire suppressants generally come in the form of retardants (ammonium phosphates) or foams (detergent-based organic chemicals). The health impacts from pesticide combustion in combination with wildfire suppressant chemicals are unknown [38]. Additionally, wildfire suppressant chemicals may leave residues on crops and in the air that livestock, farmworkers, and neighboring communities breathe.

Photo credit: Pixabay

In 2019, Cal/OSHA passed an emergency regulation to protect outdoor workers from wildfire smoke [39]. The policy stipulates that during wildfire smoke events, when the air quality index (AQI) is 151 or greater, employers are required to provide NIOSH-approved respirators for voluntary use by employees, and supply information on the proper use and limitations of respirators, and the negative health impacts of wildfire smoke exposure, among other topics.

In our study, which collected data prior to the Cal/OSHA wildfire smoke regulation, agricultural employers reported limited knowledge about the health effects of wildfire smoke exposure and how to obtain air quality information [3], unlike with heat illness. Employers had neither safety procedures to withstand wildfire smoke-induced poor air quality, nor discussions on the use of respirators or masks as potential protective equipment. Similarly, farmworkers reported experiencing occupational wildfire smoke exposure and having limited knowledge of appropriate safety precautions. As with heat illness, supervisors were discussed as having the greatest impact on the safety culture of the workplace, and farmworkers reported continuing work despite unsafe conditions due to economic need. We expect the new Cal/OSHA regulation to raise awareness of adverse wildfire smoke-induced health effects and promote strategies to reduce exposures of outdoor workers.

Discussion

Our qualitative study to identify existing and anticipated occupational health and safety hazards in California agriculture discovered varying levels of awareness of extreme weather events and potential strategies to reduce negative health impacts [3]. Additionally, important implications from workplace power dynamics emerged and should be considered during the development of occupational regulations, workplace policies, and climate change adaptation strategies. Specifically, farmworkers' economic dependence on employers and fear, due to a hostile political climate toward immigrants, reduce their ability to advocate for their own safety.

Of the three case studies discussed – drought, heat, and wildfire – agricultural employers and farmworkers are most aware of the negative health impacts of heat and the associated strategies to reduce risk. Despite widespread knowledge of the Cal/OSHA heat standard and health effects of heat stress, workers continue to experience heat illnesses, even on Cal/OSHA compliant farms. *Recommendation*: Research should evaluate the current standard guidelines and determine if modifications would reduce risk. For example, is the acclimatization period long enough? Are the temperature guidelines sufficient? What workplace policies would enable farmworkers to successfully advocate for the water, rest, and shade they need on high heat days. Additional evaluation of the impact of a piece-rate pay structure on adverse health outcomes should be conducted.

The new California state regulation to protect outdoor workers from wildfire smoke should be systematically evaluated from the outset. The regulation creates an opportunity to evaluate a state-wide policy intervention for wildfire smoke exposure. Farmworkers are a vulnerable population working long hours in physically demanding settings. Wildfire smoke compounds their health risks. This regulation is expected to raise awareness of poor air quality during wildfire events and reduce exposure for workers through respirator use, more frequent breaks, and work relocation when possible. *Recommendation*: Research should examine whether respirators provide adequate protection in the real-world agricultural working environment.

Do workers opt to wear respirators? If not, why not? How are employers adopting the new requirements and training their supervisors and workers? State and federal regulations are important components in protecting workers from occupational health and safety risks created and/or exacerbated by extreme weather; ongoing policy evaluations are critical to maintain and improve the intended outcomes.

Education and training are important risk-reducing strategies. All California employers are required to have an injury and illness prevention program (IIPP), which promotes workplace safety and health through the identification of potential workplace hazards and the planning of prevention methods. The California Worker Occupational Safety and Health Training and Education Program (WOSHTEP) provides IIPP training tailored to agricultural employers in English and Spanish [40]. In these trainings, employers gain skills in identifying the Cal/OSHA illness and injury prevention requirements, identify successful elements of an IIPP, and learn to create workplace hazard maps, in a participatory and collaborative setting. *Recommendation:* California employers should be encouraged to use their IIPP as a framework to consider and incorporate extreme weather-induced hazard prevention measures into existing workplace health and safety programs.

Our ongoing work demonstrates the importance of involving agricultural communities, worker advocates, and industry stakeholders in occupational health and safety efforts. *Recommendation:* Workplace safety messages should be tailored for the target audience (i.e., agricultural employers and farmworkers should not be given the same materials). Effective communication to employers should emphasize their obligation and responsibilities as employers with clear, concise, and practical messages. Farmworkers should be informed about their rights as workers and about strategies to take individual safety precautions. Though the majority of farmworkers in California speak Spanish, translation of existing English resources is not sufficient. Safety information should be culturally tailored, with limited text, and a clear message. When possible, agricultural employers and farmworkers benefit from in-person and interactive safety trainings that explain "why" a topic matters and "how" it affects them. As governments devise and implement climate change adaptation strategies relevant to the agricultural industry, they must recognize existing relationships of trust and paths of communication. For example, we find targeting agricultural employers and farmworker gatekeepers to be an effective approach to disseminate health and safety information and resources. Insurance companies, commodity groups, and grower associations are trusted entities by farmers and other agricultural employers, while community organizations, community workers, and *promotores,* or lay health leaders, are well regarded by farmworkers.

Conclusion

Extreme weather events are predictably unpredictable and thus present challenges to governments, communities, employers, and workers to prepare and respond with human health and safety at the forefront. Agriculture is increasingly being viewed as part of the climate change solution. Farmers are contributing to climate change mitigation and adaptation efforts through the development of climate-resilient farms, the reduction of emissions, and the use of agricultural lands as carbon sinks, among other undertakings. At the state level, policies and programs aim to expand resource conservation districts, and promote water conservation, efficient irrigation and soil health management practices, dairy farm greenhouse gas reduction, farmland conservation, best practice sharing among farms, and climate innovation [41]. While efforts that promote sustainability can be inherently beneficial for health and safety (e.g., reduced pesticide use), all farm management adaptations should be considered through a lens of worker health and safety and assess whether new hazards emerge with practice changes.

In this chapter, we demonstrated the importance of considering the health impacts of climate change on agricultural workers, a particularly vulnerable workforce. Sustaining a healthy and productive agricultural workforce is an important part of climate change adaptation. The concurrence of climate change-fueled events like extreme heat, wildfires, and drought pose a unique risk to outdoor workers in California. It is challenging to separate the effects of individual climate-driven events to best protect workers. For instance, though protecting outdoor workers from wildfire smoke exposure or during a heat wave is a recognized climate-related challenge, currently proposed solutions put workers at risk of losing their employment. Additionally, though many mental health impacts of climate-driven events are known, the combined effects of stress and anxiety related to job security, immigration concerns, and occupational hazards has not been studied. Similarly, the frequent occurrence of extreme weather events may result in yet unstudied chronic health impacts for agricultural workers. A multipronged approach is needed to adequately protect agricultural workers amidst a changing climate. This approach should include farmworkers, agricultural employers, community organizations, and government, and be supported by state and federal regulations, educational opporstunities, technological innovations, and creative solutions.

Acknowledgments The concepts described in this chapter were facilitated through grant support from the Centers for Disease Control and Prevention, National Institute for Occupational Safety and Health (CDC-NIOSH) Cooperative Agreements # U01 OH010969 and Grant # U54 OH007550.

References

1. Intergovernmental Panel on Climate Change (IPCC). Global warming of 1.5 °C. An IPCC special report [Internet]; 2018. Available from https://www.ipcc.ch/sr15/
2. Bulliten of the American Meterological Society. Explaining extreme events from a climate perspective. 2018;99(12). Available from https://www.ametsoc.org/ams/index.cfm/publications/bulletin-of-the-american-meteorological-society-bams/explaining-extreme-events-from-a-climate-perspective/

3. Riden HE, Giacinto R, Wadsworth G, Rainwater J, Andrews T, Pinkerton KE. Wildfire smoke exposure: awareness and safety responses in the agricultural workplace. J Agromedicine. 2020:1–9. https://doi.org/10.1080/1059924X.2020.1725699.
4. Schulte PA, Chun H. Climate change and occupational safety and health: establishing a preliminary framework. J Occup Environ Hyg [Internet]. 2009;6(9):542–54. Available from https://doi.org/10.1080/15459620903066008.
5. Schulte PA, Bhattacharya A, Butler CR, Chun HK, Jacklitsch B, Jacobs T, et al. Advancing the framework for considering the effects of climate change on worker safety and health. J Occup Environ Hyg [Internet]. 2016;13(11):847–65. Available from https://doi.org/10.1080/15459624.2016.1179388.
6. California Department of Public Health. Climate change and health equity issue brief [Internet]; 2019 [cited 2019 Aug 24]. Available from https://www.cdph.ca.gov/Programs/OHE/CDPHDocument Library/CCHEP-General/CDPH_CC-and-Health-Equity-Issue-Brief.pdf
7. California Department of Food and Agriculture. California agricultural statistics review [Internet]; 2018 [cited 2019 Aug 24]. Available from https://www.cdfa.ca.gov/statistics/
8. USDA Economic Research Service. Farm labor [Internet]; 2019 [cited 2019 Aug 25]. Available from https://www.ers.usda.gov/topics/farm-economy/farm-labor/
9. Martin PL, Hooker B, Akhtar M, Stockton M. How many workers are employed in California agriculture? Calif Agric [Internet]. 2017;71(1):30–34. Available from https://doi.org/10.3733/ca.2016a0011
10. Martin P, Hooker B, Stockton M. Employment and earnings of California farmworkers in 2015. Calif Agric [Internet]. 2017;72(2):107–13. Available from http://calag.ucanr.edu/Archive/?article=ca.2017a0043
11. Carroll D. California crop worker characteristics: preliminary 2015–2016 findings from the National Agricultural Workers Survey [Internet]. UC Davis Gifford Center for Population Studies presentation; 2017. Available from https://gifford.ucdavis.edu/events/
12. Courville M, Wadsworth G, Schenker M. We just have to continue working: farmworker self-care and heat-related illness. J Agric Food Syst Community Dev [Internet]. 2016;6(2). Available from https://foodsystemsjournal.org/index.php/fsj/article/view/453
13. Mitchell DC, Castro J, Armitage TL, Tancredi DJ, Bennett DH, Schenker MB. Physical activity and common tasks of California farm workers: California Heat Illness Prevention Study (CHIPS). J Occup Environ Hyg [Internet]. 2018;15(12):857–69. Available from https://doi.org/10.1080/15459624.2018.1519319.
14. National Institute for Occupational Safety and Health. Agricultural safety [Internet]. [cited 2019 Aug 24]. Available from https://www.cdc.gov/niosh/topics/aginjury/default.html
15. McCurdy SA, Carroll DJ. Agricultural injury. Am J Ind Med [Internet]. 2000;38(4):463–80. Available from https://doi.org/10.1002/1097-0274(200010)38:4%3C463::AID-AJIM13%3E3.0.CO.
16. McCurdy SA, Xiao H, Hennessy-Burt TE, Stoecklin-Marois MT, Tancredi DJ, Bennett DH, et al. Agricultural injury in California hispanic farm workers: MICASA follow-up survey. J Agromedicine [Internet]. 2013;18(1):39–49. Available from https://doi.org/10.1080/1059924X.2012.743380
17. Census of Fatal Occupational Injuries Program. Fatal occupational injuries in California: 2013–2017 [Internet]; 2018. Available from https://www.dir.ca.gov/dosh/cfoi/CFOI_2017/Fatalities-Report-2013-2017.pdf
18. Schenker MB, McCurdy SA, Riden HE, Villarejo D. Improving the health of agricultural workers and their families in California: current status and policy recommendations. Univ Calif Glob Heal Inst. 2015:1–26. Available at ucghi.universityofcalifornia.edu/about-us/publications.aspx
19. Villarejo D, McCurdy SA, Bade B, Samuels S, Lighthall D, Williams III D. The health of California's immigrant hired farmworkers. Am J Ind Med [Internet]. 2010;53(4):387–97. Available from https://doi.org/10.1002/ajim.20796.
20. California Department of Public Health. Obesity in California: the weight of the state, 2000–2014 [Internet]; 2016. Available from https://www.cdph.ca.gov/Programs/CCDPHP/DCDIC/NEOPB/CDPHDocument Library/RES_ObesityReport20002014.pdf

21. American Public Health Association. Improving working conditions for U.S. farmworkers and food production workers [Internet]; 2017 [cited 2019 Aug 24]. Available from https://apha.org/policies-and-advocacy/public-health-policy-statements/policy-database/2018/01/18/improving-working-conditions
22. Coumou D, Rahmstorf S. A decade of weather extremes. Nat Clim Chang [Internet]. 2012;2:491. Available from https://doi.org/10.1038/nclimate1452
23. Swain DL, Langenbrunner B, Neelin JD, Hall A. Increasing precipitation volatility in twenty-first-century California. Nat Clim Chang [Internet]. 2018;8(5):427–33. Available from https://doi.org/10.1038/s41558-018-0140-y.
24. Diffenbaugh NS, Swain DL, Touma D. Anthropogenic warming has increased drought risk in California. Proc Natl Acad Sci [Internet]. 2015;112(13):3931 LP–36. Available from http://www.pnas.org/content/112/13/3931.abstract
25. Motha R. Chapter 30: the impact of extreme weather events on agriculture in the United States; 2011. Publications from the USDA-ARS/UNL Faculty 1331.
26. Pathak TB, Maskey ML, Dahlberg JA, Kearns F, Bali KM, Zaccaria D. Climate change trends and impacts on California agriculture: a detailed review. Agronomy 2018;8:25. https://doi.org/10.3390/agronomy8030025.
27. Omland Ø. Exposure and respiratory health in farming in temperate zones – a review of the literature. Ann Agric Environ Med [Internet]. 2002;9(2):119–36. Available from http://europepmc.org/abstract/MED/12498578
28. World Health Organization. Guidelines on the definition and monitoring of extreme weather and climate events [Internet]; 2015. Available from https://www.wmo.int/pages/prog/wcp/ccl/opace/opace2/documents/DraftversionoftheGuidelinesontheDefinitionandMonitoringofExtremeWeatherandClimateEvents.pdf
29. Wuebbles DJ, Fahey DW, Hibbard KA, DeAngelo B, Doherty S, Hayhoe K, Horton R, Kossin JP, Taylor PC, Waple AM, CPW. Executive summary. In: Climate science special report: fourth national climate assessment, vol. I [Internet]. Washington, DC; 2017. 12–34 p. Available from https://science2017.globalchange.gov/chapter/executive-summary/
30. Balbus JM, Malina C. Identifying vulnerable subpopulations for climate change health effects in the United States. J Occup Environ Med [Internet]. 2009;51(1). Available from https://journals.lww.com/joem/Fulltext/2009/01000/Identifying_Vulnerable_Subpopulations_for_Climate.6.aspx
31. Lundgren K, Kuklane K, Gao C, Holmer I. Effects of heat stress on working populations when facing climate change. Ind Health. 2013;51(1):3–15.
32. Centers for Disease Control and Prevention. Heat-related deaths among crop workers – United States, 1992–2006 [Internet]; 2008. Available from https://www.cdc.gov/mmwr/preview/mmwrhtml/mm5724a1.htm
33. California Department of Occupational Safety and Health. Heat illness prevention [Internet]; 2006. Available from https://www.dir.ca.gov/title8/3395.html
34. Mitchell DC, Castro J, Armitage TL, Vega-Arroyo AJ, Moyce SC, Tancredi DJ, et al. Recruitment, methods, and descriptive results of a physiologic assessment of latino farmworkers: the California heat illness prevention study. J Occup Environ Med [Internet]. 2017;59(7):649–58. Available from https://www.ncbi.nlm.nih.gov/pubmed/28414703
35. Cascio WE. Wildland fire smoke and human health. Sci Total Environ [Internet]. 2018;624:586–95. Available from http://www.sciencedirect.com/science/article/pii/S004896971733512X
36. Reid CE, Jerrett M, Petersen ML, Pfister GG, Morefield PE, Tager IB, et al. Spatiotemporal prediction of fine particulate matter during the 2008 Northern California wildfires using machine learning. Environ Sci Technol [Internet]. 2015;49(6):3887–96. Available from https://doi.org/10.1021/es505846r.
37. Black C, Tesfaigzi Y, Bassein JA, Miller LA. Wildfire smoke exposure and human health: significant gaps in research for a growing public health issue. Environ Toxicol Pharmacol [Internet]. 2017;55:186–95. Available from http://www.sciencedirect.com/science/article/pii/S1382668917302478

38. Carratt S, Flayer C, Kossack M, Last J. Pesticides, wildfire suppression chemicals, and California wildfires: a human health perspective. Curr Top Toxicol [Internet]. 2017;13:1–12. Available from https://escholarship.org/uc/item/7rh1s9z8
39. California Occupational Safety and Health Standards Board. California title 8, Division 1, Chapter 4, Subchapter 7. General industry safety orders, group 16. Control of hazardous substances, Article 107. Dust, fumes, mists, vapors and gases. Section 5141.1 [Internet]; 2019. Available from https://www.dir.ca.gov/oshsb/documents/Protection-from-Wildfire-Smoke-Emergency-apprvdtxt.pdf
40. California Department of Industrial Relations. The Worker Occupational Safety and Health Training and Education Program (WOSHTEP) [Internet]. [cited 2019 Aug 26]. Available from https://www.dir.ca.gov/chswc/WOSHTEP/iipp/
41. California Natural Resources Agency. Safeguarding California plan: 2018 update [Internet]; 2018. Available from http://resources.ca.gov/docs/climate/safeguarding/update2018/safeguarding-california-plan-2018-update.pdf

Chapter 17
Household Air Pollution from Cookstoves: Impacts on Health and Climate

William J. Martin II, Tara Ramanathan, and Veerabhadran Ramanathan

Household air pollution (HAP) from cooking fires is a global problem that occurs in mostly low- and middle-income countries (LMIC) and contributes to major health and environmental risks for nearly three billion people on the planet [1–3]. HAP is a result of incomplete combustion of solid fuels such as biomass and coal that is typically used for cooking, heating, and lighting in homes of those living at the bottom of the energy ladder. Biomass fuels consist of wood, crop residues, charcoal, or dung. Exposure to HAP causes almost three million deaths annually [2, 3]. In addition, the consumption of these solid fuels causes regional environmental degradation through deforestation and the household emissions at scale represent a sizeable fraction of the outdoor air pollution in villages and cities [4]. Furthermore, some of these emissions, such as black carbon, are short-lived climate forcers that contribute to global warming and, if significantly reduced, can potentially have a mitigating impact on global temperatures in this century [5, 6]. HAP is both a major health risk for the poorest people on the planet and a major risk for global climate change. Thus, its remedies which are possible today offer the unique opportunity to improve the health and quality of life of the world's poor and at the same time provide hope that the global warming trends can be mitigated by reducing the impact of the short-lived climate forcers.

W. J. Martin II (✉)
College of Public health, The Ohio state University, Columbus, OH, USA
e-mail: Martin.3047@osu.edu

T. Ramanathan
Nexleaf Analytics, Los Angeles, CA, USA

V. Ramanathan
Scripps Institution of Oceanography, University of California at San Diego, La Jolla, CA, USA

K. E. Pinkerton, W. N. Rom (eds.), *Climate Change and Global Public Health*, Respiratory Medicine, https://doi.org/10.1007/978-3-030-54746-2_17

Cooking Fires and the Role of Women

The use of cooking fires goes back to the origins of our species and likely contributes to our evolutionary success as an intelligent species through improved nutrition [7]. Many of us harbor pleasant memories of camping fires and perceive cooking even in primitive sites as a warm and nurturing experience. These pleasant memories of domestic fires are also important to acknowledge and understand for families living in low-resource settings where traditional cooking methods have been practiced for generations with cooking of food that has preferred consistencies and flavors. These traditions and preferences can act as barriers to accepting the changes associated with newer and safer cooking methods and also potentially reduce the perception of personal risk for adverse health effects including mortality associated with traditional methods of cooking and exposure to HAP.

The traditions and cultural practices associated with domestic cooking also reinforce patterns of behavior that often contribute to defining the role of women in a social and familial context. Cooking is not only a duty that falls almost exclusively to women, they are also responsible for the fuel gathering, a form of drudgery that occupies significant time in their daily routine and places the women and accompanying children at considerable personal risk if they must walk miles from their villages to gather fuel [8]. Thus, because of the role of women in cooking and fuel-gathering and because women are often living in patriarchal societies, they are at the center of this environmental issue. Therefore, it is essential to find new ways to support women, husbands, and other family members as they move from traditional domestic cooking methods to safer and more efficient cookstoves as the acceptance of the families to these changes are the key to the success of implementing effective and sustainable solutions to reduce HAP.

Proposed interventions to reduce HAP require the successful adoption and use of new stove or fuel technology, and also invariably come at a financial cost. The ability of women to have a voice in the family decisions and use of scarce domestic resources are necessary for reducing HAP and require fundamental changes in social and cultural roles. We emphasize this message early in this chapter lest the new and increasingly affordable technologies to reduce HAP suggest that the health and environmental risks are easily managed and implemented; they are not. Failure is always more likely than success. This is well demonstrated in a TED Lecture by David Damberger, a member of Engineers Without Borders, who articulates the need in any development enterprise to carefully evaluate the long-term outcome of any intervention, as most will fail [9]. Current efforts of large-scale implementation of improved cookstove technology targeting the world's poor require the involvement of women and other family members to achieve success. One example of an effective community-based approach to engage women and families in purchase and implementation of cookstoves at scale is the million plus membership organization, Self Employed Women's Association (SEWA) [10]. In South Asia, SEWA empowers women and families to invest in cleaner cooking solutions and supports these efforts through novel group-based negotiations with manufacturers to reduce stove costs and by using microfinances loans that are essentially paid with fuel savings from the more efficient stoves. However, implementation programs that empower women and families will still require constant evaluation and oversight to be certain that expected health, environmental, economic, and climate benefits are in fact realized.

Common Challenges to Clean Cooking Solutions

The billions of people who use solid fuels for cooking or heating typically use a variation of a three-stone fire with fuel being pushed into the fire gradually from the sides or, if affordable, the use of a primitive stove that provides the basic needs of cooking [11]. If a stove exists, it is often without a chimney or flue as they typically require detailed construction and maintenance to function properly. Over time, efforts for adequate removal of cookstove emissions are often not sustainable and emissions are simply released into the household [12]. This is how almost half the planet lives.

For decades, nongovernmental organizations (NGOs), local and multinational manufacturers, development agencies, host country governments, and foundations have struggled with improving the quality of cookstoves in LMIC, which is where the majority of the world's poor live. Some of these efforts have had substantial success such as in China [13] or with more limited success as in the case of India [14]. In the majority of implementation studies around the world, there has been little study of the impact of "improved cookstove" programs on health or environment. Most implementations are often conducted at such a small scale and in such different cultural settings that benefits are assumed and comparisons across programs are difficult.

There are two types of cookstove efficiencies that impact health and the environment: (1) fuel efficiency and (2) combustion efficiency. Fuel efficiency reflects the amount of fuel required to achieve a specific task, such as a controlled water boiling test [15]. Fuel efficiency is critically important to households as stoves with improved fuel efficiency save the family fuel costs and time lost in fuel gathering. Reduction in time required to gather fuel is important for both women and children because time saved could be redirected to enhance educational and economic growth. Improved fuel efficiency will reduce the quantity of solid fuel burned and thus the quantity of CO_2 released from cooking fires. The second type of efficiency relates to the efficiency of combustion itself and is necessary for reducing particulate matter (PM) and other pollutants that impair health. $PM_{2.5}$ is that a fraction of aerosol particles that is smaller than 2.5 μm and poses special risks to human subjects due to its access to the lower respiratory tract and alveolar structures of the lung, where gas exchange occurs [16, 17]. Many outdoor air quality standards rely on $PM_{2.5}$ and $PM_{10.0}$ to reflect the risk of these air pollutant fractions to human health. Carbon monoxide (CO) is also a very dangerous pollutant, especially with the use of charcoal as a fuel [18]. Improving the combustion efficiency of a stove is key to reducing harmful emissions such as $PM_{2.5}$ and CO. Black carbon is part of the $PM_{2.5}$ fraction and is reflected as "soot" to the observer. Successful reduction of these pollutants will reduce human exposures and improve human health. There can be considerable differences in combustion and efficiency between stove testing sites and the household setting related to many factors including choice of fuel, ventilation, location of stoves, and human behavior.

Another critical component to reducing exposures to household members is to understand how human behavior or cultural traditions may impact the level of exposures. For example, the solid fuels collected (or purchased) for the stove must be sufficiently dry and combustible to perform the cooking task to achieve the reduced levels of emissions. Sometimes people will collect, out of necessity, anything that burns easily such as leaves or crop residues; these fuel sources contain excessive moisture and, when burned in even the most advanced stoves, will result in a very smoky indoor environment. In addition, there are special challenges during the transition from a traditional fire to the use of an improved stove. Many families, due to high cooking needs, use a mix of both primitive and improved cookstoves concurrently, resulting in a phenomenon known as "stove stacking" [19]. As stove stacking almost always results in persisting high levels of HAP, it is important to quantify household usage with devices such as stove use monitors (SUM) to assess the households' transition (or not) to cleaner cooking solutions [19]. At the beginning, a new improved stove or fuel may be used exclusively, but for a variety of reasons, stove stacking with traditional fires occurs over time, resulting in increasing HAP [19]. Documenting the exclusive use of improved cookstoves over time is essential to achieving sustainable benefits. Implementation at scale requires thoughtful interaction and participation with families and communities with the sensitivity to cultural traditions as well as quantitative methods to measure stove use to ensure the adoption of the new technologies and realization of the benefits. In this common scenario, despite the presence of new cookstove technology, there is a minimal impact in reducing the level of HAP. Improving the efficiency of stoves or fuels offers the potential for multiple benefits to both households and the global environment.

Globalization of Clean Cooking Solutions

In the past several years, there are increasing efforts to develop better coordination of implementation efforts and to develop a common knowledge base about the principles of stove efficiencies, affordability, and successful implementation. One of the best examples has been the Partnership for Clean Indoor Air (PCIA) led by the U.S. Environmental Protection Agency (EPA) from 2002 to 2012 with more than 500 members including NGOs, manufacturers, governments, academic institutions, and others [20]. The PCIA focused much of its attention on improving the understanding of what an efficient and clean burning stove is and created an international forum of partners to advance the science and to facilitate the effective implementation of clean cooking solutions.

In an effort to enhance resources and to expand partnerships to address the complex global problem of HAP, the EPA, US Department of State, NIH and CDC, and others developed a new public–private partnership with the United Nations Foundation in 2010 to launch the Global Alliance for Clean Cookstoves [1], which updated its name to Clean Cooking Alliance in 2018. The stated mission of the Alliance is to "improve health, protect the climate and environment, empower women, and help families save time and money by creating a thriving global market for cleaner, more modern household cooking solutions" [21].

Building on the success of the PCIA, the Alliance developed hundreds of partners to help meet its mission and goals including other governments around the world, multinational companies, foundations, and NGOs. The Alliance now provides a forum for major implementation and testing of new technology to reduce HAP and its health and atmospheric impacts that will use ongoing research and evaluation to validate whether such impacts occur at the scale expected [22]. This ambitious effort has been a "game-changer" in bringing recognition and resources to address this global threat to human health and the environment.

Cookstove Performance Standards and Testing Centers

There are multiple sites today where stoves can be tested for fuel and combustion efficiencies. The U.S. EPA offers rigorous stove testing at its facility in Research Triangle Park, North Carolina, USA, to determine emission patterns under controlled conditions [23]. In addition, Aprovecho Research Center in Cottage Grove Oregon offers similar testing but also offers a portable stove testing lab that can be used anywhere in the world [14]. Similarly, Berkeley Air in Berkeley California offers state-of-the-art testing of stoves that complement a number of technologies related to HAP and stove use including exposure-monitoring devices [24]. Despite these excellent reference laboratories, there was no global standardization of methods for stove performance until recently. The diversity of cookstoves on the market and the absence of widespread certified testing centers create confusion to consumers, NGOs, and the governments that wish to address the problem of HAP, and to do so requires standardization of stove testing. In 2012, there was an International Working Agreement for setting standards for cookstove performance from a meeting of Alliance members and other stakeholders hosted by the International Organization for Standardization (ISO) in The Hague [25, 26]. This was a major advancement as both companies manufacturing stoves and consumers buying stoves can be guided by internationally accepted standards that permit a "clean cookstove" to undergo validated testing and be marketed with a stated energy efficiency rating (Tier 1–4) [26].

Health Impacts of Household Air Pollution

Pollution is the number one environmental cause of death in the world, due largely to the overwhelming impact of ambient and household air pollution [27]. Ambient and household air pollution share many of the same products of incomplete combustion, although typically the household levels of these pollutants are of much higher concentration [28, 29]. These deaths are primarily from respiratory conditions including acute lower respiratory tract infection (ALRI) in children under age 5, chronic obstructive pulmonary disease (COPD), and lung cancer, as well as cardiovascular diseases including ischemic heart disease and stroke [30]. The lung cancer risks are almost exclusively related to coal use for cooking and heating in China [31], although some lung cancer risk is also attributable to biomass HAP exposure [3].

A trans-US government workshop held in 2011 addressed the state of the science of health impacts from HAP and offered a number of recommendations for future research related to health risks [32]. The findings included additional health risks from a small number of studies of HAP that may require replication but also include human health risks related to what we know from outdoor air pollution and tobacco smoke. Some of these putative risks will require further study in populations living with HAP, but the underlying rationale for these studies based on similar exposures is strong.

Examples of such probable health risks attributable to HAP include: other respiratory diseases such as asthma or interstitial lung diseases; pregnancy outcomes such as birth weight and prematurity or perinatal complications such as sepsis and congenital impairment; infectious diseases such as acute pneumonia in older children or tuberculosis; cancers from non-coal sources such as biomass; and ocular disorders such as cataracts or trachoma [32]. Of course, some health risks from indoor fires are unrelated to HAP. Burns and scalding are often under-reported and yet represent a life-changing risk for women and children that can include death [33]. Thus, stoves must not only be more efficient to promote health, they must also be tested to assure safety from risk of burns.

Vulnerable Populations to Adverse Health Outcomes with Household Air Pollution

It goes without saying that the world's poor who are dependent on solid fuel use are the ones most susceptible to the adverse impacts of HAP. That said, women and children, due to their domestic roles, spend more time in the household than men and thus have higher indoor exposure when burning solid fuels and less clean liquid fuels for cooking, heating, and lighting [32].

A number of respiratory diseases appear to be more common in women suggesting a possible sex-based increased susceptibility [34, 35]. For example, women may have increased risk for COPD than men for the same level of exposure to tobacco [36]. In addition, COPD is more common in women than men in the United States, even though tobacco use is more common in men. And since the year 2000, the number of women dying from COPD has surpassed that of men [37]. There are a variety of hypotheses that suggest why women may have more COPD than men [34]. When this predisposition is coupled with markedly higher exposure to HAP for women, the health risk is substantial.

The primary health risks to under age 5 children exposed to HAP include acute lower respiratory tract infection (ALRI) and low birth weight [32]. The 2015 GBD report notes significant decrease in ALRI-related death from HAP exposure from 905,000 (2005) to 729,000 (2015) [3], although it is unlikely the level of HAP exposure in this time frame has been significantly reduced. In part, this reduction in mortality might be due to improved distribution of various pneumonia vaccines as well as improved access to antibiotics. However, HAP-related ALRI mortality overall remains high in children under age 5, suggesting that improved reduction of the exposure will be essential to achieve better results.

It remains unclear how HAP exposure in early life increases the risk for severe ALRI. Two recent studies have shed further light on these health risks as a result of prenatal exposure to HAP. The first study found evidence for fetal thrombotic vasculopathy in placentas from women with high exposure to HAP during pregnancy that correlated with both increased $PM_{2.5}$ and CO measurements [38]. The second study noted that prenatal exposure from HAP impaired infant lung function, especially in girls, suggesting this may result in an increased risk of pneumonia [39]. Taken together, reducing HAP exposure during pregnancy should be a high priority to improve infant health.

These studies of prenatal HAP exposure [38, 39] provide further evidence for the enhanced vulnerability of the "first 1,000 days" (conception to 2 years of life) to environmental exposures that can impair infant health and potentially predict future risk of adult disease, a concept popularized by David Barker as the Developmental Origins of Health and Disease DOHaD [40]. Although host genetics likely play a role in response to HAP, we now appreciate that common environmental exposures can induce epigenetic marks that include DNA methylation, histone modification, chromatin structure, and short regulatory RNA [41]. For example, exposure to prenatal tobacco smoke has been shown to induce global and gene-specific DNA methylation patterns in buccal cells in children, suggesting that epigenetic changes occurred as a result of the prenatal smoke exposure [42]. Furthermore, there is evidence that sex hormones have a regulatory role on lung development and likely contribute to the smaller lungs in infant girls at birth compared to boys [43]. Adolescent girls are also more susceptible to the adverse impact of tobacco smoke on lung function [44]. The shared risk of an increased exposure of women and children to HAP, with the added potential for sex-based increased susceptibility to respiratory disease in girls, underscores a generational risk of disease risk for girls and women.

Currently there are limited available studies that identify other host risk factors for adverse health effects associated with exposure to HAP. There is strong evidence supporting an exposure–risk association with HAP, which identifies both women and children for the highest risk of adverse health consequences. We speculate that similar to outdoor particulate matter exposure and tobacco exposure that undefined host genetics likely contribute to the biological response to cookstove emissions. Ambient exposures can modify host epigenetic marks that could alter disease risk and should be considered in future studies of HAP exposure. The identification of both *susceptible* and *vulnerable* populations for the health effects of HAP exposure will require multidisciplinary studies integrating the quantification of environmental exposures and genetics/epigenetics marks together with other social and cultural determinants of health.

Regional Environmental Degradation

Fuel gathering is necessary for most of the world's poor to maintain a supply of fuel for cooking, heating, and lighting of their homes. It may reflect a range from walking long distances to collect wood in areas that are deforested, to picking up burnable debris along the roadside, to pilfering discarded chunks of coal, where available. As noted previously, fuel gathering long distances from the safety of the village places women and their accompanying children at risk from gender-based violence, as well as injuries from heavy lifting, animal attacks, and insect bites [8]. Progressive deforestation due to uncontrolled consumption of wood for fuel has enormous social, environmental, and climate consequences as the loss of trees directly impacts biodiversity with the loss of habitats for animals as well as loss of plant life required for a balanced ecosystem [4, 8, 38–40]. This in turn begins a cascade which can impair effective water management that can result in pooling of water that exacerbates environmental degradation as well as puts human subjects at risk for illness including infectious diarrhea and vector-borne disease such as malaria. As a "picture is worth a thousand words," there are several aerial photographs of national boundaries around the world that reflect differing environmental policies between countries that exist in nearly identical geographic circumstances. One such example is the island of Hispaniola in the Caribbean, which is home to both the Dominican Republic and Haiti (Fig. 17.1). Haiti relies almost entirely on charcoal as its primary energy source for residential use of solid fuels and the environmental consequences are self-evident, placing the country at major risk for repeated flooding and with a loss of its once-rich biodiversity.

Contribution of HAP to Outdoor Air Pollution

The contribution of HAP to the level and composition of outdoor air pollution remains poorly characterized. However, given the global prevalence of households that use solid fuels as the primary source of household energy needs and the

Fig. 17.1 Island of Hispaniola demonstrating the impact of deforestation in Haiti compared with Dominican Republic (DR). Haiti is the poorest country in the Western Hemisphere and shares the Island of Hispaniola with its neighbor, the DR. The population of Haiti relies on household fuel principally in the form of charcoal. There has been virtually no formal governmental policy in Haiti to protect its forests as fuel needs have increased over the past decades. The resulting deforestation results in a marked visual difference apparent in this NASA satellite photograph of the island with Haiti appearing largely barren and the DR that has federal policies regarding forest management, demonstrating a significant retention of its forests and biodiversity. http://earthobservatory.nasa.gov/IOTD/view.php?id=5352

extremely high level of HAP, it has been well established that HAP significantly contributes to outdoor air pollution. For example, one remarkable historical event is the London smog of 1952 that resulted in 12,000 excess deaths and was attributed, in part, to HAP from the myriad homes that relied on residential burning of coal [45]. Over India, about 39–59% of ambient $PM_{2.5}$ is due to residential energy sources such as cooking with coal and biomass, lighting with kerosene, and heating with solid biomass or coal (i.e. HAP) [46, 47]. Globally, about 31% of outdoor $PM_{2.5}$ is due to residential energy sources. It is recognized that black carbon is an important component of HAP. The relative contribution of HAP as a source of black carbon in outdoor air pollution, when compared to industrial emissions, can be as much as 50% or more in several south Asian countries such as India [48, 49]. Current global efforts to replace traditional cookstoves provide an opportunity to better understand the contribution of household incomplete fuel combustion on external environment.

The Surya Project, described later in this chapter, offers the first such opportunity to address this issue. Interventions on household stoves on a large-scale could have the potential co-benefits of improved indoor environment and reduce emissions that may impact outdoor air pollution.

Role of Black Carbon and Other Short-Lived Climate Forcers

Rapid and meaningful progress on slowing global warming is achievable if we recognize that global warming is caused by two different types of pollutants. The first is the long-lived carbon dioxide released by fossil fuel combustion, which stays in the atmosphere for a century to thousands of years. Most climate policies have focused on CO_2, but it will take decades and trillions of dollars to reduce emissions significantly. The world cannot afford to lose such decades. The planet has already warmed by more than 1 °C, and the resulting symptoms are being perceived in rising sea levels, melting mountain glaciers including in the Himalayas and the Alps, large-scale retreat of the Arctic sea ice and warming of the ocean waters penetrating to a depth of 1000 meters or more, and such extreme weather as droughts, floods, and heat waves. Worse, humans have already dumped enough greenhouse gases into the atmosphere to warm the planet by more than 2 °C [50]. This warming could cross the 1.5 ° C threshold by 2030 [51]. Even if we were to replace half of all fossil fuel use with renewables, the warming will continue to increase for decades, because roughly half of the CO_2 molecules persist for a century or more once released.

Fortunately, the world can get out of this seemingly hopeless predicament by broadening its focus to the *second* type of pollutants. Roughly half of total global warming is due to the release of four of these: dark soot particles called black carbon; and the gases methane, lower atmospheric ozone, and the halocarbons (HCFCS and HFCs). These pollutants stay in the atmosphere for only weeks to a few decades and hence are referred to as short-lived climate forcers. Cutting these short-lived climate warming pollutants' levels in half, which is feasible with current technologies – as UNEP's Report on Black Carbon and Ozone has recently demonstrated [52] – would quickly reduce the warming trend by 50% [53] and give the world two to four decades for the effects of CO_2 reductions to take hold. In addition, such measures can save 0.7–4.7 million lives annually and protect more than 100 million tons of crops from air pollution related damages [54]. The effects will also be quickly realized. For example, if we were to eliminate black carbon emissions by diesel vehicles today, their warming effect would disappear within weeks to a month. The cost of such reductions would not cripple economies; for example, between 1989 and 2007, California reduced its black carbon emissions by as much as 50%.

Black carbon and ozone in the atmosphere have major regional climate effects, including melting the Himalayan glaciers and decreasing the monsoon rainfall over South Asia [52, 55, 56]. In addition, both these climate warming agents lead to melting of arctic sea ice [52]. China and India have a common interest in cutting the black carbon and ozone that is melting their shared glaciers, killing millions and destroying millions of tons of crops. The United States and Europe share common interest in the Arctic where black carbon along with other short-lived pollutants are responsible for almost half of the melting ice. Modest steps that attack these short-lived climate forcers, with fast and measurable responses, are the best way to jump-start the stalled climate mitigation actions.

Improved Cookstoves or Fuels as Interventions to Reduce Health Impacts

As the majority of HAP is from cooking fires, it is reasonable to pursue interventions with more efficient stoves and fuels that will result in dramatic reductions in emissions and in exposures to family members. The challenge to date has been that although many "improved stoves" have demonstrated improved fuel efficiency with expected savings in fuels from 30% to 50%, exposure reductions have been more modest. The RESPIRE study from Guatemala suggests that exposures may need to be reduced by 50–90% to reduce the risk of pneumonia in young children [57]. These findings were the result of a controlled trial with improved built-in stoves with added chimneys that physically replaced the traditional stoves, thereby removing the risk that the families might continue to use the traditional stoves as well. Participants in the study were trained in the proper use and maintenance of the stoves and chimneys and community workers and investigators were available to monitor the intervention as well as the exposure assessments. Thus, multiple factors reinforced the correct use of the intervention to achieve the results of dramatic exposure reduction.

Since the RESPIRE trial, there have been significant efforts to scale up improved cookstove interventions for thousands of households to determine if the same health benefits can be achieved. The results to date suggest that in large-scale studies in Malawi and Nepal, using improved biomass cookstoves, there is no significant reduction in ALRI among children [58, 59]. Some problems with larger studies relate to insufficient resources for the same level of exposure assessment or detailed household follow-up to ensure proper stove use and to avoid stove stacking, as might occur more easily in smaller studies. Stove stacking persists in several field studies in different parts of the world, suggesting that improved stoves and fuels do not meet all the needs of households in LMIC [60–62]. Therefore, the benefits from achieving substantial reduction in HAP from solid fuels using improved cookstoves that might approach the WHO indoor air quality guideline are modest [63, 64].

It is even more daunting to consider how to best achieve improved health outcomes from the implementation of cookstoves that are sold in local markets but do not have the support systems in place, as may occur in a controlled trial that reinforces proper stove adoption and use. NGOs or government programs working closely with communities can develop village-level training and educational programs to provide many of the same support systems, if well planned and implemented. There are many, perhaps thousands, of cookstove types available at local markets in LMIC. Examples of many of these stoves have been tested for various performance measures including fuel and combustion efficiencies (Fig. 17.2) [9, 15, 62]. Typical "rocket stoves" achieve reasonable fuel efficiency with reductions in fuel use of about 30%, but the exposure reductions will be far less than the 50–90% noted in the RESPIRE trial to achieve risk reduction for acute pneumonia. Additions of fans to the rocket stoves, the so-called "fan stoves," offer greater efficiencies for both fuel use and emissions [15, 62]. The Philips stove was an early example of a higher quality commercially available fan stove produced at scale, but is no longer in production (Fig. 17.2). Many of the liquid fuel–based stoves, such as LPG, propane, biogas or alcohol, offer the opportunity for being ultraclean with exposure reductions greater than 90% [15, 62–64]. There are also "top loading updraft" (TLUD), natural draft, and gasifier stoves, all of which offer opportunity for marked reduction in emissions [15, 62]. More recently, an implementation study of

Fig. 17.2 Display of multiple cookstove types used around the world. This photograph shows the wide variety of cookstoves using solid fuels in LMIC including: Open "3-stone" fire, wood fuel. Berkeley Darfur, wood fuel. Envirofit G-3300, wood fuel. Onil, wood fuel. Philips HD4008, wood fuel. Philips HD4012, wood fuel. Sampada, wood fuel. StoveTec GreenFire, wood fuel. Upesi Portable, wood fuel. GERES, charcoal fuel. Gyapa, charcoal fuel. Jiko, ceramic, charcoal fuel. Jiko, metal, charcoal fuel. KCJ Standard, charcoal fuel. Kenya Uhai, charcoal fuel. StoveTec prototype, charcoal fuel. Belonio Rice Husk Gasifier, rice hull fuel. Mayon Turbo, rice hull fuel. Oorja, biomass pellet fuel. StoveTec TLUD prototype, wood pellet fuel. Jinqilin CKQ-80I, corn cob fuel. Protos, plant oil fuel. The photograph is courtesy of James Jetter, U.S. EPA, National Risk Management Research Laboratory Air Pollution Prevention and Control Division, Stove Testing Center, Research Triangle Park, North Carolina, USA

pellet-fed gasifier stoves in Rwanda showed remarkable reduction in emissions comparable to ISO Tier 4 clean fuel stoves [65]. These stoves are leased to households based on trading local biomass required for pellet production via the Inyenyeri business model to provide sustainable clean energy use in local households [65, 66].

Alternatively, solar-based stoves offer the advantage of zero emissions and no fuel costs [62]; however, there can be issues with the timing of cooking (early morning and evening) when sunlight is not available, or during rainy seasons when alternatives are needed, or, finally, the adoption of solar cooking from traditional cooking methods is too great a change for some families. Nonetheless, solar cookers are a viable alternative as the primary means of cooking or as a supplement to an improved solid fuel or LPG cookstove. An additional strategy that can extend the cooking cycle without additional energy input is heat-retention cooking [11]. This method uses devices such as a "haybox" that is insulated and houses a cooking pot recently removed from a cookstove that limits the loss of heat and permits the food to continue to cook. Such an integrated approach to cooking makes sense from both an energy usage perspective and from a health and climate perspective.

The challenge facing investigators and implementers (mostly NGOs, manufacturers, and governments) is to select cookstoves that are affordable, user-friendly, and acceptable to households and yet are sufficiently clean as to achieve dramatic reductions in both emissions and exposures. Exposure reduction of 50–90% will likely be necessary in most settings to approximate WHO air quality guidelines [64] and to significantly improve health outcomes [57]. Currently, the commercially available stoves most likely to provide both reduced emissions and exposures from the use of solid fuels are limited. As noted previously, there is a rapidly emerging class of stoves such as pellet-fed gasifiers, TLUDs, or natural draft stoves that are also available but not necessarily worldwide as yet. Commercially available charcoal stoves typically have lower PM emissions than rocket or traditional stoves but can create dangerous levels of CO as families are less aware of the dangers absent the higher PM emissions. As noted previously, liquid fuels such as LPG, propane, biogas, and alcohol offer very low emissions but the cost of fuels often represent a financial burden to families in poverty. The key to any of these strategies is to develop a monitoring and evaluation system that documents stove use and, where possible, exposure levels in and around the household. The Stove Unit Monitoring System developed by Berkeley Air offers one approach to quantitatively assess stove use for both improved and traditional stoves [60]. Personal and area exposure monitoring on a selected basis are also essential to determining whether improved stoves or fuels are delivering the impacts expected.

With the disappointing results from implementation studies of improved biomass cookstoves to reduce HAP and adverse health effects, there is increased enthusiasm to scale up clean fuels such as LPG or electricity use, "making the clean available" [67]. Embraced by the government of India, the "Give it Up" campaign asked well-to-do households to give up their LPG subsidy so that poorer households could be further subsidized to adopt LPG as a clean household fuel [68]. Tens of millions of households in India have accepted the "Give it Up" challenge to permit scaling of household LPG use. On another front, the National Institutes of Health (NIH) together with the Gates Foundation and others have launched the Household Air Pollution Investigation Network (HAPIN) to study the implementation of LPG stoves and fuels across four countries (Rwanda, India, Peru, and Guatemala) to assess technology adoption and health benefits [69]. All in all, this shift to the major implementation of clean fuels offers the best hope for reducing HAP and improving health outcomes in the next decades.

The Surya Project as a Model of Potential Interventions to Reduce Climate Impacts

Black carbon and ozone, two potent short-lived climate forcers, are also great targets for developing nations because they have other known consequences apart from their health effects. They contribute to global warming (about 25–50% of the CO_2 warming as of 2005). In addition, they perturb regional climate in major ways. The interception of sunlight by black carbon leads to about 30–50% of the warming over the elevated Himalayan-Tibetan region (Fig. 17.3) [52, 70, 71]. The interception of sunlight by black carbon and organic carbon particles from cooking smoke

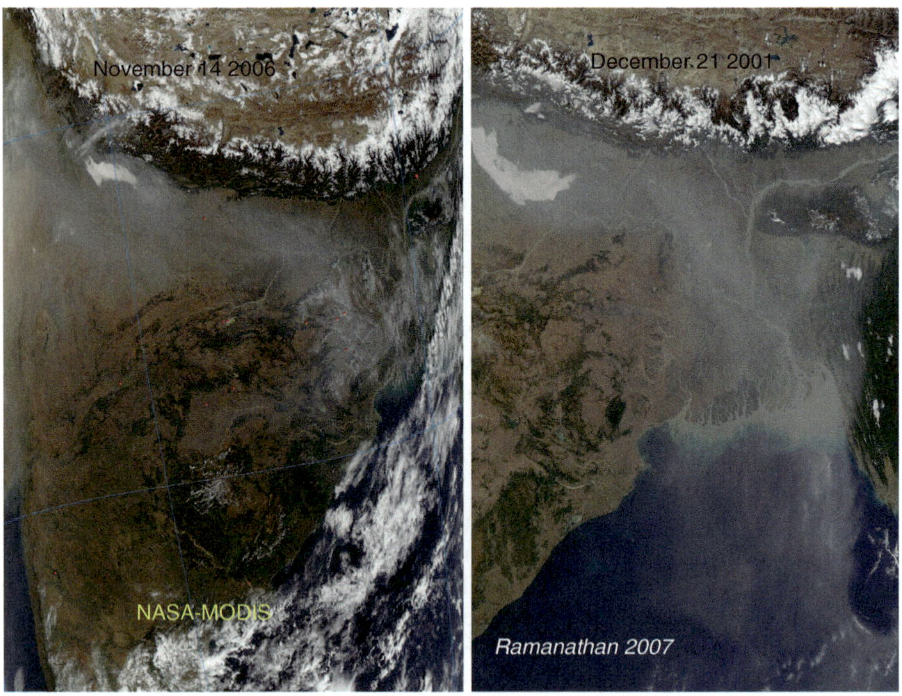

Fig. 17.3 Evidence of warming over the elevated Himalayan-Tibetan region comparing aerial photographs of 2001 with 2006. The interception of sunlight by black carbon leads to about 30–50% of the warming effect of this region with evidence for deglaciation

also weakens the monsoon circulation and reduces monsoon rainfall [52, 72, 73]. Both of these pollutants lead to widespread destruction of crops, both directly [52] and indirectly, through their effects on monsoon precipitation [74].

The world has an unprecedented opportunity to mitigate some of the disastrous effects of black carbon and ozone on climate, agriculture, water, and health with a simple act: replacing traditional cookstoves with energy-efficient and pollution-free cooking technologies. This work has already begun with international initiatives like the Clean Cooking Alliance, but challenges remain. The numerous cookstove initiatives that have taken place all over the world have demonstrated time and again that catalyzing widespread adoption of such clean cooking technologies will require innovative and affordable solutions.

This is where Project Surya, an internationally recognized cookstove project sponsored by the United Nations Environment Programme, comes in [70, 75]. Project Surya originated as a collaborative effort by three institutions: the University of California at San Diego, Nexleaf Analytics (an NGO based in Los Angeles, USA), and TERI (an NGO based in New Delhi, India). Surya is now led by Nexleaf and supported by UNEP's Climate and Clean Air Coalition, Qualcomm Wireless Reach, private foundations, and donors. Its goal is to demonstrate scientifically the environmental and health benefits of introducing clean cooking technologies and ultimately provide a rigorous evidence base for large-scale scaling of clean cooking solutions. Project Surya will use cell phones, instrument towers, and satellites, and will empower village youth to work with world-class experts in documenting the impacts.

A pilot phase was successfully completed in 2010 in several villages in one of the poorest and most polluted regions in the Indo-Gangetic plains. It has already achieved some ambitious and measurable outcomes including: documenting the connection between indoor air pollution from cooking and ambient outdoor pollution levels [49] and identifying improved cooking technologies that reduce pollution significantly [76].

In its next phase, in October 2014, Surya undertook a deployment of clean cooking in over 5000 homes. The women were provided with bank loans to purchase the stoves. One unique and revolutionary feature of this phase was to integrate the cookstoves in 456 households with state-of-the-art remote temperature sensors [77] developed by Nexleaf to quantify the usage of the stoves. The findings were published in Nature Climate Change [78]. The sensor data revealed major new insights regarding adoption: (i) It verified the feasibility of measuring the impacts of larger-scale interventions using cell phones [79]; (ii) It showed that surveys gathered from women to quantify usage were unreliable and bear no correlation with actual usage; (iii) The intervention demonstrated, for the first time, that women will purchase clean cookstoves when the affordability barrier is mitigated; (iv) The analyses showed that a purchase is not equated with sustained adoption.

While this was not the first study to show that clean cookstove adoption can decline over time [19], this was the first study to combine sensor data plus surveys to understand why that decline happens and uncovered two of the main barriers to sustained adoption: (A) usability and (B) durability. These findings led to major recommendations for the sector for truly understanding the criteria for successful adoption of new technologies: criteria 1: user-friendliness, and criteria 2: building after sales service and maintenance of the clean cookstove as part of the supply chain.

More recently, Project Surya followed up on these recommendations in a new, smaller pilot in one village in Odisha to discover if a usable and durable stove would lead to sustained adoption. During a 2-year period extending from April 2017 to April 2019, the project has witnessed sustained adoption above 90% among all households [80].

To explore if the methodology can work in multiple contexts, another parallel pilot test was launched in Abuja, Nigeria, with support from UNEP Climate and Clean Air Coalition and in partnership with local social enterprises. The pilot is taking place in 50 households and 5 villages evaluating a basket of cooking solutions including ethanol, LPG, and popular biomass cookstove models in Africa. The aim of these pilot studies is to understand factors that contribute to sustained adoption, since sustained adoption is the precursor for achieving desired climate and health impacts. With the understanding and insights gained from the series of pilot studies, Nexleaf will embark in partnerships with funders, commercial partners, and donors, on two major scale-up studies: one with 1000 households followed by a 10,000 household deployment.

Our recent data has also shown that the measured black carbon concentrations are three to five times higher than the concentrations simulated by climate models, making it all the more urgent to take action now to target it and other short-lived climate forcers [81]. Fortunately, there is a great success story to draw upon. The enormous greenhouse effect of CFC-11 and CFC-12 was discovered only in 1975 [82]. CFCs were regulated by the 1987 Montreal Protocol, because of their negative effects on stratospheric ozone, but if this had not happened, they would have added enough heat energy to warm the planet by about 1 °C or more.

Value of Co-benefits for Human Health and Climate

Improved and more efficient stoves or fuels can significantly reduce stove emissions that reduce HAP and also reduce outdoor air pollution that contributes to atmospheric changes that influence the climate [83]. Simply displacing stove emissions through a chimney or flue without improving stove or fuel efficiencies not only continues to place a family or village at risk for HAP as the pollution re-enters the home from the outside, its contribution to atmospheric pollution remains unabated. There are additional strategies needed to augment household exposure reduction. Obviously, the technology used to reduce HAP in any intervention being studied is critical to the impact on health and climate outcomes. However, the new technology must be acceptable to the user as significant reductions in HAP require exclusive use of the new stoves or fuels by the user, as opposed to shared use with the traditional means of cooking that can generate emissions that overwhelm the benefits of a new stove or fuel. There has been too little focus on the important role that human behavior and cultural traditions play in household approaches to energy use. When large-scale implementation programs with improved stoves or fuels are being conducted, there is a need to measure the impact on household and outdoor exposures, either directly or indirectly, that reflect the impact of the improved stove or fuel. Absent such measurements, the impact on human health, environment, and climate remains unknown and speculative. It is the responsibility of investigators, implementers, communities, and governments to work together to validate that major implementation programs with improved cooking solutions have the intended effects; and if not, make the necessary changes in the implementation to ensure that the health of human subjects in poverty and the health of the planet are finally realized as true co-benefits.

Summary

HAP is an exposure of poverty. The success in having a sustainable reduction in HAP requires an understanding of the traditions and culture of the family as well as the causes of poverty that place the family at the bottom of the energy ladder. An integrated approach to reducing HAP with efforts also aimed at correcting other poverty related issues is challenging but offers the hope for addressing root causes of poverty in a community setting that provides a more comprehensive and sustainable approach to improving health, the environment, and, ultimately, the global climate [84]. From one perspective, research that provides detailed exposure-responses to HAP may seem superfluous to the obvious need for poor families to breathe cleaner air at home. One can argue that we already have decades of information on the health risks from outdoor air pollution [85] or the products of incomplete combustion from tobacco smoke [86] and so further research is not needed. However, there is a compelling need to know how clean a stove or fuel must be to significantly reduce health risks, so that with proper use, major implementation of such new technology may reasonably provide the intended benefits for improved health, the regional environment, and the global climate. The alternative of providing regional electrification or widespread use of clean fuels such as LPG, once deemed unrealistic for the world's poor, is seemingly more possible in the near future. Addressing the key scientific and economic gaps related to HAP may soon provide sufficiently clean household air for families living in poverty, such that diseases are prevented, a healthier lifestyle is promoted, and a reduction in global warming trends buys more time for a planet in peril from climate change.

Acknowledgments We would like to thank James Jetter of the U.S. EPA of the National Risk Management Research Laboratory, Air Pollution Prevention and Control Division, Stove Testing Center, Research Triangle Park, North Carolina, USA for his helpful review and comments of the chapter as well as his contribution of the photograph in Fig.17.2.

References

1. Martin WJ II, Glass RI, Balbus JM, Collins FS. A major environmental cause of death. Science. 2011;334(6053):180.
2. Bonjour S, Adair-Rohani H, Wolf J, Bruce NG, Mehta S, Pruss-Ustun A, Lahiff M, Rehfuess EA, Mishra V, Smith KR. Solid fuel use for household cooking: country and regional estimates for 1980–2010. Environ Health Perspect. 2013;121:784–90.
3. Forouzanfar MH, Afshin A, Alexander LT, Anderson HR, Bhutta ZA, Biryukov S, et al. Global, regional, and national comparative risk assessment of 29 behavioural, environmental and occupational, and metabolic risks or clusters of risks, 1990–2015: a systematic analysis for the global burden of disease study 2015. Lancet. 2016;388:1659–724.
4. The International Bank for Reconstruction and Development. Household cookstoves, environment, health and climate change: a new look at an old problem; 2011. http://documents.worldbank.org/curated/en/732691468177236006/pdf/632170WP0House00Box0361508B0PUBLIC0.pdf
5. US Climate Change Science Program. Climate projections based on emissions scenarios for long-lived and short-lived radiatively active gases and aerosols; 2008. http://www.climatescience.gov/Library/sap/sap3-2/final-report/#finalreport
6. Lacey FG, Henze DK, Lee CJ, van Donkelaar A, Martin RV. Transient climate and ambient health impacts due to national solid fuel cookstove emissions. PNAS. 2017;6:1269–74. https://doi.org/10.1073/pnas.
7. Calvin WH. The ascent of mind: ice age climates and the evolution of intelligence. New York: Bantam Books; 1990.
8. Patrick E. Sexual violence and firewood collection in Darfur. Forced Migr Rev. 2007;27:40–1. https://www.fmreview.org/sites/fmr/files/FMRdownloads/en/sexualviolence/patrick.pdf. Accessed 7 June 2019.
9. Damberger D. TED lecture: what happens when an NGO admits failure?; 2011. http://www.ted.com/talks/david_damberger_what_happens_when_an_ngo_admits_failure.html
10. Chen M, Bonner C, Carre F. Organizing informal workers: benefits, challenges and successes. UNDP Human Development Report Office; 2015. http://hdr.undp.org/en/content/organizing-informal-workers-benefits-challenges-and-successes
11. Winiarski L, Aprovecho Research Center, Partnership for Clean Indoor Air (PCIA), Shell Foundation. Design principles for wood burning cook stoves; 2005. aprovecho.org/?paybox_id=77
12. Ramakrishna J, Durgaprasad MB, Smith KR. Cooking in India: the impact of improved stoves on indoor air quality. Environ Int. 1989;15:341–52. https://doi.org/10.1016/0160-4120(89)90047-0.
13. Sinton JE, Smith KR, Peabody JW, Yaping L, Xiliang Z, Edwards R. An assessment of programs to promote improved household stoves in China. Energy Sustain Dev. 2004;8:33–52. https://doi.org/10.1016/S0973-0826(08)60465-2.
14. Venkataraman C, Sagar AD, Habib G, Lam N, Smith KR. The Indian national initiative for advanced biomass cookstoves: the benefits of clean combustion. Energy Sustain Dev. 2010;14:63–72. https://doi.org/10.1016/j.esd.2010.04.005.
15. MacCarty N, Still D, Ogle D. Fuel use and emissions performance of fifty cooking stoves in the laboratory and related benchmarks of performance. Energy Sustain Dev. 2010;14:161–71. https://doi.org/10.1016/j.esd.2010.06.002.
16. US Environmental Protection Agency. $PM_{2.5}$ NAAQS implementation; 2008. http://www.epa.gov/ttnnaaqs/pm/pm25_index.html. Accessed 27 Jan 2012.
17. Ostro B, Lipsett M, Reynolds P, Goldberg D, Hertz A, Garcia C, et al. Long-term exposure to constituents of fine particulate air pollution and mortality: results from the California teachers study. Environ Health Perspect. 2010;118:363–9. https://doi.org/10.1289/ehp.0901181.
18. Kituyi E, Marufu L, Wandiga SO, Jumba IO, Andreae MO, Helas G. Carbon monoxide and nitric oxide from biofuel fires in Kenya. Energy Convers Manag. 2001;42:1517–42. https://doi.org/10.1016/S0196-8904(00)00158-8.

19. Pillarisetti A, Vaswani M, Jack D, Balakrishnan K, Bates MN, Arora NK, Smith KR. Patterns of stove usage after introduction of an advanced cookstove: the long-term application of household sensors. Environ Sci Technol. 2014;48:14525–33. https://doi.org/10.1021/es504624c.
20. Better burning, better breathing: cleaner stoves. Science and research at the U.S. Environmental Protection Agency EPA progress report 2010; 2010. https://www.epa.gov/sites/production/files/2013-12/documents/annual-report-2010.pdf. Accessed 3 June 2019.
21. Clean Cooking Alliance. https://www.cleancookingalliance.org/about/our-mission/index.html. Accessed 7 June 2019.
22. Clean Cooking Alliance International Testing Centers. https://www.cleancookingalliance.org/technology-and-fuels/testing/centers.html. Accessed 7 June 2019.
23. Jetter JJ, Kariher P. Solid-fuel household cook stoves: characterization of performance and emissions. Biomass Bioenergy. 2009;33(2):294–305. https://doi.org/10.1016/j.biombioe.2008.05.014.
24. Berkeley Air Monitoring Group. Stove performance inventory report; 2012. http://berkeleyair.com/wp-content/publications/SPT_Inventory_Report_v3_0.pdf. Accessed 7 June 2019.
25. International Organization for Standardization. Clean cookstoves and clean cooking solutions – harmonized laboratory test protocols – part 3: voluntary performance targets for cookstoves based on laboratory testing. https://www.iso.org/standard/73935.html. Accessed 7 June 2019.
26. Jetter J, Zhao Y, Smith KR, Khan B, Yelverton T, DeCarlo P, Hayes MD. Pollutant emissions and energy efficiency under controlled conditions for household biomass cookstoves and implications for metrics useful in setting international test standards. Environ Sci Technol. 2012;46:10827–34. https://doi.org/10.1021/es301693f.
27. Landringan PJ, Fuller R, Acosta NJR, Adeyi O, Arnold R, Basu N, et al. The Lancet commission on pollution and health. Lancet. 2018;391:462–512.
28. Balakrishnan K, Sambandam S, Ramaswamy P, Mehta S, Smith KR. Exposure assessment for respirable particulates associated with household fuel use in rural districts of Andhra Pradesh, India. J Expo Anal Environ Epidemiol. 2004;14(Suppl 1):S14–25. https://doi.org/10.1038/sj.jea.7500354.
29. The Health Effects Institute. Outdoor air pollution and health in the developing countries of Asia: a comprehensive review. Special report 18; 2010. https://www.healtheffects.org/system/files/SR18AsianLitReview.pdf. Accessed 7 June 2019.
30. World Health Organization. Burden of disease from the joint effects of household and ambient air pollution for 2012; 2012. https://www.who.int/phe/health_topics/outdoorair/databases/AP_jointeffect_BoD_results_March2014.pdf
31. Hosgood HD 3rd, Boffetta P, Greenland S, Lee YC, McLaughlin J, Seow A, et al. In-home coal and wood use and lung cancer risk: a pooled analysis of the international lung cancer consortium. Environ Health Perspect. 2010;118(12):1743–7. https://doi.org/10.1289/ehp.1002217.
32. Martin WJ, Glass RI, Araj H, Balbus J, Collins FS, Curtis S, et al. Household air pollution in low-and middle-income countries: health risks and research priorities. PLoS. 2013;10:e1001455. https://doi.org/10.1371/journal.pmed.1001455.
33. Peck MD, Kruger GE, van der Merwe AE, Godakumbura W, Ahuja RB. Burns and fires from non-electric domestic appliances in low- and middle-income countries. Part I. The scope of the problem. Burns. 2008;34(3):303–11. https://doi.org/10.1016/j.burns.2007.08.014.
34. Pinkerton KE, Harbaugh M, Han MK, Jourdan Le Saux C, Van Winkle LS, Martin WJ 2nd, et al. Women and lung disease. Sex differences and global health disparities. Am J Respir Crit Care Med. 2015;192:11–6. https://doi.org/10.1164/rccm.201409-1740PP.
35. Aryal S, Diaz-Guzman E, Mannino DM. Influence of sex on chronic obstructive pulmonary disease risk and treatment outcomes. Int J COPD. 2014;9:1145–54. https://doi.org/10.2147/COPD.S54476.
36. Gan WQ, Paul Man SF, Postma DS, Camp P, Sin DD. Female smokers beyond the perimenopausal period are at increased risk of chronic obstructive pulmonary disease: a systematic review and meta-analysis. Respir Res. 2006;7:52. https://doi.org/10.1186/1465-9921-7-52.

37. American Lung Association. Taking her breath away: the rise of COPD in women; 2013. https://www.lung.org/assets/documents/research/rise-of-copd-in-women-full.pdf. Accessed 5 June 2019.
38. Wylie BJ, Matechi E, Kishashu Y, Fawzi W, Premji Z, Coull BA, et al. Placental pathology associated with household air pollution in a cohort of pregnant women from Dar es Salaam. Tanzania Environ Health Perspect. 2017;125:134–40. https://doi.org/10.1289/EHP256.
39. Lee AG, Kaali S, Quinn A, Delimini R, Burkart K, Opoku-Mensah J, et al. Prenatal household air pollution is associated with impaired infant lung function with sex-specific effects. Evidence from GRAPHS, a cluster randomized cookstove intervention trial. Am J Respir Crit Care Med. 2019;199:738–46. https://doi.org/10.1164/rccm.201804-0694OC.
40. Barker DJP. The developmental origins of adult disease. J Am Coll Nutr. 2004;6:588S–95S.
41. Hanson M, Godfrey KM, Lillycrop KA, Burdge GC, Gluckman PD. Developmental plasticity and developmental origins of non-communicable disease: theoretical considerations and epigenetic mechanisms. Prog Biophys Mol Biol. 2011;106(1):272–80. https://doi.org/10.1016/j.pbiomolbio.2010.12.008.
42. Breton CV, Byun HM, Wenten M, Pan F, Yang A, Gilliland FD. Prenatal tobacco smoke exposure affects global and gene-specific DNA methylation. Am J Respir Crit Care Med. 2009;180:462–7. https://doi.org/10.1164/rccm.200901-0135OC.
43. Carey MA, Card JW, Voltz JW, Arbes SJ Jr, Germolec DR, Korach KS, Zeldin DC. It's all about sex: male-female differences in lung development and disease. Trends Endocrinol Metab. 2007;18:308–13. https://doi.org/10.1016/j.tem.2007.08.003.
44. Gold DR, Wang X, Wypij D, Speizer FE, Ware JH, Dockery DW. Effects of cigarette smoking on lung function in adolescent boys and girls. N Engl J Med. 1996;335:931–7. https://doi.org/10.1056/NEJM199609263351304.
45. Bell ML, Davis DL. Reassessment of the lethal London fog of 1952: novel indicators of acute and chronic consequences of acute exposure to air pollution. Environ Health Perspect. 2001;109(SUPPL. 3):389–94. https://doi.org/10.1289/ehp.01109s3389.
46. Sharma S, Nunez L, Ramanathan V. Atmospheric brown clouds. Oxford research encyclopedia of environmental science; 2016. https://doi.org/10.1093/acrefore/9780199389414.013.47.
47. Lelieveld J, Evans JS, Fnais M, Giannadaki D, Pozzer A. The contribution of outdoor air pollution sources to premature mortality on a global scale. Nature. 2015;525(7569):367.
48. Gustafsson Ö, Kruså M, Zencak Z, Sheesley RJ, Granat L, Engström E, Praveen PS, Rao PS, Leck C, Rodhe H. Brown clouds over South Asia: biomass or fossil fuel combustion? Science. 2009;323(5913):495–8. https://doi.org/10.1126/science.1164857.
49. Rehman IH, Ahmed T, Praveen PS, Kar A, Ramanathan V. Black carbon emissions from biomass and fossil fuels in rural India. Atmos Chem Phys. 2011;11(14):7289–99. https://doi.org/10.5194/acpd-11-10845-2011.
50. Ramanathan V, Feng Y. On avoiding dangerous anthropogenic interference with the climate system: formidable challenges ahead. Proc Natl Acad Sci U S A. 2008;105:14245–50. https://doi.org/10.1073/pnas.0803838105.
51. Xu Y, Ramanathan V, Victor DG. Global warming will happen faster than we think. Nature. 2018;564:30–2. https://doi.org/10.1038/d41586-018-07586-5.
52. United Nations Environment Programme. Integrated assessment of black carbon and tropospheric ozone: summary for decision makers; 2011. https://www.ccacoalition.org/en/resources/integrated-assessment-black-carbon-and-tropospheric-ozone. Accessed 7 June 2019.
53. Ramanathan V, Xu Y. The Copenhagen accord for limiting global warming: criteria, constraints, and available avenues. Proc Natl Acad Sci U S A. 2010;107:8055–62. https://doi.org/10.1073/pnas.1002293107.
54. Shindell D, Kuylenstierna JCI, Vignati E, van Dingenen R, Amann M, Klimont Z, et al. Simultaneously mitigating near-term climate change and improving human health and food security. Science. 2012;335:183–9. https://doi.org/10.1126/science.1210026.
55. Ramanathan V, Carmichael G. Global and regional climate changes due to black carbon. Nat Geosci. 2008;1(4):221–7.

56. Pontifical Academy of Sciences. Fate of mountain glaciers in the anthropocene. A report by the working group commissioned by the Pontifical Academy of Sciences; 2011. http://www.casinapioiv.va/content/dam/accademia/pdf/glaciers.pdf. Accessed 7 June 2019.
57. Smith KR, McCracken JP, Weber MW, Hubbard A, Jenny A, Thompson LM, et al. Effect of reduction in household air pollution on childhood pneumonia in Guatemala (RESPIRE): a randomised controlled trial. Lancet. 2011;378:1717–26. https://doi.org/10.1016/S0140-6736(11)60921-5.
58. Mortimer K, Ndamala CB, Naunje AW, Malava J, Katundu C, Weston W, et al. A cleaner burning biomass-fuelled cookstove intervention to prevent pneumonia in children under 5 years old in rural Malawi (the cooking and pneumonia study): a cluster randomised controlled trial. 2017;389:P167–75. https://doi.org/10.1016/S0140-6736(16)32507-7.
59. Tielsch JM, Katz J, Khatry SK, Shrestha L, Breysse P, Zeger S, et al. Effect of an improved biomass stove on acute lower respiratory infections in young children in rural Nepal: a cluster-randomised, step-wedge trial. Lancet Glob Health. 2016;4(Special Issue):S19. https://doi.org/10.1016/S2214-109X(16)30024-9.
60. Balakrishnan K, Sambandam S, Ghosh S, Mukhopadhyay K, Vaswani M, Arora NK, et al. Household air pollution exposures of pregnant women receiving advanced combustion cookstoves in India: implications for intervention. Ann Glob Health. 2015;81:375–85. https://doi.org/10.1016/j.aogh.2015.08.009.
61. Gould CF, Schlesinger S, Toasa AO, Thurber M, Waters WF, Graham JP, Jack DW. Government policy, clean fuel access, and persistent fuel stacking in Ecuador. Energy Sustain Dev. 2018;46:11–122. https://doi.org/10.1016/j.esd.2018.05.009.
62. Serrano-Medrano M, Garcia-Bustamante C, Berrueta VM, Martinez-Bravo R, Ruiz-Garcia VM, Ghilardi A, Masera O. Promoting LPG, clean woodburning cookstoves or both? Climate change mitigation implications of integrated household energy transition scenarios in rural Mexico. Environ Res Lett. 2018;13 https://doi.org/10.1088/1748-9326/aad5b8.
63. Pope D, Bruce N, Dherani M, Jagoe K, Rehfuess E. Real-life effectiveness of 'improved' stoves and clean fuels in reducing $PM_{2.5}$ and CO: systematic review and meta-analysis. Environ Int. 2017;101:7–18. https://doi.org/10.1016/j.envint.2017.01.012.
64. Bruce N, Pope D, Rehfuess E, Balakrishnan K, Adair-Rohani H, Dora C. WHO indoor air quality guidelines on household fuel combustion: strategy implications of new evidence on interventions and exposure-risk functions. Atmos Environ. 2015;106:451–7. https://doi.org/10.1016/j.atmosenv.2014.08.064.
65. Jagger P, Das I. Implementation and scale-up of a biomass pellet and improved cookstove enterprise in Rwanda. Energy Sustain Dev. 2018;46:32–41. https://doi.org/10.1016/j.esd.2018.06.005.
66. Champion WM, Grieshop AP. Pellet-fed gasifier stoves approach gas-stove like performance during in-home use in Rwanda. Environ Sci Technol. 2019; https://doi.org/10.1021/acs.est.9b00009.
67. Smith KR, Sagar A. Making the clean available: escaping India's chulha trap. Energy Policy. 2014;75:410–4. https://doi.org/10.1016/j.enpol.2014.09.024.
68. Smith KR, Sagar AD. LPG subsidy: analysing the 'give it up' scheme. The Economic Times; 2016. https://economictimes.indiatimes.com/blogs/et-commentary/lpg-subsidy-analysing-the-give-it-up-scheme/. Accessed 7 June 2019.
69. National Institutes of Health. Household air pollution investigation network (HAPIN). https://commonfund.nih.gov/globalhealth/hapinresources. Accessed 7 June 2019.
70. Ramanathan V, Balakrishnan, K. Project surya: reduction of air pollution and global warming by cooking with renewable sources – a controlled and practical experiment in rural India: a white paper; 2007. http://www.indiaenvironmentportal.org.in/files/Surya-WhitePaper.pdf. Accessed 7 June 2019.
71. Flanner MG, Zender CS, Hess PG, Mahowald NM, Painter TH, Ramanathan V, Rasch PJ. Springtime warming and reduced snow cover from carbonaceous particles. Atmos Chem Phy Discuss. 2008;8:19819–59. https://doi.org/10.5194/acp-9-2481-2009.

72. Ramanathan V, Chung C, Kim D, Bettge T, Buja L, Kiehl JT, et al. Atmospheric brown clouds: impacts on South Asian climate and hydrological cycle. Proc Natl Acad Sci U S A. 2005;102:5326–33. https://doi.org/10.1073/pnas.0500656102.
73. Meehl GA, Arblaster JM, Collins WD. Effects of black carbon aerosols on the Indian monsoon. J Clim. 2008;21:2869–82. https://doi.org/10.1175/2007JCLI1777.1.
74. Auffhammer M, Ramanathan V, Vincent JR. Integrated model shows that atmospheric brown clouds and greenhouse gases have reduced rice harvests in India. Proc Natl Acad Sci U S A. 2006;103:19668–72. https://doi.org/10.1073/pnas.0609584104.
75. Ramanathan V, Ramanathan, N. An unprecedented opportunity. Our Planet. 2011:28–29. http://www.igsd.org/documents/Ramanathan.pdf. Accessed 7 June 2019.
76. Kar A, Siva P, Suresh R, Rehman IH, Singh L, Singh VK, et al. Real-time assessment of black carbon pollution in Indian households due to traditional and improved biomass cook stoves. Environ Sci Technol. 2012;46:2993–3000. https://doi.org/10.1021/es203388g.
77. Graham EA, Patange O, Lukac M, Singh L, Kar A, Rehman IH, Ramanathan N. Laboratory demonstration and field verification of a wireless cookstove sensing system (WiCS) for determining cooking duration and fuel consumption. Energy Sustain Dev. 2014;23:59–67. https://doi.org/10.1016/j.esd.2014.08.001.
78. Ramanathan T, Ramanathan N, Mohanty J, Rehman IH, Graham E, Ramanathan V. Wireless sensors linked to climate financing for globally affordable clean cooking. Nat Clim Chang. 2017;7:44–7. https://doi.org/10.1038/nclimate3141.
79. Ramanathan N, Lukac M, Ahmed T, Kar A, Praveen PS, Honles T, et al. A cellphone based system for large-scale monitoring of black carbon. Atmos Environ. 2011;45:4481–7. https://doi.org/10.1016/j.atmosenv.2011.05.030.
80. Nexleaf Analytics and Tata Trusts. Beyond monitoring and evaluation: tracking improved cookstove adoption continuously and over time to achieve lasting success; 2018. https://nexleaf.org/reports/joint-learning-series/beyond-monitoring-and-evaluation.pdf. Accessed 7 June 2019.
81. Praveen PS, Ahmed T, Kar A, Rehman IH, Ramanathan V. Link between local scale BC emissions and large scale atmospheric solar absorption. Atmos Chem Phys. 2011;11:21319–61. https://doi.org/10.5194/acp-12-1173-2012.
82. Ramanathan V. Greenhouse effect due to chlorofluorocarbons: climatic implications. Science. 1975;190:50–2. https://doi.org/10.1126/science.190.4209.50.
83. Anenburg SC, Balakrishnan K, Jetter J, Masera O, Mehta S, Moss J, Ramanthan V. Cleaner cooking solutions to achieve health, climate, and economic cobenefits. Environ Sci Technol. 2013;47:3944–52. https://doi.org/10.1021/es304942e.
84. Bodereau PN. Peruvian highlands, fume-free. Science. 2011;334:157. https://doi.org/10.1126/science.1212526.
85. Pope CA 3rd, Burnett RT, Thun MJ, Calle EE, Krewski D, Ito K, Thurston GD. Lung cancer, cardiopulmonary mortality, and long-term exposure to fine particulate air pollution. JAMA. 2002;287:1132–41. https://doi.org/10.1001/jama.287.9.1132.
86. U.S. Department of Health and Human Services. How tobacco smoke causes disease. The biology and behavioral basis for smoking-attributable disease: a report of the surgeon general. U.S. Department of Health and Human Services CfDCaP, National Center for Chronic Disease Prevention and Health Promotion, Office on Smoking and Health, editors. Atlanta: U.S. Department of Health and Human Services, Centers for Disease Control and Prevention, National Center for Chronic Disease Prevention and Health Promotion, Office on Smoking and Health; 2010.

Chapter 18
Biomass Fuel and Lung Diseases: An Indian Perspective

Rajendra Prasad, Rajiv Garg, and Nikhil Gupta

Introduction

Biomass fuels are the primary source of domestic energy for about half of the world's population [1]. Biomass fuel consists of firewood, dung cakes, agricultural crop residues (straw, grass, shrubs, etc.), coal fuels, and kerosene. Together, they supply 75% of the domestic energy in India. The rest of the country relies on cleaner fuels, namely, liquefied petroleum gas (LPG) and natural gas [2]. The biomass fuels and coal are sources of high-level indoor air pollution as these are used for cooking and heating on traditional stoves or open fires, which results in incomplete combustion and heavy smoke production.

Biomass Fuel Use in India

It is estimated that about 32% of the total primary energy use in the country is derived from biomass and three-quarters of Indian households use biomass fuel as the primary means for domestic cooking [3]. Ninety percent of the rural households and 32% of the urban households cook their meals on a biomass stove. Only 25% of the cooking is done with cleaner gases. Ninety percent of the households using biomass

R. Prasad (✉)
Department of Pulmonary Medicine, Era's Lucknow Medical College & Hospital, Era University, Lucknow, India

R. Garg
Department of Respiratory Medicine, K.G's. Medical University UP, Lucknow, India

N. Gupta
Department of General Medicine, Dr. Ram Manohar Lohia Institute of Medical Sciences, Lucknow, India

K. E. Pinkerton, W. N. Rom (eds.), *Climate Change and Global Public Health*, Respiratory Medicine, https://doi.org/10.1007/978-3-030-54746-2_18

fuels cook on an open fire. There are wide variations between the rural and urban households regarding the specific kind of biomass fuel used. In rural India, 62% of the households use firewood and 14% cook with dung cakes while 13% use straw, shrubs, grass, and agricultural crop residues to fire their stoves. In urban India, 22% of the households use firewood, 8% use kerosene, and the rest use cleaner fuels like LPG or natural gas [2]. According to the World Health Organization (WHO) estimate in 2010, more than half of the Indian population (58%) depends on solid fuels for domestic purposes [4]. It can also be stated that 75% of the rural households reported firewood as their primary cooking fuel as compared to only 22% of the urban households. It is apparent that factors such as affordability, awareness, ease of availability, cooking space constraints, social customs, and demographics (e.g., working women) play a significant role in the choice of fuel in urban locality [5].

Morbidity and Mortality

Globally, almost two million deaths per year are attributable to solid fuel use with more than 99% of these occurring in developing countries [3]. Household air pollution accounts for about 3.4–4 million deaths every year [6]. The number of DALYs (disability-adjusted life years) attributable to indoor air pollution from solid fuel use for all causes accounts to 40 million. India's figures are very alarming. With a yearly death toll of 662,000 attributed to biomass fuel exposure, India tops the list of the South Asian region [2]. Biomass fuel accounts for 5–6% of the national burden of disease [6]. It has been estimated that indoor air pollution from solid fuel use in all developing countries accounted for about 1.6 million deaths annually in 2004 and about 500,000 in India in 2010, suggesting a serious impact on health [7, 8].

Emissions from a Biomass Stove and Exposure-Determining Factors

Biomass fuel combustion results in production of numerous physical and chemical products which affect the health of the lungs. When firewood is burnt, the combustion efficiency is far less than 100% [9]. The cooking stoves using biomass wastes 74% of the carbon as dissipated heat and only 18% is used for real cooking [10]. Burning of biomass fuels emits toxic fumes into the air which is a mix of small solid particles, carbon monoxide, polyorganic and polycyclic aromatic hydrocarbons (PAH), and formaldehyde.

Small solid particles are particulate matter having a size less than 10 microns (PM10), and particles having a size less than 2.5 microns are more hazardous for health as they can penetrate the lungs [11]. The concentration of indoor particles less than 10 microns (PM10) measured over 24 hours in Indian solid-fuel-using households is over 2000 $\mu g/m^3$ compared to 30 $\mu g/m^3$ in the USA [11].

The carbon monoxide produced during burning of various biomass fuels produces various short-term health effects like dizziness, headache, nausea, feeling of weakness, etc., and the long-term exposure can be similar to carbon monoxide from cigarette smoke which can lead to heart disease and fetal development anomaly [12, 13].

PAHs include a large class of compounds released during the incomplete combustion of organic matter [14]. Benzopyrene is one of the most important carcinogens of this group. PAHs are fluorine, pyrene, chrysene, benzanthracene, benzofluoranthene, benzopyrene, dibenzanthracene, benzoperylene, and indeno pyrene. All these PAHs except the first three have been classified as possible carcinogens [15]. Formaldehyde is well recognized to be an acute irritant, and long-term exposure can cause a reduction in vital capacity and chronic bronchitis. In an epidemiological study in the UK, significantly excess mortality from lung cancer was observed in workers exposed to high levels of formaldehyde [16].

The overall pathological effect of biomass smoke can be taken as mutagen, immune system suppressant, severe irritant, blood poison, inflammatory agent, CNS depressant, cilia toxin, endocrine disruptors, and neurotoxin. They have also been firmly established as human carcinogens. Several toxic inorganic chemicals are known to cause asphyxiation, stillbirth, infant death, heart disease, and severe acute and chronic lung disease. Many mechanisms of cell injury are still unexplained.

Architecture of the House and Biomass Smoke Exposure

The level of exposure to these toxic fumes from a biomass stove varies widely with the house architecture and household composition. Quantitative exposure assessments in various households have been conducted in different parts of India for development of exposure-response relationships. The climatic and cultural variations between the northern and southern Indian regions have influenced the outcome significantly. Cooking areas in many Indian households tend to be poorly ventilated, and about one-half of all households do not have a separate kitchen. Most of the households lack a chimney or any other ventilatory measures. One study conducted in Porur, Chennai, reported that 36% of the households used biomass fuels for cooking in indoor kitchens without partitions, 30% in separate kitchens inside the house, 19% in separate kitchens outside the house, and 16% in outdoor kitchens [17]. The personal exposure of cooks to the respirable particles in biomass smoke was not significantly different between indoor kitchens with or without partitions and separate kitchens outside the house but was significantly different from exposures of cooks using open outdoor kitchens as dispersion of emissions is greater outdoors as compared to indoors and therefore cooks cooking in open outdoors experience lower exposures compared to those in enclosed kitchens. Households with kitchens without partitions experienced the highest levels of living area concentrations as compared to other types. It was also observed that young children and the elderly who mostly occupy the living room are exposed to higher levels of smoke in unpartitioned indoor kitchens. Among non-cooks in households using solid fuels, women not involved in cooking and men with outdoor jobs have the lowest exposures, while women involved in assisting the cook and men staying home have the highest exposure. There seems to be no significance for the cooking duration, the number of meals cooked, outdoor area measurements, or the presence or absence of chimneys [10, 17].

Respiratory Health Effects of Biomass Fuels

Many respiratory diseases have been found to be associated with the exposure of biomass fuels.

In a study conducted in Nigeria in 2016 using a modified Medical Research Council scoring system [18], biomass-exposed women were more likely to have cough (OR: 2.89 95% CI 1.25–6.72) compared with men. No difference in the occurrence of wheeze, sputum, or dyspnea with a score greater than 2 was observed. Women had lower SF-12 (short form-12) physical functioning score (OR: 1.73 95% CI –3.17, 0.29) compared with men, but there were no differences in their lung function parameters [18].

Balcan et al. concluded that the onset of biomass use at an early age and longer duration of biomass exposure is associated with obstructive airway diseases [19]. Several studies have established the association of obstructive airway diseases with biomass fuel exposure [19–22].

The strength of association varies for such diseases like acute lower respiratory tract infections (ALRI), chronic obstructive pulmonary disease, lung cancer, pulmonary tuberculosis, asthma, and interstitial lung diseases. The evidences relating to their strength of association for ALRI in children <5 years is strong (relative risk 2.3, CI 1.9–2.7), for COPD in women more than 30 years age is strong (relative risk 3.2, CI 2.3–4.8), for lung cancer with coal smoke exposure the strength of association is also strong in women ≥30 years (relative risk 1.9, CI 1.1–3.5). For tuberculosis and asthma, the strength of association is moderate, the relative risk being 1.5 (CI 1.0–2.4) and 1.2 (CI 1.0–1.5), respectively [23]. There are also studies including meta-analyses depicting an association between solid fuel and risk of common respiratory disease from India as summarized in Table 18.1.

Acute Lower Respiratory Infection (ALRI) in Children Under 5 Years of Age

Globally, around 2.38 million children died from lower respiratory tract infection in 2016 [22]. Acute lower respiratory infection contributes to 13% of deaths and 11% of the national burden of diseases [23]. This is one of the major diseases associated

Table 18.1 Major Indian studies depicting association between current solid fuel use relative to cleaner burning fuel or electricity and risk of common respiratory diseases

Respiratory disease	Authors	Study type	Outcome	Odds ratio/ incidence risk ratio (95% CI)	References
Tuberculosis	Gupta et al. (1997)	Case-control India	Clinical pulmonary	2.54 (1.07–6.04)	[28]
	Mishra et al. (1999)	Cross-sectional: India (National Family Health Survey)	Self-reported	2.58 (1.98–3.37)	[35]
	Shetty et al. (2006)	Case-control India	Clinical pulmonary	3.26 (1.25–8.46)	[41]
	Mageshwari U et al. (2008)	Case-control India	Clinical pulmonary	0.22 (0.12–0.41)	[42]
	Kolappan et al. (2009)	Case-control India	Clinical pulmonary	2.9 (1.8–4.7)	[35]
	Behera D et al. (2010)	Case-control India	Clinical pulmonary	0.60 (0.22–1.63)	[43]
	Lakshmi et al. (2012)	Case-control India	Clinical pulmonary	2.33 (1.18–4.59)	[36]
	Sehgal et al. (2014)	Meta-analysis	Pooled OR	0.3 (0.2–0.4)	[54]
ALRI	Mishra et al. (2005)	Cross-sectional survey	Self-reported symptoms	1.58 (1.28–1.95)	[55]
	Dherani et al. (2008)	Meta-analysis	24 studies for calculation of OR	1.78 (1.45–2.18)	[44]
	Ramaswamy P et al. (2011)	Longitudinal cohort	Clinical symptoms and estimation of the incidence risk ratio among children from households using biomass fuels relative to cleaner fuels	1.33 (1.02–1.73)	[45]
Lung cancer	Gupta D et al. (2001)	Case-control India	Clinical, radiological, and histopathological assessment	1.52 (0.33–6.98)	[46]
	Behera D et al. (2005)	Case-control India	Clinical, radiological, and histopathological assessment	3.59 (1.07–11.97)	[31]

(continued)

Table 18.1 (continued)

Respiratory disease	Authors	Study type	Outcome	Odds ratio/ incidence risk ratio (95% CI)	References
	Sapkota A et al. (2008)	Case-control India	Clinical, radiological, and histopathological assessment	3.76 (1.64–8.63)	[32]
	Hosgood HD et al. (2011)	Meta-analysis	25 studies for estimation of OR	2.15 (1.61–2.89)	[47]
COPD	Behera D et al. (1991)	Descriptive study	Clinical assessment	3.04 (2.15–4.31)	[48]
	Qureshi et al. (1994)	Case-control India	Clinical assessment	2.10 (1.50–2.94)	[49]
	Kurmi OP et al. (2010)	Meta-analysis	12 studies for estimation of OR	2.80 (1.85–4.0)	[50]
Chronic bronchitis	Sehgal et al.	Meta-analysis	Pooled OR	2.37 (1.59–3.54)	[54]

with the indoor air quality. There are many studies to date that show various respiratory symptoms (coughing, wheezing, etc.) to be associated with solid fuel smoke exposures. However, none of them provide sufficient evidence to calculate the odds ratio. A host of odds ratios ranging from 1.9 to 2.7 have been worked out [24]. These ratios pertain to children with ALRI younger than 5 years only. Other factors might strongly influence ALRI incidence like housing type, location of cooking, and other cultural practices [25]. Some of the studies carried out in India have reported no association between the use of biomass fuels and ALRI in children. In a case-control study in children under 5 years of age in southern Kerala, India, where children with severe pneumonia (ascertained by WHO criteria) were compared with those having non-severe ALRI attending outpatient department, cooking fuel was not a severe risk factor for severe ALRI [26]. Also Sharma et al. in a cross-sectional study involving 642 infants dwelling in urban slums of Delhi and using wood and kerosene, respectively, did not find a significant difference in the prevalence of ALRI and the fuel types [25].

Chronic Obstructive Pulmonary Disease (COPD)

COPD contributes to 20% of DALYs due to chronic respiratory diseases, and its prevalence has increased from 3.3% in 1990 to 4.4% in 2016. About 25.8% of COPD DALYs were due to household pollution [6]. The incidence of chronic cor pulmonale is similar in both men and women. This is despite the fact that only 10% of women are smokers compared to 75% of men. Another point to note is that chronic cor pulmonale occurs 10–15 years earlier in women compared to men [27]. A relative risk of 2–4 has been arrived at for biomass fuel exposure in various Indian studies [23]. Despite the progress made in highlighting the association between biomass fuel exposure and COPD, many shortcomings still exist. Smoking is an important confounding variable for COPD and particularly so when men are included in the analyses. Another major confounding factor is age. The risk for COPD increases with age, and many age-matched studies have provided insufficient quantitative evidence to develop an odds ratio (OR). The overall risk of COPD in women exposed to biomass fuel has been estimated as 3.2 (95% CI = 2.3–4.8) [24]. There is much less evidence available about the impact on men, but the risk seems to be lower with an OR of 1.8 (95% CI = 1.0–3.2). This may be attributed to the lower exposure to biomass fumes in men [28].

Lung Cancer

Lung cancer in women is a well-demonstrated outcome of cooking with open coal stoves in China [29]. Indian women generally have low lung cancer rates [30]. This may be in a way attributed to the minimal use of coal for cooking in Indian households. Nevertheless, a few studies from India have suggested an association with lung cancer even after adjusting for active and passive smoking. An odds ratio of 3.59 (95% CI = 1.07–11.97) has been worked out [31]. In conclusion, it may be inferred that there is a general lack of epidemiological evidence relating lung cancer with biomass fuel exposure. The limited cases reported have been linked with exposure to coal fires [32].

Tuberculosis

Tuberculosis (TB) is a major public health problem in India. Out of all the cases of tuberculosis worldwide (about ten million in 2017), about one-fourth are reported from India. It is estimated that 4.5 lakh deaths occurred in 2016 due to TB in India [33]. There is a strong association between the use of biomass fuel and pulmonary TB. A high risk of pulmonary TB exists in those using wood and cow dung cake as cooking fuel [34]. It is suggested that lowered immune defense mechanisms of the lungs may be the reason for disease presentation. Biomass fuel poses a higher risk (969/100,000) of TB compared to cleaner fuels (378/100,000). It is believed that 51% of active TB in age group more than 20 years is attributable to cooking smoke from biomass fuels [35].

A study done from the northern part of India among adult women having sputum positive pulmonary tuberculosis as cases and age plus area of living matched controls revealed that the OR for biomass fuel compared with LPG was 2.33 (CI 1.18–4.59). Adjustment for confounding factors (education, type of kitchen, smoking tobacco, and TB in family member) and interaction between cooking fuel and smoker in family revealed an OR of 3.14 (CI 1.15–8.56) [36]. Other countries in

Southeast Asia have also documented an association between tuberculosis and biomass fuel exposure. Rabbani U et al. carried out a study in Pakistan in 2017 and confirmed that current users of biomass fuel were at higher risk of pulmonary TB (adjusted matched odds ratio [mOR] = 3.0; 95% CI = 1.1–4.9) compared with non-users. In comparison with former biomass users (women not using biomass for >10 years), recent biomass users (women who switched from biomass to nonbiomass ≤10 years ago), and current (lifetime) users were at a higher risk in a dose-response manner (adjusted mOR = 2.8, 95% CI = 0.9–8.2, and adjusted mOR = 3.9, 95% CI = 1.4–10.7, respectively) [37]. Given the importance of TB in India, because it is both prevalent and likely to increase with HIV epidemic, these findings need to be followed up with more detailed studies.

Pneumoconiosis and Interstitial Lung Diseases (ILD)

Pneumoconiosis has been reported from Ladakh, a hilly terrain in the northernmost part of India [38]. This place is completely devoid of industries or mines. Yet cases have been reported of diseases resembling miner's pneumoconiosis. Another factor considered responsible for the development of this respiratory morbidity is the exposure to dust from dust storms. In spring, dust storms blanket the villages in fine dust. The practice of not allowing the wood to burn quickly and smoldering for longer duration to conserve fuel adds to the high level of respirable particles indoors. Low oxygen levels or some other factor associated with high altitude may be an important contributory factor in causation of pneumoconiosis because it has been reported that the miners working at high altitude are more prone to develop pneumoconiosis than their counterparts exposed to the same levels of dust and working in the mines at normal altitude [39]. The causal role of biomass fuel exposure is however not established [38]. Similarly, a few case reports linking ILD and biomass fuel exposure have been documented. But here too, the veracity of the association is still debatable [40].

Biomass Fuel Exposure and Lung Functions

There are only three studies throwing light on the effect on lung functions from use of domestic cooking fuels including biomass fuel from India. Among this, one study included children from north India which revealed forced vital capacity (FVC) and FEV1 (forced expiratory volume in 1 sec) lowest in boys, whose households used biomass fuel ($p < 0.05$) and PEFR (peak expiratory flow rate) and FEF (forced expiratory flow) 25% and 50% also lowest in boys with their homes using kerosene as fuels. All these were the best for LPG fuel [51]. The other two studies carried out from the northern part of India revealed different outcomes; the study carried out by D. Behera et al. in 3318 rural nonsmoking women using different domestic cooking fuels like biomass fuel, liquefied petroleum gas, kerosene, and mixed revealed biomass fuel users had FVC values less than 75% predicted whereas in other groups it was more than 75% of predicted, though less than 80% of the predicted values. The absolute values of all the three parameters (FVC, FEV1, and PEFR) of lung functions were the lowest in the biomass and mixed fuel users [52]. The second study was comparative study to see the lung functions of healthy nonsmoking women who used either biomass or liquefied petroleum gas (LPG) as their sole cooking fuel. The effects of passive smoking, ventilation, overcrowding, and cooking index were also taken into account. The results of this study revealed no statistically significant differences in lung functions in the two groups except for the PEFR, which was significantly lower ($p < 0.01$) in women using biomass. No correlation was observed between different variables and pulmonary functions. The step-wise multivariate linear regression analysis showed no correlation between cooking fuel and pulmonary functions. The authors concluded that the absence of the expected adverse effects of biomass on pulmonary functions was possibly due to better ventilation in the kitchens of subjects in the biomass group compared to previous studies [53]. These studies indicate that the lung functions are adversely affected by the use of biomass as domestic cooking fuel in comparison to cleaner fuels but have a linear relationship with the duration of cooking, overcrowding, and poor ventilation and have a negative correlation with better ventilated kitchens.

Conclusion

Biomass fuel exposures contribute heavily to the burden of diseases in India. Despite heterogeneity of the published literature, the available evidence suggests significant associations with diseases such as ALRI and COPD. Other diseases, where solid fuel smoke is thought to play a role, need further study to establish the association. There exist strong associations that lung function is adversely affected with the use of biomass fuel in domestic cooking fuel compared to cleaner fuels like biogas, liquefied petroleum gas (LPG), compressed natural gas (CNG), etc. It has also been demonstrated that better ventilated kitchens improve pulmonary function. Therefore, exposure reduction strategy should be adopted. It is also believed that selection of the strategies to withdraw or reduce the exposure is very challenging and probably need of the hour for country like ours. Exposure reduction strategy will require various considerations at personal as well as cultural and economic aspects. Solutions will involve the efforts at the level of development, resources, technical capacity, domestic needs of energy, sustainability of the considered sources of energy, and the protection of the environment. Substantial improvement can be achieved by health education and cultural modification and interventions like modification of stove design and switching over to cleaner fuels or other high-efficiency low-emission fuels for cooking. In India, the government has prioritized and understands the need for cleaner fuel and has implemented a scheme whereby LPG connections are distributed to safeguard the health of women and children who are exposed to household smoke. This is highly desired considering the mammoth risk solid fuels pose in rural India. It is hoped that in future studies on biomass exposure, the associated morbidities and means for its prevention will be given high priority.

References

1. World Health Organization. The world health report: 2004: changing history. http://www.who.int/whr/2004/en/-. Accessed 12 Feb 2012.
2. International Institute of Population Sciences (IIPS). National family health survey (MCH and family planning): India 2005–06. Bombay: International Institute of Population Sciences; 2007.
3. Overview of biomass power sector in India. Available at https://biomasspower.gov.in. Updated on 20 May 2016.
4. World Health Organization. The world health statistics: 2012. http://www.who.int/whois/whostat/2012/en/-. Accessed 24 Sept 2012.
5. Government of India, Ministry of Statistics and program, New Delhi, India. Service sector implementation, NSSO Enterprises (extending trade) and consumer expenditure. National Sample Survey report, 63rd round 2006–2007.
6. Salvi S, et al. The burden of chronic respiratory diseases and their heterogeneity across the states of India: the Global Burden of Disease study 1990–2016. Lancet Glob Health. 2018;6:e1363–74.
7. Smith KR, Mehta S, Maeusezahl-Feuz M. Indoor smoke from household solid fuels. In: Ezzati M, Rodgers AD, Lopez AD, et al., editors. Comparative quantification of health risks: global and regional burden of disease due to selected major risk factors, vol. 2. Geneva: World Health Organization; 2004. p. 1435–93.
8. Wilkinson P, Smith KR, Davies M, et al. Household energy, series on the impact on public health of strategies to reduce greenhouse gases. Lancet. 2009;374:1917–29.
9. Smith KR. Biomass, air pollution, and health; a global review. New York: Plenum Press; 1987.
10. Balakrishnan K, Sankar S, Padmavathi R, Smith KR. Respirable particulate levels in rural households of Andhra Pradesh, India: daily concentrations and exposures. J Environ Stud Policy. 2002;5:87–97.
11. Smith KR. Indoor air pollution in developing countries: Growing evidence of its role in the global disease burden. In: Ikeda K, Iwata T, editors. Indoor air'96: proceedings of 7th international conference on indoor air and climate. Tokyo: Institute of Public Health; 1996. p. 33.
12. World Health Organization. International programme on chemical safety; environmental health criteria 202: selected non-heterocyclic, polycyclic aromatic hydrocarbons, vol. 3. Geneva: WHO; 1998. p. 63.
13. The Health Consequences of Smoking – Cardiovascular Diseases. A report of the surgeon general. US Department of Health and Human Services, Rockville, MD; 1983.
14. Perera FP, Hemmlinki K, Young TL, Brenner D, Kelly G, Santella RM. Detection of polyaromatic hydrocarbon-DNA adducts in white blood cells of foundry workers. Cancer Res. 1988;48:2288–91.
15. Schwarz-Miller J, Rom WN, Brandt-Rauf PW. Polycyclic aromatic hydrocarbons. In: Rom RN, editor. Environmental and occupational health, vol. 2. London: Little Brown; 1991. p. 873.
16. Acheson ED, Barnes HR, Gardner MJ, Osmond C, Pannet B, Taylor CP. Formaldehyde process, workers and lung cancer. Lancet. 1984;i:1066–7.
17. Balakrishnan K, Parikh J, Sankar S, Padmavathi R, Srividya K, Venugopal V, et al. Daily average exposures to respirable particulate matter from combustion of biomass fuels in rural households of Southern India. Environ Health Perspect. 2002;110:1069–75.
18. Obaseki D, Adeniyi B, Erhabor G, Adefuye B. Biomass exposure and its association with respiratory symptoms, quality of life, and lung function: a population study in Ife, Nigeria. Eur Respir J. 2016;48(suppl 60):PA4296.

19. Balcan B, Akan S, Ceyhan B. Effect of biomass exposure on lung function. Eur Respir J. 2015;46:PA352.
20. Capistrano SJ, van Reyk D, Chen H, Oliver BG. Evidence of biomass smoke exposure as a causative factor for the development of COPD. Toxics. 2017;5(4):36.
21. Sana A, Somda SMA, Meda N, Bouland C. Chronic obstructive pulmonary disease associated with biomass fuel use in women: a systematic review and meta-analysis. BMJ Open Respir Res. 2018;5(1):e000246.
22. Troeger C. Estimates of the global, regional, and national morbidity, mortality, and aetiologies of lower respiratory infections in 195 countries, 1990–2016: a systematic analysis for the global burden of disease study 2016. Lancet Infect Dis. 2018;18:1191–210.
23. Desai MA, Mehta S, Smith KR. Indoor smoke from solid fuels: assessing the environmental burden of disease at national and local levels, Environmental burden of disease series no. 4. Geneva: WHO protection of the human environment; 2004.
24. Smith KR. National burden of disease in India from indoor air pollution. Proc Natl Acad Sci. 2000;97:13286–93.
25. Sharma S, Sethi GR, Rohtagi A, Chaudhary A, Shankar R, Bapna JS, Joshi V, Sapir DG. Indoor air quality and acute lower respiratory infection in Indian urban slums. Environ Health Perspect. 1998;106(5):291–7.
26. Shah N, Ramankutty V, Pramila PG, Sathy N. Risk factors for severe pneumonia in children in South Kerala: a hospital based case-control study. J Trop Paediatr. 1994;40:201–6.
27. Padmavati S, Pathak SN. Chronic cor pulmonale in Delhi. Circulation. 1959;20:343–52.
28. Gupta B, Mathur N, Mahendra P, Srivastava A, Swaroop V, Agnihotri M. A study of the household environmental risk factors pertaining to respiratory disease. Energy Environ Rev. 1997;13:61–7.
29. Smith KR, Liu Y. In the epidemiology of lung cancer, Samet J, editor (Dekker, New York Edn.), vol. 3; 1994. p. 151–84.
30. Alberg AJ, Samet JM. Epidemiology of lung cancer. Chest. 2003;123:21S.
31. Behera D, Balamugesh T. Indoor air pollution as a risk factor for lung cancer in women. JAPI. 2005;3:190–2.
32. Sapkota A, Gajalakshmi V, Jetli DH, Chowdhary SR, Dikshit RP, Brennan P, Hashib M, Boffetta P. Indoor air pollution from solid fuels and risk of hypopharyngeal/laryngeal and lung cancers: a multicentric case–control study from India. Int J Epidemiol. 2008;2:1–8.
33. http://www.tbfacts.org/.
34. Kolappan C, Subramani R. Association between biomass fuel and pulmonary tuberculosis: a nested case-control study. Thorax. 2009;64:705–8.
35. Misra VK, Ratherford RD, Smith KR. Biomass cooking fuels and prevalence of tuberculosis in India. Int J Infect Dis. 1999;3:119–29.
36. Lakshmi PV, Virdi NK, Thakur JS, Smith KR, Bates MN, Kumar R. Biomass fuel and risk of tuberculosis: a case-control study from Northern India. J Epidemiol Community Health. 2010; https://doi.org/10.1136/jech.2010.115840.
37. Rabbani U, Sahito U, Nafees AA, Kazi A, Fatmi Z. Pulmonary tuberculosis is associated with biomass fuel use among rural women in Pakistan: an age- and residence-matched case-control study. Asia Pac J Public Health. 2017;29(3):211–8.
38. Saiyed HN, Sharma YK, Sadhu HG, Norboo T, Patel PD, Patel TS, Venkaiah K, Kashyap SK. Non-occupational pneumoconiosis at high altitude villages in Central Ladakh. Br J Ind Med. 1991;48:825–9.
39. Odinaev FI. The characteristics of development and course of pneumoconiosis under the conditions of a mountain climate. Grig Tr Prof Zabol. 1992;7:13–4. (Russian).
40. Dhar SN, Pathania AGS. Bronchitis due to biomass fuel burning: 'Gujjar lung' an extreme effect. Semin Respir Med. 1991;12:69–74.
41. Shetty N, Shemko M, Vaz M, D'Souza G. An epidemiological evaluation of risk factors for tuberculosis in South India: a matched case control study. Int J Tuberc Lung Dis. 2006;10:80–6.
42. Magheswari U, Johnson P, Ramaswamy P, Balakrishnan K, Jenny A, Bates M, et al. Exposure to biomass fuel smoke and tuberculosis-a case-control study in India. Epidemiology. 2007;18:s122.

43. Behera D, Aggarwal G. Domestic cooking fuel exposure and tuberculosis in Indian women. Indian J Chest Dis Allied Sci. 2010;2:139–43.
44. Dherani M, Pope D, Mascarenhas M, Smith KR, Weber M, Bruce N. Indoor air pollution from unprocessed solid fuel use and pneumonia risk in children aged under five years: a systematic review and meta-analysis. Bull WHO. 2008;86:390–8.
45. Ramaswamy P, Balakrishnan K, Ghosh S, et al. Indoor air pollution due to biomass fuel combustion and acute respiratory infection in children under 5 in Trichy District of rural Tamilnadu. India Epidemiol. 2011;22:s104.
46. Gupta D, Boffetta P, Gaborieau V, Jindal SK. Risk factors of lung cancer in Chandigarh, India. Indian J Med Res. 2001;113:142–50.
47. Hosgood HD III, Wei H, Sapkota A, Choudhury I, Bruce N, Smith KR, et al. Household coal use and lung cancer: systematic review and meta-analysis of case-control studies, with an emphasis on geographic variation. Int J Epidemiol. 2011;40:1–10.
48. Behera D, Jindal SK. Respiratory symptoms in Indian women using domestic cooking fuels. Chest. 1991;100:385–8.
49. Qureshi KA. Domestic smoke pollution and prevalence of chronic bronchitis/asthma in a rural area of Kashmir. Indian J Chest Dis Allied Sci. 1994;36:61–72.
50. Kurmi OP, Semple S, Simkhada P, Smith WCS, Ayres JG. COPD and chronic bronchitis risk of indoor air pollution from solid fuel: a systematic review and meta-analysis. Thorax. 2010;65:221–8.
51. Behera D, Sood P, Singh S. Passive smoking, domestic fuels and lung functions in north Indian children. Indian J Chest Dis Allied Sci. 1998;40(2):89–98.
52. Behera D, Jindal SK, Malhotra HS. Ventilatory function in nonsmoking rural Indian women using different cooking fuels. Respiration. 1994;61(2):89–92.
53. Reddy TS, Guleria R, Sinha S, Sharma SK, Pande JN. Domestic cooking fuel and lung functions in healthy non-smoking women. Indian J Chest Dis Allied Sci. 2004;46(2):85–90.
54. Sehgal M, Rizwan SA, Krishnan A. Disease burden due to biomass cooking-fuel-related household air pollution among women in India. Glob Health Action. 2014;7:25326.
55. Mishra V, Retherford RD, Smith KR. Cooking smoke and tobacco smoke as risk factors for stillbirth. Int J Environ Health Res. 2005;15:397–410.

Chapter 19
Climate Change and Women's Health: Risks and Opportunities

Cecilia J. Sorensen and John Balbus

Introduction

Many of the health risks that are likely to be amplified by climate change show gender differentials. Climate change impacts human health through a variety of exposures, including extreme heat, diminished air quality, and extreme weather events, as well as through ecological changes that alter vector-borne disease, reduce water quality, and decrease food security [1]. The health risks associated with these exposures are mediated through physiologic, cultural, and socioeconomic factors, which have unique features for every region and every individual. Globally, a total of 1.3 billion people in the developing world live below the poverty line, of which 70% are women [2].

As noted by the United Nations Framework Convention on Climate Change (UNFCCC) [3], women, especially those in poverty, face higher risks and experience a greater burden of health-related climate change impacts, making climate change a risk multiplier for gender-based health disparities worldwide. Climate-related disasters such as droughts, floods, and hurricanes kill more women than men; and the gender-gap effects on life expectancy following disasters tend to be more drastic in women, especially those with lower socioeconomic status [4]. Other impacts, such as food insecurity and infectious and waterborne diseases, also show

C. J. Sorensen (✉)
Department of Emergency Medicine, University of Colorado School of Medicine, Aurora, CO, USA

University of Colorado School of Public Health, Aurora, CO, USA
e-mail: Cecilia.Sorensen@ucdenver.edu

J. Balbus
National Institute of Environmental Health Sciences, Bethesda, MD, USA
e-mail: John.Balbus@nih.gov

K. E. Pinkerton, W. N. Rom (eds.), *Climate Change and Global Public Health*, Respiratory Medicine, https://doi.org/10.1007/978-3-030-54746-2_19

important gender differences. Yet while the interactions between poverty, gender-based social discrimination, and climate change threaten to amplify gender-based health disparities, women's social roles and potential for agency afford opportunities for promoting solutions to sustainability, disaster risk reduction, and health threats. This chapter will outline the major differences in health impacts due to climate exposures and highlight opportunities to promote gender equity in climate-compatible development, disaster risk reduction, and adaptation and mitigation planning at international and local levels.

Heat Impacts

Globally, ambient temperatures are projected to rise on average by 2–4 °C by 2100 [5] with an associated increase in the incidence of extreme heat waves. There is clear evidence that exposure to more frequent and intense heat waves is increasing, with an estimated 125 million additional adults exposed to heat waves between 2000 and 2016 [5]. Exposure to extreme heat is associated with significant cardiovascular and respiratory morbidity and mortality, and the risk of negative health impacts rises with only moderate increases in seasonal temperature [6]. Mechanistically, heat activates inflammatory pathways [7], perturbs blood coagulability [8], and alters central nervous system regulation of the heart [7], worsening cardiovascular disease and respiratory disease [9]. Although robust, gender-disaggregated surveillance data is lacking, several studies have demonstrated that women are more at risk of dying in heat waves, especially older women [10, 11].

Women differ from men in their physiologic compensation to elevated temperatures, which contributes to their biologic vulnerability. They dissipate less heat by sweating, have a higher working metabolic rate, and have thicker subcutaneous fat which decreases radiative cooling [12]. Social and cultural factors also contribute to poor outcomes and include: poor access to healthcare and cooling facilities due to personal safety concerns and a lack of access to personal transportation; culturally prescribed heavy clothing garments that limit evaporative cooling; and a lack of awareness of women's vulnerabilities to heat among local, national, and global decision makers and health care personnel.

Pregnancy also contributes to vulnerability. Prolonged exposure to high temperatures is associated with stillbirth, congenital birth defects, and pre-term delivery – regardless of maternal ethnicity or age, with younger mothers having an even higher risk of negative outcomes [13–16]. High ambient temperatures are also linked to pregnancy complications, such as gestational hypertension, pre-eclampsia [17], and poor neonatal outcomes [18]. Additionally, heat is teratogenic at crucial stages of development. It also increases the production of vasoactive substances, increases blood viscosity, and affects endothelial cell function which may alter placental blood flow and increase propensity for hypertensive crises and stillbirth [13].

Impacts of Poor Air Quality

Poor air quality is extensively linked to climate change and has far-reaching impacts on health. The combustion of fossil fuels directly increases ground-level ozone (O_3) and fine particulate matter ($PM_{2.5}$), and increasing ambient temperatures accelerate the formation of O_3. There is robust evidence suggesting that climate change worsens ozone pollution through increased temperatures and drives $PM_{2.5}$ generation through increased frequency and intensity of wildfires, as well as from desertification and resulting dust generation [9, 19]. Climatologic and meteorological factors, such as temperature, precipitation, cloud cover, and wind velocity, can influence the concentrations of these pollutants in ambient air and thus impact health on a regional level.

Mechanistically, inhaled particles can react with neural receptors resulting in alterations in normal functioning of the autonomic nervous system, can generate oxidative stress in alveolar-capillary cells resulting in local and systemic inflammation, and can cross the alveolar membrane, resulting in endothelial injury within cardiovascular system and prothrombic changes in blood proteins [20]. Ultimately, these pathophysiologic changes can result in respiratory and cardiovascular dysfunction and can lead to premature death [21] and exacerbation of underlying cardiovascular and respiratory disease in vulnerable populations [22, 23]. Increases in mean $PM_2.5$ are linked with mortality as well as health care system utilization.

Gender significantly affects the ultimate health impacts of poor air quality. Evidence suggests that women are more likely to experience fatal coronary heart disease as a result of exposure to ambient PM [24]. Similarly, studies have shown that the degree of artery wall thickening in relationship to ambient levels of $PM_{2.5}$ is more significant in women than men [25], potentially placing women at higher risk of all types of macro- and microvascular disease. In terms of respiratory disease, several experimental studies of pulmonary deposition of inhaled particles have shown that particle deposition characteristics differ between the genders, with women experiencing greater deposition that ultimately translates to greater health risks [26]. There is also evidence that women are more sensitive to airborne pollution than are males due to average lower red blood cell numbers, which increases sensitivity to toxicologic influences of air pollutants, as evidenced by a greater production of oxidation products in response to PM2.5 exposure [27].

Poor air quality is also associated with adverse reproductive outcomes including congenital birth defects [28], stillbirths, prematurity, and intrauterine growth restriction [29–33]. It is apparent that there is a critical period of development when the timing of exposure to airborne toxins can be even more important than the overall dose. Fetuses, in particular, are highly susceptible to a variety of toxins because of their higher rates of cell proliferation, changing metabolic capabilities, and physiologic immaturity [30]. An adverse intrauterine environment, as indicated by growth restriction and prematurity, has life-long impacts, including increased neonatal, childhood, and adult morbidity and mortality [34].

Furthermore, exposure to poor air quality threatens maternal health. Evidence suggests maternal exposure to ambient air pollution is associated with an increased risk of hypertensive disorders, including preeclampsia [35], placenta previa and accreta [36] placental abruption [37], and gestational diabetes [38]. These obstetric complications carry grave consequences for women with limited health care resources.

The use of biomass for household cooking and heating also contributes significantly to ambient air pollution [39], accounting for nearly 25% of the deaths attributable to PM [40]. Additionally, black carbon and methane emitted by inefficient stove combustion are powerful greenhouse gases [41]. In 2016, over 2.5 billion people were exposed to poor household air quality, mostly in low- and middle-income countries. For example, 43% of India's population and over 30% of China's population lack electricity and use traditional biomass fuels in the domestic setting [40]. Women, due to their traditional roles in child care and food preparation, spend more time in the home and therefore are disproportionately exposed. Health impacts from the inefficient burning of biomass and miscellaneous waste plastics include pneumonia, chronic obstructive pulmonary disease (COPD), lung cancer, stroke, and ischemic heart disease. Additionally, biomass fuel gathering increases the risk of musculoskeletal damage and consumes significant time and energy for women and children, thus limiting other activities of livelihood, such as income generation and education. In certain settings, women are additionally at risk of injury and violence during harvesting activities [41].

Extreme Weather Events and Disasters

According to a recent Lancet report, the frequency of weather-related disasters – including hurricanes, flooding, and wildfires – increased by 46% from 2007 to 2016 [9]. For many parts of the world, the IPCC projects future increases in extreme precipitation events, and increases in the severity of coastal hurricanes and flooding, coupled with intensified drought in other regions [5]. Extreme weather events and disasters pose unique health and safety risks to women worldwide [42]. In a study of 4605 natural disasters, authors found that disasters shorten women's life expectancy significantly more than men's [43]. Additionally, many women live in social conditions which constrain their mobility, behavior, education, decision-making power, and resources access, which may further exacerbate health risks.

In gender-disaggregated studies in low-middle income countries, women have been found to be more likely to die in cyclones and floods [4]. For example, in 1991 when cyclones in Bangladesh killed 140,000 people, 90% of the victims were women [44], and in 2008 when cyclone Nargis hit Myanmar, 61% of the 130,000 deaths were likewise women [4]. The combination of social inequalities in terms of access to basic social goods, culturally prescribed roles and biologic vulnerabilities may explain this disparity [45]. The gender difference in mortality has been found to be larger when women are from a lower socioeconomic status in a particular region [4]. Other research suggests that cultural factors contribute to vulnerability when women are homebound caring for children and elderly while waiting for relatives to return from a disaster-related evacuation. Underlying poor literacy and education may also play a role. If public warnings do not take into account women's access to information and the possibility that homebound women in remote areas only speak a minority language, women will be unable to appropriately take steps to safeguard their lives [24].

Physically, women of all ages are more calorie-deficient than men, leading to poor physical health and vulnerability to resource shortages ensuing from catastrophes [43]. Additionally, poor baseline nutritional status and physical health may prevent escape and survival in the acute phase of a disaster [46–48]. Pregnant women are a particularly vulnerable population, and those giving birth in the time period following disasters have been found to have an increased risk of complications including preeclampsia, uterine bleeding, and low birthweight infants [49]. Additionally, obstetric health care may not be available to women if they are displaced, resulting in poor prenatal care and cascading obstetric complications.

In the aftermath of climate driven disasters, women and girls – especially the elderly or those living in lower socioeconomic circumstances – are at higher risk of physical, sexual, and domestic violence [24, 42]. Women may be separated from family, friends, and other support systems and may avoid using shelters for fear of abuse. Infrastructure conditions in refugee camps may expose women and girls to sexual violence due to inconvenient and insecure hygiene facilities with deficient closing mechanisms [43]. Furthermore, poor, single, elderly women, adolescent girls, and women with disabilities are often at greatest risk because they have fewer personal, family, economic, and educational resources from which to draw protection, assistance, and support. These same risk factors correlate with a comparatively higher risk for mood disorders, such as depression, traumatic disorders, and anxiety, in the aftermath of disasters [50]. In the recovery phase, women also suffer disproportionate job loss and stagnant personal economic recovery. For many, there is limited access to resources in social networks, control over land and other economic resources, and access to safe housing; and often the best and only jobs available are in construction and rebuilding efforts, which are traditionally male-dominated fields [51].

Food Insecurity and Malnutrition

Under changing climatic conditions, many geographic regions are experiencing both increases in extreme precipitation as well as decreases in seasonal rainfall, with the net result being extended periods of drought [5]. Variable precipitation combined with rising seasonal temperatures can have profound implications for crop, livestock, and fishery yields and result in food insecurity and economic instability. Additionally, because of increased growth, range, and duration of growth of pests and weeds, pesticide use is expected to increase, including use in areas where specific pesticides were not previously necessary.

Women are inherently sensitive to the effects of food insecurity and resulting nutritional deficiencies due to increased needs during menstruation, pregnancy, and nursing. Furthermore, nutritional scarcity can be intensified by cultural practices that prioritize food provision to children and adult males. Research has found that in some regions, girls' nutrition status suffers most during periods of nutritional scarcity and rising food prices and that drought is more strongly associated with deaths among girls than boys [52]. Poor nutritional status with resulting anemia is highly prevalent among women and children in low- and middle-income countries [53]. Micronutrient deficiencies are associated with cognitive impairments, including poor attention span, diminished working memory, emotional and behavioral regulatory issues, and impaired sensory perception which lead to poor educational outcomes [54]. Additionally, maternal under nutrition has profound effects on neonatal development and is associated with intrauterine growth restriction, pregnancy complications, and perinatal mortality [55]. According to the Food and Agriculture Organization (FAO), in places where iron deficiency anemia is prevalent, the risk of women dying during childbirth is increased by as much as 20% [56].

In developing regions, women are the primary agricultural producers, responsible for the provision of 60–80% of all food [57]. Thus, their livelihoods, as well as their nutritional status, are threatened when changing climatic conditions prevent successful agricultural yields. Prevalent cultural norms compound these hazards to the wellbeing of women. Despite the fact that they produce most food, less than 10% of female farmers are landowners, and barely 2% of owners have proper paperwork for their land [57]. Therefore, women suffer on account of their relative lack of control over farmlands, as well as their lack of access to crop insurance to overcome the losses incurred by environmental change.

Water Scarcity and Waterborne Disease

Hydrologic factors related to climate change, such as the warming of the oceans and increased frequency and intensity of heavy downpours and droughts, alter marine and fresh water resources in a manner which affects the presence and number of many disease-causing organisms [58]. Climate-induced changes in water quality and availability as well as local temperature are closely linked to the spatial and temporal distribution of waterborne diseases. Cholera, shigella, salmonella, campylobacter, noroviruses, enteroviruses, rotaviruses, cryptosporidium, and giardia are all climate-sensitive and show variable abundance in relationship to temperature, rainfall, and distribution of host or reservoir species [10]. For example, ocean warming and reduced salinity has resulted in a steady increase in the proportion of coastal habitats suitable for outbreaks of Vibrio infections, which result in an estimated 80,000 illness and 100 deaths in the United States each year [59]. Similarly, campylobacter infections show seasonal differences in transmission rates, with warmer winters correlated with increased transmission [60]. Shifting rainfall patterns, increased rates of evaporation, and population growth are expected to increase the number of people living in water-stressed basins from about 1.5 billion in 1990 to 3–6 billion by 2050 [61].

Water scarcity forces people to drink from sources that may be biologically and toxicologically contaminated. Traditionally, women have the household role of providing water for the family and there is an increased risk of contracting waterborne diseases among primary water handlers [12, 62]. Diarrheal disease is one of the primary clinical manifestations of waterborne disease, and more than 90% of cases are attributable to a lack of access to safe water and sanitation [4]. Chronic diarrheal disease leads to insidious health impacts with long-term impacts. For example, diarrheal disease can impair growth and cognitive development, cause malnutrition and anemia. and increase susceptibility to other infectious agents [18]. Additionally, diarrheal disease may impair productivity and lead to severe illness, which may necessitate the need for medical attention and ultimately and result in lost work days [63]. Furthermore, these negative impacts are incurred by the household regardless of which family member is sick, through the costs of health care and loss of income when caring for a sick child or relative [63].

Water scarcity also equates to more time spent harvesting water and less time spent on other activities of livelihood. For example, it is estimated that during the dry season in water stressed areas of low- and middle-income countries, 30% or more of a woman's daily energy expenditure is spent harvesting water [4]. Additionally, the manual labor involved in water harvesting places women and female children at risk for cumulative damage to the spine and neck, leading to chronic skeletal pain. Traveling long distances for water also increases exposure to heat stress and heat stroke [64]. A lack of clean water and proper sanitation infrastructure also poses serious health challenges to women, especially during reproductive times [62]. Poverty exasperates the health impacts of water scarcity on women. In urban areas, due to a lack of ownership of water pipes, poorer people often have to pay higher prices for water. In such areas, a lack of access to water has been linked to higher mortality rates for women [65].

Vector-Borne Disease

As a result of alterations in temperature and precipitation, the geographic range and abundance of disease vectors is changing, exposing more people to tick-borne and mosquito-borne illnesses [66, 67]. Although the prevalence of VBDs is strongly influenced by climate, behavioral and physiologic factors determine disease burden on a local level. For example, the recent increase of dengue in certain regions is likely due to permissive ecologic conditions which favor mosquito development coupled with population expansion, unplanned urbanization, deteriorating basic sanitary conditions, and inadequate water supply and waste management systems on a local level [68].

Men and women have a different risk of acquiring VBDs because they occupy different environments throughout the day and have different biologic risks.

Pregnant women are a notably vulnerable population. They are at heightened risk for contact with vectors due to increased time spent around the home near domestic standing water. Additionally, physiologic changes during pregnancy increase vulnerability. Higher CO_2 production, a chemoattractant for mosquitos, and increased peripheral blood flow and skin temperature together increase biting risk. Furthermore, hormonally induced changes in immunologic function may suppress host defenses, resulting in higher intensity of viremia and parasitemia [56, 69]. Henceforth, studies have found that pregnant women have a risk of severe malaria that is three times as high as that of non-pregnant women [70]. Malaria infection during pregnancy results in anemia and diminished trans-placental nutrient transport from placental parasite sequestration, resulting in intrauterine growth restriction and increased vulnerability of the mother to hemorrhagic complications of delivery [71].

Other VBDs carry different pregnancy complications. Dengue virus, which is showing an increase in severity and distribution in India, is associated with increased risk of cesarean delivery, pre-eclampsia, and intrauterine growth restriction; and vertical transmission has been documented [72]. Zika virus, also transmitted by the aedes mosquito, is an emergent climate-linked infectious disease with devastating fetal impacts including microcephaly, central nervous system malformations, and impaired cognitive development [73].

Migration Issues

There is increasing evidence that environmentally induced forced migrants ("environmental refugees") are a rapidly growing demographic of displaced people [74]. Increasing natural resource scarcity and over-exploitation of resources, coupled with unprecedented population growth and unsustainable international trade patterns, are some of the antecedents of this slow disaster. Climate change compounds this situation in several ways. Increasing drought, heat, and depletion of fertile soils as a result of erosion further reduce agricultural potential and simultaneously reduce access to clean water, making regions uninhabitable for subsistence dwellers. Extreme weather events, such as hurricanes and extreme precipitation, which result in coastal and inland flooding, can destroy homes and livelihoods. Additionally, sea-level rise threatens millions of residents who dwell in coastal areas, who must now seek permanent alternative domicile and livelihood. Forced migration, primarily into urban areas, can result in worsening living conditions brought on by overcrowding, unemployment, poor sanitation, and rising social tension over limited resources. These factors result in a myriad of health impacts, including strained access to health care, violence, poor mental health, increases in the spread of communicable diseases, and personal security risks.

According to a United Nations High Commission on Refugees report, 80% of refugees in the world are women and children. In the chaos of migration, when family, community, and institutional security and protection break down, women and girls are at a higher risk of sexual violence, sexual exploitation, and trafficking [42].

This situation is exacerbated by a lack of data on gender-based violence and discrimination. Few studies have thoroughly investigated this pattern. Often, these victims do not have the social resources to seek redress through the legal system of the host country or the country from which they are escaping. Additionally, domestic violence has also been shown to increase in migration situations, when psychological stress is high and substance abuse increases [75].

When regions become uninhabitable from environmental change, individuals with skills and education generally migrate toward economically viable commercial centers to seek employment. In many parts of the world, women have less education and thus fewer employment opportunities and thus are unable to migrate, becoming "trapped" in resource-depleted areas with poor access to reliable food, water, and vocational opportunities. Women who are "trapped" have a heightened risk of communicable disease. Several studies have demonstrated that when partners travel, women are at a higher risk of HIV/AIDS infections as well as other sexually transmitted infections for a variety of reasons. A dearth of healthcare and stigma compounds these health risks [76]. For women who do migrate, their opportunities for success are often burdened by increasing strain from caregiver roles. When men who are fathers migrate, there is little expectation that children will join them. When women migrate, they bring children and elders with them, meaning they are still faced with the pressures of "working two jobs" while trying to establish a professional presence in a new context. Additionally, when women migrate with false documents, they are more likely to stay in employment situations where they are treated unfairly (Table 19.1) [76].

International Policy Response

Climate-compatible development strives to promote strategies and goals that integrate both the threats and the opportunities of a changing climate. In doing so, climate-compatible development aims to not only lower greenhouse gas emissions, but also build resilience and promote economic development, prosperity, and equality [86]. On the international policy stage, climate change, poverty, and gender inequality are increasingly recognized as global problems. Achieving the integration of policies, data and programs necessary to embed gender equality and women's health into climate compatible development, however, has proven challenging.

Recently, some advances have been made within the United Nations Framework Convention on Climate Change (UNFCCC). At its twenty-third session, the Conference of the Parties (COP) established a gender action plan to support the implementation of gender-related decisions in the UNFCCC process. The plan aims to advance women's equal participation and promote the mainstreaming of a gender perspective in the work of the Parties, through the implementation of activities grouped in five priority areas: (1) capacity-building, knowledge-sharing, and communication; (2) gender balance, participation, and women's leadership; (3) coherence; (4) gender-responsive implementation; and (5) monitoring and reporting.

Table 19.1 Climate-related health impacts on women and relevant physiologic, cultural, and socioeconomic risk factors

Exposure pathway	Gender disparities in health impacts	Physiologic and biologic vulnerabilities	Cultural and socioeconomic vulnerabilities
Increasing frequency of extreme heat events and rising average seasonal temperatures	Disproportionate heat-related morbidity and mortality	Women have a higher working metabolic rate, reduced heat dissipation through sweating, and decreased effective radiative cooling [12]	Poor access to healthcare and cooling facilities due to personal safety concerns and lack of access to personal transportation Lack of awareness of women's vulnerabilities to heat among local, national, and even global decision makers and health care personnel Dearth of gender-disaggregated heat-related health data, unknown critical exposure windows Culturally prescribed heavy clothing garments
	Adverse reproductive outcomes: pre-term delivery [77], congenital defects [16], gestational hypertension, and pre-eclampsia [17]	Heat increases the production of vasoactive substances, increases blood viscosity, and affects endothelial cell function which may alter placental blood flow and increase propensity for hypertensive crises and stillbirth [13] Hyperthermia is teratogenic, disrupting the normal sequence of gene activity during organogenesis [16]	
Combustion of fossil fuels increased O_3 and $PM_{2.5}$, resulting in poor indoor and outdoor air quality	Respiratory and cardiovascular disease [40]	In some studies, women are more likely to experience fatal coronary heart disease as a result of exposure to ambient PM [24, 78, 79] Women experience greater deposition of inhaled particles in their lungs [40]	Household biomass burning produces carbon monoxide, hydrocarbons, and particulate matter and accounts for nearly 24% of ambient air pollution from $PM_{2.5}$ [80] Women spend the maximum amount of time in the home and thus are disproportionately affected Biomass fuel gathering increases the risk of musculoskeletal damage and consumes significant time and energy for women and children; in certain settings, women are at risk of injury and violence during harvesting activities [41]
	Adverse reproductive outcomes: stillbirth, intrauterine growth restriction, and congenital defects [29, 30, 81] Obstetric complications	Poor air quality is also associated with adverse reproductive outcomes including stillbirths, prematurity, and intrauterine growth restriction [29–33] Air pollutants (e.g., CO) can cross the placenta and impact fetal growth during crucial developmental windows [28] Maternal exposure to ambient air pollution is associated with an increased risk of hypertensive disorders, including preeclampsia [35], placenta previa and accreta [36] placental abruption [37], and gestational diabetes [38]	

Increasing frequency of climate-related disasters, including hurricanes, flooding, and wildfires [9]	Women suffer disproportionate mortality during natural disasters [42] Female survivors suffer decreased life expectancy [82] Women and girls are at high risk of physical and sexual violence, especially those belonging to marginalized sectors of society [24, 83] Women are at higher risk for mood disorders such as depression and anxiety after disasters [50] Women giving birth in the time period following disasters have an increased risk of complications including preeclampsia, bleeding, and low birthweight infants [49]	Women of all ages are more calorie-deficient than men, leading to poor physical health and vulnerability to resource shortages ensuing from catastrophes [43] Women of all ages are more calorie-deficient than men, leading to poor physical health and vulnerability to resource shortages ensuing from catastrophes [46–48]	Women have unequal access to basic social goods and mortality is worsened when women have a lower socioeconomic status [4, 45] Women are often homebound caring for children and elderly while waiting for relatives to return prior to evacuation Poor, single, elderly women, adolescent girls, and women with disabilities are often at greatest risk for abuse because they have fewer personal, family, economic, and educational resources from which to draw protection, assistance, and support Women suffer disproportionate job loss and stagnant personal economic recovery following disasters [51] Poor access to obstetric care during and after disasters

(continued)

Table 19.1 (continued)

Exposure pathway	Gender disparities in health impacts	Physiologic and biologic vulnerabilities	Cultural and socioeconomic vulnerabilities
Shifting rainfall and temperature patterns impair crop, livestock, and fishery yields, contributing to food insecurity	Women suffer higher rates of macro- and micro-nutrient deficiencies Women suffer higher rates of anemia which is associated with cognitive impairments including poor attention span, diminished working memory, emotional regulatory issues, and impaired sensory perception [54] Malnutrition causes negative effects on neonatal outcomes including intrauterine growth restriction and perinatal mortality [56]	Women are inherently sensitive to the effects of food insecurity and resulting nutritional deficiencies due to increased needs during menstruation, pregnancy, and nursing	Nutritional scarcity can be intensified by cultural practices that prioritize food provision to children and adult males In developing regions, women produce between 60% and 80% of all food – livelihoods, as well as nutritional status, are threatened when climatic conditions prevent agricultural yields [84] Less than 10% of female farmers are landowners, and barely 2% have proper paperwork for their land [84] Therefore, women suffer on account of their relative lack of control over farmlands and nutritional security
Shifting rainfall and increased rates of evaporation lead to water insecurity and risk of waterborne disease	Water scarcity forces provision from sources that may be biologically and toxicologically contaminated, resulting in bacterial, viral, and protozoan infections as well as toxic exposures [12] Traveling long distances to procure water increases exposure to heat [64] Lack of access to water and sanitation creates unsafe conditions for women, especially during reproductive times [62]	Dehydration in pregnancy results in decreased uterine blood flow and is associated with pre-term labor [77] Infection in pregnancy leads to poor maternal and neonatal outcomes	Traditionally, women have the household role of providing water for the family. Water scarcity equates to more time spent harvesting water and less time spent on other activities of livelihood such as economic gain In some regions carrying water may use up to 85% of a woman's daily energy intake [12] Traveling long distances to collect water places women at risk for physical abuse and harm

Changes in temperature, precipitation, and ecology are altering the geographic distribution of vector-borne diseases	Exposure to mosquito-borne illnesses poses health threats to pregnant women who are exceptionally vulnerable Pregnant women have a risk of severe malaria that is three times as high as that of non-pregnant women [70] Zika virus carries devastating fetal impacts including microcephaly, CNS malformations, and impaired cognitive development [73] Dengue virus is associated with increased risk of cesarean delivery, eclampsia, and growth restriction [72]	Pregnant women have increased susceptibility to mosquito-transmitted diseases due to higher CO2 production, a chemoattractant for mosquitos, and increased peripheral blood flow, the heat from which allows mosquitos to locate hosts Hormonally induced changes in immunologic function during pregnancy lead to decreased immune response, which manifests as higher intensity of viremia and parasitemia [56, 69] Infection during pregnancy can result in anemia and diminished trans-placental nutrient transport, resulting in intrauterine growth restriction and increased vulnerability of the mother to hemorrhagic complications of delivery [71]	Women spend more time around the house performing domestic tasks, which places them in close proximity to domestic standing water and mosquito breeding sites Lack of access to pre-natal obstetric care and assisted deliveries places women with infections at risk of post-partum hemorrhage and poor maternal outcomes, including death
Climate-induced environmental change drives human migration and/or results in "trapped" populations	Women are more likely to undergo short-term migration (vs long-distance migration) which is often excluded from migration analysis [85] Women are more likely to be displaced by drought [85]	Lack of basic sanitation and health services compound health issues for refuges and migrants [76] Forced migration as a result of environmental changes or disasters is physiologically and mentally stressful, leading to poor health outcomes [76] Women whose partners travel frequently are at higher risk of HIV infection [76]	Marriage is a key driver of internal migration for women Low education is a risk factor for migration, and globally women suffer disproportionately from lack of education Women have fewer employment opportunities and thus are unable to migrate into economically viable and less environmentally vulnerable regions, thus becoming "trapped" Women who do migrate are still faced with the direct pressure of caring for children [76] Migration places women at risk for human trafficking [76]

UNFCCC decision 21/CP.22 (2017) additionally calls on the Parties to incorporate a gendered perspective in all elements of mitigation, adaptation, capacity, technology, and finance. Although this framework sets the stage for action, systematic integrative procedures are lacking, as are indicators to monitor progress.

The Sustainable Development Goals (SDGs) contain separate targets for poverty, gender equality, sustainability, and climate action. Opportunities to interconnect these separate targets through sub-targets and indicators that bridge sectors were largely lost during the development of the SDGs. For example, while there are energy-related indicators in the health goal (related to household use of biomass fuels), there are no health-related indicators in the energy or climate goals. Disaggregation and failure to explicitly link health with these other goal areas can lead to discordant efforts, inefficiencies, and communication barriers between involved agencies tasked with solving these problems [87].

United Nations International Strategy for Disaster Reduction (UNISDR) adopted the *Sendai Framework* in 2015 to establish common goals and standards for disaster risk reduction. This document formalizes climate change as a disaster-risk multiplier for women and recognizes women as important stakeholders in risk-reduction [1]. Furthermore, it calls on adopters to '*prepare, review and periodically update preparedness policies, plans and programmes with the involvement of all relevant institutions, considering climate change scenarios and their impact, and to facilitate the participation of all sectors and stakeholders* [1].' Strong accountability is fundamental to the Framework, which contains thirty-eight indicators to track progress in implementing the seven targets. These targets in combination aim to reduce disaster mortality and damage to critical infrastructure and economy through increased multi-hazard early warning systems, improved national and local mitigation strategies, and enhanced international cooperation. The framework also incorporates the related dimensions of SDG 1 (No poverty), 11 (Sustainable cities and communities), and 13 (Climate action). The Sendai Framework Monitor will also function as a management tool to help countries develop disaster risk reduction strategies, make risk-informed policy decisions, and allocate resources to prevent new disaster risks. It will also promote disaggregated data collection with formal bi-annual reporting via the Sendai Framework Monitoring Process.

Local Solutions: Building from the Ground Up

Although national and international policy frameworks are needed to support large-scale changes in the approach to sustainable development and climate adaptation and mitigation, local action and gradual social change inherently facilitate the integration of climate-compatible development solutions that are sustainable as well as regionally culturally and socially appropriate. Worldwide, women occupy many important spaces throughout society, including but not limited to primary food producers, guardians of natural resources, water harvesters, educators and raisers of children, caregivers for elderly, community leaders, technical and professional leaders, and political leaders. Harnessing this existing social capital and building on these roles is a starting point for meaningful local action. Women clearly play a vital role in the societal response to climate change, and their participation at all levels has been shown to result in greater responsiveness to citizen's needs and often increases cooperation across party and ethnic lines [24, 47]. Several concepts have been identified that can guide adaptation and mitigation interventions to result in gender-sensitive, climate-compatible development (Table 19.2) [87]. These include:

Table 19.2 Examples of gender-based solutions to promote health resilience to climate change at the local level

Climate-related exposure pathway	Gender-based solutions
Heat exposure	Increase knowledge among healthcare providers Provide air conditioning in maternal wards (shown to decrease intensive care need in neonatal period) [18] Increase access to pre-natal care in heat vulnerable geographic areas Implement heat early warning systems with educational messages targeted at women Collect and disseminate gender-disaggregated public health data Consider the detrimental effects of urban heat islands, especially in regions with poor access
Poor air quality	Promote domestic technologies appropriate to the needs of women and give them proper training Improve access to clean burning cook stoves – shown to reduce exposure to carbon monoxide, hydrocarbons, and particulate matter and decrease health risks [88] Consider women's transportation needs during urban planning Consider that air pollution increases maternal stress that impacts fetal outcomes and aggressively reach PM2.5 targets within the environments of reproductive-aged women [77]
Extreme-weather events	Train women to be leaders in community-level disaster risk reduction Provide gender sensitive emergency shelters that proactively safeguard women Plan for and provide emergency obstetric and gynecologic care early in the course of disasters Increase the availability of gender-disaggregated disaster-related health data Increase gender-specific public health messaging before, during, and after disasters Provide gender-sensitive psychological services in the aftermath of disasters Create economic recovery plans that provide vocational training for the female workforce
Food insecurity and malnutrition	Involve women in monitoring the effects of climate change Empower through women-centered climate-resilient farming models that encourage and assist women to gain cultivation rights and simultaneously provide skills and training to implement resilience building practices Enable community-based reintroduction of nutrient-dense, locally available wild edibles into the regular diets Strengthen nutritional interventions in reproductive-aged women

(continued)

Table 19.2 (continued)

Climate-related exposure pathway	Gender-based solutions
Water insecurity and waterborne disease	Use a gender approach when diagnosing and planning communities' freshwater requirements Increase accessibility to affordable home water filters Increase public investment in water infrastructure in high-risk areas such as urban slums Engage local female leaders and female heads of household in local, regional, and national sanitation projects to promote culturally acceptable infrastructure development that ensures women have safe and private access to hygienic facilities and clean water Promote water-saving practices that account for the different uses of water for women
Vector-borne diseases	Increase access to women's reproductive health care Train women in vulnerable regions to become health care providers Collect gender-disaggregated health data Develop vector-borne surveillance systems and early warning systems that permit effective and efficient pre-positioning of resources including bed-nets and insecticides
Forced migration	Plan and manage migration in order to reduce the chance of later humanitarian emergencies. Ease populations out of situations of vulnerability and capitalize on opportunities afforded to the individual by migration (e.g., moving populations away from flood zones into areas of safety and prosperity) Actively manage essential natural resources in climate "hotspots" to prevent forced migration/trapped populations Build urban infrastructure that is sustainable and flexible to accommodate influxes in rural migrants Actively deploy resources to protect women's personal safety and provide reproductive health services in refugee camps and informal settlement

Ensure Participation Women have crucial societal roles as educators, caregivers, holders of knowledge, and agents of social change. These roles enable women to effectively design and implement culturally acceptable interventions where they are needed most. Recognizing the importance of these roles, women should be empowered as key stakeholders at the outset of any project, with the understanding that combining scientific data and community knowledge will yield better results.

Prioritize Education Education regarding the gender-specific health threats of climate change is needed within public health, policy, medicine, and general education. Additionally, investment in skills and capacity building among women will foster leadership and strengthen resilience.

Improve Data Collecting high-quality, gender-disaggregated data will enable better understanding of gender-climate-health associations and allow for predictive modeling that can inform community-based interventions.

Enhance Pre-Planning A comprehensive assessment of women's assets and vulnerabilities is foundational to any adaptation or development project including, as examples, disaster risk reduction, transportation, finance, communication, water management, technology transfer, agriculture, and health. Such assessments not only provide a more in-depth understanding of the effects of climate change but they can also reveal the political, physical, and socioeconomic reasons why individuals suffer disproportionately, thus creating the opportunity for effective intervention.

Re-Define Success Women's health outcomes and economic prosperity can serve as surrogate markers for development, disaster risk reduction, and climate adaptation. Thus, these markers are well suited as indicators for project and policy success. Similarly, regions with poor health outcomes should be identified as "hot spots" for current and future vulnerability to climate change.

Improve Multisector Coordination Developing mechanisms for reporting and regular analysis of gender dimensions using common indicators within all sectors will increase transparency and cooperation in achieving this cross-sectoral goal.

The following series of case studies highlight the examples of successful implementation of climate-resilient health solutions on local and regional levels.

Empowering Clean Energy Solutions

Improved cook stoves have been developed and deployed in many low- and middle-income countries with a dual objective of reducing health *and* environmental risks associated with biomass combustion. However, despite enormous potential health benefits, their widespread dissemination and use has been slow. For a variety of behavioral, cultural, and policy related reasons, simply making stoves available on the market is not enough to achieve consistent household adoption [89].

Most improved cook stoves cost approximately US$60, which for individuals living with less than $1/day is prohibitive [90]. Additionally, stoves require frequent maintenance. Project Surya is piloting novel strategies to overcome this financial barrier and incentivize use among female heads of household in Nigeria and India. Through the Surya initiative, women receive personal loans from rural banks to make the initial investment. Then, the use of each stove is monitored by a wireless sensor that registers stove usage which is translated into tons of carbon mitigated. Women are compensated US$6 for each tCO_2 which equates to roughly US$32 per stove per year. The quarterly returns are distributed directly into an electronic bank account, established in the woman's name, which can be accessed through a mobile phone. Additionally, women are being trained to fix stoves, enabling entrepreneurship and female business leadership. Sustained engagement with five hamlets within the Indo-Gangetic plains resulted in a high degree of clean technology adoption and a subsequent reduction in black carbon emissions by 40%. These and similar pilot programs have demonstrated substantial decreases in lower respiratory infection in children and chronic obstructive pulmonary disease and ischemic heart disease in adults [88].

Leading Community Resilience

Women face a disproportionate burden of negative health impacts in the wake of disasters, which are increasing in frequency and threatening vulnerable communities, especially in the African continent. As a result of disasters, many urban areas are seeing rapid expansion of informal sprawling settlements, where the poorest residents live in inadequate sheet metal houses without access to basic resources such as water, sewage treatment, or electricity. Residents of these communities, especially young women, face the dangers of worsened poverty, crime, communicable diseases, and domestic, physical, and sexual abuse. Additionally, young women may assume the role of heading households and suffer from poor access to education as a result of domestic duties and endemic poor educational opportunity.

In the township of Potchefstroom, in the North West Province of South Africa, the Girls in Risk Reduction Leadership (G.I.R.R.L.) project aims to reduce the vulnerability of marginalized young women using practical capacity building initiatives to increase individual and community resilience to disasters [91]. Girls are selected based on the recommendations of school officials and local ward leaders and participates and enrolled in an intensive 2-month training program, with instruction in personal and public health, fire safety, counseling, and disaster planning. Graduates from the program then work with the local disaster coordination office to help design and coordinate plans to reduce the impact of disasters and extreme events on the community. By empowering marginalized girls to be leaders in disaster risk reduction, the traditional roles of women are expanded and recognized [24].

Climate-Resilient Farming

The district of Marathwada, India, is an agriculturally rich region that is highly susceptible to droughts. According to a study by the Indian Institute of Tropical Meteorology, between 1870 and 2015 the region faced 22 droughts, including five instances of two consecutive drought years, the most recent of which were in 2014–15 and 2015–16 [92, 93]. Successive drought years can not only lead to damaging the agricultural economy but also to harming the health of the communities, where there are high baseline rates of anemia and malnutrition [94].

Through focus groups with the local female farmers, led by Swayam Shikshan Prayog – an organization that works for women's empowerment – it was discovered that although women are extensively involved in farming, decisions related to crop selection, cultivation, and consumption rest exclusively with the male counterparts. Women expressed concerns that male counterparts were committed to growing single-strain cash crops like soya, cotton, sugarcane, which come with expensive chemical inputs. Women shared that they felt stressed and suffered from health ailments because of the increased burden of providing food for the family and high consumption of chemically contaminated food, which they believed contributed to poor health.

Given the regions' propensity for drought and food insecurity and understanding that climate change may worsen this pattern, SSP initiated a women-centered climate-resilient farming model. This approach repositions women as decision makers and bearers of knowledge, enabling them to make informed decisions related to what to grow, what to consume, and how much to sell. This model encourages women to gain cultivation rights by starting with a small section of family land, initially ranging to one-half to one acre. Women are then given multiple trainings and participate in groups centered around cultivating food crops, including cereals, pulses, oil seeds, vegetables, and animal fodder. The trainings focus on resilience-building practices including the use of bio fertilizers, preservation and exchange of local seeds, increasing crop diversity, increasing number of crop cycles, choice of drought-resistance and water efficient crops, water conserving irrigation techniques, tree plantation, and more. The adoption of the model has resulted in improved food and nutrition security of the households, reduced the cost of cultivation, increased productivity, and rendered other social, economic, and environmental benefits.

Conclusions

Women, especially those in low- and middle-income countries, will suffer disproportional negative health-related impacts from climate change. While gender has been increasingly factored in to climate change projects and policy, progress has still been slow to reduce gender-based health disparities and to mobilize women as a vast social resource for climate change mitigation, adaptation, and disaster risk reduction and management. Without full participation and contribution of women in decision-making and leadership, real climate mitigation and resilience building cannot be achieved. Moving forward, targeted local action coupled with compliance with the monitoring processes advocated by the UNFCCC, the SDGs, and the Sendai Framework are critical to address the complex interactions between poverty, gender-based social discrimination, and climate change that threaten to amplify gender-based health disparities. Additionally, high-level political engagement with the implementation of the UN landmark agreements [95] is necessary to assure policies and programs move beyond traditional separations of health, gender, and environment and embrace proactive and gender-based solutions that protect women's health and mobilize their vast social potential.

References

1. Crimmins A, Balbus J, Gamble J, Beard C, Bell J, Dodgen D, Eisen R, Fann N, Hawkins M, Herring S. The impacts of climate change on human health in the United States: a scientific assessment. Washington, DC: Global Change Research Program; 2016.
2. United Nations Division for the Advancement of Women (DAW). Environmental Management and the Mitigation of Natural Disasters: a Gender Perspective. Report of the Expert Group Meeting Ankara, Turkey, 6 – 9 November 2001.
3. The world health report 2002: reducing risks, promoting healthy life.World Health Organization; 2002.
4. Behera D, Chakrabarti T, Khanduja KL. Effect of exposure to domestic cooking fuels on bronchial asthma. Indian J Chest Dis Allied Sci. 2001;43(1):27–31.
5. Pachauri RK, Allen MR, Barros VR, Broome J, Cramer W, Christ R, Church JA, Clarke L, Dahe Q, Dasgupta P. Climate change 2014: synthesis report. Contribution of working groups I, II and III to the fifth assessment report of the Intergovernmental Panel on Climate Change, IPCC; 2014.
6. Phung D, Thai PK, Guo Y, Morawska L, Rutherford S, Chu C. Ambient temperature and risk of cardiovascular hospitalization: an updated systematic review and meta-analysis. Sci Total Environ. 2016;550:1084–102.
7. Basu R, Pearson D, Malig B, Broadwin R, Green R. The effect of high ambient temperature on emergency room visits. Epidemiology. 2012;23(6):813–20.
8. Bouchama A, Knochel JP. Heat stroke. N Engl J Med. 2002;346(25):1978–88.
9. Watts N, Amann M, Ayeb-Karlsson S, Belesova K, Bouley T, Boykoff M, Byass P, Cai W, Campbell-Lendrum D, Chambers J. The Lancet Countdown on health and climate change: from 25 years of inaction to a global transformation for public health. Lancet. 2017;391(10120):581–630.
10. Kovats RS, Hajat S. Heat stress and public health: a critical review. Annu Rev Public Health. 2008;29:41–55.
11. Sarofim MC, Saha S, Hawkins M, Mills D, Hess J, Horton R, Kinney P, Schwartz J, Juliana AS. Ch. 2: temperature-related death and illness. Washington, DC: US Global Change Research Program; 2016.
12. Duncan K. Global climate change, air pollution, and women's health. WIT Trans Ecol Environ. 2006;99:631–33.
13. Ha S, Liu D, Zhu Y, Kim SS, Sherman S, Grantz KL, Mendola P. Ambient temperature and stillbirth: a multi-center retrospective cohort study. Environ Health Perspect. 2017;125(6):067011.
14. Basu R, Sarovar V, Malig BJ. Association between high ambient temperature and risk of stillbirth in California. Am J Epidemiol. 2016;183(10):894–901.
15. Strand LB, Barnett AG, Tong S. The influence of season and ambient temperature on birth outcomes: a review of the epidemiological literature. Environ Res. 2011;111(3):451–62.
16. Van Zutphen AR, Lin S, Fletcher BA, Hwang S-A. A population-based case–control study of extreme summer temperature and birth defects. Environ Health Perspect. 2012;120(10):1443.
17. Makhseed MA, Musini VM, Ahmed MA, Monem RA. Influence of seasonal variation on pregnancy-induced hypertension and/or preeclampsia. Aust N Z J Obstet Gynaecol. 1999;39(2):196–9.
18. Kakkad K, Barzaga ML, Wallenstein S, Azhar GS, Sheffield PE. Neonates in Ahmedabad, India, during the 2010 heat wave: a climate change adaptation study. J Environ Public Health. 2014;2014:946875.
19. Nolte CG, Dolwick PD, Fann N, Horowitz LW, Naik V, Pinder RW, Spero TL, Winner DA, Ziska LH. Air quality. In: Reidmiller DR, Avery CW, Easterling DR, Kunkel KE, Lewis KLM, Maycock TK, Stewart BC, editors. Impacts, risks, and adaptation in the United States: fourth national climate assessment, vol. II. Washington, DC: U.S. Global Change Research Program; 2018. p. 512–38. https://doi.org/10.7930/NCA4.2018.CH13.

20. Newby DE, Mannucci PM, Tell GS, Baccarelli AA, Brook RD, Donaldson K, Forastiere F, Franchini M, Franco OH, Graham I. Expert position paper on air pollution and cardiovascular disease. Eur Heart J. 2014;36(2):83–93.
21. Gamble J, Balbus J, Berger M, Bouye K, Campbell V, Chief K, Conlon K, Crimmins A, Flanagan B, Gonzalez-Maddux C. Ch. 9: populations of concern. Washington, DC: US Global Change Research Program; 2016.
22. Bell ML, McDermott A, Zeger SL, Samet JM, Dominici F. Ozone and short-term mortality in 95 US urban communities, 1987–2000. JAMA. 2004;292(19):2372–8.
23. Jerrett M, Burnett RT, Pope CA III, Ito K, Thurston G, Krewski D, Shi Y, Calle E, Thun M. Long-term ozone exposure and mortality. N Engl J Med. 2009;360(11):1085–95.
24. Chen LH, Knutsen SF, Shavlik D, Beeson WL, Petersen F, Ghamsary M, Abbey D. The association between fatal coronary heart disease and ambient particulate air pollution: are females at greater risk? Environ Health Perspect. 2005;113(12):1723.
25. Künzli N, Jerrett M, Mack WJ, Beckerman B, LaBree L, Gilliland F, Thomas D, Peters J, Hodis HN. Ambient air pollution and atherosclerosis in Los Angeles. Environ Health Perspect. 2005;113(2):201.
26. Kim CS, Hu S. Regional deposition of inhaled particles in human lungs: comparison between men and women. J Appl Physiol. 1998;84(6):1834–44.
27. Sørensen M, Daneshvar B, Hansen M, Dragsted LO, Hertel O, Knudsen L, Loft S. Personal PM2. 5 exposure and markers of oxidative stress in blood. Environ Health Perspect. 2003;111(2):161.
28. Liu C-B, Hong X-R, Shi M, Chen X-Q, Huang H-J, Chen J-H, Yang K, Chen S-Q, Chen H-Q, Kan H-D. Effects of prenatal [... formula...] exposure on fetal cardiovascular malformations in Fuzhou, China: a retrospective case–control study. Environ Health Perspect. 2017;125(5):057001.
29. Glinianaia SV, Rankin J, Bell R, Pless-Mulloli T, Howel D. Particulate air pollution and fetal health: a systematic review of the epidemiologic evidence. Epidemiology. 2004;15(1):36–45.
30. Šrám RJ, Binková B, Dejmek J, Bobak M. Ambient air pollution and pregnancy outcomes: a review of the literature. Environ Health Perspect. 2005;113(4):375.
31. Ritz B, Yu F, Fruin S, Chapa G, Shaw GM, Harris JA. Ambient air pollution and risk of birth defects in Southern California. Am J Epidemiol. 2002;155(1):17–25.
32. Li X, Huang S, Jiao A, Yang X, Yun J, Wang Y, Xue X, Chu Y, Liu F, Liu Y. Association between ambient fine particulate matter and preterm birth or term low birth weight: an updated systematic review and meta-analysis. Environ Pollut. 2017;227:596–605.
33. Zhu X, Liu Y, Chen Y, Yao C, Che Z, Cao J. Maternal exposure to fine particulate matter (PM 2.5) and pregnancy outcomes: a meta-analysis. Environ Sci Pollut Res. 2015;22(5):3383–96.
34. Melody SM, Ford J, Wills K, Venn A, Johnston FH. Maternal exposure to short-to medium-term outdoor air pollution and obstetric and neonatal outcomes: a systematic review. Environ Pollut. 2018;244:915–25.
35. Pedersen M, Stayner L, Slama R, Sørensen M, Figueras F, Nieuwenhuijsen MJ, Raaschou-Nielsen O, Dadvand P. Ambient air pollution and pregnancy-induced hypertensive disorders: a systematic review and meta-analysis. Hypertension. 2014;64(3):494–500.
36. Michikawa T, Morokuma S, Yamazaki S, Fukushima K, Kato K, Nitta H. Exposure to air pollutants during the early weeks of pregnancy, and placenta praevia and placenta accreta in the western part of Japan. Environ Int. 2016;92:464–70.
37. Michikawa T, Morokuma S, Yamazaki S, Fukushima K, Kato K, Nitta H. Air pollutant exposure within a few days of delivery and placental abruption in Japan. Epidemiology. 2017;28(2):190–6.
38. Lu M-C, Wang P, Cheng T-J, Yang C-P, Yan Y-H. Association of temporal distribution of fine particulate matter with glucose homeostasis during pregnancy in women of Chiayi City, Taiwan. Environ Res. 2017;152:81–7.
39. Chafe ZA, Brauer M, Klimont Z, Van Dingenen R, Mehta S, Rao S, Riahi K, Dentener F, Smith KR. Household cooking with solid fuels contributes to ambient PM2. 5 air pollution and the burden of disease. Environ Health Perspect. 2014;122(12):1314.

40. Beggs PJ, Bambrick HJ. Is the global rise of asthma an early impact of anthropogenic climate change? Cien Saude Colet. 2006;11(3):745–52.
41. World Health Organization. Household air pollution and health; 2018. Available at https://www.who.int/news-room/fact-sheets/detail/household-air-pollution-and-health
42. International Federation of the Red Cross and Red Crescent. World disasters report; 2007. Available at https://www.ifrc.org/PageFiles/99876/WDR2007-English.pdf
43. Rahman MS. Climate change, disaster and gender vulnerability: a study on two divisions of Bangladesh. Am J Hum Ecol. 2013;2(2):72–82.
44. Aguilar L, Araujo A, Quesada-Aguilar A. Gender and climate change. San José: International Union for Conservation of Nature; 2004.
45. Moosa CS, Tuana N. Mapping a research agenda concerning gender and climate change: a review of the literature. Hypatia. 2014;29(3):677–94.
46. Chowdhury AMR, Bhuyia AU, Choudhury AY, Sen R. The Bangladesh cyclone of 1991: why so many people died. Disasters. 1993;17(4):291–304.
47. Dankelman I. Gender, climate change and human security: lessons from Bangladesh, Ghana and Senegal. Newcastle upon Tyne: Gender and Disaster Network; 2008.
48. Cannon T, Twigg J, Rowell J. Social vulnerability, sustainable livelihoods and disasters. Londres: DFID; 2003.
49. Tong VT, Zotti ME, Hsia J. Impact of the Red River catastrophic flood on women giving birth in North Dakota, 1994–2000. Matern Child Health J. 2011;15(3):281–8.
50. Norris FH, Friedman MJ, Watson PJ, Byrne CM, Diaz E, Kaniasty K. 60,000 disaster victims speak: part I. An empirical review of the empirical literature, 1981–2001. Psychiatry. 2002;65(3):207–39.
51. Tobin-Gurley J, Peek L, Loomis J. Displaced single mothers in the aftermath of Hurricane Katrina: resource needs and resource acquisition. Int J Mass Emerg Disasters. 2010;28(2):170–206.
52. United Nations Development Programme. Conceptual overview. Gender mainstreaming: a key driver of development in environment and energy; 2007. Available at http://www.energyandenvironment.undp.org/undp/indexAction.cfm?module=Library&action=GetFile&DocumentAttachmentID=2315
53. World Bank Group. World development indicators; 2018. Available at http://databank.worldbank.org/data/home.aspx
54. Jáuregui-Lobera I. Iron deficiency and cognitive functions. Neuropsychiatr Dis Treat. 2014;10:2087.
55. Food and Agriculture Organization of the United Nations. The state of food insecurity in the world, 2013: the multiple dimensions of food security; 2013. Available at http://www.fao.org/3/a-i3434e.pdf
56. Lindsay S, Ansell J, Selman C, Cox V, Hamilton K, Walraven G. Effect of pregnancy on exposure to malaria mosquitoes. Lancet. 2000;355(9219):1972.
57. German Development Institute. Drought adaptation and resilience in developing countries; 2017. Available at http://www.die-gdi.de/en/briefing-paper/article/drought-adaptation-andresiliencein-developing-countries/
58. Trtanj J, Jantarasami L, Brunkard J, Collier T, Jacobs J, Lipp E, McLellan S, Moore S, Paerl H, Ravenscroft J. Ch. 6: climate impacts on water-related illness. In: The impacts of climate change on human health in the United States: a scientific assessment. Washington, DC: US Global Change Research Program; 2016. p. 157–88.
59. Watts N, Amann M, Ayeb-karlsson S, Belesova K, Bouley T, Boykoff M, Byass P, Cai W, Depoux A, Dominguez-salas P. Review the Lancet Countdown on health and climate change: from 25 years of inaction to a global transformation for public health. Lancet. 2018;391(10120):581–630.
60. Kovats RS, Edwards SJ, Charron D, Cowden J, D'Souza RM, Ebi KL, Gauci C, Gerner-Smidt P, Hajat S, Hales S. Climate variability and campylobacter infection: an international study. Int J Biometeorol. 2005;49(4):207–14.

61. Arnell NW. Climate change and global water resources: SRES emissions and socio-economic scenarios. Glob Environ Chang. 2004;14(1):31–52.
62. Birch EL, Meleis A, Wachter S. The urban water transition: why we must address the new reality of urbanization, women, water, and sanitation in sustainable development. wH2O J Gend Water. 2012;1(1):1.
63. Kourtis AP, Read JS, Jamieson DJ. Pregnancy and infection. N Engl J Med. 2014;370(23):2211–8.
64. Shiva V, Jalees K. Water & women: a report by research foundation for science, technology, and ecology for national commission for women. Navdanya/RFSTE; 2005.
65. Sandys E. Women and water: women 2000 and beyond. New York: United Nations Division for the Advancement of Women Department of Economic and Social Affairs; 2005.
66. Beard CB, Eisen R, Barker C, Garofalo J, Hahn M, Hayden M, Monaghan A, Ogden N, Schramm P. Ch. 5: vectorborne diseases. Washington, DC: US Global Change Research Program; 2016.
67. Monaghan AJ, Sampson K, Steinhoff D, Ernst K, Ebi K, Jones B, Hayden M. The potential impacts of 21st century climatic and population changes on human exposure to the virus vector mosquito Aedes aegypti. Clim Chang. 2018;146(3–4):487–500.
68. Gupta B, Reddy BN. Fight against dengue in India: progresses and challenges. Parasitol Res. 2013;112(4):1367–78.
69. Mbonye AK, Neema S, Magnussen P. Preventing malaria in pregnancy: a study of perceptions and policy implications in Mukono district, Uganda. Health Policy Plan. 2005;21(1):17–26.
70. Rijken MJ, McGready R, Boel ME, Poespoprodjo R, Singh N, Syafruddin D, Rogerson S, Nosten F. Malaria in pregnancy in the Asia-Pacific region. Lancet Infect Dis. 2012;12(1):75–88.
71. Steketee RW, Wirima JJ, Hightower AW, Slutsker L, Heymann DL, Breman JG. The effect of malaria and malaria prevention in pregnancy on offspring birthweight, prematurity, and intrauterine growth retardation in rural Malawi. Am J Trop Med Hyg. 1996;55(1_Suppl):33–41.
72. Pouliot SH, Xiong X, Harville E, Paz-Soldan V, Tomashek KM, Breart G, Buekens P. Maternal dengue and pregnancy outcomes: a systematic review. Obstet Gynecol Surv. 2010;65(2):107–18.
73. Petersen LR, Jamieson DJ, Powers AM, Honein MA. Zika virus. N Engl J Med. 2016;374(16):1552–63.
74. Rechkemmer A, O'Connor A, Rai A, Decker Sparks JL, Mudliar P, Shultz JM. A complex social-ecological disaster: environmentally induced forced migration. Disaster Health. 2016;3(4):112–20.
75. Bartlett S. Climate change and urban children: impacts and implications for adaptation in low- and middle-income countries. Environ Urban. 2008;20(2):501–19.
76. Adanu RM, Johnson TR. Migration and women's health. Int J Gynecol Obstet. 2009;106(2):179–81.
77. Kuehn L, McCormick S. Heat exposure and maternal health in the face of climate change. Int J Environ Res Public Health. 2017;14(8):853.
78. Kim H, Kim J, Kim S, Kang SH, Kim HJ, Kim H, Heo J, Yi SM, Kim K, Youn TJ. Cardiovascular effects of long-term exposure to air pollution: a population-based study with 900 845 person-years of follow-up. J Am Heart Assoc. 2017;6(11):e007170.
79. Wang Y, Shi L, Lee M, Liu P, Di Q, Zanobetti A, Schwartz JD. Long-term exposure to PM2. 5 and mortality among older adults in the southeastern US. Epidemiology (Cambridge, MA). 2017;28(2):207.
80. Grindheim G, Toska K, Estensen ME, Rosseland LA. Changes in pulmonary function during pregnancy: a longitudinal cohort study. BJOG. 2012;119(1):94–101. https://doi.org/10.1111/j.1471-0528.2011.03158.x.
81. Pope DP, Mishra V, Thompson L, Siddiqui AR, Rehfuess EA, Weber M, Bruce NG. Risk of low birth weight and stillbirth associated with indoor air pollution from solid fuel use in developing countries. Epidemiol Rev. 2010;32(1):70–81.
82. Neumayer E, Plümper T. The gendered nature of natural disasters: the impact of catastrophic events on the gender gap in life expectancy, 1981–2002. Ann Assoc Am Geogr. 2007;97(3):551–66.

83. Chan EYY, Murray V. What are the health research needs for the Sendai framework? Lancet. 2017;390(10106):e35–6. https://doi.org/10.1016/S0140-6736(17)31670-7.
84. Liu S, Zhou Y, Wang X, Wang D, Lu J, Zheng J, Zhong N, Ran P. Biomass fuels are the probable risk factor for chronic obstructive pulmonary disease in rural South China. Thorax. 2007;62(10):889–97. https://doi.org/10.1136/thx.2006.061457.
85. Clarke L, Blanchard K, Maini R, Radu A, Eltinay N, Zaidi Z, Murray V. Knowing what we know – reflections on the development of technical guidance for loss data for the Sendai framework for disaster risk reduction. PLoS Curr. 2018;10 https://doi.org/10.1371/currents.dis.537bd80d1037a2ffde67d66c604d2a78.
86. Mitchell T, Maxwell S. Defining climate compatible development. London: Climate and Development Knowledge Network; 2010.
87. Sorensen C, Murray V, Lemery J, Balbus J. Climate change and women's health: impacts and policy directions. PLoS Med. 2018;15(7):e1002603.
88. Wilkinson P, Smith KR, Davies M, Adair H, Armstrong BG, Barrett M, Bruce N, Haines A, Hamilton I, Oreszczyn T. Public health benefits of strategies to reduce greenhouse-gas emissions: household energy. Lancet. 2009;374(9705):1917–29.
89. Sorensen C, Saunik S, Sehgal M, Tewary A, Govindan M, Lemery J, Balbus J. Climate change and women's health: impacts and opportunities in India. GeoHealth. 2018;2(10):283–97.
90. Ramanathan T, Ramanathan N, Mohanty J, Rehman IH, Graham E, Ramanathan V. Wireless sensors linked to climate financing for globally affordable clean cooking. Nat Clim Chang. 2017;7(1):44.
91. International Strategy for Disaster Reduction. Gender perspectives: integrating disaster risk reduction and climate change adaptation. Geneva: ISDR; 2008. Available at https://www.unisdr.org/we/inform/publications/3391
92. Seetharaman G. Why drought-prone Marathwada needs to look beyond the immediate monsoons. The Economic Times; 2017, April 2.
93. Kulkarni A, Gadgil S, Patwardhan S. Monsoon variability, the 2015 Marathwada drought and rainfed agriculture. Curr Sci. 2016;111(7):1182–93.
94. Government of India. National family health survey. Maharashtra state factsheet. 2015–2016. Ministry of Health and Family Welfare; 2015. Available at http://rchiips.org/NFHS/pdf/NFHS4/MH_FactSheet.pdf
95. Parties to the United Nations Framework Convention on Climate Change 2015 Paris Agreement. Available at https://unfccc.int/process-and-meetings/the-paris-agreement/the-paris-agreement

Chapter 20
The Impact of Climate Change on Public Health in Small Island States and Caribbean Countries

Muge Akpinar-Elci and Hugh Sealy

Introduction

The effects of climate change (CC) on human health are a growing global issue. Small Island states are more susceptible to CC impacts related to public health, food security, natural resources, and the fragile open economies of small developing countries. In the small island states, CC exacerbates the impacts of many diseases such as heat stress, asthma, and vector, food, and waterborne diseases. Rising sea levels threaten the health and well-being of small island states, reducing resources, while increasing temperatures and acidity negatively impact sea life in general, catastrophically impacting coral reefs. If the world does not act quickly enough to identify and implement solutions for mitigation and adaptation for CC, then the small islands will suffer more in the near future.

In this chapter, we will summarize the current knowledge on the physical, socio-economic, and health effects of CC from the small island states' perspective. We also share our experience as small islanders. The final section outlines the chapter's conclusions.

M. Akpinar-Elci (✉)
Old Dominion University, Norfolk, VA, USA
e-mail: makpinar@odu.edu

H. Sealy
The University of the West Indies, Cave Hill, Barbados, West Indies
e-mail: hugh.sealy@cavehill.uwi.edu

K. E. Pinkerton, W. N. Rom (eds.), *Climate Change and Global Public Health*, Respiratory Medicine, https://doi.org/10.1007/978-3-030-54746-2_20

The Physical Context: Current and Expected (by 2100) Physical Effects of Climate Change on Small Island States

Tropical or sub-tropical small islands exhibit variable climates but can be generally characterized by distinct wet and dry seasons. They are also characteristically threatened by periodic extreme weather events such as cyclones. The majority of the data cited below is the Intergovernmental Panel on Climate Change (IPCC), a body of government-appointed scientists with the mandate to analyze and interpret the most recent climate data. The IPCC published its fourth Assessment Report (AR4) in 2007 [1] and its fifth Assessment Report (AR5) in 2013 [2]. In 2018, the IPCC published a special report on the impacts of global warming of 1.5° [3], which was lobbied for intensely by the Alliance of Small Island States (AOSIS) during the negotiations leading up to and beyond the Paris Agreement in 2015.

Atmospheric Temperatures

According to AR5, *"The period from 1983 to 2012 was likely the warmest 30-year period of the last 1400 years in the Northern Hemisphere."* Mean atmospheric surface temperatures have risen by approximately 1 °C since 1750. *"Human activities are estimated to have caused approximately 1.0 °C of global warming above pre-industrial levels, with a likely range of 0.8–1.2 °C. Global warming is likely to reach 1.5 °C between 2030 and 2052 if it continues to increase at the current rate (high confidence)"* [3].

The above is a quote from the latest IPCC report (2018). The governments of the small islands have argued consistently that 2° of warming was too much. "*1.5 to stay alive*" has been the rallying call of small islanders. The governments represented within CARICOM (Caribbean Community consisting of 15 member countries) welcomed the release of the 1.5 Report and in a ministerial statement said "*Note with grave concern the findings of the IPCC 1.5 °C Special Report that climate-related risks for natural and human systems including health, livelihoods, food security, water supply, human security, and economic growth are significantly higher for global warming of 1.5 °C than at the present warming levels of 1.0 °C above pre-industrial levels. Particularly worrisome for Small Island Developing States (SIDS) is the finding that 70–90% of tropical coral reefs will be lost at 1.5 °C of warming and 99% of tropical coral reefs will be lost at 2 °C of warming"* [4].

From the above statement, it is evident that the Caribbean region is cognizant of the risks of climate change to the health of its citizens and is concerned. The models used in the AR4 provided a range of predicted increases (1.1–6.4 °C) for average atmospheric temperatures near the earth's surface (over land and over water) until the end of the twenty-first century. In AR5 it was revealed that GHG emissions increased by 1.3% per year during the period 1970–2000 and then increased by 2.2% per year between 2000 and 2010. The global recession of 2007–2008 provided only a temporary respite. AR5 predicted a 2.5–7.8 °C temperature increase by 2100 if the world continues on its existing trajectory.

In 2018, UN Environment published its latest emissions gap report [5]. It showed that global GHG emissions are yet to peak and in 2017 (including emissions from land use change) reached 53.5 gigatons of carbon dioxide equivalent ($GtCO_2e$). The UNEP 2018 report has determined that to limit global warming to less than 2 °C would require triple the effort and to limit warming to less than 1.5° would require a global effort 5 times greater (a reduction of 32 $GtCO_2e$ per year) than what countries have currently promised [5]. Unless emissions are reduced by 55% below 2017 levels by 2030, it may no longer be possible to contain warming to less than 1.5°. This is catastrophic news for small islands.

The temperature increases are not and will not be the same around the globe. More warming is occurring at the poles than at the equator. More warming of the atmosphere is occurring over land than over water, although greater than 80% of the increased heat is being absorbed by the oceans. This implies that for tropical small island states the increase in atmospheric temperatures may be different to the global average [6].

Data analysis by Trenberth et al. indicates consistent but non-linear warming trends in all small islands regions (Caribbean, Mediterranean, Indian Ocean, and Pacific) during the period 1901–2004 [7]. Ocean surface and island air temperatures have increased by between 0.6 and 1.0 °C with decadal increases between 0.3 and 0.5 °C in the Pacific as compared to 0–0.5 °C increases per decade in the Caribbean, Indian Ocean, and Mediterranean islands between 1971 and 2004. The seven Atmospheric Ocean General Circulation Models used in the IPCC AR4 all predict increases in surface air temperatures for all island regions. Surface air temperatures are predicted to be at least 2.5 °C higher than 1990 levels by 2100 in the South Pacific [8].

Sea-Level Rise and Ocean Acidification

Perhaps the greatest existential threats posed by CC to small islands and low-lying coastal states are those of sea-level rise (SLR) and ocean acidification. Currently, the majority of the SLR is predicted to be from thermal expansion, with temperature increases already evident at depths greater than 3000 m [9].

The amount of SLR that islands will experience by 2100 will be dependent upon the degree of further warming that will occur, with non-linear responses and tipping points potentially occurring. For example, loss of the Greenland and West Antarctica ice sheets would significantly increase the magnitude and public health threat of SLR to small islands. If temperature increases are between 3 and 5 °C by the end of the century, as was the case during the last interglacial period about 125,000 years ago, an associated SLR of 4–6 m could be expected. Pulwarty et al. claim that sea levels rose in the Caribbean on average 10 centimeters during the twentieth century [10] (Fig. 20.1).

The IPCC 2018 predicts that sea levels will rise between 0.26 and 0.77 m by 2100 under a 1.5° scenario, 0.1 m less than in the 2° scenario [3]. Irreversible loss of the Greenland ice sheet could be triggered at around 1.5–2.0° of warming. The

Fig. 20.1 A cliff on the north coast of Barbados

least reported, but perhaps the most insidious, physical impact of CC with potential far-reaching economic impacts on Small Island Developing States (SIDS) is the acidification of the world's oceans. The lowering of the pH is directly related to the concentration of carbon dioxide dissolved in the oceans. The concentration of carbon dioxide in the oceans is directly related to the concentration of carbon dioxide in the atmosphere. The acidity of the oceans restricts the storage of carbon by coral species and affects all calciferous marine species (Fig. 20.2).

It has been predicted that if the carbon dioxide levels stabilize in the atmosphere at 450 ppm (which is the CO_2 target associated with the <2 °C temperature target being negotiated under the United Nations Framework Convention on Climate Change-UNFCCC) coral reefs will cease to grow. Atmospheric stabilization at 550 ppm or greater (which is possible under current global emission pathways) would result in the dissolution of all existing corals [11]. The loss of the coral reefs would result in a loss of fisheries, irreversible damage to a very diverse ecosystem, loss of mechanical protection from waves, and loss of replenishment sand for beaches. Ultimately, most if not all sand beaches would be lost with resulting socio-economic impacts.

Hydrology and Storm Events

According to Pulwarty et al., if countries were ranked for the number of disaster events per unit area, small islands would occupy 19 out of the top 20 natural disaster prone regions in the world [10]. The following quote is from the Human Development Report 2011: *"Of the 10 countries suffering the greatest number of natural disasters per capita from 1970 to 2010, 6 were small island developing states"* [12].

Fig. 20.2 A fringing reef on the west coast of Barbados

Countries like Barbados have argued that despite their very high Human Development Index, their vulnerability to disasters should always be considered when assessing the economic status of small islands and determining eligibility for concessionary developmental financing. SIDS have consistently argued that loss and insurance mechanisms for SIDS need to be constructed specifically to address this vulnerability. Haiti, perhaps the most unfortunate of all small islands when it comes to disasters, will not recover for perhaps decades from the earthquake in January 2010, which killed over 300,000 people.

According to the Economic Commission for Latin America and the Caribbean (ECLAC, 2018) [13] hurricanes Irma and Maria caused US$5.4 billion in losses in Anguilla, The Bahamas, British Virgin Islands (BVI), Sint Maarten, and Turks and Caicos Islands in 2017, mostly in the tourism and housing sectors. According to a study [14] published in the New England Journal of Medicine, the toll from Maria has been estimated at 4645 excess deaths, mostly as a result of delayed health care in the 3 months immediately following the passage of the hurricane, although the official count of immediate deaths was 64.

The AR4 and AR5 could not detect a trend in the number of cyclones but did determine that the intensity of tropical cyclones had increased in the North Atlantic since 1970 [9]. Pulwarty et al. were unable to predict with any high degree of certainty how precipitation patterns will vary in the Caribbean as a result of CC [10]. There is some indication that the southern Caribbean and Central America region is drying, with predicted 20% less precipitation by 2100 [10]. The general indicators of ongoing CC effects for Caribbean islands predict shorter wet seasons, more intense rainfall events, longer dry seasons, and increased periods of drought.

It is likely that fresh water resource management will become a constraint in further economic development among some of the more water stressed islands (e.g.,

Barbados, Antigua & Barbuda). The use of desalination, wastewater reclamation, and rainwater harvesting are likely to become more prominent, with associated public health implications. There are few studies on the potential impacts of the changes in hydrological patterns on agriculture and food security in SIDS; however, it has been noted that the primary crops in the Caribbean – sugar, bananas, and cocoa – are not particularly drought tolerant.

Ecosystem Structural Changes

As discussed earlier, as a result of CC there will be changes to precipitation patterns and SLR will result in changes to coastal morphology and bathymetry. Both inland and coastal ecosystems are likely to be impacted structurally. It now appears inevitable, despite global mitigation pathways chosen in the future, that most small islands will face at least 0.5–1.0 m SLR before the end of the twenty-first century.

Entire coastlines will become submerged. Mangroves, seagrass beds, and coral reefs will be affected. Built infrastructure will be destroyed (airports, seaports, oil terminals, roads, hotels, schools, electricity generating plants, sewerage systems) and will have to be relocated inland with the possible concomitant deterioration of greenfield sites inland. Pressures on land use, including previously protected habitats, will become even more intense. It is noted that terrestrial watersheds in Caribbean countries have already lost, on average, 90% of their primary vegetative cover [10].

The Socio-Economic Situation: Potential Socio-Economic Effects of CC on Caribbean SIDS

CC is an existential issue for many low-lying SIDS and atolls (e.g., Kiribati, Tuvalu, the Maldives). However, most SIDS have already begun to suffer from the socio-economic impacts of CC as a result of extreme weather effects, and loss of agricultural yields due to prolonged droughts or frequent flooding. Some of the socio-economic impacts may be indirectly caused by the responses of others to CC. For example, the introduction of a "carbon tax" on airline travel into and out of Europe is likely to affect tourism revenues in the Caribbean.

Energy, Water, and Waste Management

Energy, water, and waste management/sanitation are three key crosscutting issues to be considered when any country is trying to achieve sustainable development. Access to abundant, reliable, and affordable sources of energy solves problems associated with water management. Sufficient sources of energy allow for desalination and pumping of the desalinated water to areas in demand.

Over the past decade, increases in the cost of fossil fuels (oil reached US$148/barrel in June 2008) and global efforts to mitigate CC have led to considerable interest in the deployment of renewable energy and energy efficiency technologies worldwide. More recently, reports have indicated that investments in clean energy sources have consistently exceeded US$300 billion annually over the years ranging from 2014 to 2018 [15].

However, investments in renewable energy have been limited in most small islands, despite the abundance of renewable energy sources (e.g., wind, solar, oceans) [16]. Solar energy, which can be utilized for water heating, drying, and cooling, among other things, has been described as easy to install at relatively low costs. Although this source of renewable energy has the potential to greatly positively impact islands, this technology has failed to prevail because of related maintenance, associated costs, residents' acceptance, and government support [16]. Like solar energy, wind energy is a promising source of alternative energy but faces hurdles in implementation. Among the issues associated with wind power, land ownership of wind power plants has remained a prominent issue when it comes wind energy developments. In addition to these notable barriers in renewable technology adoption, perhaps two of the reasons for the lack of market penetration and continued reliance on imported fossil fuels are (i) the lack of interest of the private sector, including the carbon markets, due to the smallness of the individual islands and (ii) the lack of public capital to invest in renewable energy technologies.

Island states have prioritized adaptation rather than mitigation, perhaps conscious that island states can make very little contribution numerically to the global CC mitigation effort. However in 2019, the government of Barbados announced its intention to become fossil fuel free by 2030 if it can receive assistance from institutions and partners like the Green Climate Fund [17].

There are potential public health impacts, both adverse and beneficial, of renewable energy and energy efficiency interventions that may be implemented as a result of CC. Examples include inter alia: disposal of the mercury contained with compact fluorescent light bulbs or disposal of batteries from electric vehicles, potential contamination of groundwater from geothermal operations, expanded habitats of infectious organisms or vectors resulting from hydroelectric projects, noise pollution from wind turbines, improved indoor air quality resulting from the use of modern efficient cooking stoves, and of course mitigation of CC, perhaps the greatest public health threat faced by mankind [18]. Public Health agencies responsible for environmental health management and monitoring will need to be cognizant of these potential impacts.

Among the pressing issues concerned with energy supply and availability in SIDS, water resource management is likely to become increasingly problematic for many SIDS. SIDS are particularly vulnerable to stresses on water resources because of the close interconnections between land and sea [19]. SIDS are inherently vulnerable to changes in precipitation patterns and to saltwater intrusion into freshwater aquifers as a result of SLR. This vulnerability is exacerbated by the lack of adequate water storage capacity (above and below ground) due to SIDS' small landmasses. If, as it is predicted for the Caribbean, the rainy season is shorter but individual rainfall events become more intense, and sea levels rise by up to 1 m, considerable strain will be placed on water utility companies to maintain adequate supplies of fresh clean water for households and businesses.

It is likely that the costs of water supply and distribution will significantly increase as utility companies and individual businesses turn to desalination, rainwater harvesting, and wastewater reclamation. The increased costs of water will impact agriculture and other sectors of the economy. Governments in severely water stressed islands will have to prioritize the provision of minimum quantities of fresh clean water to their citizens to maintain public health standards. Significant investment will be required to augment aboveground water storage capacity and to reduce leakage in aging existing water distribution infrastructure. CC will place additional technical and financial strain on water utilities that are already challenged by freshwater resource constraints, particularly in Caribbean SIDS. The provision of adequate amounts of clean, potable water for cooking and sanitation is a prerequisite for good public health.

Waste management at both the household and country level in small islands will also be impacted by CC. Although landfilling is among the least desirable waste management practices, it has persisted as the primary or only means of waste disposal in many SIDS due to lack of appropriate waste facilities, infrastructure, and limitations in land availability for ultimate waste disposal. Given the anticipated challenges in land space as a result of CC, valuable land space will become increasingly precious and landfilling will no longer be sustainable [20]. Coastal landfills (e.g., Perseverance in Grenada) will have to be sealed and relocated inland. Furthermore, the inter-relationship between waste management and disaster management will need to be appreciated by public health planners. For example, hurricane and storm events create significant quantities of inorganic and organic waste that can swiftly become public health and environmental hazards post event if not adequately managed. Improper disposal of waste in watercourses and drains can greatly exacerbate the effects of flood events. It is recommended that the public health implications associated with CC be continually identified and addressed in integrated water resource management plans, integrated solid waste management plans, and disaster management plans of small island states.

Agriculture and Fisheries: Food Security

According to the Director of the Caribbean Council, the Caribbean food import bill in 2009 may have been as high as US$5 billion [21]. Food imports have been projected to increase to US$8–10 billion by 2020 [22]. Even without CC the Caribbean SIDS have faced food security issues. The Caribbean's agricultural model is geared toward export crops – sugar cane, bananas, cocoa, nutmeg, and spices. These crops are grown and exported to earn vital foreign exchange to then allow for the import of amongst other things, the foodstuffs to feed resident populations and visiting tourists.

The yields from the major export crops of the region, sugar and bananas, are highly vulnerable to the changes in precipitation patterns that are a result of CC. One of the greatest economic impacts of Hurricane Ivan on Grenada in 2004 was the destruction of the nutmeg trees on the plantations and small farms. Recovery has taken years (a nutmeg tree matures in ~5 years), with nutmeg exports in 2011 (~350 tons) being still just a fraction (<15%) of production pre-Ivan (2500 tons/year) [23, 24].

Coastal-marine systems in SIDS, particularly those of the Caribbean, have suffered as a result of CC. Rising sea levels, ocean acidification, erosion, coral bleaching, and declines in marine fisheries are among some of the current adverse outcomes that have been linked to CC [25]. Nurse, of the University of the West Indies, concludes that the impacts of CC on Caribbean fisheries are likely to continue to be generally negative. One such example cited by Nurse of this adverse relationship is the role of increasing sea surface temperatures and persistent warm phases of the El Niño Southern Oscillation as implicatory factors associated with increased coral bleaching in Caribbean waters [26].

In addition to the impacts of increases in temperatures, ocean acidification will continue to negatively impact coral reefs and all calciferous marine species, including the "conch," which is part of the traditional diet of the Caribbean. The loss of coral habitats has the potential to negatively impact various types of marine life and adversely affect reef fisheries. Nurse noted that at the time of publication, the fisheries sector employed ~200,000 people was responsible for 10% of the protein intake and generated over US$5 billion in annual revenues for the countries within the Caribbean Community (CARICOM) [26].

Several studies have been conducted on the impact of CC on coastal upwelling of zooplankton [27]. There is some evidence that the migration patterns of pelagic species may vary according to the CC induced changes in the productivity of zooplankton [28]. There is evidence of fish moving closer to the poles as the oceans warm [29]. However, there is little specific data for the Caribbean.

Tourism

Tourism is a major industry contributor to the national economy in many small islands. According to World Travel and Tourism Council data, tourism contributes to 15.2% of the Caribbean's GDP and 13.8% of employment [30]. However, CC has a direct and significant impact on tourism. Sea-level rise, beach erosion, bleaching of coral reefs, change in rainfall, lost natural resources and biodiversity, and severe hurricanes decrease the touristic attractiveness of small islands.

The Maldives are among the many small island states adversely impacted by CC. Of particular concern for this country is SLR, anticipated to dramatically affect land and tourism revenue [31]. In 2017, tourism contributed 76.6% of GDP in the Maldives [32]. This same year, a weekend of extreme weather damaged households and properties in 62 islands across the Maldives. This extreme weather was a consequence of a cyclone that formed off the southeastern coast of Sri Lanka. The cyclone was responsible for heavy rainfall that resulted in flooding on 36 islands, strong winds that caused structural damage and felled trees on 22 islands, and 14 incidents attributable to rough seas, including the capsizing of a cargo boat [33].

According to the UNFCC Report, “In Barbados, 70% of the hotels are located within 250 m of the high-water mark. This suggests that many hotels are almost exclusively within the 1 in 500 and 1 in 100 inundation zones, placing them at risk of major structural damage” [34]. CC affects the water resources; thus, shortage of water or the emergence of vector-borne diseases may also cause a negative impact on tourism in small islands. Related with CC warmer weather in the north might also decrease the number of tourists in the tropical regions. Uyarra et al. studied the effects of CC on tourism in small islands and 654 tourists from Bonaire and Barbados participated their study. Their results concluded “CC might have a significant impact on Caribbean tourism economy through alteration of environmental features important to destination selection” [35].

The negative impact of CC on the tourism industry may cause unemployment, financial crises, rising external debt, and rising incidence of poverty and political instability in small island states.

Infrastructure and Population Displacement

In the Pacific and Caribbean islands, large populations and infrastructures are located in coastal areas more vulnerable to CCs. Severe hurricanes easily destroy buildings, damage infrastructure, disrupt public services, and cause billions of dollars in damage.

Hurricane Ivan landed in Grenada in 2004 and is a perfect example of small-island vulnerability [36]. Grenada’s socio-economic infrastructure such as housing (90% of damaged), utilities, touristic facilities (90% of damaged), and agricultural production (90% of nutmeg trees-main agricultural product) were destroyed in less than 8 hours during this category four hurricane. According to the IPCC report “Prior to Hurricane Ivan, Grenada was on course to experience an economic growth rate of approximately 5.7% per annum but negative growth of around −1.4% per annum is now forecast” [1]. In the future, CC may create more intense and frequent hurricanes; therefore island communities will have less time to recover.

In September of 2017, hurricane Irma, a category 5 storm, hit the eastern Caribbean, leveling Barbuda, damaging 95% of its buildings, threatening the Virgin Islands, Dominican Republic, Haiti, Cuba, and lashing against Puerto Rico [37]. As a consequence of hurricane Irma, about two-thirds of Puerto Rico were without power, and more than 56,000 people were without potable water [38]. Two weeks after hurricane Irma, hurricane Maria made landfall in Puerto Rico, causing an estimated US$90 billion in damages. Destruction from the storm included damages to the roads, interruptions to water, electricity, and telecommunication networks. Damages from the storm displaced thousands of persons, forcing residents to leave their homes and seek shelter elsewhere in Puerto Rico and the mainland United States [14]. In November of 2017, it was estimated that more than 200,000 people from Puerto Rico relocated to Miami, Orlando, and Tampa, Florida [39].

As exemplified by the events resulting from hurricanes Irma and Maria, increased vulnerability and economic devastation from extreme events cause extensive migration from small island states to metropolitan countries. According to Docquier and Marfouk, the Caribbean and Pacific regions are the most affected regions from skilled migration [40]. Currently, the Caribbean region has the highest emigration rates in the world; around 12% of the labor force has migrated to other countries [41]. In 2002, there were approximately 750,000 refugees from the Americas and the Caribbean and 900,000 refugees from East Asia and the Pacific [42]. Large migrations have the potential to adversely impact the health of affected communities. The quality of life of immigrants does not improve upon arrival into a new country. Immigrants are increasingly vulnerable to health difficulties in a new place [42]. According to McMichael's review "displacements can cause varied health risks: under-nutrition, exposures to infectious diseases, conflict situations, mental health problems, and altered health – related behaviors such as alcohol consumption, tobacco smoking, and transactional sex" [43]. At the same time population displacement, especially the loss of skilled workers, impacts economic growth and social stability of the small island states negatively [44].

Potential Public Health Consequences Related with Climate Change in the Caribbean

The health impacts of CC are complex and comprehensive; the real health burden is rarely recognized. According to the estimation of the World Health Organization (WHO), 200,000 deaths attributable to climate-related health problems such as crop failure and malnutrition, diarrheal disease, malaria, and flooding occur in the world's low-income countries [43].

Many small islands are located in tropical zones, which have climates already suitable for heat stress; asthma; vector-, food-, and waterborne diseases; and morbidity/mortality from extreme weather events. Incidence and prevalence of chronic diseases are increasing in the Caribbean region with unclear reasons. As stated by The United States Agency for International Development(USAID) 2009 report: "The burden of disease associated with non-communicable chronic diseases (NCDs) is greater than the burden of disease associated with communicable diseases or injuries in Latin America and the Caribbean (LAC); however, much less attention has been given to NCDs. In LAC, approximately 50% of all years of life lost are related to NCDs" [45].

Because of poor public health practices and inadequate infrastructure, these problems are already escalating in small island states [46]. Changing climate conditions will increase these health-related problems and burdens [47]. Before talking about an effective adaptation action, we need to understand the consequences of CC on health in the small island states.

As we discussed in the first part of this chapter, average annual temperatures in the Pacific Islands have increased by about 0.25 °C and in the Caribbean have increased by more than 0.5 °C approaching 1 °C over the last 100 years [48]. Continued temperature rise will be a risk to human societies and cause heat-related health problems among the small island communities. Mortality, morbidity, and hospital admissions show that death rates increase during extreme heat [43]. Patients with cardiopulmonary problems, outside workers, elderly, and the very young can be especially vulnerable to extreme heat. Remember that depending on culture and infrastructure (housing), some communities are more vulnerable than others [49]. Exposure to extreme heat can result in heat stroke, sunburn, heat exhaustion, heat cramps, heat rashes, and dehydration [50].

Rising temperatures, changing rainfall patterns, and precipitation increase the rate of vector-borne and waterborne diseases. Costello et al. stated in their paper "*Schistosomiasis, fascioliasis, alveolar echinococcosis, leishmaniasis, lyme borreliosis*, tick-borne encephalitis, and hantavirus infections are all projected to increase as a result of global CC" [51]. Malaria, dengue fever, zika, chikungunya, *filariasis*, and *schistosomiasis* have existed in tropical small island states; however, they are increasing because of changing climate conditions, poor public health practices, inadequate infrastructure, and poor waste management practices [47]. Vector reproduction, parasite maturation, and bite frequency mostly rise with temperature; as a result, malaria, tick-borne encephalitis, and dengue fever are expected to become more prevalent.

Over the past decade dengue, chikungunya, and zika, isolated in either African or Asian countries emerged in previously unaffected areas including the Caribbean [52]. This emergence is attributable to the presence of competent vectors, like the disease-carrying Aedes aegypti mosquitos who thrive in subtropical climates and quickly adapt to environmental disturbances [53]. Among these vector-borne diseases, dengue fever has been recognized to be especially sensitive to climate conditions. Rawlins et al. reported that the incidence of dengue fever rises during the warm years of the El Niño Southern Oscillation (ENSO) in the Caribbean [54]. An outbreak of dengue fever in Fiji simultaneously occurred with increased temperatures during the El Niño and the cost of the outbreak was US$3–six million [46].

Vector-borne diseases may not be the only infectious disease caused by CC impact. CC and warm weather causes increases in pathogen microorganism development and survival rates, disease transmission, and host susceptibility. For example, when ocean temperatures rise, cholera risk might be increased because of higher plankton activity (algal blooms) that supplies nutrients for Vibrio cholerae [55]. Increased rainfalls and flooding may cause *leptospirosis* or *cryptosporidiosis* outbreaks [51]. In the WHO Synthesis Workshop on Climate Variability, CC and Health in Small-Island States report it was stated that "the Epidemiology Centre and the Water and Sewage Authority of Trinidad and Tobago found that 18.6% of samples of potable water taken after heavy rainfall events were positive for *Cryptosporidium*" [46].

In the small island states, fresh water resources are predicted to reduce in relation to increased demand, decreased rainfall, and saltwater invasion due to hurricanes and SLR [6]. Singh et al. also showed that "the incidence of diarrheal diseases is associated with annual average temperature and negatively associated with water availability in the Pacific" [48]. Thus, rising temperatures and decreasing water resources related with CC may increase outbreak of diarrheal and other infectious diseases and negatively impact the quality of life and the economy of small island states. However, rising temperatures, decreased water resources, and hurricanes also cause a loss of agricultural productivity and seriously affect food security among island communities. SLR, rising temperatures, and acidification of the oceans will lead to a loss of mangroves and coral reefs, and reduced fish stocks and warm ocean temperatures cause fish populations to move to higher latitudes which will also affect food security of the islands [36]. This food insecurity will affect livelihoods in coastal populations and result in malnutrition. For example, during extreme drought, micronutrient deficiencies were found in pregnant women in Fiji. Therefore, CC can exacerbate under-nutrition and starvation.

In addition to these climate related problems, there are growing health concerns that come with the increasingly frequent and severe extreme weather events that occur annually. In 2001–2002 there were more than 50 deaths related to storms and hurricanes in the Caribbean [46]. Of particular concern are extreme events like hurricanes and cyclones which often cause flooding leaving people in affected areas vulnerable to waterborne and foodborne diseases, injuries, and adverse mental health outcomes. As a result of extreme weather events, individuals are likely to have difficulties maintaining safe and healthy living conditions [56, 57]. Previously conducted research has established that people displaced from their homes as a result of flooding are at increased likelihoods to experience depression, anxiety, and posttraumatic stress disorder. A study from Guyana reported that participants that reported previous flooding events had trouble concentrating and participated less in social events as a result of emotional problems [57]. A semi-structured focus group study conducted in 2015 among Caribbean health-care providers found that mental illness was perceived as an increasing health concern in the Caribbean [58].

CC also contributes to air quality problems; higher temperatures and/or humidity impact the frequency of smog events, seasonality of pollens, spores, and formation of various air pollutants [43]. Sunlight and high temperatures combine with nitrogen oxides and volatile organic compounds to increase ground-level ozone, which can damage respiratory systems. This effect may cause an increase in respiratory disorders, especially asthma and other chronic lung diseases [9]. CC may affect the concentration of particulate matter (PM) pollution in the air by affecting natural or "biogenic" sources of PM such as wildfires and dust from dry soils [59]. Forest fires in Indonesia occur annually and increase significantly related with a strong El Niño. The Indonesian island of Sumatra faced massive forest fires caused by El Niño–driven droughts and caused an increase in respiratory illnesses and allergy symptoms among islanders [47].

Published data from the Caribbean region stated that chronic respiratory diseases are a significant public health problem in the Caribbean [60–62]; research has reported smoking, allergy, infection, tropical climate, diesel exposure, charcoal smoke, mite, and Sahara dust as risk factors for asthma in this region [63]. One of the studies showed that climatic variables are associated with seasonal acute asthma admissions in emergency rooms in Trinidad [64]. Monteil et al. also reported Sahara dust as a risk factor for asthma in the Caribbean [65, 66]. In our recent study, we also found hospital visits due to asthma attack were correlated with Sahara dust exposure and the monthly mean rainfall level ($p < 0.05$) [67].

Hurricanes, heavy precipitation, and flooding create environments that are conducive for mold, mildew, and other bio aerosols that have the potential to negatively impact respiratory health [63]. Considering that the region is prone to tropical rain, high humidity, hurricanes, and flooding, mold should be considered as an important respiratory risk factor in the Caribbean. In one of our other community-based studies, we found that the flooding caused asthma like symptoms among the occupants of water damp buildings in Guyana after a 2008 flood [68]. This study found objective evidence of dampness and mold in 32.8% of the households.

Conclusion/Recommendations

CC is already physically affecting small island states in the Caribbean and worldwide. Due to the longevity of carbon in the atmosphere, further physical effects are inevitable such as significant SLR no matter what global mitigation efforts are taken in the near future. Changes in rainfall patterns also appear to be inevitable. A coordinated adaptive response will be required involving, inter alia, economists, engineers, physical planners, and public health professionals.

To manage the health effects of CC, we need to understand the consequences of CC on health and the solution for adaptation. Therefore, building awareness and expanding knowledge though regional-based research will be an important step for developing adaptation and prevention strategies. Establishing effective monitoring, early warning, and data management systems is critical for the management of the health effects of CC.

References

1. IPCC. Climate change 2007: synthesis report contribution of working groups I, II, and III to the fourth assessment report of the IPCC 2007. Geneva: IPCC; 2007. 104pp.
2. IPCC. 2013: climate change 2013: the physical science basis. Contribution of working group I to the fifth assessment report of the intergovernmental panel on climate change. Cambridge/New York: Cambridge University Press; 2013.
3. IPCC. 2018: summary for policymakers. In: Global warming of 1.5°C. An IPCC special report on the impacts of global warming of 1.5°C above pre-industrial levels and related global greenhouse gas emission pathways, in the context of strengthening the global response to the threat of climate change, sustainable development, and efforts to eradicate poverty. Geneva: World Meteorological Organization; 2018.
4. Regional statement on the intergovernmental panel on climate change (IPCC) special report on the impacts of global warming of 1.5 °C above pre-industrial levels (1.5 °C special

report); 2018 [cited 2019]. Available from https://caricom.org/media-center/communications/press-releases/regional-statement-on-the-intergovernmental-panel-on-climate-change-ipcc-special-report-on-the-impacts-of-global-warming-of-1.5-c-above-pre-industrial-levels-1.5-c-special-report

5. United Nations Environment Programme. Emissions gap report 2018. Nairobi; 2018.
6. IPCC. 2007: climate change 2007: impacts, adaptation and vulnerability. Contribution of working group II to the fourth assessment report of the intergovernmental panel on climate change. Cambridge: Cambridge University Press; 2007. 976pp.
7. Trenberth KE, Jones PD. Observations: surface and atmospheric climate change. Climate change 2007: The physical science basis contribution of working group I to the fourth assessment report of the IPCC 2007. Cambridge/New York: Cambridge University Press. 2007:996pp. https://www.ipcc.ch/site/assets/uploads/2018/02/ar4-wg1-chapter3-1.pdf.
8. Lal M. Climate change in small island developing countries of the South Pacific. Fijian Studies. 2004;2(1):15–31.
9. IPCC, 2007: Climate Change 2007: The physical science basis. contribution of working group i to the fourth assessment report of the intergovernmental panel on climate change. Solomon S, Qin D, Manning M, Chen Z, Marquis M, Averyt KB, Tignor M and Miller HL, (eds.). Cambridge University Press, Cambridge, United Kingdom and New York, NY, USA, 996.
10. Pulwarty RS, Nurse LA, Trotz UO. Caribbean islands in a changing climate. Environ Sci Policy Sustain Dev. 2010;52(6):16–27.
11. AOSIS Technical Negotiating Team. AOSIS frequently asked questions-technical document 12; 2009.
12. UNDP. Human development report, 2011. Sustainability and equity: a better future for all. Human development report. Palgrave Macmillan; 2011. p. 224.
13. ECLAC subregional headquarters for the Caribbean, Irma and Maria by numbers in the Magazine of the Caribbean Development and Cooperation Committee; 2018.
14. Kishore N, et al. Mortality in Puerto Rico after Hurricane Maria. N Engl J Med. 2018;379(2):162–70.
15. Bloomberg New Energy Finance. Clean energy investment exceeded $300 billion once again in 2018; 2019 [cited 2019, January 24]. Available from https://about.bnef.com/blog/clean-energy-investment-exceeded-300-billion-2018/
16. Kuang Y, et al. A review of renewable energy utilization in islands. Renew Sust Energ Rev. 2016;59:504–13.
17. Caribbean360. Barbados government signs agreement with green climate fund; 2018 [cited 2019]. Available from http://www.caribbean360.com/news/barbados-government-signs-agreement-with-green-climate-fund
18. Smith KR, et al. Energy and human health. Annu Rev Public Health. 2013;34:159–88.
19. Global Water Partnership. Integrated water resources management in the Caribbean: the challenges facing small island developing states. Stockholm; 2014.
20. Mohee R, et al. Current status of solid waste management in small island developing states: a review. Waste Manag. 2015;43:539–49.
21. Jessop D. Agriculture and food security; 2009. Available from http://www.bbc.co.uk/caribbean/news/story/2009/05/090505_jessop_caribbean_agriculture.shtml
22. Food and Agriculture Organization on the United Nations. State of food insecurity in the CARICOM Carribean meeting the 2015 hunger targets: taking stock of uneven targets. Bridgetown; 2015.
23. Caribbean360. Grenada nutmeg farmers receive Christmas bonus; 2011. Available from http://www.caribbean360.com/business/grenada-nutmeg-farmers-receive-christmas-bonus#axzz1kZpRZFLL.%20Accessed%20January%2027,%202012
24. Anon. Grenada celebrates rare, massive nutmeg production; 2011. Available from https://repeatingislands.com/2011/04/29/grenada-celebrates-rare-massive-nutmeg-production/
25. Pittman J, et al. Governance fit for climate change in a Caribbean coastal-marine context. Mar Policy. 2015;51:486–98.

26. Nurse LA. The implications of global climate change for fisheries management in the Caribbean. Clim Dev. 2011;3(3):228–41.
27. Wiafe G, et al. Impact of climate change on long-term zooplankton biomass in the upwelling region of the Gulf of Guinea. ICES J Mar Sci. 2008;65(3):318–24.
28. Rijnsdorp AD, et al. Resolving the effect of climate change on fish populations. ICES J Mar Sci. 2009;66(7):1570–83.
29. Rose G. On distributional responses of North Atlantic fish to climate change. ICES J Mar Sci. 2005;62(7):1360–74.
30. World Travel & Tourism Council. Caribbean recovery. 2018. Available from https://www.wttc.org/priorities/crisis-preparedness/caribbean-recovery/
31. Bigano A, Hamilton JM, Tol RSJ. The impact of climate change on domestic and international tourism: a simulation study. Integr Assess. 2007;7(1):25–49.
32. Anon. Maldives – contribution of travel and tourism to GDP as a share of GDP; 2018. Available from https://knoema.com/atlas/Maldives/topics/Tourism/Travel-and-Tourism-Total-Contribution-to-GDP/Contribution-of-travel-and-tourism-to-GDP-percent-of-GDP
33. Maldives Independent and Maldives: damage to 60 islands after extreme weather; 2017. Available from https://reliefweb.int/report/maldives/maldives-damage-60-islands-after-extreme-weather
34. UNEP (United Nation Environment Programme). Climate change in the Caribbean and the challenge of adaptation. Panama City: United Nations Environment Programme Regional Office for Latin America and the Caribbean; 2008.
35. Uyarra MC, et al. Island-specific preferences of tourists for environmental features: implications of climate change for tourism-dependent states. Environ Conserv. 2005;32(1):11–9.
36. Nurse L, Rawleston M. Adaptation to global climate change: an urgent requirement for small island developing states. Rev Eur Community Int Environ Law. 2005;14(2):100–7.
37. The New York Times. The storm reaches Puerto Rico: 'there is nothing like this'. The New York Times; 2017.
38. Johnson A, et al. Hurricane Irma skirts Puerto Rico, leaves 1 million without power; 2017. Available from https://www.nbcnews.com/storyline/hurricane-irma/hurricane-irma-skirts-puerto-rico-lashing-it-powerful-winds-flooding-n799086
39. Sesin C. Over 200,000 Puerto Ricans have arrived in Florida since Hurricane Maria; 2017. Available from https://www.nbcnews.com/news/latino/over-200-000-puerto-ricans-have-arrived-florida-hurricane-maria-n825111
40. Docquier F, Marfoulk A. International migration by educational attainment (1990–2000) – release 1.1; 2005. Avaliable from https://www.researchgate.net/profile/Frederic_Docquier/publication/228739789_International_Migration_by_Educational_Attainment_1990-2000-Release_11/links/0912f50c8bb81ead17000000/International-Migration-by-Educational-Attainment-1990-2000-Release-11.pdf.
41. Mishra P. Emigration and brain drain: evidence from the Caribbean. Washington, DC: International Monetary Fund; 2006.
42. Thomas SL, Thomas SD. Displacement and health. Br Med Bull. 2004;69:115–27.
43. McMichael AJ, et al. Climate change and human health. Geneva: WHO; 1996. p. 51–2.
44. Thomas-Hope E. Skilled labour migration from developing countries: study on the Caribbean region. International migration papers. Geneva: International Labour Organization; 2002.
45. Anderson GF, et al. Non-communicable chronic diseases in Latin America and the Caribbean. The LAC Bureau of the U.S. Agency for International Development; 2009. Avaliable from http://www.archive.healthycaribbean.org/publications/documents/NCD-in-LAC-USAID.pdf.
46. World Health Organization (WHO). Synthesis workshop on climate variability, climate change and health in small-island states. Geneva; 2004.
47. Ebi KL, Lewis ND, Corvalan C. Climate variability and change and their potential health effects in small island states: information for adaptation planning in the health sector. Environ Health Perspect. 2006;114(12):1957–63.

48. Singh B. Climate changes in the greater and southern Caribbean. Int J Climatol. 1997;17(10):1093–114.
49. Kovats RS, Hajat S. Heat stress and public health: a critical review. Annu Rev Public Health. 2008;29:41–55.
50. Centers for Disease Control and Prevention and National Institute for Occupational Safety and Health (NIOSH). Heat stress; 2018, June 6. Available from https://www.cdc.gov/niosh/topics/heatstress/.
51. Costello A, et al. Managing the health effects of climate change: Lancet and University College London Institute for Global Health Commission. Lancet. 2009;373(9676):1693–733.
52. Cao-Lormeau VM. Tropical Islands as new hubs for emerging arboviruses. Emerg Infect Dis. 2016;22(5):913–5.
53. Centers for Disease Control and Prevention, National Center for Emerging and Zoonotic Infectious Diseases, and Division of Vector-Borne Diseases. Dengue entomology/ecology; 2019, January 15. Available from https://www.cdc.gov/dengue/entomologyecology/index.html
54. Rawlins S, et al. The impact of climate change/variability events on the occurrence of dengue fever in parts of the Caribbean: a retrospective study for the period 1980–2002. West Indian Med J. 2004;53:54.
55. Patz JA, et al. Impact of regional climate change on human health. Nature. 2005;438(7066):310–7.
56. Akpinar-Elci M. Dr. Muge Akpinar-Elci: Hurricane devastation transcends its season. The Virginian Pilot; 2017.
57. Akpinar-Elci M, Rose S, Kekeh M. Well-being and mental health impact of household flooding in Guyana, the Caribbean. Mar Technol Soc J. 2018;52(2):18–22.
58. Macpherson CC, Akpinar-Elci M. Caribbean heat threatens health, well-being and the future of humanity. Public Health Ethics. 2015;8(2):196–208.
59. UnitedStatesEnvironmentalProtectionAgency.Climateadaptation–particulatematterandhealth. Available from https://www.epa.gov/arc-x/climate-adaptation-particulate-matter-and-health
60. Bayona M, et al. Prognostic factors of severe asthma in Puerto Rico. P R Health Sci J. 2002;21(3):213–9.
61. Gyan K, et al. African dust clouds are associated with increased paediatric asthma accident and emergency admissions on the Caribbean island of Trinidad. Int J Biometeorol. 2005;49(6):371–6.
62. Homa DM, Mannino DM, Lara M. Asthma mortality in U.S. Hispanics of Mexican, Puerto Rican, and Cuban heritage, 1990–1995. Am J Respir Crit Care Med. 2000;161(2 Pt 1):504–9.
63. Akpinar-Elci M, Olayinka O. Chapter 21: The impact of climate change and air pollution on the caribbean. In Akhtar & Palagiano C editors. Climate Change and Air Pollution. Springer Publication, New York. ISBN: 978-3-319-61346-8, 2017.
64. Ivey MA, Simeon DT, Monteil MA. Climatic variables are associated with seasonal acute asthma admissions to accident and emergency room facilities in Trinidad, West Indies. Clin Exp Allergy. 2003;33(11):1526–30.
65. Monteil MA. Saharan dust clouds and human health in the English-speaking Caribbean: what we know and don't know. Environ Geochem Health. 2008;30(4):339–43.
66. Monteil MA, Antoine R. African dust and asthma in the Caribbean: medical and statistical perspectives. Int J Biometeorol. 2009;53(5):379–81; author reply 383–5.
67. Martin F, et al. Climate change, sahara dust and the emergency room visits due to asthma in Grenada, The Caribbean. Am J Respir Crit Care Med. 2011;183(1 Meeting Abstracts):A3752.
68. Rose S, Akpinar-Elci M. An Evaluation of the prevalence of respiratory and general symptoms among occupants of households affected by flooding in guyana: the case of Cove & John. The University of Guyana Faculty of Natural Sciences First International Conference on Sustainable Development, 2013 (Conference proceeding) Avaliable from https://jare.org.gy/an-evaluation-of-the-prevalence-of-respiratory-and-general-symptoms-among-occupants-of-householdsaffected-by-flooding-in-guyana-the-case-of-cove-john/.

Chapter 21
Global Climate Change, Desertification, and Its Consequences in Turkey and the Middle East

Hasan Bayram and Ayşe Bilge Öztürk

Climate has been changed due to increases in the average global surface temperature of the earth from preindustrial period to present times. All areas of the world are expected to be affected by the consequences of climate change; however, the Middle East countries including Turkey seem to feel these effects more severe because of the long hot seasons they live and their limited natural reserves of water. Turkey is located in the Mediterranean macroclimatic zone that lies between the temperate and the subtropical zones at western parts, allowing the country to have widely diverse regional and/or seasonal variations ranging from extremely cold winters to very hot dry summers. Due to climate change impacts, widespread increases in summer temperatures are expected to be recorded in the future. Summer temperatures have been increasing mostly in the western and southwestern parts of Turkey. Also, winter precipitation in the western parts of the Turkey has been decreased significantly in the last five decades [1]. According to United Nations Environment Program (UNEP), most of the areas in Turkey are under desertification and/or high potential for desertification, and only small parts of the areas in Turkey are non-risky places [2].

Climate models are also predicting a hotter, drier, and less predictable climate for Middle East region. The region is expected to get hotter across all seasons; models predict an increase of 2.5–3.7 °C in summer and 2.0–3.1 °C in winter [3]. By the end of this century, this region is expected to have an increased mean temperature about 3–5 °C and a 20% decline in precipitation. Most of the region is expected to remain as very hot deserts under climate change scenarios. According to United Nations Development Program (UNDP) Human Development Report 2015, the Middle East is considered as one of the most water-stressed regions of the world after Africa [4]. The Middle East countries including Iraq, Iran, Israel, Jordan,

H. Bayram (✉) · A. B. Öztürk
Department of Pulmonary Medicine, Koc University School of Medicine, Koc University Research Center for Translational Medicine (KUTTAM), İstanbul, Turkey
e-mail: habayram@ku.edu.tr

K. E. Pinkerton, W. N. Rom (eds.), *Climate Change and Global Public Health*, Respiratory Medicine, https://doi.org/10.1007/978-3-030-54746-2_21

Lebanon, Syria, and Saudi Arabia are also under the threat of desertification [5]. Increased temperature is expected to cause greater seasonal variability, more severe weather events, and significant sea level rises. Furthermore, Mediterranean region is expected to shift 300–500 km northward if a 1.5 °C warming will occur, which would mean that Mediterranean ecosystem would become desert [6].

In this chapter, we review the published papers and the governmental and non-governmental reports on global climate changes including changes in temperature, greenhouse gas emissions, desertification and their consequences on sandstorms, water use, and loss of biodiversity in Middle East countries including Turkey. The impact of such changes on human health will also be reviewed in the view of the limited number of published studies and reports referring this region.

Greenhouse Gas Emissions

Turkey's energy need and demand are increasing over the years. At the end of 2017, Turkey's population has increased to 80.8 million, with an annual growth rate of 1.24% [7] and global warming has emerged with the contribution of greenhouse gases as a result of an increase in use of fossil fuels. The country's demand for general energy and electricity has increased by an annual rate of 3.7% and 7.2% for the period of 1990–2004, respectively [1]. In 2004, the ratios for coal, biomass, oil and natural gas, hydro-geothermal and wind electricity, and other renewable sources in the total energy production were as 43%, 23%, 12%, 17%, and 5%, respectively [1]. Turkey's carbon dioxide (CO_2) emission has increased by 98% between 1998 and 2009. Although the country's CO_2 emission was 20.59 million tons in the year 1990, it reached to 30.90 million tons in 2004. According to the estimates in 2000, 34% of CO_2 emission was produced by electricity generation, 32% by industry, 17% by transportation, and 16% by other sectors. However, by the year 2020, it is estimated that 41% of CO_2 emission will be produced by the generation of electricity, 33% by industry, 13% by transportation, and 13% by other sectors [8].

It is expected that the demand for electric energy in Turkey will be about 580 billion kWh by 2020 [9]. The rate of coal use as an energy supply was 24% in 2006, and it will be 36% in 2020. Due to a significant increase in coal use, CO_2emissions are expected to reach nearly 600 Mt. in 2020, which will be three times more of 2004 levels [9]. According to an autoregressive integrated moving average (ARIMA) model's results, the diesel fuel consumption in Turkey is predicted to be over four million tons in 2020 in the agricultural sector with the average growth rate of diesel consumption of 2.17% per year [10].

When Turkey is compared with other countries with respect to basic CO_2 indicators, it is ranked 23rd in total CO_2 emissions, 75th in CO_2 emissions per capita, 60th in the ratio of CO_2 emissions to the gross domestic product (GDP), and 55th in the ratio of CO_2 emissions to the GDP, measured on the basis of purchasing power parity [5]. Turkey signed the Kyoto Protocol on February 17th, 2009 [11]. However,

CO_2 intensity in Turkey still remains higher than the Western developed nations' average after the protocol. Turkey lies in a sunny belt between 36 and 42 N latitudes. The yearly average solar radiation is 3.6 kWh/m^2/day and average sunshine duration is 2640 h, corresponding to 30% of the year. In Turkey, the solar energy has a technical potential of 8.8 million tons of oil equivalent (Mtoe) electricity generation and 26.4 Mtoe heating capacity [8]. Total gross hydropower potential and total energy production capacity of Turkey are nearly 50 GW and 112 TWh/yr., respectively. The South Eastern Anatolian Project (GAP) on rivers Euphrates and Tigris has a capacity of 27 billion kWh of electricity [9]. Although, Turkey has hydropower and sun energy potential, 66% of Turkey's energy consumption is still based on fossil fuels [1].

Solar energy has the potential to equip the Middle East with centuries of sustainable, clean electricity [12]. It has been reported that the Middle East receives 3000–3500 h of sunshine per year, with more than 5.0 kW/m^2 of solar energy per day, and that average solar radiation is about 19.23 M joules per square meter in Iran. In Israel, over 700,000 households are reported to have solar water heaters [12]. As a region, the Middle East produces a tiny fraction of global emissions (less than 1% of the world total), but on per capita basis, Israel's emissions (11.8 metric tons per capita) exceed the European average (10.05 tons) [3]. The amounts of CO_2 emissions of Jordon, Syria, and Iraq are 4.9, 3.3, and 4.1 metric tons per capita, respectively [3]. However, the 88% growth of CO_2 emissions in the Middle East was the third largest in the world in 1990–2004 and more than 3 times faster than the world average; most of that growth came from fuel combustion [6].

Climate Change

There have been widespread increases in summer temperatures in Turkey [1] (Fig. 21.1). These increases are mostly recorded in the western and southwestern parts of Turkey [1]. A recent study using the regional climate model, Providing Regional Climates for Impacts Studies (PRECIS), suggests that the average temperature in 2071–2100 will be 4–5 °C higher for coastal regions and 5–6 °C higher for inland Turkey comparing to the average for 1961–1990, respectively [13].

Furthermore, winter precipitation in the western provinces of Turkey has decreased significantly in the last five decades [1]. Although precipitation has decreased along the Aegean and Mediterranean coasts, it has increased along the Black Sea coast of Turkey (Fig. 21.2). The rainfall is also expected to be 40% less in the West and 5% less in the East and the Eastern Black Sea Regions, respectively [13]. On the other hand, high mountains in Turkey started to lose their glaciers, large lakes have become smaller, and shallow lakes have vanished [14]. According to projections, nearly 20% of the surface water will be lost by the year 2030. By the year 2050 and 2100, the percentage of water loss is expected to increase up to 35% and more than 50%, respectively [15]. In the analysis using the data for years

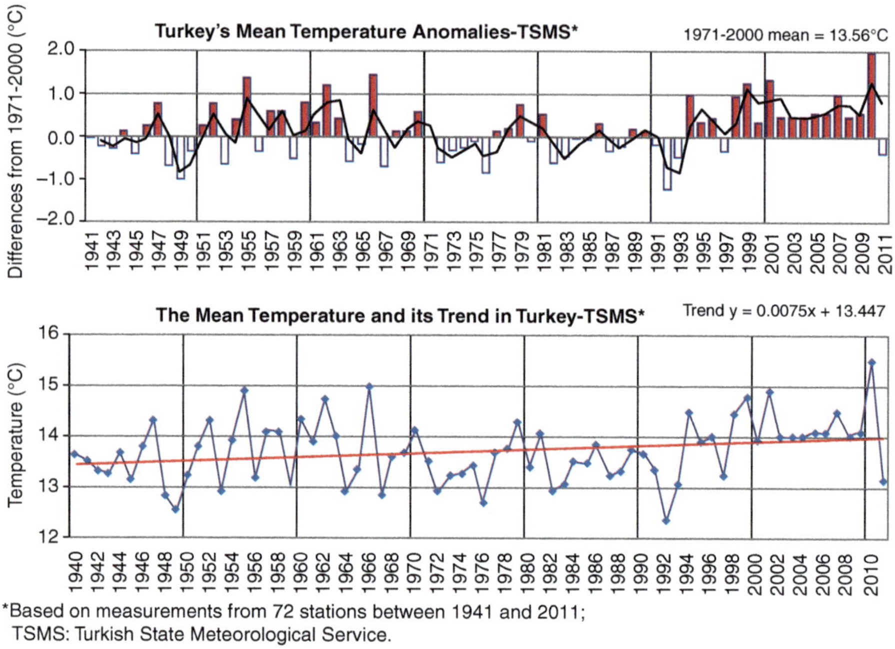

Fig. 21.1 Mean annual temperature trend in Turkey (°C) (1941–2007). Trend Y = 0.0064x + 13.474; $R^2 = 0.0422$. (Reproduced from the reference [13])

1960–2010 in Turkey, it was observed that the number of summer days, hot days and the nights increased, and the number of cool days and nights decreased, respectively. A precipitation of 564 mm with 13% precipitation rate was observed in 2013 [16–18]. Total annual precipitation trends tend to increase in the north of the country and decrease in the Southeast Anatolia, Mediterranean, and Aegean regions, respectively [16–18]. The amount of rainfall in Turkey is expected to decrease at 30% ratio in the Mediterranean and Aegean coasts and at 20% ratio in the Black Sea region, respectively [19]. A global increase in temperature and decrease in rainfall may lead to desertification in Southeast, Central Anatolia, Aegean, and Mediterranean regions, and 86.5% of the land is expected to be vulnerable to desertification in Turkey [16–18].

The Intergovernmental Panel on Climate Change (IPPC) estimates an increase in temperature in the Middle East up to 2 °C in the next 15–20 years and over 4 °C for the end of the century [6]. For example, the main climate change scenarios projected for Israel by the year 2100 include a mean temperature increase of 1.6–1.8 °C, a reduction in precipitation by −8 to −4%, an increase in evapotranspiration by 10%, and a sea-level rise of 12–88 cm [20]. According to reports from Iran, temperature has risen between 2.5 and 5 °C on average with the increase in minimum temperature being more widespread [21]. It has been reported that southwestern parts of the Caspian Sea, northwest and west of Iran, have experienced the highest

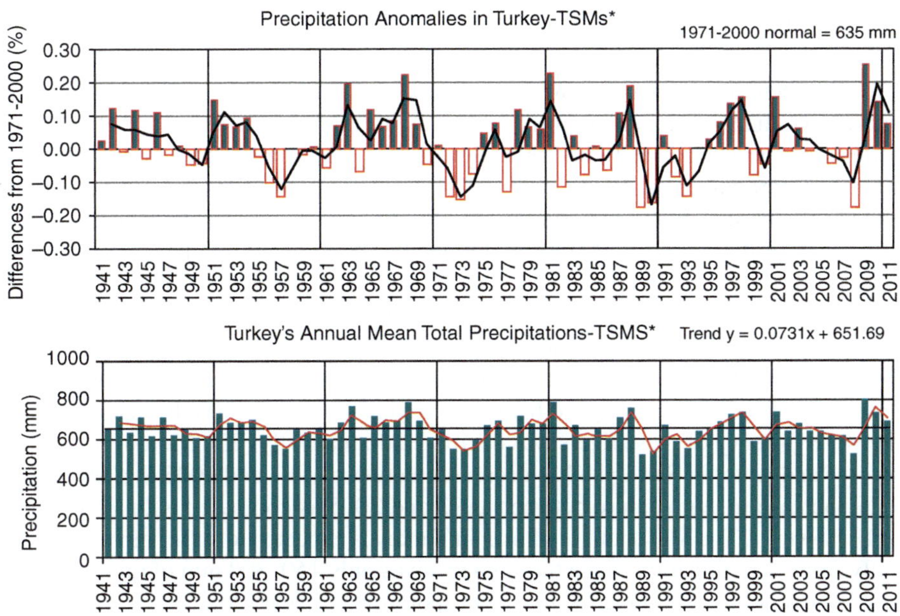

*Based on measurements from 72 stations between 1941 and 2011;
TSMS: Turkish State Meteorological Service.

Fig. 21.2 Annual mean precipitation and its trend in Turkey (1941–2007). Trend Y = −0.2917x + 656.92; $R^2 = 0.0079$. (Reproduced from the reference [13])

rate of reduction in the amount of their annual precipitation [21]. On the basis of climate change scenarios from Saudi Arabia, the average warming in the country for the year 2041 will be higher than the global average, and the highest warming (2.2–2.7 °C) is expected to occur during summer in the northwestern regions. The precipitation is also expected to decrease in the entire Kingdom from December to June [22]. According to similar climate change scenarios, the average warming in Syria for the year 2041 will be higher than the global average. The greatest increase (2.0–2.1 °C) will be expected to occur in the northwest and southeast regions of the country [23]. The IPCC projections indicate that the anticipated increase in surface temperature and reduction in rainfall will result in extreme desertification in the Middle East region [6]. It is also expected that these changes will result in a global increase in sea levels, which are expected to rise between 0.1 and 0.3 m by 2050 [3].

In the Middle East, total available water resources are 262.9 billion cubic meter (Bcm) [6]. The water deficit is likely to increase from 28.3 Bcm in the year 2000 to 75.4 Bcm in 2030. According to projections, a temperature increase of 5 °C will reduce the snow cover from 170,000 to 33,000 km² in the upland section of Euphrates and Tigris watersheds. This is expected to reduce the discharge of the Euphrates and Tigris rivers. An increase in the temperature of Jordan by 2–4 °C is expected to reduce the flow of Azraq River by 12–40% [6].

Desertification

Climatic factors that may lead to desertification in Turkey were investigated by the analysis of the spatial and temporal variations of the precipitation and aridity index series, for the period of 1930–1993. Severe and widespread dry conditions have occurred, particularly in 1973, 1977, 1984, 1989, and 1990. Southeastern Anatolia and the continental interiors of Turkey have been affected by desertification processes as a result of deterioration in the climatic factors. Significant trends from normal to drier conditions in annual precipitation and winter precipitation and towards dry subhumid or semi-arid climatic conditions have been climatic factors that lead to desertification in the Mediterranean and Aegean regions of Turkey [24]. Climatic change impacts were also investigated in the Büyük Menderes and Gediz River basins, and rivers' runoff trend was analyzed between the year 1960 and 2000. It was found that the water potency of these rivers was decreased dramatically [25]. Moreover, the salt reserve and water in Salt Lake has decreased between 1987 and 2005 as a result of a 1 °C increase in temperature between 1993 and 2005 as compared with 1970–1992 [14]. However, The Mesopotamia Basin in Turkey is expected to suffer more dramatically from desertification, since this area receives only 150–300 mm of rainfall annually but experiences 1500–2500 mm of evaporation per year [26].

In addition to changes in climate, the factors that lead to loss of land (i.e., erosion), deforestation, and soil pollution contribute to desertification in Turkey. It is estimated that 74% of the forest land and a 68% of prime agricultural land are thought to be prone to erosion [27]. The total forest area in Turkey is about 21.2 million ha (27.2% of total land); however, 49% of this is estimated to be degraded and unproductive [28]. On the other hand, Turkey is losing 11,500 ha of her forests every year with an average of 1900 fires annually [29]. As a result of accelerated destruction of forests, steppe flora gradually has become dominant in Anatolia [30]. In Turkey, 109.124 km^2 of land is desert, and 374.441 km^2 land is in danger of desertification and de-habitation [5].

Desertification is an important threat for the whole Middle East region [5, 6, 20–23]. In Iraq, areas subject to desertification are estimated to exceed to 92% of the total surface area. Since 1981, the percentage has increased, and this was partly due to military operations, which had detrimental effects on the environment including plants and the soil [5]. Syria has 25.79 km^3 renewable freshwater potential per year, and the available freshwater amount per capita is estimated to decrease from an amount of 2.089 m^3 in 1990 to 546 m^3 in 2050. In 1955, freshwater availability as per cubic meter/inhabitant in Lebanon and Syria was 3.084 and 6501 m^3 per capita, respectively. These values were decreased to 1949 and 2.089 m^3 in 1990 for Lebanon and Syria, respectively. The estimated values for years 2025 and 2050 for Syria are thought to be 1126 and 960 m^3, whereas the corresponding figures for Lebanon are expected to be 770 and 546 m^3 for years 2025 and 2050, respectively [5]. The percentage of desertificated land ranges from 10% in Syria particularly including Syrian Badia Rangelands (Steppe zone) and the marginal zone to nearly 100% in the United Arab Emirates [5]. It is estimated that the cost of soil degradation in Syria is equivalent to about 12% of the value of the country's agricultural output. In Lebanon degradation is reported to be serious on steppe mountainous land [5].

In Iran, the level of annual precipitation has decreased in the southwestern parts of the Caspian Sea, northwest and west of the country. The amount of degradation was reported to be 1.5 million ha in the country. If the rate of desertification continues in this present trend, the amount of affected land for the year 2050 is expected to be 75 million ha in Iran [21]. Desertification is also expected to be exacerbated by climate change in Israel, particularly in the Judean Desert highlands and the northern Negev [20]. Saudi Arabia is particularly vulnerable to desertification, as about 76% of the country's territory is non-arable lands, of which 38% is made up by deserts. The yearly temperature increase is expected to be 0.8–6.0 °C in the year 2100, and as a result the rate of desertification is expected to rise in this country [22].

Consequences of Climate Change and Desertification

Sandstorms and Dust Storms

Arid lands are considered as the significant contributors of dust. The phenomenon of sand dunes is thought to be one of the most dangerous consequences of desertification, due to its negative impact on every vital aspect of life. Sand dunes lead to increased sandstorms and dust storms, increased soil salinity and water logging, and widespread rangeland degradation [5]. Sandstorms and dust storms pollute the environment and agricultural production by disrupting the physiological functions of plants, especially during pollination and inflorescence. Sandstorms blow from the dune fields in central and southern areas of the Middle East region. It has been reported that their incidence has increased during recent years, and although dust storms are reported to be most common in the central plain region in Iraq and Syria [5], they have started to affect all Middle East countries. Studies suggest that Middle East countries such as Iraq face with a severe desertification problem that jeopardizes their food security through the effects of soil salinity, water logging, loss of vegetative cover, shifting sand dunes, and severe sandstorms/dust storms [5]. It has been suggested that the introduction and expansion of rain-fed agriculture in the Syrian steppe led to environmental consequences including formation of dust, dust storms, sand accumulation on roads and railroads, and formation of sand sheets, sand hummocks, and sand dunes [5]. Furthermore, dust frequency and intensity are reported to have remarkably increased during the last few years in the eastern part of the country. The frequency and amount of sandstorms and dust storms in Turkey and Lebanon are reported to be less than in Iraq and Syria [5]. However, in recent years, Turkey, in particular the southeast parts of the country, has faced more sandstorms coming from over Syrian and Saharan deserts.

Water Use

According to estimations of population growth rate of Turkey, per capita available, water was 250 L/day in the year 2000. With the assumption that Turkey will continue to grow and develop, this amount is expected to increase to 500 L/day in 2030 [31]. The total water requirement for domestic and industrial consumption is predicted to be 25.3 and 13.2 billion m^3, respectively. Per capita of potential water resources was estimated as 3070 m^3/year in 1990, however; according to climate change scenarios, the per capita of water potential will be decreased to 700–1910 m^3/year in 2050. Gross irrigatable area in Turkey is 8.5 million ha, and the whole of this area will be irrigated by the year 2030. Water requirement for this area is estimated to be 71.5 billion m^3; however, in total consumption, the percentage of irrigation is expected to drop from 75% to 65% due to the water shortage [31] (Table 21.1). According to Falkenmark indicator, Turkey is in a position that "a country facing water stress" on the basis of per capita water. The annual amount of usable water per capita is around 1519 m^3 and it will decrease to 1125 m3/year in 2030 [18]. The water stress will be more severe in the middle and western regions of Turkey by 40%. This ratio is predicted to be between 20% and 40% in the southeast and eastern regions of Turkey [16–18].

In global-scale assessments, basins are defined as being water-stressed if they have either per capita water availability below 1000 m^3 per year. Middle East is one of the regions where water-stressed basins are located. The Arab region receives an estimated 2282 billion m^3 of rainwater each year compared to estimated 205 billion m^3/year of surface water and 35 billion m3/year of groundwater [6]. Lebanon, Syria, and southern Sudan receive as much as 1500 mm of rainfall. Reduced stream flow and groundwater recharge are expected to decrease water supply 10% by 2050 [6]. Recent estimates of water resources in the Middle East region indicate that total available natural water resources are 262.8 Bcm, of this; 226.5 Bcm is made up by surface water and 36.3 Bcm by groundwater including 11.874 Bcm of nonrenewable groundwater. Per capita renewable water resources in the region have decreased from 4000 m^3 per year (year 1950) to 1100 m^3 per year in recent years. The water deficit is expected to increase from about 28.3 Bcm for the year 2000 to 75.4 Bcm in the year 2030 due to climatic and non-climatic factors [6]. Lebanon is one of the

Table 21.1 Gross total amount and consumable water in Turkey

Surface water	Rainfall (mm)	Water amount (billion m^3/year)	Gross water potential (billion m^3/year)	Exploitable (billion m^3/year)
Turkey	643	501	186	95
From bordering countries			7	3
Groundwater			41	12
Total			234	110

From Sekercioglu CH, Anderson S, Akçay E, et al. [26], with permission
Reproduced from the reference [26]

richest countries with water in the Middle East region. The total amount of available water is 3.992 million cubic meters in Lebanon. According to studies conducted by the Food and Agriculture Organization of the Nations and by the UNDP, the irrigated area of Lebanon is expected to rise to 170.000 ha by 2015 [5]. Syria has 25.79 km^3 renewable freshwater potential per year, and the available freshwater amount per capita is predicted to decrease from 2.089 m3 (in the year 1990) to 546 m^3 in the year 2050 [5]. Availability of freshwater resources in the Arab region dropped from 921 m^3 per capita per year in 2002 to 292 m^3 per capita per year in 2011. Almost 75% of the Arab population live under the water scarcity level, and nearly half lives under extreme water scarcity level of 500 m^3 per capita per year. According to the United Nations world water development report 2015, the Middle East is one of the most water-stressed regions of the world [32].

Loss of Biodiversity

The earth is made up of an ecosystem and ecological features, which are supported by biodiversity. Higher temperatures may result in a reduction in soil fertility due to higher rates of decomposition and losses of organic matter and may adversely affect nutrient cycling. As a result, climate change is expected to cause the loss of biodiversity and undermine ecological system. Turkey is considered as one of the richest countries of Europe and the Middle East with respect to biodiversity. Turkey ranks 140th out of 163 countries in biodiversity and habitat conservation [26]. The country contains 5% of the plant species found in the continent of Europe. Studies have reported that there are 163 plant families covering 1225 types, which in turn cover about 9000 species [26]. Turkey is also reported to be rich as biodiversity with 120 mammals, 400 fishes, 469 bird species, and 130 reptiles. Turkey has 33% of endemic species of totally 9000 plant species. By factors result from climate changes, of 3504 endemic plants in Turkey, 12 are reported to be extinct, and 3492 are considered to be under threat [33].

Iranian habitat supports 8200 plant species, of which 2500 are endemic, over 500 species of birds, 160 species of mammals, and 164 species of reptiles [21]. Although no systematic review has been conducted to show linkage between climate change and biodiversity in Iran, national documents in biodiversity have addressed that climate change has a negative impact on biodiversity [21]. Before Syrian Civil War, the National Syria Strategy for biodiversity was indicating that the country has more than 3000 animal species and 3077 species of flowering plants. Syria is considered as a poor country with respect to its forests, which cover only 3% of the total land area. There has also been a decrease in the wooded areas of Jebel Abdel Aziz, Abou Rajmein, and Balaas mountains, which were in the past ecosystem rich in ecological biodiversity [23]. It has also been suggested that desertification, further exacerbated by climate change, will widen the desert barrier to be crossed by the birds and will make Israel less hospitable for the migration of the migrants. Many Red Sea species have colonized the Mediterranean Sea following migration through the Suez

Canal. With increased warming, more Red Sea immigrants are expected to colonize, reproduce, and persist in the eastern Mediterranean [20]. In conclusion, the biodiversity is expected to further deteriorate due to climate change in the Middle East region [3].

Human Health

Human health is adversely influenced by the direct and indirect effects of climate change, and preliminary research has shown climate change has potentially direct and indirect adverse impacts [34–37]. A 2015 survey by the American Thoracic Society documented that climate change had an effect on patient care, and severity of chronic illness such as asthma and chronic obstructive lung disease (COPD) [38]. Changes in pollen releases impact asthma and allergic rhinitis; heat waves may cause critical care-related diseases; climate-driven air pollution increases may lead to exacerbations of asthma and chronic obstructive pulmonary disease; desertification increases particulate matter (PM) exposures; and climate-related changes in food and water security impact infectious disease through malnutrition [34–37]. Although all countries will be affected by climate change, low-resource countries including some of the Middle East countries are expected to be more effected by climate due to low-resource countries often lacking economic resources, having a close dependence on natural systems for basic food and water provision, and suffering from inadequate housing, energy, and waste management [35, 36].

Quantifying the full impact of climate change on health is extremely difficult. This is partly because many modeling techniques are still in their infancy, but partly because impacts will depend on numerous interacting factors including other environmental trends, social resources, and preexisting health status. In the twenty-first century, the Mediterranean area is delegate to be one of the most prominent and vulnerable climate change regions that will experience a large number of extremely hot temperature events, an increase of summer heat wave frequency and duration, and increasing summer temperature variability [39]. An increase in the frequency and severity of heat waves is expected to enhance both illness and death rates. Using models that estimate climate change for the years 2020 and 2050, it is predicted that summer mortality will increase dramatically; the winter mortality will decrease slightly, even if people acclimatize to the increased warmth [40].

However, there are relatively a limited number of studies investigating effects of climate change on human health [41–49]. During the 2006 California Heat Wave, emergency visits for heat-related diseases and hospitalization were reported to have increased statewide. Children (0–4 years of age) and elderly (≥65 years of age) were found to be at the greatest risk. Emergency visits also showed significant increases for acute renal failure, cardiovascular disease, diabetes, electrolyte imbalance, and nephritis [41]. Extreme heat events are associated with exacerbations of respiratory and cardiovascular disease. Hot, humid days trigger asthma symptoms. There is also evidence that extreme heat may trigger exacerbations of congestive heart failure [35–37, 42]. Sand dust storms are associated with increased cardiopulmonary mortality and emergency room visits, asthma and COPD hospitalization, and increased risk for respiratory infections and myocardial infarction [35–37, 42].

Al Eskan disease, reported in Military Medicine in 1992, was a novel and previously unreported hyper-allergic lung condition triggered by fine sand particles less than 1 μm (0.1–0.25 μm) in diameter of the central and eastern Saudi Arabian peninsula [43]. Following the Gulf War in1990, a similar clinicopathological entity was defined as "Persian Gulf syndrome" [44]. A wide range of acute and chronic symptoms including fatigue, musculoskeletal pain, cognitive problems, respiratory symptoms, skin rashes, and diarrhea was related with the exposure to sand particles less than 1 μm in Persian Gulf syndrome [44, 45]. Moreover, recent studies have reported that sand storms are associated with increased cardiopulmonary emergency visits, hospital admissions for COPD, pneumonia, and asthma in the Middle East and the other parts of Asia [46–49]. Al B et al. have reported an association between daily levels of PM_{10} and cardiovascular mortality in Gaziantep, Turkey, where dust storms are more abundant [49].

It has been suggested that climate change may also lead to increased levels of air pollution. Lelieveld et al. reported that outdoor air pollution led to 3.3 million premature deaths globally in 2010using a global chemistry model. These numbers are expected to be double in the year 2050 [50]. According to projections made by the North American Regional Climate Change Assessment Program, an increase of 0.43 ppb in average ozone concentration is expected for the year 2040 comparing to the year 2000, and this was estimated to correspond to a 0.01% increase in mortality rate and 45.2 premature deaths in the study communities attributable to the increase in future ozone levels [51].

Warmer conditions may lead to increases in the incidence and extent of infectious diseases such as malaria, dengue fever, schistosomiasis, and yellow fever. In Istanbul, Turkey, leptospirosis cases increased at the warmer periods of April–May–June, as compared to the cooler period of January–February–March in years 2004–2006 [25]. Within the last three decades, the number of malaria cases was increased in the two periods of 1977–1987 and 1993–1998 in Turkey, and this was in parallel with increased temperature [1]. In Iran, leishmaniasis diseases showed an outbreak during the period of 1995–2005 [21]. Furthermore, leishmaniasis is an endemic disease in all regions of Syria since nineteenth century, and the World Health Organization (WHO) classified border areas of the country with Iraq and Turkey as malarial high-risk areas [23].

Other consequences of climate change are expected to be the decreases in food production and increases in the cost that could lead to the risk of widespread malnutrition and hunger in the Middle East countries. A rise in sea levels and sea temperatures could also decrease the seafood stocks. Water shortages together with the higher temperatures may increase the risk of infectious diseases such as cholera, salmonella, and dysentery [35, 36]. According to climate model scenarios, Iran will experience a maximum of 1.4 °C increase in temperature during the years 2010–2039, which is expected to increase the number of hospitalizations for diarrhea and cholera [21]. The loss of biodiversity and temperature changes may possess a risk for allergic airway diseases. Hence, a study by Metintaş M, et al. evaluated the effects of geo-climatic factors on the prevalence of allergic disease in a general adult population, and it has been demonstrated that high temperatures are associated with higher levels of allergens, higher asthma prevalence, longer pollen seasons, and diversity in pollens [52].

Conclusion

Global climate change is a serious problem and has adverse impacts on the environment and human health. However, some parts of the world such as the Middle East region suffer more from the detrimental effects of climate change. The region faces heat waves, water shortage, desertification, dust storms, loss of biodiversity, and their health consequences at a much severe scale. The resident countries, in addition to their contribution to the global combat against factors leading to climate changes, need to take local and regional adaptation and mitigation measures. Furthermore, more research is needed to understand the scale of the problem and its impacts on human health.

References

1. First national communication of Turkey on climate change. In: Apak G, Ubay B, editors. Turkish Ministry of Environment and Forestry; 2007p. 1–263. http://www.dsi.gov.tr/docs/iklim-degisikligi/iklim_degisikligi_%C4%B1_ulusal_bildirim_eng.pdf?sfvrsn=2. English.
2. United Nations Environment Programme (UNEP). World atlas of desertification. London: Edward Arnold; 1992.
3. International Institute for Sustainable Development (IISD). Rising temperatures, rising tensions. In: Brown O, Crawford A, editors. Climate change and the risk of violent conflict in Middle East; 2009. p. 1–41. http://www.iisd.org/publications/pub.aspx?pno=1130
4. United Nations Development Programme (UNDP). Human development report; 2015. http://hdr.undp.org/sites/default/files/2015_human_development_report_1.pdf
5. Haktanır K, Karaca A, Omar SM. The prospects of the impact of desertifi cation on Turkey, Lebanon, Syria and Iraq. In: Marquina A, editor. Environmental challenges in the Mediterranean 2000–2050. Dordrecht: Kluwer Academic; 2004. p. 139–54.
6. Elasha BO. Arab human development report. Mapping of climate change threats and human development impacts in the Arab region; 2010. http://www.arab-hdr.org/publications/other/ahdrps/paper0xx2-en.pdf
7. Turkish Statistical Instıtute. The results of address based population registration system; 2017. http://www.turkstat.gov.tr/HbPrint.do?id=27587
8. Budak DB. Analysis of renewable energy and its impact on rural development in Turkey. AgriPolicy enlargement network for agripolicy analysis, work package WP2: studies, deliverable D2.2 study 2, Coordinators Axel Wolz, Klaus Reinsberg, Gertrud Buchenrieder (IAMO); 2009, November.
9. Yüksel İ. Global warming and environmental benefits of hydroelectric for sustainable energy in Turkey. Renew Sust Energ Rev. 2012;16:3816–25.
10. Unakıtan G, Türkekul B. Univariate modelling of energy consumption in Turkish agriculture. Energy Sources, Part B. 2014;9:284–90.

11. Sözen A, Alp İ. Comparison of Turkey's performance of green house gas emissions and local/regional pollutants with EU countries. Energy Policy. 2009;37:5007–18.
12. Meisen P, Hunter L. Global energy network institute. Renewable energy potential of the Middle East, North Africa vs The Nuclear Development Option; 2007. http://www.geni.org/globalenergy/research/middle-east-energy-alternatives/MENA-renewable-vs-nuclear.pdf
13. Demir İ, Kılıç G, Coşkun M. PRECIS bölgesel İklim Modeli ile Türkiye için İklim Öngörüleri: HadAMP3 SRES senaryosu, IV. Atmosfer Bilimleri Sempozyumu; 2008, Bildiriler Kitabı, 365–3 (in Turkish). http://www.mgm.gov.tr/FILES/iklim/iklimongoruleri.pdf
14. Ekercin S, Örmeci C. Evaluating climate change effects on water and salt resources in Salt Lake, Turkey using multitemporal SPOT imagery. Environ Monit Assess. 2010;163:361–8.
15. The Ministry of Environment and Forest, Republic of Turkey. Devlet Su İşleri Genel Müdürlüğü. İklim Değişikliği ve Yapılan Çalışmalar 2008. http://www.dsi.gov.tr/docs/iklim-degisikligi/iklim_degisikligi_ve_yap%C4%B1lan_calismalar_ekim_2008.pdf?sfvrsn=2. In Turkish.
16. MoFWA. Climate change impacts on water resources project final report executive summary; 2016. Retrieved from MoFWA website http://iklim.ormansu.gov.tr/ckfinder/userfiles/files/Iklim_Nihai%20Rapor_Executive%20Summ_05122016.pdf. Accessed 2 Jan 2018.
17. SHW. Chapter II: land and water resources-water and DSI; 2012a. Retrieved from SHW website http://www2.dsi.gov.tr/english/water-and-dsi/files/assets/basic-html/page9.html. Accessed 12 Feb 2018.
18. Birpinar ME, Tuğaç C. Impacts of climate change on water resources of Turkey, p. 145–152. In: Gastescu P, Bretcan P (edit, 2018), Water resources and wetlands, 4th international conference water resources and wetlands; 2018 September 5–9, Tulcea (Romania). p. 312.
19. Albayrakoğlu E. Climate change and security: the case for Turkey. Gazi University Akademik Bakış J. 2011;5:59–76.
20. Climate change Israel national report under the United Nations Framework Convention on Climate Change impact, vulnerability, adaptation; 2000. http://www.bgu.ac.il/BIDR/rio/Global91-editedfinal.html
21. United Nations Development Programme. Iran Second National Communication to United Nations Framework Convention on Climate Change (UNFCCC); 2010. unfccc.int/resource/docs/natc/snc_iran.pdf
22. The United Nations Framework Convention on Climate Change (UNFCCC). Second national communication Kingdom of Saudi Arabia; 2011. http://unfccc.int/resource/docs/natc/snc_report_2011_kingdom_of_saudi_arabia.pdf
23. Meslmani Y. United Nations Framework Convention on Climate Change (UNFCCC) climate change initial national communication of Syrian Arab Republic; 2010. http://unfccc.int/files/national_reports/non-annex_i_natcom/submitted_natcom/application/pdf/syria_initial_national_communication_03feb2011.pdf
24. Türkeş M. Vulnerability of Turkey to desertification with respect to precipitation and aridity conditions. Tr J Eng Environ Sci. 1999;23:363–80.
25. Birleşmiş Milletler Kalkınma Programı İklim Değişikliği ve Türkiye. Etkiler, Sektörel Analizler, Sosyoekonomik Boyutlar; 2007. http://www.tobb.org.tr/Documents/yayinlar/iklim_degisikligiveturkiye.pdf. In Turkish.
26. Sekercioglu CH, Anderson S, Akçay E, et al. Turkey's globally important biodiversity in crisis. Biol Conserv. 2011;144:2752–69.
27. Haktanır K, Karaca A, Omar SM. The prospects of the impact of desertification on Turkey, Lebanon, Syria and Iraq. In: Marquina A (ed) Environmental challenges in the Mediterranean 2000–2050, NATO science series (series IV: earth and environmental sciences), vol 37, Dordrecht: Springer; 2004, pp. 139–54.
28. TC Orman ve Su İşleri Bakanlığı Çölleşme ve Erezyonla Mücadele Genel Müdürlüğü. United Nations Convention Combat Desertification (UNCCD). Turkish National Action Programme on Combating Desertification; 2006. Available at http://www.cem.gov.tr/erozyon/Files/faaliyetler/collesme/Belgeler/Turkey_s_National_Action_Program_on_Combating_Desertification.pdf. In Turkish and English.

29. General Directorate of Forestry. 2009 forestry statistics; 2009. http://web.ogm.gov.tr/Dkmanlar/istatistikler/OrmancilikIst2009.rar. In Turkish.
30. Camci Cetin S, Karaca A, Haktanir K, Yildiz H. Global attention to Turkey due to desertification. Environ Monit Assess. 2007;128(1–3):489–93.
31. Tahmiscioglu MS, Karaca Ö, Özdemir AD, Özgüler H. Possible effect of the global climate change on water resources and floods in Turkey. International conference on climate change and the middle east past, present and future; 2006, Istanbul, Turkey. http://www.dsi.gov.tr/docs/iklim-degisikligi/possible_effect_of_the_global_climate_change_on_water_resources_and_floods_in_turkey.pdf?sfvrsn=2
32. The United Nations world water development report 2015: water for a sustainable world. https://unesdoc.unesco.org/ark:/48223/pf0000231823
33. Kahraman A, Önder M, Ceyhan E. The importance of bioconservation and biodiversity in Turkey. IJBBB. 2012;2:95–9.
34. IPCC. Fourth assessment report: climate change; 2007. http://www.ipcc.ch/publications_and_data/publications_ipcc_fourth_assessment_report_wg3rreport_mitigation_of_climate_change.htm
35. Pinkerton KE, Rom WN, Akpinar-Elci M, et al. An official American Thoracic Society workshop report: climate change and human health. Proc Am Thorac Soc. 2012;9:3–8.
36. Pinkerton KE, Rom WN, Carlsten C, et al. Climate change and global public health. Turk Toraks Derg. 2013;14:115–22.
37. Bayram H, Bauer AK, Abdalati W, Carlsten C, Pinkerton KE, Thurston GD, Balmes JR, Takaro TK. Environment, global climate change, and cardiopulmonary health. Am J Respir Crit Care Med. 2017;195(6):718–24.
38. Sarfaty M, Bloodhart B, Ewart G, Thurston GD, Balmes JR, Guidotti TL, Maibach EW. American Thoracic Society member survey on climate change and health. Ann Am Thorac Soc. 2015;12:274–8.
39. Kuglitsch FG, Toreti A, Xoplaki E, et al. Heat wave changes in the eastern Mediterranean since 1960. Geophys Res Lett. 2010;37:1–5.
40. Kalkstein LS, Greene SC. An evaluation of climate/mortality relationships in large US cities and possible impacts of climate change. Environ Health Perspect. 1997;105(1):84–93.
41. Knowlton K, Rotkin-Ellman M, King G, et al. The 2006 California heat wave: impacts on hospitalizations and emergency department visits. Environ Health Persp. 2009;117(1):61–7.
42. Rice MB, Thurston GD, Balmes JR, Pinkerton KE. Climate change. A global threat to cardiopulmonary health. Am J Respir Crit Care Med. 2014;189(5):512–9.
43. Korényi-Both AL, Korényi-Both AL, Molnár AC, Fidelus-Gort R. Al Eskan disease: desert storm pneumonitis. Mil Med. 1992;157:452–62.
44. Chiu HF, Tiao MM, Ho SC, et al. Effects of Asian dust storm events on hospital admissions for chronic obstructive pulmonary disease in Taipei, Taiwan. Inhal Toxicol. 2008;20:777–81.
45. Korényi-Both AL, Korényi-Both AL, Juncer DJ. Al Eskan disease: Persian Gulf syndrome. Mil Med. 1997;162:1–13.
46. Coker WJ, Bhatt BM, Blatchley NF, Graham JT. Clinical findings for the first 1000 Gulf war veterans in the Ministry of Defence's medical assessment programme. BMJ. 1999;318:290–4.
47. Kantani TK, Ito I, Al-Delaimy WK, et al. Desert dust exposure is associated with increased risk of asthma hospitalization in children. Am J Respir Crit Care Med. 2010;182:1475–81.
48. Cheng MF, Ho SC, Chiu HF, et al. Consequences of exposure to Asian dust storm events on daily pneumonia hospital admissions in Taipei, Taiwan. J Toxicol Environ Health A. 2008;71:1295–9.
49. Al B, Bogan M, Zengin S, et al. Effects of dust storms and climatological factors on mortality and morbidity of cardiovascular diseases admitted to ED. Emerg Med Int. 2018;2018:3758506.
50. Lelieveld J, Evans JS, Fnais M, Giannadaki D, Pozzer A. The contribution of outdoor air pollution sources to premature mortality on a global scale. Nature. 2015;25:367–71.
51. Chang HH, Zhou J, Fuentes M. Impact of climate change on ambient ozone level and mortality in southeastern United States. Int J Environ Res Public Health. 2010;7:2866–80.
52. Metintaş M, Kurt E, PAFRAIT study group. Geo-climate effects on asthma and allergic diseases in Turkey: results of PAFRAIT study. Int J Environ Health Res. 2010; 20:189–99. http://www.gm-unccd.org/FIELD/countries/Turkey/turk00en.pdf

Chapter 22
Climate Change and the Risk of Desertification with a Focus in the United States

Huda Asif and Mehdi Mirsaeidi

Climate change has emerged as a major threat for human life on the planet. The earth is experiencing a rise of surface temperatures more than ever. In this regard, the twentieth century broke records of the past 11,000 years with a 0.6 degrees Celsius (°C) rise in global surface temperature above the mean of any century in the past [23]. Furthermore, climate change is projected to intensify at an alarming rate during the twenty-first century, raising global temperatures to 1.4–5.8 °C above the mean of past centuries [23]. This projection was particularly evident during the year 2017, when the surface temperature of earth rose 0.38–0.48 °C above the mean from 1981 to 2010 period. The significance of such a seemingly small change in global surface temperature rise can be explained by the fact that over the past 5000 years, global surface temperature rose by 4–7 °C as opposed to 0.6–0.7 rise in the past century alone [108]. That is a tenfold increase in the rate of warming as compared to that in the past 5000 years.

North America has also experienced the effects of climate change; in the past two decades, there was a rise in the number of heat waves and a downward trend in the number of cold waves as shown in Fig. 22.1 (NOAA). These patterns are in line with the fact that the average surface temperature in the United States has been rising at a rate of 0.3 °C/decade. Particularly, the year 2017 was the third-warmest year in the United States since 1895 [10].

Greenhouse gases, play a key role in rising global surface temperatures. The most notorious greenhouse gases are products from combustion of carbon-based fuels including oil, coal, and natural gas [23]. Carbon monoxide, for example, is one of the six criteria pollutants designated by the United States Environmental Protection Agency (US-EPA) and also one of the main products of combustion. Criteria pollutants are the six most common air pollutants for which United States Environmental Protection Agency has set air quality standards (US-EPA). The average concentration of carbon dioxide, another greenhouse gas, on the surface of earth

H. Asif · M. Mirsaeidi (✉)
Division of Pulmonary and Critical Care, University of Miami, Miami, FL, USA
e-mail: msm249@med.miami.edu

K. E. Pinkerton, W. N. Rom (eds.), *Climate Change and Global Public Health*, Respiratory Medicine, https://doi.org/10.1007/978-3-030-54746-2_22

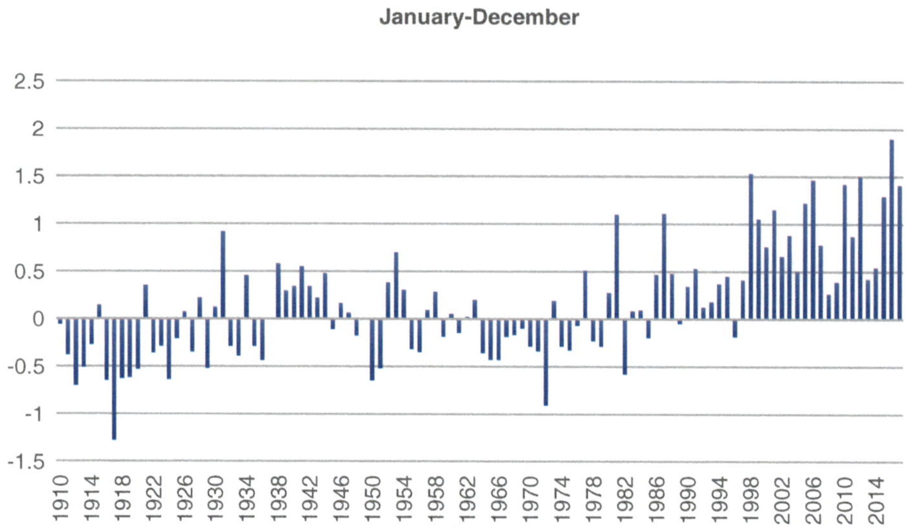

Fig. 22.1 Temperature anomalies in North America, January to December. (Adapted with modifications from [83, 84])

has risen almost four times since the 1960s. In 2017, it rose to 405 ppm, 2.2 ppm higher than in 2016, reaching highs unprecedented in modern records and ice core levels from 800,000 years ago [10].

Desertification and dust storms are known consequences of climate change, and surface temperature has a direct relation to surface soil moisture. Higher temperatures tend to extract more moisture from the soil, giving rise to drought-like conditions in the affected region [12]. With the rising global average surface temperature, drought-like conditions are projected to increase substantially by 2050. In the United States, the Rocky Mountain and Southwestern states, stretching from Wyoming and Montana down to California and Arizona, are expected to experience the most frequent droughts [54].

Droughts are closely associated with increased dust activity [71]. Increases in dust storms have been observed during droughts with the highest frequencies seen during the most severe droughts [123]. When turbulent winds blow in arid, drought-stricken regions, a large quantity of dust rises from the soil surface [74]. During such events, visibility can go below a kilometer [110] as particulate matter concentrations rise above 100 μg/m^3 [99] and the mass of atmospheric dust increases 3–ten fold in comparison to non-dust days [33].

Desert Dust Geochemistry and Its Fingerprints

It is crucial to understand dust chemical composition before studying its impact on human health. Desert dust has a unique geochemistry fingerprint based on the location of origin. Silica comprises 60% of the earth's crust [21]. Dominant

Table 22.1 Elemental composition of Asian dust from South Korea based on data from Kyung et al. [59]

Mineral	Composition (%)
SiO_2	48
Al_2O_3	12
CaO	5.18
Fe_2O_3	5
TiO_2	1

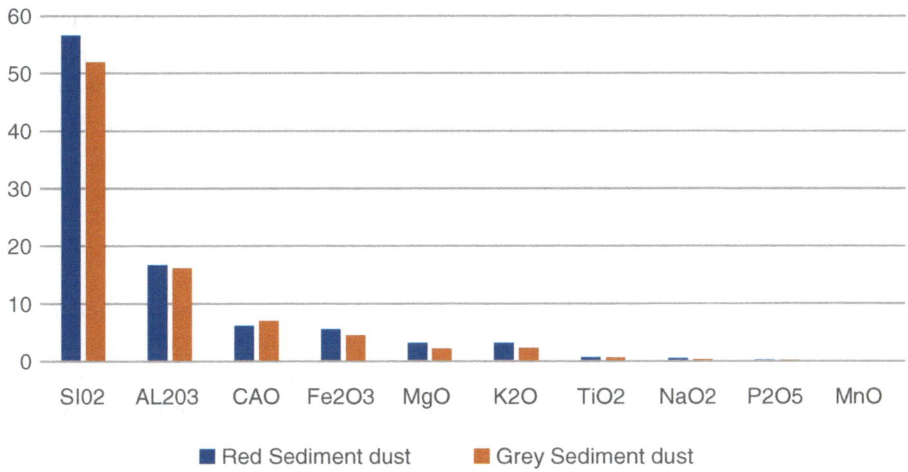

Fig. 22.2 Elemental composition of desert dust from Northeast Arizona based on data from Ghio et al. [27]

concentrations of silica, therefore, represent dust of geologic origin. Sahara Desert dust contains 80% silica, 8% calcium, and 2.5% iron [58]. Dust from Iraq contains 50% silica, 18% calcium, and 12% aluminum [40]. Asian dust also has a composition similar to the earth's crust with 60% silica [21]. Kyung et al. [59] reported that Asian dust deposited in South Korea contained 48% silica and 12% aluminum as shown in Table 22.1.

Dust composition in the southwestern United States has similar profile. Dust collected from Arizona represents a geo genic origin, with 56% silica, followed by 16% aluminum, and 7% calcium as shown in Fig. 22.2 [27]. Similarly, dust from the Mojave Desert in California shows a high (60%) concentration of silica, 14% aluminum, and 8% calcium [40].

Dust Size Definitions

The dust particles can travel over long-range distances and deposit downwind across thousands of kilometers. During this process, coarser particles tend to deposit proximal to the source while finer particles travel to deposit in more distant locations [94]. The commonly described particulate matter sizes are PM_{10}, $PM_{2.5}$, $PM_{1.0}$, and

$PM_{0.1}$, which designate the aerodynamic particle diameter in micrometers. For example, PM_{10} includes particles with an aerodynamic diameter of 10 μm or less (US-EPA). In 2013, the United States Environmental Protection Agency (EPA) revised standards for atmospheric particulate matter; the $PM_{2.5}$ limit was set to annual and daily means of 12 μg/m^3 and 35 μg/m^3, respectively, while the PM_{10} limit was set to daily mean of 150 μg/m^3. Concentrations above these limits are harmful for public health (US-EPA) [118, 119]. Health effects of different particle sizes have been described in detail in the following sections.

Particle size, chemistry, and atmospheric concentration are strongly implicated in climate change as particles suspended in air absorb or scatter solar light and shift air temperatures to variable degrees. It is therefore suggested that an increase in dust activity has a significant role in global warming [21].

Dust Geographic Distribution

Africa is the largest source of dust in the world. Drought struck North Africa in the late 1960s, chiefly involving the Sahara and the Sahel. Since then, these locations have generated more than half of the total dust mass on Earth with an estimated dust production of 517 tera-grams (equal to 5.699×10^8 Tons) annually [73]. North Africa is followed by Central Asia and China, which, generate an estimated 16% and 5%, respectively, every year [37, 73]. During 1970–80, China experienced a rapid change in land use, with increased deforestation, suboptimal farming, and overgrazing. The deserts broadened at an estimated rate of 21 km^2/yr. This gave rise to a major hub of dust generation in the perimeters of China [37]. In Central Asia, the Aral Sea is major source of dust generation. In 1960, diversion of water from the Aral Sea for agricultural purposes decreased its size from 60,000 km^2 to less than half at present. More than 33,000 km^2 of exposed sea bed has become a source of dust production, especially during dust storms [37, 87].

Australia and North America contribute an estimated 14% and 5% of the world's dust emissions, respectively [32]. In the years 1988–2011, American deserts saw a rapid intensification in dust storm activity with a 240% rise in frequency since 1990s. It is expected that such events will further increase in the coming decades due to ongoing climate change [117].

In the United States, the Great Plains (Fig. 22.3, Location 8), extending from Montana to Southern Texas, are the largest source of dust generation [30]. Arid climate, high wind speed, and erodible surface soil create a favorable setting for dust activity in the region [28]. Consequently, 60% of the wind erosion events in the United States occurs in the Great Plains. Within the Great Plains, the highest frequency of wind erosion is seen in the Southern Plains of Texas; the region sees an average 50 days of wind erosion per year, the highest in the United States, with a dust optical depth of >0.1 [30]. Dust optical depth is a measure of the dust concentration in the vertical column of atmosphere based on the degree of obstruction of the solar beam by suspended dust particles in the atmosphere. A value of 0.01

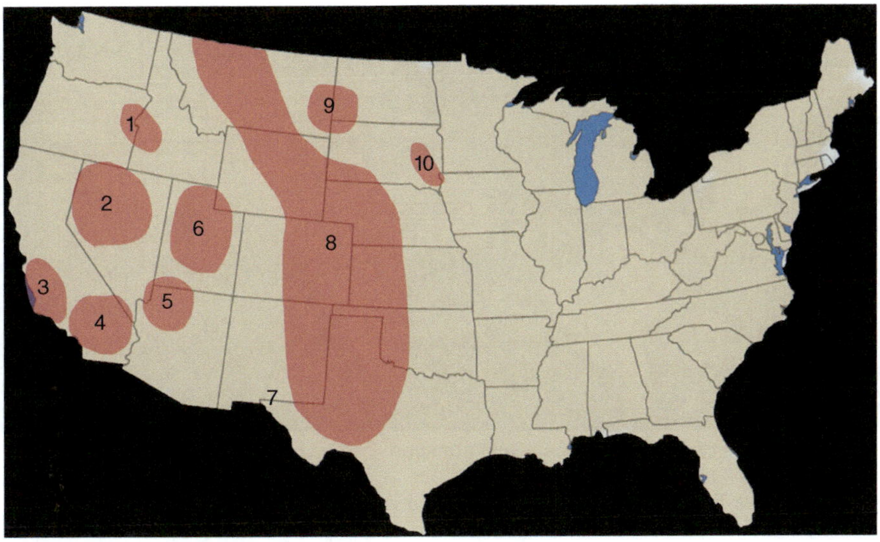

Fig. 22.3 Major locations of dust generation in the United States. Location 1 – Snake River; Location 2 – Black Rock Smoke Creek; Location 3 – San Joaquin Valley; Location 4 – Mojave Desert; Location 5 – Colorado River; Location 6 – Great Salt Lake Desert; Location 7 – Chihuahuan Desert; Location 8 – Great Plains; Location 9 – Lower Yellow Stone River; Location 10 – Big Sioux River. (Adapted from Figure 11 of Ginoux et al. [30])

indicates an extremely clean atmosphere, and a value of 0.4 indicates very hazy condition. An average aerosol optical depth for the United States is 0.1–0.15 (NOAA). The highest dust storm events take place in the Southwestern Unites States including California, Arizona, Nevada, and Washington with dust optical depth rising above 0.2 [30]. Dust storms are more frequent in this region during spring and summer with the heaviest dust activity seen in the month of July [20]. At the border of Texas and Mexico, El Paso experiences heavy dust activity during spring with dust optical depth rising above 0.2 [30]. Li et al. [61] identified 620 locations in the Southwestern United States that are hotspots for blowing dust production and frequent dust storms. These locations are situated mostly in the Southern Great Plains and partly in the Chihuahuan Desert where soil is predominantly composed of sand and silt particles. These sites consist of 26% shrub lands, 22% grasslands, 38% croplands, and 8% barren land. Some of the major dust-generating locations are shown in Fig. 22.3.

Similar to the Aral Sea in Central Asia, Owens Lake in Southern California is also a significant source of dust generation. In 1913, water levels dropped when lake water was used to serve the demand for drinking water in Los Angeles. This left behind 280 km^2 of dried lake bed rich in small-sized sediment particles. Each year, the dried bed of Lake Owens generates eight million metric tons of dust in atmosphere [37].

Apart from local sources, dust also enters the United States via long-range transport from dust events in distant locations. The main sources of dust traveling to the

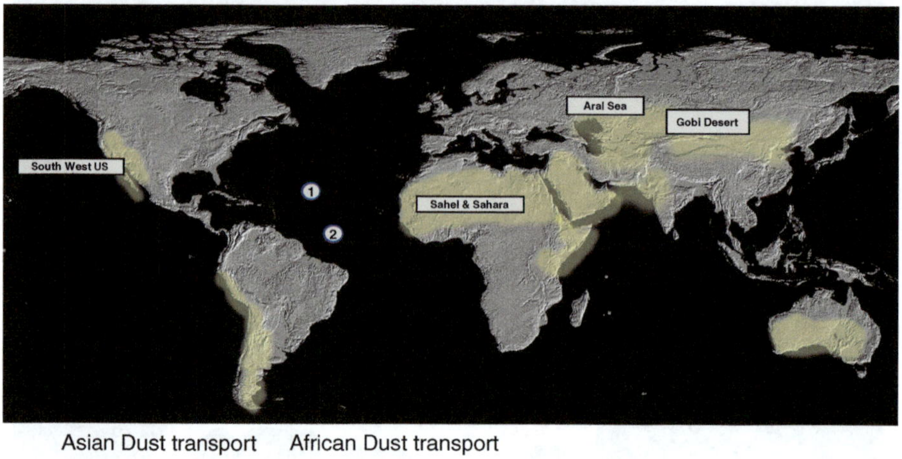

Fig. 22.4 Major locations of dust generation in the world (yellow-shaded areas) and their atmospheric pathways. Results based on Figure 3 from Kellogg and Griffin [53] and Figure 2 from Griffin [34], respectively

United States are from neighboring continental coasts. Prolonged African dust events can cross the Atlantic Ocean and reach North America and the Northern Caribbean in 3–5 days. This dust pathway (Fig. 22.4, Orange track 1) is predominant in summers between June and October, while in the winter season, from November to May, African dust travels to South America and the Southern Caribbean (Fig. 22.4, Orange track 2). Similarly, large Asian dust events travel in the eastern direction (Fig. 22.4, Green track), predominantly from February to April, and cross the Pacific Ocean to reach North America in 7–9 days [34]. In spring 2001, a large bulk of dust generated in the Gobi Desert of China traveled the entire globe exposing billions of people to its toxic effects. The dust cloud traveled east, passing through Korea and Japan, and hit the Pacific Ocean on the fifth day. The dust kept going, crossing the United States, the Atlantic, and finally settled in Europe [53].

African dust activity and concentrations in the Caribbean's Islands have been closely linked. According to a study by Kellogg and Griffin [53], increased African dust activity coincided with a significant increase in dust concentrations in Barbados. The same relationship was observed for African dust activity and culturable microorganism concentrations in Barbados. In July 2005, Griffin [34] investigated the effects of an African dust event on air quality in Florida, United States, 6500 km away from the coast of North Africa. He measured the baseline ambient particulate concentration to be 2.6×10^6 airborne particles/m^3 at a location south of Tampa Bay by a handheld laser. Ten days later, during an ongoing African dust event, the ambient particles in the same region rose to 26.1×10^6 particles/m^3. Of these recorded particles, 99% fell in the size range of $PM_{1.0}$ (0.1–0.3 μm). $PM_{1.0}$ can penetrate through the airway barrier and enter the systemic circulation [34] as discussed in detail in the following sections.

Desert Dust Effects on Health

A high atmospheric dust concentration is a significant health concern. Particularly, large dust storm events translocating over long distances negatively impact a greater portion of the population [94]. Studies have shown their wide range of effects on public health. Crooks et al. [20], for example, investigated the effects of dust storms on non-accidental mortality during the period of 1993–2005. During the study period, the highest number of dust storms occurred in the Southwest especially in Arizona and California. Daily non-accidental mortality on county-level was measured. They concluded that dust storm events were associated with an estimated 2.7% rise in daily, non-accidental mortality lagged by 0–5 days in the respective locations (95% CI: 0.4–5.1; $p = 0.023$). In Arizona, dust storms are the third largest cause of weather fatalities, resulting in 157 deaths and 1324 injuries over the past 50 years [61]. Dust affects almost all organs in humans, and effects vary from neurological to urological diseases.

Pulmonary Effects

The lungs are the first destination of inhaled dust and are, therefore, the organ most effected by dust exposure. The extent of injury is dependent on the composition and size of the dust particles. PM_{10} can lodge in the upper respiratory tract. Though it can be expectorated, it still poses significant risk of damage to epithelial cells in the upper respiratory tract [21]. It is well known that exposure to PM_{10} is directly cytotoxic to pulmonary epithelial cells as it induces oxidative stress [64] and the release of lactate dehydrogenase from cells in a dose-dependent fashion [79]. PM_{10} exposure can also cause inflammation by stimulating the release of pro-inflammatory cytokines such as Interleukin (IL)-8 from pulmonary epithelial cells [100]. One possible mechanism is PM_{10}-induced pro-inflammatory gene expression via the activation of transcription factor, Nuclear Factor Kappa B (NFκB), neutrophil influx [64], and histone acetylation at the IL-8 promotor region, which increases the expression and production of IL-8 in pulmonary epithelial cells [29]. Histone acetylation, an epigenetic modification, results in DNA unwinding and access for gene transcription activation [62]. The epigenome, a set of genome regulators [47] linking environmental variations to the genome, is subject to inheritable changes in response to environmental stimuli [62]. Oxidative stress brought about by PM_{10} can also promote fibrosis by inducing fibrogenic mediators including transforming growth factor-beta and fibronectin [59]. $PM_{2.5}$ can penetrate deeper through the bronchial epithelial layer and the alveoli, and get trapped in the deep lung layers [34]. The presence of dust particles in tissue induces macrophage-mediated activation of CD4 and CD8 T cells to upregulate their production and secretion of pro-inflammatory cytokines including Interferon-γ, IL-10, IL-17, and IL-21. Moreover, macrophages also mediate the production of granzyme-A and granzyme-B by both CD4 and CD8 T cells that play a crucial role in bronchial epithelial cell death [63]. Ultrafine $PM_{0.1}$ is much smaller than cellular structures and can enter the pulmonary submucosa and interstitium as early as 4 hours after exposure [88]. Moreover, $PM_{0.1}$ can translocate to extra-pulmonary organs such as the liver, via systemic circulation, within 24 hours of inhalational exposure [89]. In a similar fashion, $PM_{0.1}$ can also get translocated to other organs such as brain, heart, and spleen [82, 89, 113].

The most vulnerable population to pulmonary toxicity from dust exposure include the elderly with altered immune response, the young with developing pulmonary systems, and those with existing pulmonary conditions [48, 126].

Childhood pneumonia is on the rise in Karakalpaktan, Uzbekistan, with highest incidence in Central Asia. Interestingly, Karakalpaktan is located downwind of the dried part of the Aral Sea which, as described above, is a major source of dust with 23% of the total concentration composed of PM_{10}. To add to this, childhood interstitial lung disease, that is a very rare lung condition, has also been reported from Kazakhstan, which borders the Aral Sea. The incidence of interstitial lung fibrosis and bronchiectasis in school-going children is as high as 20% in some parts of Kazakhstan [3, 53, 87].

A number of studies, from East Asia to Europe, have targeted the effect of dust activity on pulmonary hospital admissions [16, 51, 52, 68, 114]. In Taiwan, Kang et al. [51, 52] found a statistically significant ($p < 0.001$) increase in the mean number of hospital admissions for pneumonia from 279 admissions/day on non-dust storm days to 292.5 admissions/day on dust storm days. There was also an increase in the mean number of pneumonia admissions on post-dust-storm days (day1–4) to 305.7 admissions/day. Possible mechanisms include PM-induced pulmonary epithelial damage, oxidative stress, and inflammation [63, 64, 89]. Increased delivery of pathogens attached to dust particles during dust storms [36], particularly in the setting of pre-existing pulmonary inflammation, might be another possible mechanism. In Rome, Italy, Mallone et al. [68] found a 2.64%–12.65% increase in respiratory mortality (95% CI = 1.18–25.42%) during dusty days with a 10.8 $\mu g/m^3$ increase in $PM_{2.5-10}$ (particles with sizes ranging between 2.5 μm and 10 μm) and a 19.8 $\mu g/m^3$ increase in PM_{10}. In the Caribbean, the incidence of asthma on the island of Barbados has increased 17 times since 1973, which coincides with the time during which long-distance transport of African dust to the Caribbean increased [53].

In the United States, Spokane, Washington, has one of the highest PM concentrations in the nation. Mar et al. [69] found a statistically significant ($p < 0.05$) association between cough and a 10 $\mu g/m^3$ increase in PM_{10}, $PM_{2.5}$, and $PM_{1.0}$ fractions, among children during 8 months follow-up. Therefore, during days with rise in ambient PM concentration, the odds of having cough on the same day (day-0) were 1.09 (95% CI = 1.02–1.16) while for next 2 days (day-1 and day-2) were 1.08 (95% CI = 1.02–1.16) and 1.10 (95% CI =1.02–1.18), respectively.

Neurological Effects

Multiple studies suggest possible associations between dust exposure and neurocognitive disorders including depression, anxiety, schizophrenia, and Alzheimer's disease [49, 91–93]. The likely mechanism of dust-induced neuropathology is via generation of reactive oxygen species, induction of pro-inflammatory cytokines, and activation of microglial cells [9]. Veronesi et al. [120] evaluated neurotoxicity in mice with pre-existing oxidative stress, following exposure to concentrated ambient particulate matter (CAPs) collected from a state park in Tuxedo, NY. In comparison to wild-type mice, those with pre-existing oxidative stress had a 29% reduction in dopaminergic neurons in the substantia nigra, a structure in the midbrain that plays an important role in movement. Loss of dopaminergic neurons in this region has been associated with Parkinson's disease. In another study, a 6-hour exposure to CAPs in BV2 cells, an immortalized C57/BL6 mouse microglial cell line, led to reduced intracellular adenosine triphosphate (ATP) levels, and induction of pro-inflammatory cytokines like IL-6 and tumor necrosis factor (TNF)-α. Deoxyribonucleic acid (DNA) microarrays of the same cells identified the stimulation of pro-inflammatory and oxidative stress–related genes. These included genes associated with the Notch-activating pathway for NFκB signaling [103]. Notch-activating pathway is an inter-cellular signaling pathway involved in the activation of NFκB which is a pro-inflammatory transcription factor [81]. Studies suggest ultrafine PM ($PM_{0.1}$) plays a significant role in neurotoxicity. It can penetrate the airway barrier upon inhalation, enter the blood circulation, and cross the blood brain barrier [9, 24]. Its large surface area enables it to induce a greater degree of oxidative stress and inflammation [95]. Ultrafine PM can also induce changes in fetal brain tissue, increasing the risk of neurodevelopmental disorders [2, 45]. Prenatal exposure to $PM_{2.5}$ and PM_{10} dust is associated with an increased incidence of autism spectrum disorder among children [50, 98, 121]. Apart from pre- and post-natal effects, multiple studies target the role of PM exposure later in life, in neurodegeneration. Jung et al. [49] found a 138% increased incidence of Alzheimer's disease for every 4.34 $\mu g/m^3$ rise in the atmospheric concentration of $PM_{2.5}$. Ailshire and Crimmins [1] investigated the association between atmospheric $PM_{2.5}$ concentrations and changes in cognitive function in 780 adults >55 years of age. They found a 1.53% increased error rate [95% confidence interval (CI) = 1.02–2.30] in cognitive assessments of individuals exposed to a $PM_{2.5}$ concentration of 13.8 $\mu g/m^3$, with cognitive decline more evident in the episodic memory component ($P < 0.05$). The 13.8 $\mu g/m^3$ concentration exceeds the US EPA's limit of 12 $\mu g/m^3$, beyond which can be harmful to public health (US-EPA). Ranft et al. [97] studied the effects of exposure to atmospheric PM_{10} in 399 women, each with the same residential address for >20 years. Women with a mean age of 74 years exposed to a PM_{10} concentration of 25–53 $\mu g/m^3$ during the study period underwent testing for mild cognitive insufficiency and olfactory function. A Consortium to Establish a Registry for Alzheimer's Disease (CERAD)-Plus neuropsychological battery was used for the former. Results revealed a statistically significant ($p < 0.01$) dose-response relationship between PM_{10} exposure and negative performance on these tests suggesting mild cognitive insufficiency.

Due to a possible role of PM in the initiation of a pro-thrombotic state [105], PM exposure is associated with increased risk of stroke [86]. Kang et al. [51, 52] investigated the effects of 46 separate dust storm events on stroke admissions in Taiwan. They concluded that there was a statistically significant ($p < 0.001$) increase in admissions on dust storm days versus non-dust storm days (239.6 versus 219.6 stroke admissions/day, respectively). This surge further intensified on the first and second post-dust storm days with 271.3 and 249.1 stroke admissions/day, respectively.

Cardiovascular Effects

The relationship between dust storms and cardiovascular health has been frequently studied in various parts of the world [32]. Impacts on cardiovascular health increase with exposure duration; immediate exposure ranging from hours to weeks puts exposed individuals at increased risk of myocardial ischemia and heart failure, while chronic exposure has been implicated in reduction in life expectancy by months to years [11]. Mallone et al. [68] found a stronger association between a 10.8 μ/m^3 rise in $PM_{2.5-10}$ concentration and cardiac mortality (9.73%; 95% CI = 4.25–15.49) during Saharan dust event days compared to dust-free days (0.86%; CI = 2.4–4.31; $p > 0.005$). The mechanism of morbidity can be explained by a range of PM effects in the body including endothelial dysfunction caused by direct damage to the basal cell membrane [105]. Also, PM exposure is associated with increased levels of mediators of coagulation [46] such as tissue factor [72] and fibrinogen [26], mediators of inflammation like C-reactive protein [17], and markers of endothelial dysfunction including but not limited to intercellular adhesion molecule-1 (ICAM-1) protein, and vascular endothelial adhesion molecule (VCAM) protein [66]. The resultant inflammation and pro-coagulatory state in the body [46] leads to accelerated atherosclerosis as shown in Fig. 22.5 [105].

Recent research suggests a possible role of epigenetics in particulate exposure-induced morbidity. According to work by Baccarelli and Ghosh [4], variation in the methylation of DNA and DNA regulatory elements can change gene expression. Baccarelli et al. [5] found a significant ($p = 0.001$) association between recent $PM_{2.5}$ exposure of 4–7 hours and hypomethylation of long interspersed nuclear element-1 (LINE-1) ($\beta = -0.13$, 95% CI). LINE-1 is one of the transposable repetitive elements that contain islands of CpG DNA sequences. LINE-1 hypomethylation also correlated with higher VCAM-1 expression in serum and high systolic, diastolic, and mean arterial blood pressures. Variation in methylation of these CpG islands can change genomic expression [4]. Increased methylation interferes with the binding of transcription factors to a gene, suppressing its expression [125]. Tarantini et al. [115] found a statistically ($p = 0.02$) significant association between

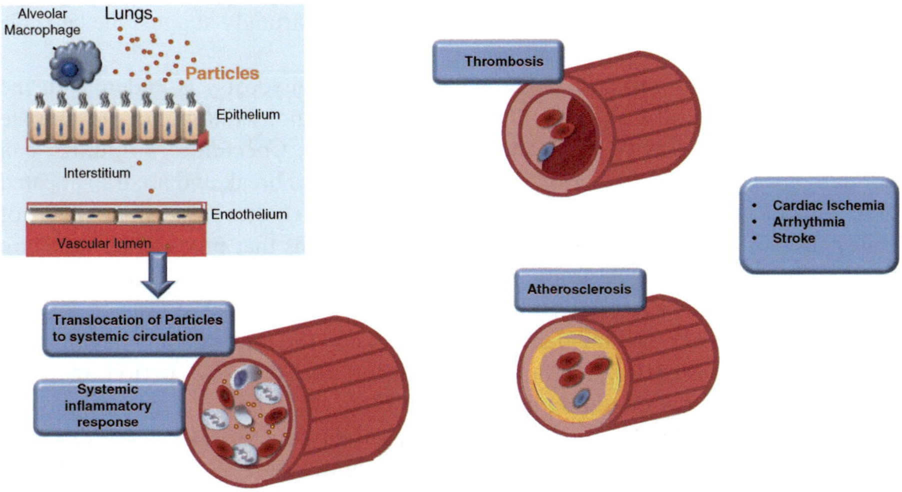

Fig. 22.5 Cardiovascular effects of dust exposure. Copied from Figure 1 by Schulz et al. [105]

short-term PM_{10} exposure and hypomethylation of promotor region of the inflammatory marker, inducible nitric oxide synthase (iNOS) as compared to pre-exposure levels [mean difference = −0.61; standard error (SE) = 0.26]. These evidences support the idea that dust exposure can cause epigenetic changes in the affected population. These epigenetic changes, with their potential to be transgenerationally inherited [41], can affect a large sum of population over time.

Infections

Dust storms transport along microorganisms with their spores leading to their distribution and propagation during dust storm events [36]. Griffin et al. [36] analyzed atmospheric concentrations of microbes in the US Virgin Islands, the Northern Caribbean, and on a cruise (From Miami, FL to Bahamas, US Virgin Islands, Haiti and Puerto Rico) during African dust events and normal days. In the two consecutive study years 2000 and 2001, they noted 5.2- and 4.9-fold increases, respectively, in atmospheric concentrations of culturable microbes during dust versus non-dust days. During this study period, the highest increase in atmospheric microbial concentrations was noted during the months of July (2000) and August (2001) during an ongoing African dust event. One of the microbes recovered during this investigation was *Pseudomonas aeruginosa*, a pathogen known to cause fatal infections in susceptible persons including those with an immunosuppressed status. Later in 2006, Kellogg and Griffin studied air microbial load prior and during dust storms in Virgin Islands. They found that background sampling days yielded 0.01 colonies of microbes/L cultured from 200 L air while African dust event days yielded 0.1

colonies/L, a ten-fold increase. These study outcomes strongly suggest the carriage of microbes with African dust transport.

In United States, valley fever (known as coccidiomycosis) is endemic in the southwestern states of California and Arizona. The region is home to arid climate and frequent dust activity [65]. The causative fungus, *Coccidioides* immitis, can normally grow in the soil. During dust storms, its spores break and are translocated with the wind as an inhalable agent [90, 116]. Multiple studies suggest incidence of valley fever is positively associated with wet conditions that provide the moisture for the fungus to propagate in the soil, followed by dry climate with dust activity which lets the spores rupture and get transported during dust storms in air and inhaled [18, 31, 122]. The incidence of meningitis, in the meningitis belt, can spike from 5 to 10 cases/100,000 population during the endemic period to 1000 cases/100,000 population during epidemics [13].

Analogous to valley fever in the United States, dust activity is a major driver of endemic meningitis in North Africa. The Meningitis Belt is a Sahelo-Sudanian region that stretches across North Africa from Uganda and the Democratic Republic of Congo, to Malawi and Mozambique, and has a high incidence of meningitis [76]. Meningitis is mainly caused by the gram-negative bacteria, *Neisseria meningitidis*. The incidence of meningitis fluctuates with the pattern of dust activity in the region [76]. It is a life-threatening disease with high morbidity and mortality that can affect over 200,000 individuals annually in the Meningitis Belt [34]. This region is the location for frequent epidemics of meningitis with seasonal spikes in endemic from November through May. This period of the year is characterized by Harmattan trade winds that are high-intensity low-humidity winds blowing toward the northeast and carry a heavy burden of dust [77]. The dry air damages the nasopharyngeal mucosal barrier and increases the risk of invasion by colonized pathogens [34]. The endemic period ends with the onset of monsoon season, when the dust settles with the rain. This further suggests an association of dusty dry weather with the outbreak of meningitis [34, 76, 77].

Dust storms can also indirectly affect the incidence of certain infections. Chronic exposure to desert dust can put individuals at higher risk of developing pneumoconiosis [27], a lung disease characterized by inflammation, coughing, and fibrosis. As mentioned above, geogenic dust carries a high silica content [21]. With chronic lung exposure to silica-containing dust, non-occupational silicosis, a type of pneumoconiosis, can develop in the exposed population [32]. This has been particularly observed in Northwest China [85] and India. The city of Ladakhi, India, home to frequent dust storms, has a 22.5% prevalence of non-occupational silicosis [102]. The inhaled silica dust upon reaching the alveoli is ingested by alveolar macrophages, resulting in damage and apoptosis of macrophages; possible mechanism is by direct or indirect generation of reactive oxygen species and oxidative stress. The activated signaling pathways induce inflammatory cytokines and apoptosis in macrophages [27]. Prolonged silica exposure leads to chronic inflammation and fibrosis [39]. Chronic silicosis is also associated with an increased risk of tuberculosis [44], an infectious disease caused by *Mycobacterium tuberculosis*, which can be found in soil and spread through the air. Mathur and Choudhary [70] found radiological silicosis among 16% of tuberculosis patients with non-occupational silica exposure in the desert of Thar, India. The possible mechanism of increased susceptibility of tuberculosis in chronic silicosis is by silica-induced necrosis of infected alveolar macrophages, releasing viable bacteria that can infect other cells [14]. Another mechanism is by silica-induced decrease in the expression of Toll-like receptor type 2 (TLR-2) on macrophages [8]. TLR-2 is a receptor found on macrophages which recognizes microbe lipoproteins and activates macrophages [8].

Long-range transport of pathogens with dust has also been implicated in the spread of avian influenza. In 2006, Kilpatrick et al. [55] studied the possible causes of the spread of the H5N1 avian influenza (Influenza A) virus. They concluded that bird migration and poultry trade patterns could not explain the H5N1 influenza outbreaks in South Korea and Japan, countries that are located downwind from Asian dust activity. In 2010, Chen et al. [15] measured ambient concentrations of Influenza A and other influenza viruses during Asian dust storm and background days in Taiwan. Results revealed increased atmospheric concentrations of Influenza A virus ($p = 0.02$) with a mean of 268 copies on dust storm days compared to 13 copies on the background days.

Health Effects of Dust Contaminated with Pesticides and Metals

Certain metals and chemicals also "hitch a ride" on dust transported over long distances. During dust events, therefore, inhalation of these substances can drastically affect pulmonary health [42]. Arsenic, for example, is a carcinogenic metal with known deleterious effects on health. Exposure to dust burdened with high arsenic concentrations can increase body levels of arsenic, leading to skin pigmentation changes, palmar and plantar hyperkeratosis, bone marrow depression, peripheral neuropathy symptoms, and skin and lung cancers [38, 111]. Subhani et al. [111] calculated the concentration of arsenic in dust in the cities of Sargodha and Lahore, Pakistan. Levels were the highest in industrial dust at 9.78 mg arsenic/kg dust followed by 7.59 mg/kg and 6.95 mg/kg in urban and rural settings, respectively. The average arsenic concentration from these three settings was 8.38 mg/kg, significantly elevated from the World Health Organization limit of 1 mg/kg. Arsenic levels similar to the levels in the three environmental settings were found in human hair and nails suggesting dust as the largest source of exposure. Similarly, in a rural area of Australia, Hinwood et al. [42] found high arsenic concentrations (32.1 mg/kg) in toenails of individuals who were exposed to a high arsenic soil concentration (>100 mg/kg) but drank arsenic-free water compared to control participants. Control participants were individuals from rural areas who drank arsenic-free water and were exposed to below set standards of soil arsenic levels (3.35 mg/kg). Soil arsenic levels of <30 mg/kg set by Australian and New Zealand Environmental Council were used for the selection of control participants.

In the dried portion of Aral Sea, nearby agricultural activity has led to increased local atmospheric pesticide concentrations. The concentration of Phosalone, an organophosphate pesticide, has been identified at a maximum of 126 mg/kg dust in Dashkhous, located in the irrigation zone close to the Aral Sea [87]. In Kazakhstan, adjacent to the Aral Sea, Bapayeva et al. [7] investigated the blood levels of cotton-growing organochlorine pesticides and compared in local female adolescents with high versus no (control) exposure [7]. High blood levels were noted in the former versus the latter for organochlorine pesticides including Lindane (18.51 ± 0.16 versus 4.05 ± 0.41 mg/l), Dieldrin (169.16 ± 3.13 versus 30.8 ± 3.7 mg/l), Dichlorodiphenyltrichloroethane, commonly known as DDT (177.78 ± 2.71 versus 109.7 ± 2.58 mg/l), and Endrin (37.57 ± 0.9 versus 4.85 ± 0.69 mg/l); all comparisons were statistically significant ($p < 0.001$). The exposed population also demonstrated a higher level of gynecologic problems (14.6%) compared to the control group (11.1%). Amenorrhea, for example, was 2.7 times more in exposed group compared to controls.

Harmful Algal Blooms and Health Impact

Harmful algal blooms are caused by uncontrolled growth of certain microbes in the sea water. The excessive production of toxins from these microbes can negatively affect organisms in the surrounding environment including humans (NOAA). Long-range transport of dust clouds can trigger the growth of the coastal marine microbes, sometimes resulting in harmful algal blooms [35]. Global dust transport is a major source of iron deposition carried mostly from North Africa downwind to Atlantic and Mediterranean oceans [67]. Iron a vital micro-nutrient for microbial growth and increases in iron concentration via long-range transport of dust clouds can trigger the growth of coastal marine microbes, resulting in harmful algal blooms [35]. The atmospheric concentration of iron in west Florida has been noted to increase from 0.1 to 0.5 nmol/kg on background days to as high as 16 nmol/kg on days of Saharan dust events. During such events, the sea surface concentration of the algal bloom bacteria such as *Trichodesmium* can rise 100-fold [60]. These algal blooms of *Trichodesmium* bacterium and *Karenia brevis* dinoflagellates can result in red tides producing marine toxins that negatively impact human health [32]. For example, humans can suffer toxic effects like abdominal upset, increased heart and respiratory rates, decreased blood pressure, and/or death upon eating fish contaminated with *Trichodesmium clupeotoxin* [35].

Karenia brevis, another red tide causing bacteria, produces a potent algal toxin, Brevetoxin, that is a heat and acid stable toxin and causes neurotoxic shellfish poisoning. Consumption of Brevetoxin-contaminated shellfish causes cell membrane depolarization and uncontrolled sodium influx. The resulting membrane excitability and cellular disruption is the mechanism by which cardiac and neurologic dysfunction is produced in affected individuals. Brevetoxin is also associated with respiratory symptoms triggering the release of the neurotransmitter, acetylcholine, from autonomic nerve endings which results in extensive mast cell degranulation, airway smooth muscle contraction, and bronchospasm [57].

Impact of Dust Exposure in US War Veterans

The Middle East is a location with frequent dust storms resulting in high ambient PM concentrations. These dust events can be frequent ranging between 20 and 50 days in a year, with the most common time for dust activity being spring and summer [78]. Engelbrecht et al. conducted sampling and analysis of atmospheric dust from 15 deployment sites in Iraq, Afghanistan, Qatar, Djibouti, and Kuwait in 2009. Results revealed consistently high $PM_{2.5}$ concentrations ranging from 35 μg/m^3 in Djibouti to 111 μg/m^3 in Tikrit, Iraq. PM_{10} concentration were also consistently high ranging from 72 μg/m^3 in Djibouti to 303 μg/m^3 in Talil, Iraq as shown in Fig. 22.6. These values exceeded the exposure limit of 15 μg/m^3 set by the US Army Center for Health Promotion and Preventive Medicine Military Exposure Guidelines [22].

Consequently, respiratory health is an emerging issue for US military personnel deployed to Iraq and Afghanistan. Troops with deployment lasting more than 30 days have a higher rate of respiratory symptoms compared to the personnel with lesser days of deployment irrespective of smoking status [106, 109]. Smith et al. demonstrated a linear dose response between respiratory symptoms duration of deployment in Army personnel. This pattern was not seen in the troops deployed to sea and air [106, 109]. In a survey conducted in 2004, 69.1% of the troops deployed to Afghanistan or Iraq presented with respiratory symptoms and 17% required medical treatment. There was also a two-fold increase seen in precombat pulmonary conditions [104].

Szema et al. reviewed 6233 cases of asthma diagnoses in Veteran Affairs records between 2004 and 2007, and among the 290 new cases of asthma, they found a 2.3% higher incidence with an odds ratio of 1.58 (95% Cl = 1.18–2.11) in those deployed

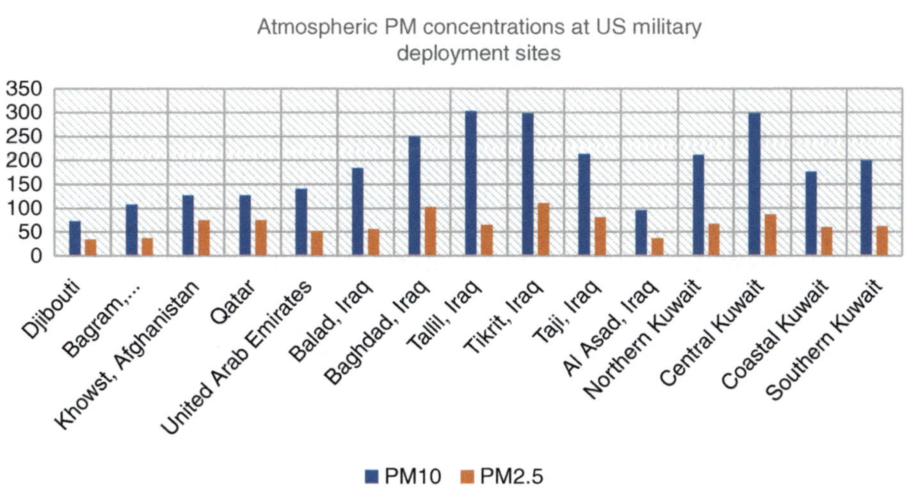

Fig. 22.6 Atmospheric PM_{10} & $PM_{2.5}$ concentrations for 15 deployment sites in the Middle East. Based on information from Engelbrecht et al. [22]

to Iraq [112]. Roop et al. [101] compared respiratory effects in asthmatics and non-asthmatics deployed to Iraq. Out of 1193 completed survey forms, 5% of participants ($n = 61$) were previously diagnosed with asthma, and 32% ($n = 375$) were smokers. Comparisons of pre- and inter-deployment effects revealed there were increased reports of wheeze by 13% ($n = 141$) in non-asthmatics and by 10% ($n = 6$) in asthmatics ($p < 0.005$), increased symptoms of cough by 29% ($n = 315$) in non-asthmatics and 20% ($n = 11$) in asthmatics ($p < 0.05$), and increased sputum production by 18% ($n = 192$) in non-asthmatics and 15% ($n = 8$) in asthmatics ($p < 0.05$). To add to this, 13% ($n = 155$) of military personnel reported new onset or aggravated dyspnea (difficult/labored breathing) during deployment, with a mean Borg dyspnea score of 2.4 ± 2. The Borg Dyspnea Scale is used to assess symptoms of dyspnea; a score of 2.4 ± 2 indicates slight dyspnea [19]. There was also a 38% increase in reported allergic rhinitis. The general performance of troops during deployment was negatively affected due to respiratory symptoms in 13% ($n = 153$) of study participants. Overall, 14% ($n = 159$) of non-asthmatics and 44% ($n = 27$) of asthmatics reported respiratory symptoms during deployment and sought medical attention. Apart from humidity and heat, these individuals were exposed to dust during the deployment period in Iraq [101].

King et al. [56] conducted a post-deployment respiratory health evaluation in war soldiers returning from Southwest Asia between the years 2005 and 2009. Out of 80 soldiers with respiratory symptoms, 49 were referred for surgical biopsy of lung tissue, 38 of whom had a biopsy suggestive of constrictive bronchiolitis, a small airway fibrotic respiratory disease that causes dry cough, shortness of breath, wheezing, and fatigue. Of these 38 individuals, 25 did not have a smoking history, and 10 had exposures to dust storms in addition to battlefield smoke, diesel exhaust, and burn pits. The observed constrictive bronchiolitis may have been associated to dust, or all mentioned exposures, collectively.

Desert Dust and Experimental Studies

A number of studies have investigated the effects of dust particles on the body. Rodriguez-Cotto et al. [100] found that African dust particles, transported to Puerto Ricco via the Atlantic Ocean, induced dose-dependent cytotoxic effects on bronchial epithelial cells in culture. Both $PM_{2.5}$ and PM_{10} extracts exposure increased concentrations of IL-6 and IL-8 in bronchial epithelial cells significantly ($p < 0.01$) with maximum increase noted in IL-8 production following exposure to PM_{10} ($p < 0.01$) (Figs. 22.7 and 22.8).

IL-6 is a pro-inflammatory cytokine with anti-inflammatory properties. It can induce an acute-phase response (a complex systemic early-defense reaction activated by stimuli such as injury, infection) by inhibiting TNF-α and IL-1. In the lungs, epithelial cells are a major source for IL-6 production [43].

IL-8, which was initially considered a product of macrophages [6], is also produced by airway epithelial cells. It is involved in neutrophil and eosinophil

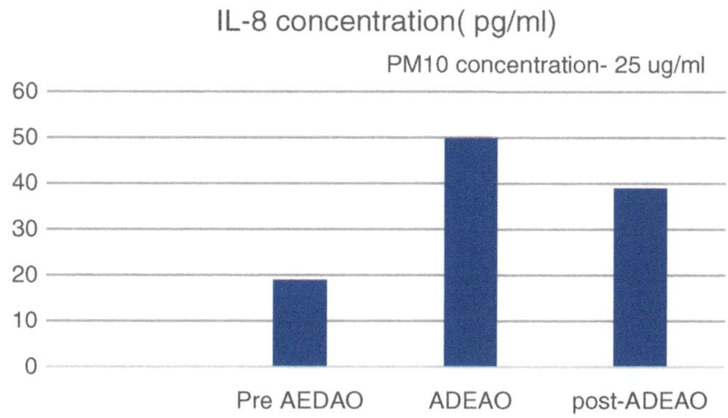

Fig. 22.7 Effects of PM_{10} dust over the Atlantic Ocean, from an African dust event, on IL-8 levels in bronchial epithelial cells. AEDAO = African dust transported to the Atlantic Ocean. Based on information in Figure 4 from Rodriguez-Cotto et al. [100]

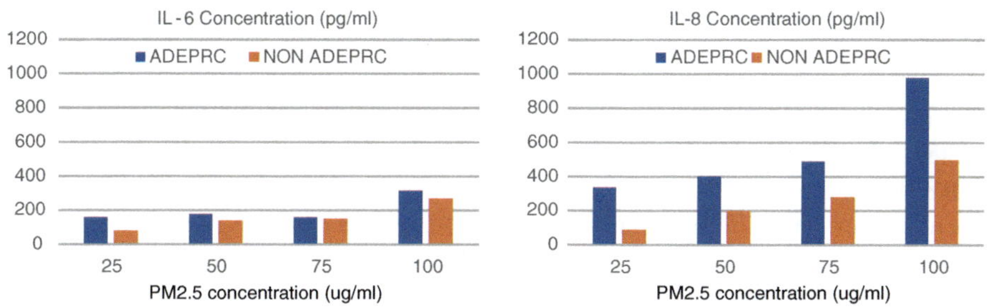

Fig. 22.8 Effects of $PM_{2.5}$ dust in Puerto Rico, from an African dust event, on IL-6 and IL-8 levels in bronchial epithelial cells. ADEPRC = African dust event in Puerto Ricco. NON-ADEPRC = Atmospheric $PM_{2.5}$ during non-dust storm days in Puerto Rico. Based on information from Rodriguez-Cotto et al. [100]

chemotaxis and recruitment [107]. It has been shown that in COPD and asthmatic patients with the evidence of neutrophilic and eosinophilic inflammation had elevated levels of IL-8 [124]. Between COPD and asthma patients, COPD patients tend to have even higher levels of IL-8 and hence higher neutrophil recruitment and subsequent tissue damage. IL-8 is negatively correlated with the FEV1/FVC ratio, a spirometric parameter that represents the portion of a patient's vital capacity that he/she is able to expire in the first second of a forced expiration. Vital capacity is the greatest volume of air that can be expelled from the lungs after taking the deepest possible breath. An increase in IL-8 levels is associated with a decrease in the FEV1/FEV ratio, a characteristic obstructive pattern [124]. It was concluded that chronic exposure to dust particles with persistent IL-8 increase and induction of neutrophils can potentially cause damage to alveolar and interstitial structures in the lungs.

Geng et al. [25] investigated the effects of exposure to $PM_{2.5}$ dust, from dust storm events in Batao city of Mongolia, on alveolar macrophages in mice as compared to $PM_{2.5}$ from non-dust days. High concentrations of $PM_{2.5}$ were associated with significant oxidative stress in alveolar macrophages with the leakage of lactate dehydrogenase ($p < 0.01$), and reduced activity of the sodium potassium ATPase pump ($p < 0.05$) in the cell membrane. Ghio et al. [27] demonstrated the inflammatory effects of American desert dust in bronchial epithelial cells. Cells were exposed to desert dust sampled from the northeast of Arizona as well as elemental silica. Following exposure to desert dust and silica separately for 24 hours, cell viability and inflammatory responses were measured. Desert dust samples–induced cytotoxic effects in bronchial epithelial cells with 0.5 units/ml increase in extracellular lactic dehydrogenase and 0.9 μM increase in Caspase-3, indicator of apoptosis. Silica exposure resulted in stronger cytotoxic response with 1.5 units/ml rise in LDH levels and 2.5 μM rise in Caspase-3. Hydrogen peroxide/peroxidase oxidant generation, as measured by the Amplex Red Fluorescence assay, was also increased in bronchial epithelial cells following exposure to each of desert dust and silica, with stronger oxidant generation seen following silica exposure. The expression of *superoxide dismutase-1 (SOD-1),* and *cyclooxygenase-2 (COX-2)* genes, which are inflammatory markers, doubled following exposure to desert dust while SOD-1 gene expression increased four-fold following exposure to elemental silica. RNA for hemeoxygenase-1, a highly sensitive and reliable marker of cellular oxidative stress [96], increased four-fold after exposure to dust and eight-fold following exposure to elemental silica. Dust and silica exposure also activated the mitogen-activated protein (MAP) kinase, p38 kinase, and extracellular signal-regulated kinase (ERK)1/2 pathways [27]. MAP kinase, p38 kinase, and ERK 1/2 pathways contribute to the activation of the inflammatory cytokines and *hemeoxygenase-1* transcriptions [80]. Consequently, pro-inflammatory mediators, including TNF-α, Interferon-γ, Interferon-1β, and IL-6 were also increased. When dust and silica separately were pharyngeally instilled in mice, bronchoalveolar lavage revealed an immunologic response which was greater to dust than to silica. There was an increase in neutrophil counts and levels of IL-1β, IL-6, macrophage inflammatory protien-2, and TNF-α ($p < 0.05$) [27]. These data suggest that dust particles are capable of causing cellular oxidative stress and inflammation, decreasing cell viability, and inducing apoptosis. The greater cytotoxic response to silica also suggests its strong role in cytotoxic response of bronchial epithelial cells to dust particles considering silica is also the dominant component of dust particles.

Future Directions

The US research community should address uncertainties on the biologic effects of dust exposure in airways, skin, and eyes. We should develop an evolving set of research priorities in terms of preventive measures for dust exposure in vulnerable populations. A national alarm system for high air-dust concentrations should be established as previously proposed [106]. There is an urgent need for new medications that attenuate dust effects in humans and animals. Local and federal governments should respond aggressively to the climate change threat and promote carbon-free industries and alternative energy sources [75].

References

1. Ailshire JA, Crimmins EM. Fine particulate matter air pollution and cognitive function among older US adults. Am J Epidemiol. 2014;180(4):359–66.
2. Allen JL, Liu X, Weston D, Prince L, Oberdorster G, Finkelstein JN, Johnston CJ, Cory-Slechta DA. Developmental exposure to concentrated ambient ultrafine particulate matter air pollution in mice results in persistent and sex-dependent behavioral neurotoxicity and glial activation. Toxicol Sci. 2014;140(1):160–78.
3. Ataniyazova O, Adrian S, Mazhitova Z, Moshammer H, Prindull G, Zetterstrom R. Continuing progressive deterioration of the environment in the Aral Sea region: disastrous effects on mother and child health. Acta Paediatr. 2001;90(5):589–91.
4. Baccarelli A, Ghosh S. Environmental exposures, epigenetics and cardiovascular disease. Curr Opin Clin Nutr Metab Care. 2012;15(4):323–9.
5. Baccarelli A, Wright RO, Bollati V, Tarantini L, Litonjua AA, Suh HH, Zanobetti A, Sparrow D, Vokonas PS, Schwartz J. Rapid DNA methylation changes after exposure to traffic particles. Am J Respir Crit Care Med. 2009;179(7):572–8.
6. Baggiolini M, Walz A, Kunkel SL. Neutrophil-activating peptide-1/interleukin 8, a novel cytokine that activates neutrophils. J Clin Invest. 1989;84(4):1045–9.
7. Bapayeva G, Issayeva R, Zhumadilova A, Nurkasimova R, Kulbayeva S, Tleuzhan R. Organochlorine pesticides and female puberty in South Kazakhstan. Reprod Toxicol. 2016;65:67–75.
8. Beamer GL, Seaver BP, Jessop F, Shepherd DM, Beamer CA. Acute exposure to crystalline silica reduces macrophage activation in response to bacterial lipoproteins. Front Immunol. 2016;7:49.
9. Block ML, Calderon-Garciduenas L. Air pollution: mechanisms of neuroinflammation and CNS disease. Trends Neurosci. 2009;32(9):506–16.
10. Blunden J, Arndt DS, Hartfield G, Sánchez-Lugo A, Scambos TA, Schreck CJ III, Stammerjohn S, Stanitski DM, Willett KM, Bissolli P. State of the climate in 2017. Bull Am Meteor Soc. 2018;99(8):S1–S310.
11. Brook RD, Rajagopalan S, Pope CA III, Brook JR, Bhatnagar A, Diez-Roux AV, Holguin F, Hong Y, Luepker RV, Mittleman MA. Particulate matter air pollution and cardiovascular disease: an update to the scientific statement from the American Heart Association. Circulation. 2010;121(21):2331–78.
12. Cai W, Cowan T, Briggs P, Raupach M. Rising temperature depletes soil moisture and exacerbates severe drought conditions across Southeast Australia. Geophys Res Lett. 2009;36(21):L21709.
13. CDC. CDC health information for international travel 2012: the yellow book; 2012.
14. Chavez-Galan L, Ramon-Luing LA, Torre-Bouscoulet L, Perez-Padilla R, Sada-Ovalle I. Pre-exposure of mycobacterium tuberculosis-infected macrophages to crystalline silica impairs control of bacterial growth by deregulating the balance between apoptosis and necrosis. PLoS One. 2013;8(11):e80971.
15. Chen PS, Tsai FT, Lin CK, Yang CY, Chan CC, Young CY, Lee CH. Ambient influenza and avian influenza virus during dust storm days and background days. Environ Health Perspect. 2010;118(9):1211–6.
16. Cheng MF, Ho SC, Chiu HF, Wu TN, Chen PS, Yang CY. Consequences of exposure to Asian dust storm events on daily pneumonia hospital admissions in Taipei, Taiwan. J Toxicol Environ Health A. 2008;71(19):1295–9.
17. Chuang KJ, Chan CC, Su TC, Lee CT, Tang CS. The effect of urban air pollution on inflammation, oxidative stress, coagulation, and autonomic dysfunction in young adults. Am J Respir Crit Care Med. 2007;176(4):370–6.
18. Comrie AC. Climate factors influencing coccidioidomycosis seasonality and outbreaks. Environ Health Perspect. 2005;113(6):688.
19. Crisafulli E, Clini EM. Measures of dyspnea in pulmonary rehabilitation. Multidiscip Respir Med. 2010;5(3):202–10.

20. Crooks JL, Cascio WE, Percy MS, Reyes J, Neas LM, Hilborn ED. The association between dust storms and daily non-accidental mortality in the United States, 1993–2005. Environ Health Perspect. 2016;124(11):1735–43.
21. Derbyshire E. Natural minerogenic dust and human health. Ambio. 2007;36(1):73–7.
22. Engelbrecht JP, McDonald EV, Gillies JA, Jayanty R, Casuccio G, Gertler AW. Characterizing mineral dusts and other aerosols from the Middle East – part 1: ambient sampling. Inhal Toxicol. 2009;21(4):297–326.
23. Forward KA, Souviney NJ, Blair RN. Desertification of the American Southwest; 2008. https://core.ac.uk/download/pdf/212970327.pdf.
24. Genc S, Zadeoglulari Z, Fuss SH, Genc K. The adverse effects of air pollution on the nervous system. J Toxicol. 2012;2012:782462.
25. Geng H, Meng Z, Zhang Q. In vitro responses of rat alveolar macrophages to particle suspensions and water-soluble components of dust storm PM2.5. Toxicol In Vitro. 2006;20(5):575–84.
26. Ghio AJ, Kim C, Devlin RB. Concentrated ambient air particles induce mild pulmonary inflammation in healthy human volunteers. Am J Respir Crit Care Med. 2000;162(3 Pt 1):981–8.
27. Ghio AJ, Kummarapurugu ST, Tong H, Soukup JM, Dailey LA, Boykin E, Ian Gilmour M, Ingram P, Roggli VL, Goldstein HL. Biological effects of desert dust in respiratory epithelial cells and a murine model. Inhal Toxicol. 2014;26(5):299–309.
28. Gillette DA, Hanson KJ. Spatial and temporal variability of dust production caused by wind erosion in the United States. J Geophys Res Atmos. 1989;94(D2):2197–206.
29. Gilmour PS, Rahman I, Donaldson K, MacNee W. Histone acetylation regulates epithelial IL-8 release mediated by oxidative stress from environmental particles. Am J Physiol Lung Cell Mol Physiol. 2003;284(3):L533–40.
30. Ginoux P, Prospero JM, Gill TE, Hsu NC, Zhao M. Global-scale attribution of anthropogenic and natural dust sources and their emission rates based on MODIS deep blue aerosol products. Rev Geophys. 2012;50(3):1–36.
31. Gorris M, Cat L, Zender C, Treseder K, Randerson J. Coccidioidomycosis dynamics in relation to climate in the southwestern United States. GeoHealth. 2018;2(1):6–24.
32. Goudie AS. Desert dust and human health disorders. Environ Int. 2014;63:101–13.
33. Gouveia N, Freitas CU d, Martins LC, Marcilio IO. Respiratory and cardiovascular hospitalizations associated with air pollution in the city of São Paulo, Brazil. Cad Saude Publica. 2006;22(12):2669–77.
34. Griffin DW. Atmospheric movement of microorganisms in clouds of desert dust and implications for human health. Clin Microbiol Rev. 2007;20(3):459–77, table of contents.
35. Griffin DW, Kellogg CA. Dust storms and their impact on ocean and human health: dust in earth's atmosphere. EcoHealth. 2004;1(3):284–95.
36. Griffin DW, Kellogg CA, Garrison VH, Lisle JT, Borden TC, Shinn EA. Atmospheric microbiology in the northern Caribbean during African dust events. Aerobiologia. 2003;19(3–4):143–57.
37. Griffin DW, Kellogg CA, Garrison VH, Shinn EA. The global transport of dust: an intercontinental river of dust, microorganisms and toxic chemicals flows through the Earth's atmosphere. Am Sci. 2002;90(3):228–35.
38. Hall AH. Chronic arsenic poisoning. Toxicol Lett. 2002;128(1–3):69–72.
39. Hamilton RF Jr, Thakur SA, Holian A. Silica binding and toxicity in alveolar macrophages. Free Radic Biol Med. 2008;44(7):1246–58.
40. Harrington AD, Schmidt MP, Szema AM, Galdanes K, Tsirka SE, Gordon T, Schoonen MAA. The role of Iraqi dust in inducing lung injury in United States soldiers-an interdisciplinary study. Geohealth. 2017;1(5):237–46.
41. Heard E, Martienssen RA. Transgenerational epigenetic inheritance: myths and mechanisms. Cell. 2014;157(1):95–109.

42. Hinwood AL, Sim MR, Jolley D, de Klerk N, Bastone EB, Gerostamoulos J, Drummer OH. Hair and toenail arsenic concentrations of residents living in areas with high environmental arsenic concentrations. Environ Health Perspect. 2003;111(2):187–93.
43. Hirano T. Interleukin 6 and its receptor: ten years later. Int Rev Immunol. 1998;16(3–4):249–84.
44. Hnizdo E, Murray J. Risk of pulmonary tuberculosis relative to silicosis and exposure to silica dust in South African gold miners. Occup Environ Med. 1998;55(7):496–502.
45. Hougaard KS, Campagnolo L, Chavatte-Palmer P, Tarrade A, Rousseau-Ralliard D, Valentino S, Park MV, de Jong WH, Wolterink G, Piersma AH, Ross BL, Hutchison GR, Hansen JS, Vogel U, Jackson P, Slama R, Pietroiusti A, Cassee FR. A perspective on the developmental toxicity of inhaled nanoparticles. Reprod Toxicol. 2015;56:118–40.
46. Huang YC, Ghio AJ, Stonehuerner J, McGee J, Carter JD, Grambow SC, Devlin RB. The role of soluble components in ambient fine particles-induced changes in human lungs and blood. Inhal Toxicol. 2003;15(4):327–42.
47. Ji H, Khurana Hershey GK. Genetic and epigenetic influence on the response to environmental particulate matter. J Allergy Clin Immunol. 2012;129(1):33–41.
48. Jimenez E, Linares C, Martinez D, Diaz J. Role of Saharan dust in the relationship between particulate matter and short-term daily mortality among the elderly in Madrid (Spain). Sci Total Environ. 2010;408(23):5729–36.
49. Jung CR, Lin YT, Hwang BF. Ozone, particulate matter, and newly diagnosed Alzheimer's disease: a population-based cohort study in Taiwan. J Alzheimers Dis. 2015;44(2):573–84.
50. Kalkbrenner AE, Windham GC, Serre ML, Akita Y, Wang X, Hoffman K, Thayer BP, Daniels JL. Particulate matter exposure, prenatal and postnatal windows of susceptibility, and autism spectrum disorders. Epidemiology. 2015;26(1):30–42.
51. Kang J-H, Keller JJ, Chen C-S, Lin H-C. Asian dust storm events are associated with an acute increase in pneumonia hospitalization. Ann Epidemiol. 2012a;22(4):257–63.
52. Kang J-H, Liu T-C, Keller J, Lin H-C. Asian dust storm events are associated with an acute increase in stroke hospitalisation. J Epidemiol Community Health. 2012b;2011:200794.
53. Kellogg CA, Griffin DW. Aerobiology and the global transport of desert dust. Trends Ecol Evol. 2006;21(11):638–44.
54. Kenneth S, Gary Y, James N, Brent B. Characterizing changes in drought risk for the United States from climate change. Environ Res Lett. 2010;5(4):044012.
55. Kilpatrick AM, Chmura AA, Gibbons DW, Fleischer RC, Marra PP, Daszak P. Predicting the global spread of H5N1 avian influenza. Proc Natl Acad Sci U S A. 2006;103(51):19368–73.
56. King MS, Eisenberg R, Newman JH, Tolle JJ, Harrell FE Jr, Nian H, Ninan M, Lambright ES, Sheller JR, Johnson JE, Miller RF. Constrictive bronchiolitis in soldiers returning from Iraq and Afghanistan. N Engl J Med. 2011;365(3):222–30.
57. Kirkpatrick B, Fleming LE, Squicciarini D, Backer LC, Clark R, Abraham W, Benson J, Cheng YS, Johnson D, Pierce R, Zaias J, Bossart GD, Baden DG. Literature review of Florida red tide: implications for human health effects. Harmful Algae. 2004;3(2):99–115.
58. Korenyi-Both AL, Korenyi-Both AL, Molnar AC, Fidelus-Gort R. Al Eskan disease: desert storm pneumonitis. Mil Med. 1992;157(9):452–62.
59. Kyung SY, Yoon JY, Kim YJ, Lee SP, Park J-W, Jeong SH. Asian dust particles induce TGF-β1 via reactive oxygen species in bronchial epithelial cells. Tuberc Respir Dis. 2012;73(2):84–92.
60. Lenes JM, Darrow BP, Cattrall C, Heil CA, Callahan M, Vargo GA, Byrne RH, Prospero JM, Bates DE, Fanning KA, Walsh JJ. Iron fertilization and the trichodesmiumresponse on the West Florida shelf. Limnol Oceanogr. 2001;46(6):1261–77.
61. Li J, Kandakji T, Lee JA, Tatarko J, Blackwell J 3rd, Gill TE, Collins JD. Blowing dust and highway safety in the southwestern United States: characteristics of dust emission "hotspots" and management implications. Sci Total Environ. 2018;621:1023–32.
62. Li J, Li WX, Bai C, Song Y. Particulate matter-induced epigenetic changes and lung cancer. Clin Respir J. 2017;11(5):539–46.

63. Ma QY, Huang DY, Zhang HJ, Wang S, Chen XF. Exposure to particulate matter 2.5 (PM2.5) induced macrophage-dependent inflammation, characterized by increased Th1/Th17 cytokine secretion and cytotoxicity. Int Immunopharmacol. 2017;50:139–45.
64. MacNee W, Donaldson K. Mechanism of lung injury caused by PM10 and ultrafine particles with special reference to COPD. Eur Respir J Suppl. 2003;40:47s–51s.
65. Maddy KT, Coccozza J. The probable geographic distribution of coccidioides immitis in Mexico. Bol Oficina Sanit Panam. 1964;57:44–54.
66. Madrigano J, Baccarelli A, Wright RO, Suh H, Sparrow D, Vokonas PS, Schwartz J. Air pollution, obesity, genes and cellular adhesion molecules. Occup Environ Med. 2010;67(5):312–7.
67. Mahowald NM, Engelstaedter S, Luo C, Sealy A, Artaxo P, Benitez-Nelson C, Bonnet S, Chen Y, Chuang PY, Cohen DD, Dulac F, Herut B, Johansen AM, Kubilay N, Losno R, Maenhaut W, Paytan A, Prospero JM, Shank LM, Siefert RL. Atmospheric iron deposition: global distribution, variability, and human perturbations. Annu Rev Mar Sci. 2009;1:245–78.
68. Mallone S, Stafoggia M, Faustini A, Gobbi GP, Marconi A, Forastiere F. Saharan dust and associations between particulate matter and daily mortality in Rome, Italy. Environ Health Perspect. 2011;119(10):1409–14.
69. Mar TF, Larson TV, Stier RA, Claiborn C, Koenig JQ. An analysis of the association between respiratory symptoms in subjects with asthma and daily air pollution in Spokane, Washington. Inhal Toxicol. 2004;16(13):809–15.
70. Mathur M, Choudhary R. Desert lung syndrome, in rural dwellers of the Thar Desert, India. J Arid Environ. 1997;35(3):559–62.
71. Middleton N. Effect of drought on dust production in the Sahel. Nature. 1985;316(6027):431.
72. Milano M, Dongiovanni P, Artoni A, Gatti S, Rosso L, Colombo F, Bollati V, Maggioni M, Mannucci PM, Bertazzi PA, Fargion S, Valenti L. Particulate matter phagocytosis induces tissue factor in differentiating macrophages. J Appl Toxicol. 2016;36(1):151–60.
73. Miller RL, Tegen I, Perlwitz J. Surface radiative forcing by soil dust aerosols and the hydrologic cycle. J Geophys Res Atmos. 2004;109(D4):1–24.
74. Miller SD, Kuciauskas AP, Liu M, Ji Q, Reid JS, Breed DW, Walker AL, Mandoos AA. Haboob dust storms of the southern Arabian Peninsula. J Geophys Res Atmos. 2008;113(D1):1–26.
75. Mirsaeidi M, Javadzadeh A, Campos M. A risky political game with climate change. Eur Respir J. 2017;50(4):1701145.
76. Molesworth AM, Cuevas LE, Connor SJ, Morse AP, Thomson MC. Environmental risk and meningitis epidemics in Africa. Emerg Infect Dis. 2003;9(10):1287–93.
77. Molesworth AM, Cuevas LE, Morse AP, Herman JR, Thomson MC. Dust clouds and spread of infection. Lancet. 2002;359(9300):81–2.
78. Mosher DE, Lachman BE, Greenberg MD, Rosen B, Nichols T. Green warriors: army environmental considerations for contingency operations from planning through post-conflict: Rand Corporation; Santa Monica, CA. 2008.
79. Naimabadi A, Ghadiri A, Idani E, Babaei AA, Alavi N, Shirmardi M, Khodadadi A, Marzouni MB, Ankali KA, Rouhizadeh A, Goudarzi G. Chemical composition of PM10 and its in vitro toxicological impacts on lung cells during the middle eastern dust (MED) storms in Ahvaz, Iran. Environ Pollut. 2016;211:316–24.
80. Nakashima K, Sato T, Shigemori S, Shimosato T, Shinkai M, Kaneko T. Regulatory role of heme oxygenase-1 in silica-induced lung injury. Respir Res. 2018;19(1):144.
81. Nakazawa M, Ishii H, Nakamura H, Yoshino SI, Fukamizu A, Nishioka K, Nakajima T. NFkappaB2 (p52) promoter activation via notch signaling pathway in rheumatoid synoviocytes. Int J Mol Med. 2001;7(1):31–5.
82. Nemmar A, Vanbilloen H, Hoylaerts MF, Hoet PH, Verbruggen A, Nemery B. Passage of intratracheally instilled ultrafine particles from the lung into the systemic circulation in hamster. Am J Respir Crit Care Med. 2001;164(9):1665–8.
83. NOAA. NOAA Global Monitoring Division, Earth system research laboratory, National Oceanic and Atmospheric Administration.
84. NOAA. NOAA National Centers for Environmental information, Climate at a Glance: North America Time Series. From https://www.ncdc.noaa.gov/cag/global/time-series/northAmerica/land/ytd/12/1880-2018

85. Norboo T, Angchuk PT, Yahya M, Kamat SR, Pooley FD, Corrin B, Kerr IH, Bruce N, Ball KP. Silicosis in a Himalayan village population: role of environmental dust. Thorax. 1991;46(5):341–3.
86. O'Donnell MJ, Fang J, Mittleman MA, Kapral MK, Wellenius GA, N. Investigators of the Registry of Canadian Stroke. Fine particulate air pollution (PM2.5) and the risk of acute ischemic stroke. Epidemiology. 2011;22(3):422–31.
87. O'Hara SL, Wiggs GF, Mamedov B, Davidson G, Hubbard RB. Exposure to airborne dust contaminated with pesticide in the Aral Sea region. Lancet. 2000;355(9204):627–8.
88. Oberdorster G, Finkelstein JN, Johnston C, Gelein R, Cox C, Baggs R, Elder AC. Acute pulmonary effects of ultrafine particles in rats and mice. Res Rep Health Eff Inst. 2000;96:5–74; disc 75–86.
89. Oberdorster G, Sharp Z, Atudorei V, Elder A, Gelein R, Lunts A, Kreyling W, Cox C. Extrapulmonary translocation of ultrafine carbon particles following whole-body inhalation exposure of rats. J Toxicol Environ Health A. 2002;65(20):1531–43.
90. Pappagianis D. Epidemiology of coccidioidomycosis. Curr Top Med Mycol. 1988;2:199–238.
91. Pedersen CB, Raaschou-Nielsen O, Hertel O, Mortensen PB. Air pollution from traffic and schizophrenia risk. Schizophr Res. 2004;66(1):83–5.
92. Perera FP, Tang D, Wang S, Vishnevetsky J, Zhang B, Diaz D, Camann D, Rauh V. Prenatal polycyclic aromatic hydrocarbon (PAH) exposure and child behavior at age 6–7 years. Environ Health Perspect. 2012;120(6):921–6.
93. Perera FP, Wang S, Vishnevetsky J, Zhang B, Cole KJ, Tang D, Rauh V, Phillips DH. Polycyclic aromatic hydrocarbons-aromatic DNA adducts in cord blood and behavior scores in New York city children. Environ Health Perspect. 2011;119(8):1176–81.
94. Perry KD, Cahill TA, Eldred RA, Dutcher DD, Gill TE. Long-range transport of north African dust to the eastern United States. J Geophys Res Atmos. 1997;102(D10):11225–38.
95. Poon HF, Calabrese V, Scapagnini G, Butterfield DA. Free radicals and brain aging. Clin Geriatr Med. 2004;20(2):329–59.
96. Poss KD, Tonegawa S. Reduced stress defense in heme oxygenase 1-deficient cells. Proc Natl Acad Sci U S A. 1997;94(20):10925–30.
97. Ranft U, Schikowski T, Sugiri D, Krutmann J, Kramer U. Long-term exposure to traffic-related particulate matter impairs cognitive function in the elderly. Environ Res. 2009;109(8):1004–11.
98. Raz R, Roberts AL, Lyall K, Hart JE, Just AC, Laden F, Weisskopf MG. Autism spectrum disorder and particulate matter air pollution before, during, and after pregnancy: a nested case-control analysis within the Nurses' health study II cohort. Environ Health Perspect. 2015;123(3):264–70.
99. Reed L, Nugent K. The health effects of dust storms in the Southwest United States. 2018;6(22):5.
100. Rodriguez-Cotto RI, Ortiz-Martinez MG, Rivera-Ramirez E, Mendez LB, Davila JC, Jimenez-Velez BD. African dust storms reaching Puerto Rican coast stimulate the secretion of IL-6 and IL-8 and cause cytotoxicity to human bronchial epithelial cells (BEAS-2B). Health (Irvine Calif). 2013;5(10b):14–28.
101. Roop SA, Niven AS, Calvin BE, Bader J, Zacher LL. The prevalence and impact of respiratory symptoms in asthmatics and nonasthmatics during deployment. Mil Med. 2007;172(12):1264–9.
102. Saiyed HN, Sharma YK, Sadhu HG, Norboo T, Patel PD, Patel TS, Venkaiah K, Kashyap SK. Non-occupational pneumoconiosis at high altitude villages in Central Ladakh. Br J Ind Med. 1991;48(12):825–9.
103. Sama P, Long TC, Hester S, Tajuba J, Parker J, Chen LC, Veronesi B. The cellular and genomic response of an immortalized microglia cell line (BV2) to concentrated ambient particulate matter. Inhal Toxicol. 2007;19(13):1079–87.
104. Sanders JW, Putnam SD, Frankart C, Frenck RW, Monteville MR, Riddle MS, Rockabrand DM, Sharp TW, Tribble DR. Impact of illness and non-combat injury during operations Iraqi freedom and enduring freedom (Afghanistan). Am J Trop Med Hyg. 2005;73(4):713–9.

105. Schulz H, Harder V, Ibald-Mulli A, Khandoga A, Koenig W, Krombach F, Radykewicz R, Stampfl A, Thorand B, Peters A. Cardiovascular effects of fine and ultrafine particles. J Aerosol Med. 2005;18(1):1–22.
106. Schweitzer MD, Calzadilla AS, Salamo O, Sharifi A, Kumar N, Holt G, Campos M, Mirsaeidi M. Lung health in era of climate change and dust storms. Environ Res. 2018;163:36–42.
107. Shute J. Interleukin-8 is a potent eosinophil chemo-attractant. Clin Exp Allergy. 1994;24(3):203–6.
108. Skilton M, Hovesepian F. The 4th industrial revolution and waste management. Waste Manag Res. 2017;35(10):997–8.
109. Smith B, Wong CA, Smith TC, Boyko EJ, Gackstetter GD, A. K. R. f. t. M. C. S. T. Margaret. Newly reported respiratory symptoms and conditions among military personnel deployed to Iraq and Afghanistan: a prospective population-based study. Am J Epidemiol. 2009;170(11):1433–42.
110. Song Z, Wang J, Wang S. Quantitative classification of northeast Asian dust events. J Geophys Res Atmos. 2007;112(D4):1–8.
111. Subhani M, Mustafa I, Alamdar A, Katsoyiannis IA, Ali N, Huang Q, Peng S, Shen H, Eqani SA. Arsenic levels from different land-use settings in Pakistan: bio-accumulation and estimation of potential human health risk via dust exposure. Ecotoxicol Environ Saf. 2015;115:187–94.
112. Szema AM, Peters MC, Weissinger KM, Gagliano CA, Chen JJ. New-onset asthma among soldiers serving in Iraq and Afghanistan. Allergy Asthma Proc. 2010;31(5):67–71.
113. Takenaka S, Karg E, Roth C, Schulz H, Ziesenis A, Heinzmann U, Schramel P, Heyder J. Pulmonary and systemic distribution of inhaled ultrafine silver particles in rats. Environ Health Perspect. 2001;109(Suppl 4):547–51.
114. Tam WW, Wong TW, Wong AH, Hui DS. Effect of dust storm events on daily emergency admissions for respiratory diseases. Respirology. 2012;17(1):143–8.
115. Tarantini L, Bonzini M, Apostoli P, Pegoraro V, Bollati V, Marinelli B, Cantone L, Rizzo G, Hou L, Schwartz J, Bertazzi PA, Baccarelli A. Effects of particulate matter on genomic DNA methylation content and iNOS promoter methylation. Environ Health Perspect. 2009;117(2):217–22.
116. Thompson GR. Strategies to combat Coccidioidomycosis: are we making any Progress? Curr Fungal Infect Rep. 2011;5(4):215.
117. Tong DQ, Wang JXL, Gill TE, Lei H, Wang B. Intensified dust storm activity and valley fever infection in the southwestern United States. Geophys Res Lett. 2017;44(9):4304–12.
118. US-EPA. United States Environmental Protection Agency. From https://19january2017snapshot.epa.gov/criteria-air-pollutants_.html
119. US-EPA United States Evironmental Protection Agency. https://19january2017snapshot.epa.gov/criteria-air-pollutants_.html.
120. Veronesi B, Makwana O, Pooler M, Chen LC. Effects of subchronic exposures to concentrated ambient particles. VII. Degeneration of dopaminergic neurons in Apo E−/−mice. Inhal Toxicol. 2005;17(4–5):235–41.
121. Volk HE, Lurmann F, Penfold B, Hertz-Picciotto I, McConnell R. Traffic-related air pollution, particulate matter, and autism. JAMA Psychiat. 2013;70(1):71–7.
122. Weaver EA, Kolivras KN. Investigating the relationship between climate and valley fever (Coccidioidomycosis). EcoHealth. 2018;15(4):840–52.
123. Wheaton E. Frequency and severity of drought and dust storms. Can J Agr Econ/Revue canadienne d'agroeconomie. 1990;38(4):695–700.
124. Yamamoto C, Yoneda T, Yoshikawa M, Fu A, Tokuyama T, Tsukaguchi K, Narita N. Airway inflammation in COPD assessed by sputum levels of interleukin-8. Chest. 1997;112(2):505–10.
125. Yang AS, Estecio MR, Doshi K, Kondo Y, Tajara EH, Issa JP. A simple method for estimating global DNA methylation using bisulfite PCR of repetitive DNA elements. Nucleic Acids Res. 2004;32(3):e38.
126. Yu HL, Chien LC, Yang CH. Asian dust storm elevates children's respiratory health risks: a spatiotemporal analysis of children's clinic visits across Taipei (Taiwan). PLoS One. 2012;7(7):e41317.

Chapter 23
Federal Programs in Climate Change and Health Research

Cecilia J. Sorensen, Caitlin Rublee, and John Balbus

Introduction

The federal government plays an integral role in supporting climate and health research. This includes funding support, conduct of research by federal scientists, and the sustained collection and curation of climate and health data. In addition to pursuing research on a wide range of science topics and applications, the various agencies also support the public through the creation of data-driven decision support tools and assessments. Because each federal agency has a different mandate and range of scientific expertise, the focus and goals of various agencies' climate change and human health research vary. For example, the National Oceanographic and Atmospheric Administration (NOAA) emphasizes the use of weather and climate forecasts and oceanographic data for public health applications, while the National Aeronautic and Space Administration (NASA) emphasizes health applications of remotely sensed data from its satellites. This chapter provides historical background, current structure, and an overview of federal science and research activities related to climate change and human health.

C. J. Sorensen (✉)
Department of Emergency Medicine, University of Colorado School of Medicine, Aurora, CO, USA

University of Colorado School of Public Health, Aurora, CO, USA
e-mail: Cecilia.Sorensen@ucdenver.edu

C. Rublee
Department of Emergency Medicine, University of Colorado School of Medicine, Aurora, CO, USA
e-mail: Caitlin.Rublee@ucdenver.edu

J. Balbus
National Institute of Environmental Health Sciences, Bethesda, MD, USA
e-mail: John.Balbus@nih.gov

K. E. Pinkerton, W. N. Rom (eds.), *Climate Change and Global Public Health*, Respiratory Medicine, https://doi.org/10.1007/978-3-030-54746-2_23

Federal Climate Change Research Prior to 1990

In 1978, Congress established the Federal Interagency Climate Program through the National Climate Act to "assist in the understanding and response to natural and human-induced climate processes and their implications" [1]. The National Climate Act required the program to conduct studies to understand the impacts of human activities on climate, to promote scientific understanding of climate change, to improve forecasts and data collection of climate processes and to encourage international cooperation in climate research [2]. The National Oceanic and Atmospheric Administration (NOAA) of the Department of Commerce coordinated the Interagency Climate Program. As a result of the climate program's initial leadership, emphasis, and design, NOAA, NASA, and National Science Foundation (NSF) became leaders of climate change science at the federal level with each agency developing its own climate change-related programs during the 1980s. While the climate program received positive evaluations from the National Academy of Sciences (NAS), many agreed that the program's limited scope and emphasis on atmospheric and climate-related science disciplines was failing to produce the range of understanding required to inform the growing demand for policy responses and "global change"-related research [2].

Prompted by increasing concerns about climate change from the international and domestic scientific communities, Congress held a series of hearings, beginning in late 1985, that spurred increasing public and legislative interest in climate change. In 1987, Congress passed the Global Climate Protection Act (P.L. 100–204), which designated the Environmental Protection Agency (EPA) and Department of State as leads for climate change policy development. Even with the passing of the Climate Protection Act of 1987, the Bush Administration had not articulated a national strategy or set of goals related to global climate change [3]. As scientific interest and public concern about climate change continued to grow, and decisive action stalled in the white house, some members of Congress became frustrated with the US government's inability to coordinate research efforts to inform climate policy decisions [4].

The Global Change Research Act of 1990 (GCRA) and the United States Global Change Research Program (USGCRP)

The Global Change Research Act of 1990 (GCRA), signed by President George H.W. Bush, in November 1990, established the GCRP with the explicit aim "to provide for development and coordination of a comprehensive and integrated United States research program, which will assist the Nation and the world to understand, assess, predict, and respond to human-induced and natural processes of global change" (P.L. 10–606). The US Global Change Research Program (GCRP) has coordinated federal research and observation on global environmental change and societal impacts since its inception. The GCRP's initial research agenda focused on developing a predictive understanding of the earth's climate [4].

Thirteen federal agencies and departments participate in the USGCRP, whose mission is "to build a knowledge base that informs human responses to climate and global change through coordinated and integrated federal programs of research, education, communication, and decision support." [5] Program activities are coordinated through interagency working groups organized around cross-disciplinary climate and global change themes [6]. These groups focus on the following program elements: Integrated Observation, Integrated Modeling, Multidisciplinary Research on the Human and Natural Components of the Earth System, Conduct Sustained Assessments, Informing Decisions, International Cooperation, Communication and Education, and Climate Change and Human Health [7]. In order to foster better integration among the sciences (biological, social, behavioral, and economic), as well as better meet societal needs, participation in the working groups extends beyond the 13 agencies formally represented in the GCRP.

The most recent strategic plan (2012) outlines four goals for GCRP coordination of federal climate change research [7]:

- *Goal 1. Advance Science*: Advance scientific knowledge of the integrated natural and human components of the Earth system.
- *Goal 2. Inform Decisions*: Provide the scientific basis to inform and enable timely decisions on adaptation and mitigation.
- *Goal 3. Conduct Sustained Assessments*: Build sustained assessment capacity that improves the Nation's ability to understand, anticipate, and respond to global change impacts and vulnerabilities.
- *Goal 4. Communicate and Educate*: Advance communications and education to broaden public understanding of global change and develop the scientific workforce of the future.

GCRP-supported research contributes to influential international reports such as the Intergovernmental Panel on Climate Change's (IPCC) climate change assessments as well as various national climate change assessments [8]. GCRP documents, program results, and plans are compiled in an annual report entitled "Our Changing Planet." [9]

Early Federal Activities Related to Climate Change and Health

Human health was not an area of strong focus for the GCRP during its first 20 years. Human health was one of several considerations within the "Human Dimensions of Climate Change," along with human behaviors that contribute to greenhouse gas emissions and other global environmental changes, the impacts of climate change on human well-being more broadly, especially economic well-being, and human processes of decision-making. While "Human Dimensions of Climate Change" was itself an expansion of scope of the GCRP promoted by National Resource Council (NRC) reports in the early 1990s [8], the NRC repeatedly called upon the GCRP to focus more attention on meeting societal needs, including human health protection. For example, in 2009, the NRC found that the GCRP's dedication to expand activities beyond climate science to research on impacts of climate change, as delineated in its 2003 strategic plan, was particularly relevant to the societal needs of 2010 and beyond [10]. At the same time, the NRC recommended that the program once again broaden its scope to better meet the needs of decision makers and stakeholders by including additional research focusing on the human dimensions of climate change [9]. In response to this guidance from the NRC, the GCRP broadened its focus from earth's systems and their functions to include comparative analysis of adaptation and mitigation strategies, capacity building, and the implication of climate change for society–environment interactions [11]. The research priorities for society–environment interactions included the following five elements: urban systems, energy systems, land use change, water resources, and human health.

In 2009, an ad hoc workgroup led by NIH and comprised of representatives from NOAA, U.S. Environmental Protection Agency, Centers for Disease Control and Prevention, U.S. Department of Agriculture, U.S. Department of State, U.S. Global Change Research Program, and the White House Office of Science and Technology Policy, assembled to draft a report outlining research needs for climate change and human health entitled, "A Human Health Perspective on Climate Change." The report, published in 2010, outlined the research priorities for climate change and human health related to 11 categories of health outcomes and exposures [12]:

- Asthma, respiratory allergies, and airway diseases
- Cancer
- Cardiovascular disease and stroke
- Foodborne disease and nutrition
- Heat-related morbidity and mortality
- Human developmental effects
- Mental health and stress-related disorders
- Neurological diseases and disorders
- Vector-borne and zoonotic diseases
- Waterborne diseases
- Weather-related morbidity and mortality

This report provided an in-depth assessment of the health impacts of climate change and laid the groundwork for interagency collaboration within the GCRP in this area. (Did it also propose a research agenda?)

Founding of the GCRP Interagency Climate Change and Human Health Group

As part of the effort to reorient the GCRP to better meet societal needs, and specifically to provide greater focus on understanding the human health implications of climate change, the GCRP chartered a new interagency working group, the Climate Change and Human Health Group (CCHHG) in December 2009. The CCHHG is intended to pilot the "end-to-end" approach to science described in the NRC report "Restructuring Federal Climate Research to Meet Societal Needs." To that end, the composition of the CCHHG includes agencies that have not traditionally participated in climate research but are translators and users of the scientific information on climate and health produced within the GCRP, like the Centers for Disease Control and Prevention (CDC) and the Department of Homeland Security (DHS). The breadth of the CCHHG's charge is reflected by the agencies serving as co-chairs of the workgroup: NIEHS, representing a research focus, CDC representing public health programmatic work, and NOAA, representing atmospheric and oceanic research as well as meteorologic, climatologic, and oceanic programmatic work and services.

Some of the roles of the CCHHG, as described in the group's charter, include:

- Coordination of federal research efforts on climate change and human health and ensuring research agendas are informed by end users of the information developed.
- Developing information serving as an information conduit between the GCRP and stakeholders on climate change and health issues
- Representing GCRP on health issues to international climate change bodies

The charter also specifies that the CCHHG will apply a "one health" concept in its work, integrating science on the health of domestic and wild animals and ecosystems with the health of humans [13]. Since its inception, the CCHHG has developed critical information resources and supported innovative programming involving multiple federal agencies and partners outside the US government. A brief description of some of these resources and activities follows.

Integrated Climate and Health Data Products: MATCH and data.gov

One of the first needs identified by health research professionals was easier access to the various types of data needed to inform climate and health decisions. This includes meteorologic observations and climate forecasts, land use and other earth observations data, and health data. There was no single location for all of these different types of data and stakeholders described challenges understanding and using data across disciplines.

In order to address these needs, the CCHHG initially developed a web portal called the Metadata Access Tool for Climate and Health, or MATCH. This required the development of definitions and ontologies for climate and health data. Using the 2010 research needs white paper as a starting place for an ontology of health effects, the CCHHG defined the part of the federal data system that was pertinent to climate and health. Over 9000 datasets were ultimately identified and metadata related to them collected and posted.

In order to give stakeholders greater access to relevant datasets, the CCHHG has allowed the MATCH portal to expire and now maintains integrated datasets through the data.gov web portal. This site allows access to dozens of relevant datasets from nine different federal agencies. Along with access to datasets, the data.gov site provides mapping tools and other useful guidance resources. It also provides another route of access to CDC's Environmental Public Health Tracking Network (see below) which curates and posts unique integrated datasets relevant to climate change and health.

Climate Resilience Toolkit

The CCHHG curates the health section of the US Climate Resilience Toolkit (https://toolkit.climate.gov/topics/human-health, last accessed Sept. 22, 2019). This portal assembles guidance documents, mapping tools, and case studies that support climate resilience efforts for the health sector. The site has six sections, including two unique resources that are built into this portal: The National Integrated Health and Health Information System (NIHHIS) provides a suite of resources and forecasts for health stakeholders concerned with the impacts of excessive heat; and the Sustainable and Climate Resilient Health Care Facilities Initiative (SCRHCFI) with a toolkit consisting of assessment checklists and other informational resources to guide the health sector towards more sustainable and resilient infrastructure. The other four sections provide curated resources on topics including air pollution, food and water safety, extreme events, and vector-borne diseases.

2016 GCRP Climate Health Assessment

In 2016, the US GCRP released a special interim report [14] as part of its expanded commitment to conducting sustained scientific assessments in between publication of the mandated National Climate Assessment reports. This report, entitled "*The Impacts of Climate Change on Human Health in the United States: A Scientific Assessment,*" comprised a thorough updated assessment of the scientific literature on the health impacts of climate change, focusing on literature published between 2007 and 2015. Innovations associated with this report included discussions of new health impacts, including nutritional and mental health impacts, as well as the first US federal estimates of future health impacts of climate change for four different health outcomes. The report is organized according to seven different types of exposure pathways, and considers climate health impacts in the context of other social and environmental determinants of health.

Workshop Report on Climate-Sensitive Infectious Disease

The CCHHG at times plays a convening role that extends beyond the GCRP, engaging other federal workgroups and committees with a common interest in understanding the interactions between climate and human disease. One such activity was a series of webinars and a workshop on enhancing the prediction of climate-sensitive infectious diseases. This activity brought together the federal workgroup on Pandemic Prediction and Forecasting Science and Technology (PPFST) as well as colleagues in the national defense and security agencies, and culminated in a report that was published in 2018 [15]. The report called for interagency collaboration in developing models and early warning systems for water- and vector-borne infectious diseases whose interannual disease activity is strongly influenced by variations in weather and climate.

International Scientific Activities on Climate Change and Health

Since its inception, the CCHHG has also helped coordinate US federal engagement with international scientific activities related to climate change and health. Most recently, the CCHHG coordinated and led US involvement in scoping an international consortium of climate science funders through the Belmont Forum [16]. This involved both assembling and coordinating the US agencies interested in participated in supporting research on climate and health through the Belmont Forum and also providing subject matter expertise to this international consortium, as most agencies participating in the forum lacked staff with specific climate and health expertise. In addition to conducting collaborative international programming, such as the Global Heat Health Information Network [17], the CCHHG has also supported review of Intergovernmental Panel on Climate Change (IPCC) Assessment and Special Reports and participated in US delegations to the United Nations Framework Convention on Climate Change (UNFCCC) meetings.

Current Federal Agency Activities in Climate Change and Health

Department of Health and Human Services (HHS)

The mission of the U.S. Department of Health and Human Services (HHS) is to enhance the health and well-being of all Americans by providing effective health and human services and by fostering sound, sustained advances in the sciences underlying medicine, public health, and social services. HHS supports a broad portfolio of research and decision support initiatives related to environmental health and the health effects of global climate change. The National Institutes of Health (NIH) and the Centers for Disease Control and Prevention (CDC) provide the focus for this effort.

HHS supports all four goal areas of the Global Change Research Program: Advance Science, Inform Decisions, Conduct Sustained Assessments, and Communicate and Educate. By conducting fundamental and applied research on the linkages between climate change and health, translating scientific advances into decision support tools for public health professionals, conducting ongoing monitoring and surveillance of climate-related health outcomes, and disseminating scientific information and engaging the public health community in two-way communication, HHS provides a model of the "end to end" science paradigm the GCRP seeks to achieve.

The NIH's National Institute of Environmental Health Sciences (NIEHS) and CDC co-chair (along with NOAA) the USGCRP's Climate Change and Human Health Interagency Working Group of the USGCRP. In addition, both NIEHS and CDC support the National Climate Assessment, which seeks to provide the scientific information that can be used by communities around the country to effectively plan for adaptation and mitigation.

The National Institutes of Health (NIH)

The NIH supports a large research portfolio relevant to the human health impacts of climate change, including research related to direct health impacts of increased temperatures and extreme weather events, the health effects of air pollution and aeroallergens, water quality and quantity, ecosystem influences on infectious disease transmission, and potential health effects of materials used in new technologies to mitigate or adapt to climate change.

At the NIH, the National Institute of Environmental Health Sciences (NIEHS) and the Fogarty International Center (FIC) co-chair the Trans-NIH Climate Change Workgroup. The Workgroup aims to coordinate and promote climate change and health research at NIH. Shortly after its founding in 2009, the Workgroup conducted a portfolio analysis of climate change research at NIH to identify gaps and prioritize research needs.

NIH Climate Change and Health Research Funding

NIH funds research through a variety of mechanisms, including focused Funding Opportunity Announcements as well as an array of investigator-initiated grant mechanisms. In 2010, the NIEHS in collaboration with ten other NIH institutes and centers, funded the Climate Change and Health: Assessing and Modeling Population Vulnerability to Climate Change grant program [18]. This program was intended to inform climate change adaptation strategies and guide public health interventions to reduce current and future harms to the most vulnerable communities. The first round of nine grantees was announced in October 2011 and two additional rounds of awards are planned through 2013 [19]. Funding for this program runs through 2014. More recent research programs in Oceans in Human Health, Aging Populations, and Global Environmental and Occupational Health have included mentions of climate change or climate-related exposures, such as heat and extreme events, in their guidance language.

The program supports the development of tools, models, and methods to better predict the health consequences of climate change and to understand the populations, both in the US and globally, which are most vulnerable to the health consequences of climate change. This research will provide better tools to decision makers involved in protecting the health of particularly vulnerable populations, including communities with low SES, the elderly, pregnant women, and other populations with increased risk.

NIEHS Climate Change and Health Program

At NIEHS, the Climate Change and Human Health Program provides leadership for a variety of NIEHS supported research and initiatives as well as trans-NIH and interdepartmental coordination of climate change and human health activities. The goals of the program include:

- Provide research on human health impacts related to climate change and adaptation
- Raise awareness and create new partnerships to advance key areas of health research and knowledge development on human health effects of climate change
- Serve as an authoritative source of information on human health effects of climate change for NIEHS stakeholders, including the public
- Represent NIEHS science in climate change research and policy activities at the NIH, HHS, federal government, and international levels

The program partners with other government agencies, the academic community, NGOs, and international organizations to identify research gaps, support ongoing investigations, and communicate health impacts of climate change to key decision makers. NIEHS is a Collaborating Centre with the World Health Organization, and its work plan includes joint activities related to the health impacts of climate and weather extremes. At the international level, the program also participates in activities related to the United Nations Framework Convention on Climate Change (UNFCCC) and Group on Earth Observations (GEO) Health and Environment Community of Practice.

In addition, the NIEHS Climate Change and Health Program has developed educational materials for use in secondary, undergraduate, graduate, and professional education, both domestically and internationally [20].

Centers for Disease Control and Prevention (CDC)

Through interdisciplinary work with local, state, and tribal health departments, nongovernmental organizations, research institutions, and other federal agencies, the CDC's Climate and Health Program oversees and implements programs targeted at empowering communities to anticipate risk and protect human health. The Climate and Health program was established with congressional funding in 2009 with a mission to lead efforts to identify populations vulnerable to climate change, to help communities prevent and adapt to the current and anticipated health impacts of climate change and to ensure effective systems are in place to detect and respond to these threats. The program's three core functions include: (1) translating climate change science to inform states, communities, and local public health departments; (2) creating decision support tools to build local capacity to respond to climate change; and (3) serving as the leader in planning for climate change-related public health impacts [21].

Building Resilience Against Climate Effects (BRACE) Framework

In order to facilitate climate preparedness and adaptation at the local level, the Climate and Health Program created the Building Resistance Against Climate Effects (BRACE) Framework. This framework is a five-step process that guides health officials through a strategic process to analyze local climate risks and implement appropriate public health interventions [22]. The process involves incorporating both short- and long-range climate projections into public health planning and response activities. In step 1, participants identify the scope of climate impacts, associated potential health outcomes and populations and locations vulnerable to these health impacts within their community [23]. In step two, participants estimate or quantify the additional burden of health outcomes associated with climate change [24]. In step three, participants identify the most suitable health interventions for the identified health impacts of greatest concern [25]. In step four, participants develop, implement, and disseminate a written adaptation plan that is regularly updated. In step 5, participants evaluate the impact of the activities undertaken. As noted, each step is accompanied by technical documents produced by the program to guide implementation.

Climate-Ready States and Cities Initiative (CRSCI)

The CDC's Climate and Health Program's Climate-Ready States and Cities initiative currently provides funding and technical assistance to 16 states and two cities around the country to implement the BRACE framework [26]. The benefits of this program are that local public health departments are empowered to gain new technical expertise to utilize meteorological and climate data in their public health planning and additionally are empowered to decide with their constituency which climate-related issues represent the most important health threats. For example, the Arizona Department of Health Services has identified the health impacts due to extreme heat, wildfires, dust storms, and poor air quality as highly relevant to current and future public health in the state. Through the BRACE program they completed the Arizona Climate and Health Adaptation Plan which addresses these threats through mobilization of community partnerships, public education, healthcare workforce education, the creation of policy, and the enforcement of laws and regulations that protect health [27].

National Environmental Public Health Tracking Network

The National Environmental Public Health Tracking Network is a tool which brings together health and environmental data from national, state and city sources to allow public health decision makers and policy makers to easily assimilate data and assess health risks in their geographic region. Recently, a new feature has been added which specifically allows users to track climate risks, including extreme heat, extreme precipitation, flood risk and health care infrastructure vulnerabilities [28]. These data can be used by federal and local policy makers to identify high-risk populations and communities, understand trends in heat-related deaths, and inform adaptation strategies.

Partnerships and Outreach

The CDC's Climate and Health Program also works with other federal agencies, nongovernmental organizations (NGO) and international governments to promote collaboration and advance implementation of CDC technical resources on climate change. For example, the program helps to facilitate the National Integrated Drought Information System as well as the National Integrated Heat Health Information System. The program also funds several NGO partners including the American Lung Association, the American Public Health Association, the National Environmental Health Association, and more. The CDC serves as a convener for workshops, establishes workgroups and provides resources and technical expertise to create communication products and informational webinars for nonprofit members.

National Aeronautics and Space Administration (NASA)

NASA plays a crucial role in global change research by supporting sustained monitoring capabilities and advancing scientific knowledge through satellite observations and development of new Earth observation systems. As of 2019, NASA had over 20 space missions collecting climate data, and providing information related to solar activity, sea level rise, atmospheric and oceanic temperatures, the state of the ozone layer, air pollution, and changes in sea ice and land ice.

Within the agency, the Applied Sciences Program, as part of the Earth Science Division, promotes efforts to discover and demonstrate innovative and practical uses of Earth observations to improve policy, business, and management decisions across five areas including: Health and Air Quality, Agriculture/Food Security, Disasters, Ecological Forecasting, and Water Resources. The program includes climate and weather-related impacts within each of these thematic areas.

Within the Applied Sciences Program, the Health and Air Quality Applications area funds applied research collaborations that use satellite and Earth observation data in novel ways to address a broad range of climate-related health threats. The program is unique in that end-users (for example, a public health department) are engaged at the onset of the project to sustainably utilize the final products of the collaboration in future decision-making. Examples include:

- Integration of Earth observation data with New York State Department of Health data to identify the threshold at which extreme heat negatively affects human health in order to inform public heat advisories issued by the National Weather Service (NWS)
- Use of NASA Earth observation data to better characterize the risk of West Nile virus in South Dakota and ultimately inform the Arbovirus Monitoring and Prediction system, which supports disease forecasting
- Use of Earth-observing satellite data to help forecast potential outbreaks of cholera in Bangladesh, Yemen, Mozambique, and in other humanitarian crisis situations
- Integration of public health data with Earth observation data to characterize the impacts of wildfire particulate matter in Colorado to inform better public health decisions

Additionally, the Health and Air Quality Applications Area partners with the international Group on Earth Observations (GEO). Through this partnership, the program is involved in applying Earth observation data for early warning systems that can inform community response to reduce health risks from infectious and vector-borne diseases. Examples include partnerships with the Myanmar Malaria Early Warning System and implementation of a geospatial surveillance and response system for vector-borne disease in the Americas. By promoting the integration of Earth observation and health data, NASA is leading some of the most cutting edge work in the field of climate-health forecasting [29].

Health and Air Quality Applied Sciences Team (HAQAST)

The primary focus of the Health and Air Quality Applied Sciences Team (HAQAST) is to facilitate the use of NASA's satellite and data products by public stakeholders in the air quality and public health communities. By forming close collaborations with academic and community partners, NASA's HAQAST team collaborates with stakeholders to identify and solve public health problems using Earth observation data. Recent examples include using satellite-derived air quality data to estimate the health burden of the 2017 California wildfires, using satellite remote sensing to derive global climate and air pollution indicators, and using satellite informed ozone measurements for estimating U.S. background ozone concentrations [30].

TOOL: SERVIR

USAID and NASA co-support SERVIR, a Regional Visualization and Monitoring System that integrates satellite and ground observation data and forecast models to help improve environmental decision making. SERVIR was developed in 2004, has activities in more than 30 countries, and has developed more than 40 custom tools in partnership with more than 200 institutions, including host country governments, universities, and nongovernmental organizations. The aim of SERVIR is to provide critical information and support services to help national, regional, and local governments, forecasters, climatologist, and other researchers track environmental changes, evaluate ecological threats, and rapidly respond to and assess damage from climatic disasters [31]. The SERVIR data products and tools provide crucial information for diverse set of activities including climate change adaptation, public health, water resource management, agricultural development, and disaster response.

Currently, SERVIR has three regional hubs: SERVIR-Africa, SERVIR-Himalaya and SERVIR-Mekong. The SERVIR-Africa hub was established in 2009 within the Regional Center for Mapping of Resources of Development in Nairobi, Kenya. Regionally, this geographic area struggles both extreme drought and heavy flooding which threatens development, human health and security. Therefore, this hub primary focuses on better hydrologic modeling of the region and delivers data products to aid in flood forecasting and flood mapping and relief [32]. The SERVIR-Himalaya hub was established in 2010 in partnership with the International Centre for Integrated Mountain Development in Kathmandu, Nepal with the aim to address challenges associated with forestry and agriculture in the setting of changes in land cover, glacial melt and poor air quality. Here, there SERVIR team assists with water resource management and in reducing the impact of climate change on water and agriculture [33]. The SERVIR-Mekong hub was launched in 2015 in partnership with Asian Disaster Preparedness Center in Bangkok, Thailand. The hub is intended to support climate resilience studies and develop early warning systems related to regional water and food security as well as monitor shifts in weather, climate, land use, and land cover [34].

U.S. Agency for International Development (USAID)

Recognizing that the adverse impacts of climate change threaten to roll back development progress in terms of reducing poverty and improving economic growth in vulnerable communities, the United States Agency for International Development instituted a mandatory climate risk management process to be implemented across all strategies, projects and activities, including those related to health [35]. Implementation began in October of 2015 for high level strategies and October 2016 for projects and activities. Under this policy, USAID project design teams are required to identify relevant climate risks, assign qualitative risk ratings, and then address those risks prior to project deployment [36]. By considering climate risks and opportunities at the outset of projects, USAID aims to protect US investments abroad, render its work more climate resilient and avoid maladaptive development efforts.

Additionally, USAID funds an initiative known as "Adaptation Thought Leadership and Assessments" or "ATLAS," to advance climate-resilient development and reduce climate-related losses in partner countries. The ATLAS initiative works across sectors, countries and regions. It aims to go beyond general climate vulnerability assessments to develop specialized decision making tools and country- or region-specific guidance for USAID partners and programs while simultaneously facilitating collaboration and capacity building [37].

Practically, the ATLAS program focuses on four major activities [38]. First, it works to improve the availability, quality, and use of weather and climate information so that decision makers and health service providers in partner countries are better equipped to manage current risks and build resilience to future climatic conditions. To facilitate sustained involvement, USAID works to build technical capacity, for example, by training health professionals in partner countries to analyze the relationships between extreme events and health and by facilitating the development of tailored information products and decision support tools such as risk and hazard maps, seasonal forecasts and vulnerability assessments. An example of such an initiative is the ATLAS-backed utilization of data available via the USAID Famine Early Warning Systems Network (FEWS NET) to create malaria early warning systems in for the health sector of Senegal [39].

Second, the ATLAS program works with partner countries to mainstream adaptation measures into governance, planning and budgeting. Often, institutional fragmentation and inadequate coordination among sectoral government agencies hinders cross-cutting climate and health integration in to policy. Therefore, the work is to identify opportunities to enhance existing policies and help stakeholders frame strategies and plans which account for climate and health risks. Examples include: improving emergency response and contingency plans to cope with increased risks associated with extreme weather events with specific attention to vulnerable groups; facilitating coordination among ministries of hydrometeorology, health, agriculture and environment by convening strategic working groups to align adaptation goals and objectives across universal goals, such as food security; and supporting improved planning around the location and construction of sanitation and health facilities, especially in high-risk coastal and flood prone areas.

Third, the ATLAS Program works with partner countries to pilot, integrate and disseminate risk-reducing public health management practices for climate-related health challenges. They provide technical assistance to countries to promote the use evidence-based practices, such as implementation of early warning systems, and simultaneously train health workers and support campaigns to educate vulnerable populations. For example, in Mozambique, the ATLAS program piloted a program to integrate climate information into health sector planning which resulted in improved understanding of how climate and weather influence malaria and diarrheal disease across the country [40].

Lastly, the ALTAS program works to mobilize finance adaptation measures from both domestic resources and international sources in partner countries where public health is constrained by insecure funding. For example, ATLAS provides technical assistance to strengthen public financial management and revenue administration, to promote integrated operations and maintenance budgets for capital investments, to conduct cost benefit analyses of adaptation options, and to support gender mainstreaming and women's participation in climate and health finance mechanisms.

TOOL: Enhancing National Climate Services (ENACTS)

In collaboration with several other partners, USAID supports the ENACTS initiative – a program which creates user-focused climate products to aid national and sub-national decision makers in multiple sectors, including health. One of ENACTS major project accomplishments is the creation of an online resource for the Ethiopian health department which depicts when conditions are conducive to the transmission of malaria at the district level. Such information is used to identify epidemic-prone districts, plan for timing of control activities and estimate the probable length of the malaria season, ultimately leading to better targeted public health interventions [41].

National Science Foundation (NSF)

As the primary funder for basic social, economic, and behavioral science research in US academic institutions, the National Science Foundation (NSF) supports research and related activities to advance our fundamental understanding of physical, chemical, biological and human systems and the interactions among them. NSF regularly collaborates with other USGCRP agencies to provide support for a range of multidisciplinary research projects and is actively engaged in a number of international partnerships related to the impacts of climate change on human health.

NSF funding of climate and health research is extensive, diverse, fragmented, and not nested under any one particular sub-agency or program. Examples of current initiatives include but are not limited to emergency preparedness and disaster response, infectious disease variability and vulnerability, heat stress and heat illness, decision making under climate uncertainty, climate impacts on coastal communities, air quality, and more [42].

Recently, NSF in conjunction with the National Institute of Environmental Health Sciences (NIEHS) awarded $30 million for new research to investigate the impact of climate change on marine and great lakes human pathogens such as aquatic toxins and harmful algae blooms [43]. Additionally, NSF recently awarded $18.7 million dollars towards natural hazards research through its Prediction of and

Resilience Against Extreme Events (PREEVENTS) program. The goal of PREEVENTS is to improve predictability and risk assessments of natural hazards, to increase resilience to these events, and to reduce their effects on human lives, societies and economies [44]. PREEVENTS also supports research to improve the understanding of processes underlying natural hazards and extreme events, including hurricanes, droughts, and heat waves.

NSF, in collaboration with the NIH, also co-funds the Ecology and Evolution of Infectious Diseases (EEID) Initiative. This multidisciplinary program supports research to understand the underlying ecological, evolutionary, and social drivers that influence the transmission dynamics of infectious disease. Research funded through this initiative is highly interdisciplinary and includes how environmental factors, such as climate change, are influencing the spread of human pathogens including zoonotic, vector-borne, or enteric pathogens [45].

NSF also supports Decision Making Under Uncertainty (DMUU) Centers, which employ researchers to utilize multiple methodologies, including forecasting and decision support, to analyze organizational decisions in settings where uncertainty is high, the risk is complex and the implications of decisions are long-term and future-oriented. Examples of projects taken on by the DMUU initiative include a robust planning and risk assessment model that shaped Louisiana's 50-year Coastal Master plan in the wake of Hurricane Katrina and Rita [46]. End users of DMUU center products include government leaders, the business community, and the public.

The NSF also funds the Dynamics of Coupled Human and Natural Systems Program, a multidisciplinary program to support teams of researchers focused on the social, natural, and physical science researching the connections between human and natural systems in the context of climate change [47].

The National Oceanic and Atmospheric Administration (NOAA)

The National Oceanic and Atmospheric Administration (NOAA), under the Department of Commerce (DOC), has a strategic climate goal of "an informed society anticipating and responding to climate and its impacts."[48] NOAA's climate change activities aim to create a predictive understanding of the changing climate and its impacts including to health, and to communicate climate information so that people can make more informed decisions in their lives, businesses, and communities. Long-term goals address climate adaptation and mitigation, being a weather-ready nation, promoting healthy oceans, and building resilient coastal communities and economies [49]. To support these objectives, NOAA provides climate predictions and services to a number of federal partners including: Environmental Protection Agency (EPA), U.S. Departments of Energy (DOE), State (DOS), Agriculture (USDA), Transportation (DOT), Interior (DOI), Health and Human Services (HHS), Homeland Security (DHS), and Defense (DOD), and the National Aeronautics and Space Administration (NASA). These collaborations help to identify climate risks and vulnerabilities, deliver climate-relevant information for decision-making, and better inform people about climate variability, change, and their impacts [50]. NOAA uses a Global Climate Observing System (GCOS) to understand key climate processes, improve modeling capabilities, and better the development and delivery of climate educational programs and information services to support climate goals including adaptation and mitigation [51].

NOAA's capabilities in linking ocean and human health as well as the agency's monitoring and prediction tools and climate science activities provide critical expertise to understand the health effects of climate change. Research further supports advancement of knowledge and application to human health. The Climate Program Office offers ten research competition topics through the Earth System Science and Modeling (ESSM), Climate and Societal Interactions (CSI), and Communication, Education and Engagement (CEE) Programs with approximately 90 new awards anticipated for 2020 [52]. The Regional Integrated Sciences and Assessments (RISA) Program works to understand context and risks, support knowledge to action networks, enhance the use of science in decision-making, and advance science policy to build capacity across the nation with public and private sector collaboration with 178 ongoing projects listed [53]. The International Research and Applications Project (IRAP) supports projects in regions such as the Caribbean, India, and Bangladesh where weather and climate challenge U.S. economic, development, scientific, and security interests. A core theme of work is integrating climate information and decision processes for regional climate resilience through translation of climate information and interdisciplinary application to enhance societal preparedness and build capacity [54]. There are six new projects being funded for fiscal years 2018–2020 [55]. The Belmont Forum also serves to support transdisciplinary research projects to investigate climate, environment, and health.

Environmental Protection Agency (EPA)

EPA's research programs aim to develop scientific information for stakeholders, policymakers, and communities to effectively manage and reduce the impacts of the changing climate on human health, ecosystems, and socioeconomic systems in the United States. EPA's research focus is informed by the Agency's mission and statutory requirements, and includes: (1) improving the scientific understanding of global change effects on air quality, water quality, ecosystems, energy, and human health in the context of other stressors; (2) assessing and developing adaptation options to effectively respond to global change risks, increase resilience of human and natural systems, and promote their sustainability; and (3) developing an understanding of the potential environmental impacts and benefits of greenhouse gas emission reduction strategies to support sustainable mitigation solutions.

EPA programs emphasize the integration of knowledge across the physical, chemical, biological, and social sciences into decision support frameworks that recognize the complex interactions between human and natural systems. Research activities include efforts to connect continental-scale temperature and precipitation changes to regional and local air quality and hydrology models to better understand the impacts of climate change on air quality and water quality, and to examine how watersheds will respond to large-scale shifts in climate, extreme events, or other global changes. This information is provided to help decision makers make climate informed policy choices and management decisions.

EPA periodically produces assessments, reports, and tools to serve as resources for decisions related to environmental issues. Some relevant examples are:

- *Assessments* have been produced on a variety of topics, such as storm water management, urban resilience, human health, and extreme heat [56].
- EPA partners with more than 40 data contributors from various government agencies, academic institutions, and other organizations to compile a key set of indicators related to the causes and effects of climate change. The indicators are published in EPA's report, *Climate Change Indicators in the United States* [57], and are designed to be a "go-to" resource for the public, scientists, analysts, decision-makers, educators, and others who can use climate change indicators as a tool for communication, environmental assessment, and informed decision-making.
- The *Climate Change Impacts and Risk Analysis (CIRA) project* [58] quantifies the physical effects and economic damages under multiple climate change scenarios.
- The *Global Climate Explorer (GCX) Integrated Climate and Land-Use Scenarios (ICLUS) Tool* [59] was developed to produce spatially explicit projections of population and land-use that are based on the Intergovernmental Panel on Climate Change's (IPCC) Special Report on Emissions Scenarios (SRES). This is part of a collection of web tools that visualize, compare, and provide access to spatial data that describe potential future environmental change. These data can serve as a starting point when assessing the vulnerability of air, water, ecosystems, and human health to climate change, land use change, and other large-scale environmental stressors.
- EPA's *Wildland Fire Research Framework* [60] outlines the Agency's wildland fire priorities. This document provides background information on wildland fire research as it relates to EPA's mission; a narrative of the existing ORD research portfolio; and a landscape of future work that is within the purview, expertise, and capacity of ORD. Additionally, this research framework outlines an approach for collaborative research activities with other federal partners.
- EPA's Environmental *Benefits Mapping and Analysis Program – Community Edition (BenMAP-CE)* [61] is an open-source Geographic Information System (GIS)-based computer tool to estimate the health and economic impacts of air quality change [62]. Benefits are estimated with health impact functions, which incorporate information about ambient air pollution levels, health effects estimates, baseline incidence rates of particular health endpoints, and the exposed population. Users can compare various climate or policy scenarios by information on air quality, demographics, economic values, and health effects or use information that is already pre-loaded. Potential changes in air pollution concentrations resulting from policy measures or changes are translated into potential health impacts.

Department of Agriculture (USDA)

The United States Department of Agriculture (USDA) aims to empower land managers, policy makers, and federal agencies with science-based knowledge to manage the risks, challenges, and opportunities posed by climate change; reduce greenhouse gas emissions; and enhance carbon sequestration. USDA includes contributions from the Agricultural Research Service (ARS), the National Institute of Food and Agriculture (NIFA), the Forest Service, Natural Resources Conservation Service (NRCS), National Agricultural Statistics Service (NASS), and Research, Education, and Economics Information System (REEIS). USDA draws upon this diversity to identify climate change challenges and priorities in continuing to meet the needs of its stakeholders, decision makers, and collaborators.

The USDA conducts research focused on understanding climate change effects on natural and managed ecosystems, developing tools and management strategies to promote adaptation, enhancing mitigation of atmospheric greenhouse gases, and providing science-based information for decision support. Research projects specifically related to extreme weather events and One Health include topics on antimicrobial resistance, water and soil quality, foodborne pathogens, and vector-borne infectious diseases [63]. The Current Research Information System (CRIS) provides project updates while the Food Safety Research Information Office (FSRIO) manages a research project database and a collection of the latest food safety publications. Other work related to sustainable use of natural resources and adaptation involves supporting collaborative projects to address soil science conservation and conservation easements on public and private lands damaged by flooding and other natural disasters.

References

1. Public Law 95–367: The National Climate Act. (15 U.S.C. 2901, 2908), 9/17/78). https://www.congress.gov/bill/95th-congress/house-bill/6669/. Accessed 27 Aug 2019.
2. Pielke RA. The policy history of the US global change research program: part I. Administrative development. Global Environ Chang. 2000;10:12. https://sciencepolicy.colorado.edu/admin/publication_files/2000.09.pdf. Accessed 27 Aug 2019.
3. GAO (General Accounting Office). Administration approach cautious pending validation of threat. Report to the chairman, subcommittee on oversight and investigations. Committee on energy and commerce, house of representatives, NSIAD-90-63, January, US GPO, WA, DC; 1990. p. 16. Accessed online http://www.gao.gov/assets/150/148577.pdf
4. Pielke RA. The policy history of the US global change research program: part II. Legislative process. Global Environ Chang. 2000;10:136. http://nome.colorado.edu/HARC/Readings/Pielke3.pdf. Accessed 1 May 2012.
5. The US Global Change Program Website. http://www.globalchange.gov/. Accessed 1 Oct 2019.
6. The US Global Change Program Website Program History. http://globalchange.gov/about/program-structure/program-history. Accessed 1 Oct 2019.

7. The US Global Change Research Program (USGCRP). USGCRP strategic plan. Washington, DC; 2012. p. 102. http://downloads.globalchange.gov/strategic-plan/2012/usgcrp-strategic-plan-2012.pdf. Accessed 1 Oct 2019.
8. Committee on the Human Dimensions of Global Change. Science priorities for the human dimensions of global change. Washington, DC: The National Academies Press; 1994. p. 496. https://www.nap.edu/catalog/9175/science-priorities-for-the-human-dimensions-of-global-change. Accessed 1 Oct 2019.
9. The US Global Change Research Program (USGCRP). Our changing planet 2010. Washington, DC; 2010. p. 7. Accessed 1 May 2012.
10. The U.S. Climate Change Science Program: vision for the program and highlights of the scientific strategic plan. https://www.globalchange.gov/browse/reports/us-climate-change-science-program-vision-program-and-highlights-scientific-strategic. Accessed 1 Oct 2019.
11. The US Global Change Research Program (USGCRP). Our changing planet 2010. Washington, DC; 2010. p. 4. Accessed 1 May 2012.
12. Portier CJ, Tart KT, Carter SR, Dilworth CH, Grambsch AE, Gohlke J, Hess J, Howard SN, Luber G, Lutz JT, Maslak T. A human health perspective on climate change: a report outlining the research needs on the human health effects of climate change. J Curr Issues Globalization. 2013;6(4):621.
13. The US Global Change Program Website. What we do – interagency crosscutting group on climate and health. http://www.globalchange.gov/what-we-do/climate-change-health. Accessed 1 Oct 2019.
14. Balbus AJ, Gamble JL, Beard CB, Bell JE, Dodgen D, Eisen RJ, Fann N, Hawkins MD, Herring SC, Jantarasami L, Mills DM, Saha S, Sarofim MC, Trtanj J, and Ziska L, Eds. United States Global Change Research Program. The impacts of climate change on human health in the united states: a scientific assessment. crimmins. U.S. Global Change Research Program, Washington, DC; 2016:312 pp. http://dx.doi.org/10.7930/J0R49NQX.
15. GCRP. Predicting climate-sensitive infectious diseases to protect public health and strengthen national security. Accessed September 22, 2019 at https://www.globalchange.gov/sites/globalchange/files/Predicting_ClimSensInfDisease_2019.pdf
16. Belmont Forum. Collaborative research actions. https://www.belmontforum.org/cras/#ceh2019. Accessed 1 Oct 2019.
17. Global Heat Health Information Network. https://www.ghhin.org/. Accessed 1 Oct 2019.
18. Ball E. NIH announces climate change and health funding. Environmental factor. NIEHS; 2010, September. http://niehs.nih.gov/news/newsletter/2010/september/spotlight-nih.cfm. Accessed 1 Oct 2019.
19. NIEHS grantees human health impacts of climate change. https://www.niehs.nih.gov/research/programs/geh/climatechange/index.cfm. Accessed 1 Oct 2019.
20. See “climate change and human health lesson plans”. Available at https://www.niehs.nih.gov/health/scied/teachers/cchh/index.cfm. Accessed 1 Oct 2019.
21. The CDC’s Climate and Health Program, About our Program. https://www.cdc.gov/climateandhealth/default.htm. Accessed online 26 Aug 2019.
22. Marinucci G, Luber G, Uejio C, Saha S, Hess J. Building resilience against climate effects – a novel framework to facilitate climate readiness in public health agencies. Int J Environ Res Public Health. 2014;11(6):6433–58.
23. Assessing health vulnerability to climate change: a guide for health departments. CDC climate and health technical report series. https://www.cdc.gov/climateandhealth/pubs/AssessingHealthVulnerabilitytoClimateChange.pdf. Accessed online 26 Aug 2019.
24. Projecting climate-related disease burden: a guide for health departments. CDC climate and health technical report series. https://www.cdc.gov/climateandhealth/docs/ProjectingClimateRelatedDiseaseBurden_508.pdf. Accessed online 26 Aug 2019.
25. Climate and health intervention assessment: evidence on public health interventions to prevent the negative health effects of climate change. CDC climate and health technical report series.

https://www.cdc.gov/climateandhealth/docs/ClimateAndHealthInterventionAssessment_508.pdf. Accessed online 26 Aug 2019.
26. The CDC's climate ready states and cities program. https://www.cdc.gov/climateandhealth/default.htm. Accessed online 26 Aug 2019.
27. Climate and Health Adaptation Plan 2017. Public health guidance for the state of Arizona. https://www.azdhs.gov/documents/preparedness/epidemiology-disease-control/extreme-weather/pubs/arizona-climate-health-adaptation-plan.pdf. Accessed online 26 Aug 2019.
28. National Environmental Public Health Tracking Network. Data explorer, climate change. https://ephtracking.cdc.gov/DataExplorer/#/. Accessed online 26 Aug 2019.
29. Health and air quality: 2018 annual summary. NASA earth science applied sciences program. https://appliedsciences.nasa.gov/system/files/sites/default/files/2018Annual%20Summary_HA.pdf. Accessed online 26 Aug 2019.
30. NASA Health and Air Quality Applied Sciences Team. Tiger Teams. Available at: https://haqast.org/tiger-teams/. Accessed August 26, 2020.
31. NASA SERVIR Overview. https://www.nasa.gov/mission_pages/servir/overview.html. Accessed online 8 Aug 2019.
32. NASA SERVIR-Eastern and Southern Africa. https://www.nasa.gov/mission_pages/servir/africa.html. Accessed online 8 Aug 2019.
33. NASA SERVIR-Himalaya. https://www.nasa.gov/mission_pages/servir/himalaya.html. Accessed online 8 Aug 2019.
34. NASA SERVIR-Mekong. https://www.nasa.gov/mission_pages/servir/mekong.html. Accessed 8 Aug 2019.
35. USAID climate risk management for projects and activities. https://www.usaid.gov/sites/default/files/documents/1868/201mal_042817.pdf. Accessed online 6 Aug 2019.
36. USAID climate risk management and screening tools. https://www.climatelinks.org/resources/climate-risk-screening-management-tool. Accessed online 26 Aug 2019.
37. USAID Adaptation Thought Leadership and Assessments (ATLAS): advancing climate-resilient development. https://www.climatelinks.org/sites/default/files/USAID%20ATLAS%20Project_Chemonics_Two%20Pager.pdf. Accessed online 8 Aug 2019.
38. USAID Building Climate Resilience in the Health Sector. https://www.climatelinks.org/sites/default/files/2019_USAID_ATLAS_Building%20Climate%20Resilience%20in%20the%20Health%20Sector.pdf. Accessed online 8 Aug 2019.
39. USAID climate change and health risks in Senegal, technical report. https://www.climatelinks.org/sites/default/files/asset/document/180327_USAID-ATLAS_Senegal%20Climate%20and%20Health_Final_rev.pdf. Accessed online 8 Aug 2019.
40. USAID climate change and health in Mozambique: impacts on diarrheal disease and malaria. https://www.climatelinks.org/resources/climate-change-and-health-mozambique-impacts-diarrheal-disease-and-malaria-0. Accessed online 8 Aug 2019.
41. Climate and Development Knowledge Network. Climate information to help combat malaria in Ethiopia. https://cdkn.org/2019/04/feature-climate-information-to-help-combat-malaria-in-ethiopia/?loclang=en_gb. Accessed online 26 Aug 2019.
42. National Science Foundation. Search results for climate and human health. https://www.nsf.gov/awardsearch/simpleSearchResult?queryText=climate+AND+human+health. Accessed online 8 Aug 2019.
43. National Science Foundation News Release. NSF, NIEHS award $30 million for new research. https://www.nsf.gov/news/news_summ.jsp?cntn_id=296579. Accessed online 8 Aug 2019.
44. National Science Foundation News Release. In wake of hurricanes, floods and wildfires, NSF awards $18.7 million in natural hazards research grants. https://www.nsf.gov/news/news_summ.jsp?cntn_id=242941. Accessed online 8 Aug 2019.
45. National Science Foundation. Ecology and Evolution of Infectious Diseases (EEID). https://www.nsf.gov/funding/pgm_summ.jsp?pims_id=5269. Accessed online 8 Aug 2019.

46. Pardee Rand Graduate School Center for Decision Making Under Uncertainty. https://www.prgs.edu/research/methods-centers/decision-making-under-uncertainty.html. Accessed online on 8 Aug 2019.
47. National Science Foundation. Dynamics of coupled natural and human systems program solicitation. https://www.nsf.gov/pubs/2018/nsf18503/nsf18503.htm. Accessed online 8 Aug 2019.
48. The US Global Change Research Program (USGCRP). The national global change research plan 2012–2021. Washington, DC; 2012. p. 110. http://downloads.globalchange.gov/strategic-plan/2012/usgcrp-strategic-plan-2012.pdf. Accessed 18 Aug 2019.
49. NOAA. NOAA's goals. https://www.performance.noaa.gov/goals/. Accessed 19 Aug 2019.
50. NOAA. Coordinating regional support for natural resource management in a changing environment. https://cpo.noaa.gov/portals/0/Docs/RISA/NaturalResourcesinChangingEnvironment_5-5.pdf. Accessed 18 Aug 2019.
51. NOAA. The Global Climate Observing System (GCOS). https://www.ncdc.noaa.gov/gosic/global-climate-observing-system-gcos. Accessed 19 Aug 2019.
52. NOAA. FY 2020 notice of funding opportunity. https://cpo.noaa.gov/Funding-Opportunities/FY-2020-Notice-of-Funding-Opportunity. Accessed 19 Aug 2019.
53. NOAA. About the regional integrated sciences and assessments program. https://cpo.noaa.gov/Meet-the-Divisions/Climate-and-Societal-Interactions/RISA/About-RISA. Accessed 19 Aug 2019.
54. NOAA. About the International Research and Applications Project (IRAP). https://cpo.noaa.gov/Meet-the-Divisions/Climate-and-Societal-Interactions/IRAP/About-IRAP. Accessed 18 Aug 2019.
55. NOAA. IRAP funded projects. https://cpo.noaa.gov/Meet-the-Divisions/Climate-and-Societal-Interactions/IRAP/Funded-Projects. Accessed 19 Aug 2019.
56. EPA. Global change assessments and research. https://www.epa.gov/risk/global-change-assessments-and-research. Accessed 19 Aug 2019.
57. EPA. Climate change indicators in the United States 2016. https://www.epa.gov/sites/production/files/2016-08/documents/climate_indicators_2016.pdf. Accessed 5 Aug 2019.
58. EPA. Climate change in the United States: benefits of global action. https://www.epa.gov/cira. Accessed 19 Aug 2019.
59. EPA. About the GCX ICLUS tool. https://www.epa.gov/gcx/about-gcx-iclus-tool. Accessed 18 Aug 2019.
60. EPA. Wildland fire research framework 2019–2022. https://www.epa.gov/sites/production/files/2019-04/documents/wildland_fire_research_framework_final-tagged.pdf. Accessed 19 Aug 2019.
61. EPA. Environmental Benefits Mapping and Analysis Program-Community Edition (BenMAP-CE). https://www.epa.gov/benmap. Accessed 19 Aug 2019.
62. EPA Environmental Benefits Mapping and Analysis Program. http://www.epa.gov/air/benmap/. Accessed 22 July 2019.
63. Freeman T, Walls I. NIFA supported activities related to climate and one health. https://nifa.usda.gov/sites/default/files/resource/NIFA-Supported-Activities-Related-to-Climate-and-One-Health.pdf. Accessed online 19 Aug 2019.

Chapter 24
Management of Climate Change Adaptation at the United States Centers for Disease Control and Prevention

Jeremy J. Hess, Gino Marinucci, Paul J. Schramm, Arie Manangan, and George Luber

The Centers for Disease Control and Prevention (CDC) is the nation's public health agency. A part of the Department for Health and Human Services, the CDC's mission is "[c]ollaboration to create the expertise, information, and tools that people and communities need to protect their health—through health promotion, prevention of disease, injury and disability, and preparedness for new health threats" [1]. In recent years, climate change has emerged as a significant potential public health threat, and the CDC has initiated a range of efforts to facilitate adaptation to climate change in the public health sector. Climate change is expected to have a wide range of health impacts [2–4], and a range of public health expertise will be required to adapt to it [5]. To facilitate leadership on the issue, CDC's climate change efforts have been housed primarily in the Climate and Health Program in the National Center for Environmental Health (NCEH), though the program collaborates closely with several intramural and extramural partners. In general the program has focused primarily on domestic efforts, in keeping with the CDC's general focus on supporting state and local public health partners. Here we provide an overview of CDC's efforts, including an overview of the Climate and Health Program, an outline of CDC's conceptual approach to the integration of climate change adaptation into public health programming, its adaptation framework Building Resilience Against

J. J. Hess
Emory Schools of Medicine and Public Health, Atlanta, GA, USA

Climate and Health Program, NCEH, CDC, Atlanta, GA, USA
e-mail: jhess@emory.edu

G. Marinucci · P. J. Schramm (✉) · A. Manangan · G. Luber
Climate and Health Program, Division of Environmental Hazards and Health Effects, National Center for Environmental Health, Centers for Disease Control and Prevention, Atlanta, GA, USA
e-mail: imw3@cdc.gov

K. E. Pinkerton, W. N. Rom (eds.), *Climate Change and Global Public Health*, Respiratory Medicine, https://doi.org/10.1007/978-3-030-54746-2_24

Climate Effects (BRACE), and activities it has supported through its Climate-Ready States and Cities Initiative (CRSCI), including recent advances in climate and health science presented at the CDC's annual science symposium on climate and health. We close with brief consideration of future adaptation needs and CDC's plans for addressing ongoing needs.

Overview of CDC'S Climate and Health Program

The CDC's Climate and Health Program (the Program) serves as the primary hub of climate change adaptation activities at CDC. The program, which is housed within the NCEH, was formed in 2006 and began receiving specific Congressional appropriations in 2009. The program seeks to identify populations most vulnerable to the impacts of climate change, anticipate future climate and associated disease trends, assure that systems are in place to detect and respond to emerging health threats, and take steps to assure that these health risks can be managed now and in the future.

In pursuit of these goals, the program serves three core functions in support of public health adaptation:

1. Translating climate science to inform public health practitioners
2. Developing decision support tools to enhance preparedness
3. Serving as a credible leader in planning for the human health impacts of a changing climate

The program works with other parts of CDC to track data on environmental conditions, disease risks, and disease occurrence related to climate change. The program also collaborates with other Federal agencies such as National Oceanic and Atmospheric Association (NOAA) and National Aeronautics and Space Administration (NASA) and has participated in both the Unites States National Climate Assessment (NCA) and the Intergovernmental Panel on Climate Change (IPCC). Finally, the bulk of the program's efforts go to supporting state and local governments in support of their climate change adaptation activities, principally through the CRSCI.

Like other efforts to anticipate and address the public health effects of climate change, the CDC Climate and Health Program was initially faced with the challenge of determining how public health should approach the problem given its broad set of projected impacts, varying time scales, and impact on complex systems, many of which are outside public health's direct control (e.g., agricultural systems and systems for maintaining critical infrastructure). From its inception the program has thus invested in efforts to clarify and define the public health threats associated with climate change and to integrate adaptation activities within existing public health programming. This investment led to a landmark publication, "Climate Change: The Public Health Response," which outlined adaptation needs using the Ten Essential Public Health Services (EPHS) framework [5], as well as a paper on using adaptive management, an iterative, modeling-based approach, to guide adaptation efforts [6]. Next we will examine the issue of adaptive management and its role in climate change adaptation in greater detail.

Adaptive Management and Its Role

Adaptive management is an iterative, cyclic approach to designing, implementing, and evaluating interventions in complex adaptive systems [7]. Such systems are typically incompletely understood and exhibit some unexpected behaviors in response to management interventions; ecosystems are a frequently cited example. Importantly, an important aspect of managing these systems is the ongoing need to learn about their behavior, particularly in response to management interventions and shifting stressors over time. Evidence indicates that adaptive management better accommodates these needs than other approaches which tend not to actively address the dynamic nature of such systems. It relies heavily on systems modeling and explicitly emphasizes learning at each stage of the process.

As codified by the National Research Council in 2004, settings in which adaptive management may be a useful approach have six major elements:

1. Management objectives that are regularly revisited and revised
2. A model of the system(s) being managed
3. A range of management choices
4. Monitoring and evaluation of outcomes
5. Mechanisms for incorporating learning into future decisions
6. A collaborative structure for stakeholder participation and learning [8]

Increasingly, adaptive management has been touted as a useful approach for managing the health effects of climate change [6, 9, 10]. Climate change is impacting a wide range of sectors and associated systems, from natural ecosystems upon which native peoples rely for food [11] to intensively managed socio-ecosystems such as urban environments in which people can be exposed to a range of climatic hazards, from heat to air pollution [12]. While many of these systems are not directly under the purview of the public health sector, in all cases public health can be considered a stakeholder (e.g., the electrical power grid, on which people rely heavily to power mechanical air conditioning, is highly pertinent to public health but managed by electrical utilities and their regulators, and emissions from fossil fuel combustion to generate electricity have significant public health impacts), and many of these systems satisfy all the criteria listed above.

Because climate change is likely to amplify stresses on certain systems essential to maintaining public health, it will be increasingly important for public health organizations to have the capacity to manage these systems as both the systems, the stressors, and management objectives evolve. In an effort to develop adaptive management expertise among its state and local public health partners, the CDC has developed a flexible approach that public health partners could choose to adapt in order to facilitate local public health adaptation to climate change entitled BRACE: Building Resilience Against Climate Effects.

Building Resilience Against Climate Effects (BRACE)

The changing climate presents a novel type of public health challenge in which assumptions based on historical climatic and meteorologic patterns and their impacts on risks for climate-sensitive health outcomes must be, at the very least, revisited. In the United States, with its federalist structure and decentralized public health system, there is a diverse arrangement of public health organizations at the state and local level, and much public health programming is locally developed and implemented. Risk assessment using anticipated future disease burden, particularly formal assessment involving projections of climate-sensitive health outcomes, is not a familiar exercise for many local public health agencies [13], and many health departments feel unready to meet the related challenges with their existing resources [14]. In a 2008 survey, health departments also indicated concern that the CDC did not have adequate expertise to facilitate their climate change preparedness efforts [13].

To address this gap in domestic public health preparedness for the health impacts of climate change, CDC has built up its climate and health expertise and initiated several programs to support state and local public health partners in building their capacity and pursuing their adaptation efforts. To ensure that states had adequate available guidance regarding climate change adaptation, CDC developed a framework entitled BRACE [15]. The BRACE framework incorporates vulnerability assessment using climate projections, modeling of projected health impacts, evidence- based evaluation of intervention options, intervention implementation, and systematic evaluation of all activities in an iterative framework that incorporates the principles of adaptive management. Once several states have implemented BRACE, the results of implementation on adaptation activities will also be evaluated.

BRACE is a five-step process that enables a health department to incorporate the best available atmospheric science into climate-health impact projections for its jurisdiction. The BRACE framework involves health departments coupling retrospectively derived response functions describing associations between weather variables and health outcomes—preferably response functions derived from data on populations within their jurisdiction—with projected atmospheric data from global circulation models. These projections are then coupled with the response functions to project future disease burdens which can be used to facilitate planning and preparedness activities.

There are already frameworks for performing vulnerability assessments related to climate change and health [16] and comparative risk assessments [17]. BRACE is not designed to supplant or supersede this guidance. Instead, BRACE was designed to present these concepts in a structure that is relatively familiar for US health departments to emphasize that the underlying process of risk assessment, identification of appropriate interventions, intervention implementation, and evaluation is similar to that used successfully in public health for decades. The main departures from a more conventional approach are in the use of climate change impact projections for risk assessment and the strong emphasis on broad stakeholder engagement, learning, modeling, and iterative decision making that are hallmarks of adaptive management.

The Five Steps of BRACE

There are five sequential steps in the BRACE Framework:

- Step 1: *Anticipating climate impacts and assessing vulnerabilities*, in which a health department identifies the scope of the most likely climate impacts, the potential health outcomes associated with those climatic changes, and the populations and locations vulnerable to these health impacts within its jurisdiction.
- Step 2: *Projecting the disease burden*, in which a health department, as best as possible, estimates or quantifies the additional burden of health outcomes due to climate change—to support prioritization and decision making.
- Step 3: *Assessing public health interventions*, in which a health department seeks to identify the most suitable health interventions for the health impacts of greatest concern.
- Step 4: *Developing and implementing a climate and health adaptation plan*, in which a health department develops and implements a health adaptation plan for climate change that addresses health impacts and gaps in critical public health functions and services, and prepares a jurisdiction to enhance its adaptive capacity.
- Step 5: *Evaluating impact and improving quality of activities*, in which a health department can evaluate the processes it has used, determine the value of utilizing the framework, and the value of climate and health activities undertaken. This step is also important for quality improvement and for incorporating refined inputs such as updated data or new information, an essential component of adaptive management.

There are some key points to consider in the implementation of the BRACE framework. First, stakeholder engagement is very important throughout the process. A targeted selection of stakeholders can add significant value to the process overall, and specific stakeholders may be particularly important at specific points in the process. For example, in step 1, where much of the emphasis is understanding climate projections, a health department may profit significantly from engagement with their state climatologist and others in the climate science community, whereas in step 3, where an assessment is being made of the appropriateness of different public health interventions, it may be appropriate to solicit input from the larger public health practitioner and affected communities.

The second key consideration is that, while the BRACE framework lays out a comprehensive, sequential approach, it is flexible in that it allows the integration of prior analysis. Steps 1 through 3 focus on providing new or enhanced information that can aid a health department when making decisions on investments and program or operational changes. At any point from step 1 to step 3, a jurisdiction may have sufficient information based on prior analyses to make decisions without undertaking parts of the step. While the BRACE framework allows for the application of prior analyses and information, it is paramount that these inputs be vetted as providing the most up-to-date, available information regarding climate-related risk.

We have laid out each step in depth below. To frame the activities a health department would undertake in each step, we start each section with framing questions that highlight the lines of inquiry driving that specific step in the process. Table 24.1 illustrates how the first three steps of BRACE have been applied to the issue of extreme heat vulnerability by the New York City Department of Health and Mental Hygiene.

Step 1: Anticipating Climate Impacts and Assessing Vulnerabilities

- *In general terms, what will the climate look like in my jurisdiction in 10, 25, and 50 years?*
- *How are the population profile and the profile of public health challenges likely to change in my jurisdiction at these intervals?*
- *How might the anticipated changes in climate interact with these demographic and other challenges to shift population health risk and place vulnerabilities?*

The goal of this first step in BRACE is to identify the range of climate impacts, associated potential health outcomes, vulnerable populations, and locations of potentially vulnerable populations within a health department's jurisdiction. In step 1 a health department works toward establishing a functional understanding of how the climate is changing in its jurisdiction, the likely associated effects on health, and the populations and systems most vulnerable to these changes. To carry out this step, health department personnel will rely on public health and medical literature, expert experience, and academic and or governmental partners with expertise in atmospheric science and modeling to gain an understanding of relevant climate health burdens and projected climatic shifts.

Step 1 is both an exploratory exercise and scoping activity. It is exploratory in the sense that health departments must first work with partners (e.g., the state climatologist) to understand how climate and health have been and are likely to be related in their jurisdictions. This entails developing an understanding of how climate and weather have historically affected population health in the health department's jurisdiction, how the climate in the region has changed to date and how it is likely to change in the future, and finally what factors have driven population vulnerability to climate-health impacts in the past. The climate-health literature, which is expanding rapidly, will likely provide insight into some of the most climate-sensitive diseases and health outcomes in a particular region as well as important factors affecting vulnerability to particular hazards and is generally where a health department should start to explore relevant climate-health relationships. The literature may not

Table 24.1 How the first three steps of BRACE have been applied to the issue of extreme heat vulnerability by the New York City Department of Health and Mental Hygiene

1. The initial step of BRACE is to assess public health vulnerabilities to climate change. To accomplish this, NYC Health utilized information in the NYC Panel on Climate Change Report (2009) detailing current and future trends in heat waves and other hazardous weather-related events. NYC Health conducted an epidemiologic analysis using vital statistics to identify subpopulations at the greatest risk for heat stroke and then mapped the distribution of heat vulnerability in the city's boroughs
2. Step 2 of BRACE involves projecting the burden of disease in a changing climate. NYC Health conducted a retrospective analysis to determine the relationship between temperature and mortality and then used global circulation model outputs to project future heat-related mortality in 2020. Their analysis showed that, all things being equal, there would be an increase in heat-related deaths
3. Step 3 of the BRACE frameworks is an assessment to determine the most effective and suitable public health interventions. NYC Health conducted a heat-health behavior survey to determine air conditioning (AC) prevalence and usage, assess behaviors of high-risk groups during hot weather, and gauge public awareness of heat warnings. They found approximately 700,000 New Yorkers were without functioning AC and approximately 550,000 were particularly vulnerable to heat illness (i.e., no functioning AC, age > 65 years old, and living with underlying chronic health conditions). About half of this population stayed at home during hot weather. Survey findings suggested that the most vulnerable populations may not understand their true risk and outreach should focus on conveying the importance of AC use and the potential lethality of both outdoor and indoor heat exposure
4. Step 4 of BRACE is the development and implementation of a climate and health adaptation plan, which is a set of public health interventions aimed to reduce the adverse health effects resulting from climate-related hazards. In an effort to prevent heat-related illness, NYC Health prepared public health messaging and materials to better convey the risk of heat stress and improved active outreach to those most vulnerable. Additionally, they implemented plans to increase access to AC to specific vulnerable areas. The health interventions put forth by NYC Health will be included in the overarching climate adaption plan for city of New York to ensure that health is an essential component
5. Step 5 of BRACE is the evaluation of the effectiveness of the climate adaptation plan and specific interventions. NYC Health is currently in the process of conducting process, outcome, and impact evaluations of its climate change adaptation planning efforts

New York City's Department of Health and Mental Hygiene (NYC Health) is a CDC Climate-Ready States and Cities Initiative (CRSCI) grantee. NYC Health has used the grant support to assess the potential increase in heat-related illness resulting from climate change. Its efforts serve as an excellent case study on how a health department can build resilience against the health effects of climate, outlined here in terms of the Building Resilience Against Climate Effects (BRACE) framework developed by the CDC's National Center for Environmental Health (NCEH)

have much specific information relevant to the health department's locale, however, so health departments will also need to solicit inputs from local partners to supplement their literature search.

Step 1 is also a scoping exercise, in that health departments must make determinations about the geographic and temporal scope of their assessments based on the intended application of climate and associated health projection information and the availability and robustness of relevant climate and health data. For example, if a jurisdiction plans to use the assessment to help inform city planning and guide decisions regarding hard infrastructure with a lifespan of at least 50 years, planners would like to avail themselves of climatic projections going at least 50 years into the future to coincide with the infrastructure lifespan, and their analysis is likely to focus in particular on historical and future extremes that may test infrastructure capacity. In contrast, assessing how vector-borne disease patterns may shift is likely to be done on a shorter time span and to focus less on extremes than on changes in means and the effect of interannual variability of temperature and precipitation on ecological conditions associated with increased disease risk.

The outcome of step 1 is typically a Climate and Health Profile Report. Such a report lays out the findings of the exercise, including the geographic and temporal scope, a summary of prevalent health concerns in the area, a list of major climate sensitive health outcomes in the region, factors affecting vulnerability historically, and an overview of how climate change is likely to affect exposures relevant to health in the region over the specified time frame. The report should also identify health impacts that may already be apparent and identify points at which other impacts are likely to manifest and highlight projected shifts in demographics that may affect population vulnerability and expected impacts on population health. Finally, the report should identify relevant infrastructure—from that in the health sector (clinics, hospitals, emergency medical services, etc.) to that in other sectors that is key to maintain public health (power plants, the electricity grid, sewage treatment, agriculture, transportation, etc.)—that may be vulnerable as the climate shifts.

Population and place vulnerability should be a theme throughout the report. Population vulnerability is relatively familiar in public health and focuses on factors that increase a population's exposure to environmental hazards or amplify an exposure's health impacts. Age, chronic health conditions, and low socioeconomic status are examples. Place vulnerability focuses on factors associated with a specific place that can increase inhabitants' vulnerability to climatic hazards, from geographic fixtures to reliance on local ecosystem services (e.g., for food and employment), as well as strong cultural place connections that could lead to adverse health impacts if ruptured. For both types of vulnerability, vulnerability factors are likely to differ by health outcome and location—age may be a significant factor for some diseases, while socioeconomic status is likely to be a major factor in others—and these vulnerability factors are not uniformly distributed.

To better characterize the distribution of vulnerability factors, health departments can include representations of their distribution in the report. One particularly useful approach entails using geographic information systems (GIS) and non-GIS-based vulnerability mapping, which incorporating demographic, risk factor, and health trend data to identify populations and locations within a jurisdiction where vulnerability is particularly high. Further analysis can be undertaken to assess infrastructure, systems, and physical features in vulnerable areas which, if compromised, may compound risk. Infrastructure and system considerations can include factors such as combined sewer systems, location of critical infrastructure such as hospitals and clinics, or vulnerability of the power grid. Physical features can include factors that can amplify exposure such as low elevation, intensity of the urban heat island, and proximity to high-traffic areas with relatively poor air quality.

Once step 1 is complete, a jurisdiction will have a rich sense of how weather and climate have historically affected population health in its area, including a sense of which populations and places are most vulnerable, and how this vulnerability and the associated health burdens are likely to shift as the climate changes. This knowledge is fundamental to the next step in BRACE: projecting future disease burdens.

Step 2: Projecting the Disease Burden

- *What is the relationship between the exposure (s) of interest and health outcomes in the recent past?*
- *What specific exposure shifts are expected as a result of climate change based on the most recent global circulation model projections?*
- *Putting together these exposure–outcome associations (health response functions) with projected climatic shifts, what is the projected burden of disease secondary to climate change in the next 10, 25, and 50 years?*

Through step 1, health departments identify the climate-sensitive health outcomes of greatest concern in their jurisdictions and consider how climate change may affect associated disease burdens over time and potential implications for the health department doing the analysis. In step 2, health departments take the next step and examine these shifting burdens more closely in an attempt to project and quantify shifting burdens associated with a changing climate.

While step 2 can be done qualitatively to yield a general impression of how climate change may affect the risk for certain outcomes, a quantitative effort is likely to be of the greater utility. A qualitative approach would, at the least, capture general trends in climatic exposures, population vulnerability, and identify associated trends in impacts to the extent possible. For instance, a region with significant projected warming and an aging population might note that extreme heat events in the region are expected to triple by 2050, that the proportion of the population over age 65 will double by that time, and that absolute risk of heat-related adverse health effects for older adults in the jurisdiction is likely to increase severalfold.

A quantitative approach entails a closer analysis of disease risks, vulnerability factors and their contributions to adverse health outcomes, changes in exposure, and relevant demographic shifts and has the potential to identify important aspects of shifting risks that might be missed in a more cursory qualitative analysis. This process has several major components, as noted in Fig. 24.1.

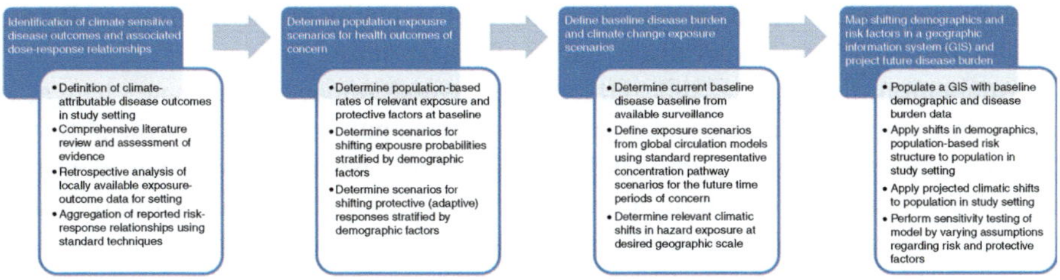

Fig. 24.1 Major steps in quantitative projection of climate-associated disease burdens

A detailed discussion of this process is outside the scope of this chapter, and there are several different studies that detail relevant methods [18–20]. Regardless of the specific approach taken, the first step is definition of the health outcomes of interest (which will have been identified in step 1 of BRACE) and the climate-health exposure pathway(s) of concern. Heat is the most commonly studied, but a wide range of outcomes are climate sensitive and may be important to study depending on the baseline burden of disease in a particular location. As noted in step 1, the chosen health outcome(s) should be relevant to the jurisdiction being studied based either on current or anticipated future disease burden, and baseline data on disease prevalence, preferably stratified by relevant demographic factors, should be available.

Several different methods have been used to project disease impacts. There is as yet no consensus regarding the most appropriate specific methods for disease projection and reporting of results, even for commonly studied exposures such as heat [21]. In general, the most commonly applied is the delta method, in which changes in the relevant climatic exposure are determined by comparing projected climatic variables (e.g., temperature, humidity, and precipitation) with historical baselines to determine the relevant shift in exposures averaged over a given period of time (e.g., an average increase of 0.7 °C in maximum temperature over June, July, and August in 2035 compared with the baseline period of 1980–2010). The shifted exposure is associated with relative risks (typically expressed as a change in relative risk per some fixed interval change in an environmental variable, e.g., an increase of emergency department visits for heat illness of 1.06 per 1 °C change in temperature above a particular threshold) derived from a comprehensive literature search and/or from retrospective analysis of locally available data for the jurisdiction. If novel associations are being evaluated, the question of whether the observed associations are indeed causal should be addressed. If possible, these exposure–outcome associations should be stratified by relevant demographic variables, e.g., age and socioeconomic status. Other strata may be relevant: for a hydrometeorological hazard such as flooding, for instance, dwelling elevation may also be a predictor of associated illness or death.

At a minimum, data required to apply the delta method include baseline disease prevalence, exposure–outcome associations for relevant climatic hazards, demographic projections for the study region, and global circulation model projections of shifts in climatic hazards in the study region for the study period. Considerations regarding data sources for projections in step 2 are listed in Table 24.2. Of note, one of the data sources listed in the table, the CDC's National Environmental Public Health Tracking Network, is discussed briefly later in this chapter.

Adaptation, i.e., activity that reduces the adverse impacts of climate change, is also important to consider, as adaptation activities in public health have the potential to limit adverse impacts significantly, though many barriers have been identified [22]. Depending on the length of study period (i.e., how far into the future health impacts are projected), projections of likely adaptations—active and passive,

Table 24.2 Common data sources used in public health climate change impact projections

Category of data required	Common data sources
Baseline disease prevalence	Public health surveillance, regional and national datasets (e.g., National Hospital Ambulatory Medical Care Survey, Healthcare Cost and Utilization Project, Nationwide Emergency Department Sample, and Behavioral Risk Factor Surveillance System)
Exposure–outcome associations	Published literature, retrospective analysis of local health outcome datasets merged with local weather and climate data from National Climatic Data Center, CDC National Environmental Public Health Tracking Network
Demographic projections	Demographic projections are available from the United States Census for the country as a whole and for individual states via the Federal-State Cooperative for Population Projections
Global circulation model projections	There are a number of climate models worldwide, and certain outputs have been made publicly available; one commonly used source is the Coupled Model Intercomparison Project (CMIP), which issues ensemble model runs for various scenarios (e.g., CMIP3, CMIP5) that are available for download

planned and unplanned—will be more or less important. If adaptations are not considered the projected disease burdens will be systematic overestimates, perhaps dramatically so if the projections are far into the future when adaptations may be widespread. There are many different adaptations to climatic exposures, some of which are passive (e.g., physiologic adaptation to heat exposure) and some of which are active (e.g., purchase, installation, and usage of mechanical air conditioning) that should be considered as part of the BRACE framework. The degree to which various adaptations may protect against exposure or dampen its impacts is not always well known but can be estimated in cases where there is no specific estimate available in the literature. In some cases, physiologic adaptation to the exposure of concern has been incorporated into exposure–outcome response functions [19]. In other cases, adaptation has been accounted for by systematically discounting estimates of future impacts [23].

Projecting disease burden is a potentially data-intensive exercise. However, once models for projecting disease burden are developed, these models can be used to guide several different types of decisions over time and can be used to engage with various stakeholders relevant to risk management decisions affecting public health. As additional information regarding exposures, adaptation options, and trends in demographics and disease burdens becomes available, the models can be updated to provide more precise estimates regarding likely future disease burdens and the cost-effectiveness of specific risk management interventions. Models can also be coupled with other efforts, such as health impact assessments aimed at characterizing climate change mitigation opportunities and associated health co-benefits (e.g., reduced emergency department visits for asthma exacerbations as a result of a shift to renewable energy sources for power generation) [24].

Step 3: Assessing Public Health Interventions

- *What are the most suitable adaptations and interventions that can be implemented to prevent or reduce anticipated increases in morbidity and mortality?*
- *What types of evidence do we have supporting particular interventions?*
- *How much morbidity and mortality might an early warning system for severe weather reliably avoid?*

Following the development of a Climate-Health Profile Report and a model for projecting the health burdens of climate change in a given jurisdiction, the next step in the BRACE framework is to identify and assess possible interventions that might be deployed to limit these anticipated impacts. This is an exercise in the evidence-based practice of public health (EBPH). While much has been written about EBPH in general, there is very little literature on EBPH and public health adaptation to climate change specifically apart from a recent publication surveying policy- relevant scientific literature in the field [25].

In general, EBPH entails problem assessment, systematic review of the public health literature to identify relevant interventions, and assessment of the identified literature to identify the interventions that have the strongest evidence of desired impacts [26, 27]. While there is abundant literature regarding the likely public health impacts of climate change (i.e., problem assessment), there is relatively little published on specific adaptations and interventions that may avoid or limit these projected impacts, even when potential exposures are considered outside the context of climate change (e.g., when strategies to protect against heat illness are considered outside of the climate change context). For instance, a recent structured review of population-level interventions to reduce the impacts of extreme heat identified only 14 studies, all of which were cross-sectional or retrospective, and the authors were unable to generate a specific impact estimate [28].

BRACE steps 1 and 2 ensure that adequate attention is paid to problem assessment, but do not provide for systematic assessment of relevant interventions. For this, a systematic literature review and accepted approach to evaluation of evidence is required. The methods for conducting systematic literature reviews and combining estimates of effect are relatively well established (see, for instance, guidelines on Preferred Reporting Items for Systematic Reviews and Meta-analyses [PRISMA] [29]), though there is not yet complete consensus regarding evaluation of evidence in public health, where experimental evidence (e.g., randomized controlled trials) is rare and it is not entirely clear when additional high-level evidence may be required [30]. In practice, public health organizations have taken an inclusive approach to evidence for public health interventions, as demonstrated by the CDC Guide to Community Preventive Services.

While experimental evidence can be particularly useful to justify more costly interventions and determine whether an outcome is causally related to the intervention, observational evidence is frequently very important in guiding day-to-day decisions that many public health officials encounter in the course of their activities. In addition to evidence available in the literature, some locales may decide it is more appropriate for them to supplement with their own evidence through analysis of locally available data to assess problems and guide interventions, a well-established approach (see Table 24.1). CDC's public health partners have also frequently cited the importance of anecdotal evidence conveyed through informal professional networks in making ad hoc decisions when little studied issues arise, such as strategies for promoting the use of cooling centers and making decisions about when to issue heat-health warnings. While considered expert opinion, such evidence is nevertheless important when formal studies have not been done and the potential harm associated with the interventions is low.

Evidence may also not be available for certain potential risks, particularly those associated with cascading failures of risk management like electrical blackouts or sewage treatment failures after extreme precipitation events. In such cases, public health officials may need to access literature outside of public health in order to identify strategies for promoting resilience across a range of linked systems upon which public health relies.

Overall, while systematic review of the literature and identification of efficacious interventions is of paramount importance, it is also clear that other forms of evidence such as observational evidence and expert opinion will also enter into deliberations regarding the interventions to pursue. As the field matures and various interventions are implemented, public health practitioners can prioritize reporting of these interventions and their effects using relevant guidelines already in the literature.

Step 4: Developing and Implementing a Climate and Health Adaptation Plan

- *What resources are required to implement the adaptations and interventions deemed suitable and feasible within the jurisdiction?*
- *How will these resources be used to implement these adaptations and interventions?*
- *Who and what needs to be mobilized to implement these adaptations and interventions?*

Having characterized climate change vulnerability in their jurisdictions, projected likely health impacts associated with climate change, and assessed the effectiveness and suitability of interventions for each of the prioritized health impacts or risk factors, health departments will be in a good position to pursue step 4, development and implementation of a climate change adaptation plan. These plans identify changes to health system functions and programs that are needed to prevent or reduce the anticipated impacts of climate change in the jurisdiction and outline steps for implementing the identified interventions.

The BRACE framework holds that plans should be comprehensive, cutting across all the essential public health functions from surveillance to regulation to outreach and education [5]. As such, the plans must be developed via both a comprehensive inward looking assessment at the health department's activities and with an outward looking engagement of stakeholders and partners to identify priorities, opportunities, and gaps in climate-sensitive disease prevention and health promotion. The planning horizon should be at least several years long, and the scope should be intersectoral with a focus on public health and the health department's role.

Climate change adaptation plans for public health are also both internal and external communication documents. To clarify internal priorities and activities, the intervention plan should clearly outline the resources required to pursue these activities, how existing activities should be modified to account for shifting risks associated with climate change, and who should be responsible for implementation. If key responsibilities lie in partnerships with other agencies, these agencies should be included, and the nature of the working partnership should be outlined explicitly. For external partners, the climate action plan should provide a vision regarding health protection in the jurisdiction and serve as an educational tool regarding ways in which partners can contribute to the overall health protection strategy.

When complete, the plan should be widely disseminated both internally and externally to all stakeholders that may have a role in executing elements of the plan. It should also identify how stakeholders can integrate adaptations into their existing functions and highlight how interventions will be evaluated and make clear the health department's commitments to communicating evaluating findings and updates to stakeholders as the adaptation plan is implemented.

Step 5: Evaluating Impact and Improving Quality of Activities

- *Did the process used to assess relevant risks, develop interventions, and engage stakeholders result in the outcomes we anticipated?*
- *Did interventions have an impact on population health outcomes?*
- *What lessons were learned from this iteration of the process?*

The final step in the BRACE framework relates to evaluating the processes from a process, outcome, and impact perspective. From a process standpoint, this step is useful for determining whether the appropriate stakeholders were involved and whether the methods of engagement resulted in the desired participation and identified the desired inputs. From an outcome standpoint, step 5 should identify the various programmatic outcomes that resulted from the activity, i.e., stronger relationships with particular stakeholders, model-building skills, increased awareness of synergies across programs, and appreciation of needed shifts in surveillance activities. From an impact standpoint, the evaluation should attempt to determine whether the interventions identified and implemented had the desired impacts on population health.

Each of these different types of evaluation—process, outcome, and impact—uses different methods and different indicators are measured to assess progress or lack thereof. Again, a comprehensive discussion of evaluation methods is outside the scope of this chapter. Health departments will have more or less resources to devote to evaluation activities and may not be able to engage each type of evaluation equally. Regardless, health departments using BRACE should have the capacity to answer the following questions at the end of their evaluation efforts:

1. Has the health department developed a reasonable estimate of future climate change health impacts?
2. Have the BRACE process enabled prioritization of health impacts and interventions?
3. Did the process result in a health department climate change adaptation plan?
4. Is climate change being considered in public health planning and implementation activities?
5. Is public health being considered in climate change planning and implementation activities?
6. Are there specific population health impact indicators that are being tracked to evaluate the interventions identified and implemented as a part of the BRACE process?
7. What aspects of the process can be improved in the next iteration?
8. What are the three top institutional learning priorities in the next round?

While evaluation is located in step 5, this is largely for ease of discussion and communication. Evaluation is in fact a central concern from the beginning of the process and is fundamental to the process of learning so central to adaptive management. Public health has a long tradition of institutional learning in response to novel threats. If the field maintains its commitment to learning it will be able to overcome many of the potential constraints and barriers to climate change adaptation in public health [22].

CDC National Environmental Public Health Tracking Network

Analysis of surveillance data is an important component of learning in public health. The CDC Environmental Public Health Tracking Network (Tracking Network) was established in 2002 to facilitate such learning in environmental health. The tracking network is administered by the NCEH and integrates health, exposure, and hazard information from various national, state, and local sources into a dynamic web-based tool that can be used to track and report environmental hazards and health problems related to them. Interested parties can query to analyze health impacts associated with environmental exposures [31]. The network involves multidisciplinary collaborations to collect, integrate, analyze, and distribute information derived from environmental hazard monitoring, human exposure surveillance, and health effects surveillance.

Among other exposures, the network tracks relevant climate and health data to help scientists and decision makers understand the connections between environmental conditions (and changes) and health impacts [32]. At this point the network focuses its climate change indicators on extreme heat with the aim of evaluating the number of heat-related deaths at the national level. Information on heat vulnerabilities, heat mortality, and temperature distribution can be used to identify patterns in extreme weather and their health effects [33]. For example, the Network can track the effects of a heat wave by aggregating and reporting the number of health conditions and reported deaths from local health departments and hospitals. These data can be used by policymakers at all levels to identify high-risk populations and communities, understand trends in heat-related deaths, and inform adaptation strategies.

The information produced by the tracking network is used by national, state, and local public health agencies to make decisions to protect human health in a timely and accurate manner hazards. For health departments interested in assessing the health effects of heat in their jurisdictions and generating exposure–outcome response functions for various heat-related exposures, the Tracking portal is the leading available tool.

Climate-Ready States and Cities Initiative

The Climate and Health Program has developed the BRACE framework as an option for all state and local health departments, and its guidance is available for all interested public health partners. To provide intensive assistance with public health adaptation to climate change in several locales, the program has developed the CRSCI. The CRSCI aims to build resilience against climate effects in communities by strengthening the capabilities of state and local health departments to deal with the challenges associated with climate change. The CRSCI is working toward this goal by providing multiple cohorts of health departments with funding and technical support. In total, health departments in 16 states and two cities have been funded through the CRSCI as illustrated in Fig. 24.2.

Funding for the Climate-Ready States and Cities is divided into two categories: (1) Assessment and Planning to Develop Climate Change Programs and (2) Building Capacity to Implement Climate Change Programs and Adaptations. Recipients of the first funding stream prepare needs assessments, gap analyses, and strategic plans to address climate change impacts on health in the short and long term using the ten EPHS framework. The second funding stream supports local health departments to implement the BRACE framework. In 2009, the first round of funding for the CRSCI provided support to eight states and two cities. Arizona, Massachusetts, New York State, North Carolina, and San Francisco received funding to assess jurisdictional capabilities and weaknesses and to plan climate change programs. Michigan, Minnesota, New York City, Oregon, and Maine received funding to build capacity to implement climate change programs and adaptations. In 2012, eight additional states received this multi-year funding: California, Wisconsin, Illinois, Vermont, New Hampshire, Rhode Island, Maryland, and Florida. All 16 of these health departments will apply the BRACE framework with appropriate amendments for each state and city, in order to determine and plan for the regionally specific effects of climate change on human health and vulnerable populations.

The CRSCI is funded via a cooperative agreement mechanism, and the state and local health departments collaborate with CDC to collectively develop a knowledge

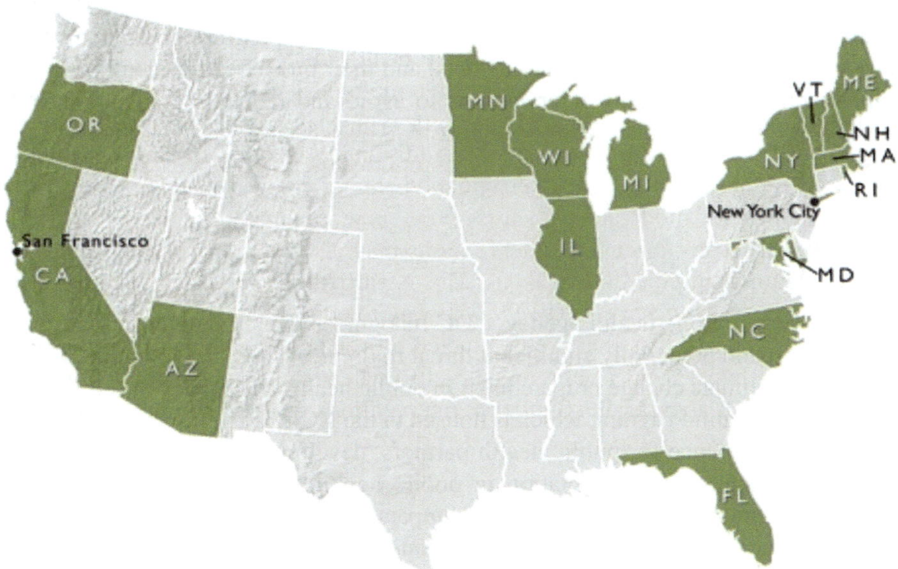

Fig. 24.2 Climate-Ready States and Cities Initiative (CRSCI)-funded locations

base regarding public health adaptation to climate change. To facilitate this process, the CDC Climate and Health Program and the grantees have regular sessions to report on progress and share findings. One such meeting is an Annual Science Symposium, where grantees and CDC scientists come together to discuss pressing public health issues related to climate change and health.

CDC Science Symposium on Climate and Health

The Climate and Health Program hosted the first Science Symposium on Climate and Health in 2011. The Symposium brought together scientists from CDC working on topics related to the health impacts of climate change. In 2012, the Symposium was expanded and co-hosted with the NOAA with participation from Health Canada and the Public Health Agency of Canada. The two-day symposium also included scientific presentations by academic institutions and state health departments.

The purpose of these symposia is to facilitate information exchange on the state of science related to climate change and to identify data, tools, and partnerships that support improved climate-related public health decision making. Presentations address the current and anticipated impact of climate, weather, and water patterns; impacts of climate patterns on marine, animal, human, and ecosystem health and safety; and climatic influence on ecological and epidemiologic factors that influence disease incidence and distribution. As CRSCI grantees move through the BRACE framework, it is expected that they will present on their progress, the models they develop to project relevant climate impacts, and their processes for identifying and implementing public health interventions to avoid and reduce the adverse health impacts of climate change.

Conclusion

As the nation's public health agency, CDC recognizes that climate change poses a multifaceted and potentially significant threat to domestic and global public health. To facilitate climate change preparedness in public health, the agency developed the Climate and Health Program, which is housed in the NCEH. The program's mission is to translate science for public health partners, develop decision support tools to facilitate climate change adaptation in public health, and to serve as a credible leader in planning for the human health impacts of a changing climate. Since its formation, the program has worked to articulate a public health approach to climate change and integrate science from public health and other sectors to facilitate public health adaptation efforts. The program has developed an adaptive management framework for public health, the BRACE framework, and is working cooperatively with several state and local health departments to pursue an evidence-based approach to climate change adaptation. As public health's expertise and experience grows, the Climate and Health Program will work to continue disseminating relevant information for the increasing number of public health practitioners focused on reducing the adverse health effects of climate change.

References

1. Centers for Disease Control and Prevention. Vision, mission, core values, and pledge. Atlanta, GA. 2011. http://www.cdc.gov/about/organization/mission.htm. Accessed 1 Feb 2012.
2. Longstreth J. Anticipated public health consequences of global climate change. Environ Health Perspect. 1991;96:139–44.
3. Patz JA, Kovats RS, Patz JA, Kovats RS. Hotspots in climate change and human health. BMJ. 2002;325(7372):1094–8.
4. Haines A, Patz JA. Health effects of climate change. JAMA. 2004;291(1):99–103.
5. Frumkin H, Hess JJ, Luber G, Malilay J, McGeehin M. Climate change: the public health response. Am J Public Health. 2008;98(3):435–45.
6. Hess J, McDowell J, Luber G. Integrating climate change adaptation into public health practice: using adaptive management to increase adaptive capacity and build resilience. Environ Health Perspect. 2012;120(2):171–9.
7. Holling C. Adaptive environmental assessment and management. New York: Wiley; 1978.
8. National Research Council. Adaptive management for water resources project planning. Washington, DC: National Academy Press; 2004.

9. Ebi K. Adaptive management to the health risks of climate change. In: Ford J, Berrang-Ford L, editors. Climate change adaptation in developed nations. New York: Springer; 2011.
10. Ebi K. Climate change and health risks: assessing and responding to them through 'adaptive management'. Health Aff. 2011;30:924–30.
11. Brubaker M, Berner J, Chavan R, Warren J. Climate change and health effects in Northwest Alaska. Glob Health Action. 2011;4.
12. Kahn ME. Urban growth and climate change. Ann Rev Resour Econom. 2009;1(1):333–50.
13. Maibach EW, Chadwick A, McBride D, Chuk M, Ebi KL, Balbus J. Climate change and local public health in the United States: preparedness, programs and perceptions of local public health department directors. PLoS One. 2008;3(7):e2838.
14. Balbus J, Ebi K, Finzer L, Malina C, Chadwick A, McBride D, Chuk M, Maibach E. Are we ready? Preparing for the public health challenges of climate change. Washington, DC: Environmental Defense Fund, National Association of County and City Health Officials, Center of Excellence in Climate Change Communication Research at George Mason University; 2008.
15. Marinucci G, Luber G. Bracing for impact: preparing a comprehensive approach to tackling climate change for public health agencies. American Public Health Association Annual Conference 2011; 11 Feb 2011; Washington, DC; 2011.
16. Ebi K, Kovats R, Menne B. An approach for assessing human health vulnerability and public health interventions to adapt to climate change. Environ Health Perspect. 2006;114(12):1930–4.
17. Campbell-Lendrum D, Woodruff R. Comparative risk assessment of the burden of disease from climate change. Environ Health Perspect. 2006;114(12):1935–41.
18. Peng R, Bobb J, Tebaldi C, McDaniel L, Bell M, Dominici F. Toward a quantitative estimate of future heat wave mortality under global climate change. Environ Health Perspect. 2011;119(5):701–6.
19. Knowlton K, Lynn B, Goldberg RA, Rosenzweig C, Hogrefe C, Rosenthal JK, Kinney PL. Projecting heat-related mortality impacts under a changing climate in the New York City region. Am J Public Health. 2007;97(11):2028–34.
20. Sheffield PE, Knowlton K, Carr JL, Kinney PL. Modeling of regional climate change effects on ground-level ozone and childhood asthma. Am J Prev Med. 2011;41(3):251–7.
21. Huang C, Barnett AG, Wang X, Vaneckova P, FitzGerald G, Tong S. Projecting future heat related mortality under climate change scenarios: a systematic review. Environ Health Perspect. 2011;119(12):1681–90.
22. Huang C, Vaneckova P, Wang X, FitzGerald G, Guo Y, Tong S. Constraints and barriers to public health adaptation to climate change. Am J Prev Med. 2011;40(2):183–90.
23. Sheridan SC, Allen MJ, Lee CC, Kalkstein LS. Future heat vulnerability in California, part II: projecting future heat-related mortality. Clim Chang. 2012:1–16.
24. Patz J, Campbell-Lendrum D, Gibbs H, Woodruff R. Health impact assessment of global climate change: expanding on comparative risk assessment approaches for policy making. Annu Rev Public Health. 2008;29(1):27–39.
25. Hosking J, Campbell-Lendrum D. How well does climate change and human health research match the demands of policymakers? A scoping review. Environ Health Perspect. 2012;120(8):1076–82.
26. Eriksson C. Learning and knowledge-production for public health: a review of approaches to evidence-based public health. Scand J Public Health. 2000;28(4):298–308.
27. Brownson RC, Fielding JE, Maylahn CM. Evidence-based public health: a fundamental concept for public health practice. Annu Rev Public Health. 2009;30:175–201.
28. Bassil KL, Cole DC. Effectiveness of public health interventions in reducing morbidity and mortality during heat episodes: a structured review. Int J Environ Res Public Health. 2010;7(3):991–1001.
29. Moher D, Liberati A, Tetzlaff J, Altman DG. Preferred reporting items for systematic reviews and meta-analyses: the PRISMA statement. PLoS Med. 2009;6(7):e1000097.

30. Petticrew M, Chalabi Z, Jones DR. To RCT or not to RCT: deciding when 'more evidence is needed' for public health policy and practice. J Epidemiol Community Health. 2012;66(5):391–6.
31. Prevention CfDCa. National environmental public health tracking network. 2012. http://ephtracking.cdc.gov/showHome.action. Accessed 5 Jan 2012.
32. Prevention CfDCa. Tracking climate change. 2012. http://ephtracking.cdc.gov/showClimateChangeTracking.action. Accessed 5 Jan 2012.
33. Prevention CfDCa. Climate change indicators available on the environmental public health tracking network. 2012. http://ephtracking.cdc.gov/showClimateChangeIndicators.action. Accessed 6 Jan 2012.

Chapter 25
Rules, Rulings, and Repeals: The Shifting State of Climate Policy at U.S. EPA

Jack Lienke

Climate policy at the U.S. Environmental Protection Agency ("EPA") is made primarily through rulemaking under the Clean Air Act. Those rules almost invariably prompt legal challenges from industry groups and anti-regulatory state attorneys general. The most significant challenges tend to work their way up to the Supreme Court, which then either deems the rule at issue a permissible exercise of EPA's authority and lets it remain on the books or remands it to the agency for revision.

The process can be likened to a game of tennis between EPA and the court. The agency serves a policy in the form of a finalized rule and the court either lets it pass (point for EPA) or judges it unlawful in some respect and returns it. But the "ball" changes with every serve—that is, a different rule is under consideration in each case. And the players change, too. Nominally, they are always "EPA" and "the Supreme Court." But the agency's policy priorities are altered—sometimes dramatically—whenever a new president is elected. And the ideological balance of the Supreme Court shifts anytime a new justice is appointed. Thus, the EPA that issues a rule may be very different from the EPA that must decide how to respond to an adverse decision from the Supreme Court regarding that rule. And the court in existence when EPA is crafting a policy may be very different from the court that ultimately sits in judgment of that policy.

The remainder of the chapter explores how this dynamic has shaped the past two decades of climate regulation at EPA. The goal is to provide an overview of each administration's highest-profile climate actions, not a comprehensive accounting of every greenhouse gas rule issue in the covered years.

J. Lienke (✉)
Institute for Policy Integrity, New York University School of Law, New York, NY, USA
e-mail: jack.lienke@nyu.edu

K. E. Pinkerton, W. N. Rom (eds.), *Climate Change and Global Public Health*, Respiratory Medicine, https://doi.org/10.1007/978-3-030-54746-2_25

Serve: Climate Policy Under the Bush EPA

Under the administration of President George W. Bush, EPA's preferred climate policy was no climate policy. Just over a year before the beginning of Bush's first term, a large coalition of environmental organizations filed a petition with the agency, arguing that it was obligated to regulate greenhouse gas emissions from new motor vehicles under Section 202 of the Clean Air Act [1]. Section 202 directs the EPA Administrator to prescribe standards limiting the emissions of "any air pollutant" emitted by motor vehicles "which in his judgment cause, or contribute to, air pollution which may reasonably be anticipated to endanger public health or welfare" (42 U.S.C. §7521(a)(1)). The petitioners noted that EPA's former general counsel Jonathan Z. Cannon had already determined in an April 1998 memorandum to the Administrator that a greenhouse gas qualified as an "air pollutant" under the Clean Air Act, which defined that term to include "any physical, chemical, biological, radioactive … substance or matter which is emitted into or otherwise enters the ambient air" (42 U.S.C. § 7602(g)). They also pointed to data showing that motor vehicles emit a variety of greenhouse gases, including more than 20 percent of the United States' carbon dioxide emissions. Finally, the petitioners cited a wealth of evidence—from sources like the Intergovernmental Panel on Climate Change (IPCC), the Centers for Disease Control and Prevention, and even EPA itself—that greenhouse gas-driven climate change endangers public health and welfare.

In September 2003, almost 4 years after the petition was filed, EPA formally denied it (68 FR 52922). The agency offered two grounds for its decision. EPA claimed, first, that it lacked legal authority to regulate greenhouse gases from motor vehicles because, contrary to Cannon's earlier finding, such gases were not air pollutants within the meaning of the Clean Air Act. Next, EPA argued that, even if it *did* have legal authority to grant the petition, it would decline to do so because regulating motor vehicles' greenhouse gas emissions under the Clean Air Act was unwise as a policy matter. On this point, the agency cited, among other things, lingering uncertainty "about the causes, extent and significance of climate change and the potential options for addressing it," and the possibility that making unilateral reductions in domestic greenhouse gas emissions would undercut the United States' negotiating position in international climate talks.

Return: The Supreme Court's Decision in *Massachusetts v. EPA*

The Supreme Court was unimpressed with EPA's reasoning. In *Massachusetts v. EPA*, 549 U.S. 497 (2007), a five-justice majority agreed with the petitioners (and Cannon, EPA's former general counsel) that greenhouse gases "fit well within the Clean Air Act's capacious definition of 'air pollutant.'" Section 202, in turn, obligated

EPA to regulate air pollutants emitted by motor vehicles whenever those pollutants posed a danger to public health or welfare. Thus, EPA could decline to regulate motor vehicles' emissions of greenhouse gases only if it determined that the gases "[did] not contribute to climate change" or provided "some reasonable explanation as to why it [could not] or [would] not exercise its discretion to determine whether they do."

EPA's policy concerns regarding the effect of domestic regulation on international climate negotiations, the court noted, had "nothing to do with whether greenhouse gas emissions contribute to climate change" and thus did "not amount to a reasoned justification for declining to form a scientific judgment." Nor could the agency "avoid its statutory obligation by noting the uncertainty surrounding various features of climate change and concluding that it would therefore be better not to regulate at this time." For uncertainty to justify continued inaction on motor vehicles' greenhouse gas emissions, it would need to be "so profound that it preclude[d] EPA from making a reasoned judgment" as to whether greenhouse gases contributed to climate change.

The court thus reversed EPA's denial of the petition and remanded the matter to the agency for further consideration.

Serve: Climate Policy Under the Obama EPA, Part 1

The decision in *Massachusetts v. EPA* did not spark meaningful action on climate change from the Bush EPA, but it did pave the way for a flurry of policymaking early in the Obama administration's first term.

Endangerment and Contribution Findings

EPA Administrator Lisa Jackson's first significant climate action was not a rule but a pair of extremely consequential findings, proposed in April 2009 and finalized the following December (74 FR 66496). In what is commonly known as the "Endangerment Finding," the Administrator concluded, based on a review of the available scientific evidence—including assessments by the U.S. Global Climate Research Program, the IPCC, and the National Research Council—that "greenhouse gases in the atmosphere may reasonably be anticipated both to *endanger* public health and to *endanger* public welfare" (emphases added). With respect to public health, Jackson noted that elevated concentrations of greenhouse gases posed "risks associated with changes in air quality, increases in temperatures, changes in extreme weather events, increases in food- and water-borne pathogens, and changes in aeroallergens" that would result from climate change. As for public welfare, the administrator cited climate-related "risks to food production and agriculture, forestry, water resources, sea level rise and coastal areas, energy, infrastructure, and settlements, and ecosystems and wildlife."

Next, in the "Contribution Finding," Administrator Jackson determined that "emissions of well-mixed greenhouse gases from the transportation sources covered under CAA section 202(a) *contribute* to the total greenhouse gas air pollution, and thus to the climate change problem, which is reasonably anticipated to endanger public health and welfare" (emphasis added). In support of this conclusion, Jackson noted that U.S. transportation sources were responsible for approximately 4 percent of global greenhouse gas emissions and over 23 percent of U.S. greenhouse gas emissions. To better illustrate the significance of this contribution, Jackson explained that emissions from these transportation sources were "larger in magnitude than the total well-mixed greenhouse gas emissions from every other individual nation with the exception of China, Russia, and India, and are the second largest emitter within the United States behind the electricity generating sector."

Mobile Source Regulation: Vehicle Emission Standards and California Waiver

Making the Endangerment and Contribution Findings left EPA obligated to establish limits on greenhouse gases from motor vehicles. In May 2009, the agency committed to developing those limits as part of a joint rulemaking process with the National Highway Traffic Safety Administration (NHTSA), in order to ensure that EPA's emission standards were "harmonized and consistent" with NHTSA's Corporate Average Fuel Economy (CAFE) standards (74 FR 24007). This interagency coordination addressed one of the policy concerns the Bush administration had raised in its petition denial: that setting greenhouse gas standards would necessarily interfere with NHTSA's CAFE program, because improving fuel economy—i.e., wringing more mileage out of each gallon of gas burned—was the most effective means of reducing vehicles' greenhouse gas emissions.

Shortly after announcing the forthcoming federal vehicle standards, EPA granted California leave to set its *own* greenhouse gas standards for vehicles—reversing a 2008 decision from the Bush administration (74 FR 3244). The Clean Air Act preempts states from setting vehicle emission standards, but California can request a waiver of this preemption if it determines that its standards will be at least as protective of public health and welfare as federal standards. EPA, in turn, is required to grant the waiver unless it concludes (1) that California's determination that its standards are at least as protective as federal standards is arbitrary and capricious, (2) that California does not "need" its standards "to meet compelling and extraordinary conditions," or (3) that California's standards are infeasible for automakers to achieve. No other state is eligible to request a waiver under the Clean Air Act, but other states do have the option of following California's standards instead of federal standards [2].

California has maintained its own standards for soot- and smog-forming vehicle emissions since the 1960s and, in 2005, requested a waiver to limit greenhouse gases as well. The Bush administration denied the request on the grounds that

California did not need greenhouse gas standards to meet compelling and extraordinary conditions (73 FR 12156). The EPA Administrator at the time, Stephen Johnson, argued that the Clean Air Act's waiver provision was intended to allow California flexibility to "address pollution problems that are local or regional," not "global climate change problems." Johnson also disputed "that the effects of climate change in California are compelling and extraordinary compared to the effects in the rest of the country."

The Obama EPA disagreed. In reversing Johnson's decision, Administrator Lisa Jackson rejected his determination that Congress intended to allow California to receive waivers only for purposes of addressing *local* pollution problems. Instead, she found that Congress "intentionally provided California the broadest possible discretion in adopting the kind of standards in its motor vehicle program that California determines are appropriate to address air pollution problems and protect the health and welfare of its citizens." Jackson also rejected the premise that California needed to demonstrate that its vulnerability to climate change was compelling and extraordinary compared to the rest of the country. As Johnson himself had acknowledged in the 2008 denial, EPA's traditional approach to waiver applications was not to ask "whether the specific standards at issue are needed to meet compelling and extraordinary conditions related to that air pollutant," but to review California's need for its motor vehicles program as a whole. In other words, so long as California needed its own standards to address compelling and extraordinary conditions related to *one* type of vehicle pollution—and no one denied that the state had a uniquely terrible smog problem—it could regulate other pollutants as well. Jackson found no statutory basis for breaking from this traditional approach in the context of greenhouse gases.

Jackson further concluded that, even if California *were* required to show that the impacts of climate change within its borders were different from those in the rest of the nation, it would still be entitled to a waiver. California, Jackson noted, had "presented evidence of a wide variety of vulnerabilities, impacts and potential impacts within California while the opponents have not demonstrated that any other state, group of states, or area within the United States would face a similar or wider range of vulnerabilities and risks." Rising temperatures would, for example, exacerbate California's already severe air quality problems, imperil its agricultural sector (the United States' largest), and reduce its already limited water resources.

For all these reasons, Jackson granted California's waiver request, allowing its greenhouse gas standards to take effect for the 2009 model year and increase in stringency through 2016. The standards were projected to reduce total greenhouse emissions from California's light-duty vehicle fleet by 18 percent in 2020 and 27 percent in 2030.

A little less than a year after EPA granted the California waiver, EPA and NHTSA finalized federal greenhouse gas and CAFE standards, which applied to light-duty vehicles in model years 2012 through 2016 (75 FR 25324). EPA's emission standards varied by vehicle size, but were, on average, projected to reduce vehicle carbon dioxide emissions from 263 grams per mile in 2012 to 225 grams per mile in

2016—equivalent to increasing average fuel economy from 30.1 miles per gallon to 35.5 miles per gallon. Relative to a business-as-usual scenario, this represented a 21 percent reduction in total U.S. light-duty vehicle emissions by 2030.

Stationary Source Regulation: Triggering and Tailoring Rules

Under a longstanding EPA interpretation of the Clean Air Act, the agency's decision to regulate greenhouse gas emissions from *mobile* sources triggered an obligation to regulate such emissions from *stationary* sources as well. Certain statutory permitting requirements applied to "major" new "sources" and "emitting facilities" (or majorly modified existing sources and emitting facilities) of "any air pollutant." EPA had, since the 1970s, read "any air pollutant" in this context to mean any air pollutant that was already regulated under some other provision of the Clean Air Act—including Section 202, the provision under which EPA had promulgated emission standards for motor vehicles.

In April 2010, EPA finalized a "Triggering Rule" in which it announced that "major" stationary sources of greenhouse gases would become subject to permitting requirements as soon as EPA's greenhouse gas emission standards for motor vehicles took effect in January 2011 (75 FR 17004). The most significant of these permitting provisions required affected stationary sources to install the "best available control technology" for all regulated pollutants.

The problem for the agency, however, was that the Clean Air Act classified sources and emitting facilities as "major" for purposes of the permitting programs if they emitted more than 100 or 250 tons per year, respectively, of a regulated pollutant. For most of the pollutants regulated by EPA, only large industrial sources like power plants, petroleum refineries, and steel plants emitted in excess of these statutory thresholds. But for greenhouse gases, millions of small businesses—like printers, furniture makers, and dry cleaners—would exceed the limits. Thus, to avoid "imposing undue costs on small sources, overwhelming the resources of permitting authorities, and severely impairing the functioning of the [permitting] programs," EPA released a "Tailoring Rule" in May 2010 (75 FR 31514).

Under the Tailoring Rule, the requirements of the permitting programs would, until July 1, 2011, apply to stationary sources' greenhouse gas emissions only if those sources both emitted more than 75,000 tons per year of carbon dioxide (or carbon dioxide equivalents) *and* emitted above the relevant statutory thresholds for some other regulated pollutant. These were known as "anyway" sources, because they would have been required to go through the permitting process even in the absence of greenhouse gas regulation. After July 2011, permitting requirements would expand to include a subset of *non*-anyway sources: new sources that emitted 100,000 tons per year or more of carbon dioxide (or carbon dioxide equivalents) and existing sources undertaking modifications that would increase their emissions by at least 75,000 tons per year of carbon dioxide (or carbon dioxide equivalent). Though these regulatory triggers were much higher than the statutory emissions thresholds, they would still capture sources representing almost 70% of the nation's total greenhouse gas emissions from stationary sources.

Return: The Supreme Court's Decision in *Utility Air Regulatory Group v. EPA*

The Obama EPA's first-term actions emerged from judicial review largely—but not entirely—unscathed. The Endangerment and Contribution Findings, the vehicle standards, and the Triggering and Tailoring Rules were all challenged in—and ultimately upheld by—the United States Court of Appeals for the District of Columbia Circuit (D.C. Circuit). States and industry groups then sought review in the Supreme Court, which opted to weigh in only on the Triggering and Tailoring Rules.

In the ensuing decision, *Utility Air Regulatory Group v. EPA*, 573 U.S. 302 (2014), a five-justice majority found that, contrary to EPA's conclusion in the Triggering Rule, stationary sources could not be obligated to undergo "major" source permitting solely by virtue of their greenhouse gas emissions. The justices declared that the Tailoring Rule's "rewriting of the statutory thresholds was impermissible," because "[a]n agency has no power to 'tailor' legislation to bureaucratic policy goals by rewriting unambiguous statutory terms." But applying statutory thresholds as written and subjecting millions of new sources to permitting requirements was also off the table, because the permitting programs were "designed to apply to, and cannot rationally be extended beyond, a relative handful of large sources capable of shouldering heavy substantive and procedural burdens." The only reasonable option, in the court's view, was to find that greenhouse gases *never* triggered permitting requirements, even though (1) the permitting programs applied to all "major" sources and emitting facilities, (2) the Clean Air Act defined "major" sources and emitting facilities as those that emitted "any air pollutant" above the 100 and 250 tons thresholds, and (3) the court in *Massachusetts* had found that greenhouse gases fit within the Act's definition of "air pollutant."

Writing for the majority, Justice Scalia explained that *Massachusetts*—a decision from which he had quite pointedly dissented—"did not hold that EPA must always regulate greenhouse gases as an 'air pollutant' everywhere that term appears in the statute, but only that EPA must 'ground its reasons for action *or inaction* in the statute.'" See the original text here: https://supreme.justia.com/cases/federal/us/573/302/. In the case of Section 202 vehicle standards, regulating greenhouse gases as air pollutants was appropriate "because nothing in the [Clean Air] Act suggested that regulating greenhouse gases under that [section] would conflict with the statutory design." But in the case of the stationary-source permitting programs, treating greenhouse gases as air pollutants that could trigger permitting requirements "would be inconsistent with—in fact, would overthrow—the Act's structure and design," because application of the statutory thresholds would affect millions of small sources that "Congress did not expect would need to undergo permitting."

But while the first portion of Scalia's opinion concluded that a stationary source's greenhouse gas emissions could not *trigger* permitting requirements, the second portion—written on behalf of a differently composed, seven-justice majority—found that EPA could nevertheless require "anyway" sources to install the best available control technology for their greenhouse gas emissions. In other words, if a source triggered permitting requirements due to its emissions of some other pollutant, EPA could require it to limit its emissions of greenhouse gases in addition to its emissions of the triggering pollutant. Scalia's opinion offered two justifications for this conclusion. First, although the Clean Air Act was somewhat ambiguous on the question of which air pollutants *triggered* the permitting process, it left no room for interpretation on the question of which pollutants were subject to control requirements once the permitting process had begun. The statute provided that the "best available control technology" was required "for each pollutant subject to regulation" under the Act. Following the promulgation of the vehicle standards, greenhouse gases undoubtedly met that bill.

Second, in contrast to treating greenhouse gas emissions as triggering pollutants, subjecting "anyway" sources to "best available control technology" requirements for their greenhouse gas emissions was not "disastrously unworkable" and would not result in a "a dramatic expansion of agency authority." Rather than "extending EPA jurisdiction over millions of previously unregulated entities," the regulation of greenhouse gas emissions from "anyway" sources would only "moderately increase[e] the demands EPA… can make of entities already subject to its regulation."

The court's decision to bless EPA regulation of "anyway" sources meant that *Utility Air Regulatory Group* had little effect on near-term greenhouse gas emissions. As Scalia's opinion acknowledged, though they are small in number, "'anyway' sources account for roughly 83% of American stationary-source greenhouse-gas emissions, compared to just 3% for the additional, non-'anyway' sources EPA sought to regulate" in the Tailoring Rule.

However limited its practical impact, *Utility Air Regulatory Group* was a significant decision because it made clear that controversy regarding the scope of EPA's authority to regulate greenhouse gas emissions was far from fully resolved. Whereas *Massachusetts* had rejected the idea that EPA could *never* regulate such emissions under the Clean Air Act, *Utility Air Regulatory Group* rejected the idea that EPA could *always* regulate such emissions under the Act. This left the agency to puzzle out which of the Clean Air Act's many provisions could and could not be used to limit greenhouse gas emissions—and in what ways those provisions could and could not be deployed. And EPA could be certain that no matter what path it chose, it was likely to be making more trips to the Supreme Court.

Serve: Climate Policy Under the Obama EPA, Round 2

In its second round of climate rulemaking under the Obama administration—begun even before the Supreme Court issued its decision in *Utility Air Regulatory Group*—EPA sought further emissions reductions from both mobile and stationary sources.

Mobile Source Regulation: More Vehicle Standards and Another California Waiver

In October 2012, the agency completed another joint rulemaking with NHTSA, which set greenhouse gas emission standards and CAFE standards for light- and medium-duty motor vehicles in model years 2017 through 2025 (77 FR 62624). As with the first set, the emission standards would grow increasingly stringent over time, ultimately requiring manufacturers to achieve an average fleet-wide emissions rate of 163 g of carbon dioxide per mile in 2025—equivalent to a fuel economy standard of 54.5 miles per gallon. EPA projected that, over the lifetimes of the vehicles sold in model years 2017 to 2025, the standards would avoid the use of approximately 4 billion barrels of oil and reduce greenhouse gas emissions by 2 billion metric tons [3].

In addition to issuing new federal standards, EPA, in January 2013, granted California a new waiver to set its own greenhouse gas standards as a component of the state's Advanced Clean Cars program for cars in model years 2015 through 2025 (78 FR 2112). The Advanced Clean Cars regulations were a "coordinated package" that included emission standards for smog-causing pollutants, emission standards for greenhouse gases, and a Zero Emission Vehicle ("ZEV") program designed to boost the market share of plug-in hybrid, fully electric, and fuel-cell vehicles to 15 percent by 2025. In approving the 2013 waiver, EPA noted that the greenhouse gas standards included in this package were "almost identical in stringency and structure" to EPA's new federal standards.

Stationary Source Regulation: Carbon Dioxide Standards for New and Existing Power Plants and Methane Standards for the Oil and Gas Sector

In August 2015, now under the leadership of Administrator Gina McCarthy, EPA finalized a pair of major rules aimed at power plants. At the time, the power sector was the nation's largest source of greenhouse gas emissions. (It is now second to the transportation sector [4].)

First, EPA promulgated a set of "new source performance standards" for carbon dioxide emissions from fossil fuel-fired electric generating units (80 FR 64510). Unlike the "best available control technology" requirement discussed above—which required an individualized determination for each individual source—new source performance standards applied nationwide to an entire category of sources. Under Section 111 of the Clean Air Act, such standards had to reflect the degree of emissions limitation achievable using the "best system of emission reduction" that the Administrator found to be "adequately demonstrated" (42 U.S.C. § 7411(a) (1)).

For coal-fired power plants, EPA determined that the best system of emission reduction was partial carbon capture and storage technology, in which some of the carbon dioxide generated by a plant is trapped before leaving the smokestack and sequestered underground, where it won't contribute to climate change. For natural gas-fired plants, EPA determined that the best system was the use of highly efficient combined-cycle generation technology, in which the waste heat that results when natural gas is burned to spin a combustion turbine is used to create steam that spins a second turbine, generating additional power.

Under expected market conditions, these new source performance standards were projected to have no effect on carbon dioxide emissions relative to a business-as-usual scenario. This was because building new coal plants was not expected to be economically desirable for power-sector firms even in the absence of the standards, due to the sustained low prices for natural gas that had resulted from the fracking-driven "shale revolution." And combined-cycle generators were already the technology of choice for firms building gas plant. In other words, the standards would have no effect on whether or how firms constructed coal plants, because no one was planning to build any coal plants anyway. And they would have no effect on whether or how firms constructed gas plants, because everyone already constructed gas plants in ways that complied with the standards.

While the new source performance standards had no effect on emissions under *expected* fuel prices, however, EPA found that if gas prices reached unlikely highs, the standards might cause firms that would have otherwise chosen to build a conventional coal plant (that is, a coal plant without carbon-capture technology) to instead build a combined-cycle gas plant. This effect would carry costs relative to a business-as-usual scenario (because running the gas plant would be more expensive than operating a conventional coal plant), but it would also yield substantial reductions in carbon dioxide and other emissions relative the business-as-usual scenario (because the gas plant would emit far less pollution than a conventional coal plant). EPA concluded that the climate and health benefits associated with these emission reductions would outweigh the associated costs. The new source performance standards could thus be viewed as a regulatory insurance policy against unexpectedly high gas prices [5].

More importantly, under the Clean Air Act, setting standards for *new* power plants was a necessary legal predicate to setting emission guidelines for *existing* power plants. EPA issued such existing-source guidelines on the same day as the new-source standards, in a rulemaking known as the Clean Power Plan (80 FR 64662).

In the Clean Power Plan, EPA defined the best system of emission reduction for carbon dioxide emissions from fossil fuel-fired electric generating units as a set of three "building blocks:

1. Improving heat rate at coal-fired plants;
2. Substituting generation from natural gas-fired plants for generation from coal-fired plants; and
3. Substituting generation from renewables (like wind and solar) for generation from both coal- *and* gas-fired plants.

Put another way, EPA determined that the best way to cut carbon dioxide emissions in the power sector was to reduce emissions from the highest-emitting sources, coal plants, both by making them more efficient and by reducing their use in favor of lower-emitting gas plants and zero-emitting renewables. The emission guidelines that EPA set based on these building blocks were projected to reduce the nation's power-sector emissions to 32 percent below 2005 levels by 2030.

In addition to limiting carbon dioxide emissions from the power sector, the Obama EPA also established new source performance standards for methane emissions from the oil and gas industry (81 FR 35824). Methane is the primary ingredient in natural gas, which as indicated above, releases significantly less carbon dioxide than coal when burned to generate electricity (about half as much per unit of energy output). But when methane escapes directly into the atmosphere, it acts as an extremely potent greenhouse gas in its own right, trapping over eighty times more heat than an equivalent amount of carbon dioxide over a 20-year period. Oil and gas operations are the United States' largest industrial emitters of methane, and EPA's performance standards sought to reduce the sector's emissions by, among other things, requiring frequent inspections of wells and related infrastructure to identify methane leaks and setting time limits for repairing those leaks.

Unlike with power plants, EPA did not pair its performance standards for *new* sources of methane with a set of emission guidelines for *existing* sources of methane in the oil and gas sector. The agency did, however, commit to issuing existing-source guidelines in the future and, as a preliminary step, issued a request for information from oil and gas companies seeking data relevant to the design of such guidelines (81 FR 35763).

Re…traction: Climate Policy Under the Trump EPA

There was no Supreme Court decision on the Obama administration's second-term climate policies. The stationary-source rules were all challenged in the D.C. Circuit, but, before any ruling was issued in those cases, Donald J. Trump was elected President of the United States. And after Trump's first EPA Administrator, Scott Pruitt, took the agency's reins in the spring of 2017, EPA asked the D.C. Circuit to hold the suits in abeyance while the agency voluntarily reconsidered the policies. Under Pruitt, EPA also opted to voluntarily reconsider its 2012 vehicle standards and the 2013 California waiver.

Ultimately, the Trump administration seems no more interested in crafting climate policy at EPA than the Bush administration was. But rather than claiming, as the Bush administration did, that it lacks authority to set *any* greenhouse limits, the Trump administration has chosen to set limits that are so weak as to be meaningless.

Mobile Sources: Freezing Vehicle Standards and Revoking the California Waiver

As discussed above, the vehicle standards EPA issued in 2012 applied to light- and medium-duty vehicles in model years 2017 through 2025. As part of that rulemaking, the agency pledged to complete a "midterm evaluation" by April 1, 2018, to ensure that the 2022 to 2025 standards remained appropriate under the Clean Air Act, taking into account factors such as changes in fuel prices, vehicle fleet mix, and technology costs. Following a lengthy technical assessment conducted jointly with NHTSA and the California Air Resources Board, EPA issued such a midterm evaluation in January 2017, concluding that the 2022 to 2025 standards were still "[f]easible at a [r]easonable [c]ost." [6]

But shortly after arriving at the agency, Trump's newly installed Administrator Pruitt announced that he was reconsidering that determination. And in the summer of 2018, Pruitt's successor, Administrator Andrew Wheeler, proposed the Safer Affordable Fuel-Efficient Vehicles Rule, which would freeze vehicle greenhouse gas standards at 2020 levels for model years 2021 through 2026 (83 FR 43986). To justify this change, Wheeler relied on economic modeling that prominent scholars have deemed "misleading" and "at odds with basic economic theory and empirical studies," which purported to show that more stringent standards would actually *harm* public health—in part by raising the cost of vehicles and inducing consumers to hold on to older, less-safe vehicles for longer periods of time [7].

Additionally, Wheeler proposed to revoke California's 2013 waiver to maintain its own greenhouse gas standards—a move that would also affect the District of Columbia and twelve other states that have opted to follow California's standards instead of federal standards. In support of this entirely unprecedented action—which is nowhere authorized in the text of the Clean Air Act—Wheeler revived the Bush-era argument that California does not need its own greenhouse gas standards to meet compelling and extraordinary conditions, because climate change is a global, rather than local or regional, problem. Wheeler further argued that "there is inadequate lead time to permit the development of technology necessary to meet" California's standards, even though the state standards are functionally identical to the federal standards that EPA deemed "feasible at a reasonable cost" in January 2017. Finally, Wheeler claimed that California's standards are preempted by the Energy Policy Conservation Act, which prohibits states from setting their own fuel economy standards. This last claim is directly at odds with two federal district court decisions finding that EPCA's preemption of state *fuel economy* standards does not prevent California from securing a Clean Air Act waiver to set its own *greenhouse gas emission* standards [2].

Stationary Sources: Weakening New Source Performance Standards and Replacing the Clean Power Plan

The Trump EPA has also taken steps to functionally—if not formally—eliminate greenhouse gas limits for the power sector. With respect to new source performance standards, EPA proposed, in December 2018, to rescind its conclusion that partial carbon and sequestration is the "best system of emission reduction" for new coal plants, claiming that such technology is not "adequately demonstrated" (80 FR 64510). Instead, EPA would deem use of the "most efficient demonstrated steam cycle" as the best system of emission reduction for coal plant, thereby justifying a much weaker standard.

Even though a new coal plant operating under the Trump standard would be permitted to emit carbon at a substantially higher rate than a plant operating under the Obama standard, the proposed change is not expected to have any near-term impact on emissions. As was the case in 2015, no firms are projected to find it economically worthwhile to build a coal plant in the foreseeable future, whatever the stringency of EPA's carbon dioxide standards [8]. But the new rule nevertheless reopens the *possibility* that new conventional coal plants could be constructed, if gas prices were to spike dramatically.

EPA has also finalized a replacement for the Clean Power Plan (84 FR 32520). Known as the Affordable Clean Energy rule, the new policy, issued in June 2019, would redefine the "best system of emission reduction" for existing power plants to include only heat rate improvements at coal-fired plants (essentially, building block 1 of the Clean Power Plan's best system).The agency concedes that this strategy will achieve less than a 1% reduction in carbon dioxide emissions relative to world in which it sets no emission limits at all [9]. Indeed, as the agency acknowledged in the Clean Power Plan, heat rate improvements, standing alone, actually have the potential to *increase* total emissions from coal plants, because, by making a plant more efficient, they render it more competitive. If the plant is dispatched more frequently as a result, its total emissions could increase, even if its per-kilowatt hour emissions *rate* has decreased.

The Trump EPA claims that is has no choice but to adopt this ineffective, potentially counterproductive, regulatory approach, because basing standards on the power sector's capacity for generation shifting exceeded the agency's authority under the Clean Air Act. The agency maintains that, under Section 111 of the Clean Air Act, a best system of emission reduction must be "limited to measures that can be applied to and at the level of the individual source," such as the installation of a new technology or the adoption of a new operational technique. But such limitation is nowhere to be found in the text of Section 111 (42 U.S.C. § 7411). Furthermore, EPA has adopted flexible, cross-source pollution-reduction strategies (like emissions trading and averaging programs) in several prior Clean Air Act rulemakings, even in the absence of express statutory authorization [10].

As for the oil and gas sector, EPA has proposed to weaken the Obama administration's new source performance standards for methane emissions by allowing less frequent inspections of regulated facilities and less prompt repairs (83 FR 52056). The agency's justification is that the monitoring frequencies required by the original standards were not cost effective, but it has conducted no new modeling to supporting this conclusion. Under the Trump administration, EPA has also withdrawn its request for information regarding methane emissions from *existing* oil and gas infrastructure, signaling that is has no intention of following through on the Obama administration's commitment to set standards for such existing sources (82 FR 12817).

Conclusion

By the time this chapter is published, challenges to all of the Trump administration's rollbacks of Obama-era climate policies will almost certainly be pending in the D.C. Circuit. It is hard to predict, though, whether the Supreme Court will ever get the opportunity to issue a decision on these rules. If the cases are not fully resolved by the end of 2020 (which is likely, given typical timelines for federal litigation), and if President Trump does not win reelection, a new, Democratic administration could ask for the challenges to be put in abeyance while it works on replacement rules—just as the Trump administration did with the Obama administration's stationary-source regulations. Finalizing such replacements could, in turn, take years—just as it has taken years for the Trump administration to develop its substitutes for Obama-era policies. And then those new rules would themselves be litigated, which could take years more. All of which is to say that it may be quite some time before EPA scores its next point in the high-stakes game of climate policy.

References

1. International Center for Technology Assessment [1999], Petition for rulemaking and collateral relief seeking the regulation of greenhouse gas emissions from new motor vehicles under [Section] 202 of the Clean Air Act.
2. Grab D, et al. No Turning Back: An Analysis of EPA's Authority to Withdraw California's Preemption Waiver Under Section 209 of the Clean Air Act. Institute for Policy Integrity; 2018 Oct.
3. U.S. EPA. Regulatory Announcement: EPA and NHTSA Finalize Historic National Program to Reduce Greenhouse Gases and Improve Fuel Economy for Cars and Trucks. U.S. Environmental Protection Agency; 2010 Apr. EPA-420-F-10-014.
4. U.S. EPA. Inventory of U.S. Greenhouse Gas Emissions and Sinks: 1990 – 2017. U.S. Environmental Protection Agency; 2019. EPA 430-R-19-001.
5. U.S. EPA. Regulatory Impact Analysis for the Final Standards of Performance for Greenhouse Gas Emissions from New, Modified, and Reconstructed Stationary Sources: Electric Utility Generating Units. US Environmental Protection Agency; 2015 Aug. EPA-452/R-15-005.

6. U.S. EPA, Final Determination on the Appropriateness of the Model Year 2022-2025 Light-Duty Vehicle Greenhouse Gas Emissions Standards under the Midterm Evaluation. U.S. Environmental Protection Agency; 2017 Jan. EPA-420-R-17-001.
7. Bento A, et al. Flawed analyses of U.S. auto fuel economy standards. Science. 2018;07: 1119–21.
8. U.S. EPA. Economic Impact Analysis for the Review of Standards of Performance for Greenhouse Gas Emissions from New, Modified, and Reconstructed Stationary Sources: Electric Utility Generating Units. U.S. Environmental Protection Agency; 2018 Dec. EPA 452/R-18-005.
9. U.S. EPA. Regulatory Impact Analysis for the Repeal of the Clean Power Plan, and the Emission Guidelines for Greenhouse Gas Emissions from Existing Electric Utility Generating Units. U.S. Environmental Protection Agency; 2019 June. EPA-452/R-19-003.
10. Revesz R, et al. Familiar Territory: A Survey of Legal Precedents for the Clean Power Plan. Environ Law Rep. 2016;46:10190–3.

Chapter 26
California's Integrated Approach to Air Quality and Climate Change

John R. Balmes

The enormity of the problem of climate change and the rapidity with which greenhouse gas (GHG) emissions are increasing requires that strategies to reduce emissions need to be highly effective and relatively easily implemented. Both short-term and long-term strategies are required to prevent a climate change tipping point. California has long been a leader in developing policies to prevent environmental degradation, especially in the area of air quality, so it should come as no surprise that the California Legislature passed and then Governor Schwarzenegger signed a landmark bill to mitigate climate change, Assembly Bill 32 (AB 32), the California Global Warming Solutions Act, in 2006 [1]. The California Air Resources Board (CARB), a state agency with authority to control air quality, was given the responsibility for implementing AB 32.

Climate change models predict that a "business as usual" (BAU) approach to reduction of atmospheric greenhouse gases (GHGs), i.e., no effort to control CO_2 emissions from combustion of fossil fuels, will result in over 2 °C increase in the annual average surface temperature relative to the 1986–2005 baseline by ~2034 [2]. A 2 °C increase in this metric is another tipping point that would mean irreversible planetary damage due to global warming [2]. The goal of the 2015 Paris climate change agreement was to organize a collective effort of the world's nations to prevent this 2 °C warming. Then California Governor Jerry Brown led the formation of an alliance of subnational governments to work to implement policies to prevent this tipping point from occurring, the Under2 Coalition [3].

With atmospheric warming comes increased air pollution. In particular, ozone increases linearly with increased temperature. The concept of a "climate gap" in ozone air quality control captures the decreased effectiveness of regulatory policies designed to reduce ozone precursors because of enhanced ozone generation with a hotter climate. This could translate into up to 30 more days per year of unhealthy

J. R. Balmes (✉)
University of California, San Francisco, CA, USA
e-mail: john.balmes@ucsf.edu

K. E. Pinkerton, W. N. Rom (eds.), *Climate Change and Global Public Health*, Respiratory Medicine, https://doi.org/10.1007/978-3-030-54746-2_26

ozone levels in the Los Angeles and San Joaquin Valley regions by 2050 [4]. Increased nitrogen dioxide (NO_2) and fine particulate matter (particles ≤2.5 μm in diameter, $PM_{2.5}$) will be emitted because of increased power generation needs for climate change adaptation, i.e., air conditioning.

It is important to keep in mind that the sources of GHGs and climate-forcing aerosols ("black carbon") are the same sources of air pollutants that harm health. In high-income jurisdictions like California, these sources are primarily found in the transportation and energy sectors. Adverse health effects from air pollution could be prevented with an integrated policy approach to reducing reliance on fossil fuels in these sectors.

Motor vehicle emissions are a major source of air pollutants that can impact health, both those that are emitted directly from tailpipes (e.g., fine and ultrafine particulate matter, black carbon, and nitrogen and sulfur oxides) and those that are generated in the atmosphere (e.g., ozone and secondary organic aerosol). Reduction of these emissions has been a major goal of the regulatory policy for over 50 years since California first mandated "smog" control technology on new automobiles sold in the state. Because California had adopted such controls before the Clean Air Act of 1970, the ability of the state to apply for a waiver to have stricter policies than those of the federal EPA was written into the law [5].

Since its inception, the CARB has continued to lead the way on the development of cleaner passenger vehicles powered by internal combustion engines with a series of technology-forcing regulations, including requirements for exhaust gas recirculation, oxidation catalysts, onboard computers, fuel injection, and oxygen sensors. As a result of these regulations, there has been a 99% reduction in passenger vehicle emissions in California since the first adoption of smog controls despite a major increase in vehicle miles traveled [6]. The CARB has also adopted aggressive regulations to reduce emissions from heavy-duty diesel trucks and buses. Combined with federal efforts to encourage development of cleaner diesel engines, these regulations have led to a major decrease in diesel exhaust particles, improving air quality and mitigating climate change. Since 1990, there has been almost a 70% reduction in diesel PM across the state [7].

In recent years, California has adopted a robust suite of climate change mitigation policies that have also been designed to achieve public health cobenefits by improving air quality. As noted above, AB 32 is a landmark legislation that gave the CARB the responsibility to develop a climate change mitigation "scoping plan" and the authority to promulgate regulatory policy to implement the plan. The primary goal of AB 32 is to reduce GHG emissions to the levels in 1990 by 2020 (a 30% reduction), and the long-range goal is a 90% reduction below the levels in 1990 by 2050. The main approaches are direct regulations, monetary and nonmonetary incentives, and a market-based mechanism to put a price on carbon emissions from multiple economic sectors. Before implementing any of these approaches, an inventory of GHG emissions was conducted. The sector in California that was the largest source of GHGs in 2016 was transportation (37%) with electric power generation close behind (34%). Industrial sources represented only 20% of GHG emissions. By

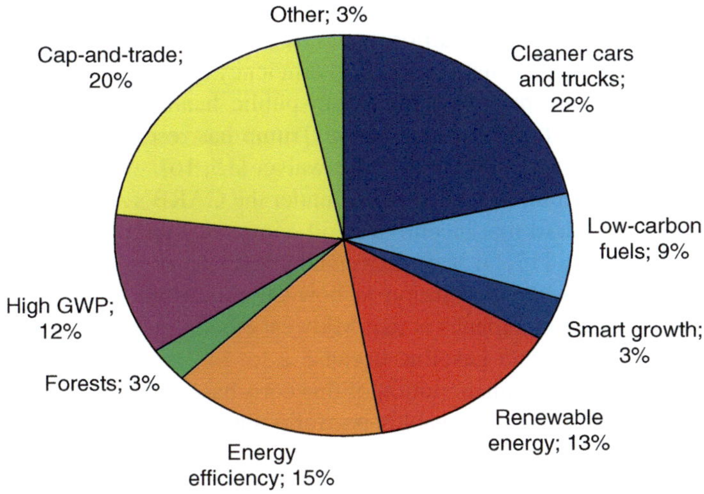

Fig. 26.1 Percentages of greenhouse gas emission reductions from components of the California's AB 32 scoping plan

2016, because of increased reliance on renewable sources, electric power had dropped to third place (19%) behind industrial sources [8].

The regulatory policies adopted under AB32 in the order of decreasing importance regarding impact on GHG emission reductions are as follows (see Fig. 26.1): advanced clean car standards (27%), renewable energy (19%), cap-and-trade (16%), low-carbon fuel standard (13%), energy efficiency (12%), and elimination of high global warming potential gases (7%) [9].

The Zero Emission Vehicle (ZEV) program was originally designed by the CARB to reduce emissions of air pollutants, but now has the additional goal of reducing GHG tailpipe emissions [10]. A total of 11 other US states have adopted the California Zero Emission Vehicle mandate. The ZEV program stimulated the development of hybrid, battery electric, and fuel cell cars that are now available for sale in California and the 11 other states. A clean vehicle rebate program has incentivized the purchase of these cars. To support recharging or hydrogen fueling of these cars, a clean fuels infrastructure building program is underway.

The CARB also promulgated a state legislatively mandated standard to reduce GHG tailpipe emissions from gasoline cars that eventually became the basis of the current US fuel efficiency (CAFE) standard [11]. Although initially aimed at increasing fuel efficiency of the US auto fleet to reduce dependence on foreign oil, the CAFE standards originally adopted in 1976 [12]also have a salutary effect on air pollution by reducing overall vehicle emissions as well as climate change by reducing carbon dioxide emissions. California requested a waiver from the US EPA to adopt stricter tailpipe emissions than those of the federal CAFE standard in 2006, but this waiver was denied by the administration of President George W. Bush [13]. However, after the election of President Obama and the recession in 2009, the US

EPA granted the requested waiver and the federal government entered into negotiations with automakers and the California Air Resources Board to adopt new CAFE standards that mandated escalating fleet fuel efficiency (54.5 mpg by 2025) [13, 14]. These standards would have benefited both public health and the environment. Unfortunately, the US EPA under President Trump has rescinded the 2025 CAFE target and is moving to rescind California's waiver [15, 16].

One of the more controversial programs under the CARB's AB32 mandate is the low-carbon fuel standard that incentivizes oil companies to produce more fuels of lower carbon intensity [17]. It is controversial because it uses a life-cycle analysis to assess carbon intensity in addition to how efficiently the fuel burns in motor vehicle engines. What this means is that Midwest ethanol has higher carbon intensity than regular California gasoline because a lot of carbon is generated in the production, refining, and transportation of this corn-based fuel.

A major success is the renewables portfolio standard that originally required California utilities to achieve 33% renewable energy generation (solar, wind, hydro, and/or geothermal) by 2020 [18]. This policy has been so successful that a new target of 100% renewable energy by 2045 has been legislatively mandated [19]. A Sustainable Freight Transport Initiative seeks to achieve both GHG emission reductions and health cobenefits through electrification of port and freight handling vehicles (e.g., cranes and forklifts) and by reductions in diesel emissions through state subsidies for new, cleaner diesel trucks and locomotives [20].

A legislative complement to AB32 is California Senate Bill 375 that mandates "Sustainable Communities Strategies" be developed by metropolitan planning organizations (MPOs) to achieve target reductions in GHG emissions set by the CARB [21]. The primary approach of the metropolitan planning organizations has been to reduce vehicle miles traveled (VMTs) by encouraging smart growth through urban infill, high-density housing along transportation corridors, investment in public transportation, and efforts to promote active commuting and walkability of neighborhoods. MPOs for all of the major population centers in the state have produced ground-breaking plans to reduce VMTs by reducing development that leads to suburban sprawl.

The most controversial of California's climate change mitigation policies is the Cap-and-Trade Program, which is expected to provide only about 16% of the state's GHG emission reductions in 2020 [22]. The Cap-and-Trade Program is a market-based mechanism that allows capped facilities to trade state-issued GHG emissions allowances, providing flexibility and reducing costs of compliance with mandated emission reductions. The "cap" limits total GHG emissions from all covered sources and declines over time to progressively reduce emissions. To prevent "leakage" of emissions from under the cap as a result of companies moving operations out of California because of the increased cost of business due to cap-and-trade, some free allowances are provided to companies judged to be placed at a competitive disadvantage, e.g., cement plants.

Critics of cap-and-trade say that it encourages heavily polluting facilities to buy allowances from clean facilities, rather than invest in emission reduction technologies, thereby continuing to impact the health of low-socioeconomic communities, where such dirty facilities are often located. For this reason, environmental justice advocates tend to support a carbon tax mechanism to put a price on carbon. Unlike a cap-and-trade program that places a cap on carbon emissions and lets the market determine price, a carbon tax leads to variable reductions in carbon emissions because it fixes the price of carbon. To address any unintended negative local consequences of the program, follow-up legislation requires that 25% of Cap-and-Trade revenue be spent on projects that benefit disadvantaged communities [23]. An innovative mapping tool, CalEnviroScreen, was developed by CalEPA to identify disadvantaged communities [24].

Because AB 32 only authorized a market mechanism to put a price on carbon through 2020, in 2017, the then Governor Brown pushed the California State Legislature to pass legislation to extend authorization of the Cap-and-Trade Program through 2030 [1]. In order to garner sufficient support for the bill (AB 398) [25], a supplementary bill was introduced and passed to address environmental justice concerns about the Cap-and-Trade Program (AB 617) [26]. This bill mandates the CARB to work with local air quality management districts and community groups to design and implement community-level air quality monitoring and emissions reduction programs, in other words, to directly address concerns about local "hot spots" of air pollution. Implementation of the AB 617-mandated programs is potentially transformative regarding air quality control policy because it moves the focus from regional background monitoring to measuring the impact of local levels.

Since 2013, ~$11 billion dollars has been generated in the Greenhouse Gas Reduction Fund from auction of CO_2 emission allowances under the Cap-and-Trade Program [22, 27]. These funds have been invested in the following major categories: sustainable communities and clean transport (including high-speed rail), energy efficiency and clean energy, and natural resources and waste diversion.

Although California is on target to achieve AB32's mandated goal of a 30% reduction in GHG emissions below the levels in 1990 by 2020, California will need to implement increasingly stringent emission control policies to meet the long-term goal of 90% reduction by 2050. In 2016 new legislation, Senate Bill 32 was passed that required statewide greenhouse gas emissions to be reduced to 40% below the level in 1990 by 2030 [28].

California's leadership on policies to advance air quality and climate change mitigation is increasingly important, given the efforts to slow or even reverse implementation of such policies at the US national level. California is a subnational jurisdiction with a population of nearly 40 million and the world's sixth largest economy. Despite the fact that California has the strictest air quality and GHG emission regulations in the United States, its economy has recently been growing faster than the nation's as a whole, and the clean energy sector is a major driver of this growth. Former Governor Jerry Brown has made it clear that California will continue to lead on climate change mitigation policy, including working together with national and subnational jurisdictions outside the United States. California alone cannot appreciably reduce global GHG emissions with even the most aggressive policies. International partnerships are necessary, and California's Cap-and-Trade Program is already linked with Quebec and hopefully will be with other jurisdictions in the future.

One of the wisest people with whom I have had the pleasure to work was David Bates, a physician-scientist who became interested in the health effects of air pollution when he had to care for patients during the worst of the London fog episodes almost 70 years ago. Shortly before he died in 2006, Dr. Bates stated in his usual clear and insightful way that "Reducing emissions of air pollutants will slow global warming, and minimizing releases of greenhouse gases will save lives that would otherwise be lost to pollution. A well-inclined government genuinely committed to 'sound' science, would recognize this and assemble a package of policy measures designed to save our lives and our future" [29].

In conclusion, the sources of air pollutants that harm health are the same as those that emit greenhouse gases and aerosols. To improve air quality and mitigate climate change, we need strong policies to move our economy away from reliance on fossil fuels.

References

1. California Environmental Protection Agency, Air Resources Board. Assembly Bill 32 Overview. [last reviewed 2014 August 5, accessed 2019 15 June]. Available from: https://www.arb.ca.gov/cc/ab32/ab32.htm.
2. IPCC Working Group. I. Summary for policymakers. In: Stocker TF, Qin D, Plattner G-K, Tignor M, Allen SK, Boschung J, Nauels A, Xia Y, Bex V, Midgley PM, editors. Climate change 2013: the physical science basis. Contribution of Working Group I to the fifth assessment report of the intergovernmental panel on climate change. New York: Cambridge University Press; 2013.
3. Under2 Coalition. Subnational global climate leadership memorandum of understanding. [updated April 2019, accessed 2019 15 June. Available from: http://under2mou.org/coalition/.
4. Millstein DE, Harley RA. Impact of climate change on photochemical air pollution in Southern California. Atmos Chem Phys. 2009;9:3745–54.
5. United States Environmental Protection Agency. Vehicle emissions California Waivers and authorizations [updated 2017 June 23., accessed 2019 15 June]. Available from: https://www.epa.gov/state-and-local-transportation/vehicle-emissions-california-waivers-and-authorizations.
6. United States Environmental Protection Agency. Progress cleaning the air and improving people's health [updated 2019 March 4, accessed 2019 15 June]. Available from.: https://www.epa.gov/clean-air-act-overview/progress-cleaning-air-and-improving-peoples-health.
7. California Environmental Protection Agency, Air Resources Board. Overview: diesel exhaust and health. [updated 2019, accessed 2019 15 June]. Available from: https://www.arb.ca.gov/research/diesel/diesel-health.htm.
8. California Environmental Protection Agency, Air Resources Board. Greenhouse gas inventory data - graphs [updated 2019, accessed 2019 15 June. https://www.arb.ca.gov/cc/inventory/data/graph/graph.htm.
9. California Environmental Protection Agency, Air Resources Board. Emissions reduction from AB32 scoping plan measures. 2010. [accessed 2019 15 June]. Available from: https://breakingenergy.com/2014/08/07/california-and-mexico-sign-climate-and-energy-agreements/.

10. California Environmental Protection Agency, Air Resources Board. Zero emission vehicle (ZEV) program. [updated 2019;accessed 2019 15 June]. Available from: http://www.arb.ca.gov/msprog/zevprog/zevprog.htm.
11. California Environmental Protection Agency, Air Resources Board. Greenhouse gas standards 2009–2016 (Pavley, Assembly Bill 1493). [updated 2014 Nov 6, accessed 2019 15 June]. Available from: https://www.arb.ca.gov/msprog/ccms/ccms_new.htm.
12. United States Transportation Department. Corporate Average Fuel Economy (CAFE) Standards[updated 2017 Aug 17, accessed 2019 15 June]. Available from.: https://www.transportation.gov/mission/sustainability/corporate-average-fuel-economy-cafe-standards.
13. United States Environmental Protection Agency. California Greenhouse Gas Waiver Request [updated 2018 Aug 10, accessed 2019 15 June]. Available from: https://www.epa.gov/regulations-emissions-vehicles-and-engines/california-greenhouse-gas-waiver-request.
14. Autos must average 54.5 mpg by 2025, new EPA standards say. Washington Post [2012 Aug 28, accessed 2019 15 June]. Available from: https://www.washingtonpost.com/national/health-science/autos-must-average-545-mpg-by-2025-new-epa-standards-are-expected-to-say/2012/08/28/2c47924a-f117-11e1-892d-bc92fee603a7_story.html?utm_term=.536832af5cba.
15. Timothy Cama and Miranda Green. Trump moves to roll back Obama emission standard. The Hill [2018 Aug 2, accessed 2019 15 June]. Available from: https://thehill.com/policy/energy-environment/400036-trump-submits-rule-to-weaken-iconic-obama-car-efficiency-standards.
16. California Environmental Protection Agency., Air Resources Board. Analysis in support of comments of the California Air Resources Board on the Safer Affordable Fuel-Efficient (SAFE) vehicles rule for model years 2021–2026 Passenger Cars and Light Trucks[2018 Oct 10, accessed 2019 15 June.] Available from https://ww2.arb.ca.gov/sites/default/files/2018-10/2018-10-26%20FINAL%20CARB%20Detailed%20Comments%20on%20SAFE%20NPRM.pdf.
17. California Environmental Protection Agency, Air Resources Board. Low carbon fuel standard. [updated 2019 May 15, accessed 2019 15 June. Available from: https://www.arb.ca.gov/fuels/lcfs/lcfs.htm.
18. California Environmental Protection Agency, Air Resources Board. Renewable electricity standard. [updated 2014 Aug 25, accessed 2019 15 June]. Available from: http://www.arb.ca.gov/energy/res/res.htm.
19. California Legislative Information. SB-100 California Renewables Portfolio Standard Program: emissions of greenhouse gases. [2018 Sept 10., accessed 2019 15 June] Available from: https://leginfo.legislature.ca.gov/faces/billNavClient.xhtml?bill_id=201720180SB100.
20. California Environmental Protection Agency, Air Resources Board. Sustainable freight initiative. [2019 February 6, accessed 2019 15 June] Available from.: https://www.arb.ca.gov/gmp/sfti/sfti.htm.
21. California Environmental Protection Agency, Air Resources Board Sustainable communities [updated 2019.; accessed 2019 15 June]. Available from: http://www.arb.ca.gov/cc/sb375/sb375.htm.
22. California Environmental Protection Agency, Air Resources Board. Cap-and-trade program. [updated 2019 June 14; accessed 2019 15 June. Available from: http://www.arb.ca.gov/cc/capandtrade/capandtrade.htm.
23. California Environmental Protection Agency. Greenhouse gas reduction investments to benefit disadvantaged communities. [updated 2019, accessed 2019 15 June]. Available from: http://www.calepa.ca.gov/envjustice/ghginvest/#sthash.rwhgv0pb.dpuf.
24. California Environmental Protection Agency, Office of Environmental Health Hazard Assessment. CalEnviroscreen version 3.0. [update 2018 June 25, accessed 2019 15 June] Available from.: https://oehha.ca.gov/calenviroscreen/report/calenviroscreen-30.

25. California Legislative Information. SB-AB-398 California Global Warming Solutions Act of 2006: market-based compliance mechanisms. [2017 July 25., accessed 2019 15 June] Available from: http://leginfo.legislature.ca.gov/faces/billNavClient.xhtml?bill_id=201720180AB398.
26. California Environmental Protection Agency, Air Resources Board Community air protection program [updated 2019.; accessed 2019 15 June]. Available from: https://ww2.arb.ca.gov/our-work/programs/community-air-protection-program.
27. California Environmental Protection Agency, Air Resources Board. California cap-and-trade program summary of proceeds to California and consigning entities. [updated June 2019, accessed 2019 15 June]. Available from: https://www.arb.ca.gov/cc/capandtrade/auction/proceeds_summary.pdf.
28. California Legislative Information. SB-32 California Global Warming Solutions Act of 2006: emissions limit. [2016 Sept 8., accessed 2019 15 June] Available from: https://leginfo.legislature.ca.gov/faces/billNavClient.xhtml?bill_id=201520160SB32.
29. Bates D. Air pollution and global warming: the seamless web. Health & Clean Air Newsletter, Spring-Summer 2005.

Chapter 27
Climate Change and Public Health Interventions

Kathryn C. Conlon and Chelsea M. Austin

Introduction

Substantial advances in our understanding of the complex relationships between climate, weather, and health have been critical to track how global climate change is impacting human health. Innovations in climate science, environmental sampling, syndromic surveillance, epidemiologic study design, and analytical methodologies have demonstrated the myriad pathways through which climate change–related exposures lead to varying degrees of injury, illness, and death throughout the world. These climate-driven harms are expected to intensify, last for longer periods of time, and impact more people than ever before. Assessing the scope and depth of the climate and health emergency by quantifying health impact is necessary for determining opportunities to prevent and reduce harm [1, 2]. Yet, there has been less explicit attention paid to the urgent need for designing, implementing, and evaluating public health interventions for climate change–related exposures [3].

Public health practitioners and researchers typically rely on reproducible evidence gathered from years' worth of analyzed data to provide informed recommendations to the public health and medical community. Randomized controlled trials are considered the gold standard for the study of interventions. However, data are often generated from more practical study designs like quasiexperimental, pre-/post-assessments, and nonrandomized control trials to test effectiveness of interventions. There are numerous challenges with climate change–related exposures that make designing interventions difficult. Perhaps, the most obvious challenge is that the

K. C. Conlon (✉)
Department of Public Health Sciences, University of California Davis, Davis, CA, USA
e-mail: kcconlon@ucdavis.edu

C. M. Austin
Department of Environmental Health Sciences, Yale School of Public Health, New Haven, CT, USA
e-mail: chelsea.austin@yale.edu

K. E. Pinkerton, W. N. Rom (eds.), *Climate Change and Global Public Health*, Respiratory Medicine, https://doi.org/10.1007/978-3-030-54746-2_27

exposures (e.g., extreme heat, drought, disease-carrying vectors) occur over varying time periods. Some exposures are extreme acute events occurring within short periods of time while others also occur as extreme events – events that deviate from an expected climate normal – but are chronic and move slowly in their outcome and resolution. Because of this, exposure misclassification is a substantial concern requiring attention at the outset of the study design. Likewise, climate change does not impact individuals equally; it is often described as a risk amplifier, meaning it magnifies existing risks amongst populations who are more susceptible to the effects of climate change [1]. Disentangling the complicated climate and health relationships is methodologically challenging, even with more straightforward ecological or observational study designs. The complex nature of our institutional, social, and interpersonal networks makes it difficult to fully assess direct impacts of climate and health interventions, particularly when many systemic inequities contribute to a given risk.

As communities become more aware of and experience the risks that climate poses to their health, calls for government action to address climate change have increased [4]. Effective government actions will invariably interrupt the climate-related drivers of adverse health impacts, while coproducing knowledge that contributes to our understanding of the complex mechanisms underlying climate change and health. Designing, implementing, and evaluating these actions or interventions are urgently needed.

This chapter first places the design, implementation, and evaluation of climate and health interventions in the public health and adaptation contexts. We present a rationale for considering public health interventions for climate change exposures as a predominantly adaptative, rather than mitigative, strategy. We then expound on what is meant by practitioner for climate and health, focusing on the intentionally broad definition of who conducts work that acts as a climate and health intervention and how that implicates different types of professionals and experts. Next, we provide examples of climate and health interventions implemented at different scales, by a variety of practitioners. Lastly, we elaborate on the role that rigorous planning in design and evaluation has in contributing toward the climate and health intervention evidence base.

The field of climate change and health interventions is nascent, but growing. Established public health practitioners with expertise in infectious disease, behavioral health, environmental health, chronic disease, and health policy, amongst others, may soon find that climate change–related issues will become more common in their respective fields because there are few areas that climate change fails to influence. Because of this, there is a role all public health practitioners and researchers may inevitably inherit in implementing climate and health interventions.

Adapting Public Health Conceptual Models for Climate and Health Interventions

Researchers define public health interventions as any action undertaken to protect health whether by assessing, maintaining, or promoting improved health status or the conditions underlying improved health status [5, 6]. When applied to specific public health issues, such as chronic disease or injury prevention, the definition takes on nuances specific to the outcomes and public health practices under study [7–9]. Popular conceptualizations of public health interventions within the USA include: the 10 Essential Public Health Services [10], which is regularly referred by the US Centers for Disease Control and Prevention (CDC); the Public Health Impact Pyramid [11]; the RE-AIM Framework [12]; and the Public Health Nurses' Intervention Wheel [13]. Despite the widening scope of interventions for public health [14], environmental health–specific frameworks elude the wider literature [15], potentially due to an ontological struggle to define and clarify the field [16].

Despite this, researchers have made quick work of defining and typifying climate adaptation in ways that have implications for health. The US Global Change Research Program (US GCRP) defines adaptation as the "adjustment in natural or human systems to a new or changing environment that exploits beneficial opportunities or moderates negative effects" [1]. Additionally, Biagini et al. [17] present a typology of adaptation actions from real-world projects that map to established public health frameworks and functions. For example, Biagini's "information" adaptations align with the third essential public health service "Inform, Education, Empower." Similar works on adaptation, such as those by Fussel et al. [18] and Smit et al. [19], describe attributes we might consider for health adaptation, including autonomous or proactive adaptation, planning horizons, and the predictability of climate hazards. Autonomous adaptation is defined as reactive, passive, or natural actions taken by the adapting entity, whereas proactive adaptation is defined as planned, intentional, or active actions. For example, an autonomous action could be staying indoors during very hot days, while a proactive adaptation for adapting to heat could be relocating to cooler parts of the region. Planning horizons refer to the time we have to think through and implement a proactive adaptation, such as the time we have to evacuate from a nearby wildfire (i.e., very little, on the order of a few minutes or hours) to the time a city planner has to green an urban heat island (i.e., much longer, often months or years). The predictability of climate hazards is defined as the certainty of future climate change and related conditions, exemplified by the predictability of hotter days (high certainty) versus changes in vector migration (lower certainty). While the body of literature describing the context and practice of climate adaptation continues to grow, the need for a similar corpus dedicated to health adaptation remains [19].

Climate and health adaptations utilize a variety of strategies, programs, and activities to interrupt climate-related hazards that impact health at any time or spatial scale (e.g., current, near-future, far-future, and individual, community, regional, global scales). The strategies, programs, and activities are adaptive in the sense that they are intervening on climate-driven predictors of health: such as housing, behavior, access to clean water, and others, to avert the worst health outcomes, change underlying conditions of risk, and improve overall quality of life. We define *climate and health interventions* as any action undertaken to protect the public's health from climate-related hazards. Here, "undertaken to" refers to the intentionality of the actors (i.e., intervention designers, implementers, and evaluators) to fulfill a public health objective. Additionally, we borrow from the lexicon of established definitions of interventions, which includes terms like assessment, maintenance, promotion, and adjustments [5, 6], and repurpose them to a climate and health context. Further, our actions must account for the complicated context in which climate and health interventions will be implemented: issues such as what we are adapting to (heat vs. sea level rise), who or what is adapting (low-income residents vs. hospital systems), and how adaptation is expected to occur (environmental [such as adding trees for shade] vs monetary [such as property buyouts for coastline dwellers]) will invariably affect the success of our adaptations [19]. Through this broad lens, climate and health interventions can span a range of applications in time (upstream vs. downstream; now vs. future), space (multiple levels of physical and social organization), and social scale (individual vs. population) necessary to protect health. Interventions might be designed and implemented today for distant outcomes; for instance, climate science education in grade school (time), institutional policies in hospitals to limit cascading failures during power safety shutoffs amidst wildfire conditions (upstream to downstream), or a national program for malarial immunization (scale). Such a lens situates climate and health interventions within a social–ecological systems perspective, whereby interventions can be implemented at all critical points in the "web of causation" [20, 21]. In Fig. 27.1, we visualize the dimensions in which climate and health interventions might occur using an adapted socioecological model.

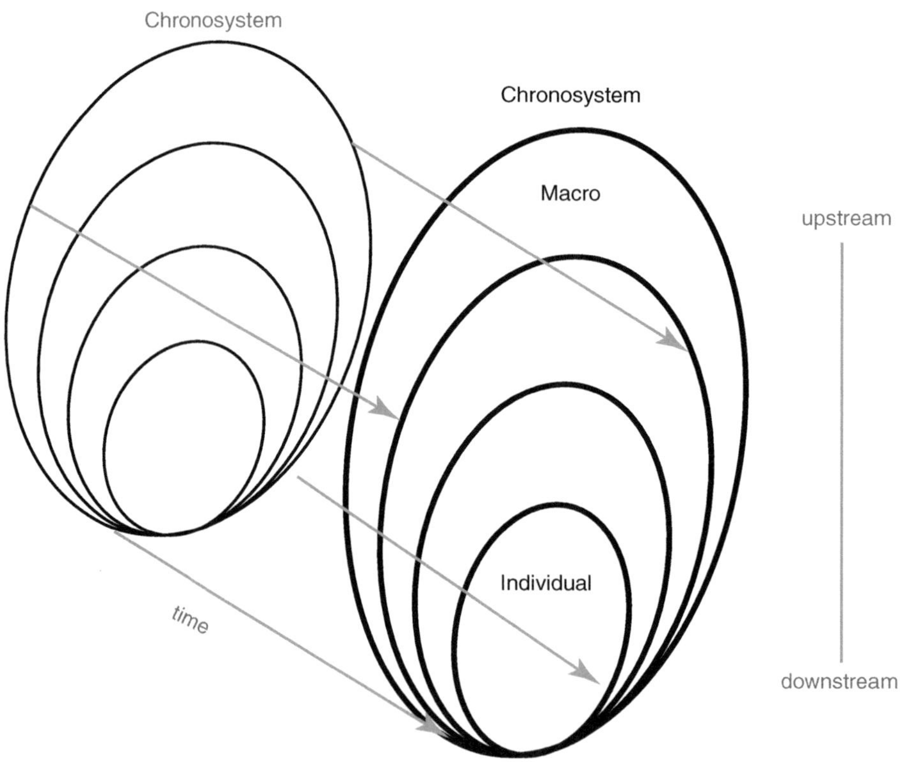

Fig. 27.1 Social ecological system for climate and health interventions. (Adapted from "Ecological systems theory," Bronfenbrenner U [22] and "Prisoners of the Proximate: Loosening the Constraints on Epidemiology in an Age of Change," McMichael, Copyright 1999 [21])

While climate and health interventions are not new conceptually or as a practice, their place within wider adaptation processes or cycles is less clear. Arguably, there are opportunities to intervene at various stages of adaptation, though they are not formally identified within published frameworks. For instance, the US GCRP, which is responsible for the National Climate Assessment (NCA) that is published every 4 years, identifies five adaptation stages in its 4th Assessment: Awareness, Assessment, Planning, Implementation, and Monitoring and Evaluation [1] . These stages conceptualize the iterative process for managing climate change risks, demonstrating the general adaptation process. At first glance, the Implementation stage, presumably, would be where we'd expect to see climate and health interventions occur because of the inherent implied action. Yet, this stage does not include a description of what counts as implementation or what practices it might consist of for the various practitioners conducting implementation. Similarly, within CDC's five-step Building Resilience Against Climate Effects Model (BRACE), intervention options are identified in Step Three and potentially implemented in Step Four, although this phase of implied work is underreported compared to the model's other steps [23]. Like the NCA example, the BRACE model also includes steps for gaining additional insight into how climate impacts health (Steps 1 and 2), which aligns with the "monitoring" and "diagnosing the problem" activities described in the Public Health Essential Services [10]. In other words, both the NCA and BRACE frameworks list stages or activities of adaptation that might be considered interventions by broader public health intervention frameworks, such as with the 10 Essential Services, and the Nurse's Intervention Wheel. Conversely, the Public Health Institute developed "A Framework for Action," providing a more comprehensive and inclusive view of the dynamic relationships between public health interventions and climate change [24]. Building from Patz and Haines' work [25], this framework illustrates intervention opportunities along various climate and health pathways. The interventions including changes to surveillance systems, education activities, engaging the public health workforce, and environmental systems changes are broad and provide a variety of roles for the public health workforce in addressing climate change and health. Figure 27.2 provides a simplified and adapted framework for climate and health interventions within a public health context, showing what climate and health interventions might look like at all stages of this framework.

In considering the wider implications of adaptation and public health practice, we begin to see climate and health interventions as a primary method of climate adaptation. Public health involves inherently upstream work for inherently downstream issues (e.g., addressing poverty or food access to decrease chronic disease). Similarly, climate and health adaptation takes an inherently chronosystemic issue – climate change – and maps the many exposure-outcome pathways through which health is eventually affected. These pathways often share a common theme with broader human, animal, and environmental problems such as infrastructure maintenance, updating building codes, managing biodiversity loss, promoting ocean health, and responsible land management.

What's more, these pathways not only elucidate the complex mechanisms through which health is harmed by climate change but also serve to illustrate an

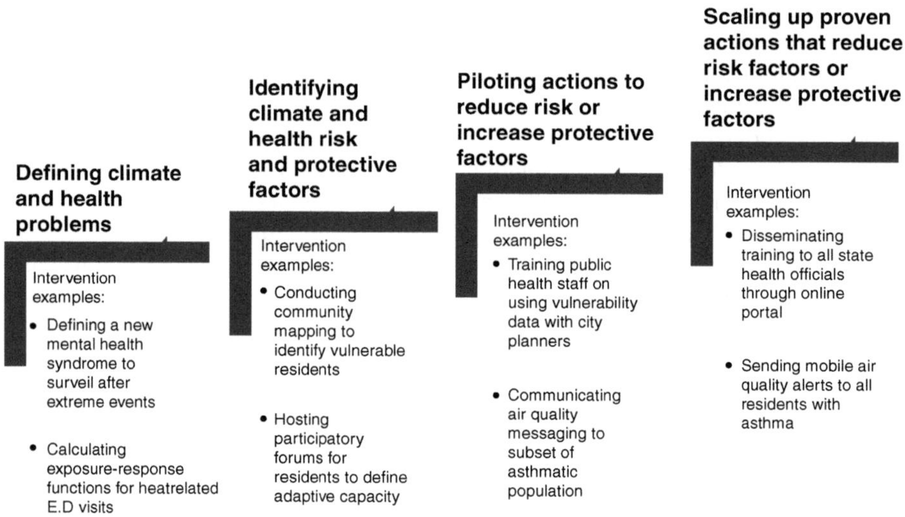

Fig. 27.2 Climate and health interventions within a public health context. (Adapted from "Public Health Policy for Preventing Violence" Mercy, et al. [26])

intervention's potential effects [27]. Furthermore, such multifactorial pathways help us illuminate cross-sectoral opportunities for cobeneficial outcomes, potentially serving other adaptation endeavors. For instance, how buildings are designed to reduce the use of energy during the hottest periods of the day may not only provide reprieve for the inhabitants' electricity bill but could also reduce the carbon footprint of the building. These aspects of climate and health interventions have many implications not only for the public health workforce, but all workforces whose functions interact with climate change and public health.

Practitioners for Climate and Health Interventions

Because the health impacts of climate change implicate numerous fields of expertise and practice, there is an argument to be made that practitioners of climate and health interventions span a broad and growing workforce. Researchers and practitioners have called on public health agencies to leverage the workforce in preventing and responding to climate-sensitive threats to health [28, 29]. After all, the public health community has a storied history of meeting urgent and burgeoning public health threats through coordinated, evidence-based preparedness and response. In 2005, Hurricane Katrina was a textbook public health natural disaster. The required expertise to respond included injury prevention and care, infectious disease surveillance, environmental health, food and water safety, chronic disease management, and maternal and child health. Despite this massive challenge, the public health community mobilized to provide urgent medical care to survivors, conduct rapid needs assessments, establish active outbreak surveillance at healthcare facilities, and develop long-term programmatic support for evacuees and survivors [30–32]. The mental health impacts studied from Hurricane Katrina, especially pertaining to displacement, have shed light on the effective approaches communities can use to maximize preparation for large-scale natural disruptions through strengthening social cohesion and connectivity [33, 34]. The nexus of research and practice is an area that public health experts operate in with relative comfort.

Like the Hurricane Katrina response, climate change–related exposures touch on all areas of public health. However, the risks posed to human health from climate change will reach fields and professional where public health does not often interact, let alone regularly collaborate; these include transportation, forestry, urban planning, waste management, utility providers, private businesses, architects, emergency managers, agrobusiness, and many more. Successful, protective climate and health interventions necessitate expertise, cooperation, and response from a wide-ranging network of practitioners and researchers. As climate and health interventions become more commonly employed and researched, the complexities that accompany each intervention will become clearer, particularly uncovering the roles that the web of experts play in employing a specific intervention. For instance, tree planting is often discussed as a climate and health intervention that is both adaptive and mitigative. Increasing tree canopy can produce shade, reducing the urban heat island, which can lower ambient temperatures and individual exposure to high temperatures [35]. Implementing a tree-planting intervention at a scale that would produce this type of effect would require input from experts ranging from urban planning, urban forestry, community developers, and policymakers. It would likely require substantial input from community members, as well as funding acquisition. Projects like Green Heart in Louisville, Kentucky have taken a multidisciplinary approach, incorporating partners from the non-governmental agencies, academia, and local government to build a public health case for understanding how the intervention could protect health [36]. These cross-sectoral representatives are practitioners of climate and health interventions. Each have key roles and responsibilities to ensure the details of an intervention are correctly employed and captured throughout the design and implementation process.

Until recently, there has been minimal, coordinated momentum amongst public health researchers and practitioners to share how climate and health adaptation is being done via intervention design, implementation, and evaluation. However, since 2009, the CDC's Climate and Health Program (CHP) has been the sole government funder for U.S. states, cities, territories, and tribal communities to address climate change–related hazards and health outcomes through a variety of public health activities, including intervention development, implementation, and evaluation [37]. CHP funded programs follow the BRACE model, which offers a roadmap for public health practitioners to systematically increase capacity to respond to climate change and health hazards. The fourth step of the model is implementing interventions. Like many public health programs and interventions, climate and health interventions are borrow and repurpose fundamental public health strategies (e.g., education, infrastructure modification, resource allocation and access, etc.) to facilitate behavior change to protect health.

One does not have to be formally labeled or employed in the public health field, funded as a climate and health practitioner, or even associated with a climate or health organization to be actively involved in a climate and health intervention. There are established networks, professional organizations, and workforce communities that have been doing climate change and health interventions but may not have been using that language. In many ways, many adaptation projects have tangentially included health considerations. EcoAdapt is an organization that provides support, training, and assistance to those interested in planning and managing for climate adaptation. Many of the resources and skills they offer touch on health considerations. EcoAdapt hosts a clearinghouse of climate adaptation activities – CAKE (Climate Adaptation Knowledge Exchange) – which can be filtered to include those with health considerations. Such tools are increasing in popularity, introducing a complementary source of information to peer-review journal publications [38]. Another growing area of interest is among medical professionals (e.g., medical doctors, nurses, educators) who recognize the threats climate change has and will continue to have on their patient populations. Organizations like the Medical Society Consortium on Climate and Health leverage their members' affiliations with medical and nursing societies to bring awareness about climate change and health to advocate for funding, education, and policy, among other community-level interventions [39].

Climate and Health Interventions in Practice

Since we view climate and health interventions with a broad lens, there is an opportunity to identify examples in numerous settings. In 2017, a collection of representatives from the CHP's funded programs and CDC scientists produced an assessment of climate and health interventions. The assessment identified three types of evidence: (1) evidence linking climate-sensitive exposure to a health outcome of interest; (2) evidence on effectiveness of interventions; and (3) evidence on the evaluation of implementation within a community. The assessment was extensive, providing an overview of fourteen exposure- and health outcome–specific interventions ranging from air pollution and respiratory illness, flooding and mold, to mental health outcomes related to drought [40]. Not surprisingly, the assessment identified few interventions that were intentionally designed to assess an exposure labelled "climate change." However, there are numerous examples of individual- and community-level climate and health interventions that could be emulated, modeled, or repurposed. Because climate-sensitive exposures are rarely uniformly defined, comparing results across interventions was unobtainable for the scope of the assessment. This is the first attempt to compile evidence pertaining to a wide array of climate and health interventions highlights the complexities and diversity of activities that we can consider intervening on climate change and health.

A common intervention for reducing the risk of heat-related illness among the most at-risk populations in hot, urban areas is to provide cooling shelters during extreme heat events. The Heat Relief Network, launched in 2005, is a well-publicized network of partnerships organized by Arizona's Maricopa County Association of Governments, municipalities, nongovernment organizations, faith-based organizations, and businesses. The Network maps locations to visit for individuals who need respite (e.g., drinking water, shelter) from the heat [41]. For this multisectoral intervention to work, roles and responsibilities for each partner must be established and tracked. How each participant engages and contributes can determine the extent to which information can be collected and built upon. While some may consider an activity like the Heat Relief Network a collaborative, public health networking program, we argue that, at its core, it – and many programs and projects like it – is a climate and health intervention. The Heat Relief Network explicitly arose as a result of an increase in homeless deaths that occurred following high nighttime temperatures in 2005 [41]. The goal of the Network is to reduce the health outcomes related to extreme heat, an exposure that is increasing in Arizona because of climate change and has been specifically identified as a concern by the State's health department. Like many climate and health interventions, this could easily be overlooked as an activity that intervenes to disrupt a climate change–related exposure on a health outcome.

In some instances, local, state, or federal policy adoption and implementation act as interventions. California's Occupational Safety and Health Administration (Cal/OSHA) passed a Heat Illness Prevention emergency regulation that was later adopted as permanent, requiring agricultural employers and facilities to provide shading, water, and heat-related illness emergency response training to workers [40]. Similarly, as mentioned in Chap. 16, Riden, the emergency regulation for California agricultural workers and wildfire smoke exposure may act as an intervention, interrupting a known climate-sensitive exposure (e.g., wildfire smoke) on health outcomes. Programs that operate more ambiguously off of existing policies, such as the Be A Buddy (BaB) Program in New York City, New York, funded by the City, have proven attractive to some communities [41]. The BaB facilitates local volunteers to educate those most at risk in the local community about the actions they need to take to become "climate prepared." In practice, this means that during extreme hot or cold temperatures, volunteers become "activated," reaching out and helping community members they've identified who may need support during those conditions. Warnings from local and national weather affiliates serve as the policies that guide the activities.

Intervention examples can also be found in the international community. For instance, in the squatter settlements of Karachi, Pakistan, diarrheal disease is one of the leading causes of death in children under 5 years of age [42]. Researchers present a multifaceted picture of risk for pediatric diarrheal disease (PDD), with varied and complicated pathways leading to climate change. In the most recent systematic review on drinking water and diarrhea risk, Gruber et al. found a positive association between the presence of bacterial pathogens in drinking water and diarrheal risk [43]. These findings are of particular importance in Karachi, Pakistan, where a high prevalence of diarrhea-causing pathogens has been measured in drinking water sources [44]. Additionally, researchers describe the significant relationship between a household's longer distances from drinking water sources and risk of diarrheal disease [45, 46]. Such distances have been positively associated with a household's decision to use outdoor taps or pipes, and nonmunicipal sources such as ponds, for drinking water in Pakistan [46]. These risk factors for diarrheal disease – unclean drinking water and distance from and use of nonmunicipal water sources – have established relationships to lower rainfall levels [47, 48], which the Karachi area has experienced in recent decades from climate change [49, 50].

To address the issue of contaminated household water in Karachi, Luby et al. devised an intervention to compare the effects of water treatment methods and handwashing practices on diarrheal disease [51]. Households were assigned to four different intervention groups: water cleaning with a bleach solution, water cleaning with a flocculent disinfectant, handwashing education only, and handwashing education and flocculent disinfectant. Control households were also assigned. The outcome measured across intervention groups was self-reported presence of diarrhea in mothers/caregivers of children or in the children themselves. The results of this study showed reductions in the prevalence of diarrheal disease for households assigned to both water treatment groups (the bleach and flocculent disinfectant), as well as the households assigned to the handwashing education and flocculent disinfectant group [51]. This example shows how intervention unidentified by its designers as being a climate and health intervention may still provide evidence for the public health community.

Building an Evidence Base of Climate and Health Interventions

As calls for more and quicker adaptation take hold of the public health conscious, so too, do calls for evidence on the effectiveness of climate and health interventions [52]. However, generating this evidence, whether through research or evaluation, is plagued by many issues [53]. Evaluators and researchers alike note the methodological complications conferred by varying, unpredictable timelines, multiscale pathways of risk and intervention, and multisectoral stakeholders who may strain consensus on the study's aims [53]. For example, study timelines might be extended given the amount of time climate-sensitive health outcomes take to develop, be observed, and measured. Extreme heat epidemiological studies, for instance, have begun to expand the study periods beyond the typical summer months (e.g., June – September) into May and October [54]. Additionally, varying or shifting climate hazards mean interventions are interrupting changing pathways, and consequently, new baselines or moderating factors (e.g., new precipitation averages) [55]. Therefore, the scales of current interventions may also introduce issues related to data collection continuity and analysis. This is particularly important where multi- or cross-scale dynamics are involved. One way to conceptualize this is when considering how evaluations that are conducted at a national level may rely on data collected at a hyper local level (e.g., sub-city) where standardized collection may not be feasible or appropriate. Similarly, interventions that target multiple systems, such as ecological and social systems, might have to account for which systems are most salient at different stages in the intervention and, thus, study design [56].

However, given the critical need for evidence, evaluators and researchers are developing methodological innovations and practical applications for adaptation interventions. Best practices for adaptation evaluation include using participatory methods to generate definitions of adaptation "success" (resilience, adaptive capacity, etc.), focusing on more upstream systems, and borrowing from the quasiexperimental and contribution analyses toolboxes to conduct studies [55, 57, 58]. For example, a health adaptation focused on food security might use open-ended questions with participants to determine nutritional needs for different climate hazards. Questions such as "how much [nutritional staple] would your household need to eat well during the next drought?" or "how would you store food during the next flood?" enable participants to set standards meaningful to their contexts as well as the evaluation. Tools such as CARE's Climate Vulnerability and Adaptation Capacity Handbook and Brooks et al.'s Tracking Adaptation and Measuring Development provide additional guidance to evaluators looking to better understand community values and adaptive thresholds [59, 60]. Additionally, focusing our learning on more upstream systems, such as education and poverty, generates evidence relevant to prevention and well-being, predictive of improved health outcomes. For instance, Striessnig et al. [61] found that education of girls and women was most strongly associated with a reduction in death during natural disasters in a study looking at macrolevel factors for disaster risk (e.g., GDP, democracy scores, the Human Development Index, etc.). Similarly, Bryan et al.'s study on predictors of adaptation by farmers found a strong link between farmers' access to safety nets, such as receipt of food or extension services, and adaptation actions like changing crop type and planting dates [62]. Such research makes a strong case for including upstream indicators, such as the social determinants of health, into our study designs and analyses.

Creating these study designs and analyses, however, requires addressing issues of shifting baselines, contribution versus attribution to outcomes, and multiscale dynamics. To do this, we pull from lessons learned in evaluation and climate adaptation in international development. In a synthesis of evaluation methods for climate adaptation, issues with time are addressed using collection methods like scenario building, quasiexperimental design, and semistructured interviews [63]. The authors define scenario building, for example, as generating a "set of possible alternative futures ranging from participatory scenarios to modeling data" [63]. This process is exemplified by Pahl-Wostl's [64] account of scenario building by officials in southern England responding to strains on water supply from climate change. In this example, officials convened stakeholders from the private (i.e., water companies) and public sectors (i.e., environmental and economic regulators) to collectively work through the dynamics and potential outcomes of expected changes in drought risk, economic growth, consumer expectations, and regulatory conditions. Using a mix of simulations and their own localized expertise, they generated a set of indicators to track several future scenarios of drought risk in southern England. According to Fisher et al. [63], this form of scenario building not only addresses shifting baselines and target-setting but is also well-suited for complex interventions. Additionally, complex interventions are studied with methods such as conducting focus groups or limiting factor analysis. Limiting factor analysis is defined by Fisher as a "[t]echnique to develop a common understanding of the key factors that must be assessed, and if necessary (and possible) managed, for project or program to be viable over the long term." [63]. As Gullison and Hardner [65] describe it, limiting factor analyses are a type of qualitative forecasting enabling practitioners and evaluators to better anticipate, manage, and study threats to success in complex, large-scale projects. Like scenario building, limiting factor analysis uses a participatory approach to move through its steps: generate a list of factors likely to impede a project's objectives, rank the status of identified factors at key points in the evaluation (e.g., baseline, annual check-ins, end of project period, etc.), and identify all the entities working to address the ranked factors including their funding source and likelihood of success. At the end of this process, practitioners are expected to better understand what factors are in their locus of control, and thus addressable, and which might moderate the project's success or viability no matter the quality of project implementation. For climate and health interventions, this process might mean anticipating, for example, limiting factors of coastline dwellers to act on behaviors (e.g., restoration of estuarine ecosystems to address sea rise) for a local communication campaign. Specifically, restoration behaviors might require a host of systems to be considered in identifying limiting factors, such as local policy, environmental practices of nearby coastal industries, and migratory habits of coastal species. Such factors could then be ranked according to their scale of influence. For instance, industrial runoff harming and, thus, modifying the coastal habitats and suitability of local vegetation and wildlife. In our evaluations, we might use these factors to improve our logic models by including new or different intervention activities. We may update our study designs through the addition of new or revised survey questions. Lastly, we may reconsider our analytical approaches by supplementing our data sources to investigate the status of limiting factors during project implementation.

Climate and health interventions require an "outside the box" approach to scoping and conducting evaluation. Given health adaptation's early stage of development [66], this also means a number of evaluation lessons have yet to be learned. This includes the possible outcome where an intervention has an unexpected – or maladaptive – impact. Thus, evaluators are charged with providing a judicious account of how they conduct their work and what they learn. This evidence will provide the critical, urgent design inputs to improve existing interventions. More importantly, this evidence will fuel the success of newer, more transformative interventions [67] for public health, necessitated by the scale and speed of our rapidly changing climate.

Conclusion

Climate change will continue to impact health in the absence of efforts to intervene on climate change–related exposures. Public health practitioners and researchers are faced with an extraordinary opportunity to contribute to building an evidence base of public health interventions for climate change. Although there are substantive methodological and evaluation challenges, there is critical need to address the evidence gap. The well-received *Project Drawdown* presents a list of thoroughly analyzed, carbon-mitigative solutions that are ranked by impact for reducing carbon pollution that contributes to global climate change [68]. The book, *Project Drawdown*, was the work of many scientists, based on thousands of peer-reviewed studies all concerned with mitigative solutions to the climate crisis. While the climate and health community is very far from having the *Project Drawdown* for climate and health interventions, it is certainly within the realm of necessity, given the pace and severity of the climate crisis we are facing. By establishing and adhering to a common set of standards for designing, implementing, and evaluating climate and health interventions, we will begin to develop the needed evidence base to identify actions that will protect health from climate change–related exposures.

References

1. US GCRP. Impacts, risks, and adaptation in the United States: fourth national climate assessment. Washington, DC: U.S. Global Change Research Program; 2018.
2. Field CB, Barros VR, Mastrandrea MD, Mach KJ, Abdrabo MAK, Adger WN, et al. Summary for policy makers: Working group 11 contribution to the fifth assessment report of the Intergovernmental panel on climate change; 2014.
3. Vogel J. Case studies: climate adaptation: the State of Practice in US Communities: Ontario Centre for climate impacts and adaptation resources; 2016.
4. Leiserowitz A, Maibach E, Rosenthal S, Kotcher J, Bergquist P, Ballew M, Goldberg M, Gustafson A. Climate change in the American mind: April 2019. New Haven: Yale Program on Climate Change Communication; 2019.
5. Rychetnik L, Frommer M, Hawe P, Shiell A. Criteria for evaluating evidence on public health interventions. J Epidemiol Community Health. 2002;56(2):119–27.
6. Fortune N, Madden R, Almborg A-H. Use of a new international classification of health interventions for capturing information on health interventions relevant to people with disabilities. Int J Environ Res Public Health. 2018;15(1):145.
7. Story M, Kaphingst KM, Robinson-O'Brien R, Glanz K. Creating healthy food and eating environments: policy and environmental approaches. Annu Rev Public Health. 2008;29:253–72.
8. Gordon RS Jr. An operational classification of disease prevention. Public Health Rep. 1983;98(2):107.
9. Geller ES, Berry TD, Ludwig TD, Evans RE, Gilmore MR, Clarke SW. A conceptual framework for developing and evaluating behavior change interventions for injury control. Health Educ Res. 1990;5(2):125–37.
10. Public Health Functions Steering Committee. Public health in America 1994. [cited 2019 April 1]. Available from: https://www.cdc.gov/publichealthgateway/publichealthservices/essentialhealthservices.html.
11. Frieden TR. A framework for public health action: the health impact pyramid. Am J Public Health. 2010;100(4):590–5.
12. Glasgow RE, Vogt TM, Boles SM. Evaluating the public health impact of health promotion interventions: the RE-AIM framework. Am J Public Health. 1999;89(9):1322–7.
13. Health MDo. Public health interventions: applications for public health nursing practice. 2019.
14. Coronado F, Koo D, Gebbie K. The public health workforce: moving forward in the 21st century. Am J Prev Med. 2014;47(5):S275–S7.

15. Lokker C, McKibbon KA, Colquhoun H, Hempel S. A scoping review of classification schemes of interventions to promote and integrate evidence into practice in healthcare. Implement Sci. 2015;10(1):27.
16. Gordon LJ. Environmental health and protection: century 21 challenges. J Environ Health. 1995;57(6):28–34.
17. Biagini B, Bierbaum R, Stults M, Dobardzic S, McNeeley SM. A typology of adaptation actions: a global look at climate adaptation actions financed through the global environment facility. Glob Environ Chang. 2014;25:97–108.
18. Füssel H-M, Klein R, Ebi K. Adaptation assessment for public health. Climate change and adaptation strategies for human health. 2006:41–62.
19. Smit B, Burton I, Klein RJT, Street R. The science of adaptation: a framework for assessment. Mitig Adapt Strateg Glob Chang. 1999;4(3-4):199–213.
20. Krieger N. Theories for social epidemiology in the 21st century: an ecosocial perspective. Int J Epidemiol. 2001;30(4):668–77.
21. McMichael AJ. Prisoners of the proximate: loosening the constraints on epidemiology in an age of change. Am J Epidemiol. 1999;149(10):887–97.
22. Bronfenbrenner U. Ecological systems theory. In: Vasta R, editor. Annals of child development. 6. London: Jessica Kingsley Publishers; 1989. p. 187–249.
23. Marinucci G, Luber G, Uejio C, Saha S, Hess J. Building resilience against climate effects—a novel framework to facilitate climate readiness in public health agencies. Int J Environ Res Public Health. 2014;11(6):6433–58.
24. Rudolph L. Climate change and health: a framework for action. In: Institute PH, editor. ; 2014.
25. Haines A, Patz JA. Health effects of climate change. (1538-3598 (electronic).
26. Mercy JA, Rosenberg ML, Powell KE, Broome CV, Roper WL. Public health policy for preventing violence. Health Aff. 1993;12(4):7–29.
27. Wise RM, Fazey I, Smith MS, Park SE, Eakin HC, Van Garderen ERMA, et al. Reconceptualising adaptation to climate change as part of pathways of change and response. Glob Environ Chang. 2014;28:325–36.
28. Watts N, Amann M, Ayeb-Karlsson S, Belesova K, Bouley T, Boykoff M, et al. The Lancet Countdown on health and climate change: from 25 years of inaction to a global transformation for public health. Lancet. 2018;391(10120):581–630.
29. Frumkin H, Hess J, Luber G, Malilay J, McGeehin M. Climate change: the public health response. Am J Public Health. 2008;98(3):435–45.
30. Mills MA, Edmondson D, Fau - Park CL, Park CL. Trauma and stress response among Hurricane Katrina evacuees. (1541-0048 (Electronic)).
31. Greenough PG, Kirsch TD. Hurricane Katrina. Public health response--assessing needs. N Engl J Med. 2005;353(15):1544–6.
32. Centers for Disease Control and Prevention. Injury and illness surveillance in hospitals and acute-care facilities after Hurricanes Katrina and Rita--New Orleans area, Louisiana, September 25-October 15, 2005. MMWR Morb Mortal Wkly Rep. 2006;55(2):35–8.
33. Pina AA, Villalta IK, Ortiz CD, Gottschall AC, Costa NM, Weems CF. Social support, discrimination, and coping as predictors of posttraumatic stress reactions in youth survivors of Hurricane Katrina. J Clin Child Adolesc Psychol. 2008;37(3):564–74.
34. Fussell E, Lowe SR. The impact of housing displacement on the mental health of low-income parents after Hurricane Katrina. Soc Sci Med. 2014;113:137–44.
35. Harlan SL, Brazel AJ, Prashad L, Stefanov WL, Larsen L. Neighborhood microclimates and vulnerability to heat stress. Soc Sci Med. 2006;63(11):2847–63.
36. Donovan GH, Michael YL, Gatziolis D, Prestemon JP, Whitsel EA. Is tree loss associated with cardiovascular-disease risk in the Women's Health Initiative? A natural experiment. Health Place. 2015;36:1–7.
37. Centers for Disease Control and Prevention. Climate and Health Program (CHP) Atlanta, GA, 2020. Available from: https://www.cdc.gov/climateandhealth/default.htm.

38. EcoAdapt. Climate Adaptation Knowledge Exchange (CAKE) 2020. Available from: https://www.cakex.org/.
39. The Medical Society Consortium on Cilmate and Health. Geoge Mason University; 2020. Available from: https://medsocietiesforclimatehealth.org/.
40. Anderson HBC, Cameron LL, Christenson M, Conlon KC, Dorevitch S, Dumas J, Eidson M,, Ferguson A GE, Hanson A, Hess JJ, Hoppe B, Horton J, Jagger M, Krueger S, Largo TW, Losurdo GM, Mack SR,, Moran C MC, Raab K, Saha S, Schramm PJ, Shipp-Hilts A, Smith SJ, Thelen M, Thie L, Walker R. BRACE Midwest, Practice. aSCo. Climate and health intervention assessment: evidence on public health interventions to prevent the negative health effects of climate change. Centers for Disease Control and Prevention: Climate and Health Program; 2017. p. https://www.cdc.gov/climateandhealth/docs/ClimateAndHealthInterventionAssessment_508.pdf.
41. Maricopa Association of Governments. Heat Relief Regional Network 2020. [Available from: https://azmag.gov/Programs/Homelessness/Heat-Relief-Regional-Network.
42. Marsh D, Husein K, Lobo M, Shah MA, Luby S. Verbal autopsy in Karachi slums: comparing single and multiple cause of child deaths. Health Policy Plan. 1995;10(4):395–403.
43. Gruber JS, Ercumen A, Colford JM Jr. Coliform bacteria as indicators of diarrheal risk in household drinking water: systematic review and meta-analysis. PLoS One. 2014;9(9):e107429.
44. Yousuf FA, Siddiqui R, Khan NA. Survey of gram negative and gram positive bacteria in drinking water supplies in Karachi, Pakistan. Microbiol Res J Int. 2014;4:592–7.
45. Nygren BL, O'Reilly CE, Rajasingham A, Omore R, Ombok M, Awuor AO, et al. The relationship between distance to water source and moderate-to-severe diarrhea in the global enterics multi-center study in Kenya, 2008–2011. Am J Trop Med Hyg. 2016;94(5):1143–9.
46. Rauf S, Bakhsh K, Hassan S, Nadeem AM, Kamran MA. Determinants of a household's choice of drinking water source in Punjab, Pakistan. Pol J Environ Stud. 2015;24(6):2751–4.
47. Burr ML, Davis AR, Zbijowski AG. Diarrhoea and the drought. Public Health. 1978;92(2):86–7.
48. Whitworth KL, Baldwin DS, Kerr JL. Drought, floods and water quality: drivers of a severe hypoxic blackwater event in a major river system (the southern Murray–Darling Basin, Australia). J Hydrol. 2012;450:190–8.
49. Hanif M, Khan AH, Adnan S. Latitudinal precipitation characteristics and trends in Pakistan. J Hydrol. 2013;492:266–72.
50. Sadiq N, Qureshi MS. Climatic variability and linear trend models for the five major cities of Pakistan. J Geograph Geol. 2010;2(1):83.
51. Luby SP, Agboatwalla M, Painter J, Altaf A, Billhimer W, Keswick B, et al. Combining drinking water treatment and hand washing for diarrhoea prevention, a cluster randomised controlled trial. Tropical Med Int Health. 2006;11(4):479–89.
52. Dilling L, Daly ME, Kenney DA, Klein R, Miller K, Ray AJ, et al. Drought in urban water systems: learning lessons for climate adaptive capacity. Clim Risk Manag. 2019;23:32–42.
53. Bours D, McGinn C, Pringle P. Monitoring & evaluation for climate change adaptation and resilience: a synthesis of tools, frameworks and approaches. 2nd ed. Oxford: SEA Change CoP, Phnom Penh, UKCIP; 2014.
54. Vanos J, Vecellio DJ, Kjellstrom T. Workplace heat exposure, health protection, and economic impacts: a case study in Canada. Am J Ind Med. 2019;62(12):1024–37.
55. Dinshaw A, Fisher S, McGray H, Rai N, Schaar J. Monitoring and evaluation of climate change adaptation. Paris: OECD Publishing; 2014.
56. Adger WN, Arnell NW, Tompkins EL. Successful adaptation to climate change across scales. Glob Environ Chang. 2005;15(2):77–86.
57. Dilling L, Prakash A, Zommers Z, Ahmad F, Singh N, de Wit S, et al. Is adaptation success a flawed concept? Nat Clim Chang. 2019;9(8):572–4.
58. Jones L. Resilience isn't the same for all: comparing subjective and objective approaches to resilience measurement. Wiley Interdiscip Rev Clim Chang. 2019;10(1):e552.
59. Brooks N, Anderson S, Burton I, Fisher S, Rai, Tellam I. An operational framework for Tracking. Adaptation and Measuring Development (TAMD). IIED Climate Change Working Paper No. 5. London, UK. 2013.

60. Daze A, Ambrose K, Ehrhart C. Climate vulnerability and capacity analysis handbook. CARE International. Atlanta, GA. 2009.
61. Striessnig E, Lutz W, Patt A. Effects of educational attainment on climate risk vulnerability. Ecol Soc. 2013;18(1):16.
62. Bryan E, Ringler C, Okoba B, Roncoli C, Silvestri S, Herrero M. Adapting agriculture to climate change in Kenya: household strategies and determinants. J Environ Manag. 2013;114:26–35.
63. Fisher S, Dinshaw A, McGray H, Rai N, Schaar J. Evaluating climate change adaptation: learning from methods in international development. N Dir Eval. 2015;2015(147):13–35.
64. Pahl-Wostl C. Participative and stakeholder-based policy design, evaluation and modeling processes. Integr Assess. 2002;3(1):3–14.
65. Gullison R, Hardner J. Using limiting factors analysis to overcome the problem of long time horizons. N Dir Eval. 2009;2009(122):19–29.
66. Bierbaum R, Smith JB, Lee A, Blair M, Carter L, Chapin FS, et al. A comprehensive review of climate adaptation in the United States: more than before, but less than needed. Mitig Adapt Strateg Glob Chang. 2013;18(3):361–406.
67. Kates RW, Travis WR, Wilbanks TJ. Transformational adaptation when incremental adaptations to climate change are insufficient. Proc Natl Acad Sci. 2012;109(19):7156.
68. Drawdown HP. The most comprehensive plan ever proposed to reverse global warming: New York, NY: Penguin; 2017.

Chapter 28
Integrating Climate Change, the Environment, and Sustainability Themes Into Professional Health Sciences Courses: A Case Study Across a University System

Tammy Nicastro, Arianne Teherani, Helene G. Margolis, and Sheri Weiser

Background and History

Climate change threatens the advancements made in global development and health over the past 50 years, placing humans at increased risk for infectious (communicable) and noncommunicable (chronic) diseases, food insecurity, diminished mental health, and forced migration (population displacement) with potential for catastrophic outcomes [1, 2]. Improvements in global public health have resulted in part from the strides made in building and stabilizing infrastructure, education, and economies in the developing world, making the deleterious effects of climate change on these sectors important for human health [2]. Efforts to mitigate climate change and environmental degradation must factor in the impacts of both on human health in order to achieve sustainable solutions [1]. Health scientists and health-care professionals must integrate and utilize knowledge of the impacts of climate change and environmental degradation on human health to advocate for global solutions.

T. Nicastro (✉)
National Center of Excellence in Women's Health, Department of Obstetrics, Gynecology, and Reproductive Sciences, University of California San Francisco, San Francisco, CA, USA
e-mail: tammy.nicastro@ucsf.edu

A. Teherani
Department of Medicine, University of California San Francisco, San Francisco, CA, USA
e-mail: Teherani@ucsf.edu

H. G. Margolis
Department of Internal Medicine, General Medicine Division, Center for Health Services Research in Primary Care, Women's Center for Health, University of California, Davis, Davis, CA, USA
e-mail: hgmargolis@UCDAVIS.EDU

S. Weiser
Department of Medicine, University of California, San Francisco, San Francisco, CA, USA
e-mail: Sheri.Weiser@ucsf.edu

K. E. Pinkerton, W. N. Rom (eds.), *Climate Change and Global Public Health*, Respiratory Medicine, https://doi.org/10.1007/978-3-030-54746-2_28

While effects may be felt most acutely in resource-poor countries, the United States and other resource-rich countries are already experiencing economic losses and health impacts due to climate change and environmental degradation [3, 4]. Heat waves, floods, hurricanes, droughts, and polar vortexes are causing economic, infrastructure, and education losses while creating health crises for the populations impacted. In addition, environmental degradation due to air and water pollution is causing increases in respiratory illnesses and a wide array of other chronic diseases and conditions, such as cardiovascular, renal, and adverse birth outcomes [2, 5–9].

Climate change is reinforcing health disparities for the most vulnerable populations in the United States and around the world [10, 11]. Health inequities will be amplified by climate change putting people of color, immigrants, people with pre-existing chronic health issues, people living in poverty, and indigenous people at the greatest risk for health threats because of limited capacity of individuals and communities to adapt to and diminish the effects of climate change [12, 13].The devasting current and projected future impacts of climate change and ecosystem degradation on human health and well-being are a call to action for the academy to deliver on its responsibility to prepare health science learners to meet the needs of the public they will serve. Currently, few health science schools provide training on the relations between climate change, ecosystem degradation, and human health [10].

This chapter illustrates how a small team developed and delivered a training system for health science faculty across the University of California's (UC) six health sciences campuses, enabling the integration of climate change, sustainability, and environmental themes into health sciences curriculum. These efforts align with a health science education initiative first characterized by the United Kingdom's Center for Sustainable Healthcare called Sustainable Healthcare Education (SHE). SHE is best described as education covering topics of impacts of climate change and ecosystem alterations on health, as well as the impact the healthcare industry is having on climate change and ecosystem alteration and ultimately, on human health and potential sustainable solutions to those negative impacts [14].

It is important to note that this initiative was implemented in a high-income country and a progressive state (i.e. California) with the world's 5th largest economy but also with notable health disparities. The need to address health disparities in California and across the US provides a platform to introduce SHE curriculum [15–17]. At the same time, California citizens and politicians have shown their political will to address climate change, sustainability, and environmental degradation through repeated ballot measures and elections of political leaders sensitive to the cause [18]. The motivation stems in part from known economic benefits of transitioning to sustainable energy sources [18]. Evidence suggests that public health workers adjust how they incorporate discussing climate change impacts on health to meet the stated goals of their funding agency or leadership of their public health office, and hence economic benefits should be part of the discussion [19].

Historically, state and federal resources have been devoted to build the nation's medical schools in exchange for the assurance that these schools serve the society's individual and community healthcare needs [20]. Social justice and health equity, which physicians are morally expected to promote, are implicit in those community healthcare needs [20]. In recent years, as less state and federal funds have been allocated to fund education, federal dollars are still funding these institutions indirectly through tax-deductible philanthropy, reinforcing the social contract that puts medical education in service of the society that has partially underwritten it [20]. During the 1970s and early 1980s, significant state and federal resources poured into developing 28 allopathic medical schools with more than half of those schools designated with a purpose to be community-based [21]. Those community-based schools, like state-funded medical schools, have a requirement to consider their educational offerings in the context of the service needs of their constituents. While medical schools and other health science schools have evolved to carry the responsibility of training students to face current and future public health threats [21], our most recent history shows that medical schools are not keeping up with training for the most imminent threats that are resulting from climate change and ecosystem degradation [10]. The public's expectations of medical schools to train for the needs of the public are but one example of the social contract between the public who supports health science schools and the academy designing the curricula of these programs. Nursing, dentistry, pharmacy, public health, and health science graduate programs also each have a social contract embedded in their mission to deliver care and support health equity for the most vulnerable [22–25].

Health science schools have long played a role in training students to explore their ethics in the context of delivering care to patients and conducting research with both human and animal subjects [26]. Ethics training for health professionals has included exploration of beliefs around what should be included in the scope of practice, how to protect private information, the difference between helping or harming, and corruption [27]. Because of the role of the healthcare sector harming the environment through an excessive and often unnecessary use of energy and production of waste, both of which harm human health, new ethical paradigms are emerging that challenge how healthcare is delivered. New models of teaching and practice that incorporate ethical questions for health practitioners and new learners around the role they play in harming or supporting the environment are warranted.

Pathways and Models for SHE Inclusion

Theories have been examined for the types of SHE content that should be incorporated into health science curricula and at what stage of training so that learners are able to become proficient in this knowledge [28–33]. Certain objectives are best taught during preclinical years and other during clinical experiences, for those health science programs with clinical training [10]. Educators should first gain an understanding of their students' knowledge and competencies around key SHE objectives. Most students have some understanding of the relationship between climate change, the environment, sustainability, and health; however, their understanding is often at a basic level and not proficient enough with this content to apply their existing knowledge toward research or clinical work [32]. Teherani conducted studies to identify specific objectives for integrating SHE themes and the expected outcomes for learners [28]. Learning objectives need to be taught throughout the learner's time on campus and in both the formal classroom or practical, applied skills setting and in nonformal academic environments. For example, if the school is attempting to teach sustainability as a responsibility of healthcare providers but the school doesn't offer recycling bins on campus or guidelines for medical waste disposal, students will have conflicted feelings about the school's commitment to sustainability. The academic institution's culture and values related to climate change and sustainability can also impact how faculty respond to the notion of incorporating new content into their courses and how students assimilate SHE themes into their thinking.

The Faculty Training Workshop

University of California Leadership

In 2013, the University of California Office of the President (UCOP) created The Carbon Neutrality Initiative (CNI) as a system-wide effort to reach carbon neutrality by 2025. The CNI is working to achieve this goal through policy changes that effect new buildings, energy consumption, and a transition to clean energy use, while engaging with faculty, students, and staff to implement new policies [34]. The CNI leverages UC's excellence in research and innovation to develop solutions that address the climate crisis and can be adapted to each of the campus' academic culture and community. Strategic communication from the CNI leadership team is a key element used within the 10-campus system (Appendix A) to learn how policy changes are impacting the campus communities and to disseminate information on CNI progress reports [35]. The CNI is organized around nine working groups that comprise UC President Janet Napolitano's Global Climate Leadership Council: Energy services; Applicable research; Campus and medical center climate action plans; Sustainability policy; Faculty engagement; Student engagement; Health sciences and services; Financial strategies; and Communications and political advocacy. These working groups represent key sectors of the UC system advising President Napolitano on how best to achieve carbon neutrality while accounting for the demands of each campus' teaching, research, and, in some cases, healthcare missions.

The Workshop Concept

In 2016, two years prior to launching this UC-wide SHE faculty training initiative, Teherani and Weiser, the investigators, participated in a program with similar components funded by President Napolitano's Global Climate Leadership Council.

This program was created to incentivize, support, and connect faculty across the 10 UC campuses who voluntarily chose to infuse existing courses with relevant climate and sustainability concepts. This program was not specific to health science and instead included faculty from engineering, urban planning, the humanities, and arts and sciences. Teherani and Weiser were the only campus team to have the responsibility of working solely with health science faculty, since their home campus, the University of California San Francisco (UCSF), is the only UC campus dedicated entirely to health. Ultimately, this exercise served as a proof-of-concept for the workshop they created to train their health sciences colleagues across the UC system in 2018. The results showed their workshop model was effective at: (1) recruiting faculty to participate; (2) providing those faculty with the tools needed to transform their courses with a focus on climate change, sustainability, and health; and (3) keeping faculty engaged through their process of infusing their course with SHE. Twenty UCSF faculty participated in this workshop, and each of them transformed at least one existing course to incorporate SHE themes. Faculty participants represented all four UCSF health professions schools – medicine, nursing, pharmacy, and dentistry – plus the UCSF graduate division which included faculty from programs such as global health sciences and medical anthropology. The faculty cohort remained committed to this effort and continued to advocate for incorporating SHE themes into health science curriculum. Examples of courses transformed include: (1) Health Policy for Pharmacists; (2) Biomaterials Science and Cast Restorations in Dentistry; (3) Principles and Methods of Epidemiology; (4) Core Curriculum in Allergy and Immunology; (5) Women's Health, the Environment, and Public Health Activism; and (6) Environmental Ethics. The success of this effort led Teherani and Weiser to conceptualize a larger initiative that would touch all UC health science campuses.

Teherani and Weiser conceptualized and proposed a faculty training initiative to the CNI leadership team, as an effort to integrate SHE into existing health science curricula across the six UC health science campuses. The workshop was designed to train two faculty leaders from each of the six health science campuses to integrate SHE themes into their courses and conduct a training workshop to advise their colleagues on how to do the same. The CNI leadership team connected this faculty training initiative's objectives with the goals of two of its working groups: (1) Health Sciences and Services and (2) Faculty Engagement and Education. The primary objectives of the health-focused initiative were: (1) To inspire and assist faculty across the health sciences in developing ideas about how to integrate climate change, sustainability, health, and carbon neutrality into existing courses; (2) Build a broad base of UC health sciences faculty engaged in UC's 2025 carbon neutrality goal who form a community of practice engaged in ongoing dialogue and sharing of educational best practices, materials, and resources on climate change and sustainable healthcare education; and (3) To develop a community of faculty facilitators on each campus that demonstrate group leadership capabilities; interdisciplinary scholarship and/or teaching; understand UC and campus-specific sustainability goals; understand the interrelated nature of climate change, sustainability, and health; and are willing to continue sharing tools and lessons gained from this initiative.

Executive Sponsorship

Our first step was securing executive sponsorship from UCSF leadership. This was important early in the project to send a clear message to faculty that these efforts were endorsed and valued by deans, the provost, and chancellor. In our case, we reached out to our chancellor, provost, and deans. In addition, partnering with UCSF leadership was an essential part of the communication plan used to reach other campuses to build a network of faculty partners. Teherani and Weiser created a model for the UC-wide initiative that replicated their proof-of-concept model by identifying two faculty leads at each of the health science campuses who could do the work they did two years prior, to train their campus colleagues.

Faculty Leads

Our second step was to identify faculty leads at each campus. In our case, this was a multistep process that included a strategic communication plan between our team, the UCSF chancellor, provost, and deans of each of UCSF's professional schools. We utilized UCSF's leadership to communicate with their counterparts at each of the other UC health science campuses, on our behalf, to help identify nominees to serve as campus leads. We described in detail the characteristics required to serve as a faculty lead for this project including being able to recruit and lead a team of faculty colleagues to attend a training workshop and then transform their course and be willing to participate in follow-up activities to measure outcomes. By working with leadership at each campus, we ensured that internal values and dynamics at each campus would be reflected in the nominees. The leadership at each campus who agreed to work with us included deans, department chairs, vice chancellors, and leaders within campus hospitals. We requested both faculty nominations from each campus' leadership network, and we requested that they circulate our project description and invite self-nominations from interested faculty. Our goal was to identify two faculty leads for the Los Angeles, Irvine, San Diego, and Davis campuses and one lead for the Riverside campus since it had recently opened its medical school. The nomination process took six months.

Each campus was unique in how faculty leads were identified, which helped us better understand the diverse campus cultures surrounding SHE themes and health. We encountered two primary obstacles while identifying faculty leads which were related to available time for nominated faculty or concern that this use of their time would not be valued by their department chair or dean. To identify the leads, we searched campus websites for key terms such as "climate change and health" or "sustainability and health," asking campus staff in the sustainability field which faculty were engaged with campus sustainability efforts, and conducted one-to-one conversations with campus department chairs to identify faculty who met our criteria. Ultimately, we identified a team of diverse, experienced, and committed faculty leads who each carried out their responsibilities to the initiative. Some key lessons learned were that sending campus-wide communications via email was not an effective mechanism to reach some of the most ideal candidates. In addition, having conversations with department chairs at each campus allowed us to understand the competing demands that were highly specific to the current priorities at each location. Having this additional information helped us shape the criteria that would best serve the needs of the initiative to recruit the right faculty for training and course transformation.

While we were identifying faculty leads within a multicampus system, the same principles apply to identifying faculty leads at one institution. First, a communication plan must be established that utilizes the appropriate leadership such that campus-wide communication can be distributed. Next, a vetting process that includes campus leadership should be established in the event more than two qualified candidates are identified. Last, plans to conduct a search using all campus networks related to SHE themes should be conducted if nominees do not emerge through leadership nomination or self-nomination processes.

Training Workshop for Faculty Leads

Our next step was to establish the content, speakers, and exercises for the one-day training at UCSF for the faculty leads from the other UC health science campuses. Before building our agenda, we established key themes we wanted represented throughout the training that emphasized the threat climate change posed to human health and the opportunity it presented to the health sciences if we organize to act strategically. In addition, we wanted to emphasize the important opportunity educators have in academic health organizations to utilize teaching, research, and clinical care to expose students to SHE themes throughout their training which can enhance students' understanding of and proficiency in SHE themes.

Agenda, Speakers, and Group Exercise

Our agenda (Appendix B) was designed to do the following: (1) bring context to the purpose of the initiative within the UC system; (2) train attendees on the importance of integrating SHE themes into health science curricula; (3) introduce methods of integrating SHE themes; (4) practice identifying content areas for integration in health science courses; and (5) train attendees on how to build a workshop at their campus to train their colleagues. We included two speakers from UCSF's leadership team, the CNI director, the co-director of the UC Global Health Institute's (UCGHI) Planetary Health Center of Expertise, a faculty member serving in an advisory role to the workshop team, and four UCSF faculty who had participated in the 2016 SHE curriculum workshop. Weiser and Teherani provided the rest of the content, and we also included some group exercises.

We invited the Executive director of the Institute for Global Health Sciences at UCSF to speak because of his experience in global health policy transformation and overall leadership in the field of global and public health. He focused his remarks in three areas: (1) the impact of climate change on global health; (2) examples of other seemingly insurmountable global health challenges that we have overcome; and (3) the power we possess to create a movement, not only as researchers and advocates but as educators. We worked with each of our speakers while they were preparing their remarks, so they aligned and reinforced key messages that would be presented throughout the workshop. We invited UCSF's Dean of the School of Medicine because of his leadership in working with underserved populations and working to eliminate health disparities locally and globally. His work connected to the important theme emphasized throughout the workshop, namely, that climate change most severely impacts the most vulnerable populations in the global south as well as on economically disadvantaged populations in the developed world. The Dean's participation also included a period for discussion that allowed faculty leads to share their concerns about issues they might face at their home campus when trying to gain support to integrate SHE themes into curricula and ask for his perspective on how to move forward. The presentation on the UCGHI Planetary Health Center of Expertise helped the attendees, who were primarily coming from human health backgrounds, put into perspective the environmental, animal, and human health interactions that will collectively impact each other and why it's important to expose learners to the planetary health perspective. The presentation given by the faculty advisor focused on California policy work related to climate change, the environment, and health, which was relevant to our group, given our publicly funded university system and the ways in which California policy affect public health. UC's CNI director presented on responsibilities and functions of the CNI and put into context how the SHE curriculum initiative would help the UC system reach carbon neutrality while instilling a culture of sustainability within this cohort of faculty and students. In addition, he presented the policies and programs implemented at each campus that were helping the UC system move closer to the goal of carbon neutrality. In our case, the system-wide CNI serves as a unifying effort that leverages the collective research, policy, and purchasing assets of our large university system. However, the idea is transferable in other smaller organizations where there is an overarching goal of reaching a sustainability goal and departments or units within the organization can be supported to design solutions to reach organization-wide goals.

Teherani and Weiser's presentations focused on climate change impacts on human health, the projections for future impacts, their previous work training faculty to integrate SHE themes into curricula, and finally, the theories and practices underlying SHE curricula integration. These presentations gave participants context for the presentations given by faculty who attended the previous workshop and had transformed their courses. Participants were further engaged through a group exercise (Appendix C) which allowed them to use the concepts presented earlier to discuss how they would incorporate SHE themes into five types of health science courses. The workshop concluded with a presentation on shifting cultures within health science academia. Drs. Weiser and Teherani discussed how students must be involved early and often throughout their training and receive messages that are solutions-based, so they can become empowered to address impacts of climate change and the sustainability practices in the healthcare systems in which they will work. In addition, Weiser's presentation highlighted how a robust research portfolio has begun to incorporate climate change and sustainability themes as they relate to expertise with infectious disease, food insecurity, and chronic illness. A student provided her insight into a transformed course taught by a faculty member who attended the 2016

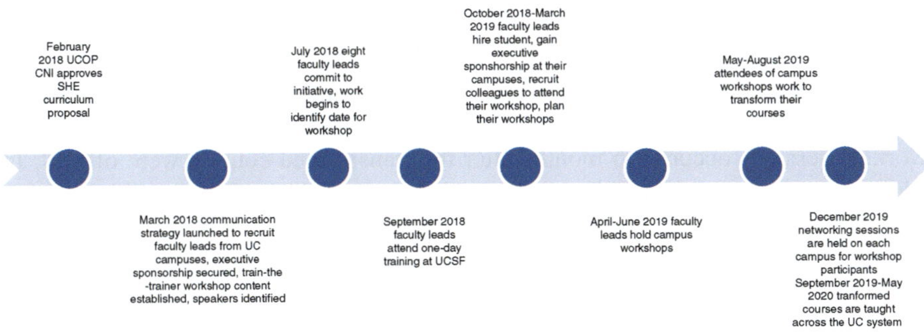

Fig. 28.1 SHE curriculum initiative timeline

workshop and discussed how engagement in the course had contributed to her master's thesis research and current research on climate change and health. The goal was to conclude the workshop with a sense of urgency and empowerment to take action.

Timelines

Our overall timeline was roughly 2 years from initiating efforts to identify faculty leads through the period when transformed courses were offered across the UC system (Fig. 28.1). The initiative was approved for funding in February of 2018 at which time we began working on the communication plan, the content, and speakers for the workshop. It took 6 months to identify the faculty leads which was a critical piece of work because it then enabled us to begin identifying a workshop date that worked for the majority. In late September, we conducted the workshop to train the faculty leads with an expectation that they would identify their workshop participants and hold their campus workshops within 7 months. After the campus workshops were completed, the faculty who attended were expected to transform their courses and deliver those courses during the soonest possible academic session. Because not all courses are offered in the fall term, this extended the period for all transformed courses to be offered, i.e., some courses were offered in the spring 2020 term. Finally, within 6 months of offering the transformed courses, faculty leads were expected to hold a networking session on their campus, providing an opportunity for faculty to share best practices on how they transformed their course and how students received the new material.

Library of Resources

A library of resources was compiled and shared on a secure cloud-based platform. The library included literature on SHE, sample agendas, slide deck presentations, and communication tools to recruit faculty participants. The password-protected platform was hosted by UCSF information technology services and allowed for any UC faculty or staff member to access the secure space.

Creating SHE Communities at Each Campus

Creating a community of faculty and other staff committed to integrating SHE themes was one of the primary objectives of the postworkshop networking session. The networking sessions occurred 6 months after the transformed courses were offered. In addition, each campus was asked to consider how it could best work within existing communication channels and resources to maintain momentum and grow the community of faculty willing to transform their courses. The idea was for each campus to develop its own system to effectively recruit and train faculty to transform their courses. At the same time, communication between all participants in this initiative was made possible by scheduled conference calls and online portals. By sharing knowledge and best practices across the UC's six health science campuses, we were able to leverage lessons learned and pathways toward success while also building networks that can be utilized throughout a faculty member's work in this arena.

Linking Deans and Vice Deans of Education

To support each of the initiative's goals, we advised each campus to seek ongoing support from their health science deans and vice deans of education. In some cases, administrative approval could be required to adjust curricula, so establishing an open dialogue with education deans can help integrate SHE themes without additional obstacles. Deans and vice deans can also play a role in establishing SHE integration as an ongoing area of focus and perhaps take on the organization of expanded faculty training to integrate SHE themes. These administrative leaders can also support the cohort of faculty engaged in SHE through a variety of ways including salary support for time required to prepare course transformation, recognition for supporting sustainability goals (or other campus goals that align with SHE), and organizing ongoing training and networking.

Evaluation Framework

The primary evaluation framework addresses whether the objectives for our workshops were met. The secondary framework consists of a realist evaluation. A realist evaluation helps decision makers and program developers learn what works, in which circumstances and for whom. The realist evaluation framework is particularly useful for the evaluation of the SHE curriculum workshop because very little to no work to date has addressed the development of faculty to educate for SHE. Some fundamental initial understanding of how faculty development in this area can and cannot be successful for a range of learners is necessary. Our evaluation of the workshops consists of both qualitative and quantitative as well as process and outcome data collection methods.

To measure the efficacy of the training workshop for faculty leads, we administered a mixed-methods survey for all participants immediately after they attended the workshop. To understand how effective our train-the-trainer model was at preparing faculty at each school to transform their courses, we administered a mixed-methods survey to all faculty who attended the campus workshops taught by the faculty leads. At the time of this publication, we are in the process of conducting interviews with faculty lead participants to better understand the success of their workshops and networking activities on influencing the community of faculty educators incorporating SHE. We will also be conducting interviews with faculty who attended a workshop at their campus, to evaluate how useful the workshop and networking sessions were at supporting faculty while integrating SHE and the impact their course had on their students. In addition, we tracked the demographics of faculty participants, names of courses transformed including the department, school, or college, how often it was offered and if it was required or an elective, and the number of students enrolled in each course. Finally, participants of the campus workshops were given a tool to reflect on the process of transforming their course and the outcomes. All quantitative and qualitative data will be analyzed and used for reporting to the CNI and various internal partners and participants (Appendix D). We also plan to submit manuscripts to peer-reviewed journals for publication.

Budgets and Forms of Support

Each campus was given a budget for a student stipend so that faculty leads could hire a student to: support the logistical planning for the workshop; support in faculty recruitment efforts; and follow-up with workshop participants for participation in the evaluation tools. Resources were distributed to each campus to support stipends for faculty leads and faculty workshop participants, as well as funds to host the workshops. Throughout the period when faculty leads and their students were planning their workshops, we provided support by hosting a monthly conference call with all faculty leads as well as individual support for any obstacles encountered. The purpose of the conference call was to work with each campus on their initial targets for recruiting support from leadership, identifying faculty participants, and shaping their workshop agenda with speakers. At the same time, faculty leads supported each other by brainstorming solutions for issues that a campus may have experienced.

Conclusion

The SHE curriculum initiative enabled more than 100 faculty to integrate climate change, sustainability, and environmental themes into their existing coursework, thereby reaching hundreds of students. Through support from the UC President's office, this initiative was able to test a faculty training model for scalability and efficacy across a 10-capus system. Gaining support from leadership at each campus and leveraging their communication channels enabled us to identify qualified faculty to serve as leaders for their campus community on SHE training and integration. It is important to understand the readiness of a campus community to adopt additional curriculum material before embarking on an initiative of this nature which requires significant time from faculty. Keeping faculty engaged throughout the process of transforming their courses is an important component in ensuring SHE themes are integrated. Bringing more SHE into health science programs will be required to meet the public health needs of the local and global populations. The train-the-trainer model can be an effective tool for grass-roots curricula transformation if financial resources, workshop materials, and localized leadership are available. Adapting this program in a less-wealthy state or low-to-middle income countries could present different opportunities and challenges from those discussed in this chapter.

Appendix A

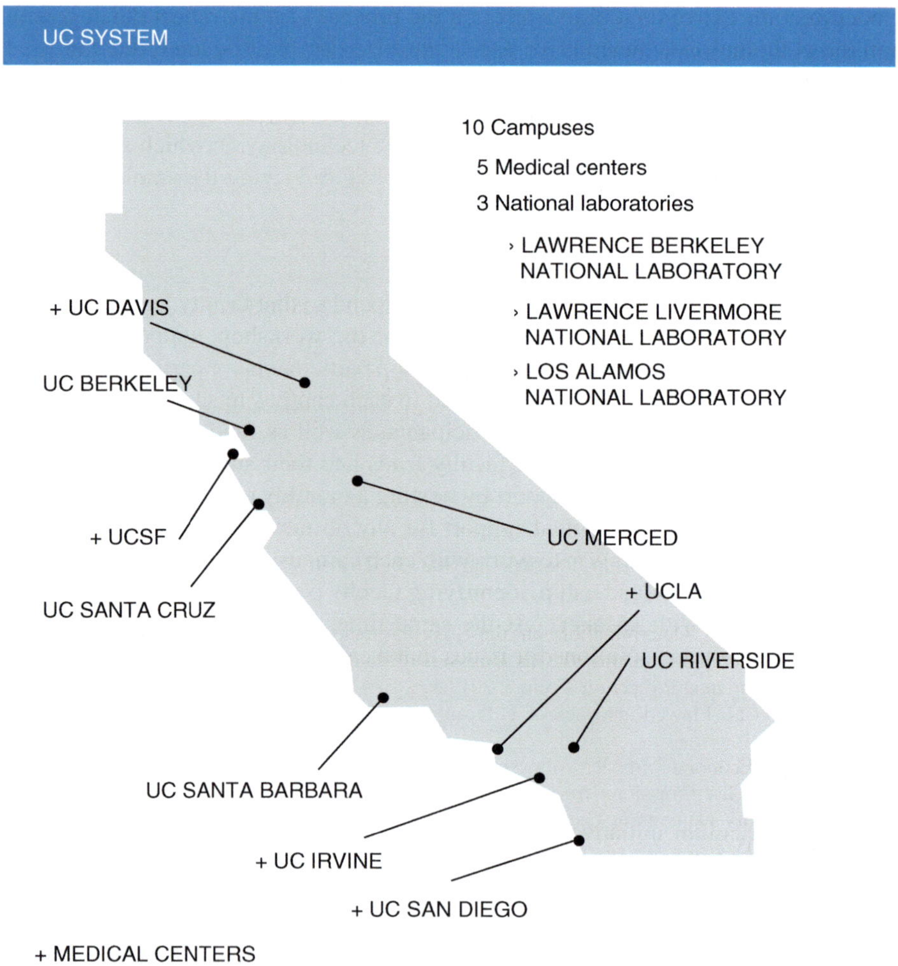

Appendix B

Agenda

Climate Change, The Environment, Sustainability, and Health Curriculum Workshop

Tuesday, September 25th, 2018

University of California San Francisco, Mission Hall room 1405

Sponsor: University of California Office of the President (UCOP), Carbon Neutrality Initiative (CNI)

Co-Leaders: Dr. Arianne Teherani, Professor, UCSF School of Medicine, and Dr. Sheri Weiser, Professor, UCSF School of Medicine

Faculty Advisor: Dr. Helene Margolis, Professor UC Davis School of Medicine

09:00–10:00 am	Breakfast buffet, arrivals, check-in
10:00–10:15 am	Opening Remarks, Dr. Jaime Sepulveda, Haile T. Debas Distinguished Professor of Global Health, and Executive Director, UCSF Institute for Global Health Sciences
10:15–10:30 am	Group Introductions
10:30–10:45 am	UCOP CNI, Matt St. Clair, UCOP Director of Sustainability
10:45–11:05 am	Introduction to Curriculum Workshop: Past and Present Workshops, Dr. Sheri Weiser
11:05–11:35 am	Climate Change and Health, California Policy Overview, Dr. Helene Margolis, UC Davis faculty and Curriculum Workshop Faculty Advisor
11:35–11:50 am	Coffee/Tea Break
11:50–12:05 pm	UC Global Health Institute (UCGHI) Planetary Health Overview, Dr. Woutrina Smith, UC Davis faculty and Co-director of UCGHI Planetary Health Center of Expertise
12:05–12:20 pm	Transforming Health Sciences Education, Dr. Arianne Teherani
12:20–1:00 pm	Group exercise: Course Transformation
1:00–2:00 pm	Lunch
2:00–2:15 pm	Lessons Learned on Course Transformation: Dr. Tom Newman, UCSF faculty, School of Medicine
2:15–2:30 pm	Lessons Learned on Course Transformation: Dr. Vincanne Adams, UCSF faculty, Graduate Division
2:30–2:45 pm	Q&A with Drs. Newman and Adams
2:45–3:00 pm	Coffee/Tea Break

3:00–3:30 pm	Importance of Climate Change, the Environment, and Sustainability for Health Science Educators and Students, Dr. Talmadge King, UCSF Dean, School of Medicine
3:30–3:45 pm	Lessons Learned on Course Transformation: Dr. Peter Chin-Hong, UCSF faculty, School of Medicine
3:45–4:00 pm	Lessons Learned on Course Transformation: Dr. James Seward, UCSF faculty, School of Medicine
4:00–4:15 pm	Q&A with Drs. Chin-Hong and Seward
4:15–4:30 pm	Online library and evaluation tools, Tammy Nicastro and Bennett Kissel
4:30–5:00 pm	Changing culture through changing education and research, Dr. Teherani and Weiser
5:00–6:30 pm	Wine reception

Appendix C

Moderator instructions:
1. *Now we will work as a group on an exercise to practice some of the concepts we've just reviewed on transforming courses.*
2. *We will spend the next 40 minutes sharing ideas on how we could incorporate climate change and sustainability into these courses.*
3. *Some questions to get us started:*
Questions:
1. *What aspect of the course is related to climate change and sustainability?*
2. *How would you teach this course with sustainability as a theme? (And which methods would you use?)*
3. *What are commonalities shared among your fields, what are differences?*

Climate Change and Sustainability Across the Curriculum

Module Exercise - University of California Courses

1. *Molecular Mechanisms of Human Diseases*
   Fundamental concepts and methodologies in modern biology, with emphasis on implications and relevance to human disease and integration of biology with mechanisms underlying disease development and applications in therapy as they apply to cancer biology, infectious disease, and modern biological approaches. (4 units)

2. *Introduction to Probability and Statistics*
   Introduction to basic principles of probability and statistical inference. Point estimation, interval estimating, and testing hypotheses, Bayesian approaches to inference. (4 units)
3. *Introduction to Sociocultural Theory*
   Seminar in the history and development of major theoretical ideas in social and cultural anthropology as applied to problems of health, illness, medicine, and medical institutions. Major concepts and problems will be illustrated through critical review of selected literature. (4 units)
4. *Bad Bugs: Perspectives on Antimicrobial Resistance*
   Bad bugs bring experts together from different disciplines to teach students the global issue of antimicrobial resistance and what their role as healthcare professionals is in addressing this public health problem. Grading will be based on attendance (as measured by mini-quizzes) as well as preparation of a brief report on an antimicrobial resistance topic of interest to the student. (1 unit).
5. *Community Health*
   This clinical course consists of population-based health initiatives in a variety of community health settings. Care of the community is emphasized. Tools for community assessment, health promotion, disease prevention, risk reduction, and rehabilitation will be applied in the clinical setting (4.5 units)

Appendix D: Evaluation Instruments, Details, and Program Objectives Addressed

Instrument	Description	Exemplar items	Objective addressed
Mixed methods survey of workshop leads (postlead workshop)	Immediately after the UCSF workshop, a mixed methods survey will be sent to faculty leads who will describe the value of the workshop in preparing them for training faculty participants on their campuses. We will also collect demographic details and teaching experience from the workshop leads.	Did the workshop provide them with a solid overview of the CNI initiative, goals of the Workshop initiative, adequate resources for recruiting for and teaching the workshop, adequate orientation to documentation needs for workshop	1,2,3
Workshop lead documentation (ongoing between lead and participant workshop)	Workshop leads will document the number and demographics of faculty who applied to or were approached to participate in the training and when relevant, factors that impacted	List of faculty recruited or applied, department and school, accepted/not. Include any accompanying documentation they submitted during the application process	1,2

(continued)

Instrument	Description	Exemplar items	Objective addressed
Mixed methods workshop participant survey (post participant workshop)	The workshop participants will complete a mixed methods survey post their campus workshop about the value of the workshop in preparing them to redesign and infuse SHE into their teaching. During this survey, participants will also provide demographic and teaching experience details as well as information about the courses they plan to infuse SHE curriculum into.	Campus, school, department Teaching experience overall Teaching experience in SHE Their expectations of what they would learn at workshop, were those expectations met, Did the workshop provide them with a solid overview of the CNI initiative, goals of the Workshop initiative, adequate resources for teaching resources, and a framing for SHE	2,3
Reflection/ documentation of courses changed by workshop participants (including a knowledge survey here based on CNI initiative and overall SHE content knowledge) (post participant networking event)	Workshop participants will complete documentation about the courses they changed including	Course name Educational modality Educational objectives of SHE content introduced Number of learners enrolled Course/content evaluation results	1
Workshop lead interviews (post networking event)	Leads will be interviewed to determine the success of their workshop/networking and the impact of their workshop/ networking on the communities of practice within workshop leads/participants and the larger institution based on the impact of their changed courses/content on their learners, fellow teachers, context, and program	Did the overall training and implementation process go well, was their local workshop successful, what were the resources they needed most to execute the workshop. Their perception of success of the workshop and networking effort over and above the evaluation data collected, whether initiative developed a community of practice, and if/how was their institution impacted as a result of this initiative. Ask about all 4 of O'Sullivan's components.	2,3

Instrument	Description	Exemplar items	Objective addressed
Workshop participant interviews (post networking event)	Participants will be interviewed to determine their retrospective view on the value of the workshop and networking event and the impact of their changed courses/content on their learners, fellow teachers, context, and program.	Their experience changing their content course, successes and barriers, perceptions of their learners, did other faculty peers engaged in the process, did they encounter any program/school/ institutional barriers to success. Ask about all 4 of O'Sullivan's components.	2,3

References

1. Costello A, Abbas M, Allen A, et al. Managing the health effects of climate change. Lancet. https://doi.org/10.1016/S0140-6736(09)60935-1.
2. Watts N, Adger WN, Agnolucci P, et al. Health and climate change: policy responses to protect public health. Lancet. 2015;386 https://doi.org/10.1016/S0140-6736(15)60854-6.
3. Patz JA, Campbell-Lendrum D, Holloway T, Foley JA. Impact of regional climate change on human health. Nature. 2005;438:310–7.
4. Cisneros R, Schweizer D, (Lee) Tarnay L, Navarro K, Veloz D, Procter CT. Climate change, forest fires, and health in California. Akhtar R, Palagiano C, Climate change and air pollution: the impact on human health in developed and developing countries . Cham Springer International Publishing, 2018: 99–130.
5. Ebi KL, Mills DM, Smith JB, Grambsch A. Climate change and human health impacts in the United States: an update on the results of the U.S. National Assessment. Environ Health Perspect. 2006;114:1318–24.
6. Ebi KL, Balbus J, Luber G, et al. Chapter 14 : Human health. In: Impacts, risks, and adaptation in the United States: the Fourth National Climate Assessment, vol. II; 2018. p. 563–94.
7. Landrigan PJ, Fuller R, Acosta NJR, et al. The Lancet Commission on pollution and health. Lancet (London, England). 2018;391:462–512.
8. Fleming E, Payne JL, Sweet WV, et al. Chapter 8 : Coastal effects. In: Impacts, risks, and adaptation in the United States: the Fourth National Climate Assessment, vol. II; 2018. p. 1092–175.
9. The National Academies. Environmental public health impacts of disasters. Washington, D.C.: National Academies Press; 2007. https://doi.org/10.17226/11840.
10. Caroline W, Perry S. It's time for medical schools to introduce climate change into their curricula. Acad Med. https://doi.org/10.1097/ACM.0000000000002368.
11. Das GM. Population, poverty, and climate change. World Bank Res Obs. 2014;29:83–108.
12. Crimmins A, Bell JE, Fann N, Hawkins M. The impacts of climate change on human health in the United States. https://doi.org/10.7930/J00P0WXS.
13. Eisenman DP, Wilhalme H, Tseng C, et al. Health & Place Heat Death Associations with the built environment, social vulnerability and their interactions with rising temperature. Health Place. 2016;41:89–99.
14. Center for Sustainable Healthcare. Sustainable Health Care Education. 2016. http://networks.sustainablehealthcare.org.uk/network/sustainable-healthcare-education.
15. McBride DL. How climate change affects children's health. J Pediatr Nurs. 2016;31:350–2.

16. Hernández D. Energy insecurity: a framework for understanding energy, the built environment, and health among vulnerable populations in the context of climate change. Am J Public Health. 2013;103:e3 2–4.
17. Kreslake JM, Price KM, Sarfaty M. Developing effective communication materials on the health effects of climate change for vulnerable groups: a mixed methods study. BMC Public Health. 2016;16(1):946. Published 2016 Sep 7. https://doi.org/10.1186/s12889-016-3546-3.
18. Banc E. Cultivating a green political landscape : lessons for climate change policy from the defeat of California ' s proposition 23. Vanderbilt Law Rev. 2013;66:399.
19. Gould S, Rudolph L. Challenges and opportunities for advancing work on climate change and public health. Int J Environ Res Public Health. 2015;12:15649–72.
20. White KL, Connelly JE. The medical school's mission and the population's health. Ann Intern Med. 2019;115:968.
21. Mullan F, Chen C, Petterson S, Kolsky G, Spagnola M, Academia and Clinic Annals Of Internal Medicine. The social mission of medical education: ranking the schools. 2019;152:804.
22. Duncan S, Rodney P. Evolving trends in nurse regulation: what are the policy impacts for nursing's social mandate ? Nurs Inq. 2015;22:27–38.
23. Isringhausen KT, VanderWielen LM, Vanderbilt AA. Addressing health care disparities and access to dental care while improving education: schools of dentistry and federally qualified health centers. J Health Care Poor Underserved. 2014;25:670–4.
24. Baverstock KA, Gargya DM, Jackson M, Stupans I. Identifying future health professionals' understanding of the determinants of health. J Nurs Educ. 2018;57:756–9.
25. Komro KA, Lang DL, Walker ER, Harper PD. Integrating structural determinants into MPH training of health promotion professionals. Am J Public Health. 2018;108:477–9.
26. Phalen RF. Core ethics for health professionals. .
27. Shoeb M, Logar T, Glass M, et al. Teaching global health ethics using simulation : interprofessional training. 2014; 94141.
28. Teherani A, Nishimura H, Apatira L, et al. Identification of core objectives for teaching sustainable healthcare education. Med Educ Online. 2017;22 https://doi.org/10.1080/10872981.2017.1386042.
29. Wellbery C, Sheffield P, Timmireddy K, Sarfaty M, Teherani A, Fallar R. It's time for medical schools to introduce climate change into their curricula. Acad Med. 2018;93:1774–7.
30. Walpole SC, Mortimer F, Inman A, Braithwaite I, Thompson T. Exploring emerging learning needs: a UK-wide consultation on environmental sustainability learning objectives for medical education. Int J Med Educ. 2015;6:191–200.
31. Thompson T, Walpole S, Braithwaite I, Inman A, Barna S, Mortimer F. Learning objectives for sustainable health care. Lancet. 2014;384:1924–5.
32. Walpole SC, Pearson D, Coad J, Barna S. What do tomorrows doctors need to learn about ecosystems? - a BEME systematic review: BEME guide No. 36. Med Teach. 2016;38:338–52.
33. Walpole SC, Vyas A, Maxwell J, et al. Building an environmentally accountable medical curriculum through international collaboration. Med Teach. 2017;39:1040–50.
34. Ramanathan V, Allison JE, Auffhammer M, et al. Executive summary of the report, bending the curve: 10 scalable solutions for carbon neutrality and climate stability. 2015. http://universityofcalifornia.edu/sites/default/files/Bending_the_Curve_F5_spreads.pdf.
35. TomKat Strategic Communications Working Group. Strategic communication to achieve carbon neutrality within the University of California: report of the TomKat Strategic Communication Working Group. 2018. https://www.nceas.ucsb.edu/files/research/projects/UC-TomKat_Strategic_Communication_MAIN-REPORT_POSTED-20180705.pdf.

Chapter 29
The Physician's Response to Climate Change

Mary B. Rice and Alexander S. Rabin

We don't always know it, but doctors treat the health effects of climate change every day. We treat the sickness of climate change when we call in prescriptions for worsening asthma and allergy symptoms triggered by longer and more potent pollen seasons. We treat climate change sickness when our patients wheeze and cough due to wildfire smoke. Though less obvious, we also treat climate change sickness when we care for the elderly patients who come into our busy emergency rooms at a higher rate on unusually warm days with life-threatening conditions such as urinary tract infections, electrolyte imbalance, kidney failure, sepsis, and exacerbation of respiratory diseases including chronic obstructive pulmonary disease [1, 2]. None of these patients report that their illness is due to heat, or the higher levels of ozone air pollution on hot days. And the doctors caring for them are busy treating the sickness, with little capacity to address how weather or air quality may drive these fragile patients into the emergency room in the first place.

Doctors are starting to connect the dots. As it is becoming increasingly obvious that climate change is not just a future threat, but is already taking place, there is growing recognition among clinicians that their patients are affected by it. For example, in a survey of the US American Thoracic Society members, 89% of whom held an M.D. or other clinical professional degree, 65% indicated that climate change is at least moderately relevant to patient care, and the majority had observed climate-related health effects in their clinical experience [3]. Surveys of other US clinical professional societies have yielded similar results [4, 5]. Outside the USA, physicians more commonly recognize how climate change affects their patients'

M. B. Rice
Harvard Medical School, Boston, MA, USA
e-mail: mrice1@bidmc.harvard.edu

A. S. Rabin (✉)
Tufts Medical Center, Boston, MA, USA
e-mail: arabin@tuftsmedicalcenter.org

K. E. Pinkerton, W. N. Rom (eds.), *Climate Change and Global Public Health*, Respiratory Medicine, https://doi.org/10.1007/978-3-030-54746-2_29

health. In one survey, 80% considered climate change at least moderately relevant to patient care [6].

Climate change affects patient care in two important ways: (1) by causing or worsening disease and (2) by interfering with the delivery of healthcare. Other chapters in the book discuss the health effects of climate change in great detail. Below, we summarize the major health effects of climate change that US physicians encounter regularly in daily practice:

- Increased wildfire activity in many parts of the country, causing high air pollution events that worsen respiratory disease, especially asthma and chronic obstructive pulmonary disease (COPD), both locally near the fire, and in more distant locations due to the spread of wildfire smoke [7, 8].
- A higher frequency of heat waves (2 or more consecutive days of extreme heat) [9] that are linked to increased hospitalization among the elderly for heat-related illness (e.g., heat stroke) and also a long list of other respiratory, infectious, and metabolic conditions that are likely poorly tolerated in the setting of thermal stress [1, 2, 10, 11]. People living in cities, particularly low-income families living in neighborhoods with large buildings and little open space, are especially vulnerable to extreme heat, because of an urban "heat island" effect [12].
- Greater ground-level ozone air pollution (smog) produced in the setting of higher temperatures, resulting in more high-ozone "episodes" [13] that trigger asthma exacerbations [14–16], arrythmia [17], and heart attacks [18].
- More unseasonably warm days during the cooler seasons and more variability in temperature, which may cause more deaths in total than extreme heat waves, and likely worsen respiratory and cardiovascular health [19–22].
- Longer pollen seasons and greater amounts of pollen production [23–25] causing more healthcare utilization for allergic disease, in terms of over-the-counter allergy medication use [26], and more emergency room and physician office visits for allergic disease [27, 28] and asthma [29–32].
- More hurricanes and coastal flooding events that disrupt communities, cause severe psychosocial stress [33], and increase residential and occupational exposures to indoor mold that can cause allergic and respiratory symptoms [34–37].

There are many more health effects of climate change. The above list highlights the more common health effects that physicians treat in clinics and hospitals across the USA. Certain high-risk groups are disproportionately affected, including asthmatic and allergic children and adults, elderly people with chronic disease, the urban poor with limited access to cooling, and those living in coastal communities and in regions affected by wildfires. However, patients in all parts of the country, with all kinds of conditions, are vulnerable to the disruption of healthcare delivery by climate change.

Modern healthcare delivery relies on complex systems to diagnose and treat patients. These systems include a large and diverse workforce, sophisticated diagnostic tools, and a web of facilities to deliver care. As doctors well know, hospital infrastructure is aging [38]. Many hospitals and clinics were built decades before the threat of rising seas, expanded flood zones, severe storms, and wildfires was known [39]. Medical centers that suffer from aging infrastructure are especially prone to damage from severe weather such as flooding or wildfires [38]. Hospitals, which are often thought of as refuges during natural disasters, may be out of reach to those most in need. Tampa General Hospital, for example, is located on an island that could be cut off from the mainland in the event of a major hurricane [40]. And in Florida's Miami-Dade County, 11% of hospitals face inundation from a Category 2 hurricane while 68% are threatened by a Category 5 hurricane [41].

As climate change increases the frequency and strength of severe weather events, direct impacts on patient care may lead to hospital closures and curtailment of emergency medical services. Indirect impacts, such as disruptions to the medical supply chain, or increased medical costs to cover lost revenue and infrastructure damage are equally damaging. Below, we provide several recent examples of healthcare delivery disruptions attributed to climate change.

Perhaps no climate-related event to date can match the devastation to a city's healthcare infrastructure as the inundation of New Orleans, Louisiana, following Hurricane Katrina in 2005. All levels of care were affected, from academic research centers to community health centers, from dialysis centers to skilled nursing facilities [39]. Charity Hospital, a large public hospital in New Orleans, and the Veterans Administration Hospital suffered catastrophic flooding that led to their permanent closure. Physicians and nursing staff fled, Tulane Medical school was relocated to Houston for a year, surgeries were postponed, and patients suffered as chronic conditions went untreated [39, 42]. A year after the storm, only three of nine acute care hospitals were operational [43]. The impact of the storm was felt far outside the city limits. As residents were displaced, surrounding cities and states struggled to care for the influx of patients [43].

The human toll of such events extends beyond the damage to hospital and clinic infrastructure. During Houston's flood that followed Hurricane Harvey in 2017, patients died from lack of access to routine care as emergency medical calls went unanswered. One woman lay stranded in a flooded neighborhood as her husband desperately awaited an evacuation. She succumbed to septic shock from an otherwise routine postoperative wound infection [44].

During climate-related disasters, medical care can be impacted nationwide due to interruptions in the supply chains of essential drugs and medical equipment. Following Hurricane Maria in 2017, small-volume normal saline manufacturing was disrupted on Puerto Rico, leading to widespread intravenous fluid shortages across the USA [45]. A scarcity of the small-volume saline bags then led to shortages of replacement products such as large-volume saline bags and lactated Ringer solution. The Food and Drug Administration scrambled to address the shortage by allowing importation of IV fluids from other countries [45].

In overstretched health systems, massive infrastructure and supply chain disruptions can cost time and money at a time when healthcare utilization peaks due to natural disasters. After a wildfire in Southern California in 2007, one analysis estimated that the associated medical costs from excess emergency department visits and hospital admissions were over $3.4 million [46]. The IV fluid shortage in 2017-2018 led to hospital staffing issues and increased the cost of care as nurses administered medications manually with a syringe rather than hanging the medication on an IV pole [47]. Damage to New York University's Langone Medical Center after Superstorm Sandy in 2012 led to a federal appropriation of $1.13 billion for recovery, the second largest payout for a single project by the Federal Emergency Management Agency [48]. As climate change accelerates the pace of such disasters, the costs to our healthcare system will continue to mount without appropriate preventative measures.

Ironically, healthcare delivery is itself also a major *cause* of climate change. Hospitals are extremely energy-intensive and use twice as much energy as office buildings, second only to food service facilities in their energy use [49]. The high energy requirements of hospitals are not surprising: hospitals are usually large buildings that are open 24 hours per day and have stringent requirements for heating, cooling, and ventilation. Hospitals also use energy-intensive equipment including imaging scanners, refrigeration, laundry, sterilization, food service, etc. and generate a huge amount of waste [50]. Inhaled volatile anesthetics used in many operating rooms, such as desflurane and isoflurane, and the carrier gas nitrous oxide, are powerful greenhouse gases that have much greater heat-trapping properties than carbon dioxide [51, 52]. The upstream manufacturing of drugs, medical devices, and other products used in healthcare is highly energy intensive [53]. Healthcare also employs the largest commuting workforce of the USA, accounting for more than 10% of the working US population [54]. Overall, healthcare delivery is responsible for approximately 10% of US greenhouse gas emissions [50]. In fact, it has been estimated that the public health impact of the air pollution from fossil fuel burning attributable to healthcare delivery causes more harm to human health than preventable medical errors [50]. While medical errors generate a great deal of attention from hospital management, the hospital's carbon footprint is generally not viewed as a problem that must be urgently addressed.

In recent years, several pioneering hospital systems, including Cleveland Clinic, Memorial Sloan Kettering and New York University Langone Medical Center, Partners Healthcare, Gunderson Health System, Rochester Regional Health, Boston Medical Center, the University of California Health, and Kaiser Permanente have demonstrated leadership in climate action by aggressively reducing energy use and/or transitioning to renewable energy sources, openly acknowledging the human health consequences of fossil fuel burning. For example, in 2017, Cleveland Clinic announced a goal of achieving carbon neutrality by 2027, through improved energy efficiency and renewable energy purchases [55]. Partners HealthCare has announced a goal of becoming carbon positive for all energy by 2025, through the purchase of low-impact hydropower and wind power, among other mechanisms [56]. Boston Medical Center has improved energy efficiency and neutralizes its carbon footprint for electricity by purchasing energy generated by a solar farm in North Carolina [56]. Many hospitals are located in densely populated urban environments, and therefore large-scale renewable energy production on hospital property is often not feasible. Therefore, each of these medical systems, in addition to improving the energy efficiency of hospital operations, has invested in off-site renewable energy production through power purchasing mechanisms.

Unfortunately, many hospital systems do not consider efforts to reduce their carbon footprint part of their overall mission to promote health, despite the clear health consequences of burning fossil fuels. This cognitive disconnect may be explained in part by who makes decisions about energy use. We argue that physicians can play a role in mission-level dialogues about the institutional carbon footprint that can help inspire carbon neutrality goals at the highest level of hospital leadership.

Making changes to hospital energy use is challenging, but convincing legislators to act boldly on climate change has proven to be even more difficult. Climate change remains a deeply polarizing issue among American voters. A 2019 Pew Research Center survey suggested that Democrats were far more likely than Republicans (67% vs. 21%) to describe climate change as a top legislative priority, even though responders across the political spectrum viewed rising healthcare costs as a top concern [57]. As physicians sit at the intersection of science, medicine, and the business of healthcare, they are in a unique position to demand change from lawmakers while calling upon all three of the modes of persuasion, as defined by Aristotle. Physicians are widely respected for their educational achievements and commitment to helping others (*ethos*) [58]. They possess the medical and scientific background to describe the public health and cost implications of climate inaction (*logos*). And importantly, they can speak to the emotional, or human, aspect of the climate crisis (*pathos*). Lawmakers listen when physicians vividly describe the consequences of an elderly person's heat stroke from an unprecedented heat wave, or the significance of a prolonged respiratory exacerbation due to wildfire smoke in a vulnerable child with asthma. Incorporating these elements in legislative testimony, and in discussions with hospital leadership, can make a powerful impact.

Organizations such as World Health Organization, Physicians for Social Responsibility, Healthcare without Harm, and the Medical Society Consortium on Climate and Health have brought together like-minded physicians, many of whom have limited advocacy experience, to speak out on behalf of climate change mitigation and resilience policies. They have also helped aggregate scientific information and policy action plans to promote physician engagement. As the following examples demonstrate, physicians are participating throughout the country and at every level to call for climate action.

At the federal level, physicians have joined legislative advocacy sessions with the help of professional medical organizations such as the American Thoracic Society [59] and the American Academy of Pediatrics [60, 61]. Working with their senators and representatives, physicians have also testified to provide context about the health benefits of policies to regulate air pollution and carbon emissions at committee hearings [62]. Through involvement in national and international medical organizations, physicians can call attention to the health impacts of climate change within their professional networks.

At the state level, there has been an increase in ambitious policy proposals to dramatically reduce fossil fuel burning for energy. Physicians have provided much-needed context when describing the health impacts of climate change and of the air pollution that is emitted when fossil fuels are burned by motor vehicles and power plants. They may help describe the impact of hospital closures from wildfires or speak about the harms of diesel pollution from school buses that impact early lung development in children [63]. There are now countless meetings and workshops throughout the country on the topic of local fossil fuel combustion and decisions about traffic and energy. Physician participation brings clinical context and scientific expertise that may otherwise be lacking among more traditional environmental activists.

There are many more examples of how physicians have acted at the city, state, and national level to protect patients from the harms of climate change. Advocating for renewable power purchase plans at hospitals, promoting clean forms of transportation, and reducing medical waste are all crucial in the fight against climate change. Personal interactions with patients who suffer from the "sickness of climate change" can help inform the public and help promote local action by connecting climate events to health outcomes. Yet, bolder proposals to achieve more aggressive reductions in greenhouse emissions are urgently needed because climate change is already exerting devastating effects from decades of unhindered carbon emissions. We physicians must seize our own unique place in society to change the conversation about climate change from politics to human well-being. The worst sickness of climate change can be prevented by taking bold action now to transition our economy away from fossil fuel burning. It is our duty as healers to protect our patients and our communities, and therefore doctors must order an end to fossil fuel burning.

References

1. Anderson GB, Dominici F, Wang Y, McCormack MC, Bell ML, Peng RD. Heat-related emergency hospitalizations for respiratory diseases in the Medicare population. Am J Respir Crit Care Med. 2013;187(10):1098–103. https://doi.org/10.1164/rccm.201211-1969OC.
2. Bobb JF, Obermeyer Z, Wang Y, Dominici F. Cause-specific risk of hospital admission related to extreme heat in older adults. JAMA. 2014;312(24):2659. https://doi.org/10.1001/jama.2014.15715.
3. Sarfaty M, Bloodhart B, Ewart G, et al. American Thoracic Society member survey on climate change and health. Ann Am Thorac Soc. 2015;12(2):274–8. https://doi.org/10.1513/AnnalsATS.201410-460BC.
4. Sarfaty M, Kreslake JM, Casale TB, Maibach EW. Views of AAAAI members on climate change and health. J Allergy Clin Immunol Pract. 2016;4(2):333–335.e26. https://doi.org/10.1016/J.JAIP.2015.09.018.
5. Sarfaty M, Mitchell M, Bloodhart B, Maibach EW. A survey of African American physicians on the health effects of climate change. Int J Environ Res Public Health. 2014;11(12):12473–85. https://doi.org/10.3390/ijerph111212473.
6. Sarfaty M, Kreslake J, Ewart G, et al. Survey of international members of the American Thoracic Society on climate change and health. Ann Am Thorac Soc. 2016;13(10):1808–13. https://doi.org/10.1513/AnnalsATS.201604-229BC.
7. Reid CE, Brauer M, Johnston FH, Jerrett M, Balmes JR, Elliott CT. Critical review of health impacts of wildfire smoke exposure. Environ Health Perspect. 2016;124(9):1334–43. https://doi.org/10.1289/ehp.1409277.
8. Delfino RJ, Brummel S, Wu J, et al. The relationship of respiratory and cardiovascular hospital admissions to the southern California wildfires of 2003. Occup Environ Med. 2009;66(3):189–97. https://doi.org/10.1136/oem.2008.041376.

9. Meehl GA, Tebaldi C. More intense, more frequent, and longer lasting heat waves in the 21st century. Science (80-). 2004;305(5686):994–7.
10. Bhaskaran K, Armstrong B, Hajat S, Haines A, Wilkinson P, Smeeth L. Heat and risk of myocardial infarction: hourly level case-crossover analysis of MINAP database. BMJ. 2012;345:e8050.
11. Dematte JE, O'Mara K, Buescher J, et al. Near-fatal heat stroke during the 1995 heat wave in Chicago. Ann Intern Med. 1998;129(3):173–81.
12. Huang G, Zhou W, Cadenasso ML. Is everyone hot in the city? Spatial pattern of land surface temperatures, land cover and neighborhood socioeconomic characteristics in Baltimore, MD. J Environ Manag. 2011;92(7):1753–9. https://doi.org/10.1016/j.jenvman.2011.02.006.
13. Shen L, Mickley LJ, Gilleland E. Impact of increasing heat waves on U.S. ozone episodes in the 2050s: results from a multimodel analysis using extreme value theory. Geophys Res Lett. 2016;43(8):4017–25. https://doi.org/10.1002/2016GL068432.
14. Babin S, Burkom H, Holtry R, et al. Medicaid patient asthma-related acute care visits and their associations with ozone and particulates in Washington, DC, from 1994-2005. Int J Environ Health Res. 2008;18(3):209–21. https://doi.org/10.1080/09603120701694091.
15. Gleason JA, Bielory L, Fagliano JA. Associations between ozone, PM2.5, and four pollen types on emergency department pediatric asthma events during the warm season in New Jersey: a case-crossover study. Environ Res. 2014;132:421–9. https://doi.org/10.1016/j.envres.2014.03.035.
16. Silverman RA, Ito K. Age-related association of fine particles and ozone with severe acute asthma in New York City. J Allergy Clin Immunol. 2010;125(2):367–373.e5. https://doi.org/10.1016/j.jaci.2009.10.061.
17. Rich DQ, Mittleman MA, Link MS, et al. Increased risk of paroxysmal atrial fibrillation episodes associated with acute increases in ambient air pollution. Environ Health Perspect. 2006;114(1):120–3.
18. Ruidavets J-B, Cournot M, Cassadou S, Giroux M, Meybeck M, Ferrières J. Ozone air pollution is associated with acute myocardial infarction. Circulation. 2005;111(5):563–9. https://doi.org/10.1161/01.CIR.0000154546.32135.6E.
19. Lee M, Nordio F, Zanobetti A, Kinney P, Vautard R, Schwartz J. Acclimatization across space and time in the effects of temperature on mortality: a time-series analysis. Environ Health. 2014;13(1):89. https://doi.org/10.1186/1476-069X-13-89.
20. Rice MB, Li W, Wilker EH, et al. Association of outdoor temperature with lung function in a temperate climate. Eur Respir J. 2019;53(1):1800612. https://doi.org/10.1183/13993003.00612-2018.
21. Mehta AJ, Kloog I, Zanobetti A, et al. Associations between changes in city and address specific temperature and QT interval--the VA Normative Aging Study. PLoS One. 2014;9(9):e106258. https://doi.org/10.1371/journal.pone.0106258.
22. Li S, Baker PJ, Jalaludin BB, Marks GB, Denison LS, Williams GM. Ambient temperature and lung function in children with asthma in Australia. Eur Respir J. 2014;43(4):1059 66. https://doi.org/10.1183/09031936.00079313.
23. Wolf J, O'Neill NR, Rogers CA, Muilenberg ML, Ziska LH. Elevated atmospheric carbon dioxide concentrations amplify Alternaria alternata sporulation and total antigen production. Environ Health Perspect. 2010;118(9):1223–8. https://doi.org/10.1289/ehp.0901867.
24. Wayne P, Foster S, Connolly J, Bazzaz F, Epstein P. Production of allergenic pollen by ragweed (Ambrosia artemisiifolia L.) is increased in CO2-enriched atmospheres. Ann Allergy Asthma Immunol. 2002;88(3):279–82. https://doi.org/10.1016/S1081-1206(10)62009-1.
25. Ziska L, Knowlton K, Rogers C, et al. Recent warming by latitude associated with increased length of ragweed pollen season in central North America. Proc Natl Acad Sci. 2011;108(10):4248–51. https://doi.org/10.1073/pnas.1014107108.
26. Sheffield PE, Weinberger KR, Ito K, et al. The association of tree pollen concentration peaks and allergy medication sales in New York city: 2003-2008. ISRN Allergy. 2011;2011:537194. https://doi.org/10.5402/2011/537194.

27. Cakmak S, Dales RE, Burnett RT, Judek S, Coates F, Brook JR. Effect of airborne allergens on emergency visits by children for conjunctivitis and rhinitis. Lancet. 2002;359(9310):947–8. https://doi.org/10.1016/S0140-6736(02)08045-5.
28. Villeneuve PJ, Doiron M-S, Stieb D, Dales R, Burnett RT, Dugandzic R. Is outdoor air pollution associated with physician visits for allergic rhinitis among the elderly in Toronto, Canada? Allergy. 2006;61(6):750–8. https://doi.org/10.1111/j.1398-9995.2006.01070.x.
29. Darrow LA, Hess J, Rogers CA, Tolbert PE, Klein M, Sarnat SE. Ambient pollen concentrations and emergency department visits for asthma and wheeze. J Allergy Clin Immunol. 2012;130(3):630–638.e4. https://doi.org/10.1016/j.jaci.2012.06.020.
30. Héguy L, Garneau M, Goldberg MS, Raphoz M, Guay F, Valois M-F. Associations between grass and weed pollen and emergency department visits for asthma among children in Montreal. Environ Res. 2008;106(2):203–11. https://doi.org/10.1016/j.envres.2007.10.005.
31. Zhong W, Levin L, Reponen T, et al. Analysis of short-term influences of ambient aeroallergens on pediatric asthma hospital visits. Sci Total Environ. 2006;370(2-3):330–6. https://doi.org/10.1016/j.scitotenv.2006.06.019.
32. Erbas B, Akram M, Dharmage SC, et al. The role of seasonal grass pollen on childhood asthma emergency department presentations. Clin Exp Allergy. 2012;42(5):799–805. https://doi.org/10.1111/j.1365-2222.2012.03995.x.
33. Evans GW. Projected behavioral impacts of global climate change. Annu Rev Psychol. 2019;70(1):449–74. https://doi.org/10.1146/annurev-psych-010418-103023.
34. Targonski PV, Persky VW, Ramekrishnan V. Effect of environmental molds on risk of death from asthma during the pollen season. J Allergy Clin Immunol. 1995;95(5 Pt 1):955–61.
35. Hernberg S, Sripaiboonkij P, Quansah R, Jaakkola JJK, Jaakkola MS. Indoor molds and lung function in healthy adults. Respir Med. 2014;108(5):677–84. https://doi.org/10.1016/j.rmed.2014.03.004.
36. Rando RJ, Lefante JJ, Freyder LM, Jones RN. Respiratory health effects associated with restoration work in post-Hurricane Katrina New Orleans. J Environ Public Health. 2012;2012:1–8. https://doi.org/10.1155/2012/462478.
37. Rando RJ, Kwon C-W, Lefante JJ. Exposures to thoracic particulate matter, endotoxin, and glucan during post-Hurricane Katrina restoration work, New Orleans 2005–2012. J Occup Environ Hyg. 2014;11(1):9–18. https://doi.org/10.1080/15459624.2013.839879.
38. King D, Beebe C, Suchomel J, Bardwell P, Donna V. A closer look at infrastructure | Health Facilities Management. Health Facilities Management.
39. Balbus J. Primary protection: enhancing health care resilience for a changing climate Acknowledgements; 2014.
40. Fears D. Sea-level rise could destroy Tampa Bay if a major hurricane hits - Washington Post. Washington Post.
41. Ebi KL, Balbus J, Luber G, et al. Chapter 14 : Human health. In: Impacts, risks, and adaptation in the United States: the Fourth National Climate Assessment, vol. II. Washington, DC; 2018. https://doi.org/10.7930/NCA4.2018.CH14.
42. Berggren RE, Curiel TJ. After the storm — health care infrastructure in post-Katrina New Orleans. N Engl J Med. 2006;354(15):1549–52. https://doi.org/10.1056/nejmp068039.
43. Rudowitz R, Rowland D, Shartzer A. Health care in New Orleans before and after Hurricane Katrina. Health Aff. 2006;25(5):w393–406. https://doi.org/10.1377/hlthaff.25.w393.
44. Fink S. Lost in the storm - The New York Times. New York Times Magazine.
45. Sacks CA, Kesselheim AS, Fralick M. The shortage of normal saline in the wake of Hurricane Maria. JAMA Intern Med. 2018;178(7):885. https://doi.org/10.1001/jamainternmed.2018.1936.
46. Kochi I, Champ PA, Loomis JB, Donovan GH. Valuing morbidity effects of wildfire smoke exposure from the 2007 Southern California wildfires. J For Econ. 2016;25:29–54. https://doi.org/10.1016/J.JFE.2016.07.002.
47. Conde X. IV bag shortages causing cost of care to increase | Wisconsin Public Radio. Wisconsin Public Radio.

48. Hartocollis A. NYU Langone Medical Center to Get $1.13 billion in storm aid - The New York Times. New York Times.
49. U.S. Energy Information Administration. commercial buildings energy consumption survey: energy usage summary. https://www.eia.gov/consumption/commercial/reports/2012/energy-usage/. Published 2012. Accessed 4 May 2019.
50. Eckelman MJ, Sherman J. Environmental impacts of the U.S. Health Care System and effects on public health. Ahmad S, ed. PLoS One. 2016;11(6):e0157014. https://doi.org/10.1371/journal.pone.0157014.
51. Ryan SM, Nielsen CJ. Global warming potential of inhaled anesthetics. Anesth Analg. 2010;111(1):92–8. https://doi.org/10.1213/ANE.0b013e3181e058d7.
52. Alexander R, Poznikoff A, Malherbe S. Greenhouse gases: the choice of volatile anesthetic does matter. Can J Anesth Can d'anesthésie. 2018;65(2):221–2. https://doi.org/10.1007/s12630-017-1006-x.
53. Eckelman MJ, Sherman JD. Estimated global disease burden from US Health Care Sector greenhouse gas emissions. Am J Public Health. 2018;108(S2):S120–2. https://doi.org/10.2105/AJPH.2017.303846.
54. Henry J. Kaiser family. Health care employment as a percent of total employment. https://www.kff.org/other/state-indicator/health-care-employment-as-total/?currentTimeframe=0&sortModel=%7B%22colId%22:%22Location%22,%22sort%22:%22asc%22%7D. Published 2017. Accessed 4 May 2019.
55. Ferenc J. Cleveland Clinic makes carbon-neutrality its newest sustainability goal. Health Facilities Management. https://www.hfmmagazine.com/articles/3210-cleveland-clinic-makes-carbon-neutrality-its-newest-sustainability-goal. Published 2017. Accessed 6 May 6 2019.
56. Ferenc J. Boston hospitals continue to set the pace for reducing emissions | Health Facilities Management. https://www.hfmmagazine.com/articles/2918-boston-hospitals-continue-to-set-the-pace-for-reducing-emissions. Accessed 6 May 2019.
57. Doherty C. Public's 2019 priorities: economy, health care, education and security all near top of list. 2019.
58. Dolton P, Marcenaro O, De Vries R, She P-W. Global teacher status. Index 2018.; 2018.
59. American Thoracic Society - ATS Advocacy.
60. American Academy of Pediatrics, American Lung Association, et al. Letter to U.S. Administrator Gina McCarthy. March 17. 2015.
61. American Academy of Pediatrics. Climate change and children's health. 2019. https://www.aap.org/en-us/advocacy-and-policy/aap-health-initiatives/climate-change/Pages/Climate-Change-and-Childrens-Health.aspx. Accessed 6 May 2019.
62. Rice M. Greenhouse gas regulations and energy costs | C-SPAN.org. C-SPAN.
63. Miller A, Kim H-J, Robinson J, Casale M. Electric buses clean transportation for healthier neighborhoods and cleaner air; 2018.

Index

A

Accumulated cyclone energy (ACE), 339
Acute febrile illness, 300
Acute lower respiratory tract infection (ALRI), 374, 375, 394, 396
Acute respiratory irritation, 220
Acute (short-term) exposure effects of particulate matter, 182
Adaptation Thought Leadership and Assessments (ATLAS), 496
Adverse birth outcomes, 244–246
Adverse health outcomes, with household air pollution, 374–376
Ae. Aegypti, 258, 277–288, 290–298, 300, 301
  climatic factors, transmission cycle
    biting behavior and extrinsic incubation period, 281, 282
    daily temperature variation affects on vectorial capacity, 282
    mechanistic dengue transmission models, 283
    observational studies, 283–285
    precipitation, 280
    relative humidity, 280
    temperature, 279–281
  populations at risk, 285–287
  regional implications
    Australia and Oceania, 290–292
    Europe, 297, 298
    Latin America, 300
    United States, 300, 301
Aedes albopictus, 276, 278, 283, 285–287, 290–292, 294–298, 300, 301
Aeroallergens, 206, 212–214
Aerosol constituents, 184
Affordable Clean Energy rule, 537
Airborne allergens, 164
Airborne pollen, 172
Air pollution, 15, 16, 164, 192–194
Air pollution co-benefit estimation, 193
Air quality, 164
Allergens, 208
Allergic conjunctivitis, 208
Allergic disease, 206
Allergic rhinitis, 171, 174, 175
Allergic shiners, 208
Alliance of Small Island States (AOSIS), 428
Anthropogenic greenhouse effect, 206
Anthropogenic greenhouse gas emissions, 5
Antibody-dependent enhancement, 276
Anxiety-mood disorders, 345
Arboviral diseases, 254
Arctic connections to global warming and health
  environmental and health policy for changing land, 97
    adapting to new Arctic, 98–100
    community health, 105–107
    food security, 104, 105
    new frontiers for development, 100–102
    threats of oil and gas development in Arctic refuge, 102–104
  future projections and feedbacks to global climate system, 94–97
  rapid environmental and societal change, 92–94
Arctic National Wildlife Refuge (ANWR), 102
Asthma, 206, 207, 212, 223, 224
ATLAS program, 496
Atmosphere-hydrosphere-land surface system, 38
Atmospheric carbon dioxide, 75

K. E. Pinkerton, W. N. Rom (eds.), *Climate Change and Global Public Health*, Respiratory Medicine, https://doi.org/10.1007/978-3-030-54746-2

Atopic disease, cardinal physical examination signs of, 212
Autonomous adaptation, 551
Autoregressive integrated moving average (ARIMA) model, 446

B
Bali road map, 82
Be a buddy (BaB) program, 557
Behavioral thermoregulation, 131
Benefits Mapping and Analysis Program–Community Edition (BenMAP-CE), 500
Benzopyrene, 393
Best system of emission reduction (BSER), 23
β-D glucan, 219
Biodiversity, loss of, 453, 454
Biofuels, 98
Biogas, 380
Biomass fuels, 369, 374, 377, 379, 381, 384, 391, 406
  architecture of house and biomass smoke exposure, 393, 394
  emissions from stove and exposure determining factors, 392, 393
  exposure and lung functions, 398, 399
  morbidity and mortality, 392
  respiratory health effects of
    acute lower respiratory infection, 394, 396
    chronic obstructive pulmonary disease (COPD), 396, 397
    interstitial lung diseases, 398
    lung cancer, 397
    pneumoconiosis, 398
    tuberculosis, 397, 398
  use in India, 391, 392
Black carbon, 4, 369, 378, 379, 382, 384
Borg Dyspnea Scale, 474
Bradford-Hill criteria, 112
Brevetoxin, 472
Building Resilience Against Climate Effects (BRACE), 505–506, 508, 509
  anticipating climate impacts and assessing vulnerabilities, 509, 510, 512, 513
  assessing public health interventions, 509, 516, 517
  climate and health adaptation plan, developing and implementing, 509, 517, 518
  evaluating impact and improving quality of activities, 509, 518, 519
  projecting disease burden, 509, 513–515
  stakeholder engagement, 509
Building resistance against climate effects (BRACE) framework, 492, 553
Business as usual (BAU) approach, 541

C
CAFE standards, 543, 544
California agricultural worker health survey (CAWHS), 357
California air resources board (CARB), 541–543, 545
California and climate changes
  biological mechanisms, 246–248
  temperature and morbidity
    temperature and adverse birth outcomes, 244–246
    temperature and hospitalizations/emergency room visits, 241–244
  temperature and mortality, 238, 239
    future projections for, 246
    heat waves and mortality, 241
    mortality displacement, 240, 241
    vulnerable subgroups, 240
California Department of Public Health (CDPH) Climate Change and Health Equity Program, 354
California global warming solutions act, 541
California's integrated approach to air quality and climate change, 541, 546
  cap-and-trade program, 544, 545
  CARB, 542, 543
  climate gap, 541
  GHG emission, 542, 544, 545
  MPOs, 544
  sustainable communities strategies, 544
  ZEV program, 543
California worker occupational safety and health training and education program (WOSHTEP), 364
Cap-and-trade program, 23, 26, 544, 545
Carbon dioxide (CO_2) emissions, 3, 75, 195, 196, 446, 447, 543
Carbon fee, 26
Carbon monoxide (CO), 371, 393, 459
Carbon tax, 26, 432
Cardiovascular disease (CVD), 139, 184
Cardiovascular effects, 468, 469
CD154 ligand, 208
CD40 Ligand, 208
CDC National Environmental Public Health Tracking Network, 519, 520

CDC Science Symposium on Climate and Health, 521, 522
CDC's Climate and Health Program, 506, 507
Centers for disease control and prevention (CDC), 342, 343, 492, 505
Chikungunya, 254, 259, 260, 264, 277, 291
Childhood pneumonia, 466
Chlorofluorocarbons (CFCs), 4
Chronic diarrheal disease, 409
Chronic diseases, 343, 344
Chronic kidney disease (CKD), 129
Chronic obstructive pulmonary disease (COPD), 139, 374, 375, 396, 397, 475
Chronic respiratory diseases, 139, 142
Chronic silicosis, 470
Chronic urticaria, 206
Classic heat stroke, 137
Clean air act, 23, 182, 196, 204, 214, 528–531, 534, 542
Clean Air Interstate Rule (CAIR), 196
Clean Cooking Alliance, 383
Clean Development Mechanism (CDM), 22
Clean Power Plan, 23, 534, 537, 538
Clean vehicle rebate program, 543
Climate adaptation, 416
Climate and health interventions
- adaptation, 550–554
- evidence base of, 558–560
- in practice, 556–558
- practitioners for, 554–556

Climate and Health Program, 505
Climate (climatic) variability, 38
Climate, definition of, 37, 38
Climate change, 38, 75–81, 505, 565
- and Women's health (*see* Women's health)
- Arctic sea ice, 79
- education for action, 85–88
- extreme weather events and health, natural disasters, 318
- floodwater, 222
- food productivity and malnutrition, 314, 315
  - fisheries, 317, 318
  - food production and drought, 315–317
- forests and forest health, impacts on, 60–63
  - diseases, 64, 66
  - insects, 64, 65
  - invasive species, 67–69
  - wildfire, 66, 67
- global and United States surface temperatures, 51–55
- healthcare facilities and medicine, accessibility to, 224
- hurricane Maria, 227–230
- hurricane Sandy, 224–227
- hurricane Katrina, 216, 217
- infectious disease, 221
- at international level
  - Bali Road Map, 82
  - COP21, 83, 84
  - Copenhagen Accord, 82, 83
  - Kyoto Protocol, 82
  - post COP21, subsequent action towards 2020 and beyond, 84
  - UNFCCC, 81
- large cohort studies, 222, 223
- mold, 218, 220
- particulate matter, 221, 222
- PTSD and asthma/smoking behavior, 223, 224
- volatile organic compounds, 220
- water resources and water quality, impacts on, 54–57
- watershed health, impacts on, 57–59

Climate change adaptation at United States, 500
- BRACE, 508, 509
  - anticipating climate impacts and assessing vulnerabilities, 509, 510, 512, 513
  - assessing public health interventions, 509, 516, 517
  - climate and health adaptation plan, developing and implementing, 509, 517, 518
  - evaluating impact and improving quality of activities, 509, 518, 519
  - projecting disease burden, 509, 513–515
  - stakeholder engagement, 509
- management and its role, 507, 508

Climate change adaptation strategies (CCAS), 124
Climate Change and Human Health Group (CCHHG), 487
- Climate Resilience Toolkit, 488
- international scientific activities on climate change and health, 489–501
- MATCH, 487, 488
- roles of, 487
- 2016 GCRP Climate Health Assessment, 488, 489
- workshop report on climate sensitive infectious disease, 489

Climate Change Impacts and Risk Analysis (CIRA) project, 500
Climate change mitigation
co-benefits assessment, framework of, 192, 193
health and air pollution benefits and costs, 192–194
monetary valuations of, 194–197
Climate change mitigation strategies (CCMS), 124
Climate change-type drought, 61, 63
Climate community, 36
Climate data, 36, 41–45
in-situ data, 42
partners, 45–49
products, 43, 44
projections, 44, 45
reanalysis, 44
scale, 42
source, 42
Climate gap, 541
Climate generation, 85–88
Climate in NYC metropolitan region, 165
epidemiologic analysis of historical data, 165, 166
future projections of, 166
health impact assessment, 167
implications, 170, 171
results, 167–169
Climate predictions, 40
Climate Protection Act, 484
Climate projections, 40, 41
Climate Ready States and Cities Initiative (CRSCI), 492, 493, 506, 520, 521
Climate Resilience Toolkit, 488
Climate scenario, 41
Climate variability, 38, 311
Greenland's ice sheet melts, 312
hurricanes, 312
hydrologic extremes, 311
sea surface temperatures, 312
vulnerable geographic regions, 312, 314
Climate-sensitive health, 111, 113, 114
Coccidioides immitis, 470
Coccidioidomycosis, 359
Combat climate change by cities and states, 27, 28
Community health, 105–107
Community mesoscale air quality (CMAQ) model, 166
Concentrated ambient particulate matter (CAPs), 467
Conference of the Parties (COP), 22, 411
Consortium to Establish a Registry for Alzheimer's Disease (CERAD)-Plus, 467
Convective heat transfer, 132
COP21, Paris Agreement, 83, 84
COP24, 84
Copenhagen Accord, 82, 83
Coral reef bleaching, 11–13
Corporate average fuel economy (CAFÉ), 23, 528
Cost of illness (COI), 194

D

DALYs (Disability Adjusted Life Years), 392
Data culture, 36, 37
Decay of fossil fuel CO_2 emissions, 3
Decision Making Under Uncertainty (DMUU) Centers, 498
Dehydration, 140
Delta method, 514
Dengue, 16, 253–259, 264, 273, 291
antibody-dependent enhancement, 276
climatic effects on entomological and viral parameters, transmission cycle, 278, 279
biting behavior and extrinsic incubation period, 281, 282
daily temperature variation affects on vectorial capacity, 282
mechanistic dengue transmission models, 283
observational studies, 283–285
precipitation, 280, 281
relative humidity, 280
temperature, 279–281
dengue virus, 275
estimated burden of, 274
global evidence consensus, 274
indirect effects of climate change on risk, 288, 289
modelled risk, 274
non-climatic risk factors, 287, 288
regional implications
Africa, 294–297
Asia, 289, 290
Australia and Oceania, 290–292
Europe, 297, 298
Latin America, 298–300
Middle East and North Africa (MENA) region, 292–294
United States, 300, 301

severe dengue, 276
societal factors, 275
symptoms, 276
vaccine, 276
vectors and transmission of, 276–278
Dengue hemorrhagic fever (DHF), 273, 276, 289
Department of Health and Human Services (HHS), 489, 490
Desert dust
effects on health, 465
geochemistry and its fingerprints, 460, 461
Desertification
in Turkey, consequences of climate change, 450
human health, 454–456
loss of biodiversity, 453, 454
sandstorms and dust storms, 451
water use, 452, 453
in United States, 459, 460
cardiovascular effects, 468, 469
desert dust, 465, 474–476
Desert dust geochemistry and its fingerprints, 460, 461
dust geographic distribution, 462–464
dust size definitions, 461, 462
harmful algal blooms and health impact, 472
impact of dust exposure in US war veterans, 473, 474
infections, 469–472
neurological effects, 467, 468
pulmonary effects, 465, 466
Diarrheal disease, 409
Dichlorodiphenyltrichloroethane, 472
Dinoflagellates, 11
Disaster risk reduction, 404, 416, 419, 420
Disseminated intravascular coagulation (DIC), 137
Diurnal temperature ranges (DTRs), 256, 282
Drought, 359, 460
Dust geographic distribution, 462–464
Dust size definitions, 461, 462
Dust storms, 451, 460, 469, 470

E
Earth Summit, 22, 81
EcoAdapt, 556
Ecology and Evolution of Infectious Diseases (EEID) Initiative, 498
Eczema, 207
El Niño Southern Oscillation (ENSO) phenomenon, 285, 291, 299, 438
End stage renal disease (ESRD), 344
Enhancing National Climate Services (ENACTS), 497
Environmental degradation, 566
Environmental Health Multiple-Determinants Model of Vulnerability (MDMv), 123, 129, 130, 145, 154
Environmental policy, 102
Environmental Protection Agency (EPA), 23, 484, 499, 500, 525
assessments, 500
BenMAP-CE, 500
CIRA project, 500
climate policy under Bush, 526
climate policy under Obama
carbon dioxide standards for new and existing power plants, 533–535
endangerment and contribution findings, 527, 528
methane standards for the oil and gas sector, 533–535
more vehicle standards and another California waiver, 533
triggering and tailoring rules, 530, 531
vehicle emission standards and California waiver, 529, 530
climate policy under Trump, 535
freezing vehicle standards and revoking California waiver, 536
weakening new source performance standards and replacing Clean Power Plan, 537, 538
Global Climate Explorer (GCX) Integrated Climate and Land-Use Scenarios (ICLUS) Tool, 500
Massachusetts v. EPA, 526, 527
research activities, 499
Utility Air Regulatory Group v. EPA, 531, 532
Wildland Fire Research Framework, 500
Environmental refugees, 410
Epigenome, 465
Essential Public Health Services (EPHS), 506
EuroHEAT project, 15
Exertional Heat Stroke, 136
Extended-Reconstructed Sea Surface Temperature Dataset (ERSST), 55
Extreme heat events (EHE), 123, 125, 144, 145, 149–154

Extreme weather events
on agricultural health and safety in california, 353, 364
and its workforce, 355, 356
climate change, 357
climate change and occupational health and safety, 354
drought, 359
farmworker occupational hazards and health status, 356, 357
heat wave, 360–362
impact, in agriculture, 358
perceptions of risk, 358
wildfire, 363
wildfires, 362, 363
definition of, 357
and disasters, 406, 407
natural disasters, 318
Extrinsic incubation period, 255, 281, 282

F
Famine Early Warning Systems Network (FEWS NET), 496
Fan stoves, 380
Federal Climate Change Research Prior to 1990, 484
Federal emergency management agency, 586
Federal programs in climate change and health research, 483
Climate Change and Human Health Group, 487
Climate Resilience Toolkit, 488
international scientific activities on climate change and health, 489–501
MATCH, 487, 488
roles of, 487
2016 GCRP Climate Health Assessment, 488, 489
workshop report on climate sensitive infectious disease, 489
current federal agency activities in climate change and health
Building Resistance Against Climate Effects Framework, 492
Climate Ready States and Cities Initiative Framework, 492, 493
Department of Health and Human Services, 489, 490
Environmental Protection Agency, 499, 500
HAQAST, 494, 495
NASA, 493, 494
National Environmental Public Health Tracking Network, 493
National Institutes of Health, 490, 491
National Oceanic and Atmospheric Administration, 498, 499
National Science Foundation, 497, 498
NIEHS, 491, 492
partnerships and outreach, 493
United States Department of Agriculture, 501
USAID, 495–497
early federal activities, climate change and health, 485, 486
Federal Climate Change Research Prior to 1990, 484
Global Change Research Act of 1990, 484, 485
The United States Global Change Research Program, 484, 485
Feed livestock, 317
Floodwater, 222
Fogarty International Center (FIC), 490
Food and Agriculture Organization (FAO), 408
Food insecurity, 17, 18, 403, 407, 408
Food production, and drought, 315–317
Food security, 104, 105, 144, 434, 435
Forced vital capacity (FVC), 398
Forecasts, 39
Forests and forest health, climate change impacts on, 60–63
diseases, 64, 66
insects, 64, 65
invasive species, 67–69
wildfire, 66, 67
Forest diseases, 64, 66
Forest insects, 64, 65
Formaldehyde, 393
Fossil fuels, 3, 433, 587, 588
Fossil–fuel combustion processes, 181
Fourth National Climate Assessment, 61
Fuel efficiency, 371

G
Gender-based solutions to promote health resilience, 417–418
General Circulation Models (GCMs), 44, 144, 287
Generalized anxiety disorder (GAD), 345
General psychological distress (GPD), 345
Genetics/epigenetics, 143
Geographic information systems (GIS), 512

Global Change Research Act of 1990 (GCRA), 484, 485
communicate and educate, 485
conduct sustained assessments, 485
inform decisions, 485
Global Circulation Models, 285
Global climate changes, 446, 456
Global Climate Explorer (GCX) Integrated Climate and Land-Use Scenarios (ICLUS) Tool, 500
Global climate models (GMCs), 254
Global Climate Observing System (GCOS), 43, 48, 499
Global dust transport, 472
Global framework on climate services (GFCS), 262
Global Historical Climatology Network (GHCN) Daily Dataset, 43
Global Historical Climatology Network-Monthly (GHCNm) dataset, 51
Global Observing Systems Information Center, 48
Global Ocean Observing System (GOOS), 43, 44, 48
Global Terrestrial Observing System (GTOS), 43, 48
Global warming, 39, 91, 144, 203, 207, 208, 340, 369, 428, 429
Arctic connections to (*see* Arctic connections to global warming and health)
climate change on biosphere
bleaching of coral reefs, 11–13
loss of Antarctic ice, 7, 8
loss of Arctic ice, 6, 7
loss of Glaciers, 8–10
mean sea level rise, 11
risks to forests, 9, 11
threat to biodiversity, 13
efforts at mitigation and adaptation
combat climate change by cities and states, 27, 28
mitigation-cap-and-trade and carbon tax, 26
mitigation-renewables, 25
policy-global, 21–23
policy-United States, 23–25
greenhouse gases, 1, 3–6
human health, consequence for
air pollution, 15, 16
heat waves, 14, 15
vector-borne diseases, 16, 17
social stability, implications for, 17
drought, forest fires, food insecurity, and migration, 17–19
hurricanes and extreme weather events, 19, 20
insurability for extreme events, 20, 21
United States military, 21
temperature, 1–6
Green energy, 148
Greenhouse gases (GHG), 1, 3–6, 144, 181, 429, 446, 447, 459, 541–545
chlorofluorocarbons, 4
CO_2, 1, 3
hydrofluorocarbons, 4
methane, 4
SO_2 aerosols, 4
Gross domestic product (GDP), 446

H

Harmful algal blooms, 472
Hayfever, 207
Hazards, 115
Head-off environmental asthma in louisiana (HEAL) study, 220
Health and Air Quality Applied Sciences Team (HAQAST), 494, 495
Health component of National Adaptation Plans (HNAPs), 116
Health consequences of climate variability, 35
Health inequities, 566
Health risks of climate change
determination, affected health, 114
Intergovernmental Panel on Climate Change, 116–119
risk management, 114, 115
traditional risk assessment approaches, 112–114, 119
vulnerability, capacity, and adaptation assessments (V&As), 115, 116
Health science schools, 567
Health sciences curriculum, 566
Heat acclimation, 124
Heat acclimatization, 124, 133–135
Heat balance, 132
Heat cramps, 136
Heat edema, 136
Heat exchange, 132
Heat exhaustion, 136
Heat-health action plans, 149
Heat illness, pathophysiology of, 130, 131
Heat-induced increased aeroallergen concentrations, 214

Heat in NYC metropolitan region, 165
epidemiologic analysis of historical data, 165, 166
future projections of, 166
health impact assessment, 167
implications, 170, 171
results, 167–169
Heat island effect, 584
Heat-related illness (HRI), 124, 136
determinants of thermoregulatory capacity
Genetics/Epigenetics, 143
race/ethnicity, 143
sex/gender, 142
heat stroke, 136, 137
pathophysiology, 137, 138
vulnerable populations, determinants of thermoregulatory capacity, 138
infants and children, 140–142
older adults, 139, 140
Heat-related morbidity/mortality (HRMM), 124, 126, 129
heat-related illness (HRI) (*see* Heat-related illness (HRI))
reduce vulnerability and incidence, strategies to, 145, 146
extreme heat event, 149–154
promote good health and access to quality healthcare, 146, 147
reduce/manage potential exposures to ambient heat and physical environmental stressors, 147–149
warm season heat preparedness and response action plans, 149–154
vulnerable populations (*see* Vulnerable populations)
Heat shock proteins (HSP), 134, 135
Heat stress, 124, 130, 131
Heat stroke, 136, 137
classic heat stroke, 137
definition of, 137
Exertional Heat Stroke, 136
Heat syncope, 136
Heat waves, 14, 15, 124, 125, 128, 137, 148, 150, 151, 153, 584
definition of, 360
extreme weather events on agricultural health and safety in california, 360–362
and morbidity, 244
and mortality, 241 (*see* Extreme heat events (EHE))
Heterogeneity, 127
Hot spots, 545
Household air pollution (HAP), 369
black carbon and other short-lived climate forcers, 378, 379
common challenges to clean cooking solutions, 371, 372
contribution to outdoor air pollution, 376–378
cooking fires and role of women, 370, 371
cookstove performance standards and testing centers, 373
cookstoves/fuels, reduce health impacts, 379–382
globalization of clean cooking solutions, 372, 373
health impacts of, 374
regional environmental degradation, 376, 377
Surya Project as model, reduce climate impacts, 382–384
value of co-benefits for human health and climate, 384, 385
vulnerable populations to adverse health outcomes with, 374–376
Human Dimensions of Climate Change, 485
Human health, 454–456
Human health co-benefits of air quality improvements
climate change mitigation
co-benefits assessment, framework of, 192, 193
health and air pollution benefits and costs, 192–194
monetary valuations of, 194–197
implications, 197
long-term exposure mortality effects of O_3, 190, 191
particulate matter, health effects of, 182
constituents, 184, 185
long-term chronic exposure, 182–184
short-term exposure, 182
tropospheric ozone, health effects of, 185–189
Human-induced climate change, 75, 164
Hurricane Harvey, 339, 585
Hurricane Irma, 339
Hurricane Katrina, 217, 218, 220, 224, 339, 343–348
Hurricane Maria, 227–230, 339, 341, 585
Hurricane Michael, 339
Hurricane Sandy, 224–227
Hurricanes, 216, 440
challenges and future of hurricane-health research, 348, 349

climate change, 340
climate variability, 312
duration, 340
mental health effects, 344, 345
exposure and post-disaster mental health, 345, 346
risk factors for post-hurricane mental health effects, 346, 347
social support and resilienceas protective factors, 347, 348
physical-health effects, 340, 341
chronic diseases, 343, 344
infectious diseases, 342, 343
injuries, 341, 342
mold-related illness, 344
mortality, 341
severity, 340
surprise, 340
Hydrofluorocarbons (HFCs), 4
Hydrologic extremes, 312
Hydrologic model, 55
Hyperendemicity, 275
Hypohydration, 140

I

Ice-albedo feedback, 95, 96
IgE allergic antibodies, 209
Indoor air pollution, 391, 392
Infections, 469–472
Infectious diseases, 221, 342, 343
Injury and illness prevention program (IIPP), 364
In-situ data, 42
Integrated Information System, 48, 49
InterAcademy Council (IAC 2010), 117
Intercellular Adhesion Molecule-1 (ICAM-1) protein, 468
Intergovernmental panel on climate change (IPCC), 5, 41, 116–119, 215, 254, 299, 428, 448, 485, 506, 526
Interleukins (IL) IL6, IL8 and IL10, 134, 474, 475
International Agency for Research on Cancer (IARC 2006), 112
International development research center (IDRC), 263
International Organization for Standardization (ISO), 373
International Research and Applications Project (IRAP), 499
Interstitial lung diseases (ILD), 398
Invasive species, 67–69

K

Karenia brevis, 472
Kigali Agreement of 2016, 4
Kyoto Protocol, 82, 446

L

Leishmaniasis diseases, 455
Liquefied petroleum gas (LPG), 381, 391, 399
Live recombinant tetravalent dengue vaccine, 276
Local or regional climate, 38
Long-term chronic exposure effects of particulate matter, 182–184
Long-term exposure mortality effects of O_3, 190, 191
Loss of biodiversity, 453, 454
Louisiana department of health and hospitals (LDHH), 342
Low-and middle-income countries (LMIC), 369, 371
Lower respiratory symptoms (LRS.), 227
Lung cancer, 397
Lyme disease, 114

M

Macrophage, 208
Major depression disorder (MDD), 345
Malnutrition, 314, 315, 455
fisheries, 317, 318
food production and drought, 315–317
women's health, 407, 408
Maternal undernutrition, 408
Mechanistic dengue transmission models, 283
Medical education, 567
Meningitis, 470
Mesoamerican nephropathy (MeN), 129
Metadata Access Tool for Climate and Health (MATCH), 487, 488
Methane, 4
Metropolitan planning organizations (MPOs), 544
Mexican immigration to california: agricultural safety and acculturation (MICASA) study, 356
Middle east countries, *see* Turkey
Mold-related illness, 218, 220, 230, 344
Montreal Protocol, 4
Mosquito-borne diseases, 254, 255
climatic drivers of, 255, 256, 258, 259
non-climatic drivers of, 259, 260
predicting climate change impacts on, 261, 262

Motor vehicle emissions, 542
Multiple (or Multi-) Determinants Models (MDM), 129
Myocardial infarctions (MI's), 182

N
National Academy of Sciences (NAS), 484
National Aeronautics and Space Administration (NASA), 493, 494, 506
National Climate Act, 484
National climate assessment (NCA), 553
National Environmental Protection Act, 103
National Environmental Public Health Tracking Network, 493
National Highway Traffic Safety Administration (NHTSA), 528
National Institute of Environmental Health Sciences (NIEHS), 490–492, 497
National Institutes of Health (NIH), 490, 491
National Integrated Health and Health Information System (NIHHIS), 488
National Integrated Heat Health Information System, 493
National Meteorological or Hydrometerological Services, 46, 48
National Morbidity and Mortality Air Pollution Study (NMMAPS), 238
National Oceanic and Atmospheric Administration (NOAA), 5, 42, 46, 51, 483, 484, 498, 499, 506
National Research Council, 527
National Science Foundation (NSF), 484, 497, 498
Natural disasters, 223
Neurological effects, 467, 468
New york climate and health project (or NYCHP), 165
Nitrogen dioxide (NO_2), 542
Nitrogen oxides (NO_x), 186
Nitrous oxide, 4
NOAA Seamless Suite of Forecasts, 37
Non-communicable chronic diseases (NCDs), 437
Non-fossil fuel energy, 24
Nongovernmental organizations (NGOs), 371, 379, 493
Notch activating pathway, 467

O
O_3 in NYC metropolitan region, 165
  epidemiologic analysis of historical data, 165, 166
  future projections of, 166
  health impact assessment, 167
  implications, 170, 171
  results, 167–169
Obesity, 140
Operational Framework for Building Climate Resilient Health Systems, 116
Organic carbon, 96
Over-the-counter allergy medication sales, 171–173
  advantages, 174, 175
  data and methods, 172, 173
Oviposition, 17
Oxidative stress, 126
Ozone, 5, 213, 378
Ozone (O_3) air pollution, 163, 181, 405
Ozone hole, 186

P
Panic disorder (PD), 345
Parameterization, 45
Paris Agreement, 83, 84
Paris Climate Agreement, 24, 27
Particulate matter (PM), health effects of, 181, 182, 221, 222, 371, 439, 454
  constituents, 184, 185
  long-term chronic exposure, 182–184
  short-term exposure, 182
Partnership for Clean Indoor Air (PCIA), 372, 373
Pediatric diarrheal disease (PDD), 557
Penn State-NCAR Mesoscale Model 5 (MM5), 166
Permafrost, 93–98, 103, 107, 108
Persian Gulf syndrome, 455
Personal protective equipment (PPE), 219
Physician's response to climate change, 583–585, 587, 588
  healthcare delivery, 586
  health effects of climate change, 584
  hospital energy use, 587
  medical costs, 585
  medical supply chain, 585
  supply chains of essential drugs and medical equipment, 585
Physiological thermoregulation, 131
Planning horizons, 551
Plantago lanceollata, 205
Pneumoconiosis, 398, 470
Pollen, 204, 206, 215, 454, 583
  disease, 206–208
  heat change plant and pollen distribution, 205

non-allergenic air pollution,
aeroallergens, 212–214
phenological phases, 205, 206
seasonal distribution of, 208
shoot growth, 205, 206
in U.S. seasonal pollen distribution, 209
water use efficiency, 205, 206
Pollens, 164
Polycyclic aromatic hydrocarbons compounds (PAH), 221, 393
Poor air quality, impacts of, 405, 406
Population vulnerability, *see* Vulnerable populations
Post COP21, 84
Post-traumatic stress disorder (PTSD), 128, 223, 224, 340, 345, 346
Potency of pollen, 204
Poverty, 381, 385
Prediction of and Resilience Against Extreme Events (PREEVENTS) program, 498
Preferred Reporting Items for Systematic Reviews and Meta-analyses (PRISMA), 516
Proactive adaptation, 551
Pseudomonas aeruginosa, 469
Public health burden, 125
Public health interventions
definition of, 550
See also Climate and health interventions
Public research university system, 565–567
budgets and forms of Support, 575
faculty training workshop
agenda, speakers, and group exercise, 571–573, 577, 578
concept, 568, 569
executive sponsorship, 570
faculty leads, 570, 571
library of resources, 573
timeline, 573
training workshop for faculty leads, 571
university of california leadership, 568
SHE
creating communities at each campus, 574
deans and vice deans of education, 574
evaluation framework, 574, 575, 579–581
pathways and models for, 567, 568
Pulmonary effects, 465, 466

Q
Quality-Adjusted Life Year (QALY), 194, 195

R
Race and ethnicity, 143, 347
Ragweed (*Artemisia* species) pollen, 205–208
Reduce emissions from deforestation and forest degradation (REDD), 23
Regional climate models (RCM), 261
Regional Climate Outlook Forums (RCOFs), 40
Regional Greenhouse Gas Initiative (RGGI), 26
Regional Integrated Sciences and Assessments (RISA) Program, 46
Representative concentration pathways (RCPs), 14, 94, 254
Respiratory health, 473

S
Sandstorms, 451
Satellites, 42
Sea surface temperature (SST), 43, 312
Self Employed Women's Association (SEWA), 370
Sendai Framework Monitoring Process, 416
SERVIR
Africa, 495
Himalaya, 495
Mekong, 495
Severe dengue, 273, 276
Shared socioeconomic pathways (SSPs), 261
Short-lived climate forcers, 378, 379
Short-term exposure effects of particulate matter, 182
Short-term exposure health effects of O_3, 187–189
Sickness of climate change, 588
Small Island Developing States (SIDS), 428, 430
Small island states, 427
physical effects of climate change on
atmospheric temperatures, 428, 429
ecosystem structural changes, 432
hydrology and storm events, 430–432
SLR and ocean acidification, 429, 430
potential public health consequences, Caribbean, 437–440
potential socio-economic effects of CC on Caribbean SIDS, 432
agriculture and fisheries, 434, 435
energy, water, and waste management, 432–434
infrastructure and population displacement, 436, 437
tourism, 435, 436
Smoking behavior, 223, 224

SO_2 aerosols, 4
Soil organic carbon, 93
Solar energy, 447
Solid fuels, 392, 394, 396, 399
South eastern anatolian project (GAP), 447
Sparse Matrix Operator Kernel Emissions Modeling System (SMOKE), 166
Special Report on Emissions Scenarios (SRES), 500
Spring pollen peaks, 171–176
Storm surge, 222
Stove Unit Monitoring System, 381
Stove use monitors (SUM), 372
Sudden infant death syndrome (SIDS), 136
Summary for Policymakers (SPM), 117
Superoxide dismutase-1 (SOD-1), 476
Superstorm Sandy, 339, 344, 346–348, 586
Surya Project as model of potential interventions to reduce climate impacts, 382–384
Sustainable and Climate Resilient Health Care Facilities Initiative (SCRHCFI), 488
Sustainable communities strategies, 544
Sustainable development, 21
Sustainable Development Goals (SDGs), 416
Sustainable healthcare education (SHE), 566, 569
  creating communities at each campus, 574
  deans and vice deans of education, 574
  evaluation framework, 574, 575, 579–581
  pathways and models for, 567, 568
Synthetic chemicals, 4
Systemic inflammatory response syndrome (SIRS), 134

T
Temporary housing units (THUs), 220
Terrestrialization, 58
The United States Global Change Research Program (USGCRP), 484, 485
Thermoregulation, 131–133
Thermotolerance, 133–135, 146
Threat to biodiversity, 13
Thunderstorm asthma, 208
Toll like receptor type 2 (TLR-2), 471
Top loading updraft (TLUD), 380, 381
Toxic diseases, 344
Traffic related air pollution (TRAP), 185
Training in tropical diseases (TDR), 263
Trichodesmium bacterium, 472
Tropical cyclones, 216
Tropospheric ozone, health effects of, 185–189
Tuberculosis (TB), 397, 398, 471
Tumor necrosis factor-alpha (TNF-α), 134
Turkey, 445
  annual mean precipitation and its trend in, 449
  climate change, 447–449
  desertification in, 450, 451
    human health, 454–456
    loss of biodiversity, 453, 454
    sandstorms and dust storms, 451
    water use, 452, 453
  greenhouse gas emissions, 446, 447
  gross total amount and consumable water in, 452
  mean annual temperature trend in, 448
12 digit Hydrologic Unit Code (HUC) resolution, 59
2016 GCRP Climate Health Assessment, 488, 489

U
U.S. Agency for International Development (USAID), 495–497
U.S. Environmental Protection Agency (USEPA), 182, 221, 372
U.S. global change research program (US GCRP), 551
United Nations Environment Program (UNEP), 21, 116, 383, 445
United Nations Framework Convention on Climate Change (UNFCCC), 81, 403, 411, 430
United Nations International Strategy for Disaster Reduction (UNISDR), 416
United States Department of Agriculture (USDA), 501
United States Environmental Protection Agency (US-EPA), 459
United States National Climate Assessment (NCA), 506
University of California leadership, 568
Urban air pollution, 212
US Global Change Research Program (GCRP), 485

V
Valley fever, 359, 470
Vector-borne diseases (VBDs), 16, 17, 438
  arboviral diseases, 254
  climate and climate change research to health policy and programming, 262–264
  malaria, 254
  mosquito-borne diseases

climatic drivers of, 255–259
non-climatic drivers of, 259, 260
predicting climate change impacts on, 261, 262
satellite imagery, 265, 266
transmission, 264
women's health, 409, 410
zika virus, 253
Vehicle miles traveled (VMTs), 544
Vibrio infections, 408
Vibrio parahaemolyticus, 113
Volatile organic compounds (VOCs), 163, 185, 219, 220, 439
Vulnerability, 124, 126, 312, 341, 431, 433
strategies to reduce, of HRMM, 145, 146
extreme heat event, 149–154
promote good health and access to quality healthcare, 146, 147
reduce/manage potential exposures to ambient heat and physical environmental stressors, 147–149
warm season heat preparedness and response action plans, 149–154
Vulnerability, capacity, and adaptation assessments (V&As), 115, 116
Vulnerable geographic regions, 312, 314
Vulnerable populations
to adverse health outcomes with household air pollution, 374–376
determinants of thermoregulatory capacity, 138
infants and children, 140–142
older adults, 139, 140
global environmental and societal challenges affecting, 143–145
multiple determinants of ambient heat health impacts
biological adaptations to heat stress and susceptibility, 130, 131
environmental health multiple-determinants model of vulnerability (MDMv), 123, 129, 130, 145, 154
heat acclimatization, 133–135
pathophysiology of heat illness, 130, 131
populations and subgroups at elevated risk, 127–129
thermoregulation, 131–133
thermotolerance, 133–135

W

Warm Season Heat Preparedness and Response Action Plans, 145
Water quality, 54–57
Water resources, 54–57
Water scarcity, 408, 409
Waterborne disease, women's health, 408, 409
Watershed Condition Classification (WCC) system, 57, 58
Watershed Condition Framework (WCF), 57
Watershed health, climate change impacts on, 57–59
Weather, definition of, 38
Wildfire, 66, 67
Wildland Fire Research Framework, 500
Willingness to Pay (WTP) methods, 194
Wolbachia bacteria, 278
Women's health
climate-related health impacts on women, 412–415
extreme weather events and disasters, 406, 407
food insecurity and malnutrition, 407, 408
heat impacts, 404
international policy response, 411, 416
local solutions, 416, 417
climate resilient farming, 420, 421
empowering clean energy solutions, 419, 420
enhance pre-planning, 419
ensure participation, 418
improve data, 418
improve multisector coordination, 419
leading community resiliance, 420
prioritize education, 418
re-define success, 419
migration issues, 410, 411
poor air quality, impacts of, 405, 406
vector-borne disease, 409, 410
water scarcity and waterborne disease, 408, 409
World Climate Research Program, 216
World Data Centers, 46
World Health Organization (WHO), 14, 149, 205, 253, 273, 276, 294, 392, 455
World Meteorological Organization, 40, 46, 48, 116, 262
World Trade Center disaster, 227

Z

Zero emission vehicle (ZEV) program, 533, 543
Zika virus, 253, 254, 277, 291

GPSR Compliance

The European Union's (EU) General Product Safety Regulation (GPSR) is a set of rules that requires consumer products to be safe and our obligations to ensure this.

If you have any concerns about our products, you can contact us on ProductSafety@springernature.com

In case Publisher is established outside the EU, the EU authorized representative is:

Springer Nature Customer Service Center GmbH
Europaplatz 3
69115 Heidelberg, Germany

Batch number: 10739796

Printed by Printforce, the Netherlands